T0215864

STRUCTURAL
Wood Design
ASD/LRFD
SECOND EDITION

STRUCTURAL Wood Design

ASD/LRFD

SECOND EDITION

Abi Aghayere
Jason Vigil

CRC Press
Taylor & Francis Group
Boca Raton London New York

CRC Press is an imprint of the
Taylor & Francis Group, an **informa** business

CRC Press
Taylor & Francis Group
6000 Broken Sound Parkway NW, Suite 300
Boca Raton, FL 33487-2742

First issued in paperback 2019

ISBN-13: 978-1-4987-4985-5 (hbk)
ISBN-13: 978-0-367-87562-6 (pbk)

Library of Congress Cataloging-in-Publication Data

Names: Aghayere, Abi O., author. | Vigil, Jason, 1974- author.
Title: Structural wood design / by Abi Aghayere and Jason Vigil.
Description: Second edition. | Boca Raton, FL : CRC Press, [2017] | Includes bibliographical references and index.
Identifiers: LCCN 2016026976 | ISBN 9781498749855 (hardback) | ISBN 9781498749862 (pdf)

Visit the Taylor & Francis Web site at
http://www.taylorandfrancis.com

and the CRC Press Web site at
http://www.crcpress.com

We thank our colleagues in the academy and in the civil, architectural and structural engineering fields who have provided valuable input and feedback on this text. In the spirit of continuous quality improvement, we always welcome your feedback.

We also thank both our families for their unflinching support that has made it possible to complete this project. Finally and most importantly, we thank our Lord and Savior, Jesus Christ, who granted us the grace, strength, and wisdom to complete this project.

Contents

Preface

This second edition covers both the allowable stress design (ASD) and the load and resistance factor design (LRFD) methods for wood and timber design in accordance with the 2015 National Design Specification (NDS) for Wood Construction. The primary audience for this book is civil and architectural engineering students, civil and construction engineering technology students, and architecture students in a typical undergraduate course in wood or timber design. This book can be used for a one-semester course in structural wood or timber design, and it sufficiently prepares the student to be able to apply the fundamentals of structural wood design to typical projects that might occur in practice. This book will also serve as a useful resource for practicing engineers, architects, and builders, and those preparing for the professional engineering licensure exams because of the practice-oriented, easy-to-follow, and detailed format that is adopted in the text, and the many practical examples applicable to typical, everyday projects that are presented.

This book provides the essentials of structural wood design from a practical perspective and helps to bridge the gap between the design of individual wood structural members and the complete design of a wood structural system, thus providing a holistic approach to structural wood design. The design of each building component is presented in a format that enables the reader to see how each component fits into the bigger picture of an entire wood building design and construction process. Structural wood details and practical example problems that realistically mirror what is found in professional design practice are presented. Other unique features of this book include the following: a discussion and description of common wood structural elements and systems that that are found in wood building structures; a complete wood building design case study; the design of wood floors for vibrations; the general analysis of shearwalls for overturning including all applicable load cases; the many realistic 3-D and 2-D drawings and illustrations in this book that enhance the reader's understanding of the concepts presented; and the easy-to-use design aids that are presented for the quick design of common structural members such as floor joists, columns, and wall studs.

This edition includes the following new features:

- An expanded student design project problem in Chapter 1.
- Several new end-of the chapter problems that mimic the design problems found in practice.
- A discussion of both the ASD and LRFD methods together with some design examples using the LRFD method for joists, girders, and axially load members.
- An expanded discussion of framing and framing systems, including metal plate connected trusses, rafters versus collar ties, and pre-engineered framing such as I-joists.
- The design of cross-laminated timber (CLT) members used in modern high-rise wood buildings is covered.

- A new chapter on practical considerations and rules of thumb for wood design is presented.
- Updated floor span charts that are useful for preliminary and final design are presented in Appendix B.

A brief description of each chapter is as follows:

Chapter 1—The reader is introduced to wood design through a discussion and description of the various wood structural elements and systems that are used in wood structures as well as the material properties of wood that affect its structural strength.

Chapter 2—The different structural loads, dead, live, snow, wind, and seismic, are discussed, and several examples are presented. This succinct treatment of structural loads gives the reader adequate information to calculate the loads acting on typical wood building structures.

Chapter 3—Both the ASD and LRFD methods for wood design are presented in this chapter. The calculation of the allowable stresses and factored stresses for both sawn lumber and glued laminated timber (glulam) in accordance with the 2015 NDS Code as well as the discussion of the various stress adjustment factors and reference design values are presented in this chapter. Glulam and other engineered wood products are introduced, and the determination of the controlling load combination in a wood building using the normalized load method is also discussed.

Chapter 4—The design using both ASD and LRFD methods and analysis of joists, beams, and girders is discussed, and several examples are presented. Continuous beams, pre-engineered wood beams and headers, weak-axis bending of top plates, and miscellaneous stresses in wood—such as stresses in notched beams and sloped rafters—are also discussed. The design of wood floors for vibrations, miscellaneous stresses in wood members, the selection of pre-engineered wood flexural members, and the design of sawn lumber decking are also discussed.

Chapter 5—The design using both ASD and LRFD methods of wood members subjected to axial and bending loads such as truss web and chord members, solid and built-up columns, and wall studs are discussed. Columns under eccentric axial loads and transverse loads are also discussed.

Chapter 6—The design of roof and floor sheathing for gravity loads and the design of the roof and floor diaphragms for lateral loads are discussed. The calculation of the forces in the diaphragm chords and drag struts is also discussed as well as the design of these axially loaded elements.

Chapter 7—The design of exterior wall sheathing for a wind load perpendicular to the face of the wall as well as the design of wood shear walls or vertical diaphragms parallel to the lateral loads is discussed. A general analysis of shear walls for overturning that takes into account all applicable lateral and gravity loads is presented.

Chapter 8—The design of connections is covered in this chapter in a simplified manner. Design examples are presented to show how the connection capacity tables in the NDS Code are used. Several practical connections and practical connection considerations are discussed.

Chapter 9—The practical considerations for the design of wood buildings are covered in this chapter. Topics such as the design of CLT used in high-rise wood buildings, rules of thumb for wood design, and fire design of heavy timber members are discussed.

Chapter 10—A complete building design case study is presented to help the reader tie together the various aspects of wood structural element design presented in the previous chapters within the context of a complete building structural system design.

This holistic and practice-oriented approach to structural wood design permeates the chapters of this text and is the hallmark of this book. The design aids presented in the appendix for the rapid design of floor joists, columns, and wall studs subjected to axial and lateral loads are utilized in this chapter.

In Appendix A, weights of common building materials are presented, and in Appendix B, practical design aids in the form of tables and charts are presented for the rapid selection of wood structural members such as joists, beams, girders, stud walls, and columns.

Abi Aghayere
Avondale, PA

Jason Vigil
Rochester, NY

Additional material is available from the CRC Web site: http://www.crcpress.com/product/isbn/9781498749855.

Chapter 1

Introduction

Wood properties, species, and grades

1.1 INTRODUCTION

Wood is a sustainable, economical, and aesthetically pleasing material that has been used for a long time in civil engineering structures. Though a large proportion of low-rise buildings in the United States and Canada are built from wood, there is a dearth of textbooks for students and practitioners alike that presents wood structural design in a holistic and easy-to-understand manner. The purpose of this book is to bridge this gap by presenting the design process for wood structures in a holistic, succinct, and practical way including a detailed discussion of the use of wood as a structural material and the analysis and design of the major structural elements in typical wood buildings.

In general, building plans and details are defined by an architect and are usually given to a structural engineer for design of structural elements and to present the design in the form of structural drawings. In this book, we take a project-based approach covering the design process that a structural engineer would go through for a typical wood-framed structure.

The intended audience for this book is students taking a course in timber or structural wood design and structural engineers and similarly qualified designers of wood or timber structures looking for a simple and practical guide for design. The reader should have a working knowledge of statics, strength of materials, structural analysis (including truss analysis), and load calculations in accordance with building codes (dead, live, snow, wind, and seismic loads). Design loads are reviewed in Chapter 2. The reader must also have available:

1. *National Design Specification for Wood Construction*, 2015 edition, ANSI/AWC (hereafter referred to as the NDS code) [1].
2. *National Design Specification Supplement: Design Values for Wood Construction*, 2015 edition, ANSI/AWC (hereafter referred to as NDS-S) [2].
3. *International Building Code*, 2015 edition, International Code Council (ICC) (hereafter referred to as the IBC) [3].
4. *Minimum Design Loads for Buildings and Other Structures*, 2010 edition, American Society of Civil Engineers (ASCE) (hereafter referred to as ASCE 7) [4].

1.1.1 The project-based approach

Wood is nature's most abundant renewable building material and a widely used structural material in the United States, where 90% of all residential buildings are of wood construction [5]. The number of building configurations and design examples that could be presented is unlimited. Some applications of wood in construction include residential buildings, retail buildings, offices, hotels, schools and colleges, healthcare and recreation facilities, senior living and retirement homes, and religious buildings. The most common wood structures are residential and multifamily dwellings as well as hotels. Residential structures are usually one to three stories in height, while multifamily and hotel structures can be up to five or six stories in height with the upper four stories framed with wood and the lower levels framed with steel or concrete. Four to five stories of wood framing is a practical upper limit for wood buildings, though a much higher number of stories are now possible for wood buildings with the introduction of cross-laminated timber. Commercial, industrial, institutional, and other structures that have higher occupancy loads and factors of safety are not typically constructed with wood, although wood may be used as a secondary structure, such as a storage mezzanine. The structures that support amusement park rides are sometimes built out of wood because of the relatively low maintenance cost of exposed wood structures and its unique ability to resist the repeated cycles of dynamic loading (fatigue) imposed on the structure by the amusement park rides. The approach taken in this text is to illustrate the design process required for each major structural element in a wood structure with proper consideration as to how the design and detailing of each element is incorporated into a typical project. In Figures 1.1 and 1.2, we identify the typical structural elements in a wood building. The elements are described in greater detail in the next section.

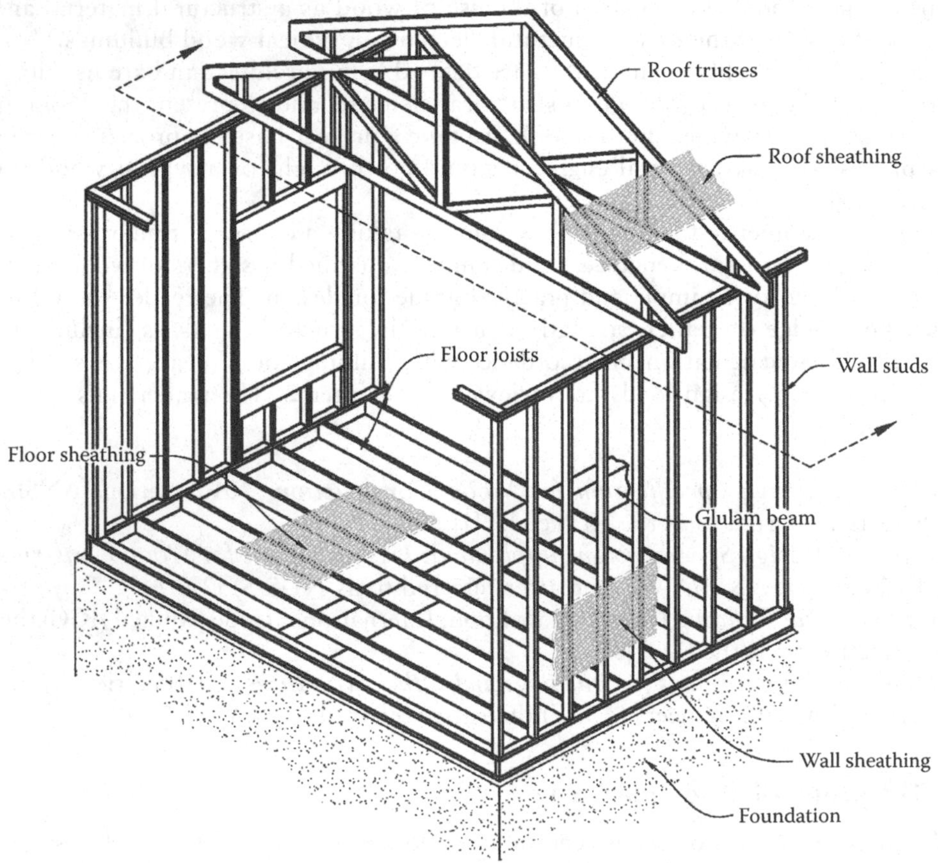

Figure 1.1 Perspective overview of a building section.

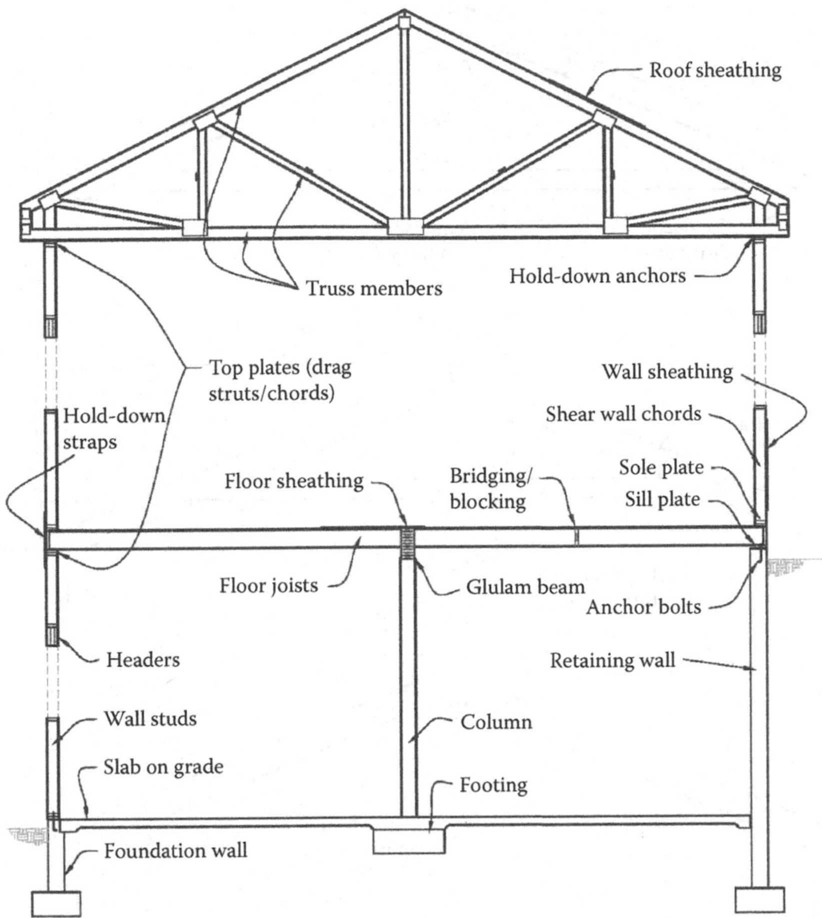

Figure 1.2 Overview of major structural elements.

1.2 TYPICAL STRUCTURAL COMPONENTS OF WOOD BUILDINGS

The majority of wood buildings in the United States are typically *platform-framed construction*, in which the vertical wall studs are built one story at a time and the floor below provides the platform to build the next level of wall that will in turn support the floor above. The walls usually span vertically between the sole or sill plates at the floor level and the top plates at the floor or roof level above. This is in contrast to the infrequently used and less economical *balloon-framed construction*, where the vertical studs are continuous for the entire height of the building and the floor framing is supported on brackets off the face of the wall studs. Platform-framed construction is the predominant method of framing for wood buildings in the United States. The typical structural elements in a wood-framed building system are described below.

Rafters (Figure 1.3): These are usually sloped sawn-dimension lumber roof beams spaced at fairly close intervals (e.g., 12, 16, or 24 in.) and carry lighter loads than those carried by the roof trusses, beams, or girders. They are usually supported by roof trusses, ridge

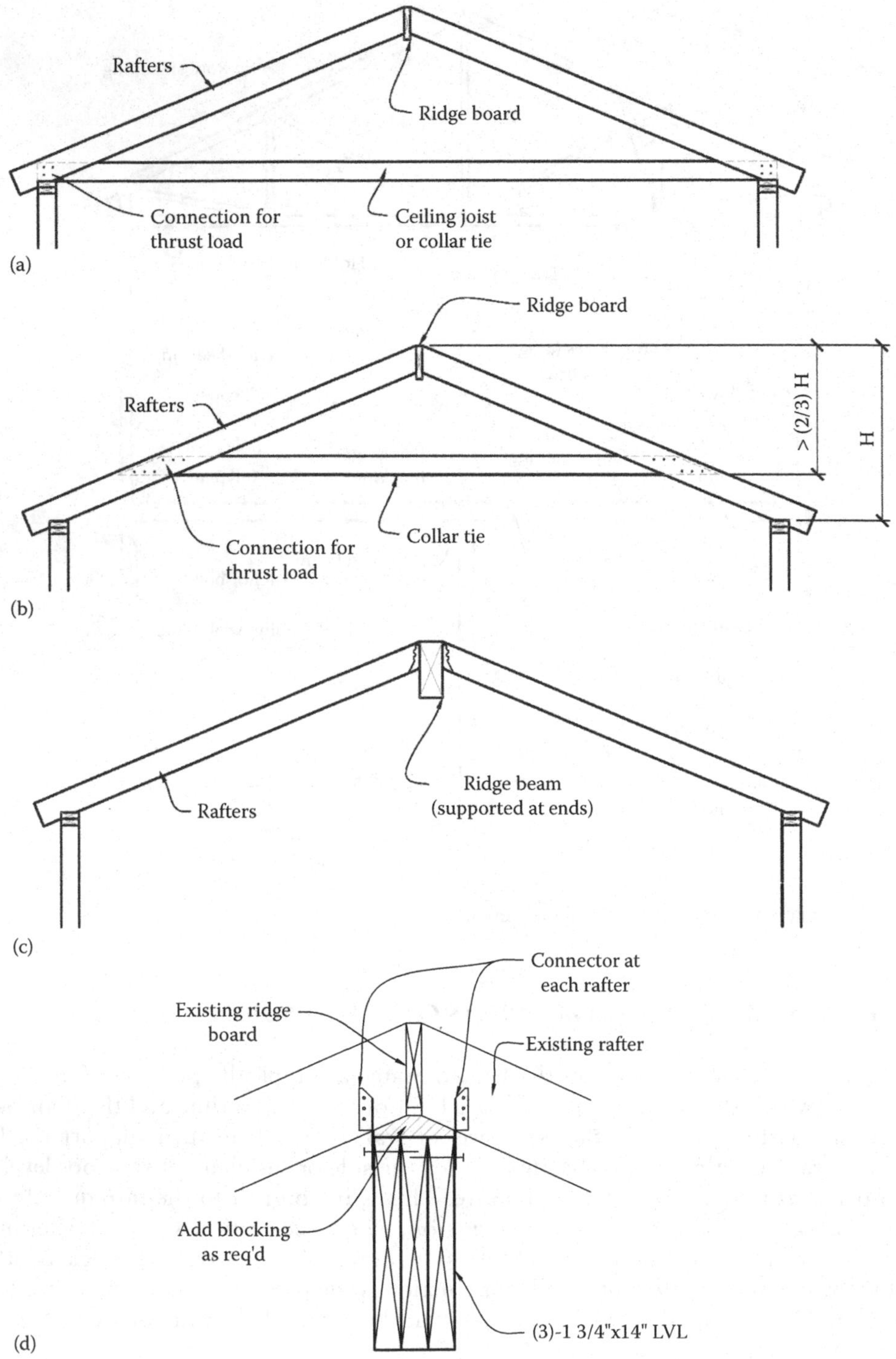

Figure 1.3 Rafter framing options: (a) rafter framing with ceiling joist, (b) rafter framing with collar tie, (c) rafters supported by a ridge beam, and (d) reinforcing at an existing ridge.

beams, hip rafters, or walls. The span of rafters is limited in practice to a maximum of 14 ft. to 18 ft., with longer spans available with the use of engineered wood members such as laminated veneer lumber (LVL) and I-joists. Rafters of varying spans that are supported by hip rafters are called *jack rafters* (see Figure 1.6). Sloped roof rafters with a nonstructural ridge, such as a 1× ridge board, require ceiling tie joists or collar ties to resist the horizontal outward thrust at the exterior walls that is due to gravity loads on the sloped rafters (Figures 1.3a and 1.3b). A ceiling joist would typically be a 2× wood member with drywall on the underside and should be provided at the same spacing as the rafters. A rafter-framed roof with ceiling tie joists acts like a three-member truss where the ceiling joists resist the tension caused by the outward horizontal thrust from the sloped rafters. The connection of the ceiling joist to the rafter has to be designed and detailed for the tension force. Rafters with collar ties (see Figure 1.3b) are used in buildings with cathedral or vaulted ceilings; the critical forces are the maximum moment in the sloped rafter and the tension in the collar tie; since there are no lateral or horizontal reactions at the ridge or at the exterior wall, the horizontal or lateral deflection of the sloped rafter at the exterior wall support should also be checked. The most effective placement of an offset collar tie is in the lower one-third of the roof height. If collar ties or ceiling joists are absent in a roof with sloped rafters, a noticeable outward lateral deflection (or kick out) of the top of the exterior walls will occur under full dead and snow or roof live loads leading to a tendency for the sloped rafters to want to flatten out. A framing scheme with just rafters and a ridge board without ceiling joists or collar ties or a ridge beam is not stable and should be avoided in practice especially in older structures.

Figure 1.3c shows rafters supported at the upper end by a ridge beam. A ridge beam is different from a ridge board in that the ridge beam is designed to support gravity load and is supported at each end. In roofs with a long ridge, the ridge beam would need to be supported by columns or a transfer beam to resist the gravity load reactions from the ridge beam.

Joists (Figure 1.4): These are sawn-lumber floor beams spaced at fairly close intervals of 12, 16, or 24 in. that support the roof or floor deck. They support lighter loads than do

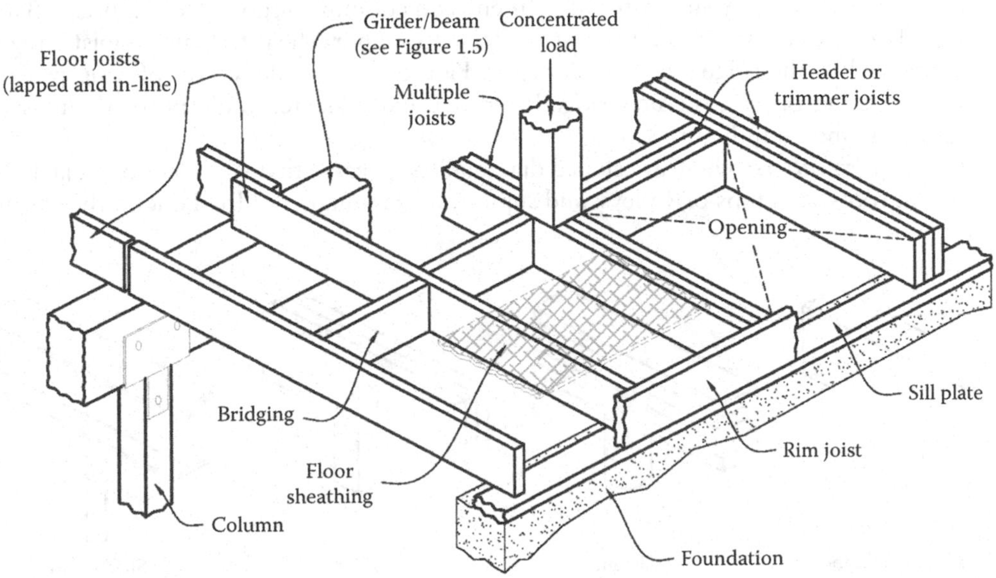

Figure 1.4 Floor framing elements.

floor beams or girders. Joists are typically supported by floor beams, walls, or girders. The spans are usually limited in practice to about 14 ft. to 18 ft. Spans greater than 20 ft usually require the use of pre-engineered products, such as I-joists that consists of flanges made from sawn lumber or structural composite lumber and webs made of plywood or oriented strand board (OSB). These prefabricated joists can vary from 10 in. to 24 in. in depth with lengths of 16 ft. to 36 ft. being commonly used [6]. Floor joists can be supported on top of the beams, either in line or lapped with other joists framing into the beam, or the joist can be supported off the side of the beams using joist hangers. In the former case, the top of the joist does not line up with the top of the beam as it does in the latter case. Lapped joists are used more commonly than in-line joists because of the ease of framing and the fact that lapped joists are not affected by the width (i.e., the smaller dimension) of the supporting beam (see Figure 1.4).

Double or triple joists: These are two or more sawn-lumber joists that are fastened together to act as one composite beam. They are used to support heavy concentrated loads or the load from a partition wall or a load-bearing wall running parallel to the span of the floor joists, in addition to the tributary floor loads. They are also used to frame around stair openings (see header and trimmer joists).

Header and trimmer joists: These are multiple-dimension lumber joists that are fastened together (e.g., double joists) and used to frame around stair openings. The trimmer joists are parallel to the long side of the floor opening and support the floor joists and the wall at the edge of the stair. The header joists support the stair stringer and floor loads and are parallel to the short side of the floor opening.

Beams and girders (Figures 1.4 and 1.5): These are horizontal elements that support heavier gravity loads than rafters and joists and are used to span longer distances. Wood beams can also be built from several joists nailed together. These members are usually made from beam and stringer (B&S) sawn lumber, glued-laminated timber (glulam), parallel strand lumber (PSL), or laminated veneer lumber (LVL). Beams and girders are usually supported by columns or multiple studs within a stud wall. For heavier applications, steel beams are used.

Ridge beams: These are roof beams at the ridge of a roof that support the sloped roof rafters. They are usually supported at their ends on columns or posts (see Figures 1.3 and 1.6). For new construction, it is cost-effective to support the rafters using joist hangers connected to the ridge beam as shown in Figure 1.3c, while for retrofitting projects with existing rafters, the new ridge beam is usually located underneath the existing rafters as shown in Figure 1.3d.

Hip and valley rafters: These are sloped diagonal roof beams that support sloped jack rafters in roofs with hips or valleys, and support a triangular roof load due to the varying

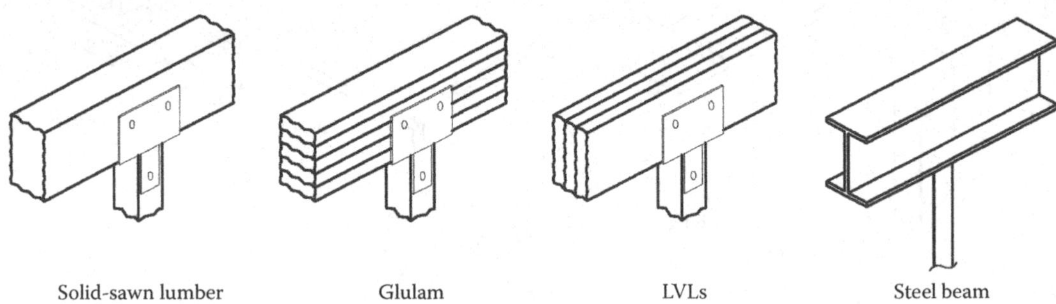

Figure 1.5 Types of beams and girders.

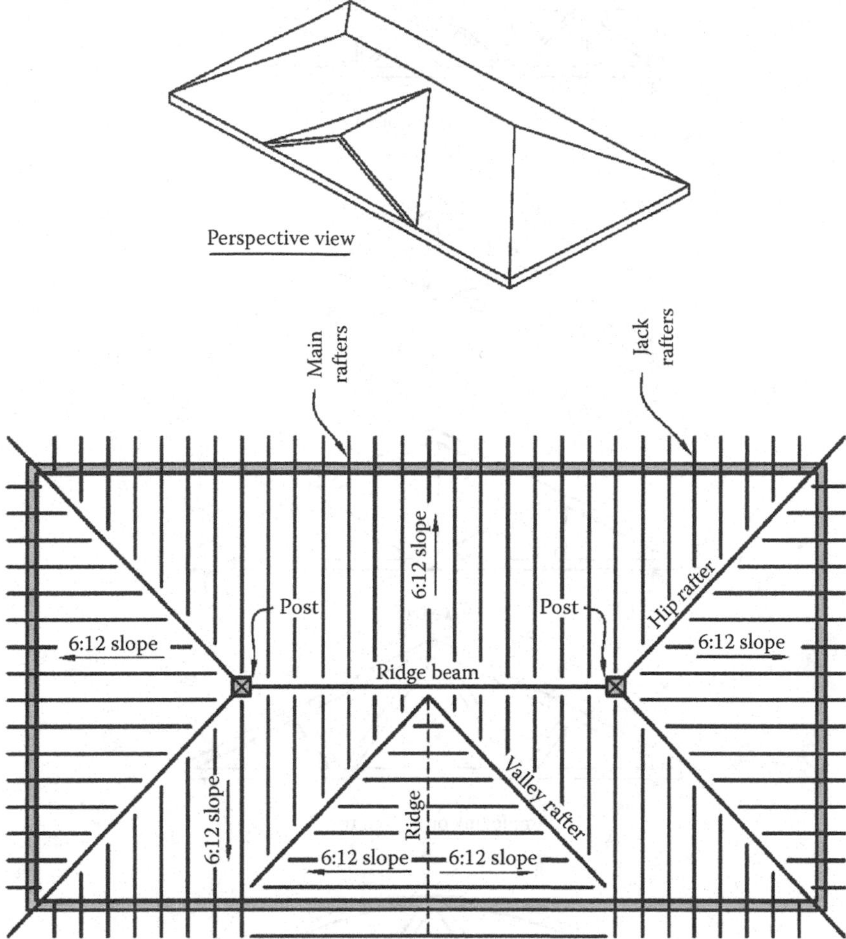

Figure 1.6 Hip and valley rafters.

spans of the jack rafters (see Figure 1.6) [7,8]. The hip rafters are simply supported at the exterior wall and on the sloped main rafter at the end of the ridge. The jack or varying span rafters are supported on the hip rafters and the exterior wall. The top of a hip rafter is usually shaped in the form of an inverted V, while the top of a valley rafter is usually V-shaped. Hip and valley rafters are designed like ridge beams.

Columns or posts: These are vertical members that resist axial compression loads and may occasionally resist additional bending loads due to lateral wind loads or the eccentricity of the gravity loads on the column. Columns or posts are usually made from post and timber (P&T) sawn lumber or glulam. Sometimes, columns or posts are built up using sawn-dimension lumber. Wood posts may also be used as the chords of shear walls, where they are subjected to axial tension or compression forces from the overturning effect of the lateral and seismic loads on the building.

Roof trusses (Figure 1.7): These are made up typically of sawn-dimension lumber top and bottom chords and web members that are subject to axial tension or compression plus bending loads. Trusses are usually spaced at not more than 48 in. on centers and are used to span long distances up to 120 ft. The trusses usually span from outside wall to outside wall. The truss configurations that are used include the Pratt truss,

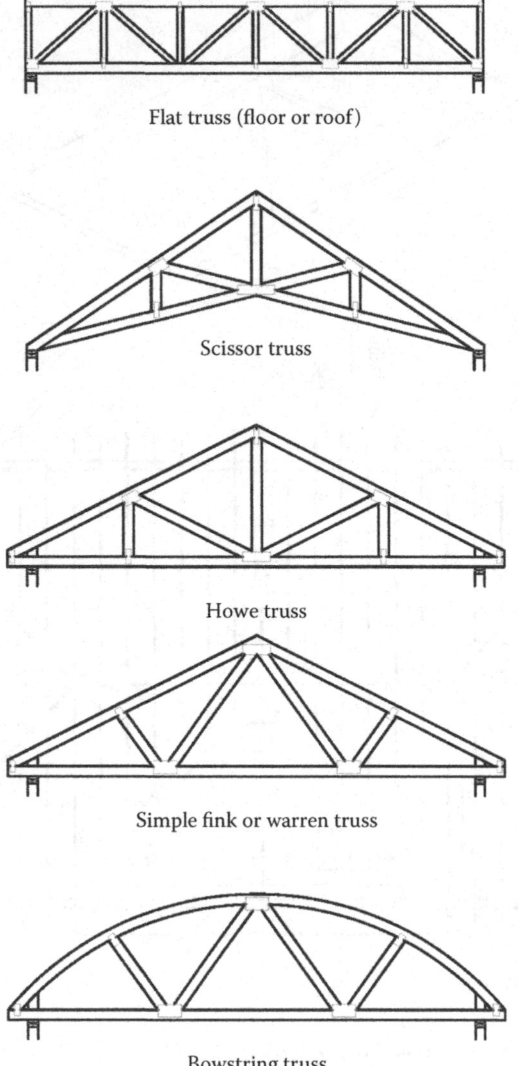

Figure 1.7 Truss profiles.

the Warren truss, the scissor truss, the Fink truss, and the bowstring truss. In building design practice, prefabricated trusses are usually specified, for economic reasons, and these are manufactured and designed by truss manufacturers rather than by the building designer. Prefabricated trusses can also be used for floor framing. These are typically used for spans where sawn lumber is not adequate. The recommended span-to-depth ratios for wood trusses are 8–10 for flat or parallel chord trusses, 6 or less for pitched or triangular roof trusses, and 6–8 for bowstring trusses [9].

Wall studs (Figure 1.8): These are axially loaded in compression and made of dimension lumber spaced at fairly close intervals (typically, 12, 16, or 24 in.). They are usually subjected to concentric axial compression loads, but exterior stud walls may also be subjected to a combined concentric axial compression load plus bending load due to wind load acting perpendicular to the wall. Wall studs may be subjected to eccentric axial load: for example, in a mezzanine floor with single-story stud and floor joists supported off the narrow face of the stud by joist hangers. Interior wall studs should, in addition to the axial load, be designed for the minimum 5 psf of interior wind pressure specified in the IBC.

Wall studs are usually tied together with plywood sheathing that is nailed to the narrow face of studs. Thus, wall studs are laterally braced by the wall sheathing for buckling about their weak axis (i.e., buckling in the plane of the wall). Stud walls also act together with plywood sheathing as part of the vertical diaphragm or shear wall to resist lateral loads acting parallel to the plane of the wall. *Jack studs* (also called *jamb* or *trimmer studs*) are the studs that support the ends of window or door headers; *king studs* are full-height studs adjacent to the jack studs; and *cripple studs* are the stubs or less-than-full-height stud members above or below a window or door opening and are usually supported by header beams. The wall frame consisting of the studs, wall sheathing, and top and bottom plates are usually built together as a unit on a flat horizontal surface and then lifted into position in the building.

Header beams (Figure 1.8): These are the beams that frame over door and window openings, supporting the dead load of the wall framing above the door or window opening as well as the dead and live loads from the roof or floor framing above. They are usually supported with beam hangers off the end chords of the shear walls or on top of jack

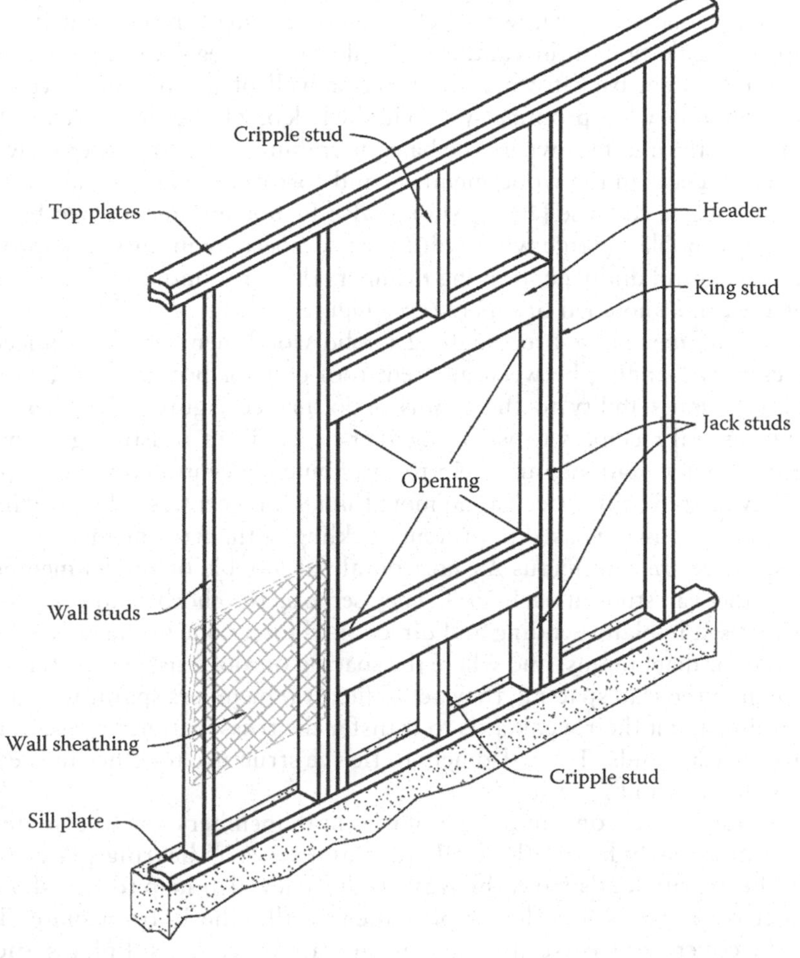

Figure 1.8 Wall framing elements.

studs adjacent to the shear wall end chords. In addition to supporting gravity loads, these header beams may also act as the chords and drag struts of the horizontal diaphragms in resisting lateral wind or seismic loads. Header beams can be made from sawn lumber, PSL, linear veneer lumber, or glued-laminated timber, or from built-up dimension lumber members nailed together. For example, a 2 × 10 double header beam implies a beam with two 2 × 10s nailed together.

Overhanging or cantilever beams (Figure 1.9): These beams consist of a back span between two supports and an overhanging or cantilever span beyond the exterior wall support below. Cantilever framing is common in multistory wood buildings especially in residential buildings where they serve as balconies at the floor levels and roof overhangs at the roof level. The typical framing of cantilevers in wood buildings is shown in Figure 1.9 for a floor balcony and an overhanging roof. In the framing of floor balconies, one design parameter is to limit the maximum deflection of the cantilever that occurs at the free end of the cantilever, and pattern live loading should be used in the analysis.

The deflection of the cantilever is dependent on the live load pattern and the ratio of the cantilever span, L_c, to the backspan, L (see Figure 1.9b). For optimum design, the length of the backspan, L, should be at least twice and preferably three times the cantilever span, L_c. For a cantilever with a pattern live load that is located only on the backspan, a negative or upward deflection of the tip of the cantilever would result. This upward deflection of the cantilever would cause a negative slope that could cause rain water to drain back toward the exterior wall of the building, which is highly undesirable. If the waterproofing system in the balcony fails, the result would be water infiltration at the interface where the balcony framing meets the exterior wall and this could lead to decay in the wood members and corrosion in any metal connectors and ultimately cause a collapse [10]. It should also be noted that roof overhangs are particularly susceptible to large wind uplift forces, especially in hurricane-prone regions. In order to control drainage, framing members that are continuous from the inside to the outside can be notched or tapered (see Figure 1.9).

Blocking or bridging: These are usually 2× solid wood members or x-braced wood or steel members spanning between adjacent roof or floor beams, joists, or wall studs, providing lateral stability to the beams or joists (see Figure 1.4). They also enable adjacent flexural members to work together as a unit in resisting gravity loads and mitigating floor vibrations, and help to distribute concentrated loads applied to the floor. They are typically spaced at no more than 8 ft on centers. The bridging or cross-bracing in roof trusses is used to prevent buckling of the truss members.

Top plates: These are continuous 2× horizontal flat (double or single) members located on top of the wall studs at each level. They serve as the chords and drag struts or collectors to resist in-plane bending and direct axial forces due to the lateral loads on the roof and floor diaphragms, and where the spacing of roof trusses rafters or floor joists do not match the stud spacing, they act as flexural members spanning between studs and bending about their weak axis to transfer the truss, rafter, or joist vertical reactions to the wall studs. They also help to tie the structure together in the horizontal plane at the roof and floor levels.

Bottom plates: These continuous 2× horizontal members or sole plates that are located immediately below the wall studs and serve as bearing plates to help distribute the gravity loads from the wall studs. They also help to transfer the lateral loads between the various levels of a shear wall. The bottom plates located on top of the concrete or masonry foundation wall are called sill plates and these are required to be pressure treated because of the presence of moisture since they are

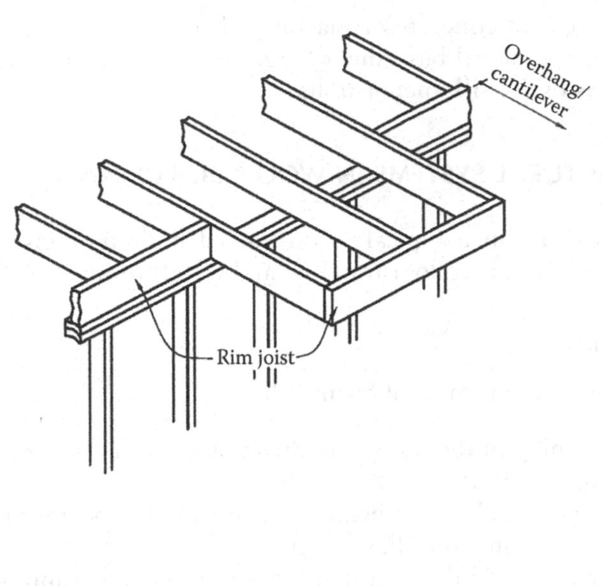

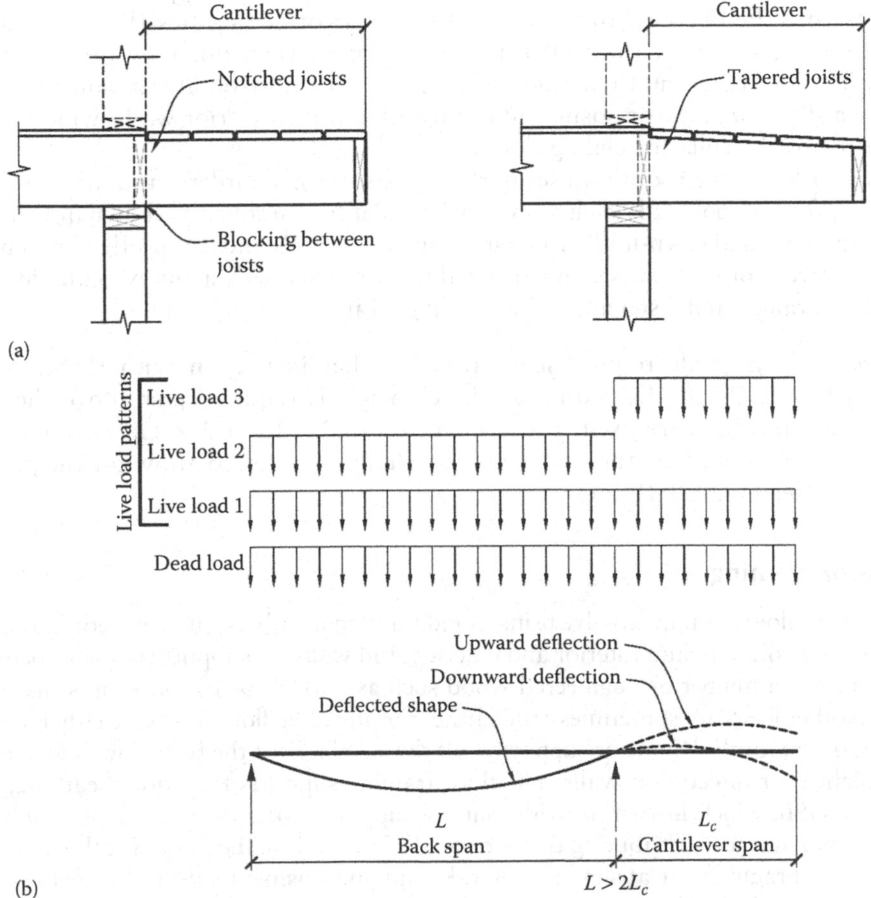

Figure 1.9 (a) Cantilever framing. (b) Pattern loading and deflected shapes in a cantilever beam.

in direct contact with concrete or masonry. They also serve as bearing plates and help to transfer the lateral base shear from the shear wall into the foundation wall below by means of the sill anchor bolts.

1.3 TYPICAL STRUCTURAL SYSTEMS IN WOOD BUILDINGS

The above-grade structure in a typical wood-framed building consists of the following structural systems: roof framing, floor framing, and wall framing.

1.3.1 Roof framing

Several schemes exist for a typical roof framing layout:

1. Roof trusses spanning in the transverse direction of the building from outside wall to outside wall (Figure 1.10a).
2. Sloped rafters supported by ridge beams and hip or valley beams or exterior walls used to form cathedral or vaulted ceilings (Figure 1.10b).
3. Sloped rafters with a 1× ridge board at the roof ridge line, supported on the exterior walls with the outward thrust resisted by collar or ceiling ties (Figure 1.10c). The intersecting rafters at the roof ridge level support each other by providing a self-equilibrating horizontal reaction at that level. This horizontal reaction results in an outward thrust at the opposite end of the rafter at the exterior walls, which has to be resisted by the collar or ceiling ties.
4. A flat or low-slope roof that uses purlins, joists, beams, girders, and interior columns to support the roof loads such as in panelized flat roof systems as shown in Figure 1.11. Purlins are small sawn-lumber members such as 2 × 4's and 2 × 6's that span between joists, rafter, or roof trusses in panelized roof systems with spans typically in the 8 ft. to 10 ft. range, and a spacing not exceeding 24 in.

The current energy code requirements are such that insulation with thickness ranging from 10 in. to 20 in. depending on the climate—is required just above the ceiling. Therefore, when rafters are used, a minimum size of 2 × 10 or 2 × 12 is often required. When trusses are used, then the heel height is usually adjusted to allow for the placement of insulation (see Figure 1.10d, e).

1.3.2 Floor framing

The options for floor framing involve using wood framing members, such as floor joists, beams, girders, interior columns, and interior and exterior stud walls, to support the floor loads. Floor joists can be sawn lumber or engineered wood such as an LVL or I-joist. Beams that transfer loads from other joists are sometimes called header beams. The floor joists are either supported by bearing on top of the beams or supported off the side faces of the beams with joist hangers, or they might bear directly on walls. The floor framing supports the floor sheathing, usually plywood or OSB, which in turn provides lateral support to the floor framing members and acts as the floor surface, distributing the floor loads. In addition, the floor sheathing acts as the horizontal diaphragm that transfers the lateral wind and seismic loads and sometimes lateral soil load to the vertical diaphragms or shear walls. Floor framing members are spaced in a way to coincide with the width of floor sheathing, which is typically a 4 ft. by 8 ft. sheet. Therefore, common spacings of floor framing are 12 in., 16 in., 19.2 in., and 24 in. Examples of floor framing layouts are shown in Figure 1.12.

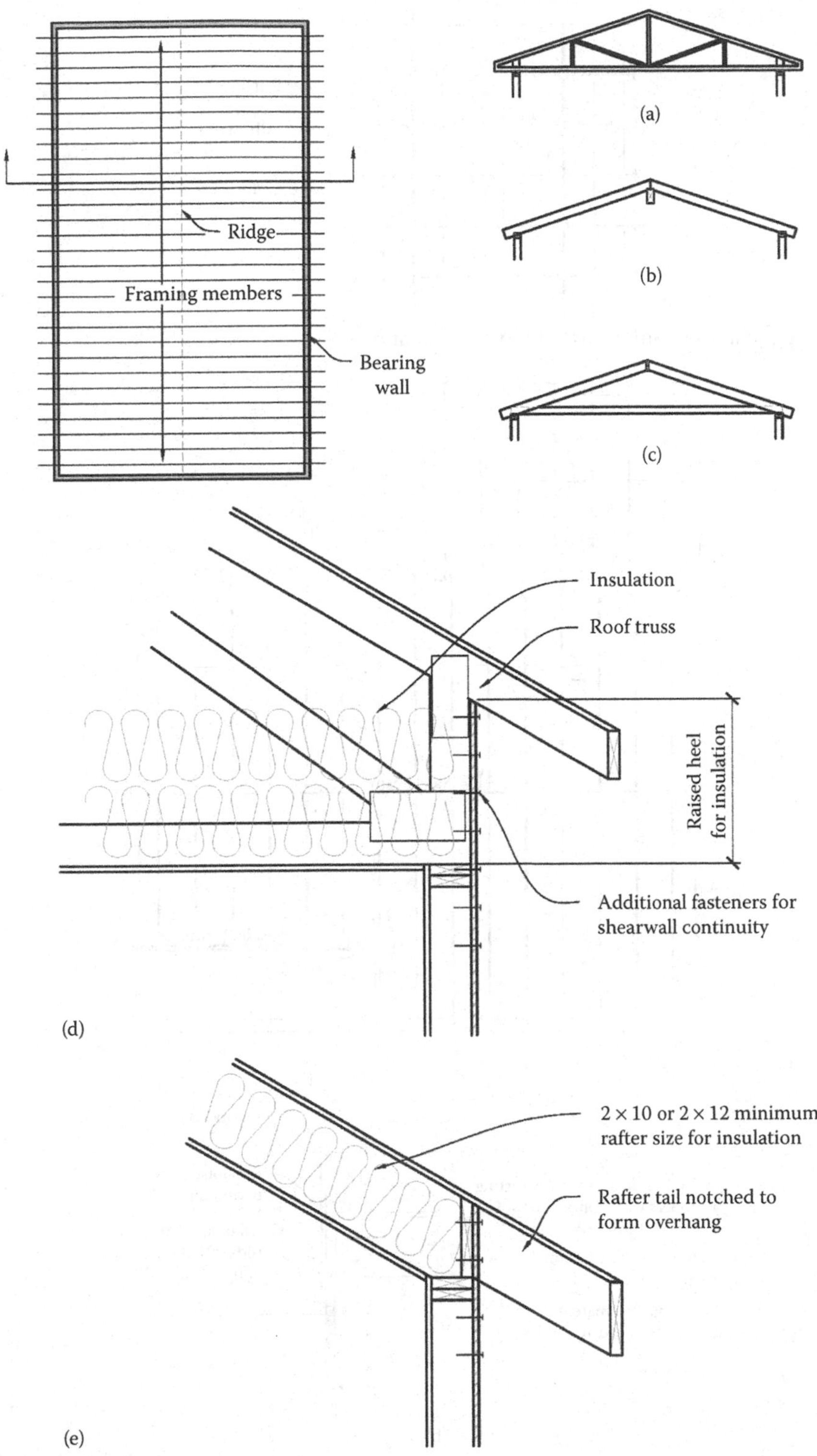

Figure 1.10 Roof framing layout and details: (a) roof truss, (b) vaulted ceiling, (c) rafter and collar tie, (d) raised heel detail truss, and (e) sloping rafter.

Girder
Purlin
Subpurlin
Column

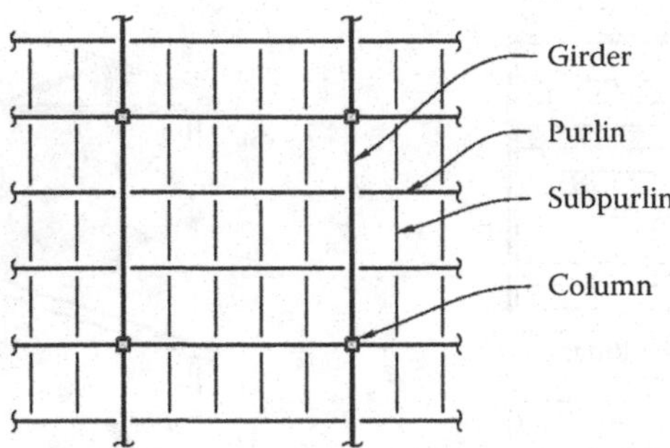

Figure 1.11 Typical roof framing layout (post and beam).

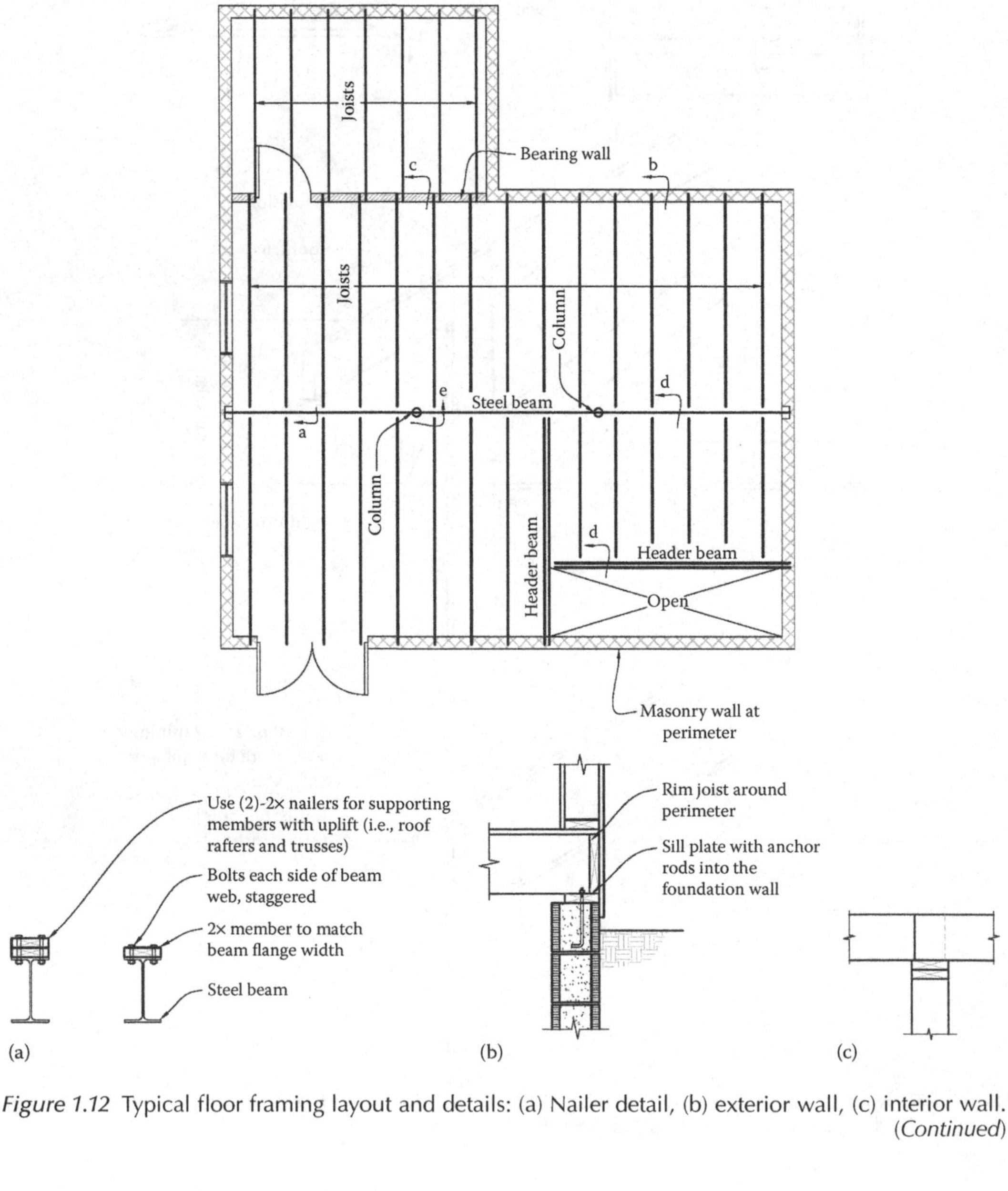

Joists
Bearing wall
c
b
Joists
Column
Steel beam
e
d
a
Column
Header beam
d
Header beam
Open
Masonry wall at perimeter
Use (2)-2× nailers for supporting members with uplift (i.e., roof rafters and trusses)
Bolts each side of beam web, staggered
2× member to match beam flange width
Steel beam
Rim joist around perimeter
Sill plate with anchor rods into the foundation wall
(a)
(b)
(c)

Figure 1.12 Typical floor framing layout and details: (a) Nailer detail, (b) exterior wall, (c) interior wall. (*Continued*)

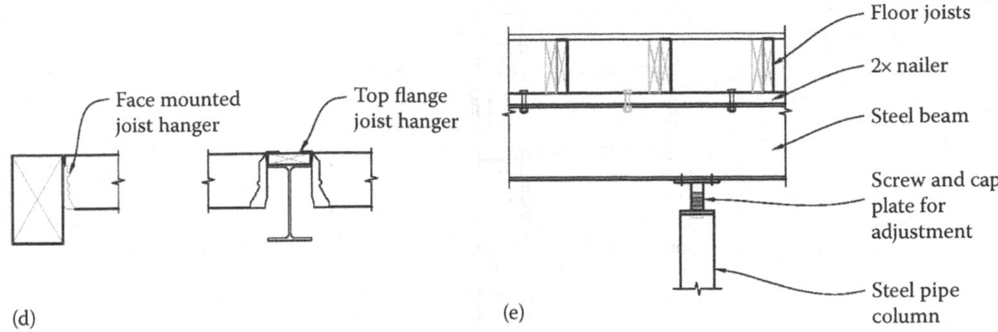

Figure 1.12 (Continued) Typical floor framing layout and details: (d) interior beam, and (e) adjustable post.

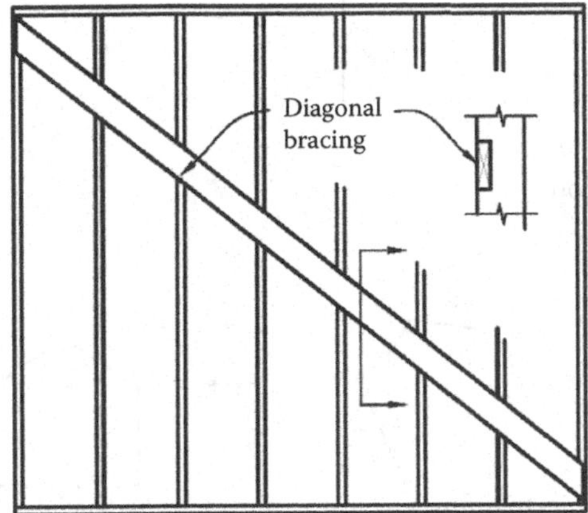

Figure 1.13 Diagonal let-in bracing.

1.3.3 Wall framing

Wall framing in wood-framed buildings consists of repetitive vertical 2 × 4 or 2 × 6 wall studs spaced at 16 or 24 in. on centers, with plywood or OSB attached to the outside face of the wall. It also consists of a top plate at the top of the wall, a sole or sill plate at the bottom of the wall, and header beams supporting loads over door and window openings. These walls support gravity loads from the roof and floor framing and resist lateral wind loads perpendicular to the face of the wall as well as acting as a shear wall to resist lateral wind or seismic loads in the plane of the wall. It may be necessary to attach sheathing to both the interior and exterior faces of the wall studs to achieve greater shear capacity in the shear wall. Occasionally, diagonal let-in bracing is used to resist lateral loads in lieu of structural sheathing, but this is not common (see Figure 1.13). A typical wall section is shown in Figure 1.14 (see also Figure 1.8).

1.3.3.1 Shear walls in wood buildings

The lateral wind and seismic forces acting on wood buildings result in sliding, overturning, and racking of a building, as illustrated in Figure 1.15. Sliding of a building is resisted by the friction between the building and the foundation walls, but in practice this friction is neglected and sill plate anchors are usually provided to resist the sliding force.

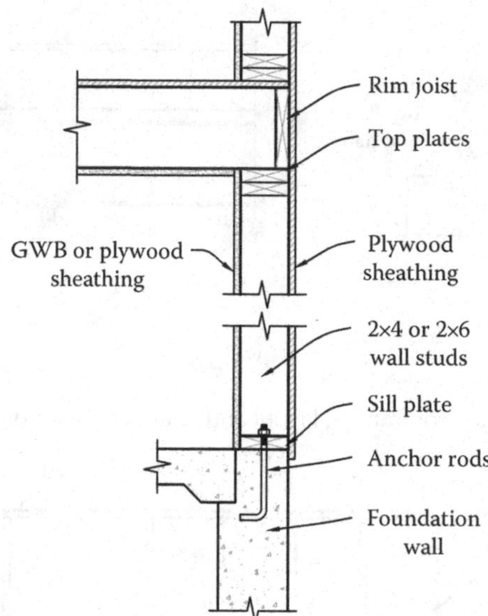

Figure 1.14 Typical wall section.

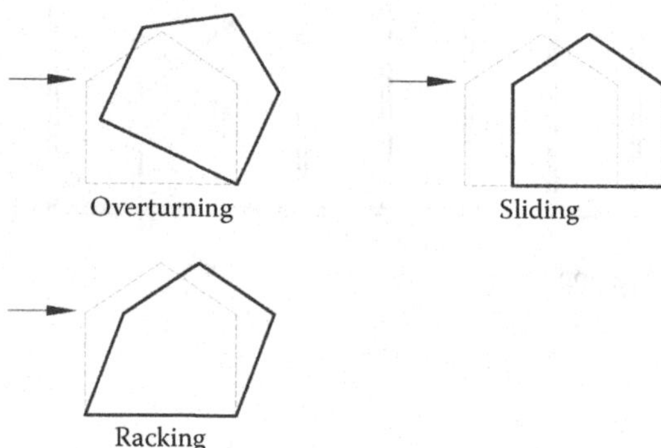

Figure 1.15 Overturning, sliding, and racking in wood buildings.

The overturning moment, which can be resolved into a downward and upward couple of forces, is resisted by the dead weight of the structure and by hold-down anchors at the end chords of the shear walls. Racking of a building is resisted by let-in diagonal braces or by plywood or OSB sheathing nailed to the wall studs acting as a shear wall.

The uplift forces due to upward vertical wind loads (or suction) on the roofs of wood buildings are resisted by the dead weight of the roof and by using toenailing or hurricane or hold-down anchors. These anchors are used to tie the roof rafters or trusses to the wall studs. The path of the uplift forces must be traced all the way down to the foundation. If a net uplift force exists in the wall studs at the ground-floor level, the sill plate anchors must be embedded deep enough into the foundation wall or grade beam to resist this uplift force, and the foundation must also be checked to ensure that it has enough dead weight, from its self-weight and the weight of soil engaged, to resist the uplift force.

1.4 WOOD STRUCTURAL PROPERTIES

Wood is a biological material and is one of the oldest structural materials in existence. It is nonhomogeneous and orthotropic, and thus its strength is affected by the direction of load relative to the direction of the grain of the wood, and it is naturally occurring and can be renewed by planting or growing new trees. Since wood is naturally occurring and nonhomogeneous, its structural properties can vary widely, and because wood is a biological material, its strength is highly dependent on environmental conditions [11]. Wood buildings have been known to be very durable, lasting hundreds of years, as evidenced by the many historic wood buildings in the United States. In this section, we discuss the properties of wood that are of importance to architects and engineers in assessing the strength of wood members and elements.

Wood fibers are composed of small, elongated, round, or rectangular tube-like cells (see Figure 1.16) with the cell walls made of cellulose, which gives the wood its load-carrying ability. The cells or fibers are oriented in the longitudinal direction of the tree log and are bound together by a glue-like material called *lignin.* The chemical composition of wood consists mainly of cellulose and lignin. The water in the cell walls is known as *bound water,* and the water in the cell cavities is known as *free water.* When wood is subjected to drying or seasoning, it loses all its free water first before it begins to lose bound water from the cell walls. It is the bound water, not the free water, that affects the shrinking or swelling of a wood member. The cells or fibers are usually oriented in the vertical direction of the tree. The strength of wood depends on the direction of the wood grain. The direction parallel to the tree trunk or longitudinal direction is referred to as the *parallel-to-grain direction*; the radial and tangential directions are both referred to as the *perpendicular-to-grain direction.*

1.4.1 Tree cross section

There are two main broad classes of trees: hardwood and softwood. This terminology is not indicative of how strong a tree is because some softwoods are actually stronger than hardwoods. *Hardwoods* have broad leaves, whereas *softwoods* have needle-like leaves and are mostly evergreen. Hardwood trees take longer to mature and grow than softwoods, are mostly tropical, and are generally more dense than softwoods. Consequently, they are more expensive and used less frequently than softwood lumber or timber in wood building construction in the United States. Examples of softwood trees include Fir, Hemlock, Pine, Redwood, and Spruce, while hardwood trees include Maple, Oak, Birch, and Basswood [12]. Softwoods constitute more than 75% of all lumber used in construction in the United States, and more than two-thirds of softwood lumber are western woods such as Douglas Fir-Larch (DF-L) and Spruce. The rest are eastern woods such as Southern Pine [13].

A typical tree cross section is shown in Figure 1.17. The growth of timber trees is indicated by an annual growth ring added each year to the outer surface of the tree trunk just

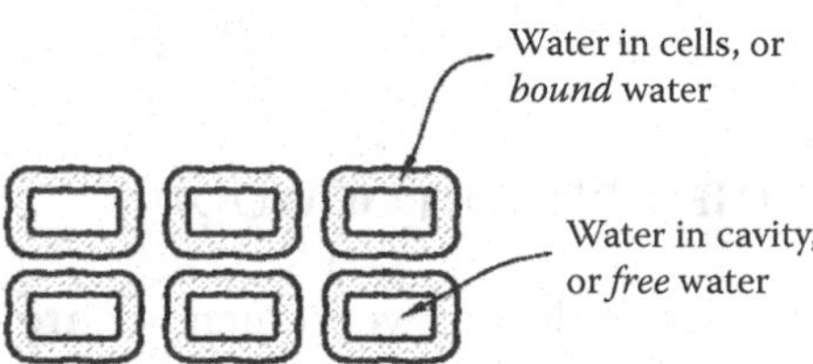

Figure 1.16 Cellular structure of wood.

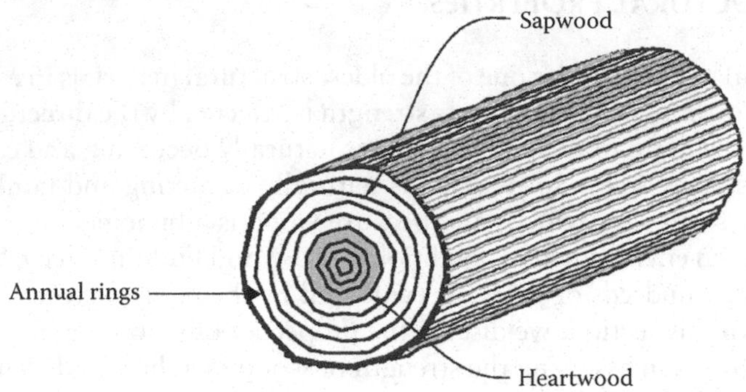

Figure 1.17 Typical tree cross section.

beneath the bark. The age of a tree can be determined from the number of annual rings in a cross section of the tree log at its base. The tree cross section shows the two main sections of the tree, the sapwood and the heartwood. *Sapwood* is light in color and may be as strong as heartwood, but it is less resistant to decay. *Heartwood* is darker and older and more resistant to decay. However, sapwood is lighter and more amenable than heartwood to pressure treatment. Heartwood is darker and functions as a mechanical support for a tree, while sapwood contains living cells for nourishment of the tree.

1.4.2 Advantages and disadvantages of wood as a structural material

Some advantages of wood as a structural material are as follows:

- Wood is renewable.
- Wood is machinable.
- Wood has a good strength-to-weight ratio.
- Wood will not rust.
- Wood is aesthetically pleasing.

The disadvantages of wood include the following:

- Wood is more combustible than other building materials.
- Wood can decay or rot and can be attacked by insects such as termites and marine borers. Moisture promotes decay and rot in wood that can lead to structural failure.
- Wood holds moisture.
- Wood is susceptible to volumetric instability (i.e., wood shrinks and warps).
- The properties of wood are highly variable and vary widely between species and even between trees of the same species. There is also variation in strength within the cross section of a tree log.

1.5 FACTORS AFFECTING THE STRENGTH OF WOOD

Several factors that affect the strength of a wood member are discussed in this section: (1) species group, (2) moisture content (MC), (3) duration of loading, (4) size and shape of the wood member, (5) defects, (6) direction of the primary stress with respect to the orientation of the wood grain, and (7) ambient temperature.

1.5.1 Species and species group

Structural softwood lumber is produced from several species of trees. Some of the species are grouped or combined together to form a *species group*, whose members are grown and harvested together, and have similar strength characteristics. There are more than 100 softwood species used in North America [14]. The NDS code's tabulated stresses for a group or combination of species (i.e., species group) were derived statistically from the results of a large number of tests to ensure that all the stresses tabulated for all species within a species group are conservative and safe. A species group is a combination of two or more species. For example, DF-L is a species group that is obtained from a combination of Douglas Fir and Western Larch species. Hem-Fir is a species group that can be obtained from a combination of Western Hemlock and White Fir. Spruce-Pine-Fir (SPF) is a species group obtained from a combination of spruce, pines, and firs commonly grown in Canada.

Structural wood members are derived from different stocks of trees, and the choice of wood species for use in design is typically a matter of economics and regional availability. For a given location, only a few species groups might be readily available. The species groups that have the highest available strengths are DF-L and Southern Pine, also called Southern Yellow Pine (SYP). Examples of widely used species groups (i.e., combinations of different wood species) of structural lumber in wood buildings include DF-L, Hem-Fir, SPF, and SYP. Each species group has a different set of tabulated design stresses in the NDS-S, and wood species within a particular species group possess similar properties and have the same grading rules.

1.5.2 Moisture content

The strength of a wood member is greatly influenced by its *moisture content*, which is defined as the percentage amount of moisture in a piece of wood. The *fiber saturation point* (FSP) is the MC at which the free water (i.e., the water in cell cavities) has been fully dissipated. Below the FSP, which is typically between 25% and 30% MC for most wood species [12], wood starts to shrink by losing water from the cell walls (i.e., the bound water). The *equilibrium moisture content* (EMC), the MC at which the moisture in a wood member has come to a balance with that in the surrounding atmosphere, is a function of the temperature and the relative humidity of the surrounding air, so the EMC will vary depending on the region and time of year. The EMC in a protected and conditioned interior environment of most wood buildings in the United States occurs typically at between 8% and 15% MC [12]. As the MC increases up to the FSP (the point where all the free water has been dissipated), the wood strength decreases, and as the MC decreases below the FSP, the wood strength increases, although this increase may be offset by some strength reduction from the shrinkage of the wood fibers. Using the oven drying method (ASTM D2016) in which samples are weighed and then oven-dried till all the moisture in the sample has dissipated, the MC of a wood member can be calculated as

$$\text{MC}(\%) = \frac{\text{weight of moist wood} - \text{weight of oven-dried wood}}{\text{weight of oven-dried wood}} \times 100$$

In practice, the oven-drying method above for calculating MC is impractical because it requires considerable time to complete the test. As an alternative, a hand-held moisture meter—though not as precise as the oven-drying method, but easy to use—can be used to measure the instantaneous MC of wood members [14].

There are two classifications of wood members based on MC: green and dry. *Green lumber* is freshly cut wood and the MC can vary from as low as 30% to as high as 200% [13]. *Dry* or

Table 1.1 Moisture content classifications for sawn lumber and glulam

	Moisture content (%)	
Lumber classification	*Sawn lumber*	*Glulam*
Dry	≤19	<16
Green	>19	≥16

seasoned lumber is wood with a MC no higher than 19% for sawn lumber and less than 16% for glulam (see Table 1.1). Wood can be seasoned by air drying or by kiln drying. Most wood members are used in dry or seasoned conditions where the wood member is protected from excessive moisture. The effect of the MC is taken into account in design by use of the moisture adjustment factor, C_M, which is discussed in Chapter 3.

1.5.2.1 Seasoning of lumber

The *seasoning* of lumber, the process of removing moisture from wood to bring the MC to an acceptable level, can be achieved through air drying or kiln drying. *Air drying* involves stacking lumber in a covered shed and allowing moisture loss or drying to take place naturally over time due to the presence of air. Fans can be used to accelerate the seasoning process. *Kiln drying* involves placing lumber pieces in an enclosure or kiln at significantly higher temperatures. During the seasoning process, the temperature has to be controlled to prevent damage to the wood members from seasoning defects such as warp, bow, twists, or crooks. Seasoned wood is recommended for building construction because of its dimensional stability. The shrinkage that occurs when unseasoned wood is used can lead to problems in the structure as the shape changes upon drying out. The amount of shrinkage in a wood member varies considerably depending on the direction of the wood grain.

1.5.3 Duration of loading

The longer the load acts on a wood member, the lower the strength of the wood member, and conversely, the shorter the load duration, the stronger the wood member. This characteristic behavior of wood has been observed in load tests of wood structural members [1]. The effect of load duration is taken into account in allowable stress design (ASD) by use of the load duration adjustment factor, C_D, and by the use of the time effect factor, λ, for load and resistance factor design (LRFD). These factors are discussed in Chapter 3.

1.5.4 Size classifications of sawn lumber

As the size of a wood member increases, the difference between the actual behavior of the member and the ideal elastic behavior assumed in deriving the design equations becomes more pronounced. For example, as the depth of a flexural member increases, the deviation from the assumed elastic properties increases and the strength of the member decreases. The various size classifications for structural sawn lumber are shown in Table 1.2, and it should be noted that for sawn lumber, the *thickness* refers to the smaller dimension of the cross section and the *width* refers to the larger dimension of the cross section. Different design stresses are given in the NDS-S for the various size classifications listed in Table 1.2.

Table 1.2 Size classifications for sawn lumber

Lumber classification	*Size*
Dimension lumber	Nominal *thickness:* 2–4 in. Nominal *width:* ≥2 in. but ≤16 in. *Examples:* 2 × 4, 2 × 6, 2 × 8, 4 × 14, 4 × 16
Beam and stringer (B&S)	Rectangular cross section Nominal *thickness:* >5 in. Nominal *width:* >2 in. + nominal thickness *Examples:* 6 × 10, 6 × 12, 8 × 14
Post and timber (P&T)	Approximately square cross section Nominal *thickness:* ≥5 in. Nominal *width:* ≤2 in. + nominal thickness *Examples:* 5 × 5, 6 × 6, 6 × 8
Decking	Nominal *thickness*: 2–4 in. Wide face applied directly in contact with framing Sometimes, tongue and grooved Used as roof or floor sheathing *Example:* 2 × 12 lumber used in a flatwise direction, 5/4 × 6, and 3 × 6

Dimension lumber, which usually comes in lengths of 8 ft. to 20 ft., is typically used for floor joists or roof rafters/trusses, and 2 × 8, 2 × 10, and 2 × 12 are the most frequently used floor joist sizes. For light-frame residential construction, dimension lumber is generally used. Beam and stringer lumber is used as floor beams or girders and as door or window headers, and post and timber lumber is used for columns or posts. Dimension lumber and many other lumber products are often supplied in length increments of 2 ft. (e.g., 10 ft., 12 ft.).

1.5.4.1 Nominal dimension versus actual size of sawn lumber

Wood members can come in dressed or undressed sizes, but most wood structural members come in dressed form. When wood is dressed on two sides, it is denoted as S2S; wood that is dressed on all four sides is denoted as S4S. Full sawn 2 × 6 lumber has an actual or nominal size of 2 × 6 in., whereas the dressed size is $1\frac{1}{2} \times 5\frac{1}{2}$ in. (Figure 1.18). The lumber size is usually called out on structural and architectural drawings using the nominal dimensions of the lumber. The reader is reminded that for a sawn lumber cross section, the thickness is the smaller dimension and the width is the larger dimension of the cross section. In this book, we assume that all wood is dressed on four sides (i.e., S4S). Section properties for wood members are given in Tables 1A and 1B of the NDS-S [2].

1.5.5 Wood defects

The various categories of defects in wood are natural and conversion defects. The nature, size, and location of defects affect the strength of a wood member because of the stress concentrations that they induce in the member. They also affect the finished appearance of the member. Some examples of natural defects are knots, shakes, splits, and fungal decay. Conversion defects occur due to unsound milling practices. Seasoning defects result from the

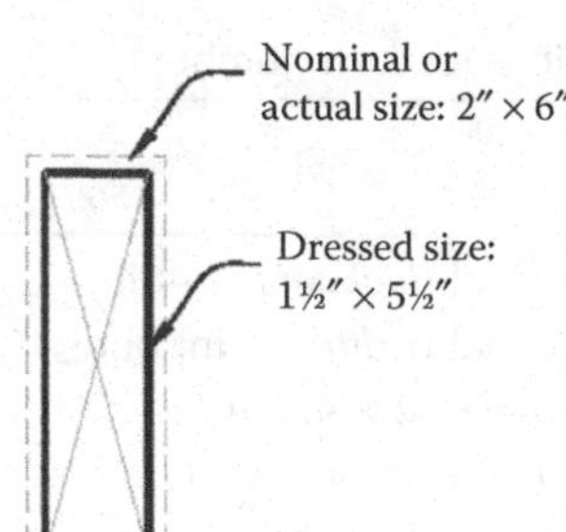

Figure 1.18 Nominal versus dressed size of a 2 × 6 sawn lumber.

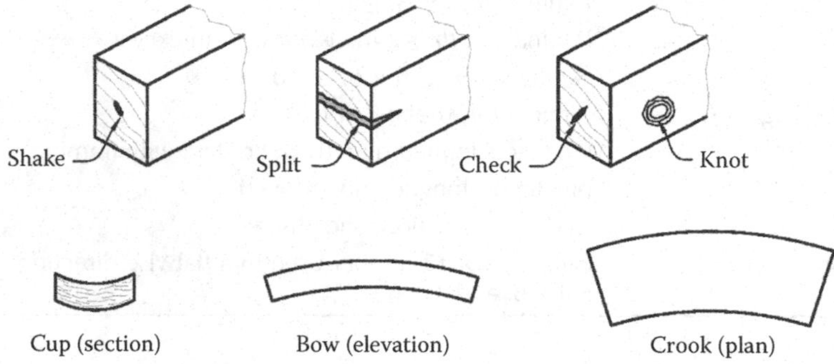

Figure 1.19 Common defects in wood.

effect of uneven or unequal drying shrinkage, examples being various types of warps, such as cupping, bowing, crooking, and twisting [13–18]. The most common types of defects in wood members are illustrated in Figure 1.19 and include the following:

- *Knots*: These are natural defects formed where the tree limbs grow out from a tree trunk. Of all the strength limiting defects in wood, knots are the most common and most significant. The strength reduction from knots arises from the following factors: a reduction in the effective size of a wood member due to the presence of the knot, stress concentrations at the knot location that lead to a reduction in capacity of the wood member due to the nonhomogeneity of the wood member around the knot; and distortion of the angle of grain of the tree trunk at the knot location due to the disrupted growth pattern of the trunk around the knot that can lead to tensile stresses perpendicular to the grain or cross-grain tension—a stress situation that should be avoided in wood members. The location of the knot along the length and across the depth of a wood member also impacts its strength. For a flexural wood member, a knot at the mid-depth (neutral axis) at mid-span is not as critical as a knot at mid-depth near the beam support. For an axially loaded wood member, a knot at mid-depth at the mid-length of the member can be critical. A knot at mid-span near the tension or compression face of a beam is critical for the bending strength of the beam [19].
- *Check*: This is the separation of the wood fibers at an angle to annual rings and occurs due to rapid drying of lumber where the interior parts of the member shrink less than the exposed outer parts, and therefore restrain the other parts of the wood member which causes the wood fibers to separate [14]. Checks do not usually penetrate the full width or thickness of the wood member.

- *Split*: This is the separation of wood fibers that occur at the end of a wood member, and extending completely through the width or thickness (i.e., the smaller cross-sectional dimension) of the member parallel to the direction of the fibers. Splits, depending on their lengths, typically affect the shear strength of beams and the buckling capacity of columns.
- *Shake*: This occurs due to separation of the wood fibers parallel to the annual rings.
- *Decay*: This is the rotting or crumbling of wood due to the presence of moisture and wood-destroying fungi that feeds on the wood fibers or cellulose. The effects of decay may be present and yet not visible on the external face of a wood member. For fungal decay to occur, the following must be present: oxygen, adequate moisture, and moderate temperatures. Fungal decay leads to a drastic reduction in the effective cross-sectional size of the member, and the progression of strength loss due to wood decay can be very rapid [20]. Decay of wood will not occur if the wood member is submerged in water and the MC is consistently below 20%, and decay will also not occur if the wood member is maintained at below freezing temperatures or well above 100°F [9].

  A fatal collapse of a wood-supported cantilevered balcony in Berkeley, California, on June 16, 2015, that claimed six lives with several others injured has been attributed to "dry rot" or fungal decay caused by moisture infiltration into the supporting cantilever wood beams near their supports; the rotting of the wood significantly reduced the capacity of the cantilevered beams leading to the collapse [21]. The use of pressure-treated wood is recommended for wood members that may be exposed to or subjected to excessive moisture such as wood-framing members in swimming pools, exterior unprotected wood members, or wood that will be in direct contact with concrete and masonry such as sill plates [9].

  Exterior wood-supported balconies and exterior balustrades as well as any exterior structural members should be properly detailed to ensure that water infiltration into these members or their supports do not occur. To prevent failures of exposed exterior structural elements and exterior structural elements that lack redundancy and are concealed by architectural finishes such the 2015 balcony collapse in Berkeley, California [10,21], the following is recommended:

  - For new buildings, use pressure-treated wood for exposed exterior structural elements and exterior structural elements that lack redundancy and are concealed by architectural finishes.
  - Enhanced coordination between the architectural and structural drawings for concealed exterior structural elements to ensure that proper waterproofing and adequate drainage are provided for these structural elements [10]. The amount of slope provided for drainage in an exterior balcony by the architect should take into account the possible upward deflection of the outer edges of the cantilevered framing members under pattern live loading as discussed in Section 1.2.
  - For existing buildings, conduct periodic inspections on exposed exterior structural elements and exterior structural elements that lack redundancy and are concealed with architectural finishes.
- *Wane*: In this defect, the corners or edges of a wood cross section lack wood material or have some of the bark of the tree as part of the cross section. This leads to a reduction in the cross-sectional area of the member which affects the structural capacity of the member.

Defects in wood members lead to a reduction in the net cross section, and their presence introduces stress concentrations in the wood member. The amount of strength reduction depends on the size and location of the defect. For example, for an axially loaded tension member, a knot anywhere in the cross section would reduce the tension capacity of the member. On the other hand, a knot at the neutral axis of the beam would not affect the bending strength but may affect the shear strength if it is located near the supports. For visually graded lumber, the grade stamp, which indicates the design stress grade assigned by the grading inspector, takes into account the number and location of defects in that member.

Other types of defects include *warping* and *compression* or *reaction wood*:

Warping results from uneven or differential drying shrinkage of wood between the radial and tangential to grain directions, causing the wood member to deviate from the horizontal or vertical plane. Examples of warping include members with a bow, cup, or crook. This defect does not affect the strength of the wood member but affects the constructability of the member. For example, if a bowed member is used as a joist or beam, there will be an initial sag or deflection in the member, depending on how it is oriented. This could affect the construction of the floor or roof in which it is used. Members that are excessively warped should not be used in wood framing.

Compression or reaction wood is caused by a tree that grows abnormally in bent shape due either to natural effects or to the effect of wind and snow loads. In a leaning tree trunk, one side of the tree cross section is subject to increased compression stresses due to the crookedness of the tree. Consequently, compression or reaction wood has nonuniform structural properties and should not be used for structural members.

1.5.6 Orientation of the wood grain

Wood is an orthotropic material with strengths that vary depending on the direction of the stress applied relative to the grain of the wood. As a result of the tubular nature of wood, three independent directions are present in a wood member: longitudinal, radial, and tangential. The variation in strength in a wood member with the direction of loading can be illustrated by a group of drinking straws glued tightly together. The group of straws will be strongest when the load is applied parallel to the length of the straws (i.e., longitudinal direction); loads applied in any other direction (i.e., radial or tangential) will crush the walls of the straws or pull apart the glue. The longitudinal direction is referred to as the *parallel-to-grain direction*, and the tangential and radial directions are both referred to as the *perpendicular-to-grain direction*. Thus, wood is strongest when the load or stress is applied in a direction parallel to the direction of the wood grain, is weakest when the stress is perpendicular to the direction of the wood grain, and has the least amount of shrinkage in the longitudinal or parallel-to-grain direction. The various axes in a wood member with respect to the grain direction are shown in Figure 1.20.

Axial or bending stress parallel to the grain: This is the strongest direction for a wood member, and examples of stresses and loads acting in this direction are illustrated in Figure 1.21a.

Axial stress perpendicular to the grain: The strength of wood in compression parallel to the grain is usually stronger than wood in compression perpendicular to the grain (see Figure 1.21b). Wood has very low strength in tension perpendicular to the grain (i.e., cross-grain tension) since only the lignin or glue is available to resist this tension force. Consequently, the NDS code does not address the loading of wood in tension perpendicular to the grain.

Stress at an angle to the grain: This case lies between the parallel-to-grain and perpendicular-to-grain directions and is illustrated in Figure 1.21c.

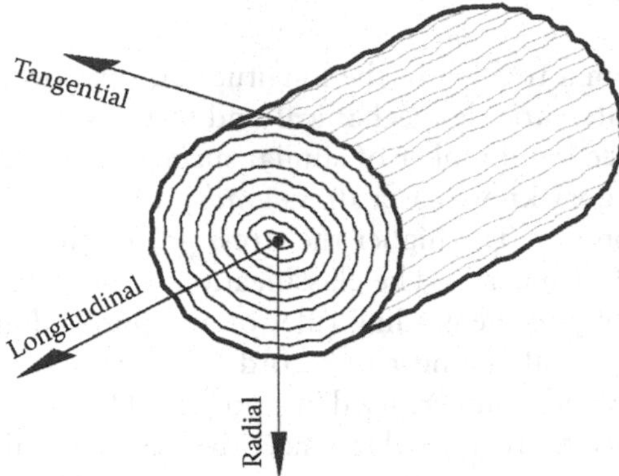

Figure 1.20 Longitudinal, radial, and tangential axes in a wood member.

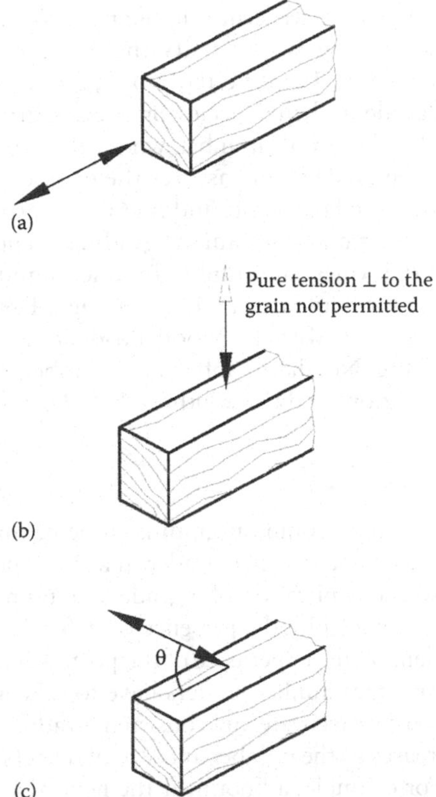

Figure 1.21 Stress applied (a) parallel to the grain, (b) perpendicular to the grain, and (c) at an angle to the grain in a wood member.

1.5.7 Ambient temperature

Wood is affected adversely by sustained temperature beyond 100°F. As the ambient temperature rises beyond 100°F, the strength of the wood member decreases. The decrease in strength is more pronounced in wood members that are subjected to both sustained high temperatures above 100°F and high MC [9]. The structural members in most insulated wood buildings have ambient temperatures of less than 100°F.

1.6 LUMBER GRADING

Lumber is usually cut from a tree log in the longitudinal direction, and because it is naturally occurring, it has quite variable mechanical and structural properties, even for members cut from the same tree log. Lumber of similar mechanical and structural properties is grouped into a single category known as a *stress grade*. This simplifies the lumber selection process and increases economy. The higher the stress grade, the stronger and, often times, more expensive the wood member is. The classification of lumber with regard to strength, usage, and defects according to the grading rules of an approved grading agency is termed *lumber grading*. The structural engineer of record (SER) is not required to carry out the grading of the structural wood members used in a project. The SER is charged with selecting wood members of appropriate stress grade to meet the design requirements.

1.6.1 Types of grading

The two types of grading systems for structural lumber are visual grading and mechanical grading. The intent of visual grading is to classify the wood members into various stress grades such as Select Structural, No. 1 and Better, No. 1, No. 2, Utility, and so on. A grade stamp indicating the stress grade and the species or species group is placed on the wood member, in addition to the MC, the mill number where the wood was produced, and the responsible grading agency. The grade stamp assures the engineer, architect, and contractor of the quality of the lumber delivered to the site and that it conforms to the contract specifications for the project. Grading rules may vary among grading agencies, but minimum grading requirements are set forth in the American Lumber Product Standard US DOC PS-20 developed by the National Institute for Standards and Technology. Examples of grading agencies in the United States [2] include the Western Wood Products Association, the West Coast Lumber Inspection Bureau, the Northern Softwood Lumber Bureau, the Northeastern Lumber Manufacturers Association, and the Southern Pine Inspection Bureau.

1.6.1.1 Visual grading

Visual grading, the oldest and most common grading system, involves visual inspection of wood members by an experienced and certified grader in accordance with established grading rules of a grading agency, and the application of a grade stamp on a wood member. In visual grading, the lumber quality is reduced by the presence of defects, and the effectiveness of the grading system is very dependent on the experience of the professional grader. Grading agencies usually have certification exams that lumber graders have to take and pass annually to maintain their certification and to ensure accurate and consistent grading of sawn lumber. The stress grade of a wood member decreases as the number and size of defects increases and as their locations become more critical. For example, a knot near the neutral axis of a wood member that will be subjected to bending will not be as critical as a knot near the tension face of the member.

1.6.1.2 Mechanical grading

Mechanical grading is a nondestructive grading system that is based on the relationship between the stiffness and strength of wood members. There are two types of machine-graded lumber used in North America [22]: machine stress-rated (MSR) lumber and machine-evaluated lumber (MEL). The mechanical grading process for MSR lumber and MEL involves the use of a "mechanical stress-rating equipment" to obtain measurements of deflections of the sawn lumber from which the stiffness is calculated. From the strength information obtained, a strength grade is assigned [14]. In addition to the nondestructive test, each piece of wood is also subjected to a visual check. The variability of MEL is higher than that of MSR lumber. The grade stamp on mechanical graded lumber includes the value of the tabulated bending stress, F_b, for single member usage and the modulus of elasticity, E, and will also include the designation "machined rated" or "MSR." The grade stamp on MEL includes a strength grade designation "M" followed by a number (e.g., M10 or M24) and includes the tabulated bending stress, the tabulated tension stress, and the modulus of elasticity. Because of the lower variability of material properties for MSR lumber, it is used in the fabrication of engineered wood products such as glulam and preengineered roof and floor trusses. A nondestructive X-ray inspection technique that measures density, in addition to a visual check, can also be used for mechanical grading of MSR lumber and MEL.

1.6.2 Stress grades

The various lumber stress grades are listed below in order of decreasing strength.

- Dense Select Structural
- Select Structural
- No. 1 and Better
- No. 1
- No. 2
- No. 3
- Stud
- Construction
- Standard
- Utility

1.6.3 Grade stamps

The use of a grade stamp on lumber assures the contractor and the engineer of record that the lumber supplied conforms to that specified in the contract documents. Lumber without a grade stamp should not be allowed on site or used in a project. A typical grading stamp on lumber might include the items shown in Figure 1.22.

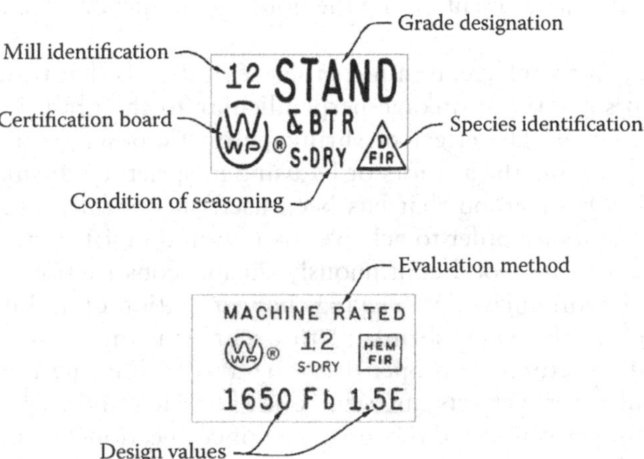

Figure 1.22 Typical grade stamp. (Courtesy of Western Woods Products Association, Portland, OR.)

1.7 SHRINKAGE OF WOOD

Shrinkage in a wood member takes place as moisture is dissipated from the member below the FSP. Wood shrinks as the MC decreases from its value at the installation of the member to the EMC, which can be as low as 8%–15% in some protected environments. Shrinkage parallel to the grain of a wood member is negligible and much less than shrinkage perpendicular to the grain or tangential to the grain. In wood members, the tangential shrinkage is on average approximately 1.5–2 times the radial shrinkage and approximately 40 times the longitudinal shrinkage [14].

Shrinkage effects in lumber can be minimized by using seasoned lumber or lumber with MC of 19% or less (KD19); KD15 (i.e., lumber with MC of 15% or less) is not always readily available.

Differential or nonuniform shrinkage is usually more critical than uniform shrinkage in wood structures. Examples of situations where differential shrinkage could result in structural distress include the following:

1. In a multistory wood building with brick cladding, the wood framing shrinks in the vertical direction (i.e., along the height of the building), whereas the brick cladding will tend to expand vertically as it absorbs water. Therefore, the connections between the brick cladding and the wood structural framing should be properly detailed to avoid inadvertently transferring structural loading to the nonstructural brick cladding due to the differential shrinkage.
2. A floor joist supported on a concrete or masonry wall at one end and a wood floor girder at the other end could be subjected to differential or nonuniform shrinkage that could cause distress in deflection-sensitive flooring like marble and terrazzo [14].
3. For wood members with two or more rows of bolts perpendicular to the direction of the wood grain, shrinkage across the width (i.e., depth) of the member causes tension stresses perpendicular to the grain (i.e., cross-grain tension) in the wood member between the bolt holes, which could lead to the splitting of the wood member parallel to the grain at the bolt location [23]. Therefore, the connection of beams to columns with bolts is less preferred than seated or saddle connections.
4. Shrinkage can adversely affect the functioning of hold-down anchors in shear walls by causing a gap between the anchor nut and the top of sill plate due to the vertical shrinkage. As a result, the shear wall (and therefore the building) will have to undergo excessive lateral displacement before the hold-down anchors can be engaged.

To reduce the effects of shrinkage, minimize the use of details that transfer loads perpendicular to the grain since the shrinkage perpendicular to the grain is much larger than that parallel to the grain. The effect of shrinkage on tie-down anchor systems can be minimized by pretensioning the anchors or by using proprietary shrinkage compensating anchor devices [24]. One method that has been used successfully to control the MC in wood during construction in order to achieve the required moisture threshold is by using portable heaters to dry the wood continuously during construction [25]. The effect of shrinkage can also be minimized by delaying the installation of architectural finishes to allow time for much of the wood shrinkage to occur. It is important to control shrinkage effects in wood structures by proper detailing and by limiting the change in MC of the member to avoid adverse effects on architectural finishes and to prevent the excessive lateral deflection of shear walls, and the loosening of connections or the splitting of wood members at connections.

Table 1.3 Shrinkage parameters

Wood species	*Width*		*Thickness*	
	a	b	a	b
Redwood, Western red cedar, Northern white cedar	3.454	0.157	2.816	0.128
Other	6.031	0.215	5.062	0.181

Source: ASTM, *Standard Practice for Establishing Allowable Properties for Visually-Graded Dimension Lumber from In-Grade Tests of Full-Size Specimens*, ASTM D 1990, ASTM International, West Conshohocken, PA, 1990.

The amount of shrinkage across the width or thickness of a wood member or element (i.e., perpendicular to the grain or to the longitudinal direction) is highly variable, but can be estimated using the following equation (adapted from ASTM D1990 [26]):

$$d_2 = d_1\left[\frac{1-(a-bM_2)/100}{1-(a-bM_1)/100}\right] \tag{1.1}$$

where:
d_1 is the initial member thickness or width at the initial MC M_1, in.
d_2 is the final member thickness or width at the final MC M_2, in.
M_1 is the MC at dimension d_1, %
M_2 is the MC at dimension d_2, %

The variables a and b are obtained from Table 1.3. The total shrinkage of a wood building detail or section is the sum of the shrinkage *perpendicular to the grain* of each wood member or element in that detail or section; longitudinal shrinkage or the shrinkage *parallel to the grain* is negligible.

1.8 DENSITY OF WOOD

The density of wood is a function of the MC of the wood and the weight of the wood substance or cellulose present in a unit volume of wood. Even though the cellulose–lignin combination in wood has a specific gravity of approximately 1.50 and is heavier than water, most wood used in construction floats because of the presence of cavities in the hollow cells of a wood member. The density of wood can vary widely between species, from as low as 20 lb/ft^3 to as high as 65 lb/ft^3 [2,13] for the most commonly used species, and is also dependent on the MC of the member; the higher the density, the higher the strength of the wood member. The density of wood up to 30% MC can be obtained from Equation 1.2 [9] as

$$\gamma_{\text{wood}} = \frac{\text{SG}_0}{1+(0.265)\text{SG}_0(\text{MC}/30\%)}\left(1+\frac{\text{MC}}{100\%}\right)(62.4\ \text{lb/ft}^3) \tag{1.2}$$

where:
SG_0 is the specific gravity of oven-dried wood (i.e., at MC = 0%)
γ_{wood} is the specific gravity of wood at the specified MC (lb/ft^3)
MC is the specified moisture content between 0% and 30%

For MC above 30%, the density of wood is calculated from Equation 1.3 as

$$\gamma_{wood} = \frac{1.3\ SG_0}{1+(0.265)SG_0}(62.4\ \text{lb/ft}^3)\left(1+\frac{MC-30\%}{30\%}\right) \tag{1.3}$$

The density of wood at a MC of 12% is typically used in design to calculate the self-weight of the wood member and dead loads [9]. In this book, a wood density of 32 lb/ft^3 is used for seasoned wood, unless otherwise noted.

1.9 UNITS OF MEASUREMENT

The United States system of units is used in this book, and accuracy to at most three significant figures is maintained in all the example problems. The standard unit of measurement for lumber in the United States is the *board foot* (bf), which is defined as the volume of 144 in.^{3} of lumber using nominal dimensions. The *Engineering News-Record*, the construction industry leading magazine, publishes the prevailing cost of lumber in the United States and Canada in units of 1000 board feet (Mbf). For example, 2 × 6 lumber that is 18 ft long is equivalent to 18 board feet or 0.018 Mbf. That is,

$$\frac{(2\ \text{in.})(6\ \text{in.})(18\ \text{ft}\times 12)}{144\ \text{in.}^3} = 18\ \text{bf or } 0.018\ \text{Mbf}$$

Example 1.1: Shrinkage in wood members

Determine the total shrinkage across (a) the width and (b) the thickness of two green 2 × 6 DF-L top plates loaded perpendicular to the grain as the MC decreases from an initial value of 30% to a final value of 15%.

Solution: For 2 × 6 sawn lumber, the actual width $d_1 = 5.5$ in. and the actual thickness = 1.5 in. The initial MC and the final EMC are $M_1 = 30$ and $M_2 = 15$, respectively.

a. *Shrinkage across the width of the two 2 × 6 top plates.* For shrinkage across the width of the top plate, the shrinkage parameters from Table 1.3 are obtained as follows:

$$a = 6.031$$

$$b = 0.215$$

From Equation 1.1, the final width d_2 is given as

$$d_2 = 5.5\left[\frac{1-[6.031-(0.215)(15)]/100}{1-[6.031-(0.215)(30)]/100}\right] = 5.32\ \text{in.}$$

Thus, the total shrinkage across the width of the two top plates = $d_1 - d_2$ = 5.5 in. − 5.32 in. = **0.18 in.**

b. *Shrinkage across the thickness of the two 2 × 6 top plates.* For shrinkage across the thickness of the top plate, the shrinkage parameters from Table 1.3 are as follows:

$$a = 5.062$$

$$b = 0.181$$

The final thickness d_2 of each top plate from Equation 1.1 is given as

$$d_2 = 1.5\left[\frac{1-[5.062-(0.181)(15)]/100}{1-[5.062-(0.181)(30)]/100}\right] = 1.46 \text{ in.}$$

The total shrinkage across the *thickness* of the two top plates will be the sum of shrinkage in each of the individual wood members:

2 top plates × $(d_1 - d_2) = (2)(1.5 \text{ in.} - 1.46 \text{ in.}) = \mathbf{0.08 \text{ in.}}$

Example 1.2: Shrinkage at framed floors

Determine the total shrinkage at each floor level for the typical wall section shown in Figure 1.23 assuming Hem-Fir wood species, and the MC decreases from an initial value of 19% to a final value of 10%. How much gap should be provided in the plywood wall sheathing to allow for shrinkage? Assume 2 × 6 sill and sole plates and 2 × 12 continuous rim joist.

Solution: For a 2 × 6 sawn lumber, the actual thickness = 1.5 in. (smaller dimension)

For a 2 × 12 sawn lumber, the actual width, d_1 = 11.25 in. (larger dimension)
The initial and final MC are $M_1 = 19$ and $M_2 = 10$.

a. *Shrinkage across the width of the 2 × 12 continuous rim joist.* The shrinkage parameters from Table 1.3 for shrinkage across the width of 2 × 12 are as follows:

$a = 6.031$

$b = 0.215$

From Equation 1.1, the final width d_2 is given as

$$d_2 = 11.25\left[\frac{1-[6.031-(0.215)(10)]/100}{1-[6.031-(0.215)(19)]/100}\right] = 11.03 \text{ in.}$$

Thus, the total shrinkage across the width of 2 × 12 is

$d_1 - d_2 = 11.25 \text{ in.} - 11.03 \text{ in.} = 0.22 \text{ in.}$

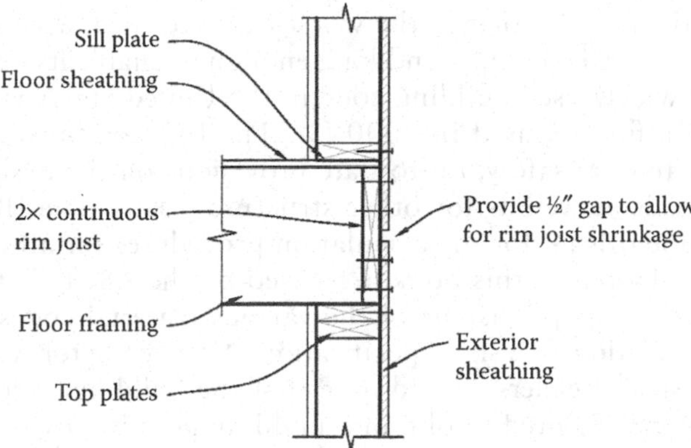

Figure 1.23 Wood shrinkage at a framed floor.

b. *Shrinkage across the thickness of the two 2 × 6 top plates and one 2 × 6 sill plate.* The shrinkage parameters from Table 1.3 for shrinkage across the thickness of the 2 × 6 plates are as follows:

$$a = 5.062$$

$$b = 0.181$$

The final thickness d_2 of each plate from Equation 1.1 is given as

$$d_2 = 1.5\left[\frac{1-[5.062-(0.181)(10)]/100}{1-[5.062-(0.181)(19)]/100}\right] = 1.475 \text{ in.}$$

The total shrinkage across the *thickness* of the two top plates and one sill plate will be the sum of the shrinkage in each of the individual wood member calculated as

$$3 \text{ plates} \times (d_1 - d_2) = 3(1.5 \text{ in.} - 1.475 \text{ in.}) = 0.075 \text{ in.}$$

Longitudinal shrinkage or shrinkage parallel to the grain in the 2 × 6 studs is negligible. Therefore, the total shrinkage per floor, which is the sum of the shrinkage of all the wood members at the floor level, is

0.075 in. + 0.22 in. = **0.3 in.** Therefore, use the ½ in. shrinkage gap.

An adequate shrinkage gap, typically about ½ in. deep, is provided in the plywood sheathing at each floor level to prevent buckling of the sheathing panels due to shrinkage. It should also be noted that for multistory wood buildings, the effects of shrinkage are even more pronounced and critical. For example, a five-story building with a typical detail as shown in Figure 1.23 will have a total accumulated vertical shrinkage of approximately five times the value calculated above!

1.10 BUILDING CODES

A *building code* is a minimum set of regulations adopted by a city or state that governs the design of building structures in that jurisdiction. The primary purpose of a building code is safety, and the intent is that in the worst-case scenario, even though a building is damaged beyond repair, it should stand long enough to enable its occupants to escape to safety. The most widely used building code in the United States is the *International Building Code* (IBC), first released in 2000 [3]. The IBC contains, among such other things as plumbing and fire safety, up-to-date provisions on the design procedures for wind and seismic loads as well as for other structural loads. The IBC now references the ASCE 7 load standards [4] for the calculation procedures for all types of structural loads. The load calculations in this book are based on the ASCE 7 standards. In addition, the IBC references the provisions of the various material codes, such as the ACI 318 for concrete, the National Design Specification (NDS) code for wood, and the AISC code for structural steel. Readers should note that the building code establishes minimum standards that are required to obtain a building permit. Owners of buildings are allowed to exceed these minimum standards if they desire, but this may increase the cost of the building.

1.11 NDS CODE AND NDS SUPPLEMENT

The primary design code for the design of wood structures in United States is the *National Design Specification* (NDS) *for Wood Construction* [1] published by the American Wood Council (AWC), in addition to the *NDS Supplement* (or NDS-S) [2], which consist of the tables listed in Table 1.4. These NDS-S tables provide reference design values (stresses and modulus of elasticity) for the various stress grades of a wood member obtained from full-scale tests on thousands of wood specimens. The NDS code was first published in 1944 and the latest (2015) edition of the NDS and NDS-S are used in this text. It should be noted that the tabulated reference design stresses are not necessarily the allowable stresses or ultimate stresses. In Chapter 3, we discuss how the reference design values are used in both the ASD method and the LRFD method to obtain the allowable stresses (ASD) and the ultimate stresses (LRFD), respectively.

1.12 STUDENT DESIGN PROJECT PROBLEM

In this section, we introduce a building design project for the student to work on throughout a typical wood design course or capstone project. Project problems are assigned to correspond with the chapter topics in this text to help the student maintain progress as each topic is covered. Several design options are indicated to be optional to allow elements of this project to be expanded at the instructor's discretion or to be used as a capstone project. Some items are not completely explained to allow the student to work through a typical design process where not all design information or assumptions are fully understood.

Overview: A wood-framed, two-story building is described below and in the given drawings. The building is located in Buffalo, New York or in an alternate location selected by the instructor. All main structural members are of wood. The National Design Specification for Wood Construction and the ASCE 7 load standard should be used.

Table 1.4 Use of NDS-S reference design values tables

NDS-S table	*Applicability*
4A	Visually graded dimension lumber (all species except Southern Pine)
4B	Visually graded Southern Pine dimension lumber
4C	Machine or mechanically graded dimension lumber
4D	Visually graded timbers (5 in. × 5 in. and larger)
4E	Visually graded decking
4F	Non–North American visually graded dimension lumber
5A	Structural glued-laminated *softwood* timber (members stressed primarily in *bending*)
5A–Expanded	Structural glued-laminated *softwood* timber combinations (members stressed primarily in *bending*)
5B	Structural glued-laminated *softwood* timber (members stressed primarily in *axial tension* or *compression*)
5C	Structural glued-laminated *hardwood* timber (members stressed primarily in *bending*)
5D	Structural glued-laminated *hardwood* timber (members stressed primarily in *axial tension* or *compression*)
6A	Treated round timber piles graded per ASTM D25
6B	Round timber piles graded per ASTM D3200

- Plan dimensions: 30'-0" × 46'-0"
- Floor-to-floor height: 11'-0"
    - Ground floor is a slab on grade, 4 in. thick
- Exterior deck: 16'-0" × 27'-0"
- Covered porch: 12'-0" × 16'-0"
- Exterior walls: 2 × 6 studs with 5/8 in. gypsum board on the inside and 5/8 in. wood shingles on plywood sheathing on the outside
- Assume all interior walls are 2 × 4 construction
- Ceilings: 5/8 in. gypsum wall board
- Second floor finish: 1 in. thick gypcrete on plywood
- Roof trusses do not bear on any interior walls
- Roofing: Asphalt shingles on plywood sheathing
- Shear walls: One on each exterior wall of the building
- Exterior deck: 5/4 in. decking boards
    - Assume wood for the covered porch is protected from the elements
    - Assume interior stair and platform bears on the first-floor slab

Material assumptions:

- DF-L for all sawn-lumber members
- Hem-Fir for the glulam girder
- Southern Pine for wood framing supporting the exterior deck
- 2600F-1.9E for any engineered lumber
    - Members labeled "girder" are to be designed as engineered lumber or multiple ply dimension lumber
    - Concrete: 28-day strength is 4000 psi
    - Masonry: $f'_m = 1500$ psi
    - Soil: 3000 psf (allowable bearing), 400 psf/ft (allowable lateral bearing)

Loading:

- Residential occupancy, regular importance
- Normal temperature and moisture conditions
- Exposure C for wind and snow loads
- Soil site class: D (seismic)
    - Handrails: Per code
    - Partitions: Use 15 psf in the floor framing design
    - Wind, seismic, and snow: Use base design values for Buffalo, NY or location selected by the instructor
    - Deflection criteria: Use a limit of $L/360$ for live loads and $L/240$ for total loads

Drawings:

May be any appropriate size. Plans should be drawn to a scale of 1/8 in. = 1'-0" and details should mainly be drawn to a scale of 3/4 in. = 1'-0". Drawings should include the following:

- *Foundation plan*: Illustrate footings, walls, and slabs; show columns and framing members that bear at the foundation/ground level.
- *Main roof plan and porch roof*: Roof framing members including ladder framing, headers, plywood pattern and fastening, roofing material.
- *Floor plan*: Floor framing members, headers, and plywood pattern and fastening, wood-column layout and glulam beam.
- *Deck plan*: Floor framing and decking, wood-column layout.

- *Transverse sections*: Indicate the various components such as roofing, roof sheathing, roof trusses, ceiling insulation, gypsum board, siding, studs, wall insulation, floor sheathing, floor joists, glulam beam, wood columns, foundation walls, finish grade outside the walls, slab-on grade, and footings.
- *Elevations*: Show all openings, stud-wall framing, and siding. Show wall sheathing attachment.
- *Trusses*: Show line diagram with member forces. Show approximate staple plate locations.
- *Shear walls*: Show applied forces and hold-down anchors and straps; show critical framing: Studs at chords, foundation connection, and sheathing attachment.
- *Specific details*:
    - Typical building wall section from foundation to roof, transverse and longitudinal walls
    - Truss to top plate (show hold-down anchor)
    - Porch: Rafter to girder and rafter to ridge
    - Joist to girder (show joist hanger)
    - Joist to bearing wall
    - Girder to column
    - Column base (interior and exterior)
    - Top plates: Lap splice connection
    - Deck: Girder to cap plate
    - Section through exterior deck handrail showing failing and post connections
    - Stair treads and stringers
    - Stringer and treads (exterior)
    - Hold-down strap (second level to first level)
    - Hold-down anchor (base of shear wall)

Members to design:

Sheathing for floor, roof, and walls: Initially, select sheathing grade, minimum thickness, span rating, and edge support requirements for vertical loads based on the IBC tables. Then check this plywood grade and thickness when designing the roof and floor diaphragms and the shear walls for lateral loads.

Roof truss: Design for the case of snow load plus dead load; use a computer analysis program for the truss analysis and check the results by a hand calculation method. Consider unbalanced loading.

Covered porch framing: Rafters, girders, and posts: Design for dead and snow loads, including snow drift and unbalanced snow loading.

Stud walls: Design for two load cases on the exterior walls:

a. Dead load plus floor live load plus snow load using the appropriate load factors for ASD or LRFD.
b. Dead load plus floor live load plus snow load plus wind load using the appropriate load factors for ASD or LRFD.

On the typical interior bearing wall, check dead plus live load using the appropriate load factors for ASD or LRFD.

Floor joists, girders, and columns: Design for floor live load plus dead load using the appropriate load factors for ASD or LRFD.

Deck framing: Joists, girders, decking, and posts: Design for dead and live loads; girder may be designed as either simple span or continuous spans; design the handrail posts, top and bottom rails, and balusters per code required loading; stringers and treads: Design for code required uniform and concentrated loads.

Roof diaphragm: Select the nailing and edge support requirements. Calculate the chord and drag strut forces and design the roof diaphragm chords and drag struts.

Floor diaphragm: Select the nailing and edge support requirements. Calculate the chord and drag strut forces and design the floor diaphragm chords and drag struts.
Shear walls: Analyze and design the most critical panel in each direction, checking the tension and compression chords; indicate whether hold-down devices are needed, and specify the vertical diaphragm nailing requirements. Note the location of the shear walls to be used.
Lintel/headers: Design lintels and headers at door and window openings for gravity loads from above. (i.e., dead plus snow loads for lintels/headers below the roof, and dead plus floor live load for lintels below the second floor).
Foundations: Design perimeter and interior wall footings for vertical loads; design exterior footings for vertical and lateral loads.

Cost estimate:
Provide a cost estimate of the main wood structural members using the table below as a guide.

Item	*Total cost* ($)
Wood columns (interior/exterior)	__________
Glulam beam	__________
Floor joists (interior)	__________
Floor joists (exterior)	__________
Girders	__________
Rafters	__________
Roof trusses	__________
Exterior decking	__________
Deck stairs	__________
Deck handrail	__________
Wall studs (ground second)	__________
Wall studs (second roof)	__________
Wall headers	__________
Floor sheathing	__________
Roof sheathing	__________
Wall sheathing (ground roof)	__________
Project total, all wood	__________

PROJECT PROBLEMS (FIGURES P 1.1–P 1.9)

P1.1 *Load calculations*:

a. Estimate the floor and roof dead loads and provide a summation table showing each item.
b. Determine the basic floor and roof live loads using ASCE 7.
c. Calculate the sloped roof snow load using ASCE 7.
d. Calculate the unbalanced snow loads on the main roof trusses and rafters; include complete loading diagrams with dimensions and intensities.
e. Calculate the unbalanced snow loads on the main roof trusses and rafters; include complete loading diagrams with dimensions and intensities.

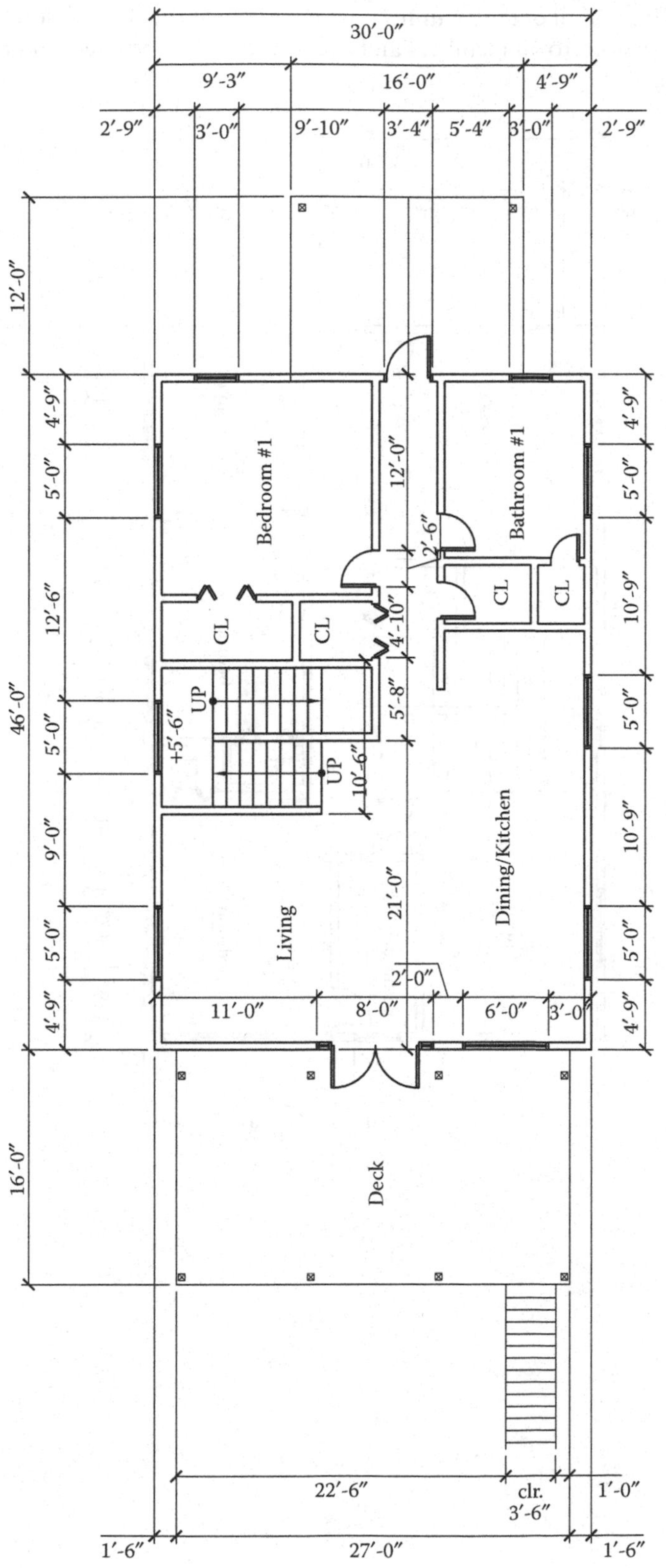

Figure P1.1 First floor plan.

P1.2 Lay out the framing for all of the main framing members at the second floor and roof levels. Show dimensions to all members and account for all overhangs and openings.

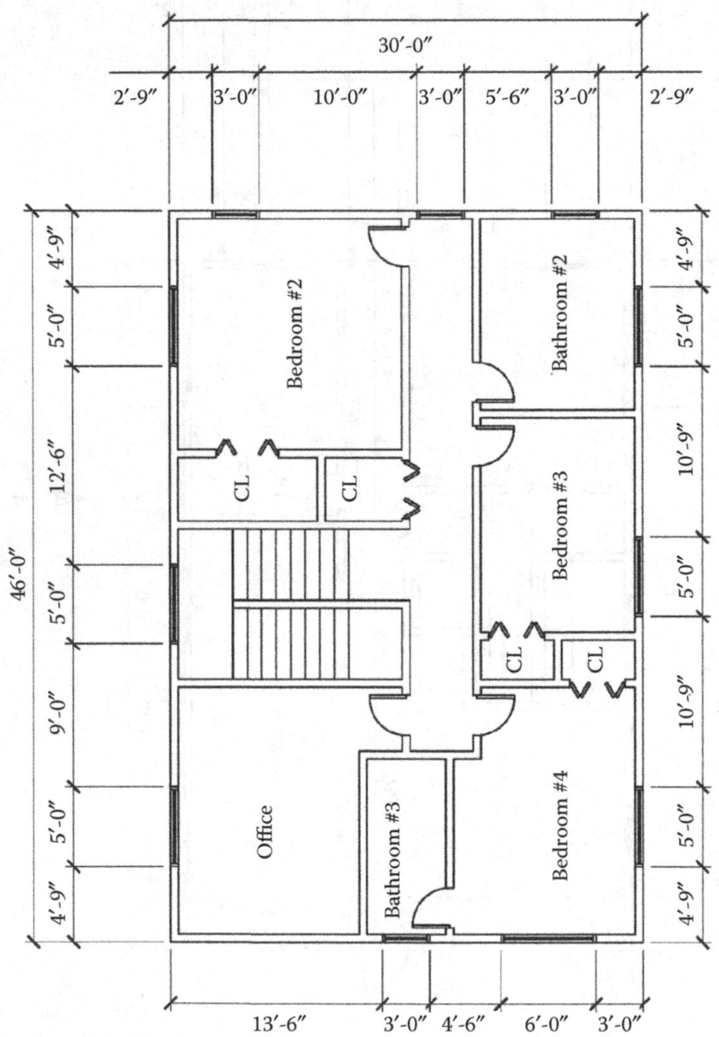

Figure P1.2 Second floor plan.

P1.3 Draw free-body diagrams showing loads, moments, and reactions for the following:

a. Floor joists
b. Floor girders and headers
c. Main roof trusses; provide output from analysis; assume 2 × 6 chords and 2 × 4 webs for the model
d. Covered porch framing members
e. Deck framing members
f. Stair stringers and treads
g. Handrail posts, rails, and balusters

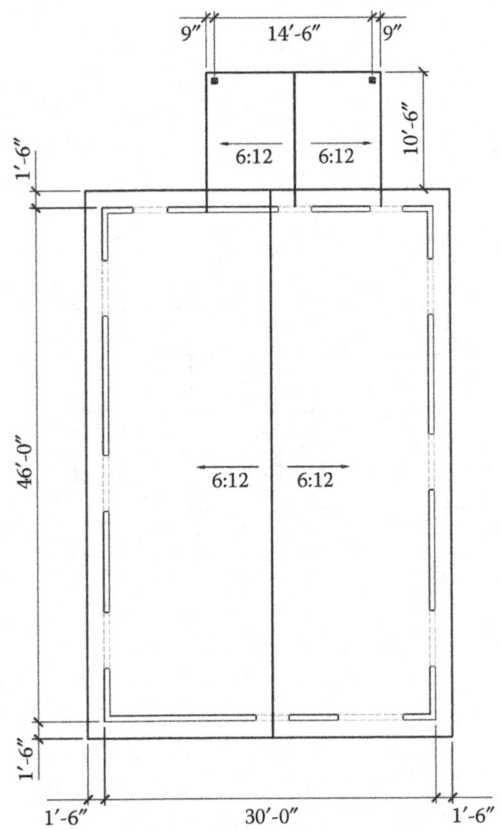

Figure P1.3 Roof plan.

P1.4 *Wind loads*:

a. Calculate the wind loads in each direction of the building for the main wind force-resisting system (ignore components and cladding).
b. Determine the wind force at each level in each direction and provide sketches showing these loads to the shear walls.
c. Determine lateral wind loads to the covered porch and provide sketches showing these loads to each post.
d. Determine the components and cladding uplift forces on the typical roof truss and porch rafter.
e. Determine the components and cladding wind load to the typical wall studs.

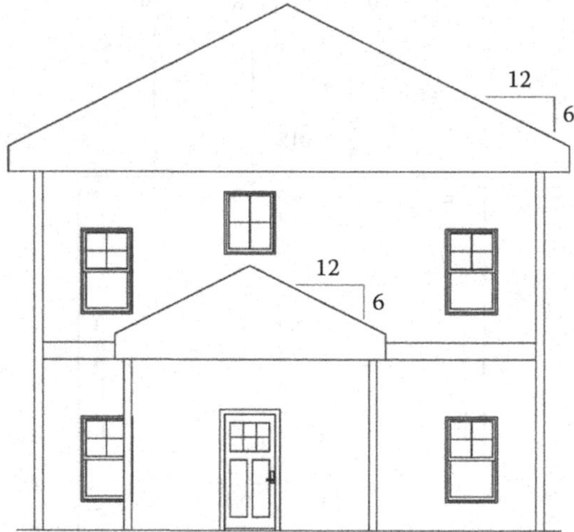

Figure P1.4 East elevation.

P1.5 *Seismic loads*:

a. Calculate the seismic base shear in each direction of the building.
b. Determine the seismic force at each level in each direction and provide sketches showing these loads to the shear walls.

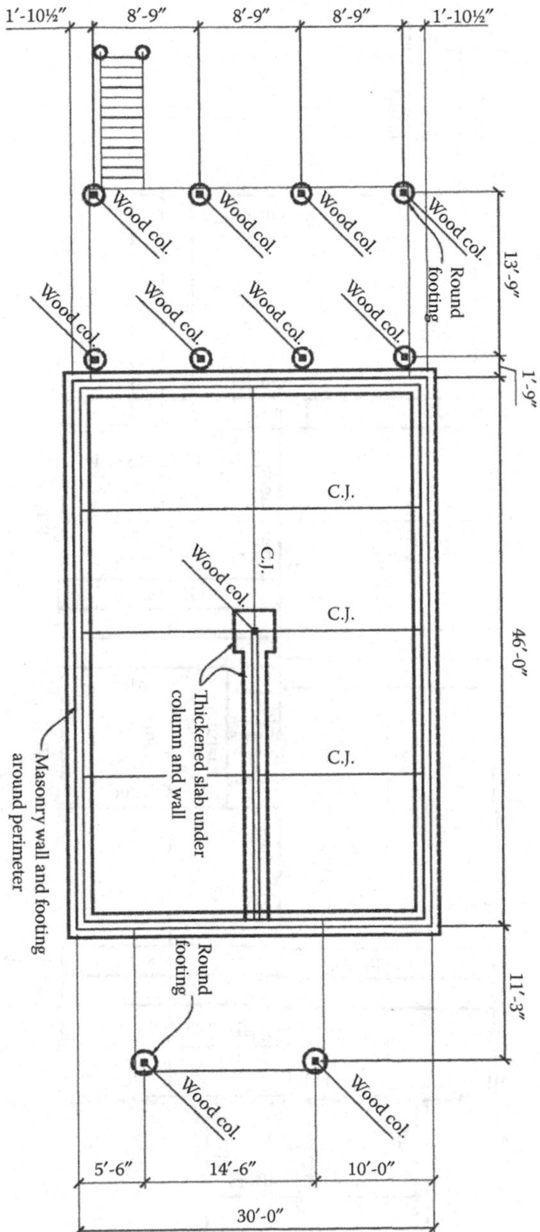

Figure P1.5 Foundation plan.

P1.6 *Floor joists*:

a. Estimate the floor joist size using the span charts in the appendix.
b. Provide calculations for bending, shear, and deflection on the typical floor joist.
c. Select a joist hanger for framing into the glulam girder.
d. Check the floor joists for vibrations.

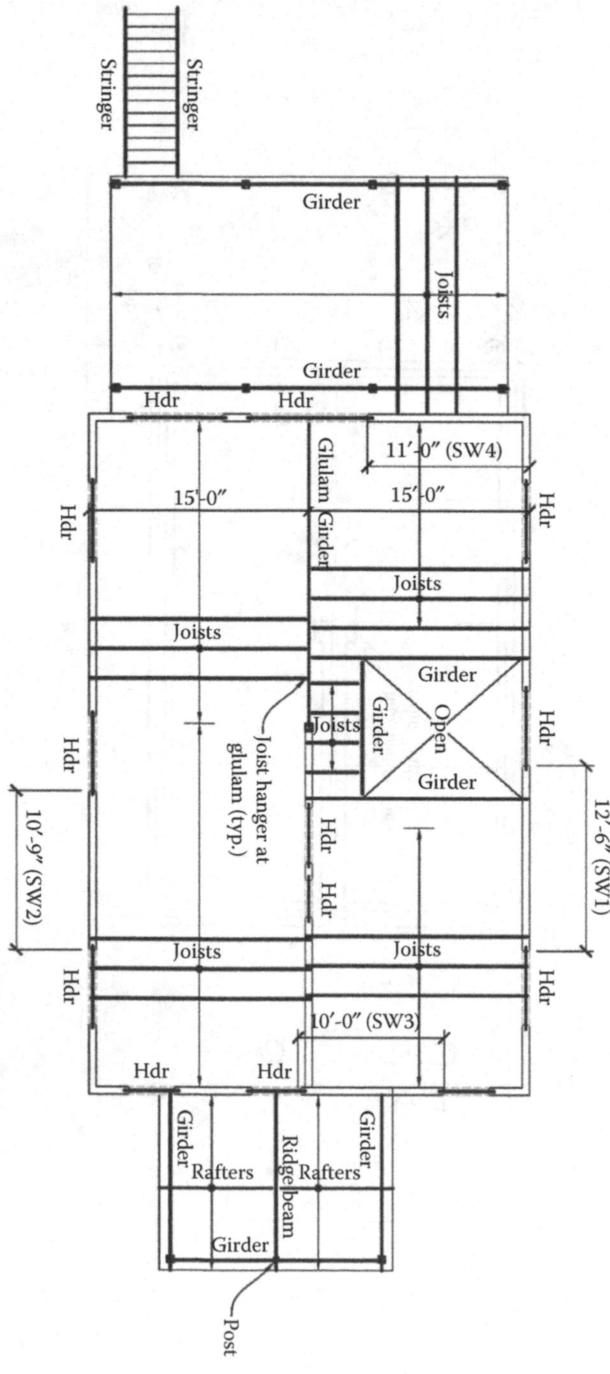

Figure P1.6 Second floor framing plan.

P1.7 *Glulam girder*:

a. Provide calculations for bending, shear, and deflection.
b. Select a connector for the girder to column connection.

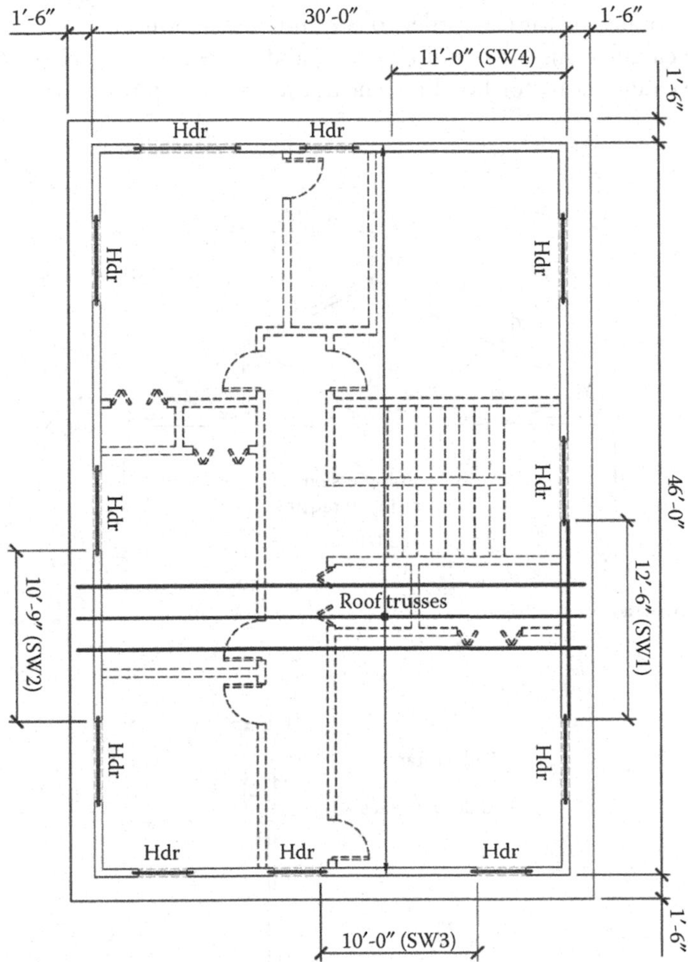

Figure P1.7 Roof framing plan.

P1.8 *Decking*:
Provide calculations for bending, shear, and deflection.

P1.9 *Deck framing*:

a. Estimate the floor joist size using the span charts in the appendix.
b. Provide calculations for bending, shear, and deflection on the typical floor joist.
c. Provide calculations for bending, shear, and deflection on the girders.
d. Select a connector for the girder to column connection.
e. Provide calculations for bending, shear, and deflection on stringers.
f. Provide calculations for bending, shear, and deflection on treads.

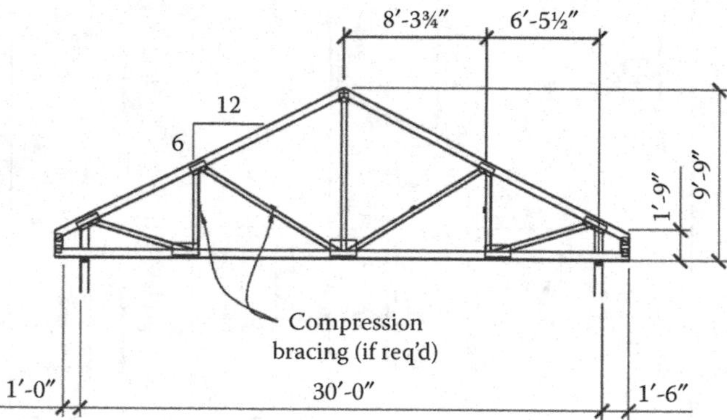

Figure P1.8 Roof truss detail.

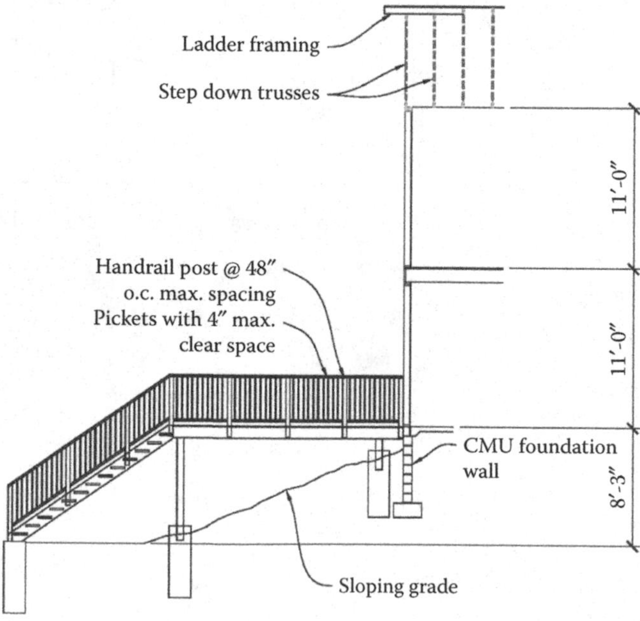

Figure P1.9 Section through deck and wall.

P1.10 *Porch framing*:

a. Estimate the rafter size using the span charts in the appendix.
b. Provide calculations for bending, shear, and deflection on the typical rafter.
c. Select a connector for the rafter to ridge and rafter to girder connection.
d. Provide calculations for bending, shear, and deflection on the ridge beam and girders.

P1.11 *Column design*:

a. Determine the axial load to each column and estimate the size using the design aids in the appendix
   i. Interior column
   ii. Covered porch column
   iii. Deck column
b. Provide calculations for the axial load capacity for each column
   i. Interior column
   ii. Covered porch column
   iii. Deck column

P1.12 *Built-up column design*:

a. Determine the axial load at the ends of headers and girders where they bear on the stud walls and provide a plan view showing these reactions for the critical load combination.
b. Design a built-up column at the locations where the axial load exceeds the capacity of a single stud.
c. Sketch the required fastening per NDS standards for the built-up columns.

P1.13 *Truss design*:

Provide analysis output of each truss member showing critical member forces. Provide free-body diagrams of the chord members showing applied loads.

a. Design the web members for the critical axial load. Provide compression bracing where needed and design the compression chord bracing.
b. Design the bottom chord for combined tension and bending for the worst-case loading.
c. Design the top chord for combined compression and bending for the worst-case loading.
d. Provide a diagram showing forces at each metal plate connection to allow for the design of the metal plates.
e. Check vertical deflections and adjust member sizes as needed.
f. Sketch the complete truss design showing the following:
   - Member axial forces for the most critical load
   - Maximum gravity and uplift reactions
   - Maximum deflections
   - Member sizes and grades
   - Compression chord bracing
   - Dimensions to all critical joints and members

P1.14 *Sheathing for gravity loads*:

a. Determine the gravity loads to the roof sheathing and select a thickness, support configuration and fastening plan. Sketch the design and fastening onto a roof framing plan.
b. Determine the gravity loads to the floor sheathing and select a thickness, support configuration and fastening plan. Sketch the design and fastening onto a floor framing plan.

P1.15 *Sheathing for lateral loads*:

a. Determine the lateral loading to the roof sheathing and select a thickness, support configuration and fastening plan for the given diaphragm shear loads. Revise the sketch from the gravity load design as necessary.
b. Determine the lateral loading to the floor sheathing and select a thickness, support configuration and fastening plan for the given diaphragm shear loads. Revise the sketch from the gravity load design as necessary and include effects from the stair opening.

P1.16 *Wall sheathing for direct wind loads*:

a. Determine the lateral (out of plane) wind load to the wall sheathing and select a thickness, support configuration and fastening plan. Sketch the design and fastening onto a wall elevation.

P1.17 *Sheathing for lateral loads*:

a. Draw full wall elevations showing shear walls and drag struts along with applied loads. Draw diagrams and calculate the maximum drag strut and shear wall forces for each wall elevation and shear wall.
b. Draw a free-body diagram of each shear wall showing all applied loads and determine the following:
   - Unit shear in each shear wall
   - Base shear force
   - Maximum compression chord force at each level
   - Maximum tension chord force at each level
c. Select a fastening pattern for each shear wall and sketch the design on a wall elevation.

P1.18 *Design the following connections and sketch each one to scale*:

a. Interior column base
b. Exterior column base
c. Tread to stringer
d. *Handrails*:
   i. Post base
   ii. Top and bottom rail to post
   iii. Baluster to rail
e. *Shear walls*:
   i. Drag struts: Top plate lap splice
   ii. Shear wall base to masonry wall
   iii. Shear wall hold down at the second floor
   iv. Shear wall hold down at the ground level

PROBLEMS

1.1 List the typical structural components of a wood building.

1.2 What is moisture content, and how does it affect the strength of a wood member?

1.3 Define the terms *equilibrium moisture content and fiber saturation point.*

1.4 Describe the various size classifications for structural lumber, and give two examples of each size classification.

1.5 List and describe factors that affect the strength of a wood member.

1.6 How and why does the duration of loading affect the strength of a wood member?

1.7 What are common defects in a wood member?

1.8 Why does the NDS code not permit the loading of wood in tension perpendicular to the grain?

1.9 Describe the two types of grading systems used for structural lumber. Which is more commonly used?

1.10 Determine the total shrinkage across the width and thickness of a green triple 2 × 4 DF-L top plate loaded perpendicular to the grain as the moisture content decreases from an initial value of 30% to a final value of 12%.

1.11 Determine the total shrinkage over the height of a two-story building that has the exterior wall cross section shown in Figure 1.24 as the moisture content decreases from an initial value of 25% to a final value of 12%.

1.12 How many board feet are there in a 4 × 16 × 36 ft long wood member? How many Mbf are in this member? Determine how many pieces of this member would amount to 4.84 Mbf (4840 bf).

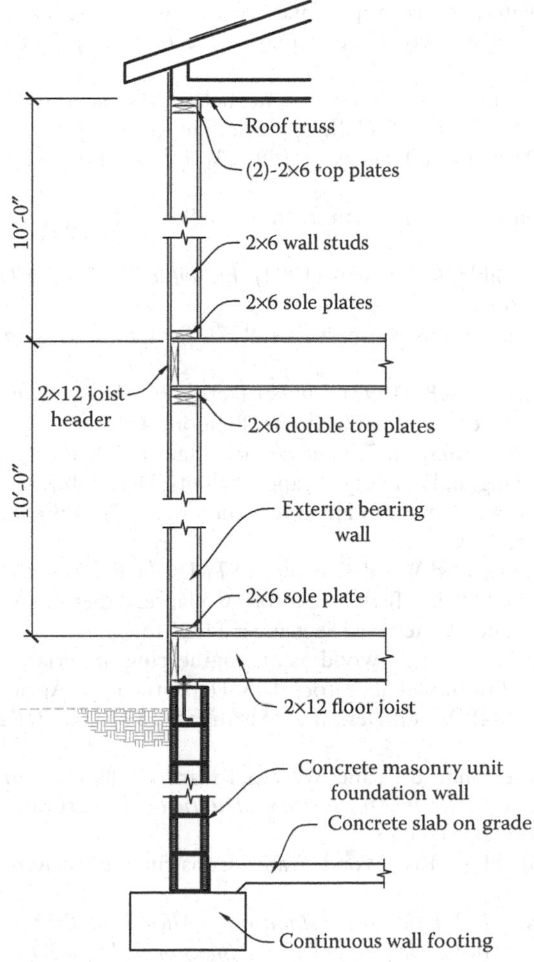

Figure 1.24 Two-story exterior wall section.

REFERENCES

1. ANSI/AF&PA (2015), *National Design Specification for Wood Construction,* American Wood Council, Leesburg, VA.
2. ANSI/AF&PA (2015), *National Design Specification Supplement: Design Values for Wood Construction,* American Wood Council, Leesburg, VA.
3. IBC (2015), *International Building Code,* International Code Council, Washington, DC.
4. ASCE (2010), *Minimum Design Loads for Buildings and Other Structures,* American Society of Civil Engineers, Reston, VA.
5. Stone, Jeffrey B. (2013), *Fire Protection in Wood Buildings—Expanding the Possibilities of Wood Design*, American Wood Council. http://www.awc.org/pdf/education/bcd/ReThinkMag-BCD200A1-DesigningForFireProtection-150801.pdf, Accessed April 17, 2016.
6. Kam-Biron, M. and Koch, L. (2014), The ABC's of Traditional and Engineered Wood Products, *STRUCTURE Magazine*, October, pp. 36–39.
7. NAHB (2000), *Residential Structural Design Guide—2000*, National Association of Home Builders Research Center, Upper Marlboro, MD.
8. Cohen, Albert H. (2002), *Introduction to Structural Design: A Beginner's Guide to Gravity Loads and Residential Wood Structural Design*, AHC, Edmonds, WA.
9. AITC (2012), *Timber Construction Manual*, 6th ed., Wiley, Hoboken, NJ.
10. Sanders, Travis P., Laurin, Geoff A., and Groess, Achim A. (2016), A Structural Engineer's Survival Guide for Waterproofed Appendages, *STRUCTURE Magazine*, April, pp. 22–27.
11. Hoyle, Robert J., Jr. (1978), *Wood Technology in the Design of Structures*, 4th ed., Mountain Press, Missoula, MT.
12. Willenbrock, Jack H., Manbeck, Harvey B., and Suchar, Michael G. (1998), *Residential Building Design and Construction*, Prentice Hall, Upper Saddle River, NJ.
13. Faherty, Keith F. and Williamson, Thomas G. (1995), *Wood Engineering and Construction,* McGraw-Hill, New York.
14. Canadian Wood Council (2011), *Introduction to Wood Design*, Canadian Wood Council, Ottawa, Canada.
15. Halperin, Don A. and Bible, G. Thomas (1994), *Principles of Timber Design for Architects and Builders,* Wiley, New York.
16. Stalnaker, Judith J. and Harris, Earnest C. (1997), *Structural Design in Wood*, Chapman & Hall, London.
17. Kim, Robert H. and Kim, Jai B. (1997), *Timber Design for the Civil and Structural Professional Engineering Exams*, Professional Publications, Belmont, CA.
18. Kermany, Abdy (1999), *Structural Timber Design*, Blackwell Science, London.
19. Anthony, Ronald W., Dugan, Kimberly D., and Anthony, Deborah J. (2016), A Grading Protocol for Structural Lumber and Timber in Historic Structures, *APT Bulletin: Journal of Preservation Technology*, 40(2), pp. 3–9.
20. Dunham, Lee (2013), Decayed Wood Structures, *STRUCTURE Magazine*, October, pp. 21–24.
21. Engineering News Record (2015), Berkeley Balcony Collapse, Other Cases Spotlight Gap in Building Oversight, ENR.com, June 26, Accessed September 14, 2015.
22. USDA (2010). Wood handbook - Wood as an engineering material, General technical report FPL-GTR-190, Forest Products Laboratory, U.S. Department of Agriculture.
23. Powell, Robert M. (2004), Wood Design for Shrinkage, *STRUCTURE Magazine*, November, pp. 24–25.
24. Nelson, Ronald F., Patel, Sharad T., and Avevalo, Ricardo (2002), *Continuous Tie-down System for Wood Panel Shear Walls in Multi-Story Structures*, Structural Engineers Association of California Convention, October 28.
25. Knight, Brian (2006), High Rise Wood Frame Construction, *STRUCTURE Magazine*, June, pp. 68–70.
26. ASTM (1990), *Standard Practice for Establishing Allowable Properties for Visually-Graded Dimension Lumber from In-Grade Tests of Full-Size Specimens*, ASTM D 1990, ASTM International, West Conshohocken, PA.

Chapter 2

Introduction to structural design loads

2.1 DESIGN LOADS

Several types of loads can act on wood buildings: dead loads, live loads, snow loads, wind loads, and seismic loads. The combinations of these loads that act on any building structure are prescribed by the relevant building code, such as the *International Building Code* (IBC) [1] or the ASCE 7 load specifications [2].

2.1.1 Load combinations

The various loads that act on a building do not act in isolation and may act on the structure simultaneously. However, these loads usually will not act on the structure simultaneously at their maximum values. The ASCE 7 load standards [2] and Section 1605 of the IBC [1] prescribe the critical combination of loads to be used for design for both the allowable stress design (ASD) method and the load and resistance factor design (LRFD) method. The basic load combinations for LRFD from Sections 2.3 and 2.4 of the ASCE 7 load standards (excluding fluid loads, F and self-restraining force, T that are zero for most building structures) are:

1. $1.4D$
2. $1.2D + 1.6\ (L + H) + 0.5\ (L_r \text{ or } S \text{ or } R)$
3. $1.2D + 1.6\ (L_r \text{ or } S \text{ or } R) + 1.6H + (f_1L \text{ or } 0.5W)$
4. $1.2D + 1.0W + f_1L + 1.6H + 0.5\ (L_r \text{ or } S \text{ or } R)$
5. $1.2D + 1.0E + f_1L + 1.6H + f_2S$
6. $0.9D + (1.0W + 1.6H)$ (D always counteracts W and H)
7. $0.9D + (1.0E + 1.6H)$ (D always counteracts E and H)

where:

f_1 = 1 for areas of public assembly with live loads that exceed 100 psf, and parking garages
= 0.5 for all other live loads

f_2 = 0.7 for roof configurations that do not shed snow off the roof structure (e.g., sawtooth roofs)
= 0.2 for all other roof configurations

Note: Where the effect of the load, H, due to lateral earth pressure, ground water pressure, or pressure of bulk materials adds to the primary variable load effect, the load factor on H shall be taken as 1.6. Where H counteracts or resists the seismic load effect, E, or the wind load effect, W, the load factor for H in load combinations 6 and 7 should be set equal to 0.9 when the load, H, is permanent (e.g., lateral soil pressure) or a load factor of zero for all other conditions, including conditions where H varies (ASCE 7, Section 2.3.2).

When designing for strength under service load conditions, the ASD load combinations (Equations 8 through 16) below should be used. The basic load combinations for ASD are as follows:

8. D
9. $D + H + L$
10. $D + H + (L_r \text{ or } S \text{ or } R)$
11. $D + H + 0.75\,(L) + 0.75\,(L_r \text{ or } S \text{ or } R)$
12. $D + H + (0.6W \text{ or } 0.7E)$
13. $D + H + 0.75\,(0.6W) + 0.75L + 0.75\,(L_r \text{ or } S \text{ or } R)$
14. $D + H + 0.75\,(0.7E) + 0.75L + 0.75\,(S)$
15. $0.6D + 0.6W + H$ (D always opposes W and H)
16. $0.6D + 0.7E + H$ (D always opposes E and H)

The ASD load combinations (Equations 8 through 16) are also used when designing for serviceability limit states such as deflections and vibrations. Where the effect of the load, H, due to lateral earth pressure, ground water pressure, or pressure of bulk materials adds to the primary variable load effect, the load factor on H shall be taken as 1.0. Where H counteracts or resists the seismic load effect, E, or the wind load effect, W, the load factor for H in load combinations 15 and 16 should be set equal to 0.6 when the load, H, is permanent (e.g., lateral soil pressure) or a load factor of zero for all other conditions, including conditions where H varies (ASCE 7, Section 2.4.1). In the above load combinations, downward loads have a positive (+) sign, while upward loads have a negative (−) sign. Load combinations 1 through 5 and 8 through 14 are used to maximize the downward acting loads, while load combinations 6 and 7 and 15 and 16 are used to maximize the uplift load or overturning effects. Therefore, in load combinations 6, 7, 15, and 16, the wind load, W, and the seismic load, E, take on *only negative or zero values*, while in all the other load combinations, W and E take on positive values. Note that E and W are calculated in the ASCE 7 load standard at their factored levels whereas all other loads (D, L, S, etc) are calculated at their service load levels. The seismic load effect, E, consists of a combination of the horizontal seismic load effect, E_h, and the vertical seismic load effect, E_v. The notations used in the above load combinations are defined below.

$E_h = \rho Q_E$
$E_v = 0.2\,S_{DS}D$
E = factored combined seismic load effect due to *horizontal* and *vertical* earthquake-induced forces
$= \rho Q_E + 0.2\,S_{DS}D$ in load combinations 5, 12, and 14
$= \rho Q_E - 0.2\,S_{DS}D$ in load combinations 7 and 16
D = the dead load
Q_E = the factored horizontal earthquake load effect due to the base shear, V (i.e., forces, reactions, moments, shears, etc. caused by the horizontal seismic force)
$0.2\,S_{DS}D$ = the factored vertical component of the earthquake force (affects mostly columns and footings)
S_{DS} = the design spectral response acceleration at short period
H = the lateral soil pressure, hydrostatic pressures, and pressure of bulk materials
L = the floor live load
L_r = the roof live load

W is the factored wind load
S is the snow load
R is the rain load
ρ is the redundancy factor

The vertical seismic load effect, E_v (= $0.2\ S_{DS}D$) can be taken as zero when S_{DS} is less than or equal to 0.125. The seismic lateral force-resisting system in both orthogonal directions of the building must be assigned a redundancy factor, ρ, the value of which can be obtained as follows from ASCE 7, Sections 12.3.4.1 and 12.3.4.2:

- For seismic design category (SDC) B or C: $\rho = 1.0$
- For SDC D, E, or F, it is conservative to assume: $\rho = 1.3$

All structural elements must be designed for the most critical of these combinations. The use of these load combinations is described in greater detail later in the book.

2.2 DEAD LOADS

Dead loads are the weights of all materials that are permanently and rigidly attached to a structure, including the self-weight of the structure, such that it will vibrate with the structure during a seismic or earthquake event (Figures 2.1 and 2.2). The dead loads can

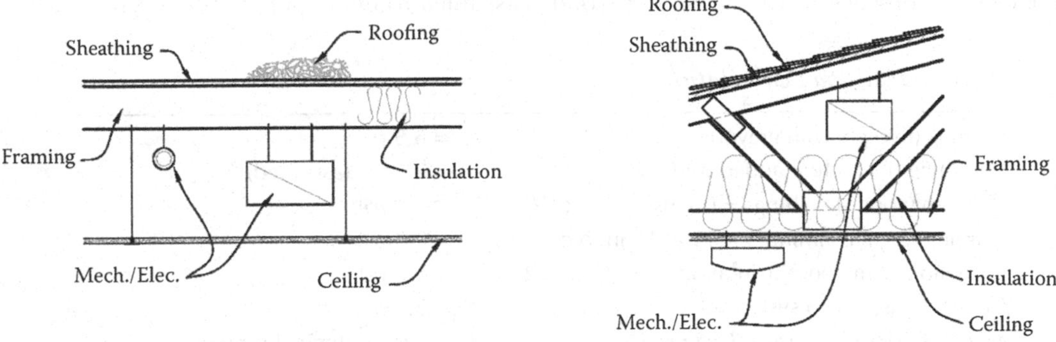

Figure 2.1 Typical roof dead loads.

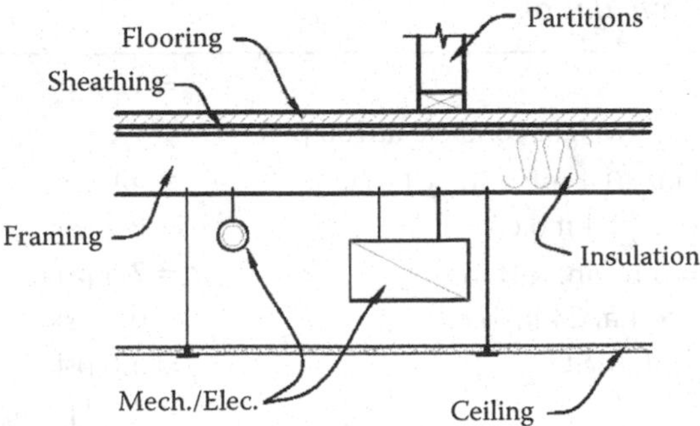

Figure 2.2 Typical floor dead loads.

be determined with more accuracy than other types of loads and are not as variable as live loads. Typical checklists for the roof and floor dead loads in wood buildings are as follows:

Typical roof dead load checklist

- Weight of roofing material
- Weight of roof sheathing or plywood
- Weight of framing
- Weight of insulation
- Weight of ceiling
- Weight of mechanical and electrical (M&E) fixtures

Typical floor dead load checklist

- Weight of flooring (i.e., the topping: hardwood, lightweight concrete, etc.)
- Weight of floor sheathing or plywood
- Weight of floor framing
- Weight of partitions (15 psf minimum, but not required when the floor live load is greater than 80 psf; see Section 4.3.2 of [2])
- Weight of ceiling
- Weight of M&E fixtures

To aid in the calculation of dead loads, weights of various building materials, such as those given in Appendix A, are typically used. The weights of framing provided in Appendix A are based on Douglas Fir-Larch with a specific gravity of 0.5 and density of 32 pcf, which is conservative for most wood buildings. The following are sample roof and floor dead-load calculations for typical wood buildings that can serve as a guide to the reader. In the calculations below, the density of wood is assumed to be 32 pcf, as stated previously.

Sample roof dead-load calculation

Roofing (five-ply with gravel)	= 6.5 psf
Reroofing (i.e., future added roof)	= 2.5 psf (assumed)
½-in. plywood sheathing (= 0.4 psf/⅛ in. × ½/⅛)	= 1.6 psf
Framing (e.g., assuming 2 × 12 at 16 in. o.c.)	= 2.8 psf
Insulation (2 in. loose insulation: 0.5 psf/in. × 2 in)	= 1.0 psf
Channel-suspended system (steel)	= 2.0 psf
Mechanical and electrical (M&E)	= 5.0 psf (typical for wood buildings)
Total roof dead load D_{roof}	= 21.4 psf ≈ **22 psf**

Sample floor dead-load calculation

Floor covering	= 12.5 psf
(e.g., assuming 1½-in. lightweight concrete at 100 pcf)	
1⅛-in. plywood sheathing (0.4 psf/⅛ in. × 1⅛/⅛)	= 3.6 psf
Framing (assuming 4 × 12 at 4 ft o.c.)	= 2.2 psf
½-in. drywall ceiling (= 5 psf/in. × ½ in.)	= 2.5 psf
Ceiling supports (say, 2 × 4 at 24 in. o.c.)	= 0.6 psf
Mechanical and electrical (M&E)	= 5.0 psf
Partition loads	= 20.0 psf (assumed)
Total floor dead load D_{floor}	= 46.4 psf ≈ **47 psf**

For buildings with floor live loads less than or equal to 80 psf, the partition dead load must be at least 15 psf (see Section 4.3.2 of [2]), while for buildings with floor live loads greater than 80 psf, no partition loads have to be considered since for such assembly occupancies, there is less likelihood that partition walls will be present.

2.2.1 Combined dead and live loads on sloped roofs

Since most wood buildings have sloped roofs, we discuss next how to combine the dead loads acting on the sloped roof surface with the live loads (i.e., snow, rain, or roof live load) acting on a horizontal projected plan area of the roof surface (Figure 2.3). Most building codes give live loads in units of pounds per square foot of the horizontal projected plan area, while the dead load of a sloped roof is in units of pounds per square foot of the sloped roof area. Therefore, to combine the dead and live loads in the same units, two approaches are possible:

1. Convert the dead load from units of pounds per square foot of sloped roof area to units of psf of horizontal projected plan area and then add to the live load, which is in units of psf of horizontal projected plan area. When using this approach, the reader should not forget the horizontal thrust or force acting at the exterior wall from the component of the dead and live loads acting parallel to the roof surface. These lateral thrusts must be considered in the design of the walls, and collar or ceiling ties should be provided to resist this lateral force.
2. Convert the live load from pounds per square foot of horizontal projected plan area to psf of sloped roof area and then add to the dead load, which is in units of psf of sloped roof area.

Option 1 is more commonly used in design practice and is adopted in this book.

Using the LRFD load combination equations presented in Section 2.1, the total factored dead plus live load, w_u, in psf of horizontal plan area will be

$$w_u = (1.2)D\frac{L_1}{L_2} + (1.6)(L_r \text{ or } S \text{ or } R) \quad \text{psf of horizontal plan area} \tag{2.1}$$

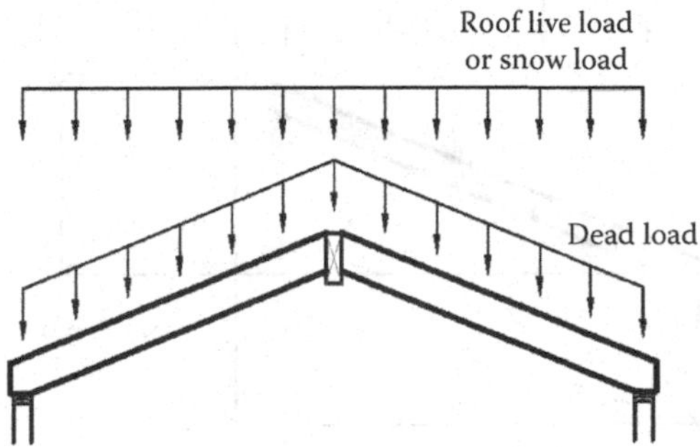

Figure 2.3 Loads on a sloped roof.

Using the ASD load combination equations presented earlier in Section 2.1, the total unfactored dead plus live load, w_s, in psf of horizontal plan area will be

$$w_s = D\frac{L_1}{L_2} + (L_r \text{ or } S \text{ or } R) \quad \text{psf of horizontal plan area} \tag{2.2}$$

where:

D is the roof dead load in psf of sloped roof area
S is the roof snow load in psf of horizontal plan area
L_r is the roof live load in psf of horizontal plan area
R is the rain load in psf of horizontal plan area (usually not critical for sloped roofs)
L_1 is the sloped length of rafter
L_2 is the horizontal projected length of rafter

2.2.1.1 Calculation of the horizontal thrust at an exterior wall

Summing moments about the exterior wall support of the rafter yields (Figure 2.4) and assuming rafters with a ridge board (i.e., no ridge beam),

$$-H_s(h) + w_s L_2 \frac{L_2}{2} = 0$$

Thus, the unfactored horizontal thrust H (without a ridge beam) becomes

$$H_s = \frac{w_s L_2 (L_2/2)}{h} \tag{2.3}$$

For rafters with a ridge beam, $V_{1s} = V_{2s} = w_s(L_2/2)$; $H_s = 0$
For rafters without a ridge beam, $V_{1s} = w_s L_2$; $V_{2s} = 0$; *and* H_s = Equation 2.3
Similarly, the factored horizontal thrust H_u without a ridge beam is calculated as

$$H_u = \frac{w_u L_2 (L_2/2)}{h} \tag{2.4}$$

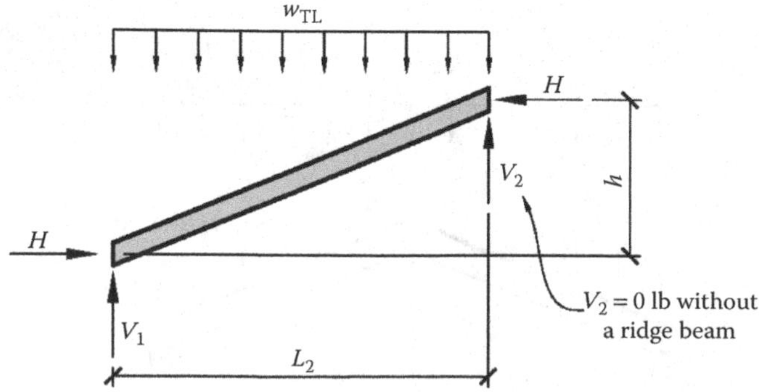

Figure 2.4 Free-body diagram of sloped rafter.

For rafters with a ridge beam, $V_{1u} = V_{2u} = w_u(L_2/2)$; $H_u = 0$

For rafters without a ridge beam, $V_{1u} = w_u L_2$; $V_{2u} = 0$; *and* H_u = Equation 2.4

2.2.2 Combined dead and live loads on stair stringers

The same Equations 2.1 and 2.2 used for calculating the total load on sloped roofs can be applied to stair stringers. Using the ASD load combinations in Section 2.1, the total unfactored load on the stair stringer is given as

$$w_{TL} = D\frac{L_1}{L_2} + L \quad \text{psf of horizontal plan area} \tag{2.5}$$

where:

D is the stair dead load in psf of sloped roof area
L is the stair live load in psf of horizontal plan area
= 100 psf except for stairs in one or two family dwellings according to IBC Table 1607.1 or ASCE 7, Table 4-1)
L_1 is the sloped length of rafter
L_2 is the horizontal projected length of rafter

Example 2.1: Design loads for a sloped roof

Given the following design parameters for a sloped roof (Figure 2.5) and assuming that the ASD method is to be used, (a) calculate the uniform total service load and the maximum shear and moment on the rafter. (b) Calculate the horizontal thrust on the exterior wall if rafters are used without a ridge beam. (c) Repeat the analysis in part (a) using the LRFD method. (d) Repeat the analysis in part (b) using the LRFD method.

- Roof dead load D = 10 psf (of sloped roof area)
- Roof snow load S = 66 psf (of horizontal plan area)
- Horizontal projected length of rafter L_2 = 18 ft
- Sloped length of rafter L_1 = 20.12 ft
- Rafter or truss spacing = 4 ft 0 in.

Solution:

a. Using the load combinations in Section 2.1, the total unfactored load (ASD method) in psf of horizontal plan area will be (see Figure 2.6)

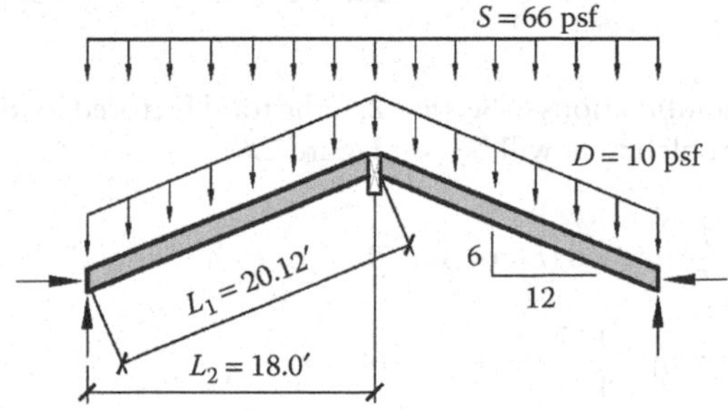

Figure 2.5 Cross section of a building with a sloped roof.

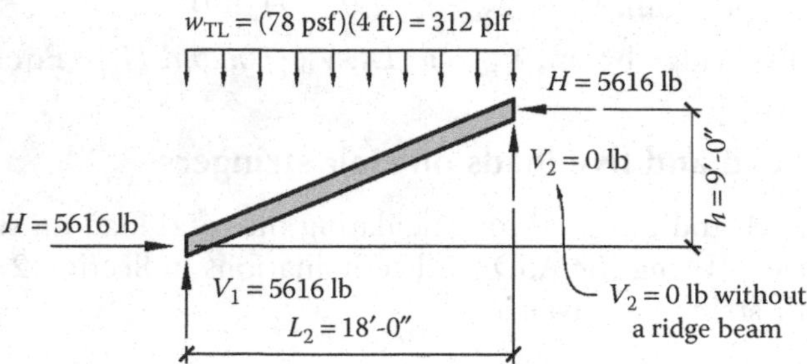

Figure 2.6 Free-body diagram of sloped rafter.

$$w_s = D\left(\frac{L_1}{L_2}\right) + (L_T \text{ or } S \text{ or } R), \quad \text{psf of horizontal plan area}$$

$$= (10\,\text{psf})\left(\frac{20.12\text{ ft}}{18\text{ ft}}\right) + 66\,\text{psf}$$

$$= \mathbf{78\ psf\ of\ horizontal\ plan\ area}$$

The total unfactored load in pounds per horizontal linear foot (lb/ft) is given as

$$w_s(\text{lb/ft}) = w_s(\text{psf}) \times \text{tributary width or spacing of rafters}$$
$$= (78\text{ psf})(4\text{ ft}) = \mathbf{312\ lb/ft}$$

The unfactored horizontal thrust H (without a ridge beam) is

$$H_s = \frac{w_s L_2 (L_2/2)}{h} = \frac{(312\text{ lb/ft})(18\text{ ft})(18\text{ ft/2})}{9\text{ ft}} = 5616\text{ lb}$$

The collar or ceiling ties must be designed to resist this horizontal thrust.

b. L_2 = 18 ft. The maximum unfactored shear force in the rafter (without a ridge beam) is

$$V_{1s} = w_s L_2 = (312\,\text{lb/ft})(18\text{ ft}) = 5616\text{ lb}$$

The maximum unfactored moment in the rafter is

$$M_{s,\max} = \frac{w_s (L_2)^2}{8} = \frac{(312)(18\text{ ft})^2}{8} = 12{,}700\text{ ft-lb} = 12.70\text{ ft-kips}$$

c. Using the load combinations in Section 2.1, the total factored load (LRFD method) in psf of horizontal plan area will be (see Figure 2.6)

$$w_u = (1.2)D\left(\frac{L_1}{L_2}\right) + (1.6)(L_T \text{ or } S \text{ or } R), \quad \text{psf of horizontal plan area}$$

$$= (1.2)(10\,\text{psf})\left(\frac{20.12\text{ ft}}{18\text{ ft}}\right) + (1.6)(66\,\text{psf})$$

$$= 119\ \text{psf of horizontal plan area}$$

The total factored load in pounds per horizontal linear foot (lb/ft) is given as

$$w_u(\text{lb/ft}) = w_u(\text{psf}) \times \text{tributary width or spacing of rafters}$$

$$= (119 \text{ psf})(4 \text{ ft}) = \mathbf{476 \text{ lb/ft}}$$

The factored horizontal thrust H_u (without a ridge beam) is

$$H_u = \frac{w_u L_2 (L_2/2)}{h} = \frac{(476 \text{ lb/ft})(18 \text{ ft})(18 \text{ ft}/2)}{9 \text{ ft}} = 8568 \text{ lb}$$

The collar or ceiling ties must be designed to resist this horizontal thrust.

d. $L_2 = 18$ ft. The maximum factored shear force in the rafter is

$$V_{1u} = w_u L_2 = (476 \text{ lb/ft})(18 \text{ ft}) = 8568 \text{ lb} \quad \text{(without a ridge beam)}$$

The maximum factored moment in the rafter is

$$M_{u,\max} = \frac{w_u (L_2)^2}{8} = \frac{(476)(18 \text{ ft})^2}{8} = 19{,}278 \text{ ft-lb} = 19.3 \text{ ft-kips}$$

Example 2.2: Design loads for stair stringers

a. Determine the total unfactored dead plus live load, shear and maximum moment on a stair stringer assuming a dead load of 50 psf and a live load of 100 psf. The plan and elevation of the stair are shown in Figure 2.7.
b. Repeat the analysis in part (a) using factored loads (LRFD method).

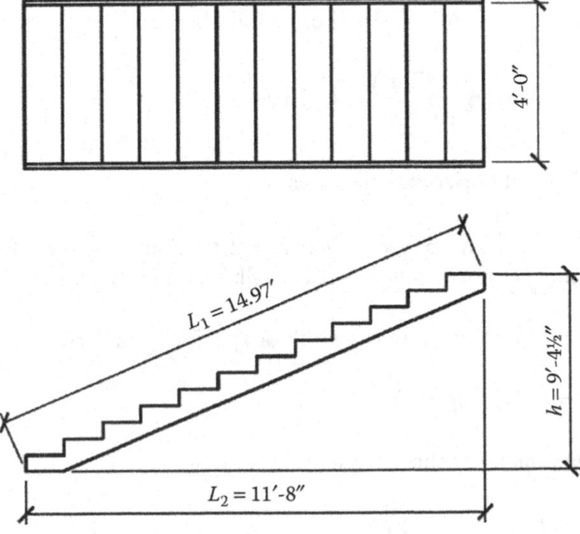

Figure 2.7 Stair plan and elevation.

Solution:

a. The stair dead load $D = 50$ psf of the sloped area and the stair live load $L = 100$ psf of the horizontal plan area. The total unfactored uniform load on the stair in psf of horizontal plan area is given as

$$w_{TL} = D\frac{L_1}{L_2} + L \quad \text{psf of horizontal plan area}$$

$$= (50\,\text{psf})\left(\frac{14.97\text{ ft}}{11.67\text{ ft}}\right) + 100\text{ psf}$$

$$= \mathbf{164\ psf} \text{ of horizontal plan area}$$

The TW for each stair stringer = 4 ft/2 = 2 ft. Thus, the total load in pounds per horizontal linear foot (lb/ft) on each stair stringer is given as

$$w_{TL}(\text{lb/ft}) = w_{TL}(\text{psf}) \times \text{tributary width or spacing of rafters}$$

$$= (164\text{ psf})(2\text{ ft}) = \mathbf{328\ lb/ft}$$

The maximum shear force in the stair stringer is

$$V_{max} = w_{TL}\frac{L_2}{2} = (328)\left(\frac{11.67\text{ ft}}{2}\right) = 1914\text{ lb}$$

The maximum moment in the stair stringer is

$$M_{max} = \frac{w_{TL}(L_2)^2}{8} = \frac{(328)(11.67\text{ ft})^2}{8} = 5584\text{ ft-lb} = 5.6\text{ ft-kips}$$

b. The total factored uniform load on the stair in psf of horizontal plan area is given as

$$w_u = (1.2)D\frac{L_1}{L_2} + (1.6)L \quad \text{psf of horizontal plan area}$$

$$= (1.2)(50\,\text{psf})\left(\frac{14.97\text{ ft}}{11.67\text{ ft}}\right) + (1.6)(100\text{ psf})$$

$$= \mathbf{237\ psf} \text{ of horizontal plan area}$$

The TW for each stair stringer = 4 ft/2 = 2 ft. Thus, the total factored uniform load in pounds per horizontal linear foot (lb/ft) on each stair stringer is given as

$$w_u(\text{lb/ft}) = w_u(\text{psf}) \times \text{tributary width or spacing of rafters}$$

$$= (237\text{ psf})(2\text{ ft}) = \mathbf{474\ lb/ft}$$

The maximum factored shear force in the stair stringer is

$$V_{u,max} = w_u\frac{L_2}{2} = (474)\left(\frac{11.67\text{ ft}}{2}\right) = 2766\text{ lb}$$

The maximum factored moment in the stair stringer is

$$M_{u,\max} = \frac{w_u(L_2)^2}{8} = \frac{(474)(11.67\text{ ft})^2}{8} = 8069\text{ ft-lb} = 8.1\text{ ft-kips}$$

2.3 TRIBUTARY WIDTHS AND AREAS

In this section, we introduce the concept of tributary widths and tributary areas (Figure 2.8). These concepts are used to determine the distribution of floor and roof loads to the various supporting structural elements. The *tributary width* (TW) of a beam or girder is defined as the width of floor or roof supported by the beam or girder, and is equal to the sum of one-half the distance to the adjacent beam on the right of the beam in question plus one-half the distance to the adjacent beam on the left of the beam in question. Thus, the TW is given as

$$\text{TW} = \frac{1}{2}\text{ (distance to adjacent beam on the right)} + \frac{1}{2}\text{ (distance to adjacent beam on the left)}$$

The *tributary area*, A_T, of a beam, girder, or column is the floor or roof area supported by the structural member. For beams and girders, the tributary area is the product of the span and the TW of the beam or girder. For columns, it is the plan area bounded by lines located at one-half the distance to the adjacent columns surrounding the column in question. It should be noted that a column does not have a TW, but only a tributary area.

2.4 LIVE LOADS

Any item not rigidly attached to a structure is usually classified as a *live load*. Live loads are short-term nonstationary gravity forces that can vary in magnitude and location. There are three main types of live loads: roof live load (L_r), snow load (S), and floor live load (L). Rain load (R) is also a roof live load but is rarely critical compared to roof live loads and snow load because of the pitched roofs in most wood buildings. These live loads are usually specified in the IBC [1]

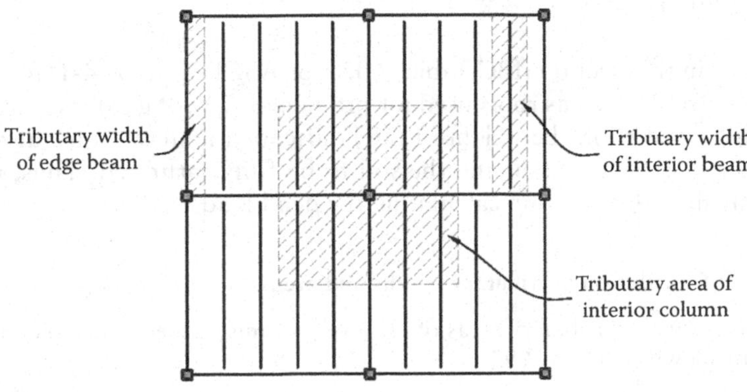

Figure 2.8 Tributary width and tributary area.

and the ASCE 7 load standards [2]. The magnitude of live loads depends on the occupancy and use of the structure and on the tributary area of the structural element under consideration. Live loads as specified in the IBC and ASCE 7 are given in units of pounds per square foot of the horizontal projected plan area, but it should be noted that the codes also specify alternative concentrated live loads (in pounds), as we will discuss later. The live loads specified in the codes are minimum values, and the owner of a building could decide to use higher live load values than those given in the code, although this will lead to a more expensive structure.

2.4.1 Roof live load

Roof live loads usually occur due to the weight of equipment and maintenance personnel during the servicing of a roof. The magnitude is a function of the roof slope and the tributary area of the structural element under consideration. The larger the tributary area, the smaller the probability that the entire tributary area, A_T, will be loaded with the maximum roof live load, resulting in a smaller design roof live load. Similarly, the steeper the roof slope, the smaller the design roof live load.

For ordinary flat, pitched, or curved roofs, the ASCE 7 load standards [2] specify the design roof live load in pounds per square foot as

$$L_r = 20R_1R_2 \tag{2.6}$$

$$12 \le L_r \le 20$$

where $R_1 = \begin{cases} 1.0 \quad \text{for } A_T \le 200 \text{ ft}^2 \\ 1.2 - 0.001A_T \quad \text{for } 200 \text{ ft}^2 < A_T < 600 \text{ ft}^2 \\ 0.6 \quad \text{for } A_T \ge 600 \text{ ft}^2 \end{cases}$

$$R_2 = \begin{cases} 1.0 \text{ for } F \le 4 \\ 1.2 - 0.05F \text{ for } 4 < F < 12 \\ 0.6 \text{ for } F \ge 12 \end{cases}$$

F is the roof slope factor
= number of inches of rise per foot for a sloped roof = 12 tan θ
= rise/span ratio multiplied by 32 for an arch or dome roof
θ is the roof slope, degrees
A_T is the tributary area, ft^2

See Table 2.2 later in this section (IBC Table 1607.1 or ASCE 7, Table 4-1) for the live loads for special purpose roofs such as roofs used for promenades, roof gardens, or assembly purposes. For a landscaped roof, the weight of the landscaped material must also be included in the dead-load calculations, assuming the soil to be fully saturated. Thus, the saturated density of the soil should be used in calculating the dead loads.

Example 2.3: Calculation of tributary widths and areas

Determine the TW and tributary areas of the joists, beams, girders, and columns in the framing plan shown in Figure 2.9.

Solution: The solution is presented in Table 2.1.

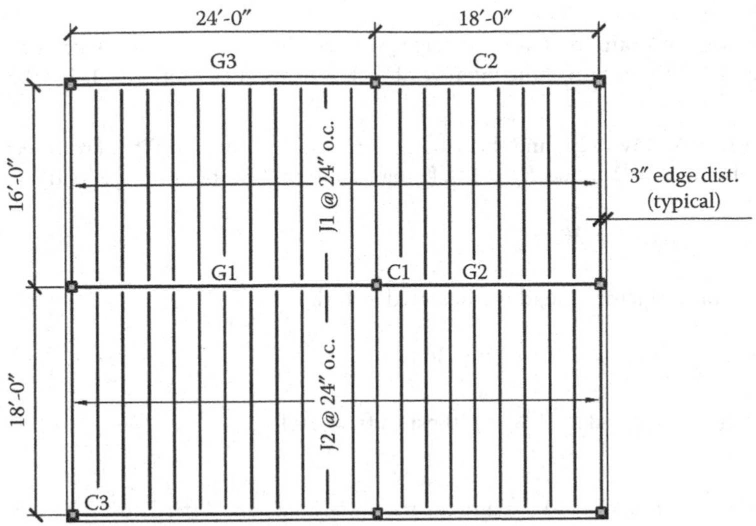

Figure 2.9 Floor framing plan (Example 2.3) and roof framing plan (Example 2.4).

Table 2.1 Tributary widths and areas of joists, beams, and columns

Structural member	*Tributary width*	*Tributary area*
Joist J1	$\frac{24\text{ in.}}{2}+\frac{24\text{ in.}}{2}=24\text{ in.}=2\text{ ft}$	$2\text{ ft}\times 16\text{ ft}=32\text{ ft}^2$
Joist J2	$\frac{24\text{ in.}}{2}+\frac{24\text{ in.}}{2}=24\text{ in.}=2\text{ ft}$	$2\text{ ft}\times 18\text{ ft}=36\text{ ft}^2$
Beam or girder G1	$\frac{16\text{ ft}}{2}+\frac{18\text{ ft}}{2}=17\text{ ft}$	$17\text{ ft}\times 24\text{ ft}=408\text{ ft}^2$
Girder G2	$\frac{16\text{ ft}}{2}+\frac{18\text{ ft}}{2}=17\text{ ft}$	$17\text{ ft}\times 18\text{ ft}=306\text{ ft}^2$
Girder G3	$\frac{16\text{ ft}}{2}+0.25\text{ ft}=8.25\text{ ft}$	$8.25\text{ ft}\times 24\text{ ft}=198\text{ ft}^2$
Column C1	–	$\left(\frac{18\text{ ft}}{2}+\frac{16\text{ ft}}{2}\right)\left(\frac{24\text{ ft}}{2}+\frac{18\text{ ft}}{2}\right)=357\text{ ft}^2$
Column C2	–	$\left(\frac{16\text{ ft}}{2}+0.25\text{ ft}\right)\left(\frac{24\text{ ft}}{2}+\frac{18\text{ ft}}{2}\right)=173.3\text{ ft}^2$
Column C3	–	$\left(\frac{18\text{ ft}}{2}+0.25\text{ ft}\right)\left(\frac{24\text{ ft}}{2}+0.25\text{ ft}\right)=113.4\text{ ft}^2$

Example 2.4: Roof load calculations

For the roof plan shown in Example 2.3, assuming a roof dead load of 10 psf and an essentially flat roof with only a roof slope of 1/4 in. per foot for drainage, determine the following loads using the ASCE 7 load combinations for ASD: (a) the unfactored uniform total load on joists J2 in lb/ft, (b) the unfactored uniform total load on girder G1 in lb/ft, (c) the total unfactored axial load (in Ibs) on column C1, (d) repeat the analysis using factored loads. Neglect the rain load R and assume that the snow load S is zero.

Solution:

Since the snow and rain load are both zero, the roof live load L_r will be critical. With a roof slope of 1/4 in. per foot, the number of inches of rise per foot, $F = 1/4 = 0.25$.

a. *Joist J2*: The TW = 2 ft and the tributary area $A_T = 36\ \text{ft}^2 < 200\ \text{ft}^2$. From Section 2.4, we obtain $R_1 = 1.0$ and $R_2 = 1.0$. Using Equation 2.6, the roof live load is given as

$$L_r = (20)(1)(1) = 20\ \text{psf}$$

The total unfactored load is calculated as follows:

$$w_s(\text{psf}) = D + L_r = 10 + 20 = 30\ \text{psf}$$

$$w_s(\text{lb/ft}) = w_s(\text{psf}) \times \text{TW} = (30\ \text{psf})(2\ \text{ft}) = 60\ \text{lb/ft}$$

b. *Girder G1*: The TW = 17 ft and the tributary area = 408 ft². Thus, $200 < A_T < 600$, and from Section 2.4, we obtain $R_1 = 1.2 - (0.001)(408) = 0.792$ and $R_2 = 1.0$. Using Equation 2.6, the roof live load is given as

$$L_r = (20)(0.792)(1) = 16\ \text{psf}$$

The total unfactored loads are calculated as follows:

$$w_s(\text{psf}) = D + L_r = 10 + 16 = 26\ \text{psf}$$

$$w_s(\text{lb/ft}) = w_s(\text{psf}) \times \text{TW} = (26\ \text{psf})(17\ \text{ft}) = 442\ \text{lb/ft}$$

c. *Column C1*: The tributary area of column C1, $A_T = 357\ \text{ft}^2$. Thus, $200 < A_T < 600$, and from Section 2.4 we obtain $R_1 = 1.2 - (0.001)(357) = 0.843$ and $R_2 = 1.0$. Using Equation 2.6, the roof live load is given as

$$L_r = (20)(0.843)(1) = 17\ \text{psf}$$

so

$$w_s(\text{psf}) = D + L_r = 10 + 17 = 27\ \text{psf}$$

and the total unfactored column axial load

$$P = (27\ \text{psf})(357\ \text{ft}^2) = 9639\ \text{lb} = 9.7\ \text{kips}$$

d. Analysis using factored loads (LRFD method)

Joist J2: The total factored load is calculated as follows:

$$w_u(\text{psf}) = 1.2D + 1.6L_r = (1.2)(10) + (1.6)(20) = 44\ \text{psf}$$

$$w_u(\text{lb/ft}) = w_u(\text{psf}) \times \text{TW} = (44\ \text{psf})(2\ \text{ft}) = 88\ \text{lb/ft}$$

Girder G1: The total factored load is calculated as follows:

$$w_u(\text{psf}) = 1.2D + 1.6L_r = (1.2)(10) + (1.6)(16) = 37.6 \text{ psf}$$

$$w_u(\text{lb/ft}) = w_u(\text{psf}) \times \text{TW} = (37.6 \text{ psf})(17 \text{ ft}) = 639 \text{ lb/ft}$$

Column C1:

$$w_u(\text{psf}) = 1.2D + 1.6L_r = (1.2)(10) + (1.6)(17) = 39.2 \text{ psf}$$

and the total factored column axial load

$$P_u = (39.2 \text{ psf})(357 \text{ ft}^2) = 13{,}994 \text{ lb} = 14 \text{ kips}$$

2.4.2 Snow load

The 50-year mean recurrence interval (MRI) ground snow loads P_g are specified in IBC Figure 1608.2 or ASCE 7, Figure 7-1, but for certain areas (designated "CS") these snow load maps call for site-specific snow load studies to establish the ground snow loads. The ground snow loads are specified in greater detail in the State Building Codes such as the *New York State Building Code* [3], which is based on the IBC, or the *Ohio State Building Code* [4], and because relatively large variations in snow loads can even occur over small geographical areas, the building codes for the particular locality or state appear to have better snow load data for the various localities within their jurisdiction compared to the snow load map given in the IBC or ASCE 7. The roof snow load is a function of the ground snow load and is typically smaller than the corresponding ground snow load because of the effect of wind at the roof level.

The roof snow load is dependent on the ground snow load P_g, the roof exposure, the roof slope, the use of the building (e.g., whether or not it is heated), and the terrain conditions at the building site, but it is unaffected by tributary area. The steeper the roof slope, the smaller the snow load because steep roofs are more likely to shed snow, and conversely, the flatter the roof slope, the larger the snow load. Depending on the type of roof, the roof snow load can be either a uniform balanced load or an unbalanced load. A *balanced snow load* is a full uniform snow load over an entire roof surface, and an *unbalanced snow load* is a nonuniform distribution of snow load over a roof surface. Only balanced snow loads are considered in this book. The various possible configurations of snow load are beyond the scope of this book, and the reader is referred to the ASCE 7 load specifications [2,5].

The design snow load on a sloped roof, P_s, is obtained using the ASCE 7 load specifications and is given as

$$P_s = C_s P_f \quad \text{psf} \tag{2.7}$$

and the flat roof snow load

$$P_f = 0.7 C_e C_t I_s P_g \quad \text{psf} \tag{2.8}$$

where:

C_e is the exposure factor from ASCE 7, Table 7-2
C_t is the thermal factor from ASCE 7, Table 7-3

C_s is the roof slope factor from ASCE 7, Figure 7-2
I_s is the importance factor from ASCE 7, Table 1.5-2 based on the Risk Category from ASCE 7, Table 1.5-2.
P_g is the ground snow load from ASCE 7, Figure 7-1 (or the governing state code)

In calculating the roof slope factor C_s, the following should be noted:

1. Slippery surface values can be used only where the roof surface is free of obstruction and sufficient space is available below the eaves to accept all the sliding snow.
2. Examples of slippery surfaces include metal, slate, and glass, and bituminous, rubber, and plastic membranes with a smooth surface.
3. Membranes with embedded aggregates, asphalt shingles, and wood shingles must not be considered slippery.

2.4.2.1 Snow drift and sliding snow loads

Snow accumulation or drifting snow could occur on the lower roof where a high roof is adjacent to a low roof or where there are large rooftop units or high parapets or other roof obstructions (Figure 2.10). These snow drift loads, which are caused by wind and are triangular in shape, must be superimposed on the uniform balanced roof snow loads. Sliding snow load occurs on low roofs where a high sloped roof is adjacent to a lower roof. The sliding snow load, which is uniform in shape, must be superimposed on the uniform balanced roof snow loads. However, the snow drift load and the sliding snow load must not be combined but must be treated separately, with the larger load used in the design of the roof members. The snow drift load and sliding snow load are superimposed separately on the balanced snow load to determine the most critical snow load case, but the snow drift loads must not be combined with sliding snow loads. The calculation of snow drift and sliding snow loads is beyond the scope of this book, and the reader is referred to Sections 7.6 and 7.7 through 7.9 of ASCE 7 [2] for the calculation procedure for unbalanced snow loads (due to wind), snow drifts on lower roofs adjacent to high roofs, snow drifts due to roof projections and roof top units greater than 15 ft in length, and sliding snow loads on lower roofs adjacent to sloped high roofs.

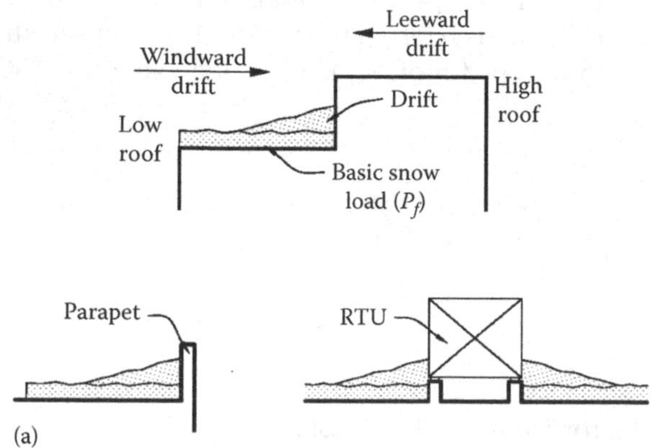

Figure 2.10 (a) Snow drift and sliding snow loads. *(Continued)*

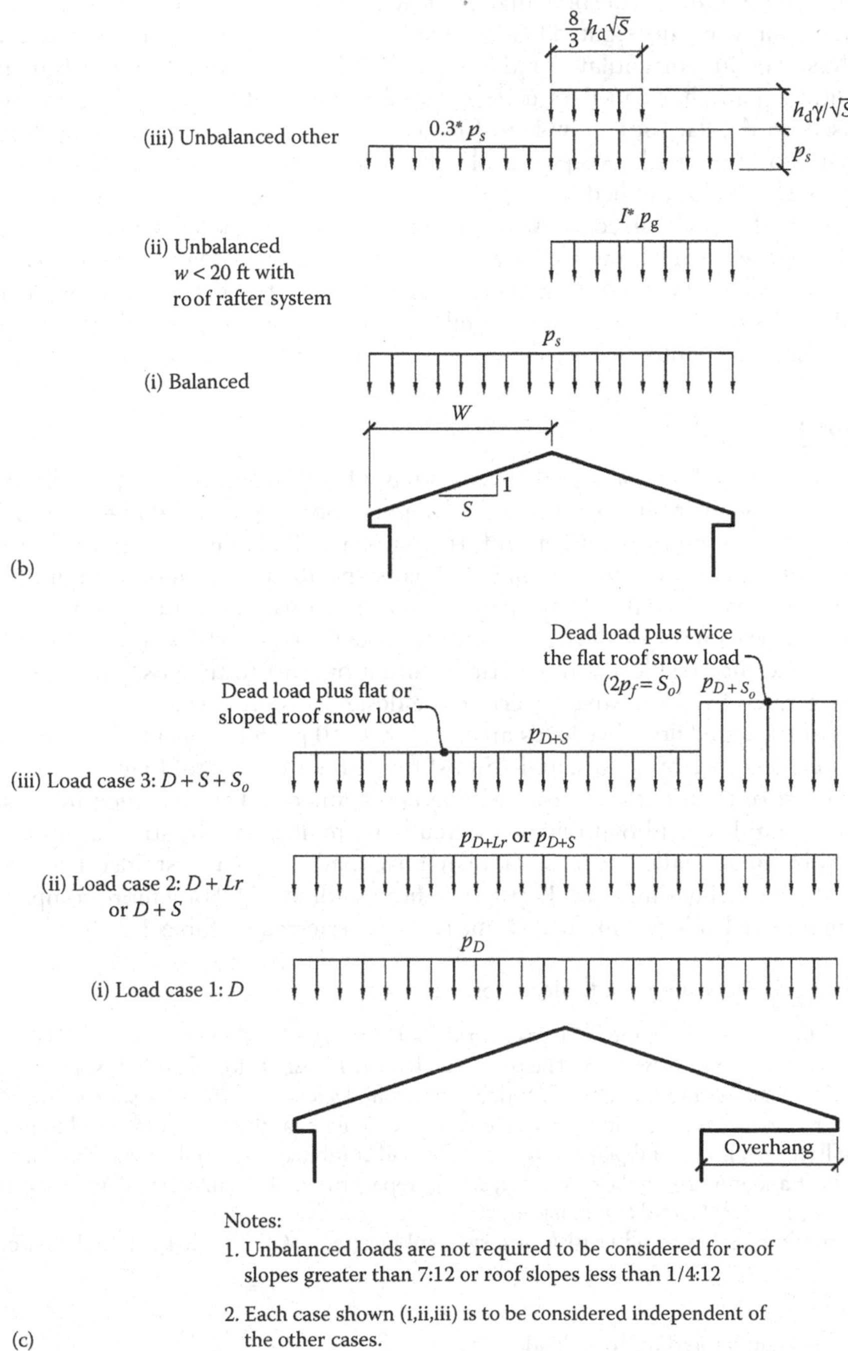

Figure 2.10 (Continued) (b) balanced and unbalanced snow loads for hip and gable roofs, and (c) loads on roof overhangs.

There are two types of warm roofs that drain water over their eaves that shall be capable of sustaining a uniformly distributed load of twice the flat roof snow load on all overhanging portions: those that are unventilated and have an R-value less than 30 ft^2 h °F/Btu and those that are ventilated and have an R-value less than 20 ft^2 h °F/Btu. The load on the overhang shall be based on the flat roof snow load for the heated portion of the roof upslope of the exterior wall. No other loads except dead loads shall be present on the roof when this uniformly distributed load is applied.

For sloped roofs, unbalanced snow loads need to be considered per ASCE 7 [2]. For sloped roofs, Figure 2.10b shows a case where unbalanced snow loading occurs when some portion of the snow moves from one side of the ridge to the other under wind loading. Figure 2.10c shows a case where snow similarly moves from one side of the ridge to the other and accumulates onto an overhang.

2.4.3 Floor live load

Floor live loads depend on the use and occupancy of the structure and the tributary area of the structural member, and are usually available from ASCE 7, Table 4-1 or IBC Table 1607.1. These gravity forces are either uniform loads specified in units of pounds per square foot of horizontal plan area or concentrated loads specified in pounds. The uniform and concentrated live loads are not to be applied simultaneously and, in design practice, the uniform loads govern most of the time, except in a few cases such as thin slabs, where the concentrated load may result in a more critical situation, due to the possibility of punching shear. Concentrated live loads also govern in the design of stair treads.

Examples of tabulated floor live loads are as follows: 40 psf for residential buildings, 50 psf for office spaces in office buildings, and 250 psf for heavy storage buildings. These tabulated live loads consist of transient and sustained load components. The sustained live load is the portion of the total live load that remains virtually permanent on the structure, such as furnishings, and is much smaller than the tabulated live loads. Typical sustained load values are 4–8 psf for residential buildings and 11 psf for office buildings [6]. For a more complete listing of the minimum live loads recommended, the reader is referred to Table 2.2.

Example 2.5: Roof snow load calculation

The building shown in Figure 2.11 has sloped roof rafters (6:12 slope) spaced 4′-0″ o.c. and is located in Lowville, New York. The roof dead load is 10 psf of sloped area. Assume a fully exposed roof and terrain category C, and use the ground snow load from the state snow map to obtain more accurate ground snow load values. Calculate (a) the total unfactored uniform load in lb/ft on a horizontal plane using the ASD load combinations, (b) the maximum unfactored shear and moment in the roof rafter, and (c) repeat the analyses in parts (a) and (b) using the factored or LRFD load combinations.

Assume no ridge beam and a Risk Category II building (see ASCE 7, Table 1.5-1 for definition).

Solution:

a. Total unfactored uniform load
   The roof slope $\boldsymbol{\theta}$ for this building is 26.6°.
   Roof live load L_r
   From Section 2.4, the roof slope factor is obtained as $F = 12\tan(26.6°) = 6$; therefore, $R_2 = 1.2 - (0.05)(6) = 0.9$.
   Since the tributary area of the rafter = 4 ft × 18 ft = 72 ft^2 < 200 ft^2, $R_1 = 1.0$.
   The roof live load will be

$$L_r = 20R_1R_2 = (20)(1.0)(0.9) = 18 \text{ psf}$$

Table 2.2 Minimum uniformly distributed and concentrated floor live loads

Occupancy	*Uniform load* (psf)	*Concentrated load* (lb)
Balconies and decks	Same as occupancy served	
Dining rooms and restaurants	100	–
Office buildings		
Lobbies and first-floor corridors	100	2000
Offices	50	2000
Corridors above first floor	80	2000
Residential (one- and two-family dwellings)	40	–
Hotels, and multifamily houses		
Private rooms and corridors serving them	40	–
Public rooms and corridors serving them	100	–
Roofs		
Ordinary flat, pitched, and curved roofs	20	–
Promenades	60	–
Gardens or assembly	100	–
Schools		
Classrooms	40	1000
Corridors above first floor	80	1000
First-floor corridors	100	1000
Stairs and exitways	100	–
One- and two-family residences only	40	–
Storage		
Light	125	–
Heavy	250	–
Stores		
Retail		
First floor	100	1000
Upper floors	75	1000
Wholesale	125	1000

Source: Adapted From International Building Code, Washington, D.C.: International Code Council, 2015, Figure 1607.1. Reproduced with permission.

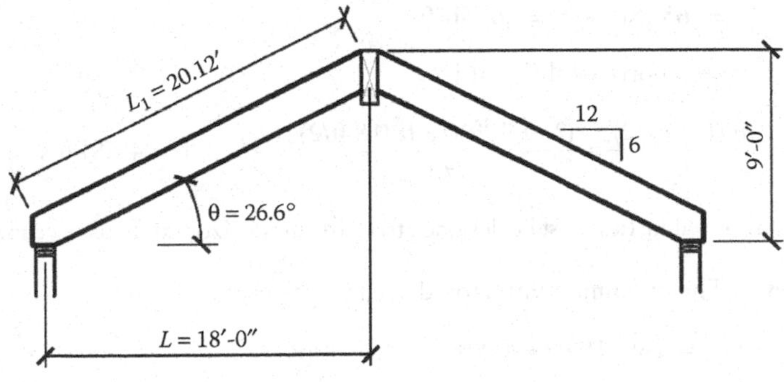

Figure 2.11 Sloped roof for Example 2.5.

Snow load
Using IBC Figure 1608.2 or ASCE 7, Figure 7-1, we find that Lowville, New York falls within the CS designated areas, where site-specific snow case studies are required. It is common practice in such cases to obtain the ground snow load from the snow map in the local or state building code. From Figure 1608.2 of the *New York State Building Code* [3], the ground snow load P_g for Lowville, New York is found to be 85 psf.

Assuming a building with a warm roof that is fully exposed and a building site with terrain category C, we obtain the coefficients as follows:

- Exposure coefficient $C_e = 0.9$ (ASCE 7, Table 7-2)
- Thermal factor $C_t = 1.0$ (ASCE 7, Table 7-3)
- Importance factor for snow, $I_s = 1.0$ (from ASCE 7, Table 1.5-2 for Risk Category II building)
- Slope factor $C_s = 1.0$ (ASCE 7, Figure 7-2 with roof slope $\theta = 26.6°$ and a warm roof)

The flat roof snow load

$$P_f = 0.7C_eC_tI_sP_g = (0.7)(0.9)(1.0)(1.0)(85) = 54 \text{ psf}$$

Thus, the design roof snow load

$$P_s = C_sP_f = (1.0)(54) = 54 \text{ psf}$$

Therefore, the snow load $S = 54$ psf

The total unfactored load on the horizontal plan area of the roof is given as

$$w_s = D\frac{L_1}{L_2} + (L_r \text{ or } S \text{ or } R) \quad \text{psf of horizontal plan area}$$

Since the roof live load L_r (18 psf) is less that the snow load S (54 psf), the snow load is more critical and will be used in calculating the total unfactored roof load

$$w_s = (10 \text{ psf})\left(\frac{20.12 \text{ ft}}{18 \text{ ft}}\right) + 54 \text{ psf}$$

$$= \mathbf{65\ psf} \text{ of horizontal plan area}$$

The total unfactored uniform load on the rafters in pounds per horizontal linear foot is given as

$$w_s(\text{lb/ft}) = w_s(\text{psf}) \times \text{tributary width or spacing of rafters}$$

$$= (65 \text{ psf})(4 \text{ ft}) = \mathbf{260\ lb/ft}$$

The unfactored horizontal thrust is

$$H_s = \frac{w_s(L_2)(L_2/2)}{h} = \frac{(260 \text{ lb/ft})(18 \text{ ft})(18 \text{ ft}/2)}{9 \text{ ft}} = 4680 \text{ lb} = 4.68 \text{kips}$$

The collar or ceiling ties must be designed to resist this horizontal thrust (see Figure 2.12).

b. $L_2 = 18$ ft. The maximum unfactored shear force in the rafter is

$$V_{1s} = w_sL_2 = (260)(18 \text{ ft}) = 4680 \text{ lb} = 4.68 \text{ kips}$$

$$V_{2s} = 0 \text{ (without a ridge beam)}$$

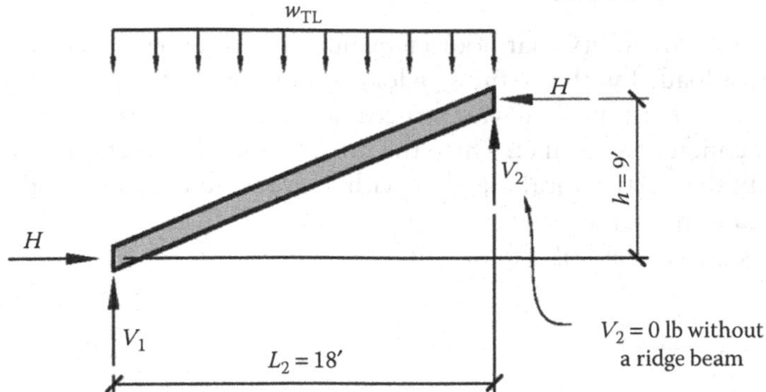

Figure 2.12 Free-body diagram of sloped rafter for Example 2.5.

The maximum unfactored moment in the rafter is

$$M_{s,\max} = \frac{w_s(L_2)^2}{8} = \frac{(260)(18\text{ ft})^2}{8}$$

$$= 10{,}530 \text{ ft-lb} = 10.53 \text{ ft-kips}$$

c. Factored or LRFD load combinations
The total factored load on the horizontal plan area of the roof is given as

$$w_u = (1.2)D\frac{L_1}{L_2} + (1.6)(L_r \text{ or } S \text{ or } R) \quad \text{psf of horizontal plan area}$$

$$= (1.2)(10\text{ psf})\left(\frac{20.12\text{ ft}}{18\text{ ft}}\right) + (1.6)(54\text{ psf})$$

$$= \mathbf{99.8\ psf} \text{ of horizontal plan area}$$

The total factored uniform load on the rafters in pounds per horizontal linear foot is given as

$$w_u(\text{lb/ft}) = w_u(\text{psf}) \times \text{tributary width or spacing of rafters}$$

$$= (99.8\text{ psf})(4\text{ ft}) = \mathbf{399\ lb/ft}$$

The factored horizontal thrust is

$$H_u = \frac{w_u(L_2)(L_2/2)}{h} = \frac{(399\text{ lb/ft})(18\text{ ft})(18\text{ ft}/2)}{9\text{ ft}} = 7182\text{ lb} = 7.18\text{kips}$$

The collar or ceiling ties must be designed to resist this horizontal thrust (see Figure 2.12). The rafter span, L_2 = 18 ft. The maximum factored shear force in the rafter is

$$V_{1u} = w_u L_2 = (399)(18\text{ ft}) = 7182\text{ lb} = 7.18\text{ kips}$$

$$V_{2u} = 0 \text{ (without a ridge beam)}$$

The maximum factored moment in the rafter is

$$M_{u,\max} = \frac{w_u(L_2)^2}{8} = \frac{(399)(18\text{ ft})^2}{8} = 16{,}160\text{ ft-lb} = 16.16\text{ ft-kips}$$

2.4.3.1 *Floor live-load reduction*

To allow for the low probability that floor elements with large tributary areas will have the entire tributary area loaded with the full live load simultaneously, the IBC and ASCE 7 load specifications allow the floor live loads for structural elements with large tributary areas to be reduced if certain conditions are met. Thus, the floor live-load reduction factor accounts for the low probability that a structural member with a large tributary area will be fully loaded over the entire area at the same time.

The reduced design live load of a floor L in psf is given as

$$L = L_0\left(0.25 + \frac{15}{\sqrt{K_{LL}A_T}}\right) \tag{2.9}$$

$\geq 0.50L_0$ for members supporting *one floor* (e.g., slabs, beams, and girders)

$\geq 0.40L_0$ for members supporting *two or more floors* (e.g., columns)

where:

L_0 is the unreduced design live load from Table 2.2 (IBC Table 1607.1 or ASCE 7, Table 4-1)

K_{LL} is the live-load element factor (see Table 4.2 of [2])
- = 4 (interior and exterior columns without cantilever slabs)
- = 3 (edge columns with cantilever slabs)
- = 2 (corner columns with cantilever slabs, edge beams without cantilever slabs, interior beams)
- = 1 (all other conditions)

A_T is the summation of the floor tributary area in ft^2 supported by the member, *excluding* the roof tributary area

- For beams and girders (including continuous beams or girders), A_T is the TW of the beam or girder multiplied by its center-to-center span between supports.
- For *one-way slabs*, A_T must be less than or equal to 1.5 (span of one-way slab)^2.
- For a member supporting more than one floor area in multistory buildings, A_T will be the summation of all the applicable floor areas supported by that member.

The IBC and ASCE 7 load specifications *do not* permit floor live-load reductions for floors satisfying any one of the following conditions:

- $K_{LL}A_T \leq 400$ ft^2.
- Floor live load $L_0 > 100$ psf.
- Floors with occupancies used for assembly purposes, such as auditoriums, stadiums, and passenger car garages, because of the high probability of overloading in such occupancies during an emergency.
- For passenger car garage floors, the live load is allowed to be reduced by 20% for members supporting two or more floors.

The following should be noted regarding the tributary area A_T used in calculating the reduced floor live load:

1. The beams or joists are usually supported by girders, which in turn are supported by columns, as indicated in our previous discussions on load paths. The tributary area A_T

for beams will usually be smaller than those for girders, and thus beams will have less floor live-load reduction than for girders. The question arises as to which A_T to use for calculating the loads on the girders.

2. For the design of the beams, use the A_T value of the beam to calculate the reduced live load that is used to calculate the moments, shears, and reactions. These load effects are used for the design of the beam and the beam-to-girder or beam-to-column connections.
3. For the girders, recalculate the beam reactions using the *A_T value of the girder.* These smaller beam reactions are used for the design of the girders only.
4. For columns, A_T is the summation of the tributary areas of all the floors with *reducible live loads* above the level at which the column load is being determined, and it excludes the roof areas.

Example 2.6: Hip roof load diagram

The hip roof shown in Figure 2.13 is subjected to a total dead plus snow load of 55 psf on the horizontal plan area. Draw the load diagrams for the typical main rafters and the hip rafter, assuming the following two options:

- Option 1: Continuous ridge board between the rafters
- Option 2: Rafters supported on a ridge beam

Solution:

The tributary widths and the load diagrams for the rafters are shown in Figures 2.14 through 2.19.

Note that for Option 1 (ridge board option) the sloped rafters supporting the hip rafters, supports a higher load than the typical main rafters.

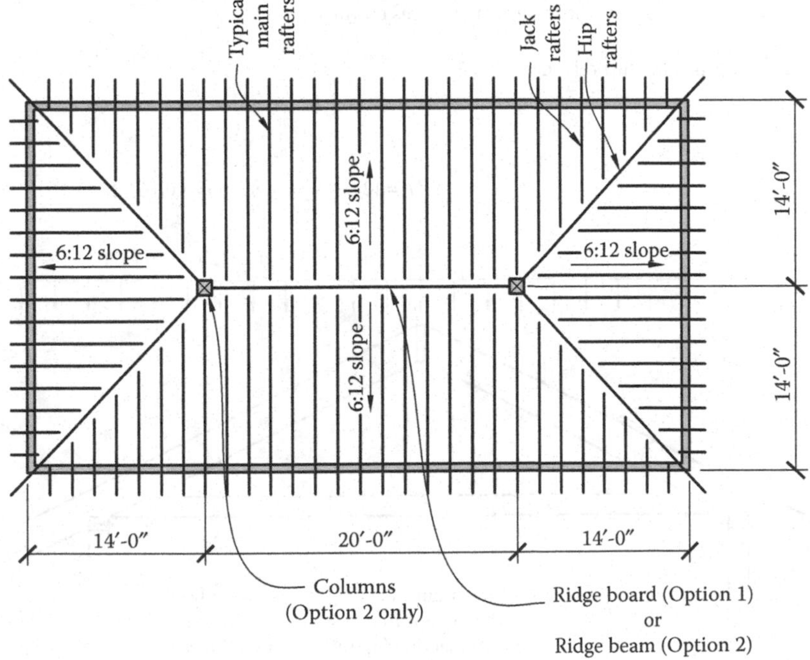

Figure 2.13 Hip roof.

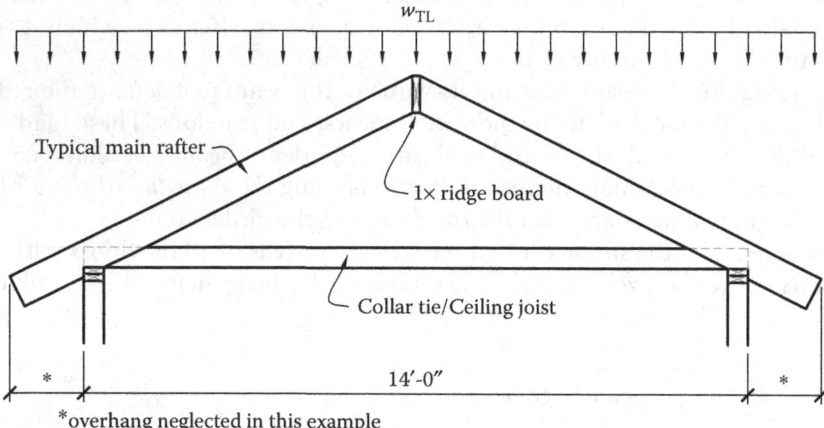

Figure 2.14 Ceiling ties and typical main rafters.

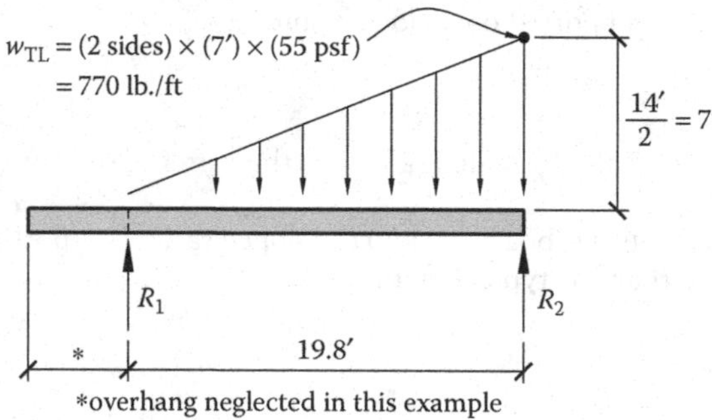

Figure 2.15 Load diagram for hip rafter.

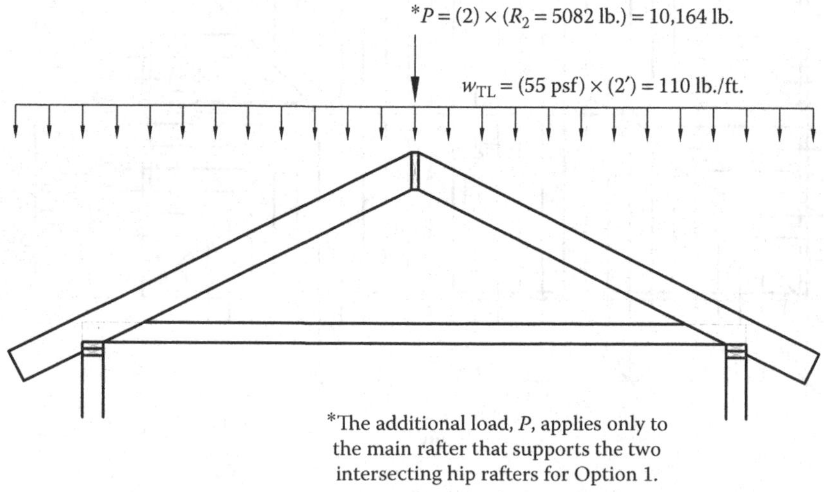

Figure 2.16 Main rafter load diagram.

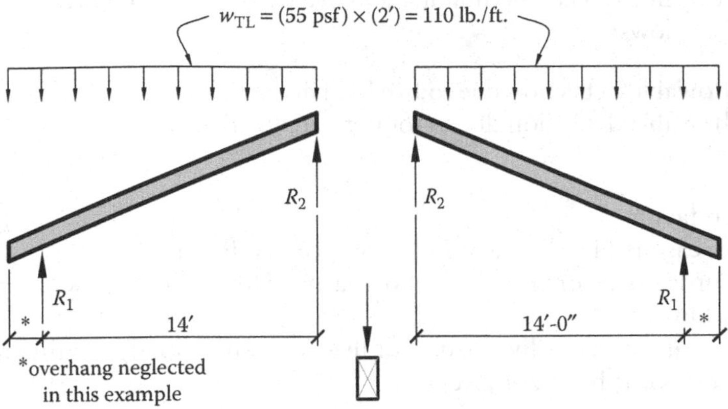

Figure 2.17 Typical rafters.

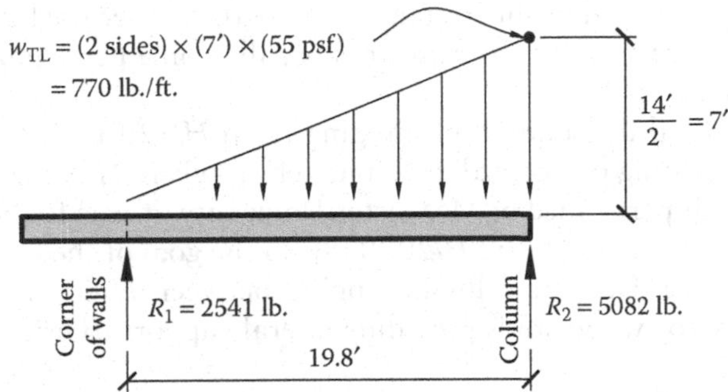

Figure 2.18 Load diagram for hip rafter (Option 2).

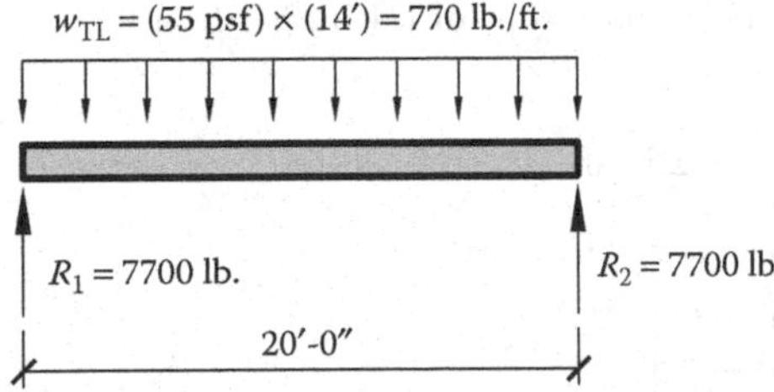

Figure 2.19 Ridge beam (Option 2).

2.5 DEFLECTION CRITERIA

The limits on deflections due to gravity loads are intended to ensure user comfort and to prevent excessive cracking of plaster ceilings, architectural partitions and any deflection sensitive flooring like stone or marble. These deflection limits are usually specified in terms of the joist, beam, or girder span, and the deflections are calculated based on elastic analysis of the structural member. The maximum allowable deflections recommended in IBC Table 1604.3 [1] are as follows:

Maximum allowable deflection due to live load $\delta_{LL} \leq L/360$
Maximum allowable deflection due to total dead plus live load $\delta_{kD+LL} \leq L/240$

where:
k is the creep factor
= 0.5 for seasoned lumber used in dry service conditions
= 1.0 for unseasoned or green lumber or seasoned lumber used in wet service conditions
D is the dead load
L is the live load (i.e., floor live load, roof live load, snow load, or rain load)
L is the span of joist, beam, or girder

For floor beams or girders that support masonry wall or glazing, the allowable total load deflection should be limited to $L/600$ or $3/8$ in., whichever is smaller, to reduce the likelihood of cracking of the masonry wall or glazing. More stringent vertical deflection limits (e.g., $L/720$) may be needed for floor members supporting stone or other deflection-sensitive floor finishes. For prefabricated members, it is common to camber the beams to help control the total deflection. The amount of the camber is usually some percentage of the dead load to prevent overcambering.

For stud walls, a lateral deflection limit ranging from $H/120$ to $H/360$ is permitted per IBC Table 1604.3 depending on the wall finish material, where H is the height of the wood stud. [7]. From research performed at McMaster University, it was found that brick veneer begins to crack at a lateral deflection of $H/2000$ [8], so the goal of the lateral deflection limit is to minimize crack size. Deflection limits ranging between $H/600$ and $H/720$ are commonly used in practice for wood studs providing lateral support to brick veneer [7,9].

Example 2.7: Column load calculation

A three-story building has columns spaced at 20 ft in both orthogonal directions and is subjected to the roof and floor loads shown below. Using a column load summation table, calculate the cumulative axial loads on a typical interior column with and without live-load reduction. Assume a roof slope of 1/4 in. per foot for drainage.

- Roof loads:
  - Dead load $D_{roof} = 25$ psf
  - Snow load $S = 35$ psf
- Second- and third-floor loads:
  - Dead load $D_{floor} = 50$ psf
  - Floor live load $L = 40$ psf

Solution:

At each level, the tributary area supported by a typical interior column is 20 ft × 20 ft = 400 ft^2.

Roof live load L_r
From Section 2.4, the roof slope factor is obtained as $F = 1/4 = 0.25$. Therefore, $R_2 = 1.0$. Since the tributary area of the column = 400 ft^2, $R_1 = 1.2 - (0.001)(400) = 0.8$. The roof live load will be

$$L_r = 20R_1R_2 = (20)(0.8)(1.0) = 16 \text{ psf} < 20 \text{ psf}$$

The reduced or design floor live loads for the second and third floors are calculated using Table 2.3.

- *Roof loads*:
  - Dead load, $D_{roof} = 25$ psf
  - Snow load, $S = 35$ psf
  - Roof live load, L_r = per code
- *Second- and third-floor loads*:
  - Dead load, $D_{floor} = 50$ psf
  - Floor live Load, $L = 40$ psf

Solution:

At each level, the tributary area, A_T, supported by a typical interior column is 20 ft × 20 ft = 400 ft^2.

Roof live load, L_r:

- For an ordinary flat roof, $L_o = 20$ psf (ASCE 7, Table 4-1).
- From Section 2.4, the roof slope factor, F, is 0.25; therefore, $R_2 = 1.0$.
- Since the tributary area, A_T, of the column = 400 ft^2, $R_1 = 1.2 - 0.001\ (400) = 0.8$
- Using Equation 2.6, the design roof live load is
- $L_r = L_o R_1 R_2 = 20\ R_1 R_2 = 20\ (0.8)\ (1.0) = 16$ psf,
- Since 12 psf $< L_r <$ 20 psf; therefore, $L_r = 16$ psf.

L_r is smaller than the snow load, S = 30 psf; therefore, the snow load, S, is more critical than the roof live load, L_r, in the applicable load combinations. All other loads, such as W, H, T, F, R, and E, are zero for the roof or floors. The applicable *LRFD* load combinations from Section 2.1 that will be used to calculate the factored column axial loads are as follows:

1. $1.4D$
2. $1.2D + 1.6L + 0.5S$
3. $1.2D + 1.6S + 0.5L$
4. $1.2D + 0.5L + 0.5S$
5. $1.2D + 0.5L + 0.2S$
6. $0.9D$
7. $0.9D$

The corresponding ASD load combinations from Section 2.1 for calculating the unfactored or ASD column axial loads are as follows:

8. D
9. $D + L$
10. $D + S$
11. $D + 0.75L + 0.75S$
12. D
13. $D + 0.75L + 0.75S$
14. $D + 0.75L + 0.75S$
15. $0.6D$
16. $0.6D$

A close examination of these load combinations will reveal that load combinations 1 and 4 through 7 are *not* the most critical or governing load combinations for the factored axial load on the column. The most critical factored load combinations are load combinations

Table 2.3 Reduced or design floor live-load calculation table

Member	*Levels supported*	*A_T (summation of floor tributary area)*	K_{LL}	*Unreduced floor live load, L_o (psf)*	*Live-load reduction factor, $0.25 + 15/\sqrt{K_{LL}A_T}$*	*Design live load, (L or S)*
Third-floor column (i.e., column below roof)	Roof only	Floor live-load reduction not applicable to roofs	–	–	–	35 psf (snow load)
Second-floor column (i.e., column below third floor)	1 floor + roof	(1 floor) (400 ft^2) = 400 ft^2	$K_{LL} = 4K_{LL}A_T = 1600$ > 400 ft^2; therefore, live-load reduction allowed	40	$\left[0.25 + \frac{15}{\sqrt{1600}}\right] = 0.625$	0.625 (40) = 25 psf ≥ 0.50 L_o
Ground- or first-floor column (i.e., column below second floor)	2 floors + roof	(2 floors) (400 ft^2) = 800 ft^2	$K_{LL} = 4K_{LL}A_T = 3200$ > 400 ft^2; therefore, live-load reduction allowed	40	$\left[0.25 + \frac{15}{\sqrt{3200}}\right] = 0.52$	0.52 (40) = 21 psf ≥ 0.40 L_o

2 ($1.2D + 1.6L + 0.5S$) and 3 ($1.2D + 1.6S + 0.5L$). A similar review of the ASD load combinations reveals that load combinations 9 ($D + L$) and 11 ($D + 0.75L + 0.75S$) are the two most critical load combinations for the calculation of the unfactored axial load on the typical interior column for this building.

The reduced or design floor live loads for the second and third floors are calculated using Table 2.3.

Using the LRFD combinations 2 (i.e., $1.2D + 1.6L + 0.5S$) and 3 ($1.2D + 1.6S + 0.5L$), the maximum factored column axial loads with and without floor live-load reductions are calculated in Table 2.4. The corresponding values for the unfactored column axial loads with and without floor live-load reductions are calculated in Table 2.5.

Therefore, the ground-floor column *with floor live-load reduction* will be designed for a cumulative *factored* axial compression load of 93.6 kips (73 kips *unfactored*), the second-floor column for a factored load of 63.4 kips (48 kips *unfactored*), and the third-floor column for a factored compression load of 34.4 kips (24 kips unfactored load).

The corresponding *factored* loads *without floor live-load* reduction are 118 kips (84.5 kips *unfactored*) for the ground-floor column, 68.6 kips (52.5 kips *unfactored*) for the second-floor column, and 34.4 kips (24 kips *unfactored*) for the third-floor column, respectively. For the ground-floor column with live-load reduction considered, there is a 21% reduction ([118–93.6]/93.6) in factored axial load and a 14% reduction ([85–73]/85) in the unfactored (ASD) axial load when live-load reduction is considered compared to when live-load reduction is not considered. The effect of floor live-load reduction on columns and their foundations is not as critical for low rise buildings as it is for taller buildings.

2.6 LATERAL LOADS

The two main types of lateral loads that act on wood buildings are wind and seismic or earthquake loads. These loads, which produce overturning, sliding, and uplift forces in the structure, are both random and dynamic in nature, but simplified procedures using equivalent static load approaches have been established in the IBC and ASCE 7 load specifications for calculating the total lateral forces acting on a building due to these loads. The wind loads specified in the ASCE 7 load specifications are *factored* wind pressures and these factored design wind pressures correspond to a 300-year mean return interval (MRI) for Risk Category I buildings, a 700-year MRI for Risk Category II buildings, and a 1700-year MRI for Risk Categories III and IV buildings. A wind load of 60% of the factored design wind load (i.e., 0.6W) should be used for allowable stress design calculations, and for lateral deflection or drift calculations. The calculation methods for wind loads are covered in Chapters 26 through 31 of the ASCE 7 load standards. Seismic design is based on a 2500-year MRI earthquake, which is an earthquake with a 2% probability of being exceeded in 50 years. For any building design, both seismic and wind loads have to be checked to determine which lateral load governs. It should be noted that even in cases where the seismic load controls the overall design of the lateral force-resisting system in a building, the wind load may still be critical for the uplift forces on the roof. We present two examples in this section to illustrate the calculation of wind and seismic lateral forces using the simplified calculation procedures of the ASCE 7 load specifications [2].

Table 2.4 Factored column load (LRFD)

Level	*Tributary area, (ft²)* A_T	*Dead load, D* (psf)	*Live load,* L_o *(S or* L_r *or R on the roof)* (psf)	*Design live load floor: L roof: S or* L_r or *R* (psf)	*Factored uniform design load at each level,* w_{u1}*; roof:* 1.2D + 0.5S *floor:* 1.2D + 1.6L (psf)	*Factored uniform design load at each level,* w_{u2}*; roof:* 1.2D + 1.6S *floor:* 1.2D + 0.5*L* (psf)	*Factored column axial load, P, at each level,* $(A_T)(w_{u1})$ or $(A_T)(w_{u2})$ (kips)	*Cumulative factored axial load,* ΣP LC 2 (kips)	*Cumulative factored axial load,* ΣP LC 3 (kips)	*Maximum cumulative factored axial load,* ΣP (kips)
***With* floor live-load reduction**										
Roof	400	25	35	35	47.5	86	19 or 34.4	19	34.4	34.4
Third floor	400	50	40	25	100	73	40 or 29	59	63.4	63.4
Second floor	400	50	40	20.8	93	70	37.3 or 28.2	93.6[a]	90.7[a]	93.6
***Without* floor live-load reduction**										
Roof	400	25	35	35	47.5	86	19 or 34.4	19	34.4	34.4
Third floor	400	50	40	40	124	80	49.6 or 32	68.6	66.4	68.6
Second floor	400	50	40	40	124	80	49.6 or 32	118	98	118

[a] For the column segment below the second floor, the reduced floor live load at the second floor level applies to all the tributary areas of all the floors supported by the column. Therefore, the total cumulative factored axial loads in the column are calculated as follows:

ΣP_{LC2} = 47.5 psf (400 ft²) + 93 psf (400 ft² + 400 ft²) = 93.6 kips, and ΣP_{LC3} = 86 psf (400 ft²) + 70 psf (400 ft² + 400 ft²) = 90.7 kips.

The maximum *factored* column loads (*with* floor live-load reduction) are as follows:

Third-story column (i.e., column below roof level) = 34.4 kips
Second-story column (i.e., column below the third floor) = 63.4 kips
First-story column (i.e., column below the second floor) = 93.6 kips

The maximum *factored* axial column loads (*without* floor live-load reduction) are as follows:

Third-story column (i.e., column below roof level) = 34.4 kips
Second-story column (i.e., column below the third floor) = 68.6 kips
First-story column (i.e., column below the second floor) = 118 kips

Table 2.5 Unfactored column axial load (ASD)

Level	*Tributary area, (A_T)* (ft²)	*Dead load, D* (psf)	*Live load, L_o (S or L_r or R on the roof)* (psf)	*Design live load roof: S or L_r or R; floor: L* (psf)	*Unfactored total design load at each level, w_{s1}; roof: D; floor: D + L* (psf)	*Unfactored total design load at each level, w_{s2}; roof: D + 0.75S (or D + S); floor: D + 0.75L* (psf)	*Unfactored column axial load at each level, P = (A_T) (w_{s1}) or (A_T) (w_{s2})* (kips)	*Cumulative unfactored axial load, ΣP_{D+L}* (kips)	*Cumulative unfactored axial load, $\Sigma P_{D+0.75L+0.75S}$* (kips)	*Maximum cumulative unfactored axial load, ΣP* (kips)
***With* floor live-load reduction**										
Roof	400	25	35	35	25	60(51.3)[a]	10 or 24(20.5)[a]	10	24.0	24
Third floor	400	50	40	25	75	69	30 or 27.5	40	48	48
Second floor	400	50	40	20.8	71	66	28.3 or 26.2	67[a]	73[a]	73
***Without* floor live-load reduction**										
Roof	400	25	35	35	25	60(51.3)[a]	10 or 24(20.5)[a]	10	24.0	24.0
Third floor	400	50	40	40	90	80	36 or 32	46	52.5	52.5
Second floor	400	50	40	40	90	80	36 or 32	82	84.5	84.5

[a] For the column segment below the second floor, the reduced floor live load at the second floor level applies to all the tributary areas of all the floors supported by the column. Therefore, the total cumulative *unfactored* reduced axial loads in the column are calculated as follows:

ΣP_{D+L} = 25 psf (400 ft²) + 71 psf (400 ft² + 400 ft²) = 67 kips, and $\Sigma P_{D+0.75L+0.75S}$ = 51.3 psf (400 ft²) + 66 psf (400 ft² + 400 ft²) = 73 kips.

The maximum *unfactored* column loads (*with* floor live-load reduction) are as follows:

Third-story column (i.e., column below roof level) = 24.0 kips
Second-story column (i.e., column below the third floor) = 48 kips
First-story column (i.e., column below the second floor) = 73 kips

The maximum unfactored axial column loads (*without* floor live-load reduction) are as follows:

Third-story column (i.e., column below roof level) = 24.0 kips
Second-story column (i.e., column below the third floor) = 52.5 kips
First-story column (i.e., column below the second floor) = 84.5 kips

2.7 WIND LOAD

There are two types of systems in a building structure for which wind loads are calculated: the *main wind force-resisting system* (MWFRS) and the *components and cladding* (C&C). The MWFRS helps to transfer the overall building lateral wind loads from the various levels of the building safely to the ground. The MWFRS in a multistory wood building consists of the roof and floor diaphragms (i.e., the horizontal diaphragms), and the shear walls (i.e., the vertical diaphragms) and other lateral load-resisting systems that are parallel to the direction of the wind load. The C&C are members that are loaded as individual components with the wind load perpendicular to these elements. Examples of C&C include walls, stud wall, cladding, and a roof deck fastener subjected to uplift wind load. The wind pressures on C&C members are usually higher than the wind pressures on the MWFRS because of local spikes in wind pressure over the small tributary areas for components and cladding. The C&C wind pressure is a function of the effective wind area, A_e, given as

$$A_e = \text{span of the C\&C member} \times \text{tributary width} \geq \left(\frac{\text{span of member}}{3}\right)^2 \tag{2.10}$$

For cladding and deck fasteners, the *effective wind area*, A_e, shall not exceed the area that is tributary to each fastener. In calculating wind pressure, positive pressures are indicated by a force "pushing" into the wall or roof surface, and negative pressures are shown "pulling" away from the wall or roof surface. The minimum design wind pressure for MWFRS and C&C is 16 psf (note that this value is for the factored load, so the value for designs using the ASD method will be 0.6 × 16 psf or 10 psf) and is applied to the vertical projected area of the wall surfaces for the MWFRS, and normal to the wall or roof surface for C&C (see ASCE 7, Figure C27.4-1).

2.7.1 Wind load calculation

The calculation of wind loads is covered in Chapters 26 through 31 of the ASCE 7 load standard. In Chapter 26, the general requirements and parameters for determining wind loads are defined and discussed. The wind hazard maps for the different risk categories and the definitions of the wind load parameters such as *open*, *partially enclosed*, and *enclosed* buildings are provided in Chapter 26. In addition, the topography factor and the building exposure categories that are dependent on the terrain roughness are discussed. Though several methods, with different conditions for their use, are available in Chapters 27 through 31 of the ASCE 7 load standard [1] for calculating the design wind loads on buildings and other structures, the method used in this text for wood buildings is the simplified method for MWFRS from Part 2 of ASCE 7, Chapter 28 and the simplified method for C&C from Part 2 of ASCE 7, Chapter 30. The minimum factored design horizontal wind pressures on the vertical wall surfaces for the MWFRS are 16 psf on the wall surface and 8 psf on the vertical projected surfaces of the roof areas. The corresponding unfactored minimum wind load for both surfaces for MWFRS is 10 psf (i.e., 0.6 W).

Components and cladding with tributary areas greater than 700 ft^2 should be designed using the provisions for MWFRS (ASCE 7, Section 30.2.3). The minimum design wind pressures on components and cladding shall be 16 psf (factored wind pressure) acting perpendicular to the surface of the C&C and in either direction. That is, the C&C should be considered for a minimum suction and a minimum factored pressure of 16 psf (ASCE 7, Section 30.2.2). The corresponding unfactored minimum wind load for C&C is 10 psf (i.e., 0.6 W). More detailed treatment of wind and seismic loads can be found in [5].

2.7.2 Exposure category

The exposure category is a measure of the terrain surface roughness and the degree of exposure or shielding of the building. There are three main exposure categories in the ASCE 7 load standard (Sections 26.7.2 and 26.7.3): exposures B, C, and D. Exposure B is the most commonly occurring exposure category—approximately 80% of all buildings fall into this category—although many engineers tend to specify exposure C for most buildings. The exposure category is determined from the ASCE 7 load standard by first establishing the type and extent of the ground surface roughness at the building site and then using that information to determine the exposure category.

2.7.3 Basic wind speed

The basic wind speed is a 3-second gust wind speed in miles per hour based on a 15%, probability of exceedance in 50 years, for Risk Category I building, 7% probability of exceedance in 50 years for Risk Category II buildings, and 2% probability of exceedance in 50 years for Risk Category III and IV buildings. These probabilities are equivalent to mean return intervals (MRI) of 300 years for Risk Category I, 700 years for Risk Category II, and 1700 years for Risk Categories III and IV. Wind speed increases as the height above the ground increases because of the reduced drag effect of terrain surface roughness at higher elevations.

The basic wind speeds at a height of 33 ft above the ground for various locations in the United States are shown in ASCE 7, Figure 26.5-1A for Risk Category II buildings and structures, Figure 26.5-1B for Risk Category III and IV buildings and structures, and Figure 26.5-1C for Risk Category I buildings and structures.

2.8 SIMPLIFIED WIND LOAD CALCULATION METHOD (ASCE 7, CHAPTER 28, PART 2)

In the simplified wind load calculation method (ASCE 7, Chapter 28, Part 2 for MWFRS in low rise buildings and Chapter 30, Part 2 for C&C), which is only applicable to low-rise buildings, the wind forces are applied perpendicular to the *vertical and the horizontal projected areas* of the building. The wind pressure diagram for the MWFRS is shown in Figure 2.20 [10]. The horizontal pressures represent the combined windward and leeward pressures with the internal pressures canceling each other out; the vertical pressures, on the other hand, include the combined effect of the external and internal pressures. The different wind pressure zones in Figure 2.20 are defined in Table 2.6 and it should be noted that for buildings with flat roofs, the simplified method yields a uniform horizontal wall pressure distribution over the entire height of the building.

The simplified procedure is applicable only if all of the following conditions are satisfied:

- Building has simple diaphragms (i.e., wind load is transferred through the roof and floor diaphragms to the vertical MWFRS).
- Building is enclosed (i.e., no large opening on any side of the building).
- Building has mean roof height that is less than or equal to the least horizontal dimension of the building.
- Building has mean roof height that is less than or equal to 60 ft.
- Building is symmetrical.
- Building has an approximately symmetrical cross section in each direction with a roof slope $\leq 45°$.

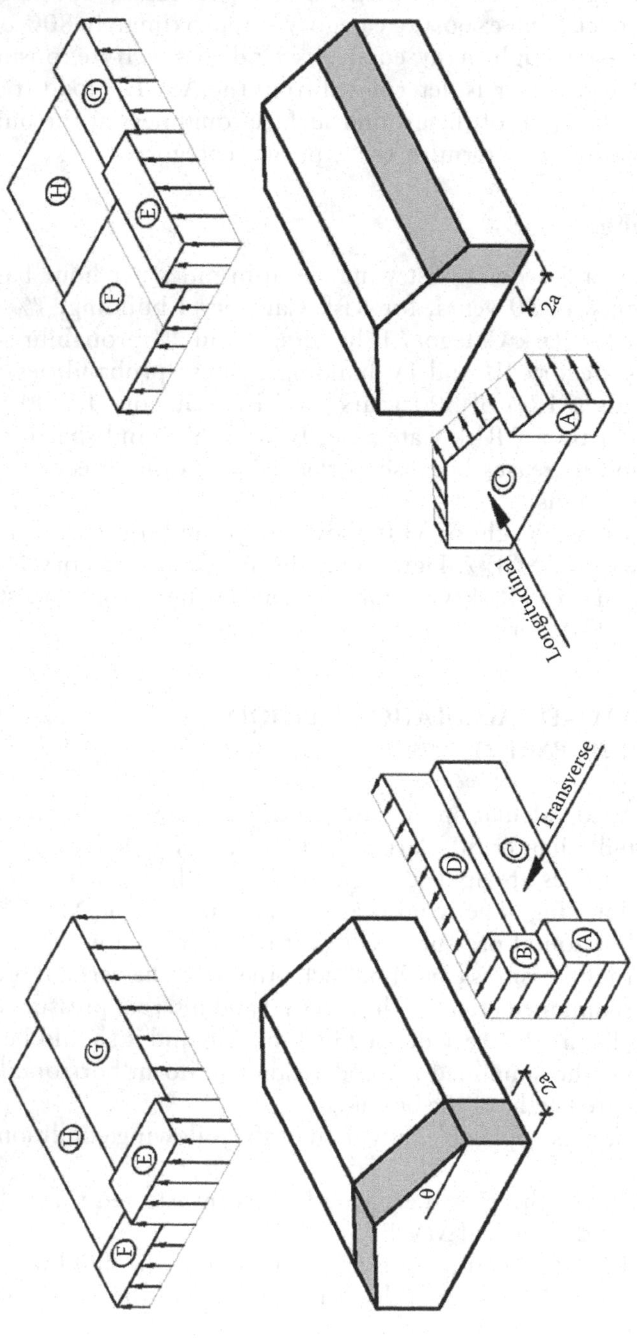

Figure 2.20 Wind pressure diagram for main wind force-resisting system (MWFRS). (From International Building Code, Washington, D.C.: International Code Council, 2003, Figure 1609.6.2.1. Reproduced with permission.)

Table 2.6 Definition of wind pressure zones—MWFRS

Zone	*Definition*
A	End-zone *horizontal* wind pressure on the *vertical* projected *wall* surface
B	End-zone *horizontal* wind pressure on the *vertical* projected *roof* surface
C	Interior-zone *horizontal* wind pressure on the *vertical* projected *wall* surface
D	Interior-zone *horizontal* wind pressure on the *vertical* projected *roof* surface
E	End-zone *vertical* wind pressure on the *windward* side of the *horizontal* projected *roof* surface
F	End-zone *vertical* wind pressure on the *leeward* side of the *horizontal* projected *roof* surface
G	Interior-zone *vertical* wind pressure on the *windward* side of the *horizontal* projected *roof* surface
H	Interior-zone *vertical* wind pressure on the *leeward* side of the *horizontal* projected *roof* surface
E_{OH}	End-zone *vertical* wind pressure on the *windward* side of the *horizontal* projected *roof overhang* surface
G_{OH}	Interior-zone *vertical* wind pressure on the *windward* side of the *horizontal* projected *roof overhang* surface

Where zone E or G falls on a roof overhang, the windward roof overhang wind pressures (i.e., E_{OH} and G_{OH}) from ASCE 7, Figure 28.6-1 should be used.

The *simplified procedure*—from ASCE 7, Chapter 28, Part 2 for MWFRS and ASCE 7, Chapter 30, Part 2 for C&C—for calculating wind loads involves the following steps:

1. Determine the applicable wind speed for the building location from the ASCE 7 load standard
2. Calculate the mean roof height and determine the wind exposure category
3. Determine the applicable horizontal and vertical wind pressures as a function of the wind speed, roof slope, zones, and effective wind area by calculating
   a. P_{s30} for MWFRS from ASCE-7, Figure 28.6-1
   b. P_{net30} for C&C from ASCE-7, Figure 30.5-1

   Note the following:
   - The tabulated values are based on an assumed exposure category of B, and a mean roof height of 30 ft. These tabulated wind pressures are to be applied to the horizontal and vertical projections of the buildings.
   - The horizontal wind pressure on the projected vertical surface area of the building is the sum of the external windward and external leeward pressures because the internal pressures cancel out each other. The resultant wind pressures are applied to one side of the building for each wind direction.
   - The wind pressures on roof overhangs are much higher than at other locations on the roof because of the external wind pressures acting on both the bottom and top exposed surfaces of the overhang.
4. Obtain the design wind pressures (P_{s30} for MWFRS and P_{net30} for C&C), as a function of the tabulated wind pressures obtained in the previous step, the applicable height and exposure adjustment factor (λ from ASCE 7, Figures 28.6-1 or 30.5-1), and the topography factor (K_{zt} from ASCE 7, Section 26.8-1). (Note that for most site conditions, K_{zt} = 1.0.). The factored design wind pressures from ASCE 7, Equations 28.6-1 and 30.5-1 are as follows:

$$P_{s30}\left(\text{MWFRS}\right) = \lambda\ K_{zt}P_{s30} \geq 16\text{ psf} \tag{2.11}$$

$$P_{\text{net30}}\left(\text{C\&C}\right) = \lambda\ K_{zt}P_{\text{net30}} \geq 16\text{ psf} \tag{2.12}$$

5. Apply the calculated wind pressures to the building as shown in ASCE 7, Figure 28.6-1 for MWFRS (end-zone width = $2a$).
   where:

$$a \leq 0.1 \times \text{least horizontal dimension of building}$$

$$\leq 0.4 \times \text{mean roof height of building}$$

$$\geq 3 \text{ ft}$$

   For a roof slope of less than 10°, the eave height should be used in lieu of the mean roof height for calculating the end-zone width, $2a$.

6. Apply the calculated wind pressures to the building walls and roof as shown in ASCE 7, Figure 30.5-1 for C&C. Note that for C&C, end-zone width = a.

2.9 SEISMIC LOAD

The earth's crust consists of several tectonic plates, many of which have fault lines, especially where the various tectonic plates interface with each other. When there is slippage along these fault lines, the energy generated creates vibrations that travel to the earth's surface and create an earthquake or seismic event. The effect of an earthquake on a structure depends on the location of that structure relative to the point at which the earthquake originated (i.e., the *epicenter*). The supporting soil for a structure plays a significant role in the magnitude of the seismic accelerations to which the structure will be subjected. The lateral force on the structure due to an earthquake is proportional to the mass and acceleration of the structure, in accordance with Newton's second law. Thus, $F = ma = W(a/g)$, where W is the weight of the structure and any components that will be accelerated at the same rate as the structure and a/g is the acceleration coefficient [11]. This basic equation is the basis for the derivation of the equations used in the IBC and ASCE 7 load standard for calculating seismic lateral forces.

The seismic loads acting on the primary system (i.e., the lateral force-resisting system) are different than those acting on the nonstructural parts and components such as architectural, mechanical, and electrical fixtures. In the ASCE 7 load standard, the seismic load calculations are based on a 2% probability of exceeding the design earthquake in 50 years for a standard occupancy building. This is equivalent to an earthquake with a mean recurrence interval (MRI) of approximately 2500 years. The seismic load calculations for the primary lateral load-resisting system are covered in ASCE 7, Chapter 12, while the seismic load calculations for the nonstructural parts and components are covered in ASCE 7, Chapter 13.

The intent of the ASCE 7 seismic design provisions is to prevent total collapse during an earthquake event while allowing limited structural damage. The inelastic action of structures during a seismic event, resulting from cracking and yielding of the members, causes an increase in the damping ratio and in the fundamental period of vibration of the structure; thus, seismic forces are reduced due to inelastic action.

The magnitude of the seismic loads acting on a building structure is a function of the type of lateral load-resisting system used in the building, the weight of the building, and the soil conditions at the building site. A seismic load calculation example for a wood building is presented in Example 2.8 using the simplified design procedure from the ASCE 7 load standard. This procedure is permitted for wood-framed structures that are three stories or less in height and that satisfy all the conditions listed in ASCE 7, Section 12.14.1.1. Most wood buildings will fall into this category. In the calculations, the building is assumed to be fixed at the base. The more in-depth equivalent lateral force analysis procedure presented in ASCE 7 is also permitted but is beyond the scope of this book.

2.9.1 Simplified design procedure for simple bearing wall or building frame systems (ASCE 7, Section 12.14)

If all the conditions listed in ASCE 7, Section 12.14.1.1 are met, the simplified procedure described below can be used. Some of the more common conditions include the following:

> Building must be in Risk Category I and II in site class A through D and not exceed three stories in height above grade. The seismic force resisting system shall be a bearing wall system or a building frame system (see ASCE 7, Table 12.14-1) and no irregularities are permitted.

For the simplified method (ASCE 7, Section 12.14.8), the factored seismic base shear and the factored lateral force at each level are given in Equations 2.13 and 2.14, respectively.

factored seismic base shear:

$$V = \frac{FS_{DS}}{R}W \tag{2.13}$$

factored seismic lateral force at level x:

$$F_x = \left(\frac{w_x}{W}\right)\left(\frac{FS_{DS}}{R}W\right) = \frac{FS_{DS}}{R}w_x \tag{2.14}$$

where:

R is the structural system response modification factor (ASCE 7, Table 12.2-1)
W_x is the portion of the total seismic dead load tributary to or assigned to level x
W is the total seismic dead load (including cladding loads) plus other loads listed below:

- 25% of floor live load for warehouses and structures used for storage of goods, wares, or merchandise (public garages and open parking structures are excluded).
- Partition load or 10 psf, whichever is greater. (Note: This only applies when an allowance for a partition load was included in the floor load calculations.)
- Total operating weight of permanent equipment. (*For practical purposes, use 50% of the mechanical room live load for schools and residential buildings, and 75% for mechanical rooms in industrial buildings.*)
- 20% of the flat roof or balanced snow load, if the flat roof snow load, P_f, exceeds 30 psf. *A higher snow load results in a tendency for the bottom part of the accumulated snow to adhere to the structure and thus contribute to the seismic load* [2].

$$W = W_{2nd} + W_{3rd} + \cdots + W_{roof} = \sum_{x=2nd}^{roof} W_x$$

$$S_{DS} = \frac{2}{3}F_a S_s$$

$F_a = 1.0$ for sites with rock*
$= 1.4$ for other soil sites

* For the simplified procedure, building sites with less than or equal to 10 ft of soil between the rock surface and the bottom of the spread footing can be considered as a rock site (ASCE 7, Section 12.14.8.1).

S_s is determined from ASCE 7, Figures 22-1 through 22-6 for the building location, but need not be larger than 1.50
$F = 1.0$ for one-story buildings
$= 1.1$ for two-story buildings
$= 1.2$ for three-story buildings

Example 2.8: Simplified wind load calculation for MWFRS

A one-story wood building 50 ft × 75 ft in plan is shown in Figure 2.21. The truss bearing (or roof datum) elevation is at 15 ft and the truss ridge is 23′-4″ above the ground-floor level. Assume that the building is enclosed and located in Rochester, New York, where the 3-second gust wind pressure is 115 mph, and the building site has a wind exposure category C. Determine the total horizontal wind force (*factored* and *unfactored*) on the main wind force-resisting system (MWFRS) in both the transverse and longitudinal directions and the gross and net vertical wind uplift pressures (*factored* and *unfactored*) on the roof (MWFRS) in both the transverse and longitudinal directions assuming a roof dead load of 15 psf. Assume a Risk Category II building.

Solution:

1. Determine the applicable 3-second gust wind speed for the building location from ASCE 7, Figure 26.5-1A. For Rochester, New York, the wind speed is 115 mph.
2. The wind exposure category is given as C (See ASCE 7, Section 26.7.3 for the descriptions of the different exposure categories).
3. Determine the applicable horizontal and vertical wind pressures as a function of the wind speed, roof slope, effective wind area, and wind load direction for MWFRS and C&C (See ASCE 7, Figure 28.6-1 for MWFRS and ASCE 7, Figure 30.5-1 for C&C):

$$\text{roof slope } \theta = \tan^{-1}\frac{23.33\text{ ft} - 15\text{ ft}}{50\text{ ft}/2} = 18.43°$$

The mean roof height = (15 ft + 23.33 ft)/2 = 19.2 ft.

MWFRS—Ps_{30} from ASCE 7, Figure 28.6-1

For the building with a roof slope **θ** of 18.43° and a basic wind speed of 115 mph, the tabulated factored horizontal wind pressures on the vertical projected area of the building (i.e., zones A, B, C, and D) are obtained by linear interpolation between the factored wind pressure values for **θ** of 0° and 20° from ASCE 7, Figure 28.6-1 as follows.

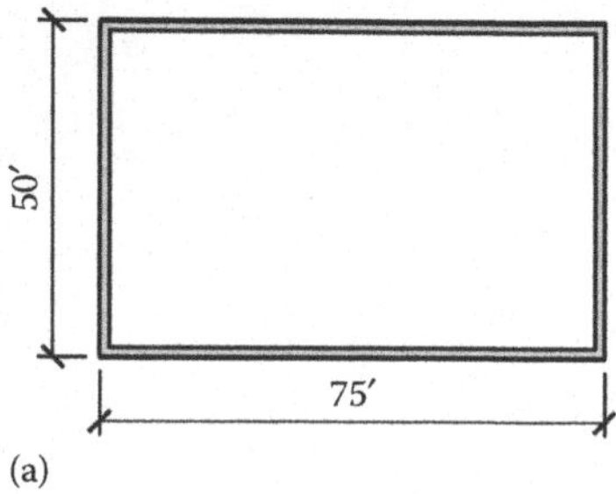

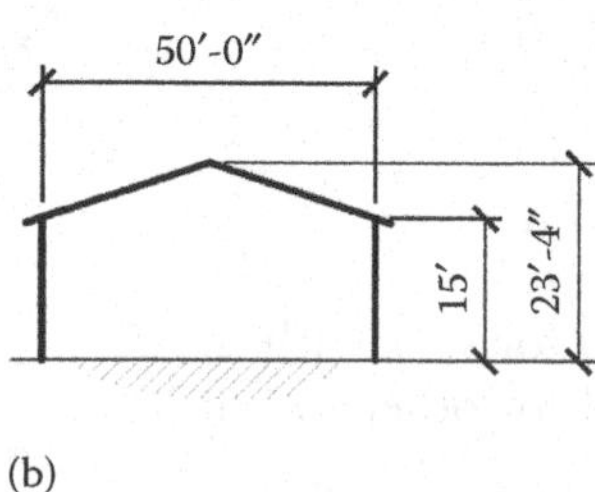

Figure 2.21 Building (a) plan view and (b) elevation.

Factored horizontal wind pressures on MWFRS—transverse wind:

$$\textit{End}\text{-zone pressure on } \textit{wall} \text{ (zone A)} = 21 \text{ psf} + \left(\frac{18.43° - 0°}{20°}\right)(29 \text{ psf} - 21 \text{ psf})$$

$$= 28.4 \text{ psf}$$

$$\textit{End}\text{-zone pressure on } \textit{roof} \text{ (zone B)} = -10.9 \text{ psf} + \left(\frac{18.43° - 0°}{20°}\right)$$

$$[(-7.7 \text{ psf}) - (-10.9 \text{ psf})] = -7.95 \text{ psf}$$

$$= 0 \text{ psf} \quad \text{(see footnote 7 in ASCE 7, Figure 28.6-1)}$$

$$\textit{Interior}\text{-zone pressure on } \textit{wall} \text{ (zone C)} = 13.9 \text{ psf} + \left(\frac{18.43° - 0°}{20°}\right)(19.4 \text{ psf} - 13.9 \text{ psf})$$

$$= 19 \text{ psf}$$

$$\textit{Interior}\text{-zone pressure on } \textit{roof} \text{ (zone D)} = -6.5 \text{ psf} + \left(\frac{18.43° - 0°}{20°}\right)[(-4.2 \text{ psf}) - (-6.5 \text{ psf})]$$

$$= -4.4 \text{ psf}$$

$$= 0 \text{ psf} \quad \text{(see footnote 7 in ASCE 7, Figure 28.6-1)}$$

Factored horizontal wind pressures on MWFRS—longitudinal wind:
In the longitudinal direction, there is no roof slope ($\boldsymbol{\theta} = 0°$) per footnote 3 in ASCE 7, Figure 28.6-1, *the factored wind pressures obtained from ASCE, Figure 28.6-1 for* $\boldsymbol{\theta} = 0°$ are as follows:

- *End*-zone pressure on *wall* (zone A) = 21 psf (for $\boldsymbol{\theta} = 0°$)
- *Interior*-zone pressure on *wall* (zone C) = 13.9 psf (for $\boldsymbol{\theta} = 0°$)

If the factored horizontal wind pressures on roofs in zones B and D (see Figure 2.20) are less than zero, a value of zero is used for the horizontal wind pressures in these zones (see footnote 7 in ASCE 7, Figure 28.6-1). In the longitudinal wind direction and for roofs with sloped rafters but no hip rafters, the triangular area between the roof datum level and the ridge is a wall surface, so only zones A and C wind pressures are applicable on this face (see Figure 2.20).

The resulting wind pressure distribution on the MWFRS is *not symmetrical,* due to the nonsymmetrical location of the end zones and the higher wind pressures acting on the end zones in ASCE 7, Figure 28.6-1. In this book, we adopt a simplified approach to calculating the resultant wind force using the average horizontal wind pressure. However, the asymmetrical nature of the wind loading on buildings and other structures and the resulting torsional effect of such loading should always be investigated to ensure that they are not critical.

For this building, the tabulated *vertical* wind pressures on the horizontal projected area of the building (i.e., roof zones E, F, G, and H) are obtained by linear interpolation between wind pressure values for $\boldsymbol{\theta}$ of 0° and $\boldsymbol{\theta}$ of 20° from the tables in ASCE 7, Figure 28.6-1 as follows:

Factored vertical wind pressures on roof—transverse wind:

End-zone pressure on windward *roof* (zone E)

$$= -25.2 \text{ psf} + \left(\frac{18.43° - 0°}{20°}\right)[-25.2 \text{ psf} - (-25.2 \text{ psf})] = -25.2 \text{ psf}$$

End-zone pressure on leeward *roof* (zone F)

$$= -14.3 \text{ psf} + \left(\frac{18.43° - 0°}{20°}\right)[-17.5 \text{ psf} - (-14.3 \text{ psf})] = -17.2 \text{ psf}$$

Interior-zone pressure on windward *roof* (zone G)

$$= -17.5 \text{ psf} + \left(\frac{18.43° - 0°}{20°}\right)[-17.5 \text{ psf} - (-17.5 \text{ psf})] = -17.5 \text{ psf}$$

Interior-zone pressure on leeward *roof* (zone H)

$$= -11.1 \text{ psf} + \left(\frac{18.43° - 0°}{20°}\right)[-13.3 \text{ psf} - (-11.1 \text{ psf})] = -13.1 \text{ psf}$$

Factored vertical wind pressures on roof—longitudinal wind:
For longitudinal wind, $\boldsymbol{\theta} = 0°$ per footnote 3 in ASCE 7, Figure 28.6-1.

- *End* zone on *windward* roof (zone E) = −25.2 psf (for $\boldsymbol{\theta} = 0°$)
- *End* zone on *leeward* roof (zone F) = −14.3 psf (for $\boldsymbol{\theta} = 0°$)
- *Interior zone* on *windward* roof (zone G) = −17.5 psf (for $\boldsymbol{\theta} = 0°$)
- *Interior* zone on *leeward* roof (zone H) = −11.1 psf (for $\boldsymbol{\theta} = 0°$)

The factored wind pressures are summarized in Figure 2.22. For the MWFRS, the end-zone width = $2a$ according to ASCE 7, Figure 28.6-1
where:

$a \leq 0.1 \times$ least horizontal dimension of building

$\leq 0.4 \times$ mean roof height of the building

$\geq 0.04 \times$ least horizontal dimension of building

≥ 3 ft

4. Multiply the tabulated factored wind pressures obtained in step 3 by the applicable height and exposure coefficients in ASCE 7, Figure 28.6-1. Tabulate the MWFRS wind loads.

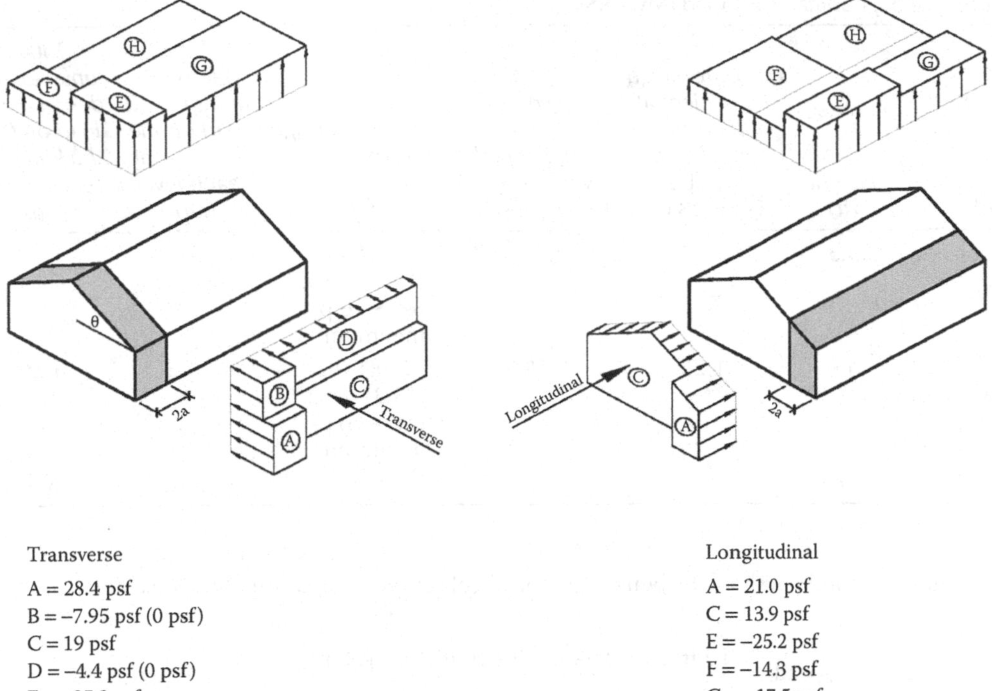

Transverse

A = 28.4 psf
B = −7.95 psf (0 psf)
C = 19 psf
D = −4.4 psf (0 psf)
E = −25.2 psf
F = −17.2 psf
G = −17.5 psf
H = −13.1 psf

Longitudinal

A = 21.0 psf
C = 13.9 psf
E = −25.2 psf
F = −14.3 psf
G = −17.5 psf
H = −11.1 psf

Figure 2.22 Summary of factored wind pressures (MWFRS).

Longitudinal wind. The end-zone width is 2*a* according to ASCE 7, Figure 28.6-1 where:

$$a \le (0.1)(50\ \text{ft}) = \mathbf{5\ ft}\ \text{(governs)}$$

$$\le (0.4)\left(\frac{15\ \text{ft} + 23.33\ \text{ft}}{2}\right) = 7.7\ \text{ft}$$

$$\ge (0.04)(50\ \text{ft}) = 2\ \text{ft}$$

$$\ge 3\ \text{ft}$$

Therefore, the edge zone = $2a$ = (2)(5 ft) = 10 ft.

Recall that for the longitudinal direction, only the horizontal wind pressures on the edge zone A and the interior zone C are applicable, and we calculated these previously in part 3 as 21 psf and 13.9 psf, respectively. For simplicity, as discussed previously, we use the average horizontal wind pressure acting on the entire building, calculated as follows:

Average *horizontal* wind pressure in the longitudinal direction (i.e., case B in ASCE 7, Figure 28.6-1), P_{s30}

$$= \frac{\left[\begin{array}{l}\text{(edge zone width)(edge zone “A” pressure)} + \text{(total width of building}\\ \quad - \text{edge zone width)(interior zone “C” pressure)}\end{array}\right]}{\text{total width of building}}$$

Table 2.7 Longitudinal wind load (MWFRS)

Level	*Height (ft)*	*Exposure/height coefficient at mean roof height*[a], λ *[ASCE 7, Figure 28.6-1]*	*Average factored horizontal wind pressure,* P_{s30} (psf)	*Factored design horizontal wind pressure,* $\lambda K_{zt} P_{s30}$	*Total factored design wind load on the building at each level* (kips)	*Total unfactored design wind load on the building at each level* (kips)
Ridge	23.33	1.29	15.3	(1.29)(1.0)(15.3) = 19.7 psf > 16 psf minimum	–	–
Roof datum (i.e., at truss bearing elevation)	15	1.29	15.3	(1.29)(1.0)(15.3) = 19.7 psf > 16 psf minimum	11.5	6.9
Base shear					11.5	6.9

[a] Mean roof height = 19.2 ft ≈ 20 ft.

where the total width of the building perpendicular to the longitudinal wind is 50 ft.

$$P_{s30\ \text{average}}(\text{wall}) = \frac{(10\ \text{ft})(21\ \text{psf}^*) + (50\ \text{ft} - 10\ \text{ft})(13.9\ \text{psf}^*)}{50\ \text{ft}}$$

$$= \mathbf{15.3\ psf} \quad \text{(i.e., factored average pressure on zones A and C)}$$

Noting that this building is in exposure category C and has a mean roof height of 19.2 ft (i.e., ~20 ft), we determine the height and exposure adjustment factor, λ, by interpolation from ASCE 7, Figure 28.6-1. The topography factor, $K_{zt} = 1.0$ (for flat terrain. See ASCE 7, Section 26.8). The factored uniform horizontal design wind pressure on the MWFRS of the building is calculated from the equation, $P = \lambda K_{zt} P_{s30} \geq 16$ psf; these values are tabulated in Table 2.7.

The total *factored* wind load on the building at the roof datum level is calculated as

$$\left[\left(\frac{1}{2}\right)(19.7\ \text{psf})(23.33\ \text{ft} - 15\ \text{ft}) + 19.7\ \text{psf}\left(\frac{15\ \text{ft}}{2}\right)\right](50\ \text{ft}) = 11{,}490\ \text{lb} = 11.5\,\text{kips}$$

The total *unfactored* wind load on the building at the roof datum level is calculated as

$$(0.6)\left[\left(\frac{1}{2}\right)(19.7\ \text{psf})(23.33\ \text{ft} - 15\ \text{ft}) + 19.7\ \text{psf}\left(\frac{15\ \text{ft}}{2}\right)\right](50\ \text{ft}) = (0.6)(11{,}490\ \text{lb}) = 6.9\,\text{kips}$$

The number ½ in the first term of the equations above account for the triangular shape of the vertical projected wall area above the roof datum (or truss bearing) level in the longitudinal direction, and for the wind pressure below the roof datum, the tributary height is 15 ft/2 as shown in the second term in the equations above. The "0.6" term in the second equation above is the load factor for wind for the ASD method as given in Section 2.1.

* See the calculated factored pressures in part 3.

Transverse wind. The end-zone width is $2a$ where:

$$a \leq (0.1)\ (50\text{ ft}) = 5\text{ ft (governs)}$$

$$\leq (0.4)\left(\frac{15\text{ ft} + 23.33\text{ ft}}{2}\right) = 7.7\text{ ft}$$

$$\geq (0.04)(50\text{ ft}) = 2\text{ ft}$$

$$\geq 3\text{ft}$$

Therefore, the edge zone = $2a$ = (2)(5 ft) = 10 ft.

Recall that for the transverse direction, the factored horizontal wind pressures on the edge zone A and the interior zone C were previously calculated in part 3 as 28.4 psf and 19 psf, respectively; the factored horizontal wind pressures in zones B and D were taken as zero. For simplicity, as discussed previously, we use the average horizontal wind pressure acting on the entire building, calculated as follows:

Average *factored* horizontal wind pressure in the transverse direction, P_{s30}

$$P_{s30} = \frac{\left[\begin{array}{l}\text{(edge zone width)(edge zone "A" pressure)} + \text{(total length of building}\\ \quad - \text{edge zone width)(interior zone "C" pressure)}\end{array}\right]}{\text{total length of building}}$$

where the total length of the building perpendicular to the transverse wind is 75 ft.

$$P_{s30\ \text{average}}(\text{wall}) = \frac{(10\text{ ft})(28.4\text{ psf}^{*}) + (75\text{ ft} - 10\text{ ft})(19\text{ psf}^{*})}{75\text{ ft}}$$

$$= 20.3\text{ psf} \quad \text{(i.e., factored average pressure on zones A and C)}$$

$$P_{s30\ \text{average}}(\mathit{roof}) = 0\text{ psf}^{*}\ \text{(per footnote 7 in ASCE 7, Figure 28.6-1)}$$

$$\text{(i.e., factored average pressure on zones B and D)}$$

Using wind exposure "C" as given (refer to ASCE 7, Section 26.7.3 for the description of the different exposure categories), the horizontal wind forces on the MWFRS are as given in Table 2.8.

The total *factored* wind load on the building at the roof datum level in the transverse direction is calculated as

$$\left[(8\text{ psf})(23.33\text{ ft} - 15\text{ ft}) + (26.2\text{ psf})\left(\frac{15\text{ ft}}{2}\right)\right](75\text{ ft}) = 19{,}736\text{ lb} = 19.7\text{ kips}$$

The total *unfactored* wind load (for ASD) on the building at the roof datum level in the transverse direction is calculated as

$$(0.6)\left[(8\text{ psf})(23.33\text{ ft} - 15\text{ ft}) + (26.2\text{ psf})\left(\frac{15\text{ ft}}{2}\right)\right](75\text{ ft}) = 11{,}841\text{ lb} = 11.8\text{ kips}$$

The "0.6" factor in the above equation is the ASD wind load factor that converts the factored wind load to the service or unfactored load.

* See part 3 for the factored horizontal wind pressures in the transverse direction.

Table 2.8 Transverse wind load (MWFRS)

Level	*Height* (ft)	*Exposure/ height coefficient at mean roof height*[a]*, λ* [ASCE 7, Figure 28.6-1]	*Factored average horizontal wind pressure, P_{s30}* (psf)	*Factored design horizontal wind pressure, $\lambda K_{zt} P_{s30}$*	*Total factored wind load on the building at each level* (kips)	*Total unfactored wind load on the building at each level* (kips)
Ridge	23.33	1.29	0[b]	(1.29)(1.0)(0) = 0 psf < 8 psf minimum, therefore use 8 psf[c]	–	–
Roof datum (i.e., at truss bearing elevation)	15	1.29	20.3	(1.29)(1.0)(20.3) = 26.2 psf > 16 psf minimum	19.7	11.8
Base shear					19.7	11.8

[a] Mean roof height = 19.2 ft ≈ 20 ft.
[b] The pressure on the vertical projected area of the roof surface is taken as zero if the wind pressure obtained is negative, per footnote 7 in ASCE 7, Figure 28.6-1. (See wind pressures for zones B and D for MWFRS.)
[c] See ASCE 7, C28.4.4 and Figure C27.4-1.

Note that for the transverse wind direction, the rectangular vertical projected area of the roof surface above the roof datum (or truss bearing) level is assigned to the roof datum level in addition to the one-half of the wall area between the roof datum level and the ground-floor level.

5. Calculate the net *vertical* wind pressures on the roof for the main wind force-resisting system (MWFRS).

The roof wind uplift diagram (Figure 2.23) shows the calculated vertical uplift wind pressures on various zones of the roof. These are obtained by multiplying the tabulated vertical wind pressures by the height/exposure adjustment factor. The wind exposure category is "C" and

$$\text{mean roof height} = \frac{15\text{ ft} + 23.33\text{ ft}}{2} = 19.2\text{ ft} \approx 20\text{ ft}$$

The height/exposure adjustment coefficient can be obtained by linear interpolation from ASCE 7, Figure 28.6-1 as 1.29.

Factored vertical wind pressures, W, on roof—longitudinal wind ($\lambda K_{zt} P_{s30}$):
Using the factored vertical wind pressures (P_{s30}) in the longitudinal direction from part 3, we calculate the factored design pressures as follows:

- *End* zone on *windward* roof (zone E) = (1.29)(1.0)(−25.2 psf) = −32.5 psf
- *End* zone on *leeward* roof (zone F) = (1.29)(1.0)(−14.3 psf) = −18.4 psf
- *Interior* zone on *windward* roof (zone G) = (1.29)(1.0)(−17.5 psf) = −22.6 psf
- *Interior* zone on *leeward* roof (zone H) = (1.29)(1.0)(−11.1 psf) = −14.3 psf* (use −16 psf)

* According to ASCE 7, Section 30.2.2 [2], the design wind pressure for components and cladding must be greater than or equal to 16 psf acting in either direction normal to the surface of the C&C.

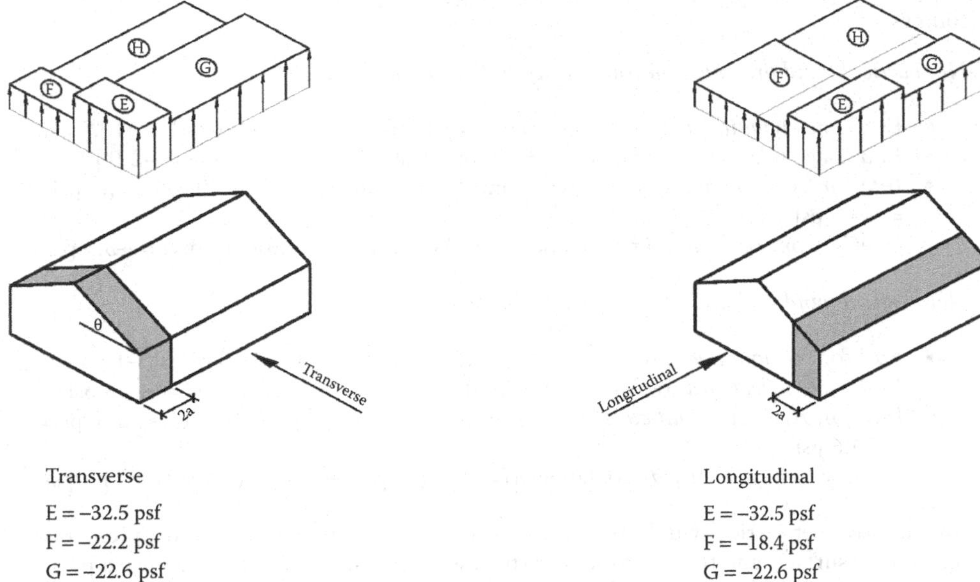

Figure 2.23 Roof uplift diagram (factored wind pressures).

Factored vertical wind pressures, W, on roof—transverse wind ($\lambda K_{zt} P_{s30}$):
Using the factored vertical wind pressures (P_{s30}) in the transverse direction from part 3, we calculate the factored design pressures as follows:

- *End* zone on *windward* roof (zone E) = (1.29)(1.0)(−25.2 psf) = −32.5 psf
- *End* zone on *leeward* roof (zone F) = (1.29)(1.0)(−17.2 psf) = −22.2 psf
- *Interior zone* on *windward* roof (zone G) = (1.29)(1.0)(−17.5 psf) = −22.6 psf
- *Interior zone* on *leeward* roof (zone H) = (1.29)(1.0)(−13.1 psf) = −16.9 psf

(Negative vertical wind pressures indicate uplift or suction on the roof, and positive values indicate downward pressure on the roof.) Using the given roof dead load of 15 psf, the net wind uplift pressures on the roof are calculated using the factored uplift wind pressures above and using the controlling ASCE 7 load combination (i.e., $0.9D + W$ *for LRFD and* $0.6D + 0.6W$ for ASD) from Section 2.1. The net factored roof pressures using the factored or LRFD load combinations are calculated below:

Net vertical wind pressures on roof—longitudinal wind (LRFD method):

- *End* zone on *windward* roof (zone E) = (0.9)(15 psf) + (1.0)(−32.5 psf) = −19 psf
- *End* zone on *leeward* roof (zone F) = (0.9)(15 psf) + (1.0)(−18.4 psf) = −4.9 psf
- *Interior* zone on *windward* roof (zone G) = (0.9)(15 psf) + (1.0)(−22.6 psf) = −9.1 psf
- *Interior zone* on *leeward* roof (zone H) = (0.9)(15 psf) + (1.0)(−16 psf) = −2.5 psf

Net vertical wind pressures on roof—transverse wind (LRFD method):

- *End* zone on *windward* roof (zone E) = (0.9)(15 psf) + (1.0)(−32.5 psf) = −19 psf
- *End* zone on *leeward* roof (zone F) = (0.9)(15 psf) + (1.0)(−22.2 psf) = −8.7 psf
- *Interior* zone on *windward* roof (zone G) = (0.9)(15 psf) – (1.0)(−22.6 psf) = −9.1 psf
- *Interior* zone on *leeward* roof (zone H) = (0.9)(15 psf) – (1.0)(−16.9 psf) = −3.4 psf

The net unfactored roof pressures using the ASD load combinations are calculated as follows.

Net vertical wind pressures on roof—longitudinal wind (ASD method):

- *End* zone on *windward* roof (zone E) = (0.6)(15 psf) + (0.6)(−32.5 psf) = −10.5 psf
- *End* zone on *leeward* roof (zone F) = (0.6)(15 psf) + (0.6)(=18.4 psf) = −2.0 psf
- *Interior* zone on *windward* roof (zone G) = (0.6)(15 psf) + (0.6)(−22.6 psf) = −4.6 psf
- *Interior zone* on *leeward* roof (zone H) = (0.6)(15 psf) + (0.6)(−16 psf) = +3.9 psf

Net vertical wind pressures on roof—transverse wind (ASD method):

- *End* zone on *windward* roof (zone E) = (0.6)(15 psf) + (0.6)(−32.5 psf) = −10.5 psf
- *End* zone on *leeward* roof (zone F) = (0.6)(15 psf) + (0.6)(−22.2 psf) = −4.3 psf
- *Interior* zone on *windward* roof (zone G) = (0.6)(15 psf) + (0.6)(−22.6 psf) = −4.6 psf
- *Interior* zone on *leeward* roof (zone H) = (0.6)(15 psf) + (0.6)(−16.9 psf) = −1.1 psf

The negative net vertical wind pressures indicate uplift or suction or upward pressures on the roof surface, and the net positive vertical wind pressures indicate downward pressures on the roof surface. Note that for roof zone H in this example, the LRFD load combination yields a more critical roof pressure (i.e., a suction of −3.4 psf) than the ASD load combination that gives a net downward pressure of +3.9 psf.

Example 2.9: Seismic load calculation

For the two-story wood-framed structure shown in Figure 2.24, calculate the factored seismic base shear *V*, in kips, and the factored lateral seismic load at each level, in kips, using the simplified analysis procedure. Assume the following unfactored design loads and seismic design parameters:

- Floor dead load = 30 psf
- Roof dead load = 20 psf
- Exterior walls = 15 psf
- Snow load P_f = 35 psf
- Site class = D, since site conditions are unknown (ASCE 7, Chapter 20)
- S_S = 0.25 (ASCE 7, Figure 22-1)
- S_1 = 0.07 (ASCE 7, Figure 22-2)
- *R* = 6.5 (light-framed wood walls sheathed with wood structural panels rated for shear resistance, from ASCE 7, Table 12.14-1 for the simplified design procedure)
- *F* = 1.1 for a two-story building (ASCE 7, Section 12.14.8.1)

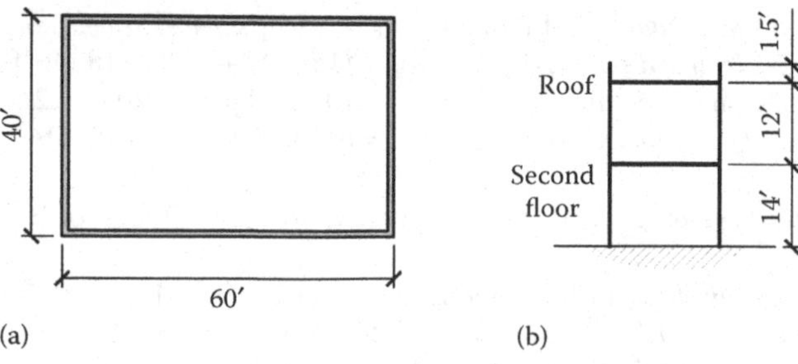

Figure 2.24 Building (a) plan view and (b) elevation.

Table 2.9 Building weights for seismic load calculation

Level	*Area*	*Tributary height*	*Weight of floor*	*Weight of walls*	W_{total}
Roof	(60 ft)(40 ft) = **2400 ft²**	(12 ft/2) + 1.5 ft = **7.5 ft**	(2400 ft²)[20 psf + (0.2 × 35 psf)] = **64.8 kips**	(7.5 ft)(15 psf) × (2) (60 ft + 40 ft) = **22.5 kips**	64.8 kips + 22.5 kips = **87.3 kips**
Second floor	(60 ft)(40 ft) = **2400 ft²**	(12 ft/2) + (14 ft/2) = **13 ft**	(2400 ft²)(30 psf) = **72.0 kips**	(13 ft)(15 psf) × (2) (60 ft + 40 ft) = **39.0 kips**	72.0 kips + 39.0 kips = **111 kips**
				ΣW = 87.3 kips + 111 kips = **198.3 kips**	

Solution:

First, calculate the weight tributary W for each level including the weight of the floor structure, the weight of the walls, and a portion of the snow load. Where the flat roof snow load P_f is greater than 30 psf, 20% of the flat roof snow load is to be included in W for the roof (ASCE 7, Section 12.14.8.1). See Table 2.9 for the calculations.

Seismic variables:

- $F_a = 1.6$ (ASCE 7, Table 11.4-1)
- $F_v = 2.4$ (ASCE 7, Table 11.4-2)
- $S_{MS} = F_a S_S = (1.6)(0.25) = 0.40$ (ASCE 7, Equation 11.4-1)
- $S_{M1} = F_v S_1 = (2.4)(0.07) = 0.168$ (ASCE 7, Equation 11.4-2)
- $S_{DS} = (2/3)S_{MS} = (2/3)(0.40) = \mathbf{0.267}$ (ASCE 7, Equation 11.4-3)
- $S_{D1} = (2/3)S_{M1} = (2/3)(0.168) = \mathbf{0.112}$ (ASCE 7, Equation 11.4-4)

Factored base shear (simplified analysis procedure):

$$V = \frac{FS_{DS}W}{R} \quad \text{(ASCE 7, Equation 12.14-11)}$$

$$= \frac{(1.1)(0.267)(198.3)}{6.5} = \mathbf{8.96\ kips}$$

Factored force at each level. The factored lateral force at each level of the building is calculated as follows:

$$F_x = \frac{FS_{DS}W_x}{R} \quad \text{(ASCE 7, Equations 12.14-11 and 12.14-12)}$$

$$F_R = \frac{(1.1)(0.267)(87.3)}{6.5} = \mathbf{3.94\ kips}$$

$$F_2 = \frac{(1.1)(0.267)(111)}{6.5} = \mathbf{5.02\ kips}$$

The seismic loads are summarized in Figure 2.25.

Note: A flat roof was assumed in this example, but for buildings with a pitched roof, the weight of the roof and walls that are assigned to the roof level are assumed to act at the roof datum level, which is equivalent to the top elevation of the exterior walls.

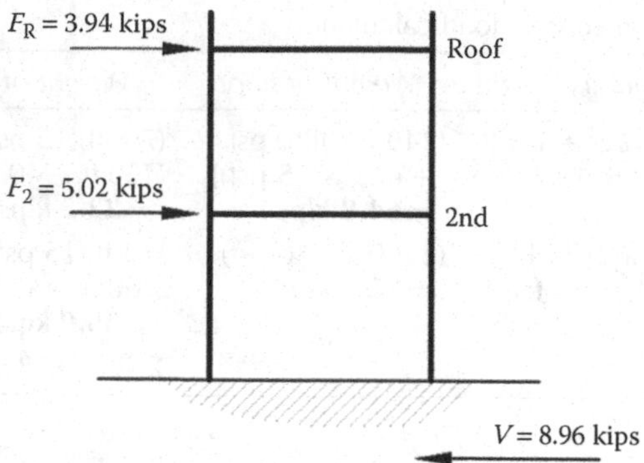

Figure 2.25 Summary of factored seismic loads.

The structural system factor R used in this example corresponds to that of a plywood panel. Where sheathing of different materials is used on both sides of the studs in a shear wall, the R value for the wall system will be the smaller of the R values of the two sheathing materials, if sheathing on both wall faces is used to resist the lateral load, or the R value of the stronger sheathing material if the weaker sheathing is neglected and only the stronger sheathing material is used to resist the lateral load. One example of this case is the use of plywood sheathing ($R = 6.5$) on the exterior face of a building and gypsum wall board ($R = 2$) on the interior face of the building.

PROBLEMS

2.1 Calculate the total uniformly distributed roof dead load in psf of the horizontal plan area for a sloped roof with the design parameters given below.
2 × 8 rafters at 24 in. o.c.
Asphalt shingles on 1/2 in. plywood sheathing
6-in. insulation (fiberglass)
Suspended ceiling
Roof slope: 6:12
Mechanical and electrical (i.e., ducts, plumbing, etc.) = 5 psf

2.2 Given the following design parameters for a sloped roof, calculate the uniform total load and the maximum shear and moment on the rafter. Calculate the horizontal thrust on the exterior wall if rafters are used.
Roof dead load $D = 20$ psf (of sloped roof area)
Roof snow load $S = 40$ psf (of horizontal plan area)
Horizontal projected length of rafter $L_2 = 14$ ft
Roof slope: 4:12 rafter or truss spacing = 4′-0″

2.3 Determine the tributary widths and tributary areas of the 2× joists, 4× beams, glulam girders, and columns in the panelized roof framing plan shown in Figure 2.26. Assuming a roof dead load of 20 psf and an essentially flat roof with a roof slope of 1/4 in. per foot for drainage, determine the following loads using the IBC load combinations. Neglect the rain load R and assume that the snow load S is zero. Calculate the following:

a. The uniform total load on the typical 2× roof joist, in lb/ft
b. The uniform total load on the typical Glulam roof girder, in lb/ft

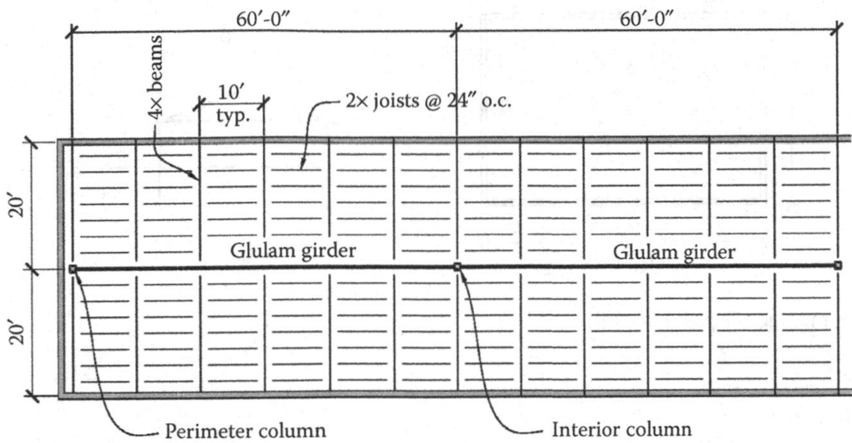

Figure 2.26 Roof framing plan—Problem 2.3.

c. The total axial load on the typical interior column, in lb
d. The total axial load on the typical perimeter column, in lb

2.4 A building has sloped roof rafters (5:12 slope) spaced at 2′-0″ on centers and is located in Hartford, Connecticut. The roof dead load is 22 psf of the sloped area. Assume a fully exposed roof with terrain category C, and use the ground snow load from the IBC or ASCE 7 snow map. The rafter horizontal span is 14 ft.

a. Calculate the total uniform load in lb/ft on a horizontal plane using the IBC.
b. Calculate the maximum shear and moment in the roof rafter.

2.5 A three-story building has columns spaced at 18 ft in both orthogonal directions and is subjected to the roof and floor loads shown below. Using a column load summation table, calculate the cumulative axial loads on a typical interior column with and without a live-load reduction. Assume a roof slope of 1/4 in. per foot for drainage.

Roof loads:

Dead load $D_{roof} = 20$ psf
Snow load $S = 40$ psf

Second- and third-floor loads:

Dead load, $D_{floor} = 40$ psf
Floor live load $L = 50$ psf

2.6 A two-story wood-framed structure 36 ft × 75 ft in plan is shown in Figure 2.27 with the following information given. The floor-to-floor height is 10 ft, the truss bearing (or roof datum) elevation is at 20 ft, and the truss ridge is 28′-4″ above the ground-floor level. The building is enclosed and is located in Rochester, New York on a site with a category C exposure. Assume the following additional design parameters:

Floor dead load = 30 psf
Roof dead load = 20 psf
Exterior walls =10 psf
Snow load $P_f = 40$ psf
Site class = D
Importance $I_e = 1.0$
$S_S = 0.25$

36′

75′

Plan view

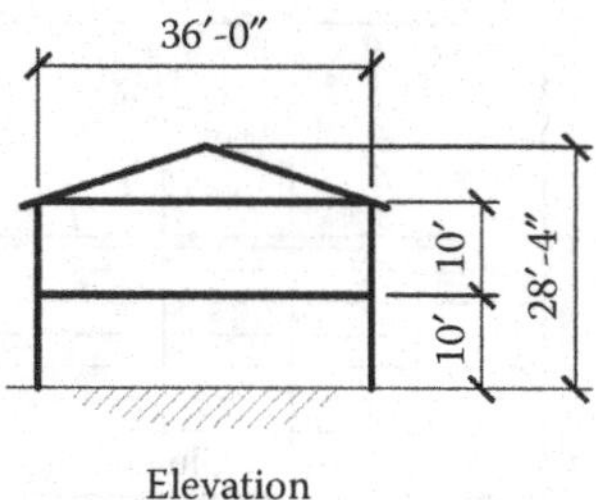

Figure 2.27 Details for Problem 2.6.

$S_1 = 0.07$
$R = 6.5$

Calculate the following:

a. The total horizontal factored wind force on the main wind force-resisting system (MWFRS) in both the transverse and longitudinal directions
b. The factored gross and net vertical wind uplift pressures on the roof (MWFRS) in both the transverse and longitudinal directions
c. The factored total seismic base shear *V*, in kips, using the simplified analysis procedure
d. The lateral seismic load at each level, in kips

2.7 For the framing plan shown in Figure 2.28:

a. Determine the floor dead load in psf
b. Determine the dead and live loads to J-1 and G-1 in plf
c. Determine the maximum moment and shear in J-1 and G-1
d. Determine the maximum service load in kips to C-1

2.8 A wood joist measures $1\frac{7}{8}'' \times 12\frac{1}{2}''$ and spans 17 ft and is spaced 16″ on center. It is subject to a uniform loading of 15 psf (dead) and 40 psf (live). Determine the following by means of a spreadsheet:

a. Area, section modulus, and moment of inertia of the wood member
b. Self-weight in plf assuming G = 0.5
c. Maximum moment and shear in the joist
d. Maximum bending stress ($f_b = M/S$ or $f_b = Mc/I$)

2.9 A single wood girder spans 40 ft and is supported at each end and at midspan. It is subjected to a uniformly distributed load of 500 plf. Determine the following for two framing scenarios:

a. Maximum moment and shear; draw both diagrams. Use either an analysis program or a design aid (hand calculations are not intended).
b. Maximum moment and shear (and diagrams) assuming that there are two simple span beams spanning 20 ft.
c. Explain the differences between the two framing scenarios and discuss the merits of using each one.

2.10 See truss design diagram and refer to Chapter 9 section on engineered wood trusses (Figure 2.29).

a. List each truss member in a table (B1, B2, T1, T2, W1 to W6) and list the following: size, length, species, grade, density, weight.

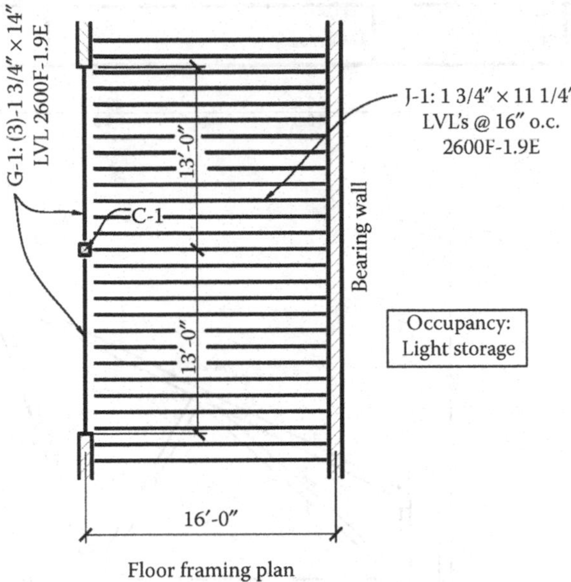

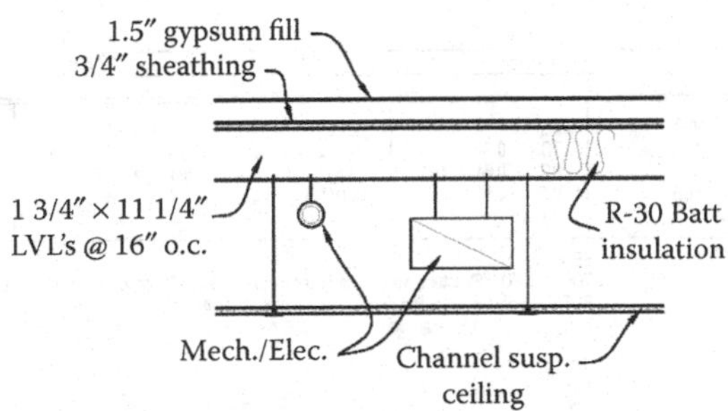

Figure 2.28 Floor framing plan—Problem 2.7.

b. Calculate the total weight of the truss using the table in (a).
c. Draw a free-body diagram of the truss and indicate the uniformly distributed loads to the top and bottom chords in pounds per lineal foot (plf) and indicate the supports.
d. What are the maximum compression loads to W2, W5, and W6 and what is the purpose of the single row of bracing at midpoint?

2.11 Given the following loads and loading diagram (Figure 2.30):

Uniform load, w	*Concentrated load, P*
$D = 500$ plf	$D = 11$ k
$L = 800$ plf	$S = 15$ k
$S = 600$ plf	$W = +12$ k or -12 k
Beam length = 25 ft	$E = +8$ k or -8 k

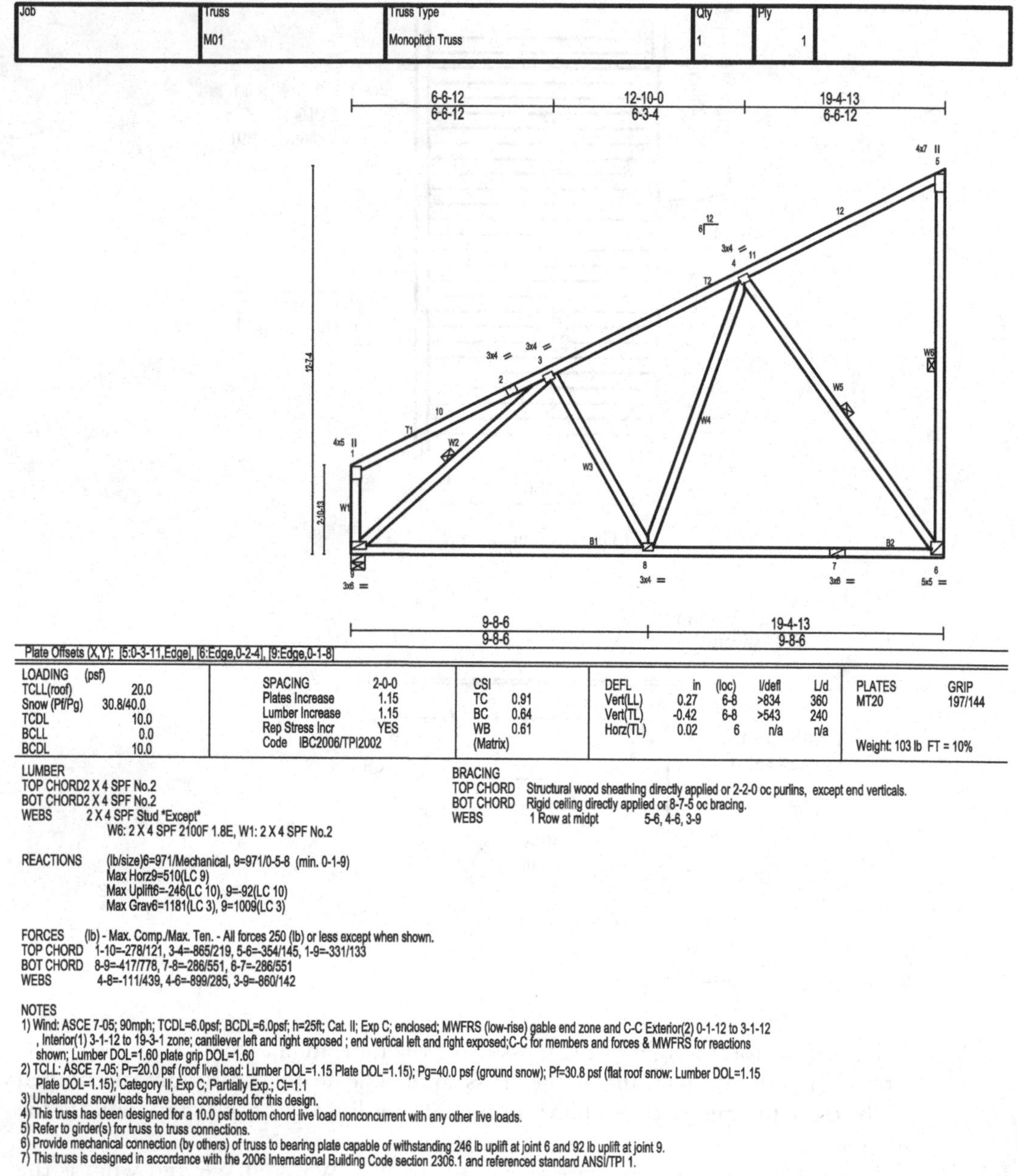

Job
Truss
M01
Truss Type
Monopitch Truss
Qty
1
Ply
1
6-6-12
6-6-12
12-10-0
6-3-4
19-4-13
6-6-12
12-7-4
2-10-13
9-8-6
9-8-6
19-4-13
9-8-6
Plate Offsets (X,Y): [5:0-3-11,Edge], [6:Edge,0-2-4], [9:Edge,0-1-8]
LOADING (psf)
TCLL(roof) 20.0
Snow (Pf/Pg) 30.8/40.0
TCDL 10.0
BCLL 0.0
BCDL 10.0
SPACING 2-0-0
Plates Increase 1.15
Lumber Increase 1.15
Rep Stress Incr YES
Code IBC2006/TPI2002
CSI
TC 0.91
BC 0.64
WB 0.61
(Matrix)
DEFL in (loc) l/defl L/d
Vert(LL) 0.27 6-8 >834 360
Vert(TL) -0.42 6-8 >543 240
Horz(TL) 0.02 6 n/a n/a
PLATES GRIP
MT20 197/144
Weight: 103 lb FT = 10%
LUMBER
TOP CHORD2 X 4 SPF No.2
BOT CHORD2 X 4 SPF No.2
WEBS 2 X 4 SPF Stud *Except*
W6: 2 X 4 SPF 2100F 1.8E, W1: 2 X 4 SPF No.2
BRACING
TOP CHORD Structural wood sheathing directly applied or 2-2-0 oc purlins, except end verticals.
BOT CHORD Rigid ceiling directly applied or 8-7-5 oc bracing.
WEBS 1 Row at midpt 5-6, 4-6, 3-9
REACTIONS (lb/size)6=971/Mechanical, 9=971/0-5-8 (min. 0-1-9)
Max Horz9=510(LC 9)
Max Uplift6=-246(LC 10), 9=-92(LC 10)
Max Grav6=1181(LC 3), 9=1009(LC 3)
FORCES (lb) - Max. Comp./Max. Ten. - All forces 250 (lb) or less except when shown.
TOP CHORD 1-10=-278/121, 3-4=-865/219, 5-6=-354/145, 1-9=-331/133
BOT CHORD 8-9=-417/778, 7-8=-286/551, 6-7=-286/551
WEBS 4-8=-111/439, 4-6=-899/285, 3-9=-860/142
NOTES
1) Wind: ASCE 7-05; 90mph; TCDL=6.0psf; BCDL=6.0psf; h=25ft; Cat. II; Exp C; enclosed; MWFRS (low-rise) gable end zone and C-C Exterior(2) 0-1-12 to 3-1-12 , Interior(1) 3-1-12 to 19-3-1 zone; cantilever left and right exposed ; end vertical left and right exposed;C-C for members and forces & MWFRS for reactions shown; Lumber DOL=1.60 plate grip DOL=1.60
2) TCLL: ASCE 7-05; Pr=20.0 psf (roof live load: Lumber DOL=1.15 Plate DOL=1.15); Pg=40.0 psf (ground snow); Pf=30.8 psf (flat roof snow: Lumber DOL=1.15 Plate DOL=1.15); Category II; Exp C; Partially Exp.; Ct=1.1
3) Unbalanced snow loads have been considered for this design.
4) This truss has been designed for a 10.0 psf bottom chord live load nonconcurrent with any other live loads.
5) Refer to girder(s) for truss to truss connections.
6) Provide mechanical connection (by others) of truss to bearing plate capable of withstanding 246 lb uplift at joint 6 and 92 lb uplift at joint 9.
7) This truss is designed in accordance with the 2006 International Building Code section 2306.1 and referenced standard ANSI/TPI 1.

Figure 2.29 Truss design diagram—Problem 2.10.

Do the following:

a. Describe a practical framing scenario where these loads could all occur as shown
b. Determine the maximum moment for each individual load effect (D, L, S, W, E)
c. Develop a spreadsheet to determine the worst-case bending moments for the code-required load combinations

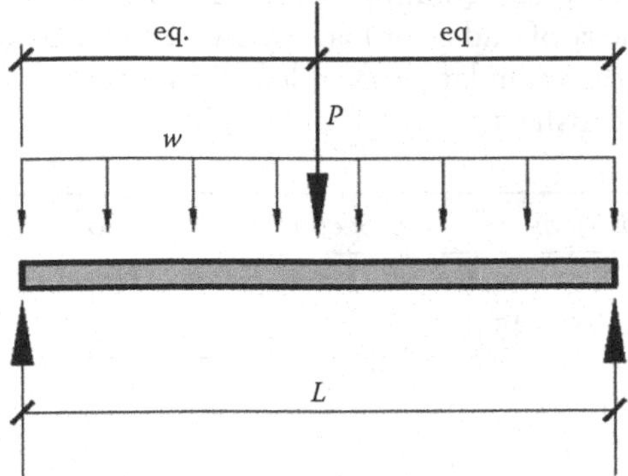

Figure 2.30 Beam loading for Problem 2.11.

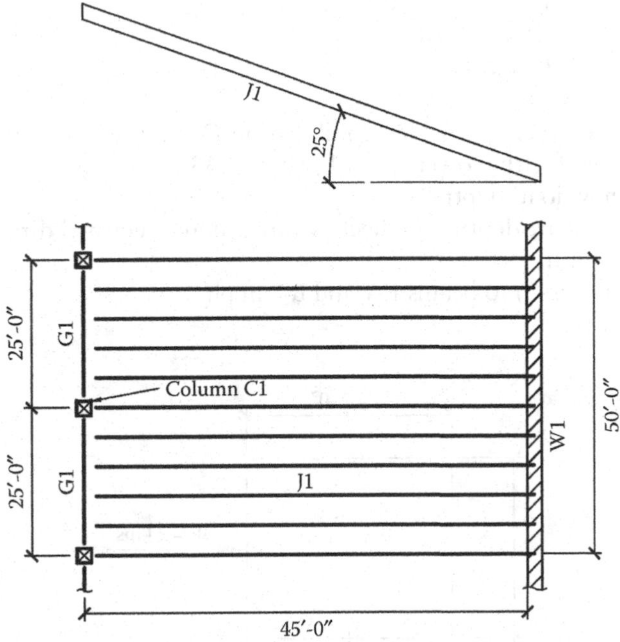

Figure 2.31 Roof framing plan for Problem 2.12.

2.12 Given the framing plan below and assuming a roof dead load of 25 psf and a 25° roof slope, determine the following using the IBC factored load combinations. Neglect the rain load, *R* and assume the snow load, *S* is zero (Figure 2.31):

a. Determine the tributary areas of B1, G1, C1, and W1
b. The uniform dead and roof live load and the factored loads on B1 in plf
c. The uniform dead and roof live load on G1 and the factored loads in plf (assume G1 is uniformly loaded)
d. The total factored axial load on column C1, in kips
e. The total factored uniform load on W1 in plf (assume trib. length of 50 ft)

2.13 A three-story building has columns spaced at 25 ft in both orthogonal directions, and is subjected to the roof and floor loads shown below. Using a column load summation table, calculate the cumulative axial loads on a typical interior column. Develop this table using a spreadsheet.

Roof loads	*Second- and third-floor loads*
Dead, $D = 20$ psf	Dead, $D = 60$ psf
Snow, $S = 45$ psf	Live, $L = 100$ psf

All other loads are 0

2.14 Using only the loads shown and the weight of the concrete footing only ($\gamma_{conc} = 150$ pcf), determine the required square footing size, B × B using the appropriate load combination to keep the footing from overturning about point A (Figure 2.32).

2.15 Given the following:
Flat roof snow load, $P_f = 42$ psf
Ground snow load, $P_g = 60$ psf
Total roof DL = 25 psf
Ignore roof live load; consider load combination $D + S$ only
Find the following for the lower roof of Figure 2.33:

a. Balanced snow load depth, h_b
b. Critical snow drift depth, h_d (check windward and leeward directions)
c. Snow drift width, w
d. Total load (D and S) to beams B-1 and B-2 in plf

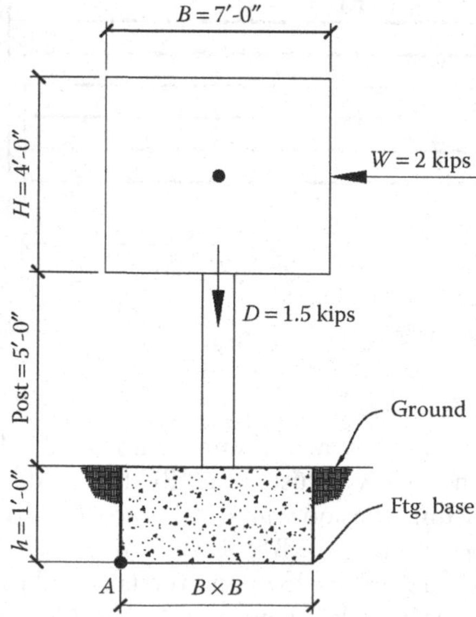

Figure 2.32 Details for Problem 2.14.

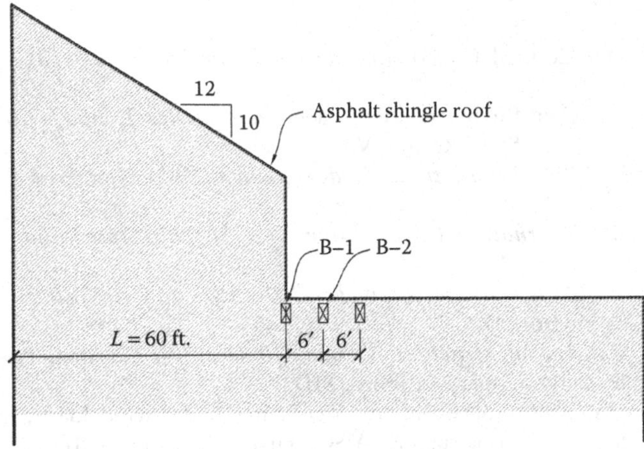

Figure 2.33 Details for Problem 2.15.

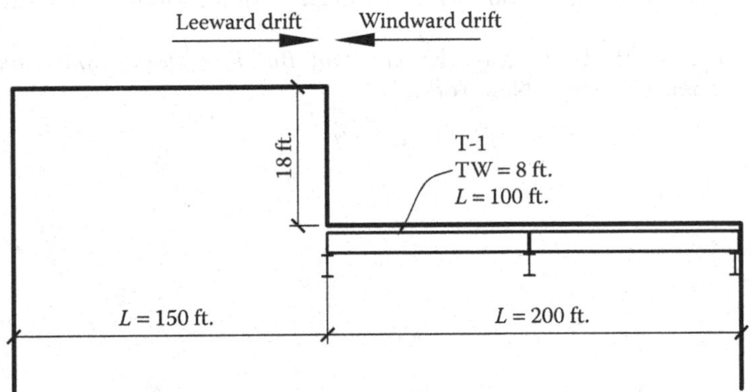

Figure 2.34 Details for Problem 2.16.

2.16 Refer to Figure 2.34

Location—Pottersville, NY; elevation is 1500 ft

Total roof DL = 20 psf

Ignore roof live load; consider load combination $D + S$ only

Use normal temperature and exposure conditions

Determine the following:

a. Flat roof snow load in psf
b. Depth and width of the leeward drift and windward drifts, which one controls the design of T-1?
c. Determine the depth of the balanced snow load and controlling drift snow load
d. Draw a free-body diagram of T-1 showing the service dead and snow loads in plf
e. Determine the maximum factored moment and shear in T-1 (analysis to be done using a computer program or by hand).

REFERENCES

1. International Codes Council (2015), *International Building Code (IBC)*—2015, ICC, Falls Church, VA.
2. American Society of Civil Engineers (2010), *Minimum Design Loads for Buildings and Other Structures (ASCE 7-10)*, ASCE, Reston, VA.
3. State of New York (2002), *International Codes Council (ICC), New York State Building Code,* Albany, NY.
4. State of Ohio (2005), *International Codes Council (ICC), Ohio State Building Code,* Columbus, OH.
5. Fanella, David A. (2012), *Structural Loads—2012 IBC and ASCE/SEI 7-10*, International Code Council, Washington, DC.
6. NAHB (2000), *Residential Structural Design Guide-2000,* National Association of Home Builders Research Center, Upper Marlboro, MD.
7. Griffis, Lawrence G. (1993), Serviceability limit states under wind loads, *Engineering Journal*, American Institute of Steel Construction (AISC), First Quarter, pp. 1–16.
8. Partain, Jason A. (2012), Design Considerations for Sawn Lumber Wood Studs, *STRUCTURE Magazine*, May.
9. Sprague, Harold O. (2008), Deflection Limits for Wood Studs Backing Brick Veneer, *STRUCTURE Magazine*, May.
10. International Codes Council (2006), *International Building Code (IBC)*—2006, ICC, Falls Church, VA.
11. Green, Norman B. (1981), *Earthquake Resistant Building Design and Construction*, Van Nostrand Reinhold Company, New York.

Chapter 3

Design method for sawn lumber and glued-laminated timber

There are two design methods in the National Design Specification (NDS) that can be used for the design of wood structures: the allowable stress design (ASD) method and the load and resistance factor design (LRFD) method. The choice of which method to use in design is the prerogative of the structural designer but the ASD method is currently more commonly used.

3.1 ALLOWABLE STRESS DESIGN METHOD

The ASD *method* has been widely used for many years for the structural design of wood structures in the United States and is still used by many practicing engineers. The method is based on stipulating a maximum allowable stress under service load conditions that incorporates a factor of safety. The factor of safety is the ratio of the maximum load or stress a structural member can support before it fails relative to the allowable load permitted by the code, and it varies between 1.5 and 2.0 for most building structures. This allowable stress must be greater than or equal to the applied stress that is calculated on the basis of an elastic structural analysis.

3.2 LOAD AND RESISTANCE FACTOR DESIGN METHOD

The LRFD *method* was first introduced into wood design in the 2005 NDS code [1]. In this method, the load effects are multiplied by various load factors, which are usually greater than 1.0 for other loads except wind and seismic loads. Recall from Chapter 2 that the wind and seismic loads (W and E) as calculated in the ASCE 7 load standard are already at their factored load levels, therefore the load factors for wind and seismic loads in the LRFD method is 1.0, to adjust the loads and load effects at the ultimate limit state. The load effects are usually determined from an elastic structural analysis. The factored resistance of the structural member or element is determined by multiplying the nominal strength or theoretical strength of the member by a strength reduction or resistance factor, which is usually less than 1.0. The factored load effect must be less than or equal to the factored resistance for the structural member to be adequate at the ultimate limit state. The member is also checked at the serviceability limit state using unfactored loads. The LRFD method accounts better than does the ASD method for the variability of loads in assigning the various load factors. In both the ASD and LRFD methods, the same reference design values from the NDS Supplement (NDS-S) [2] are used to calculate the allowable stresses (ASD) and the adjusted design value (LRFD). The applied load and load effects and the factored load and load effects are calculated using the ASD load combinations and the LRFD load combinations, respectively, previously discussed in Section 2.1.

3.3 ALLOWABLE STRESS DESIGN METHOD IN WOOD DESIGN

As stated earlier, the ASD method follows the approach that the unfactored applied stress calculated using the ASD load combinations from Section 2.1 must be less than or equal to the code-specified allowable stress in order for the structure or structural member to be deemed adequate. The unfactored applied stresses are denoted by a lower case f and the allowable stresses are denoted by an uppercase F. For example, the unfactored tension stress applied parallel to the grain f_t (= P/A) must be less than or equal to the allowable tension stress parallel to the grain F_t', where

$$F_t' = F_{t,\text{NDS-S}} \text{ (product of all applicable adjustment or } C \text{ factors)}$$

Similarly, the unfactored bending stress applied f_b (= M/S_X) must be less than or equal to the allowable bending stress F_b', where

$$F_b' = F_{b,\text{NDS-S}} \text{ (product of all applicable adjustment or } C \text{ factors)}$$

In general, the allowable stress due to unfactored loads is given as

$$F' = F_{\text{NDS-S}} \text{ (product of all applicable adjustment or } C \text{ factors)} \tag{3.1}$$

The applicable adjustment factors for ASD are given in Tables 3.1 and 3.2. The tabulated reference design values $F_{\text{NDS-S}}$ given in the NDS-S [2] assume normal load duration and dry-use conditions. Note that the load duration factor, C_D, applies only to the ASD method.

3.4 LOAD AND RESISTANCE FACTOR DESIGN METHOD IN WOOD DESIGN

The LRFD method for wood design in the NDS code follows a similar approach to the ASD method presented in the previous section, but some of the applicable adjustment factors are different for the LRFD method than for the ASD method. In the LRFD method, the factored applied stress calculated using the LRFD load combinations from Section 2.1 must be less than or equal to the LRFD adjusted design stress.

The factored applied stresses are denoted by a lower case f_u and the LRFD adjusted design stresses are denoted by an uppercase F_r'. For a structural member to be deemed adequate, the factored applied stresses, f_u, must be less than or equal to the LRFD adjusted design stresses, F_r'.

In general, the LRFD adjusted design stress or design values are given as

$$F_r' = F_{\text{NDS-S}} \text{ (product of all applicable adjustment or } C\text{-factors)}(K_F)(\phi)(\lambda) \tag{3.2}$$

where:

K_F = format conversion factor to convert the ASD-related reference design values in the NDS-S to LRFD

ϕ = resistance factor

λ = time effect factor that is dependent on the load combination

Note that the time effect factor, λ, does not apply to modulus of elasticity (both E' and $E'_{\min}$) and compression stress perpendicular to the grain (see Tables 3.3 and 3.4).

These three factors are applicable to the LRFD method only. The applicable adjustment or C factors and the K_F and ϕ factors for LRFD are shown in Tables 3.3 and 3.4. The time effect factor, λ, is obtained from NDS Table N3 and is presented here in Table 3.5. The tabulated

Table 3.1 Applicability of adjustment factors for sawn lumber (ASD)

	Load duration factor (ASD only)	*Wet service factor*	*Temperature factor*	*Beam stability factor*	*Size factor*	*Flat-use factor*	*Incising factor*	*Repetitive member factor*	*Column stability factor*	*Buckling stiffness factor*	*Bearing area factor*
$F_b' = F_b \times$	C_D	C_M	C_t	C_L	C_F	C_{fu}	C_i	C_r	–	–	–
$F_t' = F_t \times$	C_D	C_M	C_t	–	C_F	–	C_i	–	–	–	–
$F_v' = F_v \times$	C_D	C_M	C_t	–	–	–	C_i	–	–	–	–
$F_c' = F_c \times$	C_D	C_M	C_t	–	C_F	–	C_i	–	C_P	–	–
$F_{c\perp}' = F_{c\perp} \times$	–	C_M	C_t	–	–	–	C_i	–	–	–	C_b
$E' = E \times$	–	C_M	C_t	–	–	–	C_i	–	–	–	–
$E_{min}' = E_{min} \times$	–	C_M	C_t	–	–	–	C_i	–	–	C_T	–

Source: ANSI/AWC, *National Design Specification for Wood Construction*, Table 4.3.1. Courtesy of the American Wood Council, Leesburg, VA, 2015.

Table 3.2 Applicability of adjustment factors for glued laminated timber (ASD)

	Load duration factor	*Wet service factor*	*Temperature factor*	*Beam stability factor*[a]	*Volume factor*[a]	*Flat-use factor*	*Curvature factor*	*Stress interaction factor*	*Shear reduction factor*	*Column stability factor*	*Bearing area factor*
$F_b' = F_b \times$	C_D	C_M	C_t	C_L	C_V	C_{fu}	C_c	C_I	–	–	–
$F_t' = F_t \times$	C_D	C_M	C_t	–	–	–	–	–	–	–	–
$F_v' = F_v \times$	C_D	C_M	C_t	–	–	–	–	–	C_{vr}	–	–
$F_{rt}' = F_{rt} \times$	C_D	C_M	C_t	–	–	–	–	–	–	–	–
$F_c' = F_c \times$	C_D	C_M	C_t	–	–	–	–	–	–	C_P	–
$F_{c\perp}' = F_{c\perp} \times$	–	C_M	C_t	–	–	–	–	–	–	–	C_b
$E' = E \times$	–	C_M	C_t	–	–	–	–	–	–	–	–
$E_{min}' = E_{min} \times$	–	C_M	C_t	–	–	–	–	–	–	–	–

Source: ANSI/AWC, *National Design Specification for Wood Construction*, Table 5.3.1. Courtesy of the American Wood Council, Leesburg, VA, 2015.

[a] The beam stability factor C_L should not be applied simultaneously with the volume factor C_V; the lesser of these two values applies (see NDS Section 5.3.6).

Table 3.3 Applicability of adjustment factors for sawn lumber (LRFD)[a]

	Wet service factor	*Temperature factor*	*Beam stability factor*	*Size factor*	*Flat-use factor*	*Incising factor*	*Repetitive member factor*	*Column stability factor*	*Buckling stiffness factor*	*Bearing area factor*	*K_F, format conversion factor (LRFD only)*	*ϕ, resistance factor*	*λ, time effect factor*
$F_b' = F_b \times$	C_M	C_t	C_L	C_F	C_{fu}	C_i	C_r	–	–	–	2.54	0.85	λ
$F_t' = F_t \times$	C_M	C_t	–	C_F	–	C_i	–	–	–	–	2.70	0.80	λ
$F_v' = F_v \times$	C_M	C_t	–	–	–	C_i	–	–	–	–	2.88	0.75	λ
$F_c' = F_c \times$	C_M	C_t	–	C_F	–	C_i	–	C_P	–	–	2.40	0.90	λ
$F_{c\perp}' = F_{c\perp} \times$	C_M	C_t	–	–	–	C_i	–	–	–	C_b	1.67	0.90	–
$E' = E \times$	C_M	C_t	–	–	–	C_i	–	–	–	–	–	–	–
$E_{min}' = E_{min} \times$	C_M	C_t	–	–	–	C_i	–	–	C_T	–	1.76	0.85	–

Source: ANSI/AWC, *National Design Specification for Wood Construction,* Table 4.3.1. Courtesy of the American Wood Council, Leesburg, VA, 2015.

[a] For the design of all connections, the format conversion factor, K_F, and the resistance factor, ϕ, for all connection design values are 3.32 and 0.65, respectively.

Table 3.4 Applicability of adjustment factors for glued-laminated timber (LRFD)[a,b]

	Wet service factor	Temperature factor	Beam stability factor[a]	Volume factor[a]	Flat use factor	Curvature factor	Stress interaction factor	Shear reduction factor	Column stability factor	Bearing area factor	K_F, format conversion factor (LRFD only)	ϕ, resistance factor	λ, time effect factor
$F'_b = F_b \times$	C_M	C_t	C_L	C_V	C_{fu}	C_c	C_I	–	–	–	2.54	0.85	λ
$F'_t = F_t \times$	C_M	C_t	–	–	–	–	–	–	–	–	2.70	0.80	λ
$F'_v = F_v \times$	C_M	C_t	–	–	–	–	–	C_{vr}	–	–	2.88	0.75	λ
$F'_{rt} = F_{rt} \times$	C_M	C_t	–	–	–	–	–	–	–	–	2.88	0.75	λ
$F'_c = F_c \times$	C_M	C_t	–	–	–	–	–	–	C_P	–	2.40	0.90	λ
$F'_{c\perp} = F_{c\perp} \times$	C_M	C_t	–	–	–	–	–	–	–	C_b	1.67	0.90	–
$E' = E \times$	C_M	C_t	–	–	–	–	–	–	–	–	–	–	–
$E'_{min} = E_{min} \times$	C_M	C_t	–	–	–	–	–	–	–	–	1.76	0.85	–

Source: ANSI/AWC, *National Design Specification for Wood Construction,* Table 5.3.1. Courtesy of the American Wood Council, Leesburg, VA, 2015.

[a] The beam stability factor C_L should not be applied simultaneously with the volume factor C_V; the lesser of these two values applies (see NDS Section 5.3.6).

[b] For the design of all connections, the format conversion factor, K_F, and the resistance factor, ϕ, for all connection design values are 3.32 and 0.65, respectively.

reference design values $F_{NDS\text{-}S}$ or $E_{NDS\text{-}S}$ given in the NDS-S [2] assume dry-use conditions and are based on the ASD method. The conversion factor, K_F, is used to convert these ASD reference design values to LRFD reference design values.

3.4.1 NDS tabulated design stresses

The NDS tabulated reference design values for timbers are based on a 5% exclusion value of the clear-wood (i.e., wood members without knots) strength values with a reduction factor or strength ratio of less than 1.0 to allow for the effects of strength-reducing characteristics such as knot size, slope of grain, splits, checks, and shakes. A 5% exclusion value implies that 5 of every 100 pieces of wood are likely to have lower strength than the values tabulated in NDS-S. Only the elastic modulus and compression stress perpendicular to the grain are based on average values. The clear-wood (i.e., without knots) strength properties were established from laboratory tests on small straight-grained wood specimens where the wood members are loaded from zero to failure within a 5-minute duration. The *In-Grade tests*, which commenced in 1978, included some 70,000 full-size *dimension lumber* specimens of different species group obtained from several mills in the United States and Canada, which were tested to failure in bending, tension, and compression parallel to the grain. The In-Grade test program, which was carried out by the U.S. Forest Products Laboratory in collaboration with the North American Softwood Lumber industry, provided a "more scientific basis for wood engineering procedures," and the test data obtained, first published in 1991, replaced the clear-wood-based data for dimension lumber [3].

The allowable stress perpendicular to the grain is controlled by a deformation limit state and not by crushing failure. Under load perpendicular to the grain, the member continues to support increasing load because of the densification of the member. The NDS code uses a 0.04-in. deformation limit for the determination of $F'_{c\perp}$.

3.4.2 Stress adjustment factors

The various stress adjustment factors that are used in wood design and discussed in this chapter are listed in Table 3.5.

3.4.2.1 Load duration factor C_D

The strength of a wood member is a function of the duration of the load on the structure and this phenomenon has been observed in tests of wood structural members. This unique property of wood is accounted for in determining the strength of a wood member by the ASD method through use of the load duration factor, C_D. The term *duration of load* designates the total accumulated length of time that a load is applied during the life of the structure, and the shorter the load duration, the stronger the wood member. The load duration factor C_D converts stress values for normal load duration to design stress values for other durations of loading. The NDS code defines the normal load duration as 10 years, which means that the load is applied on the structure for a cumulative maximum duration of 10 years, and the load duration factor for this normal load duration is 1.0. The load duration factors for other types of loads are given in Table 3.6. Note that the load duration factor does not apply to the LRFD method; instead, the time effect factor, λ, which is different than the load duration factor, is used for LRFD. Whereas the controlling load duration factor is usually based on the shortest duration load in the ASD load combination, the time effect factor is based on the primary load in the LRFD load combination [4].

Table 3.5 Adjustment factors

Adjustment factor	*Description*
C_D	Load duration factor (applies to ASD only)
C_M	Wet service or moisture factor
C_F	Size factor
C_{fu}	Flat-use factor
C_t	Temperature factor
C_r	Repetitive member factor
C_i	Incising factor
C_P	Column stability factor
C_L	Beam stability factor
C_V	Volume factor (applies to *glulam* only)
C_b	Bearing area factor
C_T	Buckling stiffness factor
C_c	Curvature factor
C_I	Stress interaction factor (applies to *glulam* only)
C_{vr}	Shear reduction factor (applies to notched *glulam* only)
K_F	Format conversion factor (applies to *LRFD* only)
ϕ	Resistance factor (applies to *LRFD* only)
λ	Time effect factor (applies to *LRFD* only)

Table 3.6 Load duration factors, C_D (ASD only)

Type of load	*Cumulative load duration*	*Load duration factor* (C_D)
Dead load	Permanent	0.9
Floor live load	10 years or normal duration	1.0
Snow load	2 months	1.15
Roof live load	7 days	1.25
Construction load	7 days	1.25
Wind	10 minutes	1.6
Seismic (earthquake)	10 minutes	1.6
Impact	1 second or less	2.0

Source: ANSI/AWC, *National Design Specification for Wood Construction*, American Wood Council, Leesburg, VA, 2015; ANSI/AWC, *National Design Specification Supplement: Design Values for Wood Construction*, American Wood Council, Leesburg, VA, 2015.

For any load combination, the governing load duration factor C_D will be the value that corresponds to the shortest duration load in that load combination. Thus, the governing C_D value is equal to the largest C_D value in that load combination.

Example 3.1: (ASD): Load duration factors in ASD load combinations

Determine the load duration factor C_D for the following ASD load combinations: (a) dead load + floor live load; (b) dead load + floor live load + snow load; (c) dead load + floor live load + roof live load.

Solution:

a. *Dead load + floor live load:*

C_D for dead load = 0.9

C_D for floor live load = 1.0

Since the largest C_D value governs, $C_{D\,(D+L)} = \mathbf{1.0}$.

b. *Dead load + floor live load + snow load:*

C_D for dead load = 0.9

C_D for floor live load = 1.0

C_D for snow load = 1.15

Since the largest C_D value governs, $C_{D\,(D+L+S)} = \mathbf{1.15}$.

c. *Dead load + floor live load + roof live load:*

C_D for dead load = 0.9

C_D for floor live load = 1.0

C_D for roof live load = 1.25

Since the largest C_D value governs, $C_{D\,(D+L+Lr)} = \mathbf{1.25}$.

3.4.2.2 Moisture factor C_M

The strength of a wood member is affected by its moisture content. The higher the moisture content, the lower the wood strength. The tabulated design stresses in NDS-S apply to lumber that is surfaced-dry (S-dry) and used in dry service conditions. Examples will include S-dry interior wood members used in covered and insulated buildings. S-dry condition indicates a maximum moisture content of 19% for sawn lumber. For glued-laminated timber (glulam), the moisture factor is 1.0 when the moisture content is less than 16%. The moisture factor C_M for sawn lumber and glulam is shown in Table 3.7. Note that the moisture factor, C_M, accounts for the instantaneous and short-term effect of moisture on strength and does not account for the decay that may occur due to prolonged exposure of wood to moisture. For such cases, pressure-treated wood should be considered [4].

3.4.2.3 Size factor C_F

The size factor accounts for the effect of the structural member size on the strength of the sawn wood member. As the depth of a sawn wood member increases, there is some reduction in the strength of the wood member. The size factor C_F applies only to sawn lumber and is *not* applicable to Southern Pine (or Southern Yellow Pine), glulam, or machine stress-rated (MSR) lumber. Table 3.8 gives the size factors for the various size classifications of visually graded sawn lumber.

3.4.2.4 Repetitive member factor C_r

The repetitive factor applies only to flexural members placed in series and takes into account the redundancy in a roof, floor, or wall framing. It accounts for the fact that if certain conditions are satisfied, the failure or reduction in strength of one flexural member, in series with other adjacent members, will not necessarily lead to failure of the entire floor or wall

Table 3.7 Moisture factor C_M for sawn lumber and glulam

	Equilibrium moisture content, EMC (%)	*Moisture factor (C_M)*
Sawn lumber	≤19 (i.e., S-dry)	1.0
	>19	Use C_M values from NDS-S Tables 4A to 4F, as applicable
Glulam	<16	1.0
	≥16	Use C_M values from NDS-S Tables 5A to 5D, as applicable

Table 3.8 Size factor C_F for sawn lumber

Size classification	Size factor (C_F)
Dimension lumber, decking:	Use C_F from NDS-S Tables 4A, 4B, 4E, and 4F
Timbers: beam and stringer (B&S), and post and timbers (P&T) subject to bending	$C_F = \left(\frac{12}{d}\right)^{1/9} \leq 1.0$ where d is the actual member depth, in inches. For loads applied to B&S, see NDS-S Table 4D
Circular beams with diameter $d_c > 13.5''$:	$C_F = \left(\frac{12}{d_e}\right)^{1/9} \leq 1.0$ Where $d_e = \sqrt{\frac{\pi d_c^2}{4}}$
Square beams with side dimensions $d \geq 12''$ loaded in the plane of the diagonal:	$C_F = \left(\frac{12}{d}\right)^{1/9} \leq 1.0$

system because of the ability of the structure to redistribute load from a failed member to the adjacent members. Examples of repetitive members include roof or floor joists and rafters, built-up joists, beams, and columns. For a wood structural member to be classified as repetitive, all the following conditions must be satisfied:

- There are at least three parallel members of dimension lumber.
- The member spacing is not greater than 24 in. on centers.
- The members are connected or tied together by roof, floor, or wall sheathing (e.g., plywood).

If a member is repetitive, the repetitive member factor C_r is 1.15, and for all other cases, C_r is 1.0. The repetitive member factor C_r does not apply to timbers (i.e., beam and stringers and post and timbers) or glulam. The repetitive member factor applies to the tabulated bending stresses in NDS-S Tables 4A, 4B, 4C, and 4F only and has already been incorporated into the tabulated bending stress ($F_b C_r$) for decking given in NDS-S Table 4E.

3.4.2.5 Flat-use factor C_{fu}

Except for roof or floor decking, the NDS-S tabulated design stresses for dimension lumber apply to wood members bending about their stronger (or x–x) axis, as indicated in Figure 3.1a. For most cases, the flat-use factor C_{fu} is 1.0, but for the few situations where

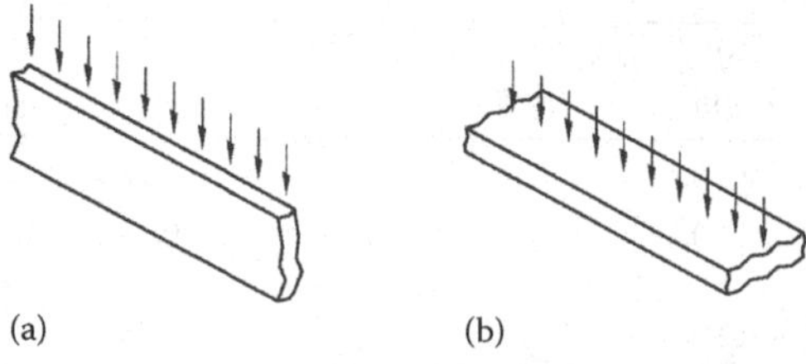

Figure 3.1 Direction of loading with respect to orthogonal axes: (a) bending about the strong axis and (b) bending about the weak axis.

wood members are stressed about their weaker (y–y) axis as indicated in Figure 3.1b, the flat-use factor C_{fu} is obtained from NDS-S Tables 4A and 4B for dimension lumber, and from NDS-S Tables 5A to 5D for glulam. An example of a situation where dimension lumber may be bent about the weaker (y–y) axis is in stair treads. The flat-use factor does not directly apply to decking because this factor has already been incorporated in the tabulated NDS-S reference design values for decking.

For glulam members, the flat-use factor is applied to the reference bending stress about the y–y or weaker axis, $F_{b,\,y-y}$. The C_{fu} values for glulam are tabulated in the adjustment factors section of NDS-S Tables 5A to 5D and are to be applied to the F_{by} (or weak axis bending) stress only. For glulam beams subjected to bending only about the strong axis, the flat-use factor will be 1.0.

3.4.2.6 Incising factor C_i

This is a factor that is used for pressure-treated wood. Some species of lumber do not easily accept pressure preservative treatment, and small incisions parallel to the grain may be cut in the wood to increase the depth of penetration of preservatives in the wood member. The incising factor applies only to dimension lumber and has a value of 1.0 for lumber that is not incised. Design values are multiplied by the factors shown in Table 3.9 when dimension lumber is incised parallel to the grain, as follows:

Maximum depth = 0.4 in.
Maximum length = 0.375 in.
Maximum incision density = 1100 per ft^2

3.4.2.7 Temperature factor C_t

Beyond a temperature of 100°F, an increase in temperature leads to a reduction in the strength of a wood member. The NDS code Table 2.3.3 lists the temperature factors C_t for various temperatures. For temperatures not greater than 100°F, the temperature factor C_t is 1.0 for wood used in dry or wet service conditions, and this applies to most insulated wood building structures. When the temperature in a wood structure is greater than 100°F, as may occur in some industrial structures, the C_t values can be obtained from NDS code Table 2.3.3.

3.4.2.8 Beam stability factor C_L

The beam stability factor accounts for the effect of lateral torsional buckling in a flexural member. When a beam is subjected to bending due to gravity loads, the top edge is usually in compression while the bottom edge is in tension. As a result of the compression stresses, the top edge of the beam will be susceptible to out-of-plane or sidesway buckling if it is not braced adequately. As the top edge moves sideways, the bottom edge, which is in tension, tends to move sideways in the opposite direction, and this causes twisting in the beam if the beam ends are restrained. This tendency to twist is called *lateral-torsional buckling,* and it should be

Table 3.9 Incising factor

Design value	*Incising factor (C_i)*
E, E_{min}	0.95
F_b, F_t, F_c, F_v	0.80
$F_{c\perp}$	1.00

Source: ANSI/AWC, *National Design Specification for Wood Construction,* American Wood Council, Leesburg, VA, 2015.

noted that if the beam ends are unrestrained, the beam will become unstable and could roll over. However, lateral restraints are usually always provided, at least at the beam supports. Generally, if the compression edge of a flexural member is continuously braced by decking or plywood sheathing, the beam will not be susceptible to lateral-torsional buckling and the beam stability factor will be 1.0 in this case.

Two methods are prescribed in the NDS code that could be used to determine the beam stability factor C_L: rule-of-thumb method and the calculation method.

Rule-of-thumb method (NDS code Section 3.3.3). The beam stability factor C_L is 1.0 for each of the following cases provided that the conditions described for each case—using nominal dimensions—are satisfied:

1. If $d/b \leq 2$, no lateral support and no blocking are required. *Examples:* 2 × 4 or 4 × 8 wood member.
2. If $2 < d/b \leq 4$, the beam ends are held in position laterally using solid end blocking. *Examples:* 2 × 6, 2 × 8, or 4 × 16 wood member.
3. If $4 < d/b \leq 5$, the compression edge of a beam is laterally supported throughout its length. Solid end blocking is not required by the NDS code, but is recommended. *Example:* 2 × 10 wood member with plywood sheathing nailed to the top face of the wood member at fairly regular intervals.
4. If $5 < d/b \leq 6$, bridging or full-depth solid blocking is provided at 8 ft intervals or less, or plywood sheathing is nailed to the top face of the wood member and blocking provided at the ends of the joist and at midspan where the joist span exceeds 12 ft or plywood sheathing nailed to the top and bottom faces of the joist. *Example:* 2 × 12.
5. If $6 < d/b \leq 7$, both edges of the beam are held in line for the entire length. *Examples:* 2 × 14 with plywood sheathing nailed to the top face; a ceiling nailed to the bottom face of the wood member.

For all five cases above, the member dimensions d and b are the nominal width (larger dimension) and thickness (smaller dimension) of the wood member, respectively.

3.4.3 Calculation method for C_L

When the nominal width (or larger dimension) d is greater than the nominal thickness (or smaller dimension) b, the beam stability factor is calculated using the following equations:

ASD:

$$C_L = \frac{1 + F_{bE}/F_b^*}{1.9} - \sqrt{\left(\frac{1 + F_{bE}/F_b^*}{1.9}\right)^2 - \frac{F_{bE}/F_b^*}{0.95}} \tag{3.3}$$

LRFD:

$$C_{L,r} = \frac{1 + F_{bE,r}/F_{b,r}^*}{1.9} - \sqrt{\left(\frac{1 + F_{bE,r}/F_{b,r}^*}{1.9}\right)^2 - \frac{F_{bE,r}/F_{b,r}^*}{0.95}} \tag{3.4}$$

where:

$F_{bE} = 1.20E'_{min}/R_B^2$ = Euler critical buckling stress for *bending* members

$E'_{min} = E_{min}C_M C_t C_i C_T$ (ASD reduced modulus of elasticity for buckling calculations for sawn lumber)

$E'_{min} = E_{min}C_M C_t$ (ASD reduced modulus of elasticity for buckling calculations for Glulam)

$F_{bx}^* = F_{bx\,\text{NDS-S}} \times$ (product of all applicable adjustable factors except C_{fu}, C_V, C_L, and C_I)

For the LRFD method, the corresponding parameters are as follows:

$F_{bE} = 1.20E'_{\min,r}/R_B^2$ = Euler critical buckling stress for *bending* members

$E'_{\min,r} = E_{\min}C_MC_tC_iC_T(1.76)(0.85)$ = LRFD reduced modulus of elasticity for buckling calculations for sawn lumber

$E'_{\min,r} = E_{\min}C_MC_t(1.76)(0.85)$ = LRFD reduced modulus of elasticity for buckling calculations for glulam

$F^*_{bx,r} = F_{bx\ \text{NDS-S}}\ K_F\phi\lambda\ \times$ (product of all applicable adjustable factors except C_D, C_{fu}, C_V, C_L, and C_I)

$R_B = \sqrt{l_e d/b^2} \le 50$

(when R_B is greater than 50, the beam width—that is, the smaller dimension, b—has to be increased and/or lateral bracing provided at close spacing to the compression edge of the beam to bring R_B within the limit of 50)

l_e = effective length obtained from Table 3.10 as a function of the unbraced length l_u

l_u = the unsupported length of the compression edge of the beam, or the distance between points of lateral support preventing rotation and/or lateral displacement of the compression edge of the beam

[l_u is the unbraced length of the compression edge of the beam in the plane (i.e., y–y axis) in which lateral torsional buckling will occur; without continuous lateral support, there is a reduction in the allowable bending stress, but in many cases,

Table 3.10 Effective length l_e for bending members

Span condition	$L_u/d < 7$	$7 \le l_u/d$: $7 \le l_u/d \le 14.3$	$7 \le l_u/d$: $14.3 < l_u/d$
Cantilever			
Uniformly distributed load	$l_e = 1.33l_u$	$l_e = 0.90l_u + 3d$	
Concentrated load at unsupported end	$l_e = 1.87l_u$	$l_e = 1.44l_u + 3d$	
Single span beam			
Uniformly distributed load	$l_e = 2.06l_u$	$l_e = 1.63l_u + 3d$	
Concentrated load at center with no intermediate lateral support	$l_e = 1.80l_u$	$l_e = 1.37l_u + 3d$	
Concentrated load at center with lateral support at center		$l_e = 1.11l_u$	
Two equal concentrated loads at 1/3 points with lateral support at 1/3 points		$l_e = 1.68l_u$	
Three equal concentrated loads at 1/4 points with lateral support at 1/4 points		$l_e = 1.54l_u$	
Four equal concentrated loads at 1/5 points with lateral support at 1/5 points		$l_e = 1.68l_u$	
Five equal concentrated loads at 1/6 points with lateral support at 1/6 points		$l_e = 1.73l_u$	
Six equal concentrated loads at 1/7 points with lateral support at 1/7 points		$l_e = 1.78l_u$	
Seven or more concentrated loads, evenly spaced, with lateral support at points of load application		$l_e = 1.84l_u$	
Equal end moments		$l_e = 1.84l_u$	
Single span or cantilever conditions not specified above	$l_e = 2.06l_u$	$l_e = 1.63l_u + 3d$	$l_e = 1.84l_u$

Source: ANSI/AWC, *National Design Specification for Wood Construction*, Table 3.3.3. Courtesy of the American Wood Council, Leesburg, VA, 2015.

continuous lateral support is provided by the plywood roof or floor sheathing, which is nailed and glued to the compression face of the joists and beams, and for these cases, the unbraced length is taken as zero.]

K_F, ϕ, λ are factors that apply to the LRFD method only (see Tables 3.3 and 3.4).

3.4.3.1 Column stability factor C_p

The column stability factor accounts for buckling stability in columns due to slenderness and is discussed more fully in Chapter 5. It is a function of the column end fixity, the column effective length, and the column cross-sectional dimensions, and it applies only to the compression stress parallel to the grain, F_c' and $F_{c,r}'$. When a column is fully or completely braced about an axis of buckling such as by an intersecting stud wall or by plywood wall sheathing, the column stability factor C_p about that axis will be equal to 1.0.

3.4.3.2 Volume factor C_V

The volume factor, which accounts for size effects in glulam members, is applicable only for bending stress calculations. When designing glulam members for bending, the smaller of the beam buckling stability factor C_L and the volume factor C_V are used to calculate the allowable bending stress. The volume factor is given as

$$C_V = \left(\frac{21}{L}\right)^{1/x}\left(\frac{12}{d}\right)^{1/x}\left(\frac{5.125}{b}\right)^{1/x} = \left(\frac{1291.5}{bdL}\right)^{1/x} \leq 1.0 \tag{3.5}$$

where:

L is the length of beam between points of zero moment, ft (conservatively, assume that L is the span of beam)
d is the depth of beam, in.
b is the width of beam, in.
x is the 20 for Southern Pine
x is the 10 for all other species

For the same size glulam, the C_V factor is higher for Southern Pine than for other species.

3.4.3.3 Bearing area factor C_b

The bearing area factor accounts for the increase in allowable stresses at interior supports. NDS code Section 3.10.4 indicates that for bearing supports not nearer than 3 in. to the end of a member and where the bearing length l_b measured parallel to the grain of the member is less than 6 in., a bearing area adjustment factor C_b could be applied, where

$$\begin{aligned} C_b &= \frac{l_b + 0.375}{l_b} && \text{for } l_b \leq 6 \text{ in.} \\ &= 1.0 && \text{for } l_b > 6 \text{ in.} \\ &= 1.0 && \text{for bearings at the ends of a member} \end{aligned} \tag{3.6}$$

This equation accounts for the increase in bearing capacity at interior reactions or interior concentrated loads (see Figure 3.2) where there are unstressed areas surrounding the stress bearing area.

3.4.3.4 Curvature factor C_c

The curvature factor applies to glulam only and is used to adjust the allowable bending stresses for curved glulam members such as arches. The curvature factor C_c applies only

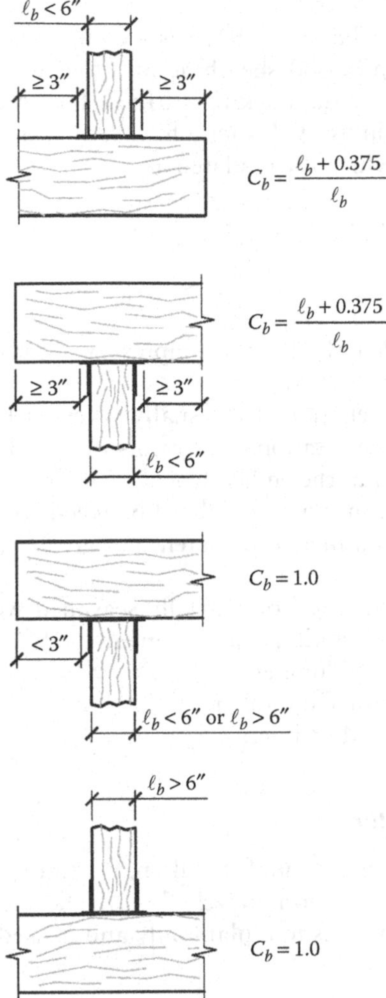

Figure 3.2 Bearing stress factors for different bearing conditions.

to the curved section of the glulam member, and for straight glulam members, C_c is 1.0. The curvature factor for a curved glulam member is given as

$$C_c = 1 - (2000)\left(\frac{t}{R}\right)^2 \tag{3.7}$$

where:

t is the thickness of laminations, in.
R is the radius of curvature of the inside face of the lamination, in.

$\frac{t}{R} \le \frac{1}{100}$ for hardwoods and Southern Pine

$\le \frac{1}{125}$ for other softwoods

3.4.3.5 Buckling stiffness factor C_T

When a 2 × 4 or smaller sawn lumber is used as a compression chord in a roof truss with a minimum of 3/8-in.-thick plywood sheathing attached to its narrow face, the stiffness against buckling is increased by the bucking stiffness factor C_T. This is accounted for by multiplying the reference modulus of elasticity for beam and column stability, E_{min}, by the buckling stiffness factor C_T, which is calculated as

$$C_T = 1 + \frac{K_m l_e}{K_T E} \tag{3.8}$$

where:

l_e is the effective column length of truss compression chord ($= K_e l$) ≤ 96 in. (for $l_e > 96$ in., use $l_e = 96$ in.)

K_e is the effective column length factor, usually taken as 1.0 for members supported at both ends, that is, the rotational restraints due to the members framing into the compression member at the ends are neglected

l is the unbraced length of the truss top chord between connections and panel points

K_m = 2300 for wood with a moisture content not greater than 19% at the time of plywood attachment
= 1200 for unseasoned wood or partially seasoned wood at the time of plywood attachment (i.e., wood with moisture content > 19%)

K_T = 0.59 for visually graded lumber
= 0.75 for machine-evaluated lumber
= 0.82 for glulams and MSR lumber

3.4.3.6 Shear reduction factor C_{vr}

The NDS-S reference shear design values for glulam, F_{vx} and F_{vy}, were derived based on tests on prismatic glulam beams subject to noncyclic loads. The shear reduction factor, C_{vr}, has a constant value of 0.72 and applies to glulam only and is used for the shear design for the following cases [4]:

- *Nonprismatic* glulam beams or girders
- Notched glulam beams
- Curved and tapered beams, arches
- Glulam beams under impact or cyclic loading or repetitive loading
- Glulam members at connections

It is not recommended to have the tension side of glulam beams tapered.

3.4.3.7 Stress interaction factor C_I

The stress interaction factor applies only to glulam and accounts for the interaction between bending stress, compressive stress, tension stress, and shear stress in tapered glulam beams and curved nonprismatic glulam beams. This factor is calculated using Equation 3.9 [4].

$$C_I = \frac{1}{\sqrt{1 + (F_b \tan\theta / F_v C_{vr})^2 + (F_b \tan^2\theta / F_{c\perp})^2}} \tag{3.9}$$

θ is the angle of taper of the beam

- For glulam beams tapered on the compression side, use either C_I or C_V, whichever is smaller, to calculate the adjusted reference design values. Also, $F_{c\perp}$ is replaced in Equation 3.9 by F_{rt}, the reference design value in radial tension.
- For glulam beams tapered on the tension side, though this is not recommended, use either C_I or C_L, whichever is smaller, to calculate the adjusted reference design values.

3.4.3.8 Format conversion factor K_F

The reference design values for different wood species and stress grades given in the NDS-S are given in an ASD format. In order to use those same reference design values for the LRFD method, the NDS code provides format conversion factors for each type of stress that are used to convert the ASD reference design values that are based on normal duration of loading to the LRFD reference design values. The format conversion factor applies to the LRFD method only [1,4].

3.4.3.9 Resistance factor ϕ

The resistance factors account for material and dimensional variability in a structural member and the consequences of failure, and it varies with the type of stress or failure mode considered. The resistance factor applies to the LRFD method only.

3.4.3.10 Time effect factor λ

The time effect factor applies to the LRFD method only and takes into account the duration of the loading on a wood member. It is dependent on the combination of loads acting on the structural member as shown in Table 3.11 and the factors were derived using a probabilistic-based approach to achieve a consistent probability of failure for different load durations [4]. The time effect factor does not apply to compression stress perpendicular to the grain, the modulus of elasticity for deflection calculations and the buckling modulus of elasticity.

Table 3.11 Time effect factor, λ (LRFD only)

Load combination	λ
$1.4D$	0.6
$1.2D + 1.6L + 0.5(L_r \text{ or } S \text{ or } R)$	0.7 when L is from storage; 0.8 when L is from occupancy; 1.25 when L is from impact[a]
$1.2D + 1.6(L_r \text{ or } S \text{ or } R) + (L \text{ or } 0.5W)$	0.8
$1.2D + 1.0W + L + 0.5(L_r \text{ or } S \text{ or } R)$	1.0
$1.2D + 1.0E + L + 0.2S$	1.0
$0.9D + 1.0W$	1.0
$0.9D + 1.0E$	1.0

Source: ANSI/AWC, *National Design Specification for Wood Construction*, American Wood Council, Leesburg, VA, 2015, Table N3.

[a] Time effect factor for design of connections or structural members pressure-treated with water-borne preservatives or fire retardant chemicals shall not be greater than 1.0.

Table 3.12 Radial tension design values, F_{rt}, for curved glulam members

Species group	*Radial tension design values, F_{rt}, for curved glulam members*	
	Type of loading	
Southern Pine	All loading conditions	$F_{rt} = (1/3)F_{vx}C_{vr}$
Douglas Fir-Larch, Douglas Fir-South, Hem-Fir, Western Woods, and Canadian Softwood species	Wind or earthquake loading	$F_{rt} = (1/3)F_{vx}C_{vr}$
	Other type of loading	$F_{rt} = 15$ psi

3.4.3.11 Radial tension stress perpendicular to grain, F_{rt}

The radial tension stress perpendicular to grain, F_{rt}, applies to curved glulam bending members and is obtained from Table 3.12 in NDS Code Section 5.2.8 [1]:

3.4.3.12 Radial compression stress perpendicular to grain, F_{rc}

The radial compression stress perpendicular to grain, F_{rc}, applies to curved glulam bending members and is taken as the reference compression stress perpendicular to grain on the side face of the glulam member, $F_{c\perp y}$ (NDS Code Section 5.2.9)

3.4.4 Procedure for calculating allowable stress or the LRFD adjusted design stress

The procedure for calculating the allowable stresses for ASD or the LRFD adjusted design stress from the tabulated NDS reference design stresses is as follows:

1. Identify the size classification of the wood member: that is, whether it is *dimension lumber*, *beam and stringer (B&S)*, or *post and timber (P&T)* or *decking*.
2. Obtain the tabulated reference design stress values from the appropriate table in the NDS-S based on the size classification of the member, the wood species, the stress grade, and the structural function or stress. Refer to Table 1.4 for the applicable NDS-S tables. Note that the reference design values already include a factor of safety for ASD.
3. Obtain the applicable adjustment factors from the NDS code adjustment factors applicability tables (see Tables 3.1 through 3.4).
4. Calculate the allowable stresses for the ASD method, $F'_{[\,]}$, or the LRFD adjusted design stresses, $F'_{[\,],r}$:

   ASD:

   $F'_{[\,]} = F_{\text{NDS-S}} \times$ (product of all applicable adjustment factors from Tables 3.1 and 3.2)

   LRFD:

   $F'_{[\,],r} = F_{\text{NDS-S}} \times$ (product of all applicable adjustment factors from Tables 3.3 and 3.4)

   - For ASD, the allowable stress must be greater than or equal to the applied stress due to the ASD load combinations from Section 2.1.
   - For LRFD, the LRFD adjusted design stress or resisting stress must be greater than or equal to the factored stress (load effects) due to the LRFD load combinations from Section 2.1.

3.4.5 Moduli of elasticity for sawn lumber

There are two moduli of elasticity design values tabulated in NDS-S for sawn lumber. The reference modulus of elasticity, E, is used for beam deflection calculations and other serviceability calculations such as floor vibrations, while the reduced modulus of elasticity, E_{min}, is used for beam and column stability calculations. E_{min} is derived from the reference modulus of elasticity E with a factor safety of 1.66 applied, in addition to accounting for the coefficient of variation in the modulus of elasticity and an adjustment factor to convert reference E values to a pure bending basis.

3.5 ENGINEERED WOOD PRODUCTS

Several proprietary engineered wood products have been developed for use in wood structures that require longer clear spans and higher load-carrying capacity than can be sustained by sawn lumber products. These engineered wood products also known as "manufactured lumber" include wood I-joists; structural composite lumber (SCL) such as laminated veneer lumber (LVL) and parallel strand lumber (PSL); wood structural panels such as plywood and oriented strand board (OSB)—these are discussed in Chapters 6 and 7; cross-laminated timber (CLT); and glued-laminated timber or glulam [5].

3.5.1 Wood I-joists

These are I-shaped proprietary prefabricated wood products made up of plywood or OSB web members bonded to sawn lumber or SCL tension and compression flanges with waterproof adhesives. They are used as floor joists or rafters with long spans (up to 36 ft) and have a much higher stiffness-to-weight ratio when compared to sawn lumber, and perform better than sawn lumber beams in minimizing floor vibrations. Holes may be cut in the webs of I-joists according to the manufacturer's guidelines to allow for the passage of mechanical duct and piping [1,5,6].

3.5.2 Manufactured lumber products

SCL are proprietary that include LVL and PSL and they have higher strength and stiffness compared to sawn lumber. LVL consists of thin wood veneers strips of 0.25 in. maximum thickness with grains that are parallel to the longitudinal direction of the member and bonded together with waterproof adhesive under heat and pressure. PSL consists of wood strands with a least dimension of 0.25 in. with grains that are parallel to the longitudinal axis of the member. The average length of the strands shall be at least 150 times the least dimension [1,5]. LVLs are typically used as beams and are available in widths of 1.75 in. and depths of up to 24 in., and lengths of up to 60 ft. Multiple LVLs may be used side by side to form a larger size beam. PSLs are typically used as beams and columns, and as headers over door openings. They are available in lengths of up to 60 ft, widths between 3.5 and 7 in., and depths of up to 18 in. [5]. The reference design values for SCL products are published by the manufacturer of the product. These products should not be modified on site by notching or drilling holes through them without coordination with the SCL manufacturer since the load capacity will be altered by these modifications.

3.5.3 Cross-laminated timber

Cross-laminated timber (CLT) panels are made from bonding several layers of sawn lumber together with waterproof adhesive with each layer orthogonal to the adjacent layer, and panel widths can vary up to 12 ft and lengths of up to 50 ft. They are used as walls and floor panels and are typically manufactured from Douglas Fir-Larch, Pine, and Spruce. The total thickness of CLT can vary from 2 to 20 in. and each laminate can vary from 5/8 to 2 in. The number of laminate layers varies from 3 to 9 and above. CLT products have so far been predominantly used in Europe though they are gradually been introduced into the North American market, and the NDS code now includes a section on CLT [1]. The use of CLT in high-rise wood buildings will be discussed in Chapter 9.

3.6 GLUED-LAMINATED TIMBER

Glulam is the acronym for glued-laminated timber, which is made from gluing together under pressure thin laminations of kiln-dried sawn lumber where the direction of the grains in all the laminations is parallel. They are used for very long spans (up to 100 ft or more) and to support very heavy loads where sawn lumber becomes impractical, and are also widely used in architecturally exposed structures because of their aesthetic appeal. Other advantages of glulam compared to sawn lumber include higher strength, better fire rating because of their relatively large size compared to sawn lumber framing, better aesthetics, better quality control because of the small size of the laminations, greater dispersion of defects in the laminations of the glulam, ready formation to curved shape for arches, and economical for long spans and large loads. Glulams are widely used in the construction of arch and long-span wood structures and are fabricated from very thin laminations of sawn lumber as shown in Figure 3.3 bonded together with adhesives that meet specifications for exterior use and are durable under high and low temperatures. They are usually end jointed and the laminations are stacked and glued together under pressure with water-resistant adhesives to produce wood members of practically any size and length. The lamination thickness varies from 1½ or 1⅜ in. for straight/slightly curved members and is ¾ in. for sharply curved members. Note that proprietary wood members such as Microlams (LVL) and Parallams (PSL) are sometimes used for floor beams, girders, and columns in lieu of glulams.

Glulam beams with large span lengths are often fabricated with an upward deflection or camber so as to reduce the final maximum downward deflection of the glulam beam measured from the horizontal roof or floor datum level. The maximum recommended upward camber is 1.5× the dead load deflection for roof beams and the dead load deflection for floor beams [5].

Glulams are available for depths as high as 72 in. or more and the standard glulam widths are 3⅛, 5⅛, 6¾, 8¾, 10¾, and 12¼ in. for western species glulam (i.e., Douglas Fir-Larch and Hem-Fir), and 3, 5, 6¾, 8½, and 10½ in. for Southern Pine glulam [2]. The laminations could be made from one piece of wood or from more than one piece of wood joined end to end to achieve a desired length or joined edge to edge to achieve a desired width of the glulam beam. Curved glulams are made by bending the laminations to the desired curved

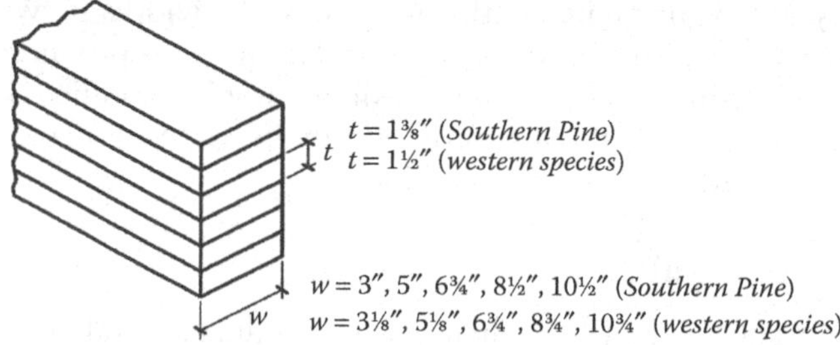

Figure 3.3 Glulam cross-section.

shape during the gluing process [6]. The most commonly used wood species for glulam are Douglas Fir-Larch, Hem-Fir, and Southern Pine [7]. Other common species used for glulam include Hem-Fir, Spruce-Pine-Fir, Eastern Spruce, and Alaska Cedar. Glulams can be manufactured in any length, limited only by transportation and handling limit requirements [6]. As noted earlier, the sizes of glulam are usually called out on plans using their actual dimensions, whereas sawn lumber is called out using nominal dimensions. Table 1C of NDS-S lists the various glulam sizes and section properties. In practice, it is usually more economical for the designer to specify the desired allowable stresses than to specify the glulam combination, hence the introduction of the stress class system in NDS-S Table 5A.

3.6.1 End joints in glulam

The most common type of end jointing system used in glulam members is the finger joint as shown in Figure 3.4. The scarf joint shown in Figure 3.4c is no longer used except for special situations or in existing buildings. Scarf joints have a maximum slope of 1:10 for tension joints and 1:5 for compression joints [5–7]. The end joints are the most highly engineered and quality controlled part of the glulam beam since that is where the beam is weakest and likely to fail. The end joints are bonded with high strength durable and waterproof adhesives and subject to rigorous qualification procedures and testing.

3.6.2 Grades of glulam

Traditionally, the two main structural grades of glulam are bending combination and axial combination glulams. The bending combination glulams are given in NDS-S Table 5A–Expanded, and the axial combination glulams are given in NDS-S Table 5B. The values in NDS-S Tables 5A and 5B are for softwood glulam, the most frequently used glulam in the United States, and NDS-S Tables 5C and 5D are for hardwood glulam.

3.6.2.1 Bending combination glulams (NDS-S Tables 5A, 5A–Expanded, and 5C)

The bending combination glulams are used predominantly to resist bending loads, and for efficiency, their outer laminations are usually made of higher quality lumber than the inner laminations. They could also be used to resist axial loads but may not be as efficient to resist axial loads as axial combination glulams. There are two types of bending combination lamination layup—balanced and unbalanced layup. In the balanced layup, which is recommended for continuous beams, the tension and compression laminations have the same tabulated design bending stress, whereas in the unbalanced layup, the tension lamination has a higher tabulated bending stress than the compression lamination. Unbalanced layup are recommended only for

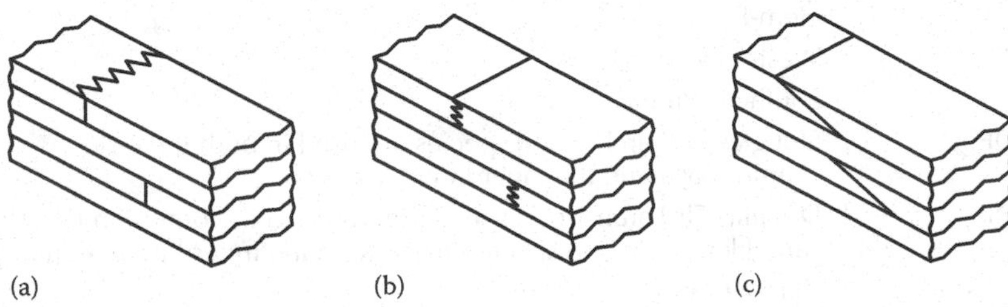

Figure 3.4 End joints in glulam: (a) vertical finger joint, (b) horizontal finger joint, and (c) scarf joint.

simply supported beams. An example of a bending combination glulam is 24F-V4 DF or 24F-V3 SP, the most commonly used glulam combinations with unbalanced layup, where the "24" refers to a tabulated reference bending stress of 2400 psi in the tension lamination, "V" refers to glulam with visually graded laminations, and "4" or "3" is the combination symbol. Another example is 20F-E3, where the "20" refers to a tabulated reference bending stress of 2000 psi, "E" refers to glulam with E-rated laminations, and "3" is the combination symbol. The most commonly used glulam with a balanced lamination layup is 24F-V8 DF or 24F-V5 SP [8].

3.6.2.2 Axial combination glulams (NDS-S Tables 5B and 5D)

Axial combination glulams are used predominantly to resist axial loads (tension or compression), and therefore all the laminations are made of the same quality or grade. They could also be used to resist bending loads but are not as efficient as bending combination glulams. One example of an axial combination glulam is 5DF, where "5" is a combination symbol and "DF" refers to the wood species used in the glulam laminations.

3.6.3 Appearance grades for glulam

Glulam is produced in four standard appearance grades [6,8] as follows (note that the reference design values for glulam as tabulated in the NDS-S are independent of these appearance grades):

- *Framing*: Used in concealed locations, for example, for headers in residential buildings.
- *Industrial*: Used in concealed applications or where aesthetics of the glulam member is not of paramount importance.
- *Architectural*: These have a smoother surface finish and are used in exposed applications where the aesthetics of the glulam member is of paramount importance such as in architecturally exposed structures.
- *Premium*: These also have a smoother surface finish and are used in architecturally exposed structures.

3.6.4 Wood species used in glulam

The wood species or species groups used in glulam laminations are shown in Table 3.13. The outer laminations of glulam are usually of higher quality lumber than the inner laminations, and often, the tension outer laminations are made of higher quality lumber than the

Table 3.13 Wood species used in glulam

Wood species symbol	*Description*
DF	Douglas Fir-Larch
DFS	Douglas Fir-South
HF	Hem-Fir
WW	Western woods
SP	Southern Pine
DF/DF	Douglas Fir-Larch wood species are used in both the outer laminations and the inner laminations
DF/HF	Douglas Fir-Larch wood species are used in the outer laminations, and Hem-Fir wood species are used in the inner laminations
AC	Alaska Cedar
SW	Softwood species

compression outer laminations. Glulam beams with tension and compression laminations of equal strength are called a *balanced lay-up* system. Glulams with different higher strength in the tension lamination compared to the compression laminations are called an *unbalanced lay-up* system. Unbalanced lay-up systems are used in simply supported beams, whereas balanced lay-up glulams should be used for continuous glulam beams. In practice, all glulams have to be grade-stamped, and unbalanced glulam beams have to be grade stamped on the top with a "TOP" designation to prevent installing the beam upside down [4]. When a glulam member will be exposed and unprotected from the environment, preservative treatment should be applied to the member, or wood species that are naturally resistant to decay—such as California Redwood or Alaska Yellow Cedar—should be used for the glulam laminations.

3.6.5 Stress class system

In the 2015 NDS-S, Table 5A provides an alternative means for specifying glulam based on the required allowable stress or stress class. This stress class approach is favored by the glulam industry because it gives glulam manufacturers flexibility in providing glulams where the required reference design stresses are specified, rather than specifying a specific glulam combination. The stress class approach greatly reduces the number of glulam combinations that the designer has to choose from, thus increasing design efficiency. In the stress class system of Table 5A, glulam is called out, for example, as 20F-1.5E, where "20" refers to the tabulated bending stress of 2000 psi (for the tension laminations stressed in tension) and "1.5E" refers to a bending modulus of elasticity of 1.5×10^6 psi obtained by mechanical grading. The intent is that future NDS-S editions will use the stress class approach in place of the bending combination glulams. The stress class approach does not apply to axial combination glulams.

3.6.5.1 NDS-S Tables 5A, 5A–Expanded, and 5B

In NDS-S Tables 5A, 5A–Expanded, and 5B, the allowable bending stress when high-quality *tension* laminations are stressed in *tension* is denoted as F_{bx}^{+}. The allowable bending stress when lower quality *compression* laminations are stressed in *tension* is denoted as F_{bx}^{-}. For continuous beams and beams with cantilevers, it is preferable and recommended to use glulam with equal tensile strengths in the compression and tension laminations (i.e., $F_{bx}^{+} = F_{bx}^{-}$). This is called a *balanced lay-up system*. For simply supported single span glulam beams, it is more efficient and practical to use an unbalanced lay-up system (i.e., glulams that have a higher strength in the tension laminations than in the compression laminations). Note that there are several values of elasticity modulus E given in NDS-S Tables 5A and 5B:

- E_x is used in deflection calculations for strong (x–x)-axis bending of the glulam member.
- $E_{x,\min}$ is used in glulam beam and column stability calculations for buckling about the x–x axis. It is used in C_L and C_p calculations for buckling about the x–x axis.
- E_y is used for deflection calculations for weak (y–y)-axis bending of the glulam member.
- $E_{y,\min}$ is used in glulam beam and column stability calculations for buckling about the y–y axis. It is used in C_L and C_p calculations for buckling about the y–y axis.
- E is used to calculate deflection calculations in axially loaded glulam members.
- $E_{\min}$ is used in glulam beam and column stability calculations in axially loaded glulam members.

Two values of compression stress perpendicular to the grain ($F_{c\perp}$) are given in NDS-S Table 5A, and the locations of these stresses in a glulam beam are indicated in Figure 3.5.

- $F_{c\perp \text{ tension face}}$ applies to the bearing stress on the outer face of the tension lamination.
- $F_{c\perp \text{ compression face}}$ applies to the bearing on the outer face of the compression lamination.

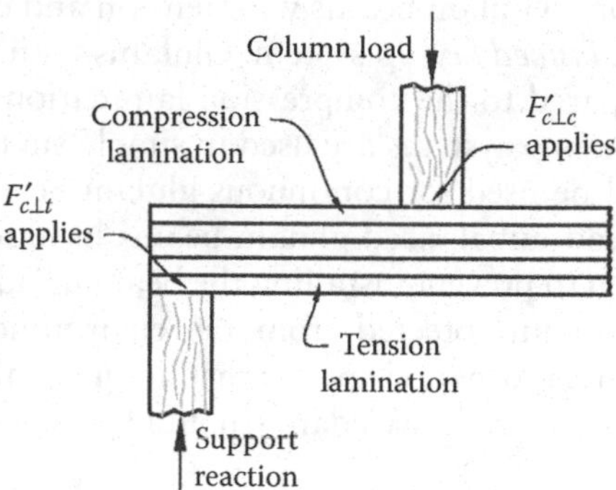

Figure 3.5 Bearing stress at supports and at interior loads.

The allowable bearing stresses on the tension and compression laminations for ASD are given as

$$F'_{c\perp t} = F_{c\perp \text{ tension face}} C_M C_t C_b$$

$$F'_{c\perp c} = F_{c\perp \text{ compression face}} C_M C_t C_b$$

The LRFD adjusted bearing stresses on the tension and compression laminations for are given as

$$F'_{c\perp t,r} = F_{c\perp \text{tension face}} C_M C_t C_b K_F \phi = F_{c\perp \text{tension face}} C_M C_t C_b (1.67)(0.90)$$

$$F'_{c\perp c,r} = F_{c\perp \text{compression face}} C_M C_t C_b K_F \phi = F_{c\perp \text{tension face}} C_M C_t C_b (1.67)(0.90)$$

3.6.6 Failure of glulam structures

In general, wood is weakest when subjected to tension perpendicular to the grain direction, and one of the most common types of failures in glulam structures is failure due to parallel-to-grain cracks occurring along the laminations caused by tension perpendicular to the grain. Failures due to tension stress perpendicular to the grain are brittle and sudden, and appropriate attention should be given to the detailing of glulam beams, girders, and trusses and their connections to avoid or minimize tension stress perpendicular to the grain direction or to reinforcing the glulam member when subjected to tension stress perpendicular to the grain [9,10]. Such reinforcement may include threaded screws or carbon fiber reinforced plastic wrap. Examples of glulam members subjected to tension perpendicular to the grain include beams with notches or holes, double tapered beams, pitched and curved beams brace members or diagonal web members of trusses with eccentric connections or members with unintended moment connections that causes tension stress perpendicular to the grain [11].

3.7 ALLOWABLE STRESS AND LRFD ADJUSTED STRESS CALCULATION EXAMPLES

In this section, we present several examples to illustrate the calculation of allowable (ASD) and LRFD adjusted stresses for sawn lumber and glulam members.

3.8 ASD LOAD COMBINATIONS AND THE GOVERNING LOAD DURATION FACTOR

The ASD load combinations were presented in Chapter 2. The distinct critical ASD load combinations for which the various structural elements in a typical wood structure need to be designed are given in Table 3.14. For the roof beams, the seismic and floor live loads are assumed to be zero, while for the floor beams, the roof live loads, snow load, seismic load, and wind load are taken as zero since floor beams are not usually subjected to these types of loads. In ASD load combinations 15 and 16 in Chapter 2, the wind and seismic loads oppose the dead load. Therefore, W and E take on negative values only in these two load combinations in order to maximize the uplift or overturning effect. These two ASD load combinations are used to include the effects of uplift or overturning in structural elements and systems due to wind and seismic load effects.

3.8.1 Normalized load method—ASD method only

From Table 3.13, it is apparent that except for floor beams, several ASD load combinations need to be considered in the design of wood members, but because of the varying load duration factors C_D, it is not possible or easy to determine by inspection the most critical ASD load combinations for the design of a particular wood member. Thus, it can become quite tedious to design for all these load combinations with differing C_D values, especially when the design is carried out manually.

To help in identifying the most critical or governing ASD load combination quickly, thus reducing the tedium in design, a normalized load method applicable only to the ASD method is described in this section. In this method, the total load for each ASD load combination is normalized with respect to the governing load duration factor C_D for that load combination. The ASD load combination that has the largest normalized load is the most critical, and the wood member is designed for the actual total load corresponding to that governing load combination. The reader is cautioned to use the actual total load, in design, not the normalized load. The normalized load method is applicable only if all the loads in the ASD load combination are of the same type and pattern. For example, the method is not valid if in the same load combination, some loads are uniform loads (given in psf or ksf) and others are concentrated loads (given in kips or lbs). When a load value is zero in a load combination, that load is neglected when determining the governing load duration factor C_D for that load combination. Note that the normalized load method may not work well for column design because C_D is a factor in the nonlinear C_p equation.

Table 3.14 Applicable design load combinations

Roof beams	*Floor beams*	*Columns*
$E = 0, L = 0, F = 0, H = 0, T = 0$[a]	$L_r = 0, S = 0, E = 0, W = 0, H = 0, F = 0, T = 0$[a]	
$D + (L_r$ or S or $R)$	$D + L$	Check all eight load combinations in Section 2.1
$D + 0.75\ (0.6W) + 0.75\ (L_r$ or S or $R)$	Design for only one load combination	
$0.6D + 0.6W$		

[a] See load combinations in Section 2.1 for definition of the various load symbols.

Example 3.2: (ASD): Allowable stresses in sawn lumber (dimension lumber)

Given a 2 × 10 No. 2 Douglas Fir-Larch wood member that is fully braced laterally, subjected to dead load and snow load and exposed to the weather under normal temperature conditions, determine all the applicable ASD *allowable* stresses.

Solution:

1. Size classification = dimension lumber. Therefore, use NDS-S Table 4A
2. From NDS-S Table 4A, we obtain the following reference design values:
   - Bending stress $F_b = 900$ psi
   - Tension stress parallel to the grain $F_t = 575$ psi
   - Horizontal shear stress parallel to the grain $F_v = 180$ psi
   - Compression stress perpendicular to the grain $F_{c\perp} = 625$ psi
   - Compression stress parallel to the grain $F_c = 1350$ psi
   - Pure bending modulus of elasticity $E = 1.6 \times 10^6$ psi
   - Minimum modulus of elasticity $E_{min} = 0.58 \times 10^6$ psi
3. Determine the applicable stress adjustment or C factors. From Section 3.1, for a dead load plus snow load ($D + S$) combination, the governing load duration factor C_D is 1.15 (i.e., the largest C_D value in the load combination governs). From the adjustment factors section of NDS-S Table 4A, noting that the wood member is exposed to the weather, we obtain the following C factors:
   - $C_M(F_b) = 1.0$
   - $C_M(F_t) = 1.0$
   - $C_M(F_v) = 0.97$
   - $C_M(F_{c\perp}) = 0.67$
   - $C_M(F_c) = 0.8$
   - $C_M(E) = 0.9$
   - $C_F(F_b) = 1.1$
   - $C_F(F_t) = 1.1$
   - $C_F(F_c) = 1.0$
   - $C_p = 1.0$ (column stability factor needs to be calculated, but assume 1.0 for this problem)
   - $C_t = 1.0$ (normal temperature condition)

   C_L, C_{fu}, C_i, C_r, C_b, $C_T = 1.0$ (assumed)

   Note:

   If $F_bC_F \leq 1150$, $C_M(F_b) = 1.0$:

   $F_bC_F = (900)(1.1) = 990 < 1150$; therefore, $C_M(F_b) = 1.0$

   If $F_cC_F \leq 750$, $C_M(F_c) = 1.0$:

   $F_cC_F = (1350)(1.0) = 1350 > 750$; therefore, $C_M(F_c) = 0.8$
4. Calculate the allowable stresses (ASD) using the adjustment factors applicability table (Table 3.1)

   Allowable bending stress:

   $$F_b' = F_bC_DC_MC_tC_LC_FC_{fu}C_iC_r$$

   $$= (900)(1.15)(1.0)(1.0)(1.0)(1.1)(1.0)(1.0)(1.0) = 1139 \text{ psi}$$

   Allowable tension stress parallel to the grain:

   $$F_t' = F_tC_DC_MC_tC_FC_i$$

   $$= (575)(1.15)(1.0)(1.0)(1.1)(1.0) = 727 \text{ psi}$$

   Allowable horizontal shear stress:

   $$F_v' = F_vC_DC_MC_tCi$$

   $$= (180)(1.15)(0.97)(1.0)(1.0) = 200 \text{ psi}$$

Allowable bearing stress or compression stress perpendicular to the grain:

$$F'_{c\perp} = F_{c\perp} C_M C_t C_i C_b$$

$$= (625)(0.67)(1.0)(1.0)(1.0) = 419 \text{ psi}$$

Allowable compression stress parallel to the grain:

$F'_c = F_c C_D C_M C_t C_F CiC_P$ (C_P needs to be calculated, but assume 1.0 for this example; this factor is discussed in Chapter 5)

$$= (1350)(1.15)(0.8)(1)(1.0)(1.0)(1.0) = 1242 \text{ psi}$$

Allowable pure bending modulus of elasticity for deflection and other serviceability calculations:

$$E' = E C_M C_t C_i$$

$$= (1.6 \times 10^6)(0.9)(1.0)(1.0) = 1.44 \times 10^6 \text{ psi}$$

Allowable minimum bending modulus of elasticity for buckling calculations:

$$E'_{\min} = E_{\min} C_M C_t C_i C_T$$

$$= (0.58 \times 10^6)(0.9)(1.0)(1.0)(1.0) = 0.52 \times 10^6 \text{ psi}$$

Example 3.3: LRFD adjusted stresses in sawn lumber (dimension lumber)

Given a 2 × 10 No. 2 Douglas Fir-Larch wood member that is fully braced laterally, subjected to dead load and snow load and exposed to the weather under normal temperature conditions, determine all the applicable *LRFD adjusted design* stresses.

Solution:

1. Size classification = dimension lumber. Therefore, use NDS-S Table 4A
2. From NDS-S Table 4A, we obtain the following reference design values:
   Bending stress F_b = 900 psi
   Tension stress parallel to the grain F_t = 575 psi
   Horizontal shear stress parallel to the grain F_v = 180 psi
   Compression stress perpendicular to the grain $F_{c\perp}$ = 625 psi
   Compression stress parallel to the grain F_c = 1350 psi
   Pure bending modulus of elasticity $E = 1.6 \times 10^6$ psi
   Minimum modulus of elasticity $E_{\min} = 0.58 \times 10^6$ psi
3. Determine the applicable stress adjustment or C factors. From Section 3.1, for $1.2D + 1.6S$ load combination, the time effect factor, λ, is 0.8 from Table 3.11. From the adjustment factors section of NDS-S Table 4A, noting that the wood member is exposed to the weather, we obtain the following *adjustment* factors:
   $C_M(F_b) = 1.0$
   $C_M(F_t) = 1.0$
   $C_M(F_v) = 0.97$
   $C_M(F_{c\perp}) = 0.67$
   $C_M(F_c) = 0.8$
   $C_M(E) = 0.9$
   $C_F(F_b) = 1.1$
   $C_F(F_t) = 1.1$
   $C_F(F_c) = 1.0$

C_p = 1.0 (column stability factor needs to be calculated, but assume 1.0 for this problem)
C_t = 1.0 (normal temperature condition)
C_L, C_{fu}, C_i, C_r, C_b, C_T = 1.0 (assumed)

Note:

If $F_bC_F \leq 1150$, $C_M(F_b) = 1.0$:
$F_bC_F = (900)(1.1) = 990 < 1150$; therefore, $C_M(F_b) = 1.0$
If $F_cC_F \leq 750$, $C_M(F_c) = 1.0$:
$F_cC_F = (1350)(1.0) = 1350 > 750$; therefore, $C_M(F_c) = 0.8$

4. Calculate the LRFD adjusted design stresses using the adjustment factors applicability table (Table 3.3)

LRFD adjusted bending stress:
From Table 3.3 for bending stress, $K_F = 2.54$, $\phi = 0.85$.
From Table 3.11, $\lambda = 0.8$

$$F'_{b,r} = F_bC_MC_tC_LC_FC_{fu}C_iC_rK_F\phi\lambda$$

$$= (900)(1.0)(1.0)(1.0)(1.1)(1.0)(1.0)(1.0)(2.54)(0.85)(0.8) = 1710 \text{ psi}$$

LRFD adjusted tension stress parallel to the grain:
From Table 3.3 for tension stress, $K_F = 2.70$, $\phi = 0.80$.
From Table 3.11, $\lambda = 0.8$

$$F'_{t,r} = F_tC_MC_tC_FC_iK_F\phi\lambda$$

$$= (575)(1.0)(1.0)(1.1)(1.0)(2.70)(0.80)(0.8) = 1093 \text{ psi}$$

LRFD adjusted horizontal shear stress:
From Table 3.3 for shear stress, $K_F = 2.88$, $\phi = 0.75$.
From Table 3.11, $\lambda = 0.8$

$$F'_{v,r} = F_vC_MC_tC_iK_F\phi\lambda$$

$$= (180)(0.97)(1.0)(1.0)(2.88)(0.75)(0.8) = 302 \text{ psi}$$

LRFD adjusted bearing stress or compression stress perpendicular to the grain:
From Table 3.3 for bearing stress perpendicular to the grain, $K_F = 1.67$, $\phi = 0.9$.

$$F'_{c\perp,r} = F_{c\perp}C_MC_tC_iC_bK_F\phi$$

$$= (625)(0.67)(1.0)(1.0)(1.0)(1.67)(0.9) = 629 \text{ psi}$$

LRFD adjusted compression stress parallel to the grain:
From Table 3.3 for compression stress parallel to the grain, $K_F = 2.40$, $\phi = 0.90$.
From Table 3.11, $\lambda = 0.8$

$F'_{c,r} = F_cC_MC_tC_FC_iC_pK_F\phi\lambda$ (C_p needs to be calculated, but assume 1.0 for this example; this factor is discussed in Chapter 5)

$$= (1350)(0.8)(1)(1.0)(1.0)(1.0)(2.40)(0.90)(0.8) = 1866 \text{ psi}$$

LRFD adjusted pure bending modulus of elasticity, E':

$$E'_{,r} = EC_MC_tC_i$$

$$= (1.6\times10^6)(0.9)(1.0)(1.0) = 1.44\times10^6 \text{ psi}$$

LRFD adjusted minimum modulus of elasticity, E'_{min}:

From Table 3.3 for minimum modulus of elasticity, $K_F = 1.76$, $\phi = 0.85$.

$$E'_{min,r} = E_{min} C_M C_t C_i C_T K_F \phi$$

$$= (0.58 \times 10^6)(0.9)(1.0)(1.0)(1.0)(1.76)(0.85) = 0.78 \times 10^6 \text{ psi}$$

Example 3.4: (ASD): Allowable stresses in sawn lumber (timbers)

Given an 8 × 20 No. 2 Douglas Fir-Larch roof girder that is fully braced for bending and subject to a dead load, snow load, and wind load combination, calculate the ASD for dry service and normal temperature conditions.

Solution:

1. Size classification is beam and stringer, so use NDS-S Table 4D. For 8 × 20 sawn lumber, $b = 7.5$ in. and $d = 19.5$ in
2. From NDS-S Table 4D, we obtain the reference design values:
   Bending stress $F_b = 875$ psi
   Tension stress $F_t = 425$ psi
   Horizontal shear stress $F_v = 170$ psi
   Bearing stress or compression stress perpendicular to the grain $F_{c\perp} = 625$ psi
   Compression stress parallel to the grain $F_c = 600$ psi
   Pure bending modulus of elasticity $E = 1.3 \times 10^6$ psi
   Minimum modulus of elasticity $E_{min} = 0.47 \times 10^6$ psi
   Note that two values are given in this NDS-S Table 4D for No. 2 Douglas Fir-Larch beam and stringer depending on the Grading Agency. Though both reference design values are the same in this case, sometimes they may be different, so it is important that the structural designer use the appropriate values corresponding to the Grading Agency specified on the structural drawings or in the project specifications.
3. Determine the applicable stress adjustment factors. From Section 3.4.2.1, for a dead load plus wind load plus snow load ($D + W + S$) combination, the governing load duration factor $C_D = 1.6$ (i.e., the largest C_D value in the load combination governs). As discussed previously, the reader should check to ensure that the local code allows a C_D value of 1.6 to be used for wind loads. From the adjustment factors section of NDS-S Table 4D, we obtain the following stress adjustment or C factors:

   $C_M = 1.0$ (dry service)

   $$C_F = \left(\frac{12}{d}\right)^{1/9} \leq 1.0$$

   $$= \left(\frac{12}{19.5}\right)^{1/9} = 0.95 < 1.0 \qquad \text{OK}$$

   $C_p = 1.0$ (column stability factor; need to calculate, but assume 1.0 for now)
   $C_t = 1.0$ (normal temperature conditions apply)
   $C_L = 1.0$ (member is fully braced laterally for bending)
   C_{fu}, C_i, C_r, $C_b = 1.0$ (Assumed)
4. Calculate the allowable stresses using the adjustment factors applicability table (Table 3.1)

   Allowable bending stress:

   $$F'_b = F_b C_D C_M C_t C_L C_F C_{fu} C_i C_r$$

   $$= (875)(1.6)(1.0)(1.0)(1.0)(0.95)(1.0)(1.0)(1.0) = 1330 \text{ psi}$$

Allowable tension stress parallel to the grain:

$$F_t' = F_t C_D C_M C_t C_F C_i$$
$$= (425)(1.6)(1.0)(1.0)(0.95)(1.0) = 646 \text{ psi}$$

Allowable horizontal shear stress:

$$F_v' = F_v C_D C_M C_t C_i$$
$$= (170)(1.6)(1.0)(1.0)(1.0) = 272 \text{ psi}$$

Allowable bearing stress or compression stress perpendicular to the grain:

$$F_{c\perp}' = F_{c\perp} C_M C_t C_i C_b$$
$$= (625)(1.0)(1.0)(1.0)(1.0) = 625 \text{ psi}$$

Allowable compression stress parallel to the grain:

$$F_c' = F_c C_D C_M C_t C_F C_i C_p$$
$$= (600)(1.6)(1)(1)(0.95)(1.0)(1.0) = 912 \text{ psi}$$

Note: C_p needs to be calculated, and this factor is discussed in Chapter 5. For this example, we assume a value of 1.0.
Allowable pure bending modulus of elasticity:

$$E' = E C_M C_t C_i$$
$$= (1.3\times10^6)(1.0)(1.0)(1.0) = 1.3\times10^6 \text{ psi}$$

Allowable minimum (buckling) modulus of elasticity:

$$E_{\min}' = E C_M C_t C_i C_T$$
$$= (0.47\times10^6)(1.0)(1.0)(1.0)(1.0) = 0.47\times10^6 \text{ psi}$$

Example 3.5: (LRFD): LRFD adjusted stresses in sawn lumber (timbers)

Given an 8 × 20 No. 2 Douglas Fir-Larch roof girder that is fully braced for bending and subject to a dead load, snow load, and wind load combination, calculate the *LRFD adjusted design* stresses for dry service and normal temperature conditions.

Solution:

1. Size classification is beam and stringer, so use NDS-S Table 4D. For 8 × 20 sawn lumber, $b = 7.5$ in. and $d = 19.5$ in
2. From NDS-S Table 4D, we obtain the reference design values:
   Bending stress $F_b = 875$ psi
   Tension stress $F_t = 425$ psi
   Horizontal shear stress $F_v = 170$ psi
   Bearing stress or compression stress perpendicular to the grain $F_{c\perp} = 625$ psi
   Compression stress parallel to the grain $F_c = 600$ psi
   Pure bending modulus of elasticity $E = 1.3 \times 10^6$ psi
   Minimum modulus of elasticity $E_{\min} = 0.47 \times 10^6$ psi
   Note that two values are given in this NDS-S Table 4D for No. 2 Douglas Fir-Larch beam and stringer depending on the Grading Agency. Though both

reference design values are the same in this case, sometimes they may be different, so it is important that the structural designer use the appropriate values corresponding to the Grading Agency specified on the structural drawings or in the specifications.

3. Determine the applicable stress adjustment factors. The applicable load combinations from Section 2.1.1 for the given applied loads are $1.2D + 1.6S + 0.5W$ and $1.2D + 1.0W + 0.5S$ for which the time effect factors, λ, are 0.8 and 1.0, respectively, from Table 3.11. The lower value of 0.8 governs. From the adjustment factors section of NDS-S Table 4D, we obtain the following stress adjustment factors:

   $C_M = 1.0$ (dry service)

$$C_F = \left(\frac{12}{d}\right)^{1/9} \leq 1.0$$

$$= \left(\frac{12}{19.5}\right)^{1/9} = 0.95 < 1.0 \qquad \textbf{OK}$$

   $C_p = 1.0$ (column stability factor; need to calculate, but assume 1.0 for now)
   $C_t = 1.0$ (normal temperature conditions apply)
   $C_L = 1.0$ (member is fully braced laterally for bending)
   C_{fu}, C_i, C_r, C_b, $C_T = 1.0$ (Assumed)

4. Calculate the LRFD adjusted design stresses using the adjustment factors applicability table (Table 3.2)

LRFD adjusted bending stress:
From Table 3.3 for bending stress, $K_F = 2.54$, $\phi = 0.85$.
From Table 3.11, $\lambda = 0.8$

$$F'_{b,r} = F_b C_M C_t C_L C_F C_{fu} C_i C_r K_F \phi \lambda$$
$$= (875)(1.0)(1.0)(1.0)(0.95)(1.0)(1.0)(1.0)(2.54)(0.85)(0.8) = 1436 \text{ psi}$$

LRFD adjusted tension stress parallel to the grain:
From Table 3.3 for tension stress, $K_F = 2.70$, $\phi = 0.80$.
From Table 3.11, $\lambda = 0.8$

$$F'_{t,r} = F_t C_M C_t C_F C_i K_F \phi \lambda$$
$$= (425)(1.0)(1.0)(0.95)(1.0)(2.70)(0.80)(0.8) = 698 \text{ psi}$$

LRFD adjusted horizontal shear stress:
From Table 3.3 for shear stress, $K_F = 2.88$, $\phi = 0.75$.
From Table 3.11, $\lambda = 0.8$

$$F'_{v,r} = F_v C_M C_t C_i K_F \phi \lambda$$
$$= (170)(1.0)(1.0)(1.0)(2.88)(0.75)(0.8) = 294 \text{ psi}$$

LRFD adjusted bearing stress or compression stress perpendicular to the grain:
From Table 3.3 for bearing stress perpendicular to the grain, $K_F = 1.67$, $\phi = 0.9$.

$$F'_{c\perp,r} = F_{c\perp} C_M C_t C_i C_b K_F \phi$$
$$= (625)(1.0)(1.0)(1.0)(1.0)(1.67)(0.9) = 939 \text{ psi}$$

LRFD adjusted compression stress parallel to the grain:
From Table 3.3 for compression stress parallel to the grain, $K_F = 2.40, \phi = 0.90$.
From Table 3.11, $\lambda = 0.8$

$F'_{c,r} = F_cC_MC_tC_FC_iC_pK_F\phi\lambda$ (C_p needs to be calculated, but assume 1.0 for this example; this factor is discussed in Chapter 5)
$= (600)(1.0)(1.0)(0.95)(1.0)(1.0)(2.40)(0.90)(0.8) = 985$ psi

LRFD adjusted pure bending modulus of elasticity, E':

$$E'_r = EC_MC_tC_i$$
$$= (1.3\times10^6)(1.0)(1.0)(1.0) = 1.3\times10^6 \text{ psi}$$

LRFD adjusted minimum modulus of elasticity, $E'_{\min,r}$:
From Table 3.3 for minimum modulus of elasticity, $K_F = 1.76, \phi = 0.85$.

$$E'_{\min,r} = E_{\min}C_MC_tC_iC_TK_F\phi$$
$$= (0.47\times10^6)(1.0)(1.0)(1.0)\,(1.0)(1.76)(0.85) = 0.70\times10^6 \text{ psi}$$

Example 3.6: (ASD): Allowable stresses in glulam

An $8\frac{3}{4}\times34\frac{1}{2}$ in. 24F-V4 DF/DF simply supported glulam roof girder spans 32 ft (Figure 3.6), is fully braced for bending, and supports a dead load (*D*), snow load (*S*), and wind load (*W*) combination.

a. Calculate the ASD for dry service and normal temperature conditions.
b. Determine the maximum notch depth permitted by the NDS code for this glulam girder and recalculate the allowable shear stress assuming a maximum notch depth per the NDS code in the tension zone at the girder supports.

Solution:
Allowable stress design (**ASD**)

1. Size classification = glulam in bending (i.e., bending combination glulam). Therefore, use NDS-S Table 5A. For this glulam beam, the width $b = 8.75$ in. and depth $d = 34.5$ in. The distance between points of zero moments $L = 32$ ft (simply supported beam).
2. From NDS-S Table 5A, we obtain the reference design values:
   Bending stress with tension laminations stressed in tension $F_{bx}^{+} = 2400$ psi
   Bending stress with compression laminations stressed in tension $F_{bx}^{-} = 1850$ psi
   Bearing stress or compression perpendicular to the grain on tension lamination, $F_{c\perp xx,t} = 650$ psi
   Bearing stress or compression perpendicular to the grain on compression laminations, $F_{c\perp xx,c} = 650$ psi

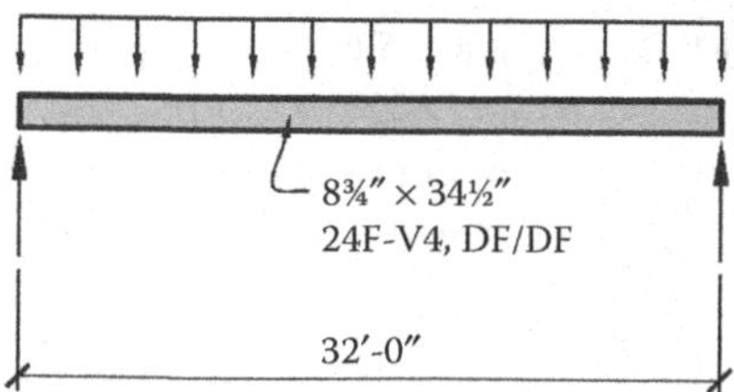

Figure 3.6 Simply supported glulam beam.

Horizontal shear stress parallel to the grain, $F_{v,xx} = 265$ psi
Pure bending modulus of elasticity $E_{xx} = 1.8 \times 10^6$ psi
Minimum modulus of elasticity $E_{y\,min} = 0.85 \times 10^6$ psi

3. Determine the applicable stress adjustment or C factors. From Section 3.1, for a dead load plus wind load plus snow load ($D + W + S$) combination, $C_D = 1.6$. (Check if the local code allows a C_D value of 1.6 to be used for wind loads.) From the adjustment factors section of NDS-S Table 5A, we obtain the following C factors:
   $C_M = 1.0$ (dry service)
   $C_t = 1.0$ (normal temperature)
   $C_L = 1.0$ (beam is fully braced)
   C_b, C_c, and $C_{fu} = 1.0$
   Shear reduction factor, $C_{vr} = 1.0$ (for glulam beam or girder **without** notches)
   Shear reduction factor, $C_{vr} = 0.72$ (for glulam beam or girder **with** notches)
   From Equation 3.5, we calculate the volume factor as

$$C_v = \left(\frac{21}{L}\right)^{1/x}\left(\frac{12}{d}\right)^{1/x}\left(\frac{5.125}{b}\right)^{1/x} = \left(\frac{1291.5}{bdL}\right)^{1/x} \leq 1.0$$

$$= \left(\frac{1291.5}{(8.75\text{ in.})(34.5\text{ in.})(32\text{ ft})}\right)^{1/10} = 0.82 < 1.0 \qquad \text{OK}$$

   Recall that in calculating the allowable bending stress for glulams, the smaller of the C_V and C_L factors are used. Since $C_V = 0.82$ is less than $C_L = 1.0$, the C_V value of 0.82 will govern for the calculation of the allowable bending stress.

4. Calculate the allowable stresses for the ASD method using the adjustment factors applicability in Table 3.4.
   Allowable bending stress with tension laminations stressed in tension:

$$F'^{+}_{bx} = \text{the smaller of } F^{+}_{bx}C_D C_M C_t C_L C_{fu} C_c C_I \quad \text{or} \quad F^{+}_{bx}C_D C_M C_t C_V C_{fu} C_c C_I$$
$$= (2400)(1.6)(1.0)(1.0)(1.0)(1.0)(1.0)(1.0) = 3840 \text{ psi}$$

   or

$$= (2400)(1.6)(1.0)(1.0)(0.82)(1.0)(1.0)(1.0) = \mathbf{3148\ psi\ (governs)}$$

   Therefore,

$$F^{+}_{bx} = 3148 \text{ psi}$$

   Allowable bending stress with compression laminations stressed in tension:

$$F'^{-}_{bx} = \text{the smaller of } F_{bx}C_D C_M C_t C_L C_{fu} C_c \quad \text{or} \quad F^{-}_{bx}C_D C_M C_t C_V C_{fu} C_c$$
$$= (1850)(1.6)(1.0)(1.0)(1.0)(1.0)(1.0) = 2960 \text{ psi}$$

   or

$$= (1850)(1.6)(1.0)(1.0)(0.82)(1.0)(1.0) = \mathbf{2427\ psi\ (governs)}$$

   Therefore,

$$F'^{-}_{bx} = 2427 \text{ psi}$$

   Allowable bearing stress or compression perpendicular to the grain in the tension lamination:

$$F'_{c\perp xx,t} = F_{c\perp xx,t}C_M C_t C_b$$
$$= (650)(1.0)(1.0)(1.0) = 650 \text{ psi}$$

Allowable bearing stress or compression perpendicular to the grain in the compression lamination:

$$F'_{c\perp xx,c} = F_{c\perp xx,c}C_MC_tC_b$$

$$= (650)(1.0)(1.0)(1.0) = 650 \text{ psi}$$

Allowable horizontal shear stress:

$$F'_{v,xx} = F'_{v,xx}C_DC_MC_tC_{vr}$$

$$= (265)(1.6)(1)(1)(1) = 424 \text{ psi}$$

Pure bending modulus of elasticity for bending about the strong x–x axis:

$$E'_{xx} = E_{xx}C_MC_t$$

$$= (1.8\times10^6)(1)(1) = 1.8\times10^6 \text{ psi}$$

The allowable minimum modulus of elasticity, $E'_{y\,\min}$ used for stability and buckling calculations:

$$E'_{y\,\min,\,r} = E_{\min}C_MC_t$$

$$= (0.85\times10^6)(1)(1) = 0.85\times10^6$$

Notched glulam girder:
Total depth of the glulam girder, d = 34.5″.

According to the NDS code Section 5.4.5, the maximum allowable depth of an unreinforced notch in the tension zone of a glulam beam or girder at the bearing support is the smaller of the following:

10% of the total depth = 0.1 (34.5 in.) = 3.45″

Or

3″ (Governs)

Therefore, the net depth of glulam girder at the notch, d_n = 34.5″ – 3″ = 31.5″
For a notched glulam beam or girder, the shear reduction factor, C_{vr} = 0.72.

The revised allowable horizontal shear stress (ASD) for a notched glulam beam at the bearing support is calculated as

$$F'_{v,\,xx} = F'_{v,\,xx}C_DC_MC_tC_{vr}$$

$$= (265)(1.6)(1)(1)(0.72) = 305 \text{ psi}$$

Example 3.7: (LRFD): LRFD adjusted stresses in glulam

An 8 3/4 × 34 1/2 in. 24F–V4 DF/DF simply supported glulam roof girder spans 32 ft (Figure 3.6), is fully braced for bending, and supports a dead load (D), snow load (S), and wind load (W) combination.

a. Calculate the LRFD adjusted stresses for dry service and normal temperature conditions.
b. What is the maximum notch depth permitted by the NDS code for this girder and recalculate the LRFD adjusted shear stress assuming a notch in the tension zone at the girder supports.

Solution:

Load and resistance factor design (**LRFD**)

1. Size classification = glulam in bending (i.e., bending combination glulam). Therefore, use NDS-S Table 5A. For this glulam beam, the width b = 8.75 in. and depth d = 34.5 in. The distance between points of zero moments L = 32 ft (simply supported beam).
2. From NDS-S Table 5A, we obtain the reference design values:
   Bending stress with tension laminations stressed in tension $F_{bx}^{+} = 2400$ psi.
   Bending stress with compression laminations stressed in tension $F_{bx}^{-} = 1850$ psi.
   Bearing stress or compression perpendicular to the grain on tension lamination, $F_{c\perp xx,t} = 650$ psi.
   Bearing stress or compression perpendicular to the grain on compression laminations, $F_{c\perp xx,c} = 650$ psi.
   Horizontal shear stress parallel to the grain, $F_{v,xx} = 265$ psi.
   Pure bending modulus of elasticity $E_{xx} = 1.8 \times 10^6$ psi.
   Minimum modulus of elasticity $E_{y\,\min} = 0.85 \times 10^6$ psi.
3. Determine the applicable stress adjustment. The applicable load combinations are $1.2D + 1.6L + 0.5W$ and $1.2D + 1.0W + 0.5S$ for which the time effect factors, λ, are 0.8 and 1.0, respectively. The lower value of 0.8 governs. From the adjustment factors section of NDS-S Table 5A, we obtain the following *adjustment* factors:
   $C_M = 1.0$ (dry service)
   $C_t = 1.0$ (normal temperature)
   $C_L = 1.0$ (beam is fully braced)
   C_b, C_c, C_I, and $C_{fu} = 1.0$
   Shear reduction factor, $C_{vr} = 1.0$ (for glulam beam or girder **without** notches)
   Shear reduction factor, $C_{vr} = 0.72$ (for glulam beam or girder **with** notches)
   From Equation 3.5, we calculate the volume factor as

$$C_v = \left(\frac{21}{L}\right)^{1/x}\left(\frac{12}{d}\right)^{1/x}\left(\frac{5.125}{b}\right)^{1/x} = \left(\frac{1291.5}{bdL}\right)^{1/x} \le 1.0$$

$$= \left(\frac{1291.5}{(8.75\text{ in.})(34.5\text{ in.})(32\text{ ft})}\right)^{1/10} = 0.82 < 1.0 \qquad \textbf{OK}$$

   Recall that in calculating the LRFD adjusted bending stress for glulams, the smaller of the C_V and C_L factors are used. Since $C_V = 0.82$ is less than $C_L = 1.0$, the C_V value of 0.82 will govern for the calculation of the ultimate bending stress.
4. Calculate the LRFD adjusted design stresses
   The LRFD adjusted design bending stress with tension laminations stressed in tension using the adjustment factors applicability in Table 3.4:
   From Table 3.4 for bending stress, $K_F = 2.54$, $\phi = 0.85$.
   From Table 3.11, $\lambda = 0.8$

$$F'_{b,r} = F_b C_M C_t C_L C_V C_{fu} C_c C_I K_F \phi \lambda$$

$$F'^{+}_{bx,r} = \text{the smaller of } F^{+}_{bx} C_M C_t C_L C_{fu} C_c C_I (K_F)(\phi)(\lambda) \quad \text{or} \quad F^{+}_{bx} C_M C_t C_V C_{fu} C_c C_I (K_F)(\phi)(\lambda)$$

$$= (2400)(1.0)(1.0)(1.0)(1.0)(1.0)(1.0)(2.54)(0.85)(0.8) = 4145 \text{ psi}$$

or

$$= (2400)(1.0)(1.0)(0.82)(1.0)(1.0)(1.0)(2.54)(0.85)(0.8)$$

$$= \textbf{3399 psi (governs)}$$

Therefore,

$$F'^{+}_{bx,r} = \mathbf{3399\ psi}$$

The LRFD adjusted bending stress with compression laminations stressed in tension:
From Table 3.4 for bending stress, $K_F = 2.54$, $\phi = 0.85$.
From Table 3.11, $\lambda = 0.8$

$$F'^{-}_{bx,r} = \text{the smaller of } F_{bx}C_M C_t C_L C_{fu} C_c C_I (K_F)(\phi)(\lambda) \quad \text{or} \quad F^{-}_{bx}C_M C_t C_V C_{fu} C_c C_I (K_F)(\phi)(\lambda)$$

$$= (1850)(1.0)(1.0)(1.0)(1.0)(1.0)(1.0)(2.54)(0.85)(0.8) = 3195 \text{ psi}$$

or

$$= (1850)(1.0)(1.0)(0.82)(1.0)(1.0)(1.0)(2.54)(0.85)(0.8)$$

$$= \mathbf{2620\ psi\ (governs)}$$

Therefore,

$$F'^{-}_{bx,r} = \mathbf{2620\ psi}$$

The LRFD adjusted bearing stress or compression perpendicular to the grain in the tension lamination:
From Table 3.4 for bearing stress perpendicular to the grain, $K_F = 1.67$, $\phi = 0.90$.

$$F'_{c\perp xx,t,r} = F_{c\perp xx,t} C_M C_t C_b (K_F)(\phi)$$

$$= (650)(1.0)(1.0)(1.0)(1.67)(0.90) = 977 \text{ psi}$$

The LRFD adjusted bearing stress or compression stress perpendicular to the grain in the compression lamination:
From Table 3.4 for bearing stress perpendicular, $K_F = 1.67$, $\phi = 0.90$.

$$F'_{c\perp xx,c,r} = F_{c\perp xx,c} C_M C_t C_b (K_F)(\phi)$$

$$= (650)(1.0)(1.0)(1.0)(1.67)(0.90) = 977 \text{ psi}$$

The LRFD adjusted horizontal shear stress (no notches):
From Table 3.4 for shear stress, $K_F = 2.88$, $\phi = 0.75$.
From Table 3.11, $\lambda = 0.8$

$$F'_{v,xx,r} = F'_{v,xx} C_M C_t C_{vr} (K_F)(\phi)(\lambda)$$

$$= (265)(1)(1)(1.0)(2.88)(0.75)(0.80) = 458 \text{ psi}$$

The LRFD adjusted pure bending modulus of elasticity for bending about the strong x–x axis:

$$E'_{xx,r} = E_{xx} C_M C_t$$

$$= (1.8\times10^6)(1)(1) = 1.8\times10^6 \text{ psi}$$

The LRFD adjusted minimum modulus of elasticity, $E'_{y\,\min,r}$, used for stability and buckling calculations:
From Table 3.4 for minimum modulus of elasticity, $K_F = 1.76$, $\phi = 0.85$.

$$E'_{y\,\min,r} = E_{\min} C_M C_t (K_F)(\phi)$$

$$= (0.85\times10^6)(1)(1)(1.76)(0.85) = 1.27\times10^6 \text{ psi}$$

Notched glulam girder:
Total depth of the glulam girder, $d = 34.5''$

According to the NDS code Section 5.4.5, the maximum allowable depth of an unreinforced notch in the tension zone of a glulam beam at the bearing support is the smaller of the following:

10% of the total depth = 0.1 (34.5 in) = 3.45″

Or

3″ (Governs)

Therefore, net depth of the glulam girder at the notch, $d_n = 34.5'' - 3'' = 31.5''$

The revised LRFD adjusted horizontal shear stress for a notched glulam girder at the bearing support:

For a notched glulam beam or girder, the shear reduction factor, C_{vr} will be 0.72; therefore, the revised LRFD adjusted horizontal shear stress is

$$F'_{v,xx,r} = F'_{v,xx} C_M C_t C_{vr} (K_F)(\phi)(\lambda)$$
$$= (265)(1)(1)(0.72)(2.88)(0.75)(0.8) = 330 \text{ psi}$$

Example 3.8: (ASD): Governing load combination (beams)—ASD only

For a roof beam subject to the following loads, determine the most critical ASD load combination using the normalized load method: $D = 10$ psf, $L_r = 16$ psf, $S = 20$ psf, $W = 5$ psf (downward). Assume that the wind load was calculated according to the ASCE 7 load standard (i.e., W is a factored load).

Dead load $D = 10$ psf
Roof live load $L_r = 16$ psf
Snow load $S = 20$ psf
Wind load $W = +5$ psf (since there is no wind uplift or suction, $W = 0$ in LRFD load combination 6).

All other loads are neglected.

Solution:

Note: Loads with zero values are not considered in determination of the load duration factor for that load combination and have been excluded from the load combinations.

Since all the loads are distributed uniformly with the same units of pounds per square foot, the load type or patterns are similar. Therefore, the *normalized load method* can be used to determine the most critical load combination. The applicable load combinations are shown in Table 3.15.

Summary:

- The governing ASD load combination is $D + S$ because it has the highest normalized load.
- The roof beam should be designed for a total unfactored uniform load of 30 psf with a load duration factor $C_D = 1.15$.

Example 3.9: (LRFD): Governing load combination (beams)—LRFD

For a roof beam subject to the following loads, determine the most critical LRFD load combination using the normalized load method: $D = 10$ psf, $L_r = 16$ psf, $S = 20$ psf, $W = 5$ psf (downward). Assume that the wind load was calculated according to the ASCE 7 load standard.

Dead load $D = 10$ psf
Roof live load $L_r = 16$ psf
Snow load $S = 20$ psf

Table 3.15 Applicable and governing ASD load combination

Load combination	*Value of load combination (w)*	*C_D factor for load combination*	*Normalized load (w/C_D)*
D	10 psf	0.9	$\frac{10}{0.9} = 11.1$
$D + L_r$	10 + 16 = 26 psf	1.25	$\frac{26}{1.25} = 20.8$
$D + S$	10 + 20 = **30 psf**	1.15	$\frac{30}{1.15} = 26.1 \Leftarrow$(governs)
$D + 0.6W$	10 + (0.6)(5) = 13 psf	1.6	$\frac{13}{1.6} = 8.13$
$D + 0.75\,(0.6W) + 0.75L_r$	10 + (0.75)[(0.6)(5) + 16] = 24.3 psf	1.6	$\frac{24.3}{1.6} = 15.2$
$D + 0.75(0.6W) + 0.75S$	10 + (0.75)[(0.6)(5) + 20] = 27.3 psf	1.6	$\frac{27.3}{1.6} = 17.1$
$0.6D + 0.6W$	(0.6)(10) + (0.6)(0)[a] = 6 psf	0.9	$\frac{6}{1.6} = 3.75$

Note: The value given in bold is the governing value.

[a] In ASD load combinations 15 and 16 given in Section 2.1, the wind load W and seismic load E are always opposed by the dead load D. Therefore, in these combinations, D takes on a positive number while W and E take on negative values only.

Wind load $W = +5$ psf (since there is no wind uplift or suction, $W = 0$ in LRFD load combination 6 [see Section 2.1.1]).

All other loads are neglected.

Solution:

Since all the loads are distributed uniformly with the same units of pounds per square foot, the load type or patterns are similar. Therefore, the *normalized load method* can be used to determine the most critical LRFD load combination. The applicable LRFD load combinations are shown in Table 3.16.

Summary:

- The governing LRFD load combination is $\mathbf{1.2D + 1.6S + 0.5W}$ because it has the highest normalized load.
- The roof beam should be designed for a total factored uniform load of **46.5 psf** with a time effect factor $\boldsymbol{\lambda} = \mathbf{0.8}$.
- For serviceability design (such as checking deflections and vibrations), an unfactored total load of 27.3 psf should be used. This is determined using the ASD load combination $D + 0.75\,(0.6W) + 0.75S$.

Example 3.10: (ASD): Governing load combination (columns)—ASD only

For a two-story column subject to the axial loads given below, determine the most critical ASD load combination using the normalized load method. Assume that the wind loads have been calculated in accordance with the ASCE 7 load standard.

Roof dead load $D_{roof} = 10$ kips
Floor dead load $D_{floor} = 10$ kips
Roof live load $L_r = 12$ kips
Floor live load $L = 20$ kips

Table 3.16 Applicable and governing LRFD load combination

LRFD load combination	*Value of load combination (w)*	*Time effect factor, λ for load combination*	*Normalized load (w/λ)*
$1.4D$	1.4(10) = 14 psf	0.6	$\frac{14}{0.6} = 23.3$
$1.2D + 1.6L + 0.5L_r$	1.2(10) + 1.6(0) + 0.5(16) = 20 psf	0.8	$\frac{20}{0.8} = 25$
$1.2D + 1.6L + 0.5S$	1.2(10) + 1.6(0) + 0.5(20) = 22 psf	0.8	$\frac{22}{0.8} = 27.5$
$1.2D + 1.6L_r + 0.5L$	1.2(10) + 1.6(16) + 0.5(0) = 37.6 psf	0.8	$\frac{37.6}{0.8} = 47$
$1.2D + 1.6L_r + 0.5W$	1.2(10) + 1.6(16) + 0.5(5) = 40.1 psf	0.8	$\frac{40.1}{0.8} = 50.1$
$1.2D + 1.6S + 0.5L$	1.2(10) + 1.6(20) + 0.5(0) = 44 psf	0.8	$\frac{44}{0.8} = 55$
$1.2D + 1.6S + 0.5W$	1.2(10) + 1.6(20) + 0.5(5) = **46.5 psf**	0.8	$\frac{46.5}{0.8} = 58.1 \Leftarrow$ (governs)
$1.2D + 1.0W + 0.5L + 0.5L_r$	1.2(10) + 1.0(5)+ 0.5(0) + 0.5(16) = 25 psf	1.0	$\frac{25}{1.0} = 25$
$1.2D + 1.0W + 0.5L + 0.5S$	1.2(10) + 1.0(5) +0.5(0) + 0.5(20) = 27 psf	1.0	$\frac{27}{1.0} = 27$
$0.9D + 1.0W$	0.9(10) + 1.0(0)[a] = 9 psf	1.0	$\frac{9}{1.0} = 9$

Note: The value given in bold is the governing value.

[a] In LRFD load combinations 6 and 7 given in Section 2.1, the wind load *W* and seismic load *E* are always opposed by the dead load *D*. Therefore, in these combinations, *D* takes on a positive number while *W* and *E* take on *negative* values only.

Snow load $S = 18$ kips
Wind load $W = \pm 10$ kips
Earthquake or seismic load $E = \pm 12$ kips

Solution:
In the load combinations, the dead load *D* is the sum of the dead loads from the all levels of the building. Thus,

$$D = D_{\text{roof}} + D_{\text{floor}} = 10 \text{ kips} + 10 \text{ kips} = 20 \text{ kips}$$

Since all the loads on this column are concentrated axial loads and therefore similar in pattern or type, the *normalized load method* can be used (Table 3.17).

Summary:

- The governing ASD load combination is dead load plus floor live load plus snow load ($D + 0.75L + 0.75S$) because it has the highest normalized load.
- The column should be designed for a total unfactored axial load P = **48.5 kips** with a load duration factor C_D = **1.15.**

Example 3.11: (LRFD): Governing load combination (columns)—LRFD

For a two-story column subject to the axial loads given below, determine the most critical LRFD load combination using the normalized load method. Assume that the wind loads have been calculated in accordance with the ASCE 7 load standard.

Table 3.17 Applicable and governing ASD load combination

ASD load combination	*Value of load combination (P)*	*C_D factor for load combination*	*Normalized load (P/C_D)*
D	20 kips	0.9	$\frac{20}{0.9} = 22.2$
$D + L$	20 + 20 = 40 kips	1.0	$\frac{40}{1.0} = 40$
$D + L_r$	20 + 12 = 32 kips	1.25	$\frac{32}{1.25} = 25.6$
$D + S$	20 + 18 = 38 kips	1.15	$\frac{38}{1.15} = 33.0$
$D + 0.6W$	20 + (0.6)(10) = 26 kips	1.6	$\frac{26}{1.6} = 16.3$
$D + 0.75L + 0.75L_r$	20 + (0.75)(20 + 12) = 44 kips	1.25	$\frac{44}{1.25} = 35.2$
$\mathbf{D + 0.75L + 0.75S}$	20 + (0.75)(20 + 18) = **48.5 kips**	**1.15**	$\frac{48}{1.15} = 42.2 \Leftarrow$(governs)
$D + 0.75\,(0.6W) + 0.75L + 0.75L_r$	20 + (0.75)[(0.6)(10) + 20 + 12] = 48.5 kips	1.6	$\frac{48.5}{1.6} = 30.3$
$D + 0.75\,(0.6W) + 0.75L + 0.75S$	20 + (0.75)[(0.6)(10) + 20 + 18] = 53 kips	1.6	$\frac{53}{1.6} = 33.1$
$D + (0.75)(0.7E) + 0.75L + 0.75S$	20 + 0.75[(0.7)(12) + 20 + 18] = 54.8 kips	1.6	$\frac{54.8}{1.6} = 34.3$
$0.6D + 0.6W$	(0.6)(20) + (0.6)(–10)[a] = 6 kips	1.6	$\frac{6}{1.6} = 3.75$
$0.6D + 0.7E$	(0.6)(20) + (0.7)(–12)[a] = 3.6 kips	1.6	$\frac{3.6}{1.6} = 2.25$

Note: The values given in bold are the governing values.

[a] In ASD load combinations 15 and 16 given in Section 2.1 the wind load W and seismic load E are always opposed by the dead load D. Therefore, in these combinations, D takes on a positive number while W and E take on negative values only.

Roof dead load $D_{roof} = 10$ kips
Floor dead load $D_{floor} = 10$ kips
Roof live load $L_r = 12$ kips
Floor live load $L = 20$ kips
Snow load $S = 18$ kips
Wind load $W = \pm 10$ kips
Earthquake or seismic load $E = \pm 12$ kips

Solution:

In the load combinations, the dead load D is the sum of the dead loads from the all levels of the building. Thus,

$$D = D_{roof} + D_{floor} = 10 \text{ kips} + 10 \text{ kips} = 20 \text{ kips}$$

Table 3.18 Applicable and governing LRFD load combination

LRFD load combination	*Value of load combination (P)*	*Time effect factor, λ for load combination*	*Normalized load (P/λ)*
$1.4D$	1.4(20) = 28 kips	0.6	$\frac{28}{0.6} = 46.7$
$1.2D + 1.6L + 0.5L_r$	1.2(20) + 1.6(20) + 0.5(12) = 62 kips	0.8	$\frac{62}{0.8} = 77.5$
$1.2D + 1.6L + 0.5S$	1.2(20) + 1.6(20) + 0.5(18) = **65 kips**	0.8	$\frac{65}{0.8} = 81.3 \Leftarrow$(governs)
$1.2D + 1.6L_r + 0.5L$	1.2(20) + 1.6(12) + 0.5(20) = 53.2 kips	0.8	$\frac{53.2}{0.8} = 66.5$
$1.2D + 1.6L_r + 0.5W$	1.2(20) + 1.6(12) + 0.5(10) = 48.2 kips	0.8	$\frac{48.2}{0.8} = 60.3$
$1.2D + 1.6S + 0.5L$	1.2(20) + 1.6(18) + 0.5(20) = 62.8 kips	0.8	$\frac{62.8}{0.8} = 78.5$
$1.2D + 1.6S + 0.5W$	1.2(20) + 1.6(18) + 0.5(10) = 57.8 kips	0.8	$\frac{57.8}{0.8} = 72.3$
$1.2D + 1.0W + 0.5L + 0.5L_r$	1.2(20) + 1.0(10) + 0.5(20) + 0.5(12) = 50 kips	1.0	$\frac{50}{1.0} = 50$
$1.2D + 1.0W + 0.5L + 0.5S$	1.2(20) + 1.0(10) + 0.5(20) + 0.5(18) = 53 kips	1.0	$\frac{53}{1.0} = 53$
$1.2D + 1.0E + 0.5L + 0.2S$	1.2(20) + 1.0(12) + 0.5(20) + 0.2(18) = 49.6 kips	1.0	$\frac{49.6}{1.0} = 49.6$
$0.9D + 1.0W$	0.9(20) + 1.0(−10)[a] = 8 kips (net downwards)	1.0	$\frac{8}{1.0} = 8$
$0.9D + 1.0E$	0.9(20) + 1.0(−12)[a] = 6 kips (net downwards)	1.0	$\frac{6}{1.0} = 6$

Note: The value given in bold is the governing value.

[a] In LRFD load combinations 6 and 7 given in Section 2.1, the wind load *W* and seismic load *E* are always opposed by the dead load *D*. Therefore, in these combinations, *D* takes on a positive number while *W* and *E* take on *negative* values only.

Since all the loads on this column are concentrated axial loads and therefore similar in pattern or type, the *normalized load method* can be used (Table 3.18).

Summary:

- The governing LRFD load combination is $\mathbf{1.2D + 1.6L + 0.5S}$ because it has the highest normalized load.
- The ground floor column should be designed for a total factored axial load of **65 kips** with a time effect factor $\boldsymbol{\lambda = 0.8}$.
- For serviceability design (such as for calculating the column foundation size), an unfactored total load of 48.5 kips should be used. This is determined using the ASD load combination $D + 0.75L + 0.75S$.

PROBLEMS

3.1 Explain why and how stress adjustment factors are used in wood design.

3.2 List the various stress adjustment factors used in the design of wood members.

3.3 Determine the load duration factor, C_D, for the following ASCE 7 ASD load combinations.

a. D
b. $D + H + L$
c. $D + H + (L_r \text{ or } S \text{ or } R)$
d. $D + H + 0.75(L) + 0.75(L_r \text{ or } S \text{ or } R)$
e. $D + H + (0.6W \text{ or } 0.7E)$
f. $D + H + 0.75\ (0.6W \text{ or } 0.7E) + 0.75L + 0.75\ (L_r \text{ or } S \text{ or } R)$
g. $D + H + 0.75\ (0.7E) + 0.75L + 0.75(S)$
h. $0.6D + 0.6W + H$
i. $0.6D + 0.7E + H$

3.4 What is glulam, and how is it manufactured?

3.5 Described briefly the various grades of glulam and describe two ways of specifying glulam.

3.6 Describe the effect of notches on glulam beams.

3.7 What is the advantage of the stress class system for specifying glulam?

3.8 Given a 2 × 12 Douglas Fir-Larch Select Structural wood member that is fully braced laterally, subject to dead load + wind load + snow load ($D + W + S$), and exposed to the weather under normal temperature conditions, determine all applicable allowable stresses for allowable stress design (ASD). Calculate the corresponding LRFD adjusted design stresses.

3.9 Given a 10 × 20 Douglas Fir-Larch Select Structural roof girder that is fully braced for bending and subject to dead load + wind load + snow load ($D + W + S$). Assuming dry service and normal temperature conditions, calculate the allowable bending stress, shear stress, modulus of elasticity, and bearing stress perpendicular to the grain. Calculate the corresponding LRFD adjusted design stresses.

3.10 A 5⅛ × 36 in. 24F-V8 DF/DF (i.e., 24F-1.8E) simply supported glulam roof girder spans 50 ft, is fully braced for bending, and supports a uniformly distributed dead load + snow load + wind load combination ($D + S + W$). Assuming dry service and normal temperature conditions, calculate the allowable bending stress, shear stress, modulus of elasticity, and bearing stress perpendicular to the grain for allowable stress design (ASD). Calculate the corresponding LRFD adjusted design stresses.

3.11 A 6¾ × 36 in. 24F-V8 DF/DF (i.e., 24F-1.8E) simply supported glulam floor girder spans 64 ft between simple supports and supports a uniformly distributed dead load + floor live load combination. The compression edge of the beam is laterally braced at 8 ft on centers. Assuming dry service and normal temperature conditions, calculate the allowable bending stress, shear stress, modulus of elasticity, and bearing stress perpendicular to the grain for allowable stress design (ASD). Calculate the corresponding LRFD adjusted design stresses.

3.12 For a roof beam subject to the following loads, determine the most critical ASD load combination using the normalized load method: $D = 20$ psf, $L_r = 20$ psf, $S = 35$ psf, and $W = 10$ psf (downward). Assume that the wind load was calculated according to the ASCE 7 load standard.

3.13 For a column subject to the axial loads given below, determine the most critical ASD load combination using the normalized load method. Assume that the wind loads have been calculated according to the ASCE 7 load standard.
Roof dead load $D_{\text{roof}} = 8$ kips
Floor dead load $D_{\text{floor}} = 15$ kips
Roof live load $L_r = 15$ kips
Floor live load $L = 26$ kips
Snow load $S = 20$ kips
Wind load $W = \pm 8$ kips
Earthquake or seismic load $E = \pm 10$ kips

3.14 Find F_b', F_t', F_v', F_c', E', and $E_{\min}'$ for the following assuming ASD:
a. 2 × 10 floor joists at 19.2″ on center in an interior condition under $D + L$ loading with continuous floor sheathing. Wood is Hem-Fir, #1.
b. 2 × 6 wall studs at 12″ o.c. in an interior condition braced at mid-height only and 12 ft tall. Wood is Spruce-Pine-Fir, No. 1. Loading is $D + L_r$
c. 6 × 16 wood beam spanning 24 ft and braced at one-third points in an interior condition under $D + S$ loading. Wood is Douglas Fir-Larch, No. 1.

3.15 Find F_b', F_t', F_v', F_c', E' and $E_{\min}'$ for the following assuming ASD:
a. 2 × 12 roof rafters at 16″ on center in an interior condition under $D + L_r$ loading with continuous roof sheathing. Wood is MSR 2400F-2.0E.
b. 2 × 8 floor joists at 16″ on center supporting a deck for an assembly occupancy in an exterior condition with continuous wood decking boards. Wood is Southern Pine, No. 1.
c. 8 × 12 wood beam spanning 20 ft and braced at midspan in an interior condition under $D + L$ loading. Wood is Hem-Fir, Select Structural.

3.16 Find F_b', F_t', F_v', F_c', E', and $E_{\min}'$ for the following assuming ASD:
a. 2 × 6 members in the bottom chord of a roof truss, with trusses spaced 24″ on center in an interior condition under $D + S$ loading with no sheathing. Use a length of 8 ft. Wood is MSR 2600F-1.6E.
b. 2 × 4 members in the top chord of a roof truss, with trusses spaced 24″ on center in an interior condition under $D - W$ loading with continuous roof sheathing. Wood is MSR Spruce-Pine-Fir, No. 1.
c. 10 × 24 wood beam spanning 40 ft and braced at on-quarter points in an interior condition under $D + L + S$ loading. Wood Douglas Fir-Larch, Select Structural.

3.17 Develop a spreadsheet to calculate C_L, beam stability factor. Submit the results for a 2 × 10, Hem-Fir, Select Structural. Plot a curve showing C_L on the vertical axis and unbraced length (0–20 ft) on the horizontal axis.

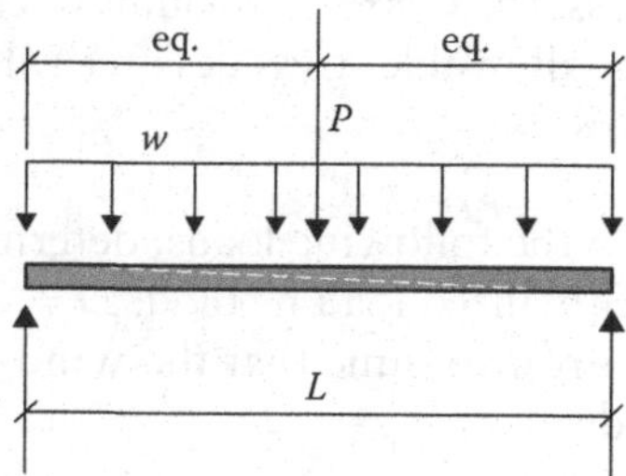

Figure 3.7 Simply supported glulam.

3.18 Develop a spreadsheet to calculate C_p, column stability factor. Submit the results for a 6 × 6, Southern Pine, No. 1. Plot a curve showing C_p on the vertical axis and unbraced length (0–16 ft on the horizontal axis).

3.19 Refer to the details in Figure 3.7 with a beam length of $L = 25$ ft.
Given loads:

Uniform load (w)	*Concentrated load (P)*
$D = 500$ plf	$D = 11$ k
$L = 800$ plf	$S = 15$ k
$S = 600$ plf	$W = +12$ k or -12 k
	$E = +8$ k or -8 k

Determine the following:

a. Describe a practical framing scenario where these loads could all occur as shown.
b. Determine the maximum moment for each individual load effect (D, L, S, W, E).
c. Develop a spreadsheet to determine the worst-case bending moments for the code-required load combinations.

3.20 A three-story building has columns spaced at 24 ft in both orthogonal directions, and is subjected to the roof and floor loads shown below. Using a column load summation table, calculate the cumulative axial loads on a typical interior column. Develop this table using a spreadsheet.

Roof loads	*Second and third floor loads*
Dead, $D = 20$ psf	Dead, $D = 60$ psf
Snow, $S = 45$ psf	Live, $L = 100$ psf

All other loads are 0.

REFERENCES

1. ANSI/AWC (2015), *National Design Specification for Wood Construction*, American Wood Council, Leesburg, VA.
2. ANSI/AWC (2015), *National Design Specification Supplement: Design Values for Wood Construction*, American Wood Council, Leesburg, VA.
3. WWPA (2005), *In-Grade Lumber Testing and Design Values for Dimension Lumber*, Tech Notes, 2005-2, Portland, Oregon.

4. AITC (2012), *Timber Construction Manual*, 6th ed., American Institute of Timber Construction, Englewood, CO.
5. Kam-Biron, Michelle and Koch, Lori (2014), The ABC's of Traditional and Engineered Wood Products, *STRUCTURE Magazine*, October.
6. ANSI/AWC (2015), *ASD/LRFD Manual for Engineered Wood Construction*, American Wood Council, Leesburg, VA.
7. Somayaji, Shan (1990), *Structural Wood Design*, West Publishing Company, St. Paul, MN.
8. DeStafano, Jim (2003), Exposed Laminated Timber Roof Structures, *STRUCTURE Magazine*, March, pp. 16–17.
9. Ayub, Mohammad (2015), Investigation of the November 13 and 14, 2014 Collapses of Two Pedestrian Bridges Under Construction at Wake Technical Community College Campus, Raleigh, NC, Report, U.S. Department of Labor, Occupational Safety and Health Administration, Directorate of Construction.
10. Jockwer, Robert (2014), *Structural Behaviour of Glued Laminated Timber Beams with Unreinforced and Reinforced Notches*, PhD Dissertation, ETH, Zurich, Switzerland.
11. Fruhwald, Eva, Serrano, Erik, Toratti, Tomi, Emilsson, Arne, and Thelandersson, Sven (2007), Design of Safe Timber Structures—How Can We Learn from Structural Failures in Concrete, Steel and Timber? Report TVBK-3053, Division of Structural Engineering, Lund Institute of Technology, Lund University, Lund, Sweden.

Chapter 4

Design and analysis of beams and girders

4.1 DESIGN OF JOISTS, BEAMS, AND GIRDERS

In this chapter we introduce procedures for the analysis and design of joists, beams, and girders. These horizontal structural members are usually rectangular in crosssection with bending about the strong (x–x) axes, and they are used in the framing of roofs and floors of wood buildings. The stresses that need to be checked in the design and analysis of these structural elements include (1) bending stresses (including beam lateral torsional buckling), (2) shear stresses, (3) deflection (live-load and long-term deflections), and (4) bearing stresses.

4.1.1 Definition of beam span

The *span* of a beam or girder is defined as the clear span face to face of supports, L_n, plus one-half the required bearing length, $l_{b,\text{ req'd}}$, at each end of the beam. Thus, the span $L = L_n + l_{b,\text{req'd}}$ (see Figure 4.1). In practice, it is conservative to assume the span of a beam as the distance between the centerline of the beam supports.

4.1.2 Layout of joists, beams, and girders

It is more economical to lay out the joists or beams to span the longer direction of the roof or floor bay with the girders spanning the short direction, since the girders are usually more heavily loaded than the joists or beams (Figure 4.2). The maximum economical bay size for wood framing using sawn-lumber joists and girders is approximately 14 × 14 ft. Longer girder spans can be achieved using glulams.

4.1.3 Design procedure

In designing beams and girders, the first step is to calculate all the loads and load effects, and then the beam size is selected to satisfy the bending, shear, deflection, and bearing stress requirements. The design procedure for sawn-lumber joists, beams, or girders and glulam beams and girders is as follows.

4.1.3.1 Check of bending stress

1. Calculate all the loads and load effects, including maximum moment, shear, and reactions. Also calculate the dead and live loads that will be used later for deflection calculations.

   Note: Because of the relatively significant self-weight of girders, an assumption must initially be made for the girder self-weight and included in calculation of the load effects for the girder. This self-weight is checked after the final girder selection is made.

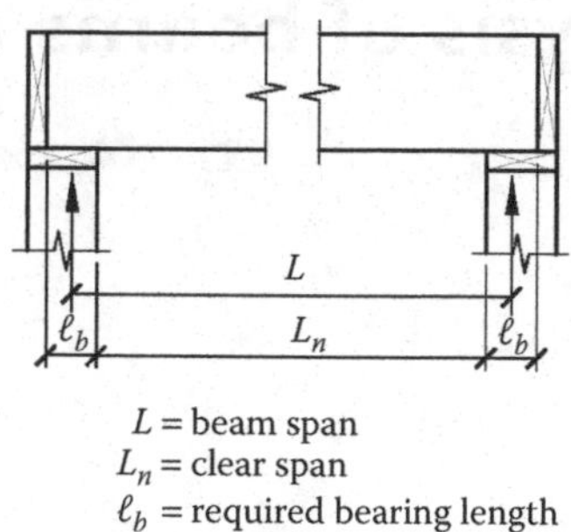

Figure 4.1 Definition of beam span.

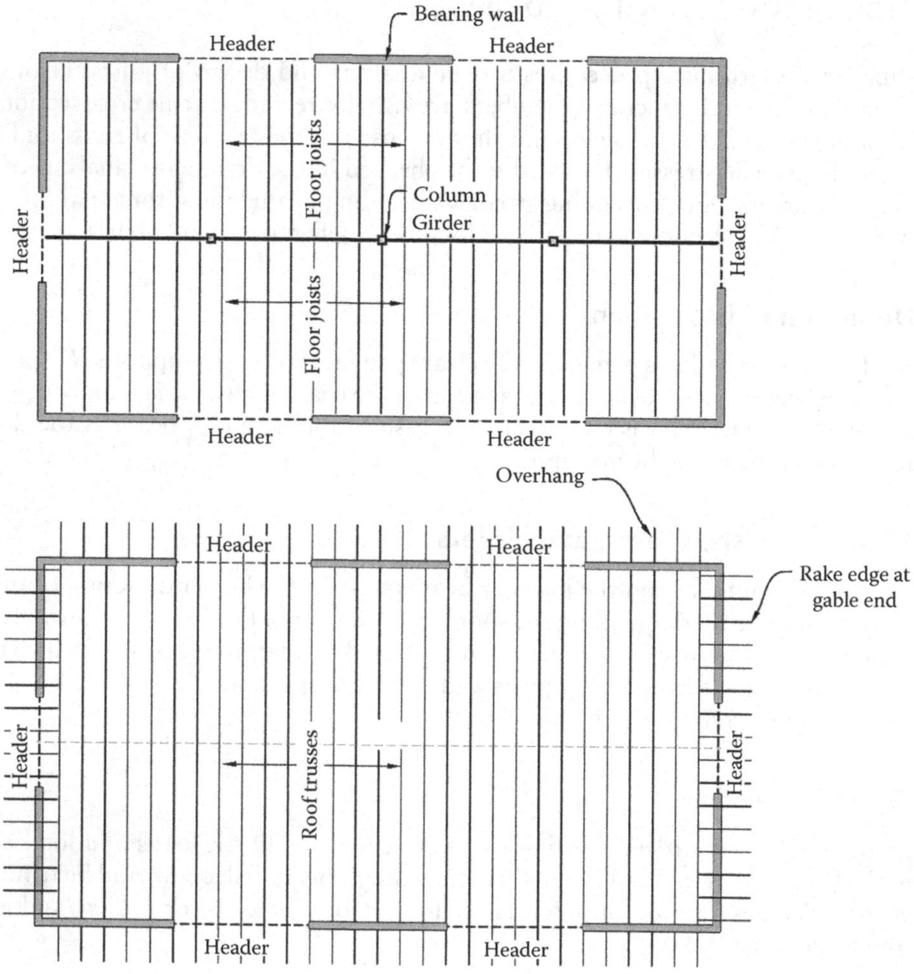

Figure 4.2 Typical roof and floor framing layout.

As previously discussed, a wood density of 32 lb/ft^3 is used in this book to calculate the self-weight of the wood member unless noted otherwise. Alternatively, the self-weight could be obtained directly from NDS-S Table 1B by interpolation between the densities given. The self-weight of the joists is usually included in the pounds per square foot roof or floor dead load.

2. Assume a stress grade (i.e., Select Structural, No. 1 and Better, No. 1, etc.) and size classification (i.e., dimension lumber, timbers, glulam, etc.), and determine the applicable NDS-S Table to be used in the design.
3. Assume initially that the allowable bending stress F_b' or factored bending strength $F_{b,r}'$ is equal to the NDS-S tabulated reference bending stress, F_b for allowable stress design (ASD) or nominal bending strength $F_{b,r}K_{F,r}\phi_b\lambda$ for load and resistance factor design (LRFD). Therefore, the required approximate section modulus of the member is given as

$$S_{xx,\text{ req'd}} \geq \frac{M_{\max}}{F_b} \text{ for ASD} \tag{4.1}$$

$$S_{xx,\text{ req'd}} \geq \frac{M_u}{F_{b,r}K_{F,r}\phi_b\lambda} \text{ for LRFD}$$

4. Using the approximate section modulus S_{xx} calculated in step 3, select a trial member size from NDS-S Table 1B (for sawn lumber), NDS-S Table 1C (for western species glulam), or NDS-S Table 1D (for Southern Pine glulam). For economy, the member size with the least area that has a section modulus at least equal to that required from step 2 should be selected.
5. For the member size selected in step 4, determine all the applicable stress adjustment or C factors, obtain the tabulated NDS-S bending stress, F_b, and calculate the actual allowable bending stress or the factored bending strength,

$$F'_b = F_b \text{ (product of applicable adjustment or C factors) for ASD} \tag{4.2}$$

$$F'_{b,r} = F_b K_{F,r}\phi_b\lambda \text{ (product of applicable adjustment or C factors) for LRFD}$$

6. Using the size selected in step 4, calculate the applied bending stress f_b *for ASD or factored bending stress,* f_{bu}, *for LRFD* and compare this to the actual available bending stress calculated in step 5.

For ASD:
If $f_b = M_{\max}/S_{xx} \leq F'_b$, the beam is adequate in bending.
If $f_b = M_{\max}/S_{xx} > F'_b$, the beam is not adequate in bending; increase the member size and repeat the design process.

For LRFD:
If $f_{bu} = M_u/S_{xx} \leq F'_{b,r}$, the beam is adequate in bending.
If $f_{bu} = M_u/S_{xx} > F'_{b,r}$, the beam is not adequate in bending; increase the member size and repeat the design process.

4.1.3.2 Check of shear stress

Wood is weaker in horizontal shear than in vertical shear because of the horizontal direction of the wood grain relative to the vertical shear force. Thus, horizontal shear is the most critical shear stress for wood members. For vertical shear, the shear force is perpendicular to the grain, whereas for horizontal shear, the shear force is parallel to the grain, and only the lignin that glues the fibers together (see Chapter 1) is available to resist this horizontal shear.

The critical section for shear can conservatively be assumed to be at the face of the beam bearing support. Where the beam support is subjected to confining compressive stresses as shown in Figure 4.3, and where there are no applied concentrated loads within a distance d from the face of the beam support, the critical section for shear can be assumed to be at a

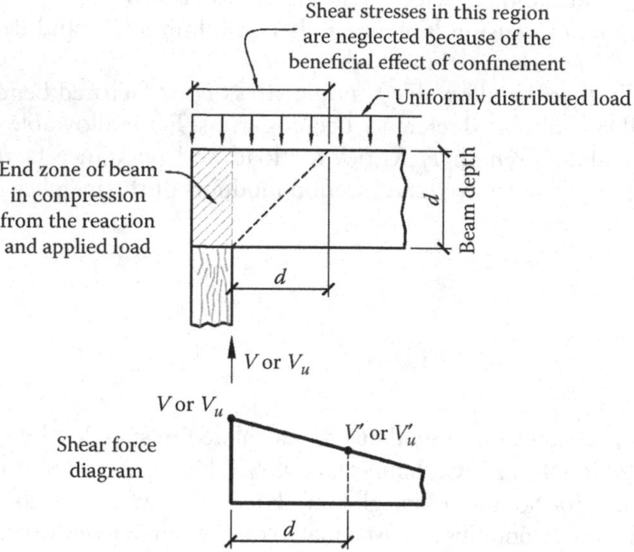

Figure 4.3 Critical section for shear.

distance d from the face of the beam support. Where a beam or joist is supported from the face of a girder on joist hangers, the critical section for shear should be taken at the face of the support.

The relationship between the shear stress applied in a rectangular wood beam at the centerline of the beam support f_v and the allowable shear stress F'_v is calculated as

$$f_v = \frac{1.5V}{A} \le F'_v \text{ for ASD} \tag{4.3}$$

$$f_{vu} = \frac{1.5V_u}{A} \le F'_{v,r} \text{ for LRFD}$$

If applicable, the reduced applied shear stress at a distance d from the face of the beam support is calculated as

$$f'_v = \frac{1.5V'}{A} \le F'_v \text{ for ASD} \tag{4.4}$$

$$f'_{vu} = \frac{1.5V'_n}{A} \le F'_{v,r} \text{ for LRFD}$$

where:

V = maximum shear at the critical section; for simplicity, this critical section can conservatively be assumed to be at the centerline of the joist, beam, or girder bearing support

V' or V_u' = applied shear at a distance d from the face of the support

F'_v = allowable shear stress = $F_v C_D C_M C_t C_i C_{vr}$ for ASD

$F'_{v,r}$ = available shear stress = $F_{v,r} C_M C_t C_i C_{vr} K_{F,v} \phi_v \lambda$ for LRFD

The C factors are discussed in Chapter 3. Where there is a concentrated load on the beam with full bearing within a distance d from the face of the beam support, the critical section for shear will be at the face of the beam support. For these cases, use Equation 4.3. The following shear design checks need to be made:

For ASD:

If f_v or $f'_v \leq F'_v$, the beam is adequate in shear.

If f_v or $f'_v > F'_v$, the beam is not adequate; therefore, increase the beam size and recheck the shear stress.

For LRFD:

If f_{vu} or $f'_{vu} \leq F'_{v,r}$, the beam is adequate in shear.

If f_{vu} or $f'_{vu} > F'_{v,r}$, the beam is not adequate; therefore, increase the beam size and recheck the shear stress.

Note: If the shear stress f_v *or* f_{vu} at the centerline of the beam support is adequate, there is no need to calculate further the shear stress f'_v *or* f_{vu} at a distance d from the face of the support since the shear stress would only decrease away from the support. However, for situations where the calculated shear stress f_v *or* f_{vu} at the centerline of the beam support is not adequate, *it may be necessary to examine the shear stress at a distance "d" away from the support or an increase in the beam section may be necessary.*

4.1.3.3 Check of deflection (calculating the required moment of inertia)

The beam or girder deflection is calculated using engineering mechanics principles and assuming linear elastic material behavior. The maximum deflection is a function of the type of loading, the span of the beam, and the allowable pure bending modulus of elasticity E' $(=E\ C_M C_t C_i)$. The formulas for moment, shear, and deflection for common loading conditions are shown in Figure 4.4 and Table 4.1.

The deflection limits from IBC Section 1604.3.6 [1] are given in Table 4.2, and k is a factor that accounts for *creep effects* in wood members, which is the continuous deflection of wood members under constant load. It should be noted that for the dead- plus live-load deflection, it is the incremental deflection that needs to be calculated and compared to the deflection limits, and this incremental dead- plus live-load deflection includes the effects of creep, but excludes the instantaneous dead-load deflection, which takes place during construction and before any of the deflection-sensitive elements such as glazing, masonry walls, or marble flooring are installed. If the deflection limits in Table 4.2 are satisfied, the beam is considered adequate for deflection; otherwise, the beam size has to be increased and the deflection rechecked. When the live load is greater than twice the dead load, the $L/360$ limit will control over $L/240$ for total loads.

In accordance with section 1604 of the IBC, deflections in wood members having a moisture content of less than 16% at the time of installation and used under dry conditions are permitted to be calculated based on $k = 0.5$. For green or unseasoned lumber or glulam used in wet service conditions, $k = 1.0$. It is conservative to assume that $k = 1.0$. In Table 4.2, L is the span of a joist, beam, or girder between two adjacent supports, Δ_{DL} is the deflection due to dead load D, and Δ_{LL} is the deflection due to live load L or (L_r or S).

There might be occasions, especially for large-span girders, such as glulam girders, where although the deflection limits are satisfied, the absolute value of these deflections may be sufficiently large (say, greater than 1 in.) as to become unsightly. It can also increase the possibility of ponding in panelized roofs. Ponding is the increase in rain load on flat roofs due to water collecting in the roof depressions caused by the beam and girder deflections. This increase in loading causes more deflections that in turn further increase the roof live load. This cycle can continue and may lead to roof failure. In these cases it may be prudent to camber the girder to negate the effect of the downward deflection. *Cambering* is the fabrication of a beam or girder with a built-in upward deflection, usually equal to *1.5 times the dead-load deflection* for glulam roof beams and *1.0 times the dead-load deflection* for floor beams with the intent that when loads are applied, the beam or girder will become essentially flat or horizontal. Cambering of the cantilevered portion of cantilevered beams is not recommended. For beams

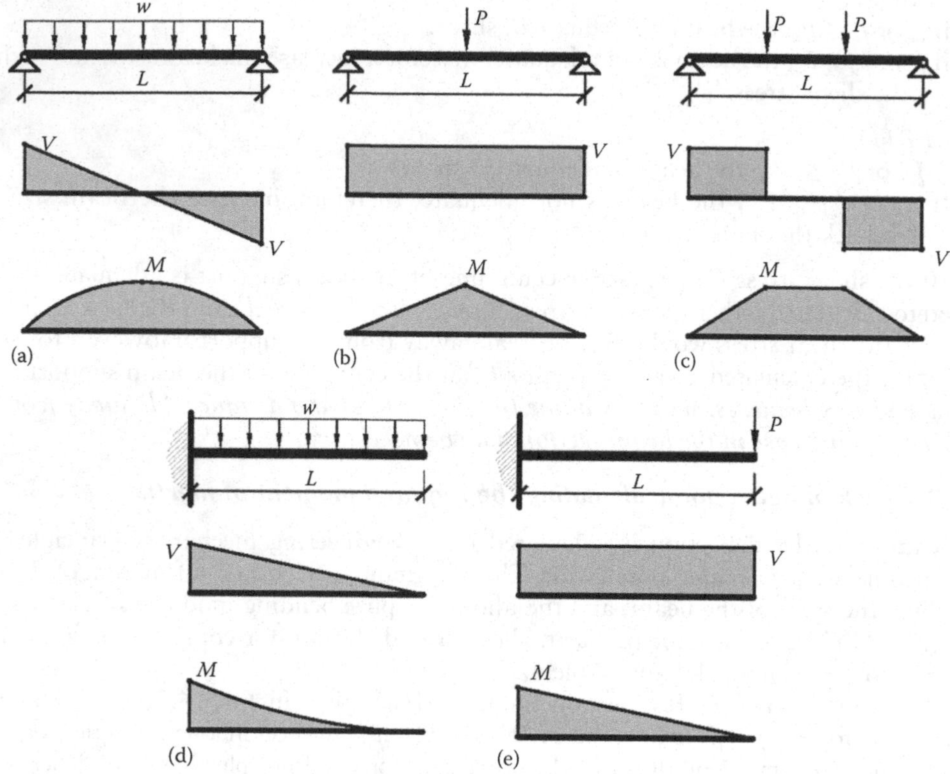

Figure 4.4 Beam load and deflection diagrams: (a) Uniformly loaded, (b) concentrated load at midspan, (c) concentrated load at 1/3 points, (d) uniformly loaded, cantilever, and (e) concentrated load at end of cantilever.

Table 4.1 Beam loading and deflection formulas

Load case	*Diagrams*	*Maximum shear*	*Maximum moment*	*Maximum deflection*
Uniformly loaded	Figure 4.4a	$V = \frac{wL}{2}$	$M = \frac{wL^2}{8}$	$\Delta = \frac{5wL^4}{384EI}$
Concentrated load at midpsan	Figure 4.4b	$V = \frac{P}{2}$	$M = \frac{PL}{4}$	$\Delta = \frac{PL^3}{48EI}$
Equal concentrated loads at one-third points	Figure 4.4c	$V = P$	$M = \frac{PL}{3}$	$\Delta = \frac{PL^3}{28EI}$
Uniformly loaded cantilever	Figure 4.4d	$V = wL$	$M = \frac{wL^2}{2}$	$\Delta = \frac{wL^4}{8EI}$
Concentrated load at the end of a cantilever	Figure 4.4e	$V = P$	$M = PL$	$\Delta = \frac{PL^3}{3EI}$

and girders in flat roofs, further analysis for possible ponding should be carried out when the beam deflection exceeds 1/2 in. under a uniform load of 5 psf [2].

The deflection limits provided in this section are intended to limit the cracking of nonstructural elements and finishes, due to large deflections, and are also intended to help control floor vibrations or bouncing floors. However, these deflection limits have been found to be unsatisfactory

Table 4.2 Deflection limits for framing members (reference IBC Section 1604)

Member		*Live load*	*Snow or wind load*	*Dead plus live load (kD + L)*
Roof	Supporting plaster ceiling	*L*/360	*L*/360	*L*/240
	Supporting nonplaster ceiling	*L*/240	*L*/240	*L*/180
	Not supporting a ceiling	*L*/180	*L*/180	*L*/120
Floor	All members	*L*/360	*L*/360	*L*/240

for controlling floor vibrations for floor joist with spans exceeding about 15 ft. Where floor joist spans exceed about 15 ft, it is recommended that a live-load deflection limit of 0.5 in. be used. It should be noted that where floor joists are supported on floor girders instead of stud walls, the floor vibration will be magnified because of the flexibility of both the joists and the girders. The girders should be designed to a smaller deflection limit than *L*/360 to control the floor vibration. To help control floor vibration, it is also recommended that the plywood sheathing be nailed as well as glued to the floor joists. Floor vibrations are discussed in more detail in Section 4.7.

In the wood industry, more stringent deflection limits are often used for engineered products such as I-joists and open-web floor trusses. The most common published limit is a live-load deflection limit of *L*/480 for floor framing members and is commonly used in residential structures, especially in areas that do not have walls to dampen floor vibrations. A limit of *L*/600 might be used for commercial floors and limits between *L*/600 and *L*/1000 can often be required for floors that support floor tiles with grouted joints. The use or occupancy of any floor space and the presence of deflection sensitive elements should always be considered before assigning a deflection limit.

An alternative approach to calculating the deflections is to calculate the required moment of inertia by equating the deflection to the appropriate deflection limit. For example, for a *uniformly loaded beam or girder,* we obtain the required moment of inertia based on the live load,

$$\Delta_{LL} = \frac{5w_{LL}L^4}{384E'I} = \frac{L}{360}$$

Solving and rearranging the equation above yields the required moment of inertia,

$$I_{req'd} = \frac{5w_{LL}L^3}{384E'}(360) \quad \text{in.}^4 \tag{4.5}$$

Similarly, the moment of inertia required based on the total load is obtained for a *uniformly loaded beam or girder* as

$$I_{req'd} = \frac{5w_{k(DL)+LL}L^3}{384E'}(240) \quad \text{in.}^4 \tag{4.6}$$

The higher moment of inertia from Equations 4.5 and 4.6 will govern. The preceding equations can be further refined for quick use in a design situation so that variables with common units are used. Equation 4.6 can be normalized as follows:

$$I_{req'd} = \frac{5w_{D+L}L^3\,(240)\left(\dfrac{1\text{ ft}}{12\text{ in.}}\right)\left(12\dfrac{\text{in.}}{\text{ft}}\right)^3}{384E'\left(\dfrac{1\text{ psi}}{1000\text{ ksi}}\right)} = \frac{0.45\,wL^3}{E'}$$

Table 4.3 Deflection formulas for design

Load case	*Maximum deflection*	*L/240*	*L/360*
Uniformly loaded	$\Delta = \frac{wL^4}{44.4E'I}$	$I_{req'd} = \frac{wL^3}{2.22E'}$	$I_{req'd} = \frac{wL^3}{1.48E'}$
Concentrated load at midspan	$\Delta = \frac{36PL^3}{E'I}$	$I_{req'd} = \frac{720PL^2}{E'}$	$I_{req'd} = \frac{1080PL^2}{E'}$
Variable moment[a]	$\Delta = \frac{180ML^2}{E'I}$	$I_{req'd} = \frac{3600ML}{E'}$	$I_{req'd} = \frac{5400ML}{E'}$

[a] Loading is based on a uniformly distributed load on a simple-span beam.

where:

w is in pounds per foot (plf)
L is in feet
E' is in ksi

For quick design, normalized deflection formulas are summarized in Table 4.3.

Δ is in inches
w is in pounds per foot (plf)
P is in kips
L is in feet
M is in kip-feet
E' is in ksi

4.1.3.4 Check of bearing stress (compression stress perpendicular to the grain)

The NDS-S tabulated design stress perpendicular to the grain, $F_{c\perp}$, is based on a deformation limit of 0.04 in. for a steel plate bearing on a wood member (Figure 4.5). The 0.04 in. deformation limit has been found, based on experience, to provide an adequate level of service in typical wood frame building construction. However, if a lower deformation limit is desired, the NDS code provides an equation for calculating the equivalent tabulated design value for the lower deformation limit of 0.02 in. based on NDS-S tabulated values:

$$F_{c\perp 0.02"} = 0.73\ F_{c\perp \text{NDS-S}} \tag{4.7}$$

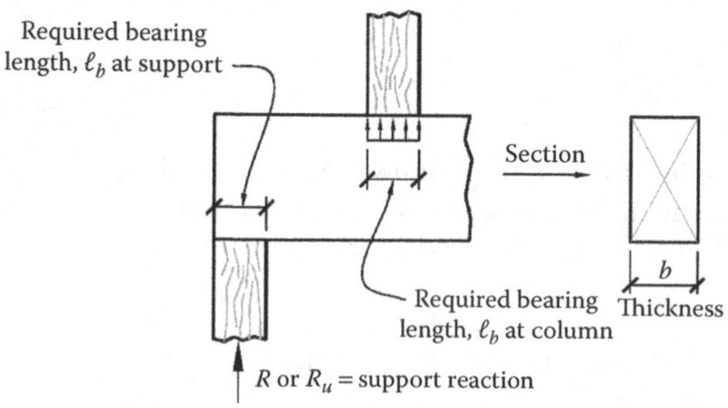

Figure 4.5 Bearing at supports and at interior loads.

The applied bearing stress or compression stress perpendicular to the grain is calculated as

$$f_{c\perp} = \frac{R}{A_{\text{bearing}}} \leq F'_{c\perp} \text{ for ASD} \tag{4.8}$$

$$f_{c\perp u} = \frac{R_u}{A_{\text{bearing}}} \leq F'_{c\perp,r} \text{ for LRFD}$$

where:

$F'_{c\perp}$ is the allowable bearing stress perpendicular to the grain (ASD)
$F'_{c\perp,r}$ is the available bearing stress perpendicular to the grain (LRFD)
R or R_u is the maximum reaction at the support
A_{bearing} is the bearing area = (thickness of beam)(bearing length) = $(b)(l_b)$
$F_{c\perp}$ is the NDS-S tabulated design stress perpendicular to the grain (ASD)
$F_{c\perp,r}$ is the NDS-S tabulated nominal design stress perpendicular to the grain (LRFD)
$f_{c\perp}$ is the applied bearing stress perpendicular to the grain
$f_{c\perp u}$ is the factored bearing stress perpendicular to the grain

The minimum required bearing length l_b is obtained by rearranging Equation 4.8. Thus,

$$l_{b,\text{ req'd}} \geq \frac{R}{bF'_{c\perp}} \text{ for ASD} \tag{4.9}$$

$$l_{b,\text{ req'd}} \geq \frac{R_u}{bF'_{c\perp,r}} \text{ for LRFD}$$

If the bearing length l_b required from Equation 4.9 is greater than the bearing length available, the designer or engineer can do one of two things: increase beam thickness b and/or use a stress grade that gives a higher value of $F'_{c\perp}$ or $F'_{c\perp,r}$. The practical recommended minimum bearing length is 1.5 in. for wood members bearing on wood and 3 in. for wood members bearing on masonry or concrete walls.

Some of the bearing details that could be used for joists, beams, or girders are shown in Figure 4.6. In Figure 4.6a, the joists from the adjacent spans butt against each other at the girder support with a tolerance gap of approximately 1/2 in. provided; thus, the bearing length available for the joist will be less than one-half the thickness b of the supporting girder. In the overlapping joist detail (Figures 4.6d and 4.6e), the bearing length available for the joist is equal to the thickness b of the girder. In Figures 4.6b and 4.6c, the joist or beam is supported off the face of the girder with proprietary joist or beam hangers. The available bearing length for this detail is given in the manufacturer's catalog for each hanger. Figure 4.6e shows the girder support at a wood column using a U-shaped column cap. The available length of bearing for the girder in this detail will depend on the length of the U-shaped column cap.

4.2 ANALYSIS OF JOISTS, BEAMS, AND GIRDERS

In the analysis of beams and girders, the beam or girder size, stress grade, and size classification are usually known, and the load is usually known or can be calculated. The intent is to determine if the existing beam or girder is adequate to resist the loads applied or to calculate the load capacity of the beam or girder. The analysis procedure for sawn-lumber joists, beams, or girders and glulam beams and girders is as follows.

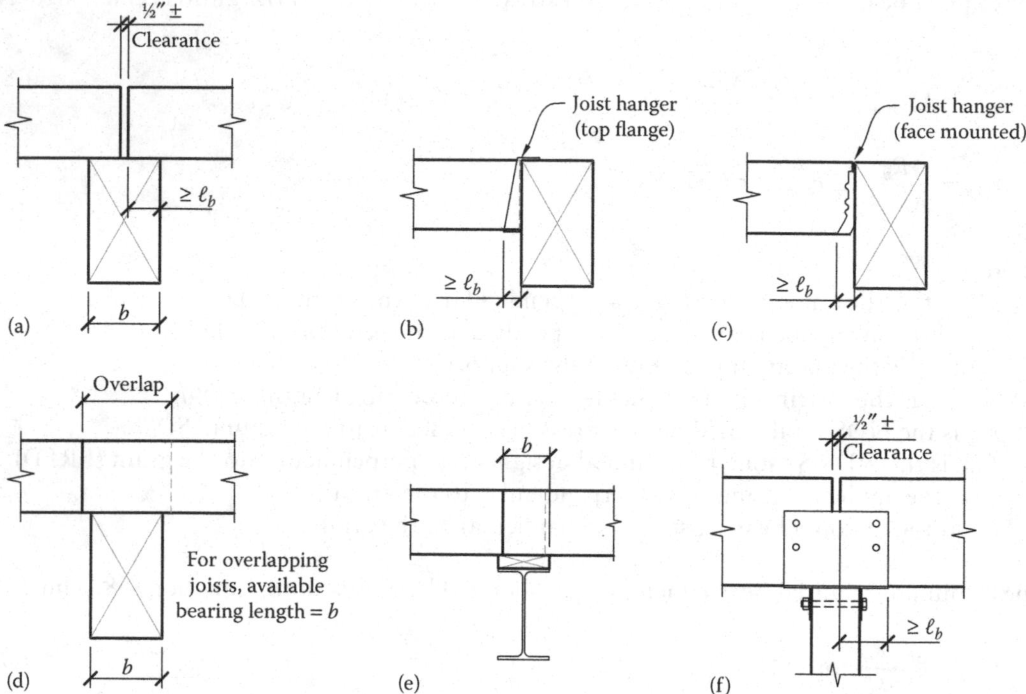

Figure 4.6 Bearing details for joist, beams, and girders.

4.2.1 Check of bending stress

1. Calculate all the loads and load effects, including the maximum moment, shear, and reactions. Also calculate the dead and live loads that will be used later for the deflection calculations. Include the self-weight of the member.
2. From the known stress grade (i.e., Select Structural, No. 1 and Better, No. 1, etc.) and size classification (i.e., dimension lumber, timbers, glulam, etc.), determine the applicable NDS-S Table to be used in the design.
3. Determine the cross-sectional area A_{xx}, the section modulus S_{xx}, and the moment of inertia I_{xx} from NDS-S Table 1B (for sawn lumber), NDS-S Table 1C (for western species glulam), or NDS-S Table 1D (for Southern Pine glulam).
4. Determine the applied bending stress f_b *or* f_{bu},

$$f_b = \frac{M_{\max}}{S_{xx}} \text{ for ASD} \tag{4.10}$$

$$f_{bu} = \frac{M_u}{S_{xx}} \text{ for LRFD}$$

5. Determine all the applicable stress adjustment factors, obtain the tabulated NDS-S bending stress F_b for ASD or nominal bending stress $F_b\,K_{F,r}\phi_b\lambda$ for LRFD, and calculate the allowable or design bending stress using Equation 4.2:
   $F_b' = F_b$ (product of applicable adjustment or C factors)
   $F_{b,r}' = F_b\,K_{F,r}\phi_b\lambda$ (product of applicable adjustment or C factors) for LRFD
6. Compare the applied stress f_b *for ASD or the factored bending* stress f_{bu} from step 4 to the allowable bending stress F_b' *for ASD or the design bending stress* $F_{b,r}'$ calculated in step 5.

For ASD:

If $f_b = M_{max}/S_{xx} \leq F'_b$, the beam is adequate in bending.

If $f_b = M_{max}/S_{xx} > F'_b$, the beam is not adequate in bending; increase the member size and repeat the design process.

For LRFD:

If $f_{bu} = M_u/S_{xx} \leq F'_{bn}$, the beam is adequate in bending.

If $f_{bu} = M_u/S_{xx} > F'_{bn}$, the beam is not adequate in bending; increase the member size and repeat the design process.

7. To determine the maximum allowable load or moment that a beam or girder can support, based on bending stress alone, equate the allowable bending stress F'_b to M_{max}/S_{xx}, and solve for the allowable maximum moment M_{max} and thus the allowable maximum load. Therefore,

$$\text{the allowable maximum moment } M_{max} = F'_b S_{xx} \quad \text{in.-lb}$$

$$= \frac{F'_b S_{xx}}{12} \quad \text{ft-lb}$$

Knowing the allowable maximum moment, we can determine the allowable maximum load based on bending stress alone, using the relationship between the maximum moment and the applied load. For example, for a *uniformly loaded beam or girder,* we obtain the applied maximum moment as

$$M_{max} = \frac{w_{TL}L^2}{8}$$

where L is the span of the beam or girder in feet. Solving and rearranging the equation above yields the maximum allowable total load:

$$w_{TL} = \frac{8M_{max}}{L^2} = \frac{(8)(F'_b S_{xx}/12)}{L^2} \quad \text{lb/ft} \tag{4.11}$$

Knowing the allowable total load w_{TL} and the applied dead load w_{DL} on the member, the allowable live load w_{LL} based on bending stress alone can be obtained using the load combination relationships (see Section 2.1.1):

$$w_{TL} = w_{DL} + w_{LL}$$

Therefore,

the allowable maximum live load $w_{LL} = w_{TL} - w_{DL}$

The allowable maximum load can be converted to units of pounds per square foot by dividing the allowable load (in lb/ft) by the tributary width of the beam or girder in feet.

4.2.2 Check of shear stress

8. Determine the applied shear stress

$$f_v = \frac{1.5V}{A} \quad \text{for ASD} \tag{4.12}$$

$$f_{vu} = \frac{1.5V_u}{A} \quad \text{for LRFD}$$

$$f'_v = \frac{1.5V'}{A} \text{ for ASD} \tag{4.13}$$

$$f'_{vu} = \frac{1.5V'_u}{A} \text{ for LRFD}$$

where V is the maximum shear at the centerline of the joist, beam, or girder bearing support, and V' is the maximum distance d from the face of the support. Use V' to calculate the applied shear stress *only* for beams or girders with no concentrated loads within a distance d from the face of the support and where the bearing support is subjected only to confining compressive stresses. (i.e., use V' only for beams or girders subject to compression at the support reactions).

9. Determine all the applicable stress adjustment or C factors, obtain the tabulated NDS-S shear stress F_v, and calculate the allowable shear stress using Equation 4.14:

$$F'_v = F_v C_D C_M C_t C_i \text{ for ASD} \tag{4.14}$$

$$F'_{v,r} = \text{available shear stress} = F_v K_{F,v} \phi_v \lambda C_M C_t C_i C_{vr} \text{ for LRFD}$$

10. Compare the applied shear stress from step 8 to the allowable or available shear stress calculated in step 9.

For ASD:

If f_v or $f'_v \leq F'_v$, the beam is adequate in shear.

If f_v or $f'_v > F'_v$, the beam is not adequate; therefore, increase the beam size and recheck the shear stress.

For LRFD:

If f_{vu} or $f'_{vu} \leq F'_{v,r}$, the beam is adequate in shear.

If f_{vu} or $f'_{vu} > F'_{v,r}$, the beam is not adequate; therefore, increase the beam size and recheck the shear stress.

11. To determine the maximum allowable load or shear that a beam or girder can support, based on shear stress alone, equate the allowable shear stress F'_v to $1.5V/A$ or $1.5V'/A$, and solve for the allowable shear V_{max} and thus the allowable maximum load. That is,

$$\frac{1.5V_{max}}{A_{xx}} = F'_v$$

Therefore,

$$\text{the allowable maximum shear } V_{max} \text{ or } V'_{max} = \frac{F'_v A_{xx}}{1.5} \text{ for ASD}$$

Similarly, the maximum factored shear that can be supported is

$$V_u \text{ or } V'_u = \frac{F'_{v,r} A}{1.5} \quad \text{for LRFD}$$

Knowing the allowable maximum shear at the centerline of the support, we can determine the allowable maximum load based on shear stress alone using the relationship between the maximum shear and the applied load. However, it should be noted that shear rarely governs the design of joists, beams, and girders.

4.2.3 Check of deflection

12. Using the appropriate deflection formulas, determine and compare the live-load and total-load deflections to the appropriate deflection limits. For a *uniformly loaded beam or girder,* we obtain

$$\text{live-load deflection } \Delta_{LL} = \frac{5w_{LL}L^4}{384E'I} \leq \frac{L}{360}$$

$$\text{total load deflection (including creep) } \Delta_{k(DL)+LL} = \frac{5w_{k(DL)+LL}L^4}{384E'I} \leq \frac{L}{240}$$

where:

$E' = EC_MC_tC_i$; the stress adjustment factors are as discussed previously, and C_i does not apply to glulam

k = creep effect

= 1.0 for green or unseasoned lumber or glulam used in wet service conditions

= 1.0 for seasoned lumber used in wet service conditions

= 0.5 for seasoned or dry lumber, glulam, and prefabricated I-joists used in dry service conditions

L = span of joist, beam, or girder, in.

$\Delta_{k(DL)+LL}$ = deflection due to the total load $k(D) + L$ (or L_r or S), in.

Δ_{LL} = deflection due to the live load L (or L_r or S), in.

$w_{k(DL)+LL}$ = total load including $k(D) + L$ (or L_r or S), lb/in.

w_{LL} = maximum live load L (or L_r or S), lb/in.

13. To determine the maximum allowable load that can be supported by a beam or girder based on deflections alone, equate the calculated deflection to the appropriate deflection limit and solve for the allowable maximum load. For example, for a *uniformly loaded beam or girder,* the live-load deflection is calculated as

$$\Delta_{LL} = \frac{5w_{LL}L^4}{384E'I} = \frac{L}{360}$$

Solving and rearranging the equation above yields the allowable maximum live load,

$$\begin{aligned} w_{LL} &= \frac{384E'I}{5L^4}\left(\frac{L}{360}\right) \quad \text{lb/in.} \\ &= \frac{384E'I}{5L^4}\left(\frac{L}{360}\right)(12) \quad \text{lb/ft} \end{aligned} \tag{4.15}$$

Similarly, for a *uniformly loaded beam or girder,* the maximum allowable total load is given as

$$\begin{aligned} w_{k(DL)+LL} &= \frac{384E'I}{5L^4}\left(\frac{L}{240}\right) \quad \text{lb/in.} \\ &= \frac{384E'I}{5L^4}\left(\frac{L}{240}\right)(12) \quad \text{lb/ft} \end{aligned} \tag{4.16}$$

Knowing that $w_{k(DL)+LL} = w_{k(DLq)} + w_{LL} = k(w_{DL}) + w_{LL}$, we can determine the allowable maximum live load w_{LL} since we know k and w_{DL}. The smaller of this value and the value calculated from Equation 4.15 will be the allowable maximum live load. This load can be converted to units of pounds per square foot by dividing this load (lb/ft) by the tributary width of the beam or girder in feet.

4.2.4 Check of bearing stress (compression stress perpendicular to the grain)

14. Determine the applied bearing stress or compression stress perpendicular to the grain and compare to the allowable or available compression stress perpendicular to the grain,

$$f_{c\perp} = \frac{R}{A_{\text{bearing}}} = \frac{R}{bl_b} \leq F'_{c\perp} \text{ for ASD} \tag{4.17}$$

$$f_{c\perp u} = \frac{R_u}{A_{\text{bearing}}} = \frac{R_u}{bl_b} \leq F'_{c\perp,r} \text{ for LRFD}$$

where:

$F'_{c\perp}$ = allowable bearing stress perpendicular to the grain (ASD)
$= F_{c\perp}C_M C_t C_b C_i$
$F'_{c\perp,r}$ = available bearing stress perpendicular to the grain (LRFD)
$= F_{c\perp}C_M C_t C_b C_i K_{F,c\perp}\phi_c\lambda$
R or R_u = maximum reaction at the support
A_{bearing} = bearing area = (thickness of beam) (bearing length) = $(b)(l_b)$
b = thickness (smaller dimension) of the beam or girder
l_b = length of bearing at the beam or girder support
$F_{c\perp}$ = NDS-S tabulated design stress perpendicular to the grain

The minimum required bearing length l_b is obtained by rearranging Equation 4.17. Thus,

$$l_{b,\text{req'd}} \geq \frac{R}{bF'_{c\perp}} \text{ for ASD} \tag{4.18}$$

$$l_{b,\text{req'd}} \geq \frac{R_u}{bF'_{c\perp,r}} \text{ for LRFD}$$

15. To determine the maximum allowable load (or reaction) that can be supported by a beam or girder, based on bearing stress alone using the ASD method, equate the applied bearing stress to the allowable bearing stress and solve for the reaction R:

$$\frac{R}{A_{\text{bearing}}} = F'_{c\perp}$$

Thus,

$$R = A_{\text{bearing}}\, F'_{c\perp} = bl_b F'_{c\perp}$$

Knowing the allowable maximum reaction R_1, we can determine the allowable maximum load based on bearing stress alone using the relationship between the maximum reaction and the applied load. For example, for a *uniformly loaded beam or girder,* we obtain

$$R_{\max} = \frac{w_{\text{TL}}L}{2}$$

where L is the span of the beam or girder in feet. Solving and rearranging the equation above yields the maximum allowable total load,

$$\begin{aligned} w_{\text{TL}} &= \frac{2R_{1,\max}}{L} = \frac{2A_{\text{bearing}}F'_{c\perp}}{L} \text{ lb/ft} \\ &= \frac{2bl_bF'_{c\perp}}{L} \text{ lb/ft} \end{aligned} \tag{4.19}$$

Knowing the allowable total load w_{TL} and the applied dead load w_{DL} on the member, the allowable live load w_{LL} based on bearing stress alone can be obtained using the load combinations in Chapter 2:

$$w_{TL} = w_{DL} + w_{LL}$$

Therefore, the allowable maximum live load based on bearing stress alone,

$$w_{LL} = w_{TL} - w_{DL}$$

Note: The actual allowable maximum loads for the beam or girder will be the smallest of the maximum loads calculated for bending stress, shear stress, deflection, and bearing stress.

4.2.5 Design examples

In this section we present design examples for glulam girders and the structural elements in stairs.

Example 4.1: Design of joist (or beams) and girders (ASD/LRFD)

A typical floor plan in the private room areas of a hotel building is shown in Figure 4.7. Design a typical (a) joist B1 and (b) girder G1 assuming the use of Douglas Fir-Larch.

ASD Solution:

(a) *Design of floor joist B1* (Figure 4.8). Joist span = 19 ft. For private room areas and corridors serving them, the floor live load L = 40 psf

Dead Load:

1-in. gypsum	= 6 psf
1-in. plywood sheathing (0.4 psf per 1/8 in. × 6)	= 2.4 psf (assumed)
Framing: 2 × 14 at 16 in. o.c.	= 3.3 psf (assumed self-weight of floor framing)
Mechanical and electrical	= 8 psf (assumed)
Partitions	= 20 psf (15 psf is the minimum value allowed; see Chapter 2)
Total dead load D	= 39.7 psf ≃ 40 psf

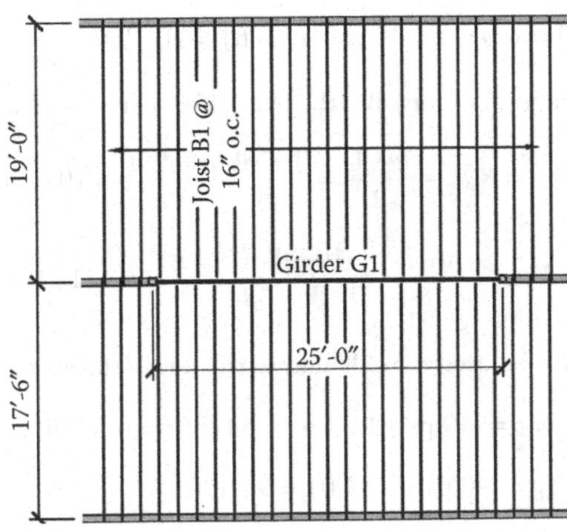

Figure 4.7 Floor plan for Example 4.1.

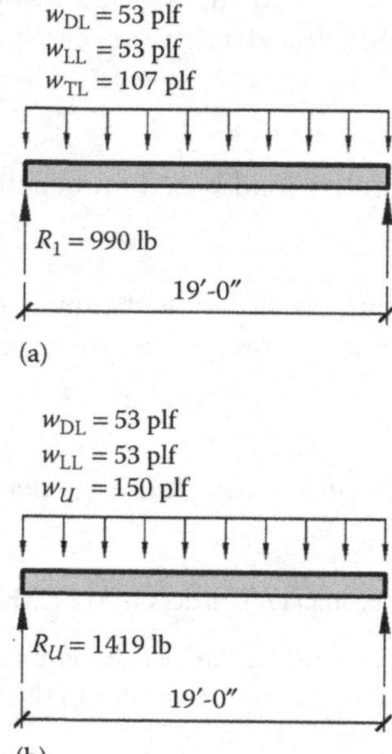

Figure 4.8 Free-body diagram of joist B1. (a) Unfactored loads/reactions and (b) factored loads/reactions.

$$\text{Tributary area of joist or beam B1} = \left(\frac{16 \text{ in.}}{12}\right)(19 \text{ ft}) = 25.3 \text{ ft}^2$$

From Section 2.4.3.1 or ASCE 7, Table 4.2, we obtain the following parameters: A_T, the tributary area = 25.3 ft² and $K_{LL} = 2$ (interior beam or girder).

$$K_{LL}A_T = (2)(25.3 \text{ ft}^2) = 50.6 \text{ ft}^2 < 400 \text{ ft}^2$$

Therefore, no live-load reduction is permitted. The tributary width (TW) of joist or beam B1 = 16 in. = 1.33 ft. The total uniform load on the joist that will be used to design the joist for bending, shear, and bearing is

$$w_{TL} = (D + L)(\text{TW}) = (40 \text{ psf} + 40 \text{ psf})(1.33 \text{ ft}) = 107 \text{ lb/ft}$$

Using the free-body diagram of joist B1, the load effects are calculated as follows:

$$\text{Maximum shear, } V_{max} = R_{max} = \frac{w_{TL}L}{2} = \frac{(107 \text{ lb/ft})(19 \text{ ft})}{2} = 1014 \text{ lb}$$

$$\text{Maximum moment } M_{max} = \frac{w_{TL}L^2}{8} = \frac{(107 \text{ lb/ft})(19 \text{ ft})^2}{8} = 4813 \text{ ft-lb} = 57{,}800 \text{ in.-lb}$$

The following loads will be used for calculating the joist deflections:

$$\text{uniform dead load } w_{DL} = (40 \text{ psf})(1.33 \text{ ft}) = 54 \text{ lb/ft} = 4.5 \text{ lb/in.}$$

$$\text{uniform live load } w_{LL} = (40 \text{ psf})(1.33 \text{ ft}) = 54 \text{ lb/ft} = 4.5 \text{ lb/in.}$$

We now proceed with the design following the steps outlined previously in this chapter.

Check of Bending Stress

1. Summary of load effects (the self-weight of joist B1 was included in the determination of the floor dead load in psf):
   maximum shear V_{max} = maximum reaction R_{max} = 1014 lb
   maximum moment M_{max} = 57,800 in.-lb
   uniform dead load is w_{DL}= 4.5 lb/in.
   uniform live load is w_{LL} = 4.5 lb/in.
2. For the stress grade, assume Select Structural, and assume that the size classification for the joist is dimension lumber. Thus, use NDS-S Table 4A.
3. From NDS-S Table 4A, we find that F_b = 1500 psi. Assume initially that $F_b' = F_b$ = 1500 psi. From Equation 4.1 the approximate section modulus of the member required is given as

$$S_{xx,\ \text{req'd}} \geq \frac{M_{max}}{F_b} = \frac{57{,}800}{1500} = 38.5 \text{ in}^3$$

4. From NDS-S Table 1B, determine the trial size with the least area that satisfies the section modulus requirement of step 3: Try a 2 × 14 DF-L Select Structural member:
   b = 1.5 in. and d = 13.25 in.
   Size is still dimension lumber as assumed in step 1. OK
   S_{xx} provided = 43.9 in.^{3} > 38.5 in.^{3} OK
   Area A provided = 19.88 in.^{2}
   I_{xx} provided = 290.8 in.^{4}
5. The NDS-S tabulated stresses are
   F_b = 1500 psi
   F_v = 180 psi
   $F_{c\perp}$ = 625 psi
   $E = 1.9 \times 10^6$ psi
   The adjustment or C factors are listed in Table 4.4. Using Equation 4.2, the allowable bending stress is
   $F_b' = F_b$ (product of applicable adjustment or C factors)
   $= F_b C_D C_M C_t C_L C_F C_r C_{fu} C_i$
   = (1500)(1.0)(1.0)(1.0)(0.9)(1.15)(1.0)(1.0)
   = **1553 psi**
6. The actual applied bending stress is

$$f_b = \frac{M_{max}}{S_{xx}} = \frac{57{,}800 \text{ lb-in.}}{43.9 \text{ in}^3} = 1316 \text{ psi}$$

   $< F_b' = 1553$ psi OK

   Therefore, the beam is adequate in bending.

Check of Shear Stress

V_{max} = 1014 lb. The beam cross-sectional area A = 19.88 in.^{2}. The applied shear stress in the wood beam at the centerline of the beam support is

$$f_v = \frac{1.5V}{A} = \frac{(1.5)(1014 \text{ lb})}{19.88 \text{ in}^2} = 77 \text{ psi}$$

The allowable shear stress is

$$F_v' = F_v C_D C_M C_t C_i = (180)(1)(1)(1)(1) = 180 \text{ psi}$$

Thus, $F_v < F_v'$; therefore, the beam is adequate in shear.

Table 4.4 Adjustment factors for joists

Adjustment factor	*Symbol*	*Value*	*Rationale for the value chosen*
Beam stability factor	C_L	1.0	The compression face is fully braced by the floor sheathing
Size factor	$C_F(F_b)$	0.9	From NDS-S Table 4A
Moisture or wet service factor	C_M	1.0	The equilibrium moisture content is $\leq$19%
Load duration factor	C_D	1.0	The largest C_D value in the load combination of dead load plus floor live load (i.e., $D + L$)
Temperature factor	C_t	1.0	Insulated building; therefore, normal temperature conditions apply
Repetitive member factor	C_r	1.15	All the three required conditions are satisfied: • The 2 × 14 trial size is dimension lumber • Spacing = 16 in. ≤ 24 in. • Plywood floor sheathing nailed to joists
Bearing stress factor	C_b	1.0	$C_b = \frac{l_b + 0.375}{l_b}$ for $l_b \leq 6$ in. $= 1.0$ for $l_b > 6$ in. = 1.0 *for bearings at the ends of a member* (see Section 3.4.3.3)
Flat use factor	C_{fu}	1.0	
Incision factor	C_i	1.0	

Note: Since the shear stress f_v at the centerline of the beam support is adequate (i.e., $<F'_v$), there is no need to calculate the shear stress f'_v further at a distance d from the face of the support since it is obvious *that* f'_v would be less than f_v, and thus will be adequate in that case.

Check of Deflection (see Table 4.5)
The allowable pure bending modulus of elasticity is

$$E' = EC_M C_t C_i = (1.9 \times 10^6)(1.0)(1.0)(1.0) = 1.9 \times 10^6 \text{ psi}$$

The moment of inertia about the strong axis is $I_{xx} = 290.8$ in.^{4}, the uniform dead load is $w_{DL} = 4.5$ lb/in., and the uniform live load is $w_{LL} = 4.5$ lb/in.

The dead-load deflection is

$$\Delta_{DL} = \frac{5wL^4}{384E'I} = \frac{(5)(4.5 \text{ lb/in.})(19 \text{ ft} \times 12)^4}{(384)(1.9 \times 10^6 \text{ psi})(290.8 \text{ in}^4)} = 0.29 \text{ in.}$$

The live-load deflection is

$$\Delta_{LL} = \frac{5wL^4}{384E'I} = \frac{(5)(4.5 \text{ lb/in})(19 \text{ ft} \times 12)^4}{(384)(1.9 \times 10^6 \text{ psi})(290.8 \text{ in}^4)} = 0.29 \text{ in.} < \frac{L}{360} \quad \text{OK}$$

Table 4.5 Joist deflection limit

Deflection	*Deflection limit*
Live-load deflection Δ_{LL}	$\frac{L}{360} = \frac{(19 \text{ ft})(12)}{360} = 0.63$ in.
Total-load deflection $\Delta_{TL} = k\Delta_{DL} + \Delta_{LL}$	$\frac{L}{240} = \frac{(19 \text{ ft})(12)}{240} = 0.95$ in.

Since seasoned wood in dry service conditions is assumed to be used in this building, the creep factor $k = 0.5$. However, it is conservative to use $k = 1.0$, which will be used here. The total dead plus floor live-load deflection is

$$\Delta_{TL} = k(\Delta_{DL}) + \Delta_{LL}$$

$$= (1.0)(0.29 \text{ in.}) + 0.29 \text{ in.} = 0.58 \text{ in.} < \frac{L}{240} \qquad \text{OK}$$

Alternatively, the required moment of inertia can be calculated using Equations 4.5 and 4.6 as follows:

$$I_{req'd} = \frac{5w_{LL}L^3}{384E'I}(360)\text{ in}^4 = \frac{(5)(4.5)(19 \text{ ft} \times 12)^3}{(384)(1.9 \times 10^6)}(360) = 132 \text{ in}^4$$

Similarly, the required moment of inertia based on the total load is obtained for a uniformly loaded beam or girder as

$$\left(\frac{(5)(9)(19x12)^3}{(384)(1.9x10)^6}\right)(240) = 175 \text{ in}^4$$

Therefore,

the required moment of inertia = 175 in.^{4} < I_{actual} = 290.8 in.^{4} OK

Alternatively, since the dead load equals the live load, a deflection limit of $L/240$ would control here. Therefore, using the applicable equation from Table 4.5:

$$I_{req'd} = \frac{wL^3}{2.22E'} = \frac{(107)(19)^3}{2.22(1900)} = 174 \text{ in.}^4$$

which agrees with the previous calculation.

Check of Bearing Stress (Compression Stress Perpendicular to the Grain)

The maximum reaction at the support $R = 1014$ lb and the thickness of a 2 × 14 sawn-lumber joist, $b = 1.5$ in. The allowable bearing stress or compression stress parallel to the grain is

$$F'_{c\perp} = F_{c\perp}C_M C_t C_b C_i = (625)(1.0)(1.0)(1.0)(1.0) = 625 \text{ psi}$$

From Equation 4.9, the minimum required bearing length l_b is

$$l_{b,\text{ req'd}} \geq \frac{R}{bF'_{c\perp}} = \frac{1014 \text{ lb}}{(1.5 \text{ in.})(625 \text{ psi})} = 1.09 \text{ in.}$$

The floor joists will be connected to the floor girder using face-mounted joist hangers with the top of the joists at the same level as the top of the girders. From a proprietary joist hanger load table (Figure 4.9) connector catalog, select a face-mounted hanger JH-214 with the following properties:

Load duration factor C_D for dead plus floor live load = 1.0 = 100%.

Therefore, the allowable load for JH-214 = 1280 lb > maximum reaction = 1014 lb.

Available bearing length = 2 in. > $L_{b,\text{ req'd}}$ = 1.09 in. OK

Use a 2 × 14 DF-L Select Structural for joist B1.

(b) *Design of floor girder G1.* Girder span = 25 ft. Since the joist selected was 2 × 14, we would expect the girder to be larger than a 2 × 14. Thus, the girder is probably going to be a timber (i.e., beam and stringer). Assume the girder self-weight = 90 lb/ft (this will be checked later). For hotels, the live load $L = 40$ psf.

Dead load:

From the design of the floor joist B1, the total dead load $D = 40$ psf.

$$\text{Tributary width of girder G1} = \frac{19 \text{ ft}}{2} + \frac{17.5 \text{ ft}}{2} = 18.25 \text{ ft}$$

Tributary area of girder G1 = (18.25 ft)(25 ft) = 456 ft^2

Joist size	Model #	Dimensions			Douglas Fir-Larch or Southern Pine allowable loads (lb)			
		W	H	B	100%	115%	125%	160% (uplift)
2 × 14	JH-214	1⁹⁄₁₆″	10″	2″	1280	1470	1600	2040
	xx	xx	xx	xx	xx	xx	xx	xx

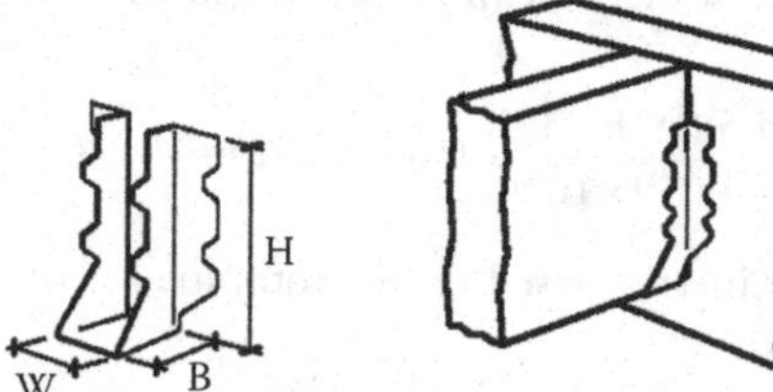

Figure 4.9 Generic load Table for joist hangers.

From ASCE 7, Table 4.2, $\tilde{A}_T$, the tributary area = 456 ft², and $K_{LL} = 2$ (interior girder).

- $K_{LL}A_T = (2)(456 \text{ ft}^2) = 912 \text{ ft}^2 > 400 \text{ ft}^2$
- Floor live load $L = 40$ psf < 100 psf
- Floor occupancy is not assembly occupancy

Since all the conditions above are satisfied, live-load reduction is permitted.

Live-load reduction. The reduced or design floor live load for the girder is calculated using Equation 2.9 as follows:

$$L = L_0\left(0.25 + \frac{15}{\sqrt{K_{LL}A_T}}\right) = (40 \text{ psf})\left[0.25 + \frac{15}{\sqrt{(2)(456)}}\right] = \mathbf{30\ psf}$$

$\geq (0.50)(40 \text{ psf}) = 20$ psf for members supporting one floor (e.g., slabs, beams, and girders)

$\geq (0.40)(40 \text{ psf}) = 16$ psf for members supporting two or more floors (e.g., columns)

The total uniform load on the girder that will be used to design for bending, shear, and bearing is obtained below using the dead load, the reduced live load calculated above, and the assumed self-weight of the girder, which will be checked later.

$$w_{TL} = (D + L)\ (\text{tributary width}) + \text{girder self-weight}$$
$$= (40 \text{ psf} + 30 \text{ psf})(18.25 \text{ ft}) + 90 \text{ lb/ft} = \mathbf{1368\ lb/ft}$$

Using the free-body diagram of girder G1 (Figure 4.10), the load effects are calculated as follows:

maximum shear (and maximum reaction R_{max} =)

$$V_{max} = \frac{w_{TL}L}{2} = \frac{(1368 \text{ lb/ft})(25 \text{ ft})}{2} = 17{,}100 \text{ lb}$$

$$\text{maximum moment } M_{max} = \frac{w_{TL}L^2}{8} = \frac{(1368 \text{ lb/ft})(25 \text{ ft})^2}{8} = 106{,}875 \text{ ft-lb} = 1{,}282{,}500 \text{ in.-lb}$$

The following loads will be used for calculating the joist deflections:

uniform dead load $w_{DL} = (40 \text{ psf})(18.25 \text{ ft}) + 90 \text{ lb/ft} = 820 \text{ lb/ft} = 68.33 \text{ lb/in.}$
uniform live load $w_{LL} = (30 \text{ psf})(18.25 \text{ ft}) = 547.5 \text{ lb/ft} = 45.63 \text{ lb/in.}$

We now proceed with the design following the steps described earlier in the chapter.

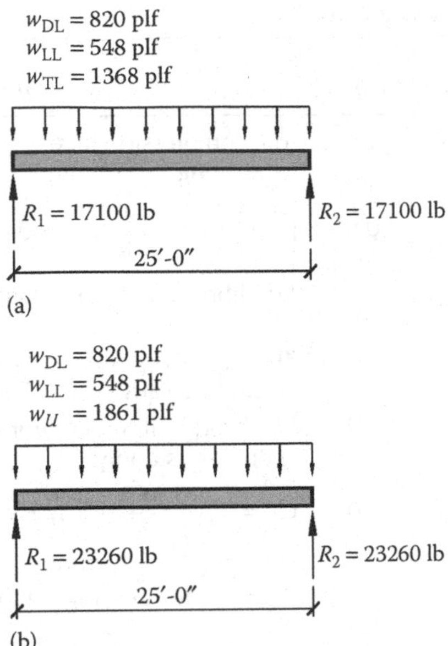

Figure 4.10 Free-body diagram of girder G1. (a) Unfactored loads/reactions and (b) factored loads/reactions.

Check of Bending Stress (ASD method)

1. Summary of load effects (the self-weight of girder G1 was assumed, but this will be checked later in the design process):
   maximum shear V_{max} and maximum reaction R_{max} = 17,100
   maximum moment M_{max} = 1,282,500 in.-lb
   uniform dead load w_{DL} = 68.33 lb/in.
   uniform live load w_{LL} = 45.63 lb/in.
2. For the stress grade, assume Select Structural, and assume that the size classification for the girder is beam and stringer (B&S), that is, timbers. Therefore, use NDS-S Table 4D.
3. From NDS-S Table 4D, we find that F_b = 1600 psi. Assume initially that $F_b' = F_b$ = 1600 psi. From Equation 4.1, the required approximate section modulus of the member is given as

$$S_{xx,\ \text{req'd}} \geq \frac{M_{\max}}{F_b} = \frac{1,282,032}{1600} = 801.3\ \text{in.}^3$$

4. From NDS-S Table 1B, determine the trial size with the least area that satisfies the section modulus requirement of step 3. Try a 10 × 24 DF-L Select Structural.
   b = 9.5 in. and d = 23.5 in.
   Size is B&S, as assumed in step 1 OK
   S_{xx} provided = 874.4 in.^{3} > 801.3 in.^{3} OK
   Area A provided = 223.3 in.^{2}
   I_{xx} provided = 10,274 in.^{4}
5. The NDS-S tabulated stresses are
   F_b = 1600 psi
   F_v = 170 psi
   $F_{c\perp}$ = 625 psi
   $E = 1.6 \times 10^6$ psi

Table 4.6 Adjustment factors for wood girder

Adjustment factor	*Symbol*	*Value*	*Rationale for the value chosen*
Beam stability factor	C_L	1.0	The compression face is fully braced by the floor sheathing
Size factor	C_F	0.93	$\left(\frac{12}{d}\right)^{1/9} = \left(\frac{12}{23\text{ in.}}\right)^{1/9} = 0.93$
Moisture or wet service factor	C_M	1.0	Equilibrium moisture content is ≤19%
Load duration factor	C_D	1.0	The largest C_D value in the load combination of dead load plus floor live load (i.e., $D + L$)
Temperature factor	C_t	1.0	Insulated building; therefore, normal temperature conditions apply
Bearing stress factor	C_b	1.0	$C_b = \frac{l_b + 0.375}{l_b}$ for $l_b \leq 6$ in. = 1.0 for $l_b > 6$ in. = 1.0 for bearings at the ends of a member
Flat use factor	C_{fu}	1.0	
Curvature factor	C_c	1.0	

The adjustment or C factors are listed in Table 4.6. Using Equation 4.2, the allowable bending stress is

$$F_b' = F_b \text{ (product of applicable adjustment or C factors)}$$

$$= F_b C_D C_M C_t C_L C_F C_r C_{fu} C_c$$

$$= (1600)(1.0)(1.0)(1.0)(0.93)\,(1.0)(1.0)(1.0)$$

$$= \mathbf{1488\ psi}$$

6. The actual applied bending stress is

$$f_b = \frac{M_{max}}{S_{xx}} = \frac{1{,}282{,}500\text{ lb-in.}}{874.4\text{ in}^3} = 1467\text{ psi} < F_b' = 1488\text{ psi} \qquad \text{OK}$$

Therefore, the beam is adequate in bending.

Check of Shear Stress

$V_{max} = 17{,}100$ lb. The beam cross-sectional area $A = 223.3$ in.^{2}. The applied shear stress in the wood beam at the centerline of the girder support is

$$f_v = \frac{1.5V}{A} = \frac{(1.5)\,(17{,}100\text{ lb})}{223.3\text{ in}^2} = 115\text{ psi}$$

The allowable shear stress is

$$F_v' = F_v C_D C_M C_t C_i = (170)(1)(1)(1)(1) = 170\text{ psi}$$

Thus, $f_v < F_v'$; therefore, the beam is adequate in shear. See Figure 4.11.

Note: Since the shear stress f_v at the centerline of the girder support is adequate (i.e., $< F_v'$), there is no need to calculate the shear stress f_v' at a distance d from the face of the support since it is obvious that f_v' would be less than f_v and thus will be adequate in that case.

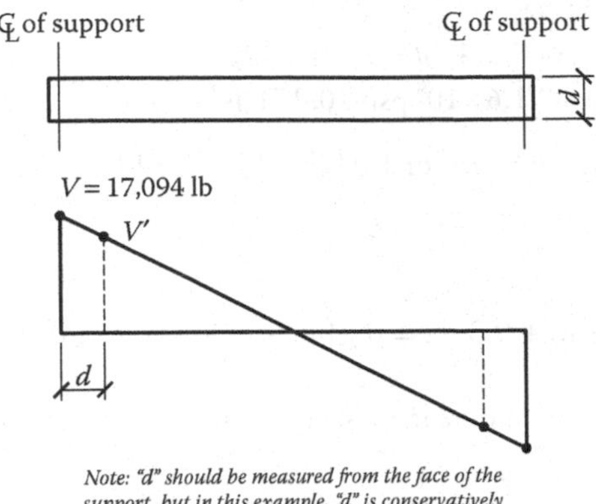

℄ of support
℄ of support
d
V = 17,094 lb
V′
d
Note: "d" should be measured from the face of the support, but in this example, "d" is conservatively measured from the centerline of support since the bearing dimensions are not yet known

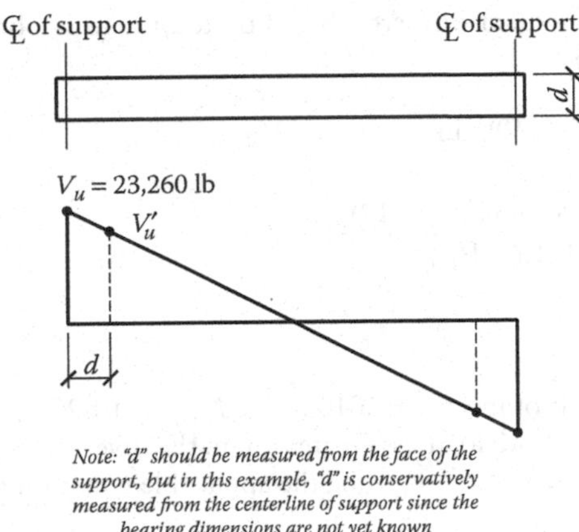

℄ of support
℄ of support
d
Vu = 23,260 lb
V′u
d
Note: "d" should be measured from the face of the support, but in this example, "d" is conservatively measured from the centerline of support since the bearing dimensions are not yet known

Figure 4.11 Shear force diagram for girder G1.

Check of Deflection (see Table 4.7)
The allowable pure bending modulus of elasticity is

$$E' = EC_M C_t C_i = (1.6 \times 10^6)(1.0)(1.0)(1.0) = 1.6 \times 10^6 \text{ psi}$$

The moment of inertia about the strong axis I_{xx} = 10,270 in.^{4}, the uniform dead load is w_{DL} = 68.33 lb/in., and the uniform live load is w_{LL} = 45.63 lb/in.

The dead-load deflection is

$$\Delta_{\text{DL}} = \frac{5wL^4}{384E'I} = \frac{(5)(68.33 \text{ lb/in.})(25 \text{ ft} \times 12)^4}{(384)(1.6 \times 10^6 \text{ psi})(10274 \text{ in}^4)} = 0.44 \text{ in.}$$

Table 4.7 Girder deflection limit

Deflection	*Deflection limit*
Live-load deflection Δ_{LL}	$\frac{L}{360} = \frac{(25 \text{ ft})(12)}{360} = 0.83$ in.
Total-load deflection $\Delta_{\text{TL}} = k\Delta_{\text{DL}} + \Delta_{\text{LL}}$	$\frac{L}{240} = \frac{(25 \text{ ft})(12)}{240} = 1.25$ in.

The live-load deflection is

$$\Delta_{LL} = \frac{5w_{LL}L^4}{384E'I} = \frac{(5)(45.63 \text{ lb/in.})(25 \text{ ft} \times 12)^4}{(384)(1.6 \times 10^6 \text{ psi})(10,274 \text{ in}^4)} = 0.30 \text{ in.} < \frac{L}{360} = 0.83 \text{ in.} \qquad \text{OK}$$

As discussed previously, a k-factor of 1.0 is used for dead-load deflections. The total-load deflection is

$$\Delta_{TL} = k\Delta_{DL} + \Delta_{LL}$$

$$= (1.0)(0.44 \text{ in.}) + 0.3 \text{ in.} = 0.74 \text{ in.} < \frac{L}{240} = 1.25 \text{ in.} \qquad \text{OK}$$

Alternatively, the required moment of inertia can be calculated using Equations 4.5 and 4.6 as follows:

$$I_{\text{req'd}} = \frac{5w_{LL}L^3}{384E'}(360) \quad \text{in}^4 = \frac{(5)(45.63)(25 \text{ ft} \times 12)^3}{(384)(1.6 \times 10^6)}(360) = 3609 \text{ in.}^4$$

Similarly, the required moment of inertia based on total load is obtained for a *uniformly loaded beam or girder* as

$$I_{\text{req'd}} = \frac{5w_{k(DL)+LL}L^3}{384E'}(240) \quad \text{in}^4$$

$$= \frac{(5)(68.33 + 45.63)(25 \text{ ft} \times 12)^3}{(384)(1.6 \times 10^6)}(240)$$

$$= 6010 \text{ in.}^4$$

Therefore,

the required moment of inertia = 6010 in.^{4} < I_{actual} = 10,270 in.^{4} OK

Alternatively, since the dead load is greater than the live load, a deflection limit of $L/240$ would control here. Therefore, using the applicable equation from Table 4.5:

$$I_{\text{req'd}} = \frac{wL^3}{2.22E'} = \frac{(820 + 547.5)(25)^3}{2.22(1600)} = 6010 \text{in.}^4$$

which agrees with the previous calculation.

Check of Bearing Stress (Compression Stress Perpendicular to the Grain)

The maximum reaction at the support R = 17,100 lb and the thickness of a 10 × 24 sawn-lumber joist b = 9.5 in. The allowable bearing stress or compression stress parallel to the grain is

$$F'_{c\perp} = F_{c\perp}C_M C_t C_b C_i = (625)(1.0)(1.0)(1.0)(1.0) = 625 \text{ psi}$$

From Equation 4.8, the minimum required bearing length l_b is

$$l_{b,\text{ req'd}} \geq \frac{R}{bF'_{c\perp}} = \frac{17,100 \text{ lb}}{(9.5 \text{ in.})(625 \text{ psi})} = 2.88 \text{ in.} \quad \text{say, 3 in.}$$

The girders will be connected to the column using a U-shaped column cap detail as shown below, and the minimum length of the column cap required is

(3 in.)(2) + 1/2-in. clearance between ends of girders = 6.5 in.

From a proprietary wood connector catalog, the maximum column cap width available for a U-shaped cap is 9.5 in. Since this is not less than the thickness (smaller dimension) of the girder, the proprietary U-shaped cap can be used. Therefore, use a U-shaped column cap similar to that shown in Figure 4.6e.

Note: The maximum column cap width available in most catalogs is 9.5 in., so where the width of a girder or beam is greater than this value, a special column cap having a minimum width equal to the thickness (smaller dimension) of the girder or beam would have to be designed.

Check of Assumed Girder Self-Weight

As discussed in Chapter 2, the density of wood is assumed in this book to be 32 lb/ft³, and this is used to calculate the actual weight of the girder selected.

Assumed girder self-weight = 90 lb/ft

Actual weight of the 10 × 24 girder selected (from NDS-S Table 1B):

$$\frac{223.3\text{ in}^2\ (32\text{ lb/ft}^3)}{(12\text{ in.})(12\text{ in.})} = 49\text{ lb/ft, which is less than the assumed value of 90 lb/ft} \qquad \text{OK}$$

Note: The actual self-weight of the girder in this example is less than the assumed self-weight. However, let us consider a fictitious case where the actual self-weight is greater than the assumed value by, say, 50%. The self-weight calculated should then be used to recalculate the design loads, and the member should be redesigned. However, it should be noted that in many cases, the self-weight of the girder is usually a small percentage of the total load on the girder, and thus the total load is not affected significantly by even large positive variations of up to a maximum of 50% in the girder self-weight.

Assume in our fictitious example that we had selected a girder size that has a self-weight of 135 lb/ft, which is 50% heavier than the 90 lb/ft, which was assumed initially in the load calculations. Therefore, the revised total load would become

$$\begin{aligned} w_{TL} &= (D + L)(\text{tributary width}) + \text{girder self-weight} \\ &= (40\text{ psf} + 30\text{ psf})(18.25\text{ ft}) + 135\text{ lb/ft} = 1413\text{ lb/ft} \end{aligned}$$

Compared to the total load of 1368 lb/ft calculated previously using the assumed self-weight of 90 lb/ft, the percentage increase in the total load and load effect for this fictitious example will be (1413 − 1368)/1368 × 100% = 3%. This is quite a small and insignificant change for a 50% increase in girder self-weight, and thus its effect can be neglected. So we find that for this girder, even a 50% increase in the self-weight of the girder would not have had a significant effect on the total load and the load effects.

Use a 10 × 24 DF-L Select Structural for girder G1.

LRFD Solution:

(a) *Design of floor joist B1* (Figure 4.8): Joist span = 19 ft. For private room areas and corridors serving them, the floor live load L = 40 psf.

Dead load:	
1-in. gypsum	= 6 psf
1-in. plywood sheathing (0.4 psf/1/8 in. × 6)	= 2.4 psf (assumed)
Framing: 2 × 14 at 16 in. o.c.	= 3.3 psf (assumed self-weight of floor framing)
Mechanical and electrical	= 8 psf (assumed)
Partitions	= 20 psf (15 psf is the minimum value allowed; see Chapter 2)
Total dead load D	= 39.7 psf = 40 psf

$$\text{Tributary area of joist or beam B1} = \left(\frac{16\text{ in.}}{12}\right)(19\text{ ft}) = 25.3\text{ ft}^2$$

From Chapter 2, we obtain the following parameters: A_T, the tributary area = 25.3 ft² and $K_{LL} = 2$ (interior beam or girder).

$K_{LL}A_T = (2)(25.3\text{ ft}^2) = 50.6\text{ ft}^2 < 400\text{ ft}^2$

Therefore, no live-load reduction is permitted. The tributary width (TW) of joist or beam B1 = 16 in. = 1.33 ft. The total factored uniform load on the joist that will be used to design the joist for bending, shear, and bearing is

$$Wu = 1.2D + 1.6L = (1.2)(40\text{ psf}) + (1.6)(40\text{ psf}) = 112\text{ psf}$$

$$w_{TL} = WuTW = (112)(1.33) = 150\text{ plf}$$

Using the free-body diagram of joist B1, the load effects are calculated as follows:

$$\text{Maximum shear, } V_U = R_U = \frac{w_U L}{2} = \frac{(150\text{ lb/ft})(19\text{ ft})}{2} = 1419\text{ lb}$$

$$\text{Maximum moment } M_U = \frac{w_U L^2}{8} = \frac{(150\text{ lb/ft})(19\text{ ft})^2}{8} = 6769\text{ ft-lb} = 81225\text{ in.-lb}$$

The following loads will be used for calculating the joist deflections:

uniform dead load $w_{DL} = (40\text{ psf})(1.33\text{ ft}) = 54\text{ lb/ft} = 4.5\text{ lb/in.}$

uniform live load $w_{LL} = (40\text{ psf})(1.33\text{ ft}) = 54\text{ lb/ft} = 4.5\text{ lb/in.}$

We now proceed with the design following the steps described previously in this chapter.

Check of Bending Stress

1. Summary of load effects (the self-weight of joist B1 was included in the determination of the floor dead load in psf):
   maximum shear V_U = maximum reaction $R_U = 1419$ lb
   maximum moment $M_{max} = 81225$ in.-lb
   uniform dead load is $w_{DL} = 4.5$ lb/in.
   uniform live load is $w_{LL} = 4.5$ lb/in.
2. For the stress grade, assume Select Structural, and assume that the size classification for the joist is dimension lumber. Thus, use NDS-S Table 4A.
3. From NDS-S Table 4A, we find that $F_b = 1500$ psi. Assume initially that $F_b' = F_b = 1500$ psi. From Equation 4.1 the approximate section modulus of the member required is given as

$$S_{xx,\text{req'd}} \geq \frac{M_{max}}{F_b K_{F,b} \phi_b \lambda} = \frac{81225}{(1500)(2.54)(0.85)(0.8)} = 31.4\text{ in}^3$$

   From Appendix N:
   $K_{F,b} = 2.54$
   $\phi_b = 0.85$
   $\lambda = 0.8$
4. From NDS-S Table 1B, determine the trial size with the least area that satisfies the section modulus requirement of step 3: Try a 2 × 14 DF-L Select Structural member.
   $b = 1.5$ in. and $d = 13.25$ in.
   Size is still dimension lumber as assumed in step 1. OK

S_{xx} provided = 43.9 in^3 > 31.4 in^3 OK
Area A provided = 19.88 in^2
I_{xx} provided = 290.8 in^4

Note that a 2 × 12 would seem to work since S_x = 31.6 in^3 > 31.4 in^3, but it is very close to the limit. As the design progresses, it will evident that a size larger than 2 × 12 will be needed.

5. The NDS-S tabulated stresses are
   $F_b = 1500$ psi
   $F_v = 180$ psi
   $F_{c\perp} = 625$ psi
   $E = 1.9 \times 10^6$ psi

   The adjustment or C factors are listed in Table 4.4. Using Equation 4.2, the design bending stress is

   $$F'_{b,r} = F_b K_{F,b}\,\phi_b \lambda \text{ (product of applicable adjustment or C factors)}$$
   $$= F_b\, K_{F,b}\,\phi_b \lambda\, C_M C_t C_L C_F C_r C_{fu} C_i$$
   $$= (1500)(2.54)(0.85)(0.8)(1.0)(1.0)(0.9)(1.15)(1.0)(1.0)$$
   $$= 2681 \text{ psi}$$

6. The factored bending stress is

   $$f_{bu} = \frac{M_U}{S_{xx}} = \frac{81225\,\text{lb}-\text{in.}}{43.9\,\text{in}^3} = 1850\,\text{psi}$$

   $$< F'_{b,r} = 2681\,\text{psi} \quad \text{OK}$$

   Therefore, the beam is adequate in bending.

Check of Shear Stress

V_{max} = 990 lb. The beam cross-sectional area A = 19.88 in^2. The factored shear stress in the wood beam at the centerline of the beam support is

$$f_{vu} = \frac{1.5V_u}{A} = \frac{(1.5)(1419\,\text{lb})}{19.88\,\text{in}^2} = 108\,\text{psi}$$

The design shear stress is

$$F'_{v,r} = F_v K_{F,v} \phi_v \lambda\, C_M C_t C_i = (180)(2.88)(0.75)(0.8)(1)(1)(1) = 311 \text{ psi}$$

From Appendix N:

$K_{F,v} = 2.88$
$\phi_v = 0.75$
$\lambda = 0.8$

Thus, $f_v < F'_v$; therefore, the beam is adequate in shear.
Note: Since the shear stress f_v at the centerline of the beam support is adequate (i.e., $< F'_v$), there is no need to calculate the shear stress f'_v further at a distance d from the face of the support since it is obvious that f'_v would be less than f_v, and thus will be adequate in that case.

Check of Deflection (see Table 4.5)
The allowable pure bending modulus of elasticity is

$$E' = EC_M C_t C_i = (1.9 \times 10^6)(1.0)(1.0)(1.0) = 1.9 \times 10^6 \text{ psi}$$

The moment of inertia about the strong axis $I_{xx} = 290.8$ in.^{4}, the uniform dead load is $w_{DL} = 4.5$ lb/in., and the uniform live load is $w_{LL} = 4.5$ lb/in.

The dead-load deflection is

$$\Delta_{DL} = \frac{5wL^4}{384E'I} = \frac{(5)(4.5 \text{ lb/in})(19 \text{ ft} \times 12)^4}{(384)(1.9 \times 10^6 \text{psi})(290.8 \text{ in}^4)} = 0.29 \text{ in.}$$

The live-load deflection is

$$\Delta_{LL} = \frac{5wL^4}{384E'I} = \frac{(5)(4.5 \text{ lb/in})(19 \text{ ft} \times 12)^4}{(384)(1.9 \times 10^6 \text{ psi})(290.8 \text{ in}^4)} = 0.29 \text{ in.} < \frac{L}{360} \text{ OK}$$

Since seasoned wood in dry service conditions is assumed to be used in this building, the creep factor $k = 0.5$. However, it is conservative to use $k = 1.0$, which will be used here. The total dead plus floor live load deflection is

$$\Delta_{TL} = k(\Delta_{DL}) + \Delta_{LL}$$

$$= (1.0)(0.29 \text{ in.}) + 0.29 \text{in.} = 0.58 \text{ in.} < \frac{L}{240} \quad \text{OK}$$

Alternatively, the required moment of inertia can be calculated using Equations 4.5 and 4.6 as follows:

$$I_{\text{req'd}} = \frac{5w_{LL}L^3}{384E'}(360) \text{ in}^4 = \frac{(5)(4.5)(19 \text{ ft} \times 12)^3}{(384)(1.6 \times 10^6)}(360) = 157 \text{ in}^4$$

Similarly, the required moment of inertia based on the total load is obtained for a uniformly loaded beam or girder as

$$I_{\text{req'd}} = \frac{5w_{k(DL)+LL}L^3}{384E'}(240)\text{in}^4 = \frac{(5)(4.5+4.5)(19\text{ft} \times 12)^3}{(384)(1.6 \times 10^6)}(240) = 209 \text{ in}^4$$

Therefore,

the required moment of inertia = 144.2 in^4 < I_{actual} = 290.8 in^4 OK

Alternatively, since the dead load equals the live load, a deflection limit of L/240 would control here. Therefore, using the applicable equation from Table 4.5:

$$I_{\text{req'd}} = \frac{wL^3}{2.22E'} = \frac{(108)(19)^3}{2.22(1600)} = 209 \text{ in}^4$$

Which agrees with the previous calculation. Note that previously a 2 × 12 was considered based on bending strength, but based on deflection since 209 in^4 > 199 in^4, a 2 × 12 would not work for deflection.

Check of Bearing Stress (Compression Stress Perpendicular to the Grain)
The maximum reaction at the support $R = 990$ lb and the thickness of a 2 × 14 sawn-lumber joist, $b = 1.5$ in. The allowable bearing stress or compression stress parallel to the grain is

$$F'_{c\perp r} = F_{c\perp}\, K_{F,c\perp}\phi_c\lambda\, C_M C_t C_b C_i = (625)(1.67)(0.9)(0.8)(1.0)(1.0)(1.0)(1.0) = 751 \text{ psi}$$

From Appendix N:

$K_{F,c\perp} = 1.67$
$\phi_v = 0.9$
$\lambda = 0.8$

From Equation 4.9, the minimum required bearing length l_b is

$$l_{b,\,\text{req'd}} = \frac{R}{bF'_{c\perp}} = \frac{1419\text{ lb}}{(1.5\text{ in.})(751\text{ psi})} = 1.26\text{ in.}$$

The floor joists will be connected to the floor girder using face-mounted joist hangers with the top of the joists at the same level as the top of the girders. From a proprietary joist hanger load table (Figure 4.9) connector catalog, select a face-mounted hanger JH-214 with the following properties:

Note that most proprietary joist hanger catalogs utilize service loads, so from the ASD solution:

$R = 1014$ lb < 1280 lb, OK

Load duration factor, C_D for dead plus floor live load = 1.0 = 100%.
Available bearing length = 2 in. > $l_{b,\,\text{req'd}} = 1.26$ in. OK
Use a 2 × 14 DF-L Select Structural for joist B1.

(b) *Design of floor girder G1.* Girder span = 25 ft. Since the joist selected was 2 × 14, we would expect the girder to be larger than a 2 × 14. Thus, the girder is probably going to be a timber (i.e., beam and stringer). Assume the girder self-weight = 90 lb/ft (this will be checked later). For hotels, the live load $L = 40$ psf.
Dead load:
From the design of the floor joist B1, total dead load, $D = 40$ psf.

$$\text{Tributary width of girder G1} = \frac{18.5\text{ ft}}{2} + \frac{18.0\text{ ft}}{2} = 18.25\text{ ft}$$

$$\text{Tributary area of girder G1} = (18.25\text{ ft})(25\text{ ft}) = 456\text{ ft}^2$$

From ASCE 7, Table 4-2, A_T, the tributary area = 456 ft², and $K_{LL} = 2$(interior girder).

- $K_{LL}A_T = (2)(456\text{ ft}^2) = 912\text{ ft}^2 > 400\text{ ft}^2$
- Floor live load $L = 40$ psf < 100 psf
- Floor occupancy is not assembly occupancy

Since all the conditions above are satisfied, live-load reduction is permitted.
Live-load reduction. The reduced or design floor live load for the girder is calculated using Equation 2.9 as follows:

$$L = L_0\left(0.25 + \frac{15}{\sqrt{K_{LL}A_T}}\right) = (40\text{ psf})\left[0.25 + \frac{15}{\sqrt{(2)(456)}}\right] = 30\text{ psf}$$

≥ (0.50)(40 psf) = 20 psf for members supporting one floor (e.g., slabs, beams, and girders)

≥ (0.40)(40 psf) = 16 psf for members supporting two or more floors (e.g., columns)

The total uniform load on the girder that will be used to design for bending, shear, and bearing is obtained below using the dead load, the reduced live load calculated above, and the assumed self-weight of the girder, which will be checked later.

w_{TL} = (D + L) (tributary width) + girder self-weight

= (40 psf + 30 psf)(18.25 ft) + 90 lb/ft = **1368 lb/ft**

Using the free-body diagram of girder G1 (Figure 4.10), the load effects are calculated as follows:

uniform dead load w_{DL} = (40 psf)(18.25 ft) + 90 lb/ft = 820 lb/ft = 68.33 lb/in.
uniform live load w_{LL} = (30 psf)(18.25 ft) = 547.5 lb/ft = 45.63 lb/in.
w_U = 1.2D + 1.6L = (1.2)(820) + (1.6)(548) = 1861 plf
maximum reaction and shear:

$$V_U = \frac{w_U L}{2} = \frac{(1861\text{ lb/ft})(25\text{ ft})}{2} = 23{,}260\text{ lb}$$

$$\text{maximum moment } M_{max} = \frac{w_{TL}L^2}{8} = \frac{(1861\text{ lb/ft})(25\text{ ft})^2}{8}$$

$$= 145{,}400\text{ ft-lb} = 1{,}744{,}500\text{ in.-lb}$$

We now proceed with the design following the steps described earlier in the chapter.

Check of Bending Stress

1. Summary of load effects (the self-weight of girder G1 was assumed, but this will be checked later in the design process):

   maximum shear V_{max} and maximum reaction R_{max} = 23260
   maximum moment M_{max} = 1,744,500 in.-lb
   uniform dead load w_{DL} = 68.33 lb/in.
   uniform live load w_{LL} = 45.63 lb/in.

2. For the stress grade, assume Select Structural, and assume that the size classification for the girder is beam and stringer (B&S), that is, timbers. Therefore, use NDS-S Table 4D.

3. From NDS-S Table 4D, we find that F_b = 1600 psi. Assume initially that $F_b' = F_b$ = 1600 psi. From Equation 4.1 the required approximate section modulus of the member is given as

$$S_{xx,\text{req'd}} \geq \frac{M_u}{F_b K_{F,b} \phi_b \lambda} = \frac{1{,}744{,}500}{(1600)(2.54)(0.85)(0.8)} = 631\text{ in}^3$$

   From Appendix N:
   $K_{F,b} = 2.54$
   $\phi_b = 0.85$
   $\lambda = 0.8$

4. From NDS-S Table 1B, determine the trial size with the least area that satisfies the section modulus requirement of step 3. Try a 10 × 22 DF-L Select Structural.
   b = 9.5 in. and d = 21.5 in.
   Size is B&S, as assumed in step 1 OK

S_{xx} provided = 731.9 in^3 > 631 in^3 OK
Area A provided = 204 in^2
I_{xx} provided = 7868 in^4

5. The NDS-S tabulated stresses are
   $F_b = 1600$ psi
   $F_v = 170$ psi
   $F_{c\perp} = 625$ psi
   $E = 1.6 \times 10^6$ psi

   The adjustment or C factors are listed in Table 4.6. Using Equation 4.2, the allowable bending stress is

   $$F'_{b,r} = F_b K_{F,b} \phi_b \lambda \text{ (product of applicable adjustment or } C \text{ factors)}$$
   $$= F_b\, K_{F,b} \phi_b \lambda\, C_M C_t C_L C_F C_r C_{fu} C_i$$
   $$= (1600)(2.54)(0.85)(0.8)(1.0)(1.0)(1.0)(1.0)(0.93)(1.0)(1.0) = 2570 \text{ psi}$$

6. The actual applied bending stress is

   $$f_b = \frac{M_{max}}{S_{xx}} = \frac{1{,}744{,}500 \text{ lb-in.}}{731.9 \text{ in}^3} = 2383 \text{ psi}$$
   $$< F'_b = 2570 \text{ psi OK}$$

Therefore, the beam is adequate in bending.

Check of Shear Stress

V_{max} = 23,260 lb. The beam cross-sectional area A = 247.3 in^2. The applied shear stress in the wood beam at the centerline of the girder support is

$$f_v = \frac{1.5\,V}{A} = \frac{(1.5)(23{,}260 \text{ lb})}{204 \text{ in}^2} = 171 \text{ psi}$$

The allowable shear stress is

$$F'_{v,r} = F_v K_{F,v}\ \phi_v \lambda\ C_M C_t C_i = (170)(2.88)(0.75)(0.8)(1)(1)(1) = 293 \text{ psi}$$

From Appendix N:
$K_{F,v} = 2.88$
$\phi_v = 0.75$
$\lambda = 0.8$

Thus, $f_v < F'_v$; therefore, the beam is adequate in shear (See Figure 4.10b).

Note: Since the shear stress f_v at the centerline of the girder support is adequate (i.e., < F'_v), there is no need to calculate the shear stress f'_v at a distance d from the face of the support since it is obvious that f'_v would be less than f_v and thus will be adequate in that case.

Check of Deflection (see Table 4.7)

The allowable pure bending modulus of elasticity is

$$E' = E C_M C_t C_i = (1.6 \times 10^6)(1.0)(1.0)(1.0) = 1.6 \times 10^6 \text{ psi}$$

The moment of inertia about the strong axis I_{xx} = 7868 in.^{4}, the uniform dead load is w_{DL} = 68.33 lb/in., and the uniform live load is w_{LL} = 45.63 lb/in.
The dead-load deflection is

$$\Delta_{DL} = \frac{5wL^4}{384E'I} = \frac{(5)(68.33 \text{ lb/in.})(25 \text{ ft} \times 12)^4}{(384)(1.6 \times 10^6 \text{ psi})(7868 \text{ in}^4)} = 0.57 \text{ in.}$$

The live-load deflection is

$$\Delta_{LL} = \frac{5w_{LL}L^4}{384E'I} = \frac{(5)(45.33\text{ lb/in.})(25\text{ ft}\times 12)^4}{(384)(1.6\times 10^6\text{ psi})(7868\text{ in}^4)} = 0.38\text{ in.} < \frac{L}{360} = 0.83\text{ in.} \quad \text{OK}$$

As discussed previously, a k-factor of 1.0 is used for dead load deflections. The total load deflection is

$$\Delta_{TL} = k\Delta_{DL} + \Delta_{LL}$$

$$= (1.0)(0.57\text{ in.}) + 0.38\text{ in.} = 0.95\text{ in.} < \frac{L}{240} = 1.25\text{ in.} \qquad \text{OK}$$

Alternatively, the required moment of inertia can be calculated using Equations 4.5 and 4.6 as follows:

$$I_{\text{req'd}} = \frac{5w_{LL}L^3}{384E'}(360)\text{in}^4 = \frac{(5)(45.63)(25\text{ ft}\times 12)^3}{(384)(1.6\times 10^6)}(360) = 3609\text{ in}^4$$

Similarly, the required moment of inertia based on total load is obtained for a *uniformly loaded beam or girder* as

$$I_{\text{req'd}} = \frac{5w_{k(DL)+LL}L^3}{384E'}(240)\text{in}^4 = \frac{(5)(68.33+45.63)(25\text{ ft}\times 12)^3}{(384)(1.6\times 10^6)}(240) = 6010\text{ in}^4$$

Therefore,

the required moment of inertia = 6010 in^4 < I_{actual} = 7868 in^4 OK

Alternatively, since the dead load is greater than the live load, a deflection limit of L/240 would control here. Therefore, using the applicable equation from Table 4.5:

$$I_{\text{req'd}} = \frac{wL^3}{2.22E'} = \frac{(820+547.5)(25)^3}{2.22(1600)} = 6010\text{ in}^4$$

Which agrees with the previous calculation.

Check of Bearing Stress (Compression Stress Perpendicular to the Grain)

The maximum reaction at the support R = 23260 lb and the thickness of a 10 × 22 sawn-lumber joist b = 9.5 in. The allowable bearing stress or compression stress parallel to the grain is

$$F'_{c\perp r} = F_{c\perp}\, K_{F,c\perp}\phi\lambda\, C_M C_t C_b C_i = (625)(1.67)(0.9)(0.8)(1.0)(1.0)(1.0)(1.0) = 751\text{ psi}$$

From Appendix N:

$K_{F,c\perp} = 1.67$
$\phi_v = 0.9$
$\lambda = 0.8$

From Equation 4.8, the minimum required bearing length l_b is

$$l_{b,\text{ req'd}} \geq \frac{R_U}{bF'_{c\perp r}} = \frac{23260\text{ lb}}{(9.5\text{ in.})(751\text{ psi})} = 3.26\text{ in.} \quad \text{say, 3.5 in.}$$

The girders will be connected to the column using a U-shaped column cap detail as shown below, and the minimum length of the column cap required is

$$(3.5\text{ in.})(2)+\frac{1}{2}\text{-in. clearance between ends of girders} = 7.5\text{ in}$$

From a proprietary wood connector catalog, the maximum column cap width available for a U-shaped cap is 9.5 in. Since this is not less than the thickness (smaller dimension) of the girder, the proprietary U-shaped cap can be used. Therefore, use a U-shaped column cap similar to that shown in Figure 4.6e.
Note: The maximum column cap width available in most catalogs is 9.5 in., so where the width of a girder or beam is greater than this value, a special column cap having a minimum width equal to the thickness (smaller dimension) of the girder or beam would have to be designed.

Check of Assumed Girder Self-Weight
As discussed in Chapter 2, the density of wood is assumed in this book to be 32 lb/ft^3, and this is used to calculate the actual weight of the girder selected.
Assumed girder self-weight = 90 lb/ft
Actual weight of the 10 × 22 girder selected (from NDS-S Table 1B):

$$\frac{247.3\text{ in}^2(31.2\text{ lb/ft}^3)}{(12\text{ in.})(12\text{ in.})} = \left(\frac{54\text{ lb/ft which is less than the assumed value of 90 lb/ft} \quad \text{OK}}{(12\text{ in.})(12\text{ in})}\right)$$

Example 4.2: Analysis of a wood beam (ASD/LRFD)

1. Determine the adequacy of a 4 × 14 Southern Pine No. 1 sawn-lumber beam to support a concentrated moving load of 1000 lb that can be located anywhere on the beam. The beam span is 15 ft, normal temperature and dry service conditions apply, and the beam is laterally braced at the supports only. Assume a wood density of 36 lb/ft^3 and a load duration factor C_D of 1.0. The available bearing length l_b is 3 in.
2. Determine the maximum concentrated live load that can be supported by this beam. Assume the creep factor, $k = 1.0$.

Solution:
Since the concentrated load is a moving load, the most critical load locations that will result in maximum moment, shear, and reaction in the beam have to be determined. For the maximum moment in this simply supported beam, the critical location of the crane load will be at the midspan of the beam, and for maximum shear the critical location of the concentrated moving load is within a distance d from the face of the support. For maximum reaction, the most critical location for the moving load is when the load is as close to the centerline of the beam support as possible.

Check of Bending Stress (ASD Method)

1. Calculate all the loads and load effects, including maximum moment, shear, and reactions. Also calculate the dead and live loads that will be used for the deflection calculations later. Include the self-weight of the member.
   a. Calculate the self-weight of the beam:

$$w_{D,\text{ beam}} = (36\text{ lb/ft}^3)(46.38\text{ in.}^2/144) = 12\text{ lb/ft}$$

   b. Calculate the maximum moment and shear. Since the concentrated load is a moving load, the maximum moment in the beam will occur when the moving

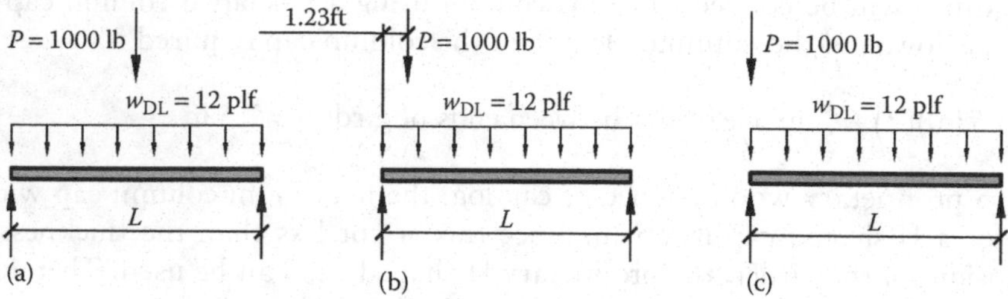

Figure 4.12 Free-body diagram of a beam for different loading conditions: (a) Maximum moment, (b) maximum shear, and (c) maximum reaction.

load is at midspan, whereas the maximum shear and reaction will occur when the moving load is at or near the support (Figure 4.12). The maximum moment is calculated as

$$M_{max} = \frac{w_{D,\,beam}L^2}{8} + \frac{PL}{4} = \frac{(12\text{ lb/ft})(15\text{ ft})^2}{8} + \frac{(1000\text{ lb})(15\text{ ft})}{4}$$

$$= 4088\text{ ft-lb} = 49{,}050\text{ in.-lb}$$

The maximum shear at the centerline of the beam support when the concentrated moving load is assumed to be at or very near the face of the beam support (i.e., within a distance d from the face of the support) is calculated as

$$V_{max} = \frac{w_{D,\,beam}L}{2} + P = \frac{(12\text{ lb/ft})(15\text{ ft})}{2} + (1000\text{ lb})\left(\frac{15\text{ ft} - 1.23\text{ ft}}{15\text{ ft}}\right) = 1008\text{ lb}$$

The maximum reaction will occur when the concentrated load is at or very near the centerline of the beam support and is calculated as

$$R_{max} = \frac{w_{D,beam}L}{2} + P = \frac{(12\text{ lb/ft})(15\text{ ft})}{2} + 1000\text{ lb} = 1090\text{ lb}$$

2. From the known stress grade (i.e., Select Structural, No. 1 and Better, No. 1, etc.) and size classification (i.e., dimension lumber, timbers, glulam, etc.), determine the applicable NDS-S Table to be used in the design. For 4 × 14 Southern Pine No. 1 dimension lumber, use NDS-S Table 4B.
3. Determine the cross-sectional area A_{xx}, the section modulus S_{xx}, and the moment of inertia I_{xx} from NDS-S Table 1B (for sawn lumber), NDS-S Table 1C (for western species glulam), or NDS-S Table 1D (for Southern Pine glulam). From NDS-S Table 1B, we obtain the section properties for 4 × 14 sawn lumber:

   $A_{xx} = 46.38$ in.^{2} ($b = 3.5$ in. and $d = 13.25$ in.)
   $I_{xx} = 678.5$ in.^{4}
   $S_{xx} = 102.4$ in.^{4}

   From NDS-S Table 4B, we obtain the tabulated stress values:

   $F_b = 1000$ psi
   $F_v = 175$ psi
   $E = 1.6 \times 10^6$ psi
   $E_{min} = 0.58 \times 10^6$ psi
   $F_{c\perp} = 565$ psi

4. Determine the applied bending stress f_b from Equation 4.10:

$$f_b = \frac{M_{max}}{S_{xx}} = \frac{49,050}{102.4} = 479 \text{ psi}$$

5. Determine all the applicable stress adjustment or C factors, obtain the tabulated NDS-S bending stress F_b, and calculate the allowable bending stress using Equation 4.2. The stress adjustment or C factors from NDS-S Table 4B are given in Table 4.8. The allowable bending stress of the beam if the beam stability coefficient C_L is 1.0 is given as

$$F_b^* = F_b C_D C_M C_t C_F C_r C_{fu} C_i = (1000)(1.0)(1.0)(1.0)(0.9)(1.0)(1.0)(1.0) = \mathbf{900 \text{ psi}}$$

The allowable pure bending modulus of elasticity E' and the bending stability modulus of elasticity E'_{min} are calculated as

$$E' = E\, C_M C_t C_i = (1.6 \times 10^6)(1.0)(1.0)(1.0) = 1.6 \times 10^6 \text{ psi}$$

$$E'_{min} = E_{min}\, C_M C_t C_i C_T = (0.66 \times 10^6)(1.0)(1.0)(1.0)(1.0) = 0.66 \times 10^6 \text{ psi}$$

Calculation of the Beam Stability Factor C_L

From Equation 3.3 the beam stability factor is calculated as follows. The unsupported length of the compression edge of the beam or distance between points of lateral support preventing rotation and/or lateral displacement of the compression edge of the beam is

$$l_u = 15 \text{ ft} = 180 \text{ in.}$$

$$\frac{l_u}{d} = \frac{180 \text{ in.}}{13.25 \text{ in.}} = 13.6$$

Table 4.8 Beam adjustment factors

Adjustment factor	*Symbol*	*Value*	*Rationale for the value chosen*
Beam stability factor	C_L	0.968	See calculations below
Size factor	$C_F(F_b)$	0.90	See adjustment factors section of NDS-S Table 4B
Moisture or wet service factor	C_M	1.0	Equilibrium moisture content is ≤19%
Load duration factor	C_D	1.0	The largest C_D value in the load combination of dead load plus floor live load (i.e., $D + L$)
Temperature factor	C_t	1.0	Insulated building, therefore, normal temperature conditions apply
Repetitive member factor	C_r	1.0	Does not satisfy all three requirements needed for a member to be classified as repetitive
Incision factor	C_r	1.0	1.0 for wood that is not incised, even if pressure treated
Buckling stiffness factor	C_T	1.0	1.0 except for a 2 × 4 truss top chords with plywood sheathing attached
Bearing area factor	C_b	1.0	$C_b = \frac{l_b + 0.375}{l_b}$ for $l_b \leq 6$ in. = 1.0 for $l_b > 6$ in. = 1.0 for bearings at the ends of a member
Flat use factor	C_{fu}	1.0	
Incising factor	C_i	1.0	

Using the l_u/d value, the effective length of the beam of the beam is obtained from Table 3.9 (or NDS code Table 3.3.3) as

$$l_e = 1.63l_u + 3d = (1.63)(180\text{ in.}) + (3)(13.25\text{ in.}) = 333\text{ in.}$$

$$R_B = \sqrt{\frac{l_e d}{b^2}} = \sqrt{\frac{(333)(13.25\text{ in.})}{(3.5\text{ in.})^2}} = 19 < 50 \qquad \text{OK}$$

$$F_{bE} = \frac{1.20E'_{\min}}{R_B^2} = \frac{(1.20)(0.66\times10^6)}{(19)^2} = 2194\text{ psi}$$

$$\frac{F_{bE}}{F_b^*} = \frac{2194\text{ psi}}{900\text{ psi}} = 2.44$$

From Equation 3.3 the beam stability factor is calculated as

$$C_L = \frac{1+F_{bE}/F_b^*}{1.9} - \sqrt{\left(\frac{1+F_{bE}/F_b^*}{1.9}\right)^2 - \frac{F_{bE}/F_b^*}{0.95}}$$

$$= \frac{1+2.44}{1.9} - \sqrt{\left(\frac{1+2.44}{1.9}\right)^2 - \frac{2.44}{0.95}} = 0.968$$

The allowable bending stress in the beam is calculated as

$$F_b' = F_b C_D C_M C_t C_L C_F C_r = F_b^* C_L = (900)(0.968) = \mathbf{871\text{ psi}}$$

6. Compare the applied stress f_b from step 4 to the allowable bending stress F_b' calculated in step 5. The applied bending stress (see step 4) $f_b = 479$ psi $< F_b' = 871$ psi. Therefore, the beam is adequate in bending.
7. To determine the maximum allowable load or moment, the beam or girder can support based on bending stress alone, equate the allowable bending stress F_b' to $M_{\max}/S_{xx}$, and solve for the allowable maximum moment $M_{\max}$ and thus the allowable maximum load. The applied maximum moment is given as

$$M_{\max} = \frac{w_{D,\text{ beam}}L^2}{8} + \frac{PL}{4}$$

The allowable maximum moment is given as

$$M_{\max} = F_b' S_{xx} \quad \text{in.-lb}$$

$$= \frac{F_b' S_{xx}}{12} \quad \text{ft-lb}$$

Equating the $M_{\max}$ equations and solving for the maximum allowable concentrated load yields

$$\frac{w_{D,\text{ beam}}L^2}{8} + \frac{PL}{4} = \frac{F_b' S_{xx}}{12} \quad \text{ft-lb}$$

Therefore,

$$P_{LL} = \left(\frac{4}{L}\right)\left(\frac{F_b' S_{xx}}{12} - \frac{w_{D,\text{ beam}}L^2}{8}\right) = \left(\frac{4}{15\text{ ft}}\right)\left[\frac{(871\text{ psi})(102.4\text{ in.}^3)}{12} - \frac{(12\text{ lb/ft})(15\text{ ft})^2}{8}\right]\text{lb}$$

$$= \mathbf{1892\text{ lb}}$$

Since the concentrated load is a live load, the maximum allowable moving load P_{LL} (based on bending stress alone) is 1892 lb. This assumes that the only dead load on the beam is the uniform self-weight of the beam, as specified in the problem statement.

Check of Shear Stress

8. Determine the applied shear stress f_v or f'_v, if applicable. In this problem, since the concentrated load could be located within a distance d from the face of the beam, only the shear stress f_v calculated using the shear at the centerline of the beam support V is applicable. Therefore, from Equation 4.12 we obtain the applied shear stress as

$$f_v = \frac{1.5V}{A} = \frac{(1.5)(1090)}{46.38} = 52.9 \text{ psi}$$

9. Determine all the applicable stress adjustment or C factors, obtain the tabulated NDS-S shear stress F_v, and calculate the allowable shear stress using Table 3.1. The allowable shear stress from Equation 4.14 is

$$F'_v = F_v C_D C_M C_t C_i = (175 \text{ psi})(1.0)(1.0)(1.0)(1.0) = 175 \text{ psi}$$

10. Compare the applied shear stress f_v or f'_v from step 8 to the allowable shear stress F'_v calculated in step 9. Since $f_v = 52.9$ psi $< F'_v = 175$ psi, the beam is adequate in shear.
11. To determine the maximum allowable load or shear that can be supported by the beam or girder based on shear stress alone, equate the allowable shear stress F'_v to $1.5V/A$ or $1.5V'/A$, and solve for the allowable shear V_{max} and thus the allowable maximum load. That is,

$$\frac{1.5V_{max}}{A_{xx}} = F'_v$$

The allowable maximum shear

$$V_{max} = \frac{F'_v A_{xx}}{1.5} \text{ lb}$$

Knowing the allowable maximum shear at the centerline of the support, we can determine the allowable maximum load based on shear stress alone, using the relationship between maximum shear and the applied load, while recalling that the concentrated live load is a moving load. For this case, the maximum shear due to applied loads was obtained in step 2 as

$$V_{max} = \frac{w_{D,\text{beam}} L}{2} + P\left(\frac{15 \text{ ft} - 1.23 \text{ ft}}{15 \text{ ft}}\right) = \frac{F'_v A_{xx}}{1.5}$$

Thus, the maximum allowable total concentrated load (based on shear stress alone) is

$$0.918P_{LL} = \frac{F'_v A_{xx}}{1.5} - \frac{w_{D,\text{beam}} L}{2} \text{ lb}$$

$$= \frac{(175 \text{ psi})(46.38 \text{ in}^2)}{1.5} - \frac{(12 \text{ lb/ft})(15 \text{ ft})}{2}$$

$$P_{LL} = 5796 \text{ lb}$$

Since this is greater than the value obtained from step 7, this value will not govern because the beam will fail first in bending at the lower load $P_{LL} = 1892$ lb before the concentrated load reaches the value of 5796 lb. As mentioned previously, shear rarely governs the design of wood joists, beams, and girders.

Check of Deflection

12. Using the appropriate deflection formula for a beam with a uniformly distributed dead load (due to self-weight) plus a concentrated crane (live) load, determine and compare the live-load and total-load deflections to the appropriate deflection limits. Since the maximum deflection will take place at midspan when the concentrated live load is at the midspan, we obtain the deflections as

$$\text{live-load deflection } \Delta_{LL} = \frac{P_{LL}L^3}{48E'I} = \frac{(1000 \text{ lb})(15 \text{ ft} \times 12 \text{ in./ft})^3}{(48)(1.6 \times 10^6 \text{ psi})(678.5 \text{ in.}^4)}$$

$$= 0.110 \text{ in.}$$

$$< \frac{L}{360} = 0.50 \text{ in.} \qquad \text{OK}$$

Total-load deflection

$$\Delta_{k(DL)+LL} = \frac{5w_{k(DL)}L^4}{384E'I} + \frac{P_{LL}L^3}{48E'I}$$

$$= \frac{(5)(12/12 \text{ lb/in.})(15 \text{ ft} \times 12 \text{ in./ft})^4}{(384)(1.6 \times 10^6 \text{ psi})(678.5 \text{ in.}^4)} + \frac{(1000 \text{ lb})(15 \text{ ft} \times 12 \text{ in./ft})^3}{(48)(1.6 \times 10^6 \text{ psi})(678.5 \text{ in.}^4)}$$

$$= 0.12 \text{ in.}$$

$$\le \frac{L}{240} = \frac{(15 \text{ ft})(12 \text{ in./ft})}{240} = 0.75 \text{ in.} \qquad \text{OK}$$

13. To determine the maximum allowable load that can be supported by the beam or girder based on deflections alone, equate the calculated deflection to the appropriate deflection limit and solve for the allowable maximum load. That is,

$$\Delta_{LL} = \frac{P_{LL}L^3}{48E'I} = \frac{L}{360}$$

Therefore, the allowable maximum live load

$$P_{LL} = \frac{48E'I}{L^3}\left(\frac{L}{360}\right) = \frac{(48)(1.6 \times 10^6)(678.5 \text{ in.}^4)}{(15 \text{ ft} \times 12 \text{ in./ft})^3}\left(\frac{15 \text{ ft} \times 12 \text{ in./ft}}{360}\right) = 4464 \text{ lb}$$

The maximum total-load deflection

$$\Delta_{k(DL)+LL} = \frac{5w_{k(DL)}L^4}{348E'I} + \frac{P_{LL}L^3}{48E'I} = \frac{L}{240}$$

Therefore, the allowable maximum live load

$$P_{LL} = \frac{48E'I}{L^3}\left(\frac{L}{240} - \frac{5w_{k(DL)}L^4}{384E'I}\right)$$

$$= \frac{(48)(1.6 \times 10^6)(678.5 \text{ in.}^4)}{(15 \text{ ft} \times 12 \text{ in./ft})^3}\left[\frac{(15 \text{ ft} \times 12 \text{ in./ft})}{240} - \frac{(5)(12/12 \text{ lb/in.})(15 \text{ ft} \times 12 \text{ in./ft})^4}{(384)(1.6 \times 10^6)(678.5 \text{ in.}^4)}\right]$$

$$= 6585 \text{ lb}$$

The smaller P_{LL} value of 4464 lb governs as far as deflections are concerned. However, this value is still less than the allowable live load calculated in step 7 based on bending stress, so the value of 1892 lb (calculated in step 7) remains the governing allowable maximum live load for this beam.

Check of Bearing Stress (Compression Stress Perpendicular to the Grain)

14. Determine the bearing stress or compression stress applied perpendicular to the grain and compare to the allowable compression stress perpendicular to the grain. From Equation 4.17 the bearing stress applied is

$$f_{c\perp} = \frac{R_{\max}}{A_{\text{bearing}}} = \frac{R_{\max}}{bl_b} = \frac{1090 \text{ lb}}{(3.5 \text{ in.})(3 \text{ in.})} = 104 \text{ psi}$$

where b = 3.5 in. and l_b (given in the problem statement) = 3 in. The allowable bearing stress or compression stress perpendicular to the grain is

$$F'_{c\perp} = F_{c\perp}C_M C_t C_b C_i = (565 \text{ psi})(1.0)(1.0)(1.0)(1.0) = 565 \text{ psi} > f_{c\perp} = 104 \text{ psi} \qquad \text{OK}$$

Similar to shear, the beam reaction rarely controls the design strength of a joist or member, and as such will not be checked. The actual allowable maximum loads for the beam or girder will be the smallest of the maximum loads calculated for bending stress (step 7), shear stress (step 11), and deflection (step 13). Therefore, the allowable maximum live load

P_{LL} (from step 7) = **1892 lb**

Thus, the bending stress criterion controls the allowable maximum live load in this beam.

A 4 × 14 Southern Pine No. 1 is adequate; the maximum live load $P_{LL}(\max)$ = 1892 lb.

LRFD Solution:

1. Determine the adequacy of a 4 × 14 Southern Pine No. 1 sawn-lumber beam to support a concentrated moving load of 1000 lb that can be located anywhere on the beam. The beam span is 15 ft, normal temperature and dry service conditions apply, and the beam is laterally braced at the supports only. Assume a wood density of 36 lb/ft^3 and a time effect factor λ of 0.8. The available bearing length is 3 in.
2. Determine the maximum concentrated live load that can be supported by this beam. Assume the creep factor, k = 1.0.

Solution: Since the concentrated load is a moving load, the most critical load locations that will result in maximum moment, shear, and reaction in the beam have to be determined. For the maximum moment in this simply supported beam, the critical location of the crane load will be at the midspan of the beam, and for maximum shear the critical location of the concentrated moving load is within a distance d from the face of the support. For maximum reaction, the most critical location for the moving load is when the load is as close to the centerline of the beam support as possible.

Check of Bending Stress

1. Calculate all the loads and load effects, including maximum moment, shear, and reactions. Also calculate the dead and live loads that will be used for the deflection calculations later. Include the self-weight of the member.
   a. Calculate the self-weight of the beam:

   $$w_{D,beam} = (36 \text{ lb/ft}^3)(46.38 \text{ in}^2/144) = 12 \text{ lb/ft}$$

   b. Calculate the maximum moment and shear. Since the concentrated load is a moving load, the maximum moment in the beam will occur when the moving load is at midspan, whereas the maximum shear and reaction will occur when the moving load is at or near the support (Figure 4.12b). The maximum moment is calculated as

Use the load combination 1.2D + 1.6L

$$M_U = \frac{1.2w_{D,\text{beam}}L^2}{8} + \frac{1.6PL}{4} = \frac{(1.2)(12\ \text{lb/ft})(15\ \text{ft})^2}{8} + \frac{(1.6)(1000\ \text{lb})(15\ \text{ft})}{4}$$

$$= 6405\ \text{ft-lb} = 76860\ \text{in.-lb}$$

The maximum shear at the centerline of the beam support when the concentrated moving load is assumed to be at or very near the face of the beam support (i.e., within a distance d from the face of the support) is calculated as

$$V_U = \frac{1.2w_{D,\text{beam}}L}{2} + 1.6P = \frac{(1.2)(12\ \text{lb/ft})(15\ \text{ft})}{2} + (1.6)(1000\ \text{lb})\left(\frac{15\ \text{ft}\ \ 1.23\ \text{ft}}{15\ \text{ft}}\right) = 1577\ \text{lb}$$

The maximum reaction will occur when the concentrated load is at or very near the centerline of the beam support and is calculated as

$$R_U = \frac{1.2w_{D,\text{beam}}L}{2} + 1.6P = \frac{(1.2)(12\ \text{lb/ft})(15\ \text{ft})}{2} + (1.6)(1000\ \text{lb}) = 1708\ \text{lb}$$

2. From the known stress grade (i.e., Select Structural, No. 1 and Better, No. 1, etc.) and size classification (i.e., dimension lumber, timbers, glulam, etc.), determine the applicable NDS-S table to be used in the design. For 4 × 14 Southern Pine No. 1 dimension lumber, use NDS-S Table 4B.
3. Determine the cross-sectional area A_{xx}, the section modulus S_{xx}, and the moment of inertia I_{xx} from NDS-S Table 1B (for sawn lumber), NDS-S Table 1C (for western species glulam), or NDS-S Table 1D (for Southern Pine glulam). From NDS-S Table 1B, we obtain the section properties for 4 × 14 sawn lumber:

$A_{xx} = 46.38\ \text{in}^2$ ($b = 3.5$ in. and $d = 13.25$ in.)

$I_{xx} = 678.5\ \text{in}^4$

$S_{xx} = 102.4\ \text{in}^4$

From NDS-S Table 4B, we obtain the tabulated stress values:

$F_b = 1000$ psi

$F_v = 175$ psi

$E = 1.6 \times 10^6$ psi

$E_{\min} = 0.58 \times 10^6$ psi

$F_{c\perp} = 565$ psi

4. Determine the factored bending stress f_{bu} from Equation 4.10:

$$f_{bu} = \frac{M_U}{S_{xx}} = \frac{76860}{102.4} = 751 \text{ psi}$$

5. Determine all the applicable stress adjustment or C factors, obtain the tabulated NDS-S bending stress F_b, and calculate the allowable bending stress using Equation 4.2. The stress adjustment or C factors from NDS-S Table 4B are given in Table 4.8. The available bending stress of the beam if the beam stability coefficient C_L is 1.0 is given as

$$F^*_{b,r} = K_{F,b}\phi_b\lambda\, F_b C_M C_t C_F C_r$$
$$= (2.54)(0.85)(0.8)(1000)(1.0)(1.0)(0.9)(1.0) = \mathbf{1554\ psi}$$

From Appendix N:

$$K_{F,b} = 2.54$$

$$\phi_b = 0.85$$

$$\lambda = 0.8$$

The allowable pure bending modulus of elasticity E' and the bending stability modulus of elasticity $E'_{\min}$ are calculated as

$$E' = EC_M C_t C_i = (1.6 \times 10^6)(1.0)(1.0)(1.0) = 1.6 \times 10^6 \text{ psi}$$

$$E'_{\min,r} = E_{\min}\, C_M C_t C_i C_T\, K_{F,E}\phi_{st}$$

$$= (0.66 \times 10^6)(1.0)(1.0)(1.0)(1.0)(1.76)(0.85) = 0.987 \times 10^6 \text{ psi}$$

$$K_{F,E} = 1.76$$

$$\phi_{st} = 0.85$$

Calculation of the Beam Stability Factor C_L

From Equation 3.2 the beam stability factor is calculated as follows. The unsupported length of the compression edge of the beam or distance between points of lateral support preventing rotation and/or lateral displacement of the compression edge of the beam is

$$l_u = 15 \text{ ft} = 180 \text{ in.}$$

$$\frac{l_u}{d} = \frac{180 \text{ in.}}{13.25 \text{ in.}} = 13.6$$

Using the l_u/d value, the effective length of the beam of the beam is obtained from Table 3.10 (or NDS code Table 3.3.3) as

$$l_e = 1.63l_u + 3d = (1.63)(180\text{ in.}) + (3)(13.25\text{ in.}) = 333\text{ in.}$$

$$R_B = \sqrt{\frac{l_e d}{b^2}} = \sqrt{\frac{(333)(13.25\text{ in.})}{(3.5\text{ in.})^2}} = 19 < 50 \quad \text{OK}$$

$$F_{bE,r} = \frac{1.20E'_{\min,r}}{R_B^2} = \frac{(1.20)(0.987\times 10^6)}{(19)^2} = 3288\text{ psi}$$

$$\frac{F_{bE,r}}{F_{b,r}^*} = \frac{3288\text{ psi}}{1554\text{ psi}} = 2.12$$

From Equation 3.2 the beam stability factor is calculated as

$$C_L = \frac{1+F_{bE}/F_{b,r}^*}{1.9} - \sqrt{\left(\frac{1+F_{bE}/F_{b,r}^*}{1.9}\right)^2 - \frac{F_{bE}/F_{b,r}^*}{0.95}}$$

$$= \frac{1+2.12}{1.9} - \sqrt{\left(\frac{1+2.12}{1.9}\right)^2 - \frac{2.12}{0.95}} = 0.968$$

The allowable bending stress in the beam is calculated as

$$F'_{b,r} = F_b K_{F,b} \phi_b \lambda\, C_M C_t C_L C_F C_r C_{fu} C_i$$

$$= (900)(2.54)(0.85)(0.8)(1.0)(1.0)(0.968)(0.9)(1.0)(1.0)(1.0) = \mathbf{1354\ psi}$$

From Appendix N:

$$K_{F,b} = 2.54$$

$$\phi_b = 0.85$$

$$\lambda = 0.8$$

6. Compare the factored stress f_{bu} from step 4 to the available bending stress $F'_{b,r}$ calculated in step 5. The factored bending stress (see step 4) $f_{bu} = 751\text{ psi} < F'_{b,r} = 1354\text{ psi}$. Therefore, the beam is adequate in bending.
7. To determine the maximum moment the beam can support based on bending stress alone, equate the available bending stress $F'_{b,r}$ to M_u/S_{xx}, and solve for the maximum moment and thus the maximum load. The factored maximum moment is given as

$$M_{U\max} = \frac{1.2w_{D,\text{beam}}L^2}{8} + \frac{1.6PL}{4}$$

The allowable maximum moment is given as

$$M_{U\max} = F'_{b,r}S_{xx}\text{ in.-lb}$$

$$= \frac{F'_{b,r}S_{xx}}{12}\text{ ft-lb}$$

Equating the M_{Umax} equations and solving for the maximum concentrated load yields

$$\frac{1.2w_{D,\text{beam}}L^2}{8}+\frac{1.6PL}{4}=\frac{F'_{b,r}S_{xx}}{12}\ \text{ft-lb}$$

Therefore,

$$P_{LL}=\left(\frac{4}{1.6L}\right)\left(\frac{F'_{b,r}S_{xx}}{12}-\frac{1.2w_{D,\text{beam}}L^2}{8}\right)=\left(\frac{4}{(1.6)(15\,\text{ft})}\right)$$

$$\left[\frac{(1354\text{ psi})(102.4\text{ in}^3)}{12}-\frac{(1.2)(12\text{ lb/ft})(15\text{ ft})^2}{8}\right]\text{lb}$$

$$=\mathbf{1858\,lb}$$

Since the concentrated load is a live load, the maximum allowable moving load P_{LL} (based on bending stress alone) is 1858 lb. This assumes that the only dead load on the beam is the uniform self-weight of the beam, as specified in the problem statement.

Check of Shear Stress

8. Determine the factored shear stress f_{vu} or f'_{vu}, if applicable. In this problem, since the concentrated load could be located within a distance d from the face of the beam, only the shear stress f_{vu} calculated using the shear at the centerline of the beam support V_u is applicable. Therefore, from Equation 4.12 we obtain the factored shear stress as

$$f_{vu}=\frac{1.5V_U}{A}=\frac{(1.5)(1708)}{46.38}=55.2\text{ psi}$$

9. Determine all the applicable stress adjustment or C factors, obtain the tabulated NDS-S shear stress F_v, and calculate the available shear stress using Table 3.3. The allowable shear stress from Equation 4.14 is

$$F'_{v,r}=F_vK_f\phi_v\lambda\,C_MC_tC_i=(175)(2.88)(0.75)(0.8)(1)(1)(1)=302\text{ psi}$$

From Appendix N:

$$K_f=2.88$$

$$\phi_v=0.75$$

$$\lambda=0.8$$

10. Compare the applied shear stress f_{vu} or f'_{vu} from step 8 to the available shear stress $F'_{v,r}$ calculated in step 9. Since $f_{vu}=55.2$ psi $< F'_{v,r}=302$ psi, the beam is adequate in shear.

11. To determine the maximum allowable load or shear that can be supported by the beam or girder based on shear stress alone, equate the allowable shear stress $F'_{v,r}$ to 1.5 $V_{u/A}$ or $1.5V'_{u/A}$, and solve for the maximum shear V_{Umax} and thus the allowable maximum load. That is,

$$\frac{1.5V_{U\max}}{A_{xx}} = F'_{v,r}$$

The maximum available shear

$$V_{U\max} = \frac{F'_{v,r}A_{xx}}{1.5}\ \text{lb}$$

Knowing the allowable maximum shear at the centerline of the support, we can determine the allowable maximum load based on shear stress alone, using the relationship between maximum shear and the applied load, while recalling that the concentrated live load is a moving load. For this case, the maximum shear due to applied loads was obtained in step 2 as

$$V_{U\max} = \frac{1.2w_{\text{D,beam}}L}{2} + 1.6P_{LL}\left(\frac{15\ \text{ft} - 1.23\ \text{ft}}{15\ \text{ft}}\right) = \frac{F'_{v,r}A_{xx}}{1.5}$$

Thus, the maximum allowable total concentrated load (based on shear stress alone) is

$$(0.918)(1.6)P_{LL} = \frac{F'_{v,r}A_{xx}}{1.5} - \frac{1.2w_{\text{D,beam}}L}{2}\ \text{lb}$$

$$= \frac{(302\ \text{psi})(46.38\ \text{in}^2)}{1.5} - \frac{(1.2)(12\ \text{lb/ft})(15\ \text{ft})}{2}$$

$$P_{\text{LL}} = 6283\ \text{lb}$$

Since this is greater than the value obtained from step 7, this value will not govern because the beam will fail first in bending at the lower load P_{LL} = 1858 lb before the concentrated load reaches the value of 6283 lb. As mentioned previously, shear rarely governs the design of wood joists, beams, and girders.

Check of Deflection

12. Using the appropriate deflection formula for a beam with a uniformly distributed dead load (due to self-weight) plus a concentrated crane (live) load, determine and compare the live load and total load deflections to the appropriate deflection limits. Since the maximum deflection will take place at midspan when the concentrated live load is at the midspan, we obtain the deflections as

$$\text{live-load deflection } \Delta_{\text{LL}} = \frac{p_{\text{LL}}L^3}{48E'I} = \frac{(1000\ \text{lb})(15\ \text{ft} \times 12\ \text{in./ft})^3}{(48)(1.6 \times 10^6\ \text{psi})(678.5\ \text{in}^4)}$$

$$= 0.110\ \text{in.}$$

$$< \frac{L}{360} = 0.50\ \text{in.}\quad \text{OK}$$

Total load deflection

$$\Delta_{k(\mathrm{DL})+\mathrm{LL}} = \frac{5w_{k(\mathrm{DL})}L^4}{384E'I} + \frac{P_{\mathrm{LL}}L^3}{48E'I}$$

$$= \frac{(5)(12/12\ \mathrm{lb/\ in})(15\ \mathrm{ft}\times 12\ \mathrm{in./ft})^4}{(384)(1.6\times 10^6\ \mathrm{psi})(678.5\ \mathrm{in}^4)} + \frac{(1000\ \mathrm{lb})(15\ \mathrm{ft}\times 12\ \mathrm{in./ft})^3}{(48)(1.6\times 10^6\ \mathrm{psi})(678.5\ \mathrm{in}^4)}$$

$$= 0.12\ \mathrm{in.}$$

$$\leq \frac{L}{240} = \frac{(15\,\mathrm{ft})(12\,\mathrm{in./ft})}{240} = 0.75\ \mathrm{in.}\quad \mathrm{OK}$$

13. To determine the maximum allowable load that can be supported by the beam or girder based on deflections alone, equate the calculated deflection to the appropriate deflection limit and solve for the allowable maximum load. That is,

$$\Delta_{\mathrm{LL}} = \frac{P_{\mathrm{LL}}L^3}{48E'I} = \frac{L}{360}$$

Therefore, the allowable maximum live load

$$P_{\mathrm{LL}} = \frac{48E'I}{L^3}\left(\frac{L}{360}\right) = \frac{(48)(1.6\times 10^6)(678.5\ \mathrm{in}^4)}{(15\ \mathrm{ft}\times 12\ \mathrm{in./ft})^3}\left(\frac{15\ \mathrm{ft}\times 12\ \mathrm{in./ft}}{360}\right) = 4464\ \mathrm{lb}$$

The maximum total load deflection

$$\Delta_{k(\mathrm{DL})+\mathrm{LL}} = \frac{5w_{k(\mathrm{DL})}L^4}{348E'I} + \frac{P_{\mathrm{LL}}L^3}{48E'I} = \frac{L}{240}$$

Therefore, the allowable maximum live load

$$P_{\mathrm{LL}} = \frac{48E'I}{L^3}\left(\frac{L}{240} - \frac{5w_{k(\mathrm{DL})}L^4}{384E'I}\right)$$

$$= \frac{(48)(1.6\times 10^6)(678.5\ \mathrm{in.}^4)}{(15\ \mathrm{ft}\times 12\ \mathrm{in./ft})^3}\left[\frac{(15\ \mathrm{ft}\times 12\ \mathrm{in./ft})}{240} - \frac{(5)(12/12\ \mathrm{lb/in.})(15\ \mathrm{ft}\times 12\ \mathrm{in./ft})^4}{(384)(1.6\times 10^6)(678.5\ \mathrm{in.}^4)}\right]$$

$$= 6585\ \mathrm{lb}$$

The smaller P_{LL} value of 4464 lb governs as far as deflections are concerned. However, this value is still less than the allowable live load calculated in step 7 based on bending stress, so the value of 1858 lb (calculated in step 7) remains the governing allowable maximum live load for this beam.

Check of Bearing Stress (Compression Stress Perpendicular to the Grain)

14. Determine the bearing stress or compression stress applied perpendicular to the grain and compare to the allowable compression stress perpendicular to the grain. From Equation 4.17 the bearing stress applied is

$$f_{c\perp U} = \frac{R_U}{A_{\mathrm{bearing}}} = \frac{R_U}{bl_b} = \frac{1708}{(3.5\ \mathrm{in.})(3\ \mathrm{in.})} = 163\ \mathrm{psi}$$

where $b = 3.5$ in. and lb (given in the problem statement) = 3 in. The available bearing stress or compression stress perpendicular to the grain is

$$F'_{c\perp r} = F_{c\perp}\, K_{F,c\perp}\phi_c\lambda\; C_M C_t C_b C_i = (565)(1.67)(0.9)(0.8)(1.0)(1.0)(1.0)(1.0) = 679 \text{ psi}$$

$$> f_{c\perp U} = 163 \text{ psi}, \qquad \text{OK}$$

From Appendix N:

$$K_{F,c\perp} = 1.67$$

$$\phi_v = 0.9$$

$$\lambda = 0.8$$

Similar to shear, the beam reaction rarely controls the design strength of a joist or member, and as such will not be checked. The actual allowable maximum loads for the beam or girder will be the smallest of the maximum loads calculated for bending stress (step 7), shear stress (step 11), and deflection (step 13). Therefore, the allowable maximum live load

$$P_{\text{LL}} \text{ (from step 7)} = 1858 \text{ lb}$$

Thus, the bending stress criterion controls the allowable maximum live load in this beam.

A 4 × 14 Southern Pine No. 1 is adequate; the maximum live load $P_{\text{LL}}(\text{max}) = 1858$ lb.

Example 4.3: Design of a glulam girder (ASD/LRFD)

The panelized roof framing system shown in Figure 4.13 consists of 2× subpurlins supported on 4× purlins that are supported on simply supported glulam girders. The total roof dead load, including the weight of the subpurlins and purlins, is 20 psf and the snow load is 35 psf. The unfactored wind load on the roof is ±10 psf. Design the glulam girder assuming the wood density of 36 lb/ft^3, dry service, and normal temperature conditions.

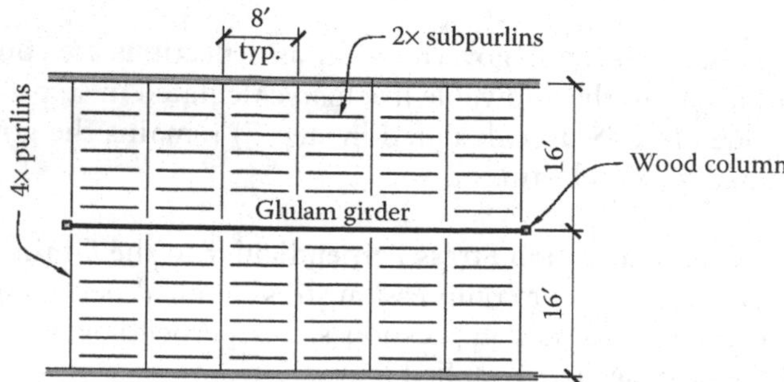

Figure 4.13 Framing plan for Example 4.3.

Solution:
The total roof dead load D is 20 psf (excludes the self-weight of girder). Assume that the girder self-weight is 100 lb/ft (this will be checked later). The snow load S is 35 psf and the wind load W is ±10 psf. All other loads, E, L, R, H, F, and $T = 0$.

$$\text{Tributary width of glulam girder} = \frac{16\text{ ft}}{2} + \frac{16\text{ ft}}{2} = 16$$

$$\text{Tributary area of glulam girder} = (16\text{ ft})(48\text{ ft}) = 768\text{ ft}^2 > 600\text{ ft}^2$$

From Section 2.4.1, $R_1 = 0.6$. Since we have a flat roof with minimum slope (1/4 in./ft) for drainage, the roof slope factor (see Section 2.4.1) $F = 0.25$; therefore, $R_2 = 1.0$. From Equation 2.6, we calculate the roof live load as

$$L_r = 20R_1R_2 = (20)(1.0)(0.6) = 12\text{ psf} < 20\text{ psf}$$

Therefore, from Section 2.1, the governing load combination for this roof girder is determined using the normalized load method. Since the loads E, L, R, H, F, and T are zero, the applicable load combinations are given in Table 4.9.

Summary: The governing load combination is $D + S$ with a load duration factor C_D of 1.15.

We assume that this girder is subjected to uniform loads because there are at least five equally spaced reaction points where the purlins frame into the girder with equal reactions, indicating that the load on the girder is actually close to being uniform. The total uniform load on the girder that will be used to design for bending, shear, and bearing is obtained below using the dead load, the snow load, and the assumed self-weight of the girder, which will be checked later.

$$w_{TL} = (D + S)\,(\text{tributary width}) + \text{girder self-weight}$$
$$= (20\text{ psf} + 35\text{ psf})(16\text{ ft}) + 100\text{ lb/ft} = \mathbf{980\text{ lb/ft}}$$

Table 4.9 Applicable and governing load combinations

Load combination	*Value of load combination (W)*	*C_D factor for load combination*	*Normalized load (W/C_D)*
D	20 psf	0.9	$\frac{20}{0.9} = 22.2$
$D + L_r$	20 + 12 = 32 psf	1.25	$\frac{32}{1.25} = 25.6$
$D + S$	20 + 35 = **55 psf**	1.15	$\frac{55}{1.15} = 47.8 \Leftarrow$ (governs)
$D + 0.75(0.6W) + 0.75L_r$	20 + (0.75)(10 + 12) = 36.5 psf	1.6	$\frac{36.5}{1.6} = 22.8$
$D + 0.75(0.6W) + 0.75S$	20 + (0.75)(10 + 35) = 53.8 psf	1.6	$\frac{53.8}{1.6} = 33.6$
$0.6D + (0.6W)$	(0.6)(20) + (−10)[a] = 2 psf	1.6	$\frac{2}{1.6} = 1.25$

Note: The value given in bold is the controlling value.

[a] In load combinations 7 and 8 given in Section 2.1, the wind load W and seismic load E are always opposed by the dead load D. Therefore, in these combinations, D takes on a positive number while W and E take on negative values only.

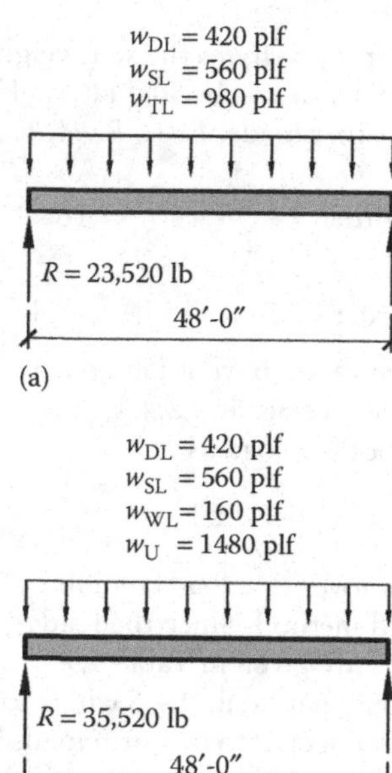

Figure 4.14 Free-body diagram of a glulam girder. (a) Unfactored load/reactions and (b) factored load/reactions.

Using the free-body diagram of the girder (Figure 4.14), the load effects are calculated as follows:

$$\text{maximum shear } V_{max} = \frac{w_{TL}L}{2} = \frac{(980 \text{ lb/ft})(48 \text{ ft})}{2} = 23,520 \text{ lb}$$

$$\text{maximum reaction } R_{max} = 23{,}520 \text{ lb}$$

$$\text{maximum moment } M_{max} = \frac{w_{TL}L^2}{8} = \frac{(980 \text{ lb/ft})(48 \text{ ft})^2}{8} = 282,240 \text{ ft-lb}$$

$$= 3{,}386{,}880 \text{ in.-lb}$$

The following loads will be used for calculating the girder deflections:

uniform dead load w_{DL}= (20 psf)(16 ft) + 100 lb/ft = 420 lb/ft = 35 lb/in.
uniform live load w_{LL} = (35 psf)(16 ft) = 560 lb/ft = 46.7 lb/in.

We now proceed with the design following the steps described earlier in the chapter.

Check of Bending Stress (Girders)

1. Summary of load effects (the self-weight of girder G1 was assumed, but this will be checked later in the design process):
   maximum shear V_{max} = 23,520 lb
   maximum reaction R_{max} = 23,520 lb
   maximum moment M_{max} = 3,386,880 in.-lb
   uniform dead load is w_{DL} = 35 lb/in.
   uniform live load is w_{LL} = 46.7 lb/in.
2. For glulam used primarily in bending, use NDS-S Table 5A–Expanded.
3. Using NDS-S Table 5A–Expanded, assume a 24F–V4 DF/DF glulam; therefore, F_{bx}^{+} = 2400 psi (tension lamination stressed in tension). Assume initially that

$F'^{+}_{bx} = F^{+}_{bx} = 2400$ psi. From Equation 4.1, the required approximate section modulus of the member is given as

$$S_{xx,\,\text{req'd}} \geq \frac{M_{\max}}{F^{+}_{bx}} = \frac{3,386,880 \text{ in.-lb}}{2400 \text{ psi}} = 1411.2 \text{ in.}^3$$

4. From NDS-S Table 1C (for western species glulam), the trial size with the least area that satisfies the section modulus requirement of step 3 is 6 3/4 in. × 36 in.:
   $b = 6.75$ in. and $d = 36$ in.
   S_{xx} provided = 1458 in.^{3} > 1411.2 in.^{3} OK
   area A provided = 243 in.^{2}
   I_{xx} provided = 26,240 in.^{4}
5. The NDS-S (Table 5A–Expanded) tabulated stresses are
   $F^{+}_{bx} = 2400$ psi (tension lamination stressed in tension)
   $F_{v,xx} = 265$ psi
   $F_{c\perp xx,\,\text{tension lam}} = 650$ psi
   $E_x = 1.8 \times 10^6$ psi
   $E_y = 1.6 \times 10^6$ psi
   $E_{y\,\min} = 0.85 \times 10^6$ psi ($E_{y,\min}$, and not $E_{x,\min}$, is used for lateral buckling of the girder about the weak axis)

The adjustment or C factors are given in Table 4.10.

From the adjustment factor applicability table for glulam (Table 3.2), we obtain the allowable bending stress of the glulam girder with C_V and C_L equal to 1.0 as

$$F'^{+}_{bx} = F^{+}_{bx} C_D C_M C_t C_{fu} C_c C_I = (2400)(1.15)(1.0)(1.0)(1.0)(1.0)(1.0) = \mathbf{2760 \text{ psi}}$$

The allowable pure bending modulus of elasticity E'_x and the bending stability modulus of elasticity $E'_{y,\min}$ are calculated as

$$E'_x = E_x C_M C_t = (1.8 \times 10^6)(1.0)(1.0) = 1.8 \times 10^6 \text{ psi}$$

$$E'_{y,\min} = E_{y,\min} C_M C_t = (0.85 \times 10^6)(1.0)(1.0) = 0.85 \times 10^6 \text{ psi}$$

Table 4.10 Adjustment factors for glulam

Adjustment factor	*Symbol*	*Value*	*Rationale for the value chosen*
Beam stability factor	C_L	0.967	See calculation below
Volume factor	C_V	0.802	See calculation below
Curvature factor	C_c	1.0	Glulam girder is straight
Flat use factor	C_{fu}	1.0	Glulam is bending about its strong *x–x* axis
Moisture or wet service factor	C_M	1.0	Equilibrium moisture content is <16%
Load duration factor	C_D	1.15	The largest C_D value in the load combination of dead load plus snow load (i.e., $D + S$)
Temperature factor	C_t	1.0	Insulated building; therefore, normal temperature conditions apply
Repetitive member factor	C_r	–	Does not apply to glulams; neglect or enter a value of 1.0
Bearing area factor	C_b	1.0	$C_b = \frac{l_b + 0.375}{l_b}$ for $l_b \leq 6$ in. = 1.0 for $l_b > 6$ in. = 1.0 for bearings at the ends of a member

Calculating the Beam Stability Factor C_L
From Equation 3.3 the beam stability factor is calculated as follows: the unsupported length of the compression edge of the beam or distance between points of lateral support preventing the rotation and/or lateral displacement of the compression edge of the beam is

l_u = 8 ft = 96 in. (i.e., the distance between lateral supports provided by the purlins)

$$\frac{l_u}{d} = \frac{96 \text{ in.}}{36 \text{ in.}} = 2.67$$

We previously assumed a uniformly loaded girder; therefore, using the l_u/d value, the effective length of the beam of the beam is obtained from Table 3.10 (or NDS code Table 3.3.3) as

$$l_e = 2.06 l_u = (2.06)(96 \text{ in.}) = 198 \text{ in.}$$

$$R_B = \sqrt{\frac{l_e d}{b^2}} = \sqrt{\frac{(198)(36 \text{ in.})}{(6.75 \text{ in.})^2}} = 12.5 \le 50 \qquad \text{OK}$$

$$F_{bE} = \frac{1.20 E'_{\min}}{R_B^2} = \frac{(1.20)(0.85 \times 10^6)}{(12.5)^2} = 6527 \text{ psi}$$

$$\frac{F_{bE}}{F_b^*} = \frac{6527 \text{ psi}}{2760 \text{ psi}} = 2.365$$

From Equation 3.3 the beam stability factor is calculated as

$$C_L = \frac{1 + F_{bE}/F_b^*}{1.9} - \sqrt{\left(\frac{1 + F_{bE}/F_b^*}{1.9}\right)^2 - \frac{F_{bE}/F_b^*}{0.95}}$$

$$= \frac{1 + 2.365}{1.9} - \sqrt{\left(\frac{1 + 2.365}{1.9}\right)^2 - \frac{2.365}{0.95}} = 0.967$$

Calculating the Volume Factor C_V

L = length of beam in feet between adjacent points of zero moment = 48 ft
d = depth of beam = 36 in.
b = width of beam, in. = 6.75 in.
x = 10 (for western species glulam)

From Equation 3.5,

$$C_V = \left(\frac{21}{L}\right)^{1/x}\left(\frac{12}{d}\right)^{1/x}\left(\frac{5.125}{b}\right)^{1/x} = \left(\frac{1291.5}{bdL}\right)^{1/x} = \left[\frac{1291.5}{(6.75 \text{ in.})(36 \text{ in.})(48 \text{ ft})}\right]^{1/10} = 0.802 \le 1.0$$

The smaller of C_V and C_L will govern and is used in the allowable bending stress calculation. Since $C_V = 0.802 < C_L = 0.97$, use $C_V = 0.802$. Using Table 3.2 (i.e., adjustment factor applicability Table for glulam), we obtain the allowable bending stress as

$$F'^{+}_{bx} = F^{+}_{bx} C_D C_M C_t C_{fu} C_c C_I (C_L \text{ or } C_V) = F^{*}_{b}(C_L \text{ or } C_V) = (2760)(0.802) = \mathbf{2213.5 \text{ psi}}$$

6. Using Equation 4.10, the actual bending stress applied is

$$f_b = \frac{M_{\max}}{S_{xx}} = \frac{3{,}386{,}880 \text{ lb-in.}}{1458 \text{ in.}^3} = 2322 \text{ psi}$$

$>$ allowable bending stress $F'_b = 2213.5$ psi **not good**

Although this section is only about 5% overstressed, it is prudent to select a larger section that will yield an applied stress that is slightly lower than the allowable stress. To do this we will need to recalculate C_V, C_L, the bending stress applied, and the allowable bending stress.

Try a 6¾ × 37½ in. 24F–V4 DF/DF glulam girder (A_{xx} = 253.1 in.[2]; S_{xx} = 1582 in.[3]; I_{xx} = 29,660 in.[4]), and the reader should verify that the recalculated values of C_V, C_L, and F'_b are as follows:

$C_V = 0.799$
$C_L = 0.964$
$F'^{+}_{bx} = 2205$ psi
$f_b = 2141 \text{ psi} < F'^{+}_{bx}$ OK

Therefore, a 6¾ × 37½ in. 24F–V4 DF/DF glulam girder is adequate in bending.

7. **Check of Shear Stress**
V_{max} = 23,520 lb. The beam cross-sectional area A = 253.1 in.[2]. The shear stress applied in the wood beam at the centerline of the girder support is

$$f_v = \frac{1.5V}{A} = \frac{(1.5)(23,520 \text{ lb})}{253.1 \text{ in}^2} = 139.4 \text{ psi}$$

Using the adjustment factor applicability table for glulam from Table 3.2, we obtain the allowable shear stress as

$$F'_v = F_v C_D C_M C_t = (265)(1.15)(1)(1) = 304 \text{ psi}$$

Thus, $f_v < F'_v$; therefore, the girder is adequate in shear. See Figure 4.15.

Note: Since the shear stress f_v at the centerline of the girder support is adequate (i.e., $< F'_v$), there is no need to further calculate the shear stress f_v' at a distance d from the face of the support since it is obvious *that* f_v' would be less than f_v and thus will also be adequate.

8. **Check of Deflection** (see Table 4.11)
The allowable pure bending modulus of elasticity for strong (i.e., x–x)-axis bending of the girder was calculated previously:

$$E'_x = E_x C_M C_t = (1.8 \times 10^6)(1.0)(1.0) = 1.8 \times 10^6 \text{ psi}$$

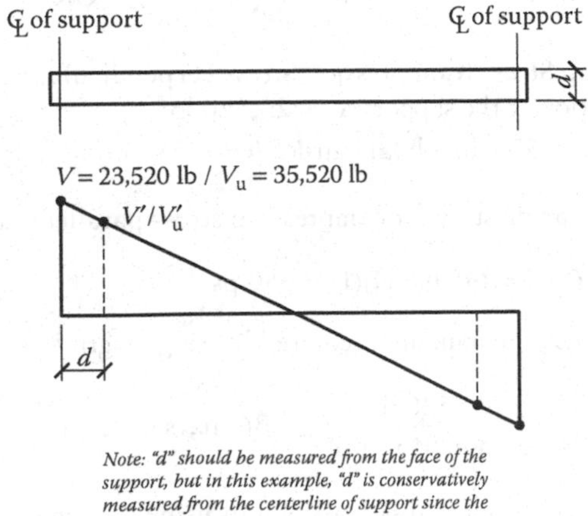
℄ of support
℄ of support
d
V = 23,520 lb / V_u = 35,520 lb
V′/V′_u
d
Note: "d" should be measured from the face of the support, but in this example, "d" is conservatively measured from the centerline of support since the bearing dimensions are not yet known

Figure 4.15 Glulam girder shear force diagram.

Table 4.11 Glulam girder deflection limits

Deflection	*Deflection limit*
Live-load deflection Δ_{LL}	$\frac{L}{360}=\frac{(48\text{ ft})(12)}{360}=1.6$ in.
Total-load deflection $\Delta_{TL}=k\Delta_{DL}+\Delta_{LL}$	$\frac{L}{240}=\frac{(48\text{ ft})(12)}{240}=2.4$ in.

The moment of inertia about the strong axis $I_{xx}=29{,}660$ in.^{4}.

uniform dead load $w_{DL}=35$ lb/in.
uniform live load $w_{DL}=46.7$ lb/in.

The dead-load deflection is

$$\Delta_{DL}=\frac{5w_{DL}L^4}{384EI}=\frac{(5)(35\text{ lb/in.})(48\text{ ft}\times 12)^4}{(384)(1.8\times 10^6\text{ psi})(29{,}660\text{ in}^4)}=0.94\text{ in.}$$

The live-load deflection is

$$\Delta_{LL}=\frac{5w_{LL}L^4}{384EI}=\frac{(5)(46.7\text{ lb/in.})(48\text{ ft}\times 12)^4}{(384)(1.8\times 10^6\text{ psi})(29{,}660\text{ in}^4)}=\mathbf{1.25\text{ in.}}<\frac{L}{360}=1.6\text{ in.}\qquad\text{OK}$$

As discussed previously, a value of $k=1.0$ will conservatively be used. The total dead plus floor live-load deflection is

$$\boldsymbol{\Delta}_{TL}=\boldsymbol{k\Delta}_{DL}+\boldsymbol{\Delta}_{LL}$$
$$=(1.0)(0.94\text{ in.})+1.25\text{ in.}=\mathbf{2.19\text{ in.}}\frac{L}{240}<=2.4\text{ in.}\qquad\text{OK}$$

Note: Although the deflections calculated are within the IBC deflection limits, the total-load deflection and the live-load deflection are on the high side (>1 in.), and therefore the beam should be cambered. The camber provided for glulam roof beams should be approximately 1.5 × the dead-load deflection (Section 4.1). With the camber provided, the net downward total deflection becomes

$$2.19\text{ in.}-(1.5)(0.94\text{ in.})=0.78\text{ in.}=\frac{L}{800}<\frac{L}{240}\qquad\text{OK}$$

9. **Check of Bearing Stress (Compression Stress Perpendicular to the Grain)**
Maximum reaction at the support $R_1=23{,}520$ lb
Width of the **6¾ × 37½ in.** glulam girder $b=6.75$ in.

The allowable bearing stress or compression stress parallel to the grain is

$$F'_{c\perp}=F_{c\perp}C_MC_tC_b=(650)(1)(1)(1)=650\text{ psi}$$

From Equation 4.9, the minimum required bearing length l_b is

$$l_{b,\text{ req'd}}\geq\frac{R}{bF'_{c\perp}}=\frac{23{,}520\text{ lb}}{(6.75\text{ in.})(650\text{ psi})}=5.36\text{ in., say 5.5 in.}$$

The girders will be connected to the column using a U-shaped column cap detail similar to that shown in Figure 4.6d, and the minimum length of the column cap required is (5.5 in.)(2) + 2 ½-in. clearance between ends of girders = 11.5 in.

10. **Check of Assumed Self-Weight of Girder**
A wood density of 36 lb/ft³ specified in this example is used to calculate the actual weight of the girder selected.
Assumed girder self-weight = 100 lb/ft
Actual weight of a 6 ¾ × 37 ½ in. glulam girder selected from NDS-S Table 1C

$$= \frac{(253.1\text{ in}^2)(36\text{ lb/ft}^3)}{(12\text{ in.})(12\text{ in.})}$$

= 63 lb/ft, which is less than the assumed value of 100 lb/ft OK

Note: Even though the assumed self-weight of the girder differs from the actual self-weight of the girder selected, this difference does not affect the results significantly, as demonstrated below, and thus can be neglected. The total load calculated previously using the assumed self-weight of 100 lb/ft was

$$w_{TL} = (D + S)(\text{tributary width}) + \text{girder self-weight}$$

$$= (20\text{ psf} + 35\text{ psf})(16\text{ ft}) + 100\text{ lb/ft} = 980\text{ lb/ft}$$

The corresponding total load using the actual self-weight of the girder selected is

$$w_{TL} = (D + S)(\text{tributary width}) + \text{girder self-weight}$$

$$= (20\text{ psf} + 35\text{ psf})(16\text{ ft}) + 63\text{ lb/ft} = 943\text{ lb/ft}$$

The difference between the assumed and actual self-weight as a percentage of the actual self-weight is

$$= \frac{100-63}{63} \times 100\% = 59\%$$

The difference between the total load calculated using the assumed self-weight and the actual total load w_{TL} is

$$\frac{980-943}{943} \times 100\% = 4\%$$

Thus, a 59% error in the self-weight of the girder yields only a 4% error in the total load, which is insignificant and can therefore be neglected, especially since we err on the safe side (i.e., the assumed self-weight was greater than the actual self-weight). Use a 6 ¾ × 37 ½ in. 24F–V4 DF/DF glulam girder.

LRFD Solution:

Example 4.4: Design of a glulam girder

The panelized roof framing system shown in Figure 4.13 consists of 2× subpurlins supported on 4× purlins that are supported on simply supported glulam girders. The total roof dead load, including the weight of the subpurlins and purlins, is 20 psf and the snow load is 35 psf. The wind load on the roof is ±10 psf Design the glulam girder assuming wood density of 36 lb/ft³, dry service and normal temperature conditions.

Solution: The total roof dead load D is 20 psf (excludes the self-weight of girder). Assume that the girder self-weight is 100 lb/ft (this will be checked later). The snow load S is 35 psf and the wind load W is ±10 psf All other loads, E, L, R, H, F, and $T = 0$.

$$\text{Tributary width of glulam girder} = \frac{16\,\text{ft}}{2} + \frac{16\,\text{ft}}{2} = 16$$

$$\text{Tributary area of glulam girder} = (16\text{ ft})(48\text{ ft}) = 768\text{ ft}^2 > 600\text{ ft}^2$$

From Section 2.4.1, $R_1 = 0.6$. Since we have a flat roof with minimum slope (1/4 in./ft) for drainage, the roof slope factor (see Section 2.4.1) $F = 0.25$; therefore, $R_2 = 1.0$. From Equation 2.6, we calculate the roof live load as (Table 4.12).

$$L_r = 20R_1R_2 = (20)(1.0)(0.6) = 12\text{ psf} < 20\text{ psf}$$

Summary: The governing load combination is 1.2D + *1.6S + 0.5W*.

We assume that this girder is subjected to uniform loads because there are at least five equally spaced reaction points where the purlins frame into the girder with equal reactions, indicating that the load on the girder is actually close to being uniform. The total uniform load on the girder that will be used to design for bending, shear, and bearing is obtained below using the dead load, the snow load, the wind load and the assumed self-weight of the girder, which will be checked later.

For deflections, the higher of snow and wind loads will be used.

Using the free-body diagram of the girder (Figure 4.14), the load effects are calculated as follows:

$$\text{uniform dead load } w_{\text{DL}} = (20\text{ psf})(16\text{ ft}) + 100\text{ lb/ft} = 420\text{ lb/ft} = 35\text{ lb/in.}$$

$$\text{uniform snow load } w_{\text{SL}} = (35\text{ psf})(16\text{ ft}) = 560\text{ lb/ft} = 46.7\text{ lb/in.}$$

$$\text{uniform wind load } w_{\text{WL}} = (10\text{ psf})(16\text{ ft}) = 160\text{ lb/ft}$$

Factored uniform load (controlling case is 85.8 psf/0.8 > 51.5/1.0, see Table 4.12):

$$w_u = 1.2w_D + 1.6w_s + 0.5w_w = (1.2)(420) + (1.6)(560) + (0.5)(160) = 1480\text{ plf}$$

$$\text{factored shear } V_U = \frac{w_{\text{TL}}L}{2} = \frac{(1480\text{ lb/ft})(48\text{ ft})}{2} = 35520\text{ lb}$$

$$\text{maximum reaction } R_U = 35520\text{ lb}$$

$$\text{maximum moment } M_U = \frac{w_U L^2}{8} = \frac{(1480\,\text{lb/ft})(48\,\text{ft})^2}{8} = 426,240\,\text{ft-lb}$$

$$= 5,115,000\text{ in.-lb}$$

We now proceed with the design following the steps described earlier in the chapter.

Table 4.12 Applicable and governing load combinations

Load combination	*Value of load combination, Wu*
1.4D	(1.4)(20) = 28 psf
1.2D + 1.6 L_r	(1.2)(20) + (1.6)(12) = 43.2 psf
1.2D + 0.5S	(1.2)(20) + (0.5)(35) = 41.5 psf
1.2D + 1.6S + 0.5W	(1.2)(20) + (1.6)(35) + (0.5)(10) = **85 psf**
1.2D + 1.0*W* + 0.5*S* 1.2*D* + 0.2*S*	(1.2)(20) + (1.0)(10) + (0.5)(35) = 51.5 psf
1.2D + 0.2*S*	(1.2)(20) + (0.2)(35) = 31 psf
0.9*D* + 1.0*W*	(0.9)(20) + (1.0)(−10)[a] = 8 psf (no net uplift)

[a] In this load combination, *W* is always opposed by the dead load *D*. Therefore, in these combinations, *D* takes on a positive number while *W* takes on negative values only.

Check of Bending Stress (Girders)

1. Summary of load effects (the self-weight of girder G1 was assumed, but this will be checked later in the design process):

   Factored maximum shear $V_U = 35520$ lb

   Factored maximum reaction $R_U = 35520$ lb

   Factored maximum moment $M_U = 5{,}115{,}000$ in.-lb

   uniform dead load is $w_{DL} = 35$ lb/in.

   uniform snow load is $w_{SL} = 46.7$ lb/in.

2. For glulam used primarily in bending, use NDS-S Table 5A-Expanded.
3. Using NDS-S Table 5A-Expanded, assume a 24F-V4 DF/DF glulam; therefore, $F^{+}_{bx} = 2400$ psi. (tension lamination stressed in tension). From Equation 4.1, the required approximate section modulus of the member is given as

$$S_{xx,\text{req'd}} \geq \frac{M_U}{F^{+}_{bx} K_{F,b} \phi_b \lambda} = \frac{5{,}115{,}000 \text{ in.-lb}}{(2400 \text{ psi})(2.54)(0.85)(0.8)} = 1233 \text{ in}^3$$

   From Appendix N:

   $K_{F,b} = 2.54$

   $\phi_b = 0.85$

   $\lambda = 0.8$

4. From NDS-S Table 1C (for western species glulam), the trial size with the least area that satisfies the section modulus requirement of step 3 is 6 ¾ in. × 34.5 in.:

   $b = 6.75$ in. and $d = 34.5$ in.

   S_{xx} provided $= 1339 \text{ in}^3 > 1233 \text{ in}^3$ OK

   area A provided $= 232.9 \text{ in}^2$

   I_{xx} provided $= 23100 \text{ in}^4$

5. The NDS-S (Table 5A-Expanded) tabulated stresses are

   $F^{+}_{bx} = 2400$ psi (tension lamination stressed in tension)

   $F_{v,xx} = 265$ psi

   $F_{c\perp xx,\text{ tension lam}} = 650$ psi

   $E_x = 1.8 \times 10^6$ psi

   $E_y = 1.6 \times 10^6$ psi

$E_{y,min} = 0.85 \times 10^6$ psi ($E_{y,min}$, and not $E_{x,min}$, is used for lateral buckling of the girder about the weak axis)

The adjustment or C factors are given in Table 4.10 except $C_D = 1.6$.

From the adjustment factor applicability table for glulam (Table 3.2), we obtain the allowable bending stress of the glulam girder with C_D, C_V and C_L equal to 1.0 as

$$F^*_{bx,r} = F^+_{bx} C_M C_t C_c \, K_{F,b} \phi_b \lambda$$

$$= (2400)(1.0)(1.0)(1.0)(2.54)(0.85)(0.8) = 4145 \text{ psi}$$

$$K_{F,b} = 2.54$$

$$\phi_b = 0.85$$

$$\lambda = 0.8$$

The allowable pure bending modulus of elasticity E'_x and the bending stability modulus of elasticity $E'_{y,min}$ are calculated as

$$E'_x = E_x C_M C_t = (1.8 \times 10^6)(1.0)(1.0) = 1.8 \times 10^6 \text{ psi}$$

$$E'_{y,min,r} = E_{y,min} C_M C_t \, K_{F,E} \phi_{st} = (0.85 \times 10^6)(1.0)(1.0)(1.76)(0.85) = 1.27 \times 10^6 \text{ psi}$$

$$K_{F,E} = 1.76$$

$$\phi_{st} = 0.85$$

Calculating the Beam Stability Factor C_L

From Equation 3.2 the beam stability factor is calculated as follows: The unsupported length of the compression edge of the beam or distance between points of lateral support preventing rotation and/or lateral displacement of the compression edge of the beam is

$l_u = 8$ ft $= 96$ in. (i.e., the distance between lateral supports provided by the purlins)

$$\frac{l_u}{d} = \frac{96 \text{ in.}}{33 \text{ in.}} = 2.91$$

We previously assumed a uniformly loaded girder; therefore, using the l_u/d value, the effective length of the beam of the beam is obtained from Table 3.10 (or NDS code Table 3.3.3) as

$$l_e = 2.06 l_u = (2.06)(96 \text{ in.}) = 198 \text{ in.}$$

$$R_B = \sqrt{\frac{l_e d}{b^2}} = \sqrt{\frac{(198)(34.5 \text{ in.})}{(6.75 \text{ in.})^2}} = 12.2 \le 50 \quad \text{OK}$$

$$F_{bE} = \frac{1.20 E'_{min}}{R_B^2} = \frac{(1.20)(1.27 \times 10^6)}{(12.2)^2} = 10190 \text{ psi}$$

$$\frac{F_{bE}}{F^*_{b,r}} = \frac{10190 \text{ psi}}{4145 \text{ psi}} = 2.46$$

From Equation 3.3 the beam stability factor is calculated as

$$C_L = \frac{1+F_{bE}/F_{b,r}^*}{1.9} - \sqrt{\left(\frac{1+F_{bE}/F_{b,r}^*}{1.9}\right)^2 - \frac{F_{bE}/F_{b,r}^*}{0.95}}$$

$$= \frac{1+2.46}{1.9} - \sqrt{\left(\frac{1+2.46}{1.9}\right)^2 - \frac{2.46}{0.95}} = 0.97$$

Calculating The Volume Factor C_V

L = Length of beam in feet between adjacent points of zero moment = 48 ft
d = Depth of beam = 34.5 in.
b = Width of beam, in. = 6.75 in.
x = 10 (for western species glulam)

From Equation 3.5,

$$C_V = \left(\frac{21}{L}\right)^{1/x}\left(\frac{12}{d}\right)^{1/x}\left(\frac{5.125}{b}\right)^{1/x} = \left(\frac{1291.5}{bdL}\right)^{1/x} = \left[\frac{1291.5}{(6.75\text{ in.})(34.5\text{ in.})(48\text{ ft})}\right]^{1/10}$$

$$= 0.806 \le 1.0$$

The smaller of C_V and C_L will govern and is used in the allowable bending stress calculation. Since $C_V = 0.809 < C_L = 0.97$, use $C_V = 0.809$. Using Table 3.4 (i.e., adjustment factor applicability table for glulam), we obtain the available bending stress as

$$F'_{b,r} = F^+_{bx}\, K_{F,b}\phi_b\lambda\, C_M C_t (C_V \text{ or } C_L) C_r C_{fu} C_c C_I$$

$$=(2400)(2.54)(0.85)(0.8)(1.0)(1.0)(0.806)(1.0)(1.0)(1.0) = \mathbf{3341\ psi}$$

6. Using Equation 4.10, the factored bending stress is

$$f_{bU} = \frac{M_U}{S_{xx}} = \frac{5{,}115{,}000\text{ lb-in.}}{1339\text{ in}^3} = 3820\text{ psi}$$

$>$ available bending stress $F'_{b,r} = 3341$ psi **NG**

Therefore, use a 6 ¾ x 37-½ in. 24F-V4 DF/DF glulam for bending with $f_{bu} = 3233$psi.

7. **Check of Shear Stress**

$V_{max} = 33{,}600$ lb. The beam cross-sectional area $A = 253$ in^2. The factored shear stress in the wood beam at the centerline of the girder support is

$$f_{vU} = \frac{1.5V}{A} = \frac{(1.5)(33{,}600\text{ lb})}{253\text{ in}^2} = 211\text{ psi}$$

Using the adjustment factor applicability table for glulam from Table 3.4, we obtain the available shear stress as

$$F'_{v,r} = F_v\, K_{F,v}\phi_v\lambda\, C_M C_t C_{vr} = (265)(2.88)(0.75)(0.8)(1)(1)(1) = 458\text{ psi}$$

From Appendix N:

$K_{F,v} = 2.88$

$\phi_v = 0.75$

$\lambda = 0.8$

Thus, $f_{vu} < F'_{v,r}$; therefore, the girder is adequate in shear (See Figure 4.15b). *Note:* Since the shear stress f_{vu} at the centerline of the girder support is adequate (i.e., $<F'_{v,r}$), there is no need to further calculate the shear stress f'_{vu} at a distance d from the face of the support since it is obvious *that* f'_{vu} would be less than f_{vu} and thus will also be adequate.

8. **Check of Deflection** (see Table 4.11)
The allowable pure bending modulus of elasticity for strong (i.e., x–x)-axis bending of the girder was calculated previously:

$$E'_x = E_x C_M C_t = (1.8 \times 10^6)(1.0)(1.0) = 1.8 \times 10^6 \text{ psi}$$

The moment of inertia about the strong axis $I_{xx} = 29{,}663$ in.^{4}

uniform dead load $w_{DL} = 35$ lb/in.

uniform live load $w_{DL} = 46.7$ lb/in.

The dead-load deflection is

$$\Delta_{DL} = \frac{5w_{DL}L^4}{384EI} = \frac{(5)(35\text{ lb/in.})(48\text{ ft}\times 12)^4}{(384)(1.8\times 10^6\text{ psi})(29{,}663\text{ in}^4)} = 0.94\text{ in}$$

The live-load deflection is

$$\Delta_{LL} = \frac{5w_{LL}L^4}{384EI} = \frac{(5)(46.7\text{ lb/in.})(48\text{ ft}\times 12)^4}{(384)(1.8\times 10^6\text{ psi})(29{,}663\text{ in}^4)} = 1.25\text{ in.} < \frac{L}{360} = 1.6\text{ in.}\quad \text{OK}$$

The total load deflection is

$$\mathbf{\Delta_{TL} = k\Delta_{DL} + \Delta_{LL}}$$

$$= (1.0)(0.94\text{ in.}) + 1.25\text{ in.} = 2.19\text{ in.} < L/240 = 2.4\text{ in.}\qquad \text{OK}$$

Note: Although the deflections calculated are within the IBC deflection limits, the total load deflection and the live-load deflection are on the high side (>1 in.), and therefore the beam should be cambered. The camber provided for glulam roof beams should be approximately 1.5 × the dead-load deflection (Section 4.1). With the camber provided, the net downward total deflection becomes

$$2.19\text{ in.} - (1.5)(0.94\text{ in.}) = 0.78\text{ in.} = \frac{L}{800} < \frac{L}{240}\quad \text{OK}$$

9. Check of Bearing Stress (Compression Stress Perpendicular to the Grain)

Maximum reaction at the support $R_U = 33{,}600$ lb

Width of 6¾ × 37½ in. glulam girder $b = 6.75$ in.

The allowable bearing stress or compression stress parallel to the grain is

$$F'_{c\perp r} = F_{c\perp}\, K_{F,c\perp}\phi_{c\perp}\lambda C_M C_t C_b C_i = (650)(1.67)(0.9)(0.8)(1.0)(1.0)(1.0)(1.0) = 781 \text{ psi}$$

From Appendix N:

$K_{F,c\perp} = 1.67$

$\phi_{c\perp} = 0.9$

$\lambda = 0.8$

From Equation 4.9, the minimum required bearing length l_b is

$$l_{b,\text{req'd}} \geq \frac{R_U}{bF'_{c\perp}} = \frac{35{,}520 \text{ lb}}{(6.75 \text{ in.})(781 \text{ psi})} = 6.73 \text{in., say } 6.75 \text{in.}$$

The girders will be connected to the column using a U-shaped column cap detail similar to that shown in Figure 4.6d, and the minimum length of column cap required is

(6.75 in.)(2) + 2½-in. clearance between ends of girders = 16 in.

10. Check of Assumed Self-Weight of Girder
A wood density of 36 lb/ft³ specified in this example is used to calculate the actual weight of the girder selected.
Assumed girder self-weight = 100 lb/ft
Actual weight of a 6¾ × 37½ in. glulam girder selected from NDS-S Table 1C

$$= \frac{(253.1 \text{ in}^2)(36 \text{ lb/ft}^3)}{(12 \text{ in.})(12 \text{ in.})}$$

= 63 lb/ft which is less than the assumed value of 100 lb/ft OK

Note: Even though the assumed self-weight of the girder differs from the actual self- weight of the girder selected, this difference does not affect the results significantly, as demonstrated below, and thus can be neglected. The total load calculated previously using the assumed self-weight of 100 lb/ft was

$$w_{TL} = (1.2D + 1.6S + 0.5W)(\text{tributary width}) + 1.2(\text{girder self weight})$$

$$= [(1.2)(20) + (1.6)(35) + (0.5)(10)] + (\text{tributary width}) + 1.2(100)$$
$$= 1480 \text{ plf}$$

The corresponding total load using the actual self-weight of the girder selected is

$$w_{TL} = (1.2D+1.6S+0.5W)(\text{tributary width})+1.2(\text{girder self weight})$$

$$= [(1.2)(20)+(1.6)(35)+(0.5)(10)]+(\text{tributary width})+1.2(63) = 1463 \text{ plf}$$

The difference between the assumed and actual self-weight as a percentage of the actual self-weight is

$$= \frac{100-63}{63} \times 100\% = 59\%$$

The difference between the total load calculated using the assumed self-weight and the actual total load w_{TL} is

$$\frac{1480-1436}{1436} \times 100\% = 3\%$$

Thus, a 59% error in the self-weight of the girder yields only a 3% error in the total load, which is insignificant and can therefore be neglected, especially since we err on the safe side (i.e., the assumed self-weight was greater than the actual self-weight).

Use a 6¾ × 37½ in. 24F-V4 DF/DF glulam girder.

Example 4.5: Design of stair treads and stringers (ASD/LRFD)

For the typical floor plan shown in Figure 4.16, the second-floor dead load is 20 psf and the live load is 40 psf. Assuming Hem-Fir wood species, design the stair treads and stair stringers (Figure 4.17).

Solution:
Some IBC requirements for stairs include the following (see also Chapter 9):

Stair risers must be ≥4 in. and ≤7 in. (7.75 in. maximum in groups R-2 and R-3).
Stair treads (excluding nosing) ≥11 in. (10 in. minimum in groups R-2 and R-3).
Minimum width of stair = 3 ft 8 in.
Minimum head room clearance = 6 ft 8 in.
The floor-to-floor height between stair landings must not exceed 12 ft.

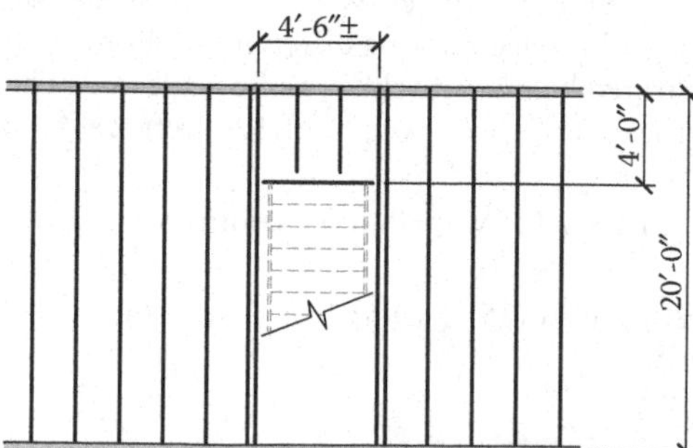

Figure 4.16 Second-floor framing plan.

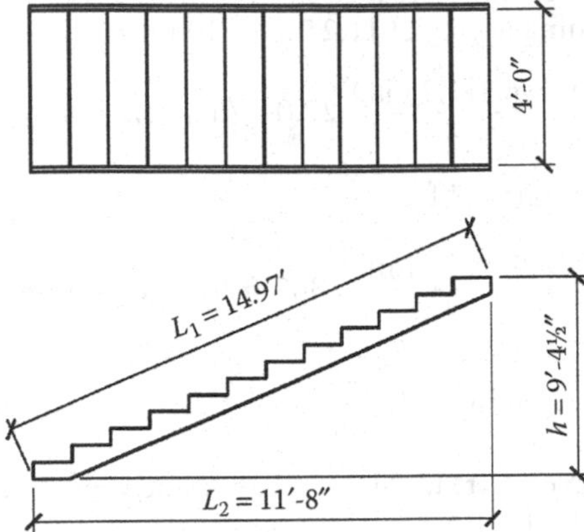

Figure 4.17 Plan and elevation of stair.

To provide comfort for the users, the following rules of thumb are commonly used:

Number of treads = number of risers – 1
$2R + T = 25$ (ensures the comfort of people using the stair)

where T is the tread width in inches and R is the riser height in inches.
Risers:

$$\text{Floor-to-floor height} = 9 \text{ ft } 4\frac{1}{2} \text{ in.} = 112.5 \text{ in.}$$

$$\text{Assuming 7-in. risers, number of risers} = \frac{112.5 \text{ in.}}{7 \text{ in.}} = 16.07$$

$$\text{Assuming 7.5-in. risers, number of risers} = \frac{112.5 \text{ in.}}{7.5 \text{ in.}} = 15$$

Use 15 risers at 7.5 in. = 112.5 in. floor-to-floor height.

Treads:

$$T + 2 \times R = 25$$

$$T = 25 - (2)(7.5 \text{ in.}) = 10 \text{ in.}$$

Number of treads = number of risers – 1 = 15 – 1 = 14

Use 15 risers at 7.5 in. and 14 treads at 10 in.
Horizontal run of stair = span of stair = 14 treads × 10 in. = 140 in. = 11.67 ft
Width of stair = 4 ft

Design of tread. The tread is bending about its weaker (y–y) axis, and it should be noted that the uniform stair live load of 100 psf will not usually govern for the design of the treads. Instead, the most critical load for which the tread should be designed is the alternative 300-lb concentrated live load in addition to the uniform tread self-weight and finishes (see ASCE 7, Table 4.2) for the alternative concentrated floor live loads.

Assume 2 × 12 for the treads (actual size = 1.5 in. × 11.25 in.). Therefore, the tread width = 11.25 in. (including nosing of 11.25 in. – 10 in. = 1.25 in.).

$$\text{Self-weight of tread} = \frac{(1.5\text{ in.})(11.25\text{ in.})}{144}(27.0\text{ lb/ft}^3) = 3.2\text{ lb/ft (density of Hem-Fir} = 27.0\text{ lb/ft}^3)$$

Span of tread = width of stair = 4 ft

$$M_{\max} = \frac{(3.2\text{ lb/ft})(4\text{ ft})^2}{8} + \frac{(300\text{ lb})(4\text{ ft})}{4} = 307\text{ ft-lb (concentrated load at midspan)}$$

$$V_{\max} = \frac{(3.2\text{ lb/ft})(4\text{ ft})}{2} + 300\text{ lb} = 307\text{ lb (concentrated load at support)}$$

Try a 2 × 12 Hem-Fir No. 1 tread. Since this is dimension-sawn lumber, use NDS-S Table 4A. From NDS-S Table 1B,

$S_{yy} = 4.219\text{ in.}^3$ (note that S_{yy} is used because the tread is bending about the weak axis)
$F_b = 975$ psi
$F_v = 150$ psi

From the adjustment or C factors section of NDS-S Table 4A, we obtain

$C_D(D + L) = 1.0$
$C_M = 1.0$ (dry service conditions)
$C_t = 1.0$ (normal temperature conditions)
$C_F = 1.0$
$C_{fu} = 1.2$ (the flat use factor is used since the 2 × 12 tread is used on the flat and is bending about the weak axis)

Bending stress:

$$\text{Applied bending stress } f_{b,yy} = \frac{M_{\max}}{S_{yy}} = \frac{(307)(12)}{4.219} = 873\text{ psi}$$

$$\text{Allowable bending stress } F_b' = F_b C_D C_M C_t C_F C_{fu} C_i C_r$$

$$= (975)(1.0)(1.0)(1.0)(1.0)(1.2)(1.0)(1.0)$$

$$= 1170\text{ psi} > 873\text{ psi} \quad \text{OK}$$

Shear stress:

$$\text{Applied shear stress } f_{v,yy} = \frac{1.5V}{A} = \frac{(1.5)(307)}{16.88} = 28\text{ psi}$$

Allowable shear stress $F_v' = F_v C_D C_M C_t C_i = (150)(1.0)(1.0)(1.0)(1.0)$ $= 150\text{ psi} > 28\text{ psi}$ OK

Use a 2 × 12 Hem-Fir No. 1 tread (nosing = 1.25 in.).

Analysis and Design of Stair Stringer

Width of stair = 4 ft
Actual size of riser = 1.5 in. × 6 in. (see Figure 4.18)
Horizontal run L_2 = 140 in. = 11.67 ft
Floor-to-floor height H = 9.375 ft

Sloped length of stair $L_1 = \sqrt{(H)^2 + (L_2)^2} = \sqrt{(9.375\text{ ft})^2 + (11.67\text{ ft})^2} = 15\text{ ft}$

Stair loads:

Treads = (14 treads × 3.2 lb/ft × 4-ft span)(6.0 in./11.25 in.)/(4 ft × 11.67 ft) = 2.1 psf

Risers = (15 risers × 3.2 lb/ft × 4-ft span)(6.0 in./11.25 in.)/(4 ft × 11.67 ft) = 2.2 psf

Two stringers (assume 3 × 16 per stringer) = (2 × 8.3 lb/ft)/4 ft = 4.2 psf

Handrails (assume 2 psf) = 2 psf

Total dead load on stair D = 10.5 psf ≈ 11 psf of sloped area

Live load on stair L = 100 psf of horizontal plan area

To convert the dead load to units of psf of the horizontal plan area, we use the method introduced in Chapter 2:

Total load on stair w_{TL} (psf on the horizontal plane) $= D\frac{L_1}{L_2} + L$ (see Chapter 2)

$$= (11\text{ psf})\left(\frac{15\text{ ft}}{11.67\text{ ft}}\right) + 100\text{ psf}$$

$$= 115\text{ psf}$$

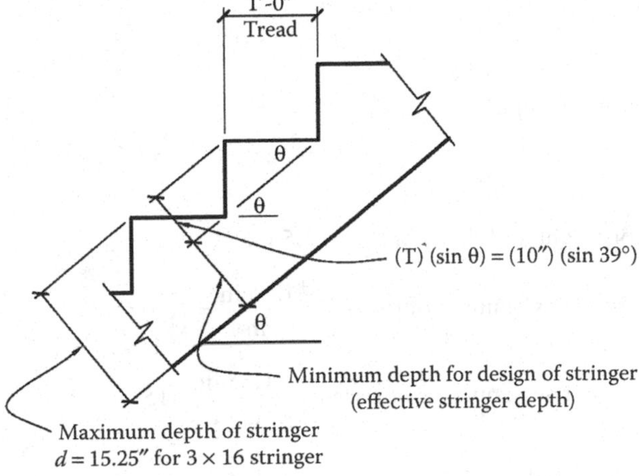

Figure 4.18 Effective depth of stair stringer.

$$\text{Tributary width per stair stringer} = \frac{4\text{ ft}}{2} = 2\text{ ft}$$

Total load on stair stringer w_{TL} (lb/ft on the horizontal plane) = (115 psf)(2 ft) = **230 lb/ft**
Horizontal span of stair stringer L_2 = 11.67 ft

The load effects on the stair stringer are calculated as follows:

$$\text{maximum moment } M_{max} = \frac{w_{TL}(L_2)^2}{8} = \frac{(230\text{ lb/ft})(11.67\text{ ft})^2}{8} = 3915\text{ ft-lb}$$

$$\text{maximum shear } V_{max} = \frac{w_{TL}L_2}{2} = \frac{(230\text{ lb/ft})(11.67\text{ ft})}{2} = 1342\text{ lb}$$

LRFD Solution:

Example 4.6: Design of Stair Treads and Stringers

For the typical floor plan shown in Figure 4.16b the second-floor dead load is 20 psf and the live load is 40 psf. Assuming Hem-Fir wood species, design the stair treads and stair stringers (Figure 4.17b).

Solution:
Some IBC requirement for stairs include the following (see also Chapter 10):

Stair risers must be > 4 in. and < 7 in. (7.75 in. maximum in groups R-2 and R-3).
Stair treads (excluding nosing) > 11 in. (10 in. minimum in groups R-2 and R-3).
Minimum width of stair = 3 ft 8 in.
Minimum head room clearance = 6 ft 8 in
The floor-to-floor height between stair landings must not exceed 12 ft.

To provide comfort for the users, the following rules of thumb are commonly used:

Number of treads = number of risers − 1
$2R + T = 25$ (ensures the comfort of people using the stair)

where:
T is the tread width in inches
R is the riser height in inches

Risers:

Floor-to-floor height = 9 ft $4\frac{1}{2}$ in. = 112.5 in.

$$\text{Assuming 7-in. risers, number of risers} = \frac{112.5\text{ in}}{7\text{ in}} = 16.07$$

$$\text{Assuming 7.5-in. risers, number of risers} = \frac{112.5\text{ in}}{7.5\text{ in}} = 15$$

Use 15 risers at 7.5 in. = 112.5 in. floor-to-floor height.

Treads:

$$T + 2 \times R = 25$$

$$T = 25 - (2)(7.5\,\text{in.}) = 10\,\text{in.}$$

$$\text{Number of treads} = \text{number of risers} - 1 = 15 - 1 = 14$$

Use 15 risers at 7.5 in. and 14 treads at 10 in.

$$\text{Horizontal run of stair} = \text{span of stair} = 14\ \text{treads} \times 10\,\text{in.} = 140\,\text{in.} = 11.67\,\text{ft}$$

$$\text{Width of stair} = 4\ \text{ft}$$

Design of tread. The tread is bending about its weaker (*y-y*) axis, and it should be noted that the uniform stair live load of 100 psf will not usually govern for the design of the treads. Instead, the most critical load for which the tread should be designed is the alternative 300-lb concentrated live load in addition to the uniform tread self-weight and finishes (see Table 2.2) for the alternative concentrated floor live loads.

Assume 2 × 12 for the treads (actual size = 1.5 in. × 11.25 in.). Therefore, the tread width = 11.25 in. (including nosing of 11.25 in.–10 in. = 1.25 in.)

$$\text{Self-weight of tread} = \frac{(1.5\,\text{in.})(11.25\,\text{in.})}{144}(26.9\ \text{lb/ft}^3) = 3.2\ \text{lb/ft} \quad (\text{density of Hem-Fir} = 27.0\ \text{pcf})$$

Span of tread = width of stair = 4 ft

Use the load combination 1.2D + 1.6L

$$M_U = \frac{1.2(3.2\,\text{lb/ft})(4\text{ft})^2}{8} + \frac{1.6(300\,\text{lb})(4\,\text{ft})}{4} = 388\,\text{ft-lb}\ (\text{concentrated load at midspan})$$

$$Vu = \frac{1.2(3.2\,\text{lb/ft})(4\text{ft})}{2} = (1.6)(300\,\text{lb}) = 488\,\text{lb}\ (\text{concentrated load at support})$$

Try a 2 × 12 Hem-Fir No. 1 tread. Since this is dimension-sawn lumber, use NDS-S Table 4A. From NDS-S Table 1B,

$S_{yy} = 4.219\ \text{in}^3$ (note that S_{yy} is used because the tread is bending about the weak axis)

$F_b = 975$ psi

$F_v = 150$ psi

From the adjustment or C factors section of NDS-S Table 4A, we obtain

$C_M = 1.0$ (dry service conditions)

$C_t = 1.0$ (normal temperature conditions)

$C_F = 1.0$

$C_{fu} = 1.2$ (the flat use factor is used since the 2×12 tread is used on the flat and is bending about the weak axis)

Bending stress:

$$\text{factored bending stress } f_{bu,yy} = \frac{M_U}{S_{yy}} = \frac{(488)(12)}{4.219} = 1388\,\text{psi}$$

Available bending stress:

$$F'_{b,r} = F_b K_{F,b} \phi_b \lambda C_M C_t C_L C_F C_r C_{fu} C_i$$

$$= (975)(2.54)(0.85)(0.8)(1.0)(1.0)(1.0)(1.2)(1.0)(1.0)$$

$$= 2020 \text{ psi} > 1388 \text{ psi, OK}$$

Shear stress:

$$\text{Factored shear stress } f_{vu,yy} = \frac{1.5V_U}{A} = \frac{(1.5)(488)}{16.88} = 44\,\text{psi}$$

Available shear stress:

$$F'_{v,r} = F_v K_{F,v} \phi_v \lambda C_M C_t C_i = (150)(2.88)(0.75)(0.8)(1)(1)(1) = 259 \text{ psi} > 44 \text{ psi, OK}$$

From Appendix N:

$K_{F,v} = 2.88$
$\phi_v = 0.75$
$\lambda = 0.8$

Use a 2 × 12 Hem-Fir No. 1 tread (nosing = 1.25 in.).

Analysis and Design of Stair Stringer

Width of stair = 4 ft
Actual size of riser = 1.5 in. × 6 in. (see Figure 4.18b)
Horizontal run $L_2 = 140\,\text{in.} = 11.67\,\text{ft}$
Floor-to-floor height $H = 9.375\,\text{ft}$

$$\text{Sloped length of stair } L_1 = \sqrt{(H)^2 + (L_2)^2} = \sqrt{(9.375\,\text{ft})^2 + (11.67\,\text{ft})^2} = 15\,\text{ft}$$

Stair loads:

Treads = (14 treads × 3.2 lb/ft × 4-ft span)/(4 ft × 11.67 ft) = 3.8 psf

Risers = (15 risers × 3.2 lb/ft × 4-ft span)/(4 ft × 11.67 ft) = 4.1 psf

Two stringers (assume 3 × 16 per stringer) = (2 × 8.3 lb/ft)/4 ft = 4.2 psf

Handrails (assume 2 psf) = 2 psf

Total dead load on stair D = 14.1 psf ≈ 14 psf of sloped area

Live load on stair L = 100 psf of horizontal plan area

To convert the dead load to units of psf of horizontal plan area, we use the method introduced in Chapter 2:

$$\text{Total load on stair } w_{TL} \text{ (psf on horizontal plane)} = D\frac{L_1}{L_2} + L \text{(see Chapter 2)}$$

$$= (14\,\text{psf})\left(\frac{15\,\text{ft}}{11.67\,\text{ft}}\right) + 100\,\text{psf}$$

$$= 118\,\text{psf}$$

$$\text{Tributary width per stair stringer} = \frac{4\,\text{ft}}{2} = 2\,\text{ft}$$

Total factored uniform load on stair stringer w_U (lb/ft on horizontal plane)

$$= [(1.2)(18) + (1.6)(100)](2\,\text{ft}) = \mathbf{363\ lb/ft}$$

Horizontal span of stair stringer L_2 = 11.67 ft
The load effects on the stair stringer are calculated as follows:

$$\text{maximum moment} = M_U = \frac{w_U(L_2)^2}{8} = \frac{(3641\text{b/ft})(11.67\,\text{ft})^2}{8} = 6182\,\text{ft-lb}$$

$$\text{maximum shear/reaction} = V_U = \frac{w_U(L_2)^2}{2} = \frac{(3641\text{b/ft})(11.67\,\text{ft})^2}{2} = 2124\,\text{ft-lb}$$

Effective depth of stringer. Typically, the stair stringers are notched to provide support for the treads and risers. As a result of this notch, the effective depth of the stringer is reduced (Figure 4.18). The effective depth of stair stringer to be used in design is

actual depth – depth to allow for the treads/risers "cutting" into the stringer for support
= width of unnotched stair stringer, d – tread width × sin θ

where

$$\theta = \text{angle of slope of stair} = \tan^{-1}\frac{H}{L_2} = \tan^{-1}\frac{9.375\,\text{ft}}{11.67\,\text{ft}} = 39°$$

The tread width (excluding the nosing) is 10 in. If we assume 3 × 16 stair stringers, the effective stringer depth is 15.25 in.–(10 in.)(sin 39°) = 9 in. Therefore, the effective stringer section = 2.5 × 9 in. (see previous discussion of the conservative use of this stringer section)

Though this effective depth, which is actually the minimum depth of the stringer, ignores the stress concentration effects at the re-entrant corners, the authors believe that the approach is conservative since the stringer depth actually varies linearly between this minimum value and a maximum value (which is equal to the full unnotched depth of the stringer) at mid-way between the re-entrant corners.

The stair stringer should then be designed as a beam (similar to the beam design examples presented previously in this chapter) for the load effects calculated previously using this effective section. The stair stringer should be designed to satisfy the following deflection limits: $L/240$ under the unfactored dead plus live load, and $L/360$ under the unfactored live load, where L is the horizontal span of the stair stringer. It is recommended that a ¼" diameter pilot hole be predrilled in the stringer at the notch locations to properly locate the notch corner and avoid overcutting beyond the corner, and to reduce the stress concentrations from a square notch. Mechanical connectors are recommended for the connection of the stringer to the supporting headers at the floor or landing levels [3].

The header and trimmer joists supporting the stair stringer at the second-floor level will be subjected to concentrated loads due to reactions from the stair stringers, in addition to the second-floor uniform loads. Once the joists have been analyzed for the maximum shears, moments, and reactions, the design will follow the same approach as was used in Example 4.1. A structural analysis software may be needed for the deflection computations for these nonuniformly loaded joists.

4.2.6 Continuous beams and girders

Thus far, we have focused on simply supported single-span beams and girders. For girders supported on three or more columns, it is more cost-effective to analyze the beam or girder as a continuous beam, subject to the limitations on the length of beams or girders that can be transported. However, the analysis of continuous beams is considerably more tedious than the analysis of simply supported single-span beams. For continuous beams, a *pattern load analysis* has to be carried out in which the live load is randomly located on the spans of the beam or girders to produce the maximum load effects. It should be noted that the dead load, on the other hand, is always located over the entire length of the girder. The girder is then analyzed for the different load cases and the absolute maximum negative or positive moment, the absolute maximum positive or negative shear, and the absolute maximum reaction are obtained; the girder is then designed for these absolute maximum values. The maximum deflection at the midspan of the endspan of the continuous beam is calculated and compared to the beam deflection limits discussed earlier. For continuous roof beams in areas with snow, the partial snow load prescribed in Section 7.5 of ASCE 7 should also be considered. Refer to Figure 4.19d for various load cases to consider for continuous beams.

4.2.7 Beams and girders with overhangs or cantilevers

These types of beams or girders are used to frame balconies or overhanging portions of buildings. It consists of the overhang or cantilever and the back span (see Figure 4.19c). It is more efficient to have the length of the back span be at least three times the length of the overhang or cantilever span. In these types of beams, the designer needs to, among other things, check the deflection of the tip of the cantilever or overhang and the uplift force at the back-span end, as these could be critical.

Similar to continuous beams, a pattern load analysis is required to obtain the maximum possible moment, shear, reaction, and deflection. The beam would then be designed for these maximum load effects using the procedure discussed previously. The partial snowload prescribed in Section 7.5 of ASCE 7 should also be considered for cantilevered roof beams in areas with snow. The deflection limits for the back span are as presented previously. However, for the cantilever or overhang, the appropriate deflection limits are presented in Table 4.13. The reader should note that the number "2" in the cantilever deflection limits below is a factor that accounts for the fact that the deflected profile of a cantilever beam is one-half that of a simply supported single-span beam (IBC 1604.3).

At the supports of continuous beams or the interior support of a cantilever girder with an overhang, the bottom face of the beam between the points of inflection near the interior support of a continuous beam (see Figure 4.19b), and between the point of inflection within

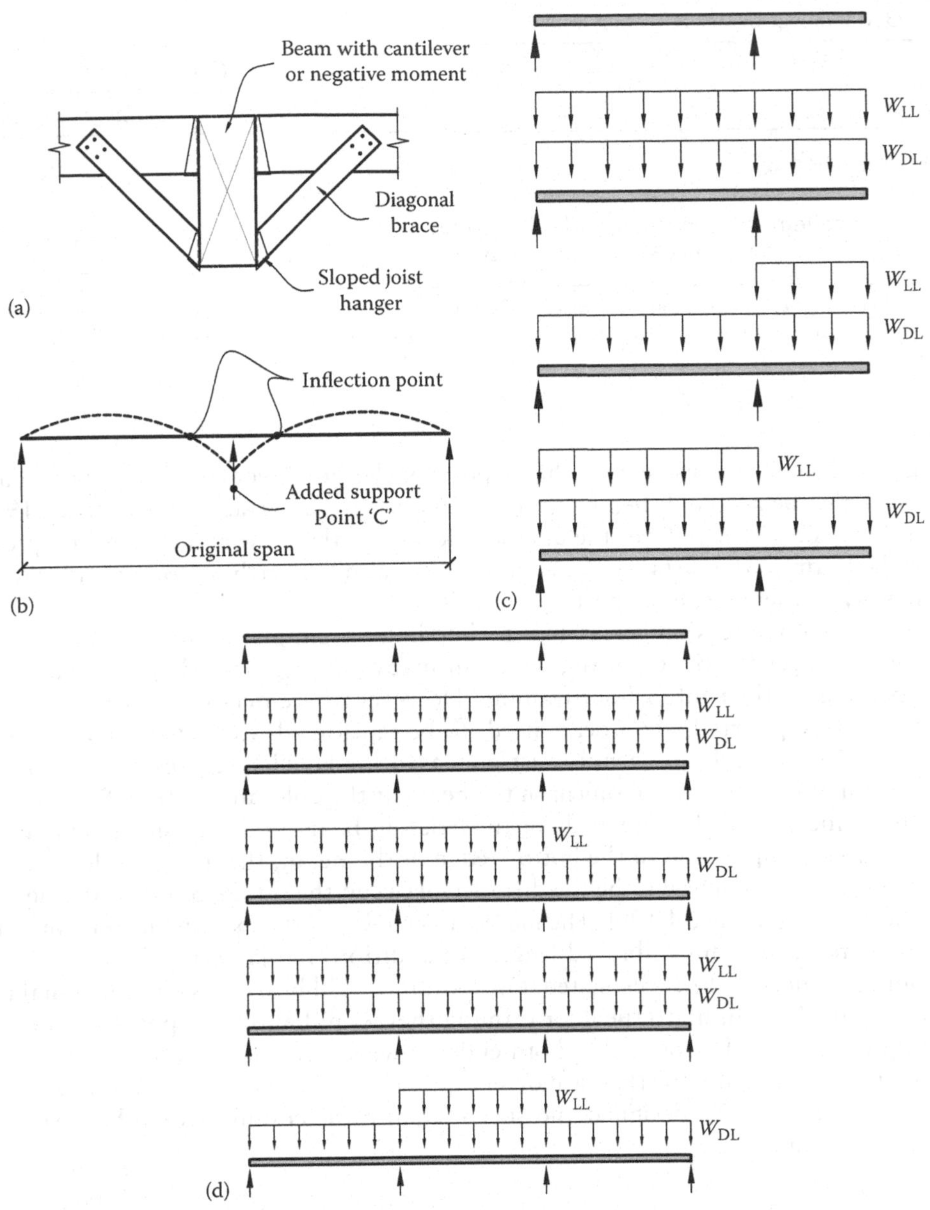

Figure 4.19 Beam lateral bracing detail. (a) Bracing detail at negative moment, (b) loading with negative moment, (c) possible load cases for a cantilever beam, and (d) possible load cases for a 3-span continuous beam.

the back-span of a cantilever beam (near the interior support) and the tip of the cantilever or overhang (see Figures 4.19c and 4.19d) is unbraced because of the negative moment, which causes compression stresses in the bottom fibers. Lateral bracing is required at the bottom of the beam at the interior support (point C) to prevent lateral instability of the column support, and for the same reason, lateral bracing is also required at the supports of continuous beams and girders. The lateral support can be provided by a member of similar depth

Table 4.13 Cantilever beam deflection limits[a]

Deflection	*Deflection limit*	
	Back span	*Overhang or cantilever*
Live-load deflection Δ_{LL}	$\frac{L_{bs}}{360}$	$\frac{2L_{cant}}{360}$
Incremental long-term deflection due to dead load plus live load (including creep effects), $\Delta_{TL} = k\Delta_{DL} + \Delta_{LL}$	$\frac{L_{bs}}{240}$	$\frac{2L_{cant}}{240}$

[a] L_{bs} is the back span of a joist, beam, or girder between two adjacent supports, L_{cant} is the cantilever or overhang span of joist, beam, or girder, Δ_{DL} is the deflection due to dead load *D*, and Δ_{LL} is the deflection due to live load *L* or (L_r or *S*).

framing into the continuous beam at the supports or the cantilever beam at point C. Lateral bracing can also be accomplished by using a diagonal kicker as shown in Figure 4.19. The diagonal brace or kicker reduces the unbraced length of the bottom face (i.e., compression zone) of the beam to the distance between the brace and the farthest point of inflection on opposite sides of the interior support.

The flexural compression stress due to bending in simply supported wood beams occurs on the top edge of the beam, which in many cases is braced against lateral torsional buckling by the roof or floor framing. However, there may be occasions where an existing simply supported wood beam needs to be reinforced with a new post or column at the midspan to enable the beam resist increased loads. This creates a two-span continuous beam with a negative moment in the beam at the column location. Since the post is usually connected to the underside of the existing beam and the negative moment at the post causes compression in the bottom edge of the beam, the unbraced length of the beam for negative moment will be the distance between the points of inflection on either side of the post (see Figure 4.19b). The moment capacity of the existing beam would have to be calculated using this unbraced length. In addition, the lateral stability of the new post would also need to be investigated. One option would be to provide a diagonal brace from the top of the column to the floor framing that would allow the post to be designed as a compression member pinned at both ends; however, if it is not possible to laterally brace the top of the column, then a moment connection at the base of the new post or column would have to be designed and specified, and the column would be designed as a cantilevered column.

4.3 SAWN-LUMBER DECKING

In lieu of using plywood sheathing (see Chapter 6) to resist gravity loads on roofs and floors, tongue-and-grooved decking may be used to span longer distances between roof trusses or floor beams. Sawn-lumber decking may also be required on floors where equipment or forklift loading produces concentrated loads for which plywood sheathing is inadequate. Decking is also aesthetically pleasing because the underside of the decking creates an attractive ceiling feature, obviating the need for additional plaster or suspended ceiling. Wood decking are manufactured as 2-in.-thick single tongue-and-groove panels at a moisture content of 15% or as 3- and 4-in.-thick double tongue-and-groove panels at a moisture content of 19%. The standard nominal sizes for wood decking are 2 × 6, 2 × 8, 3 × 6, and 4 × 6. Prior to installation, the lumber decking should be allowed to reach the moisture content of the surrounding atmosphere, to avoid problems that may result from shrinkage.

The tongue and grooves prevent differential deflections of adjacent deck panels, and each piece of deck is usually nailed through the tongue at each support. Decking consists of repetitive sawn-lumber members that are used in a flat orientation. However, the flat-use factor should not be applied in calculating the allowable stresses because this factor is already embedded in the NDS-S tabulated design values for decking in NDS-S Table 4E. Single-member and repetitive-member bending stresses are given in this Table. The different stress grades of decking include Select, Commercial Dex and Select Dex.

There are several possible decking layouts as shown in Figure 4.20. The two most popular decking patterns are (1) nonstaggered decking panels supported on only two supports with all the panel joints continuous, and (2) staggered decking panels supported on at least two supports with continuous panel joints parallel to the longer dimension of the panels. It should be noted that tongue-and-groove lumber decking, although efficient for supporting gravity loads, is not very efficient as horizontal diaphragms in resisting lateral loads from wind or seismic forces. It possesses very low diaphragm shear capacity in the direction

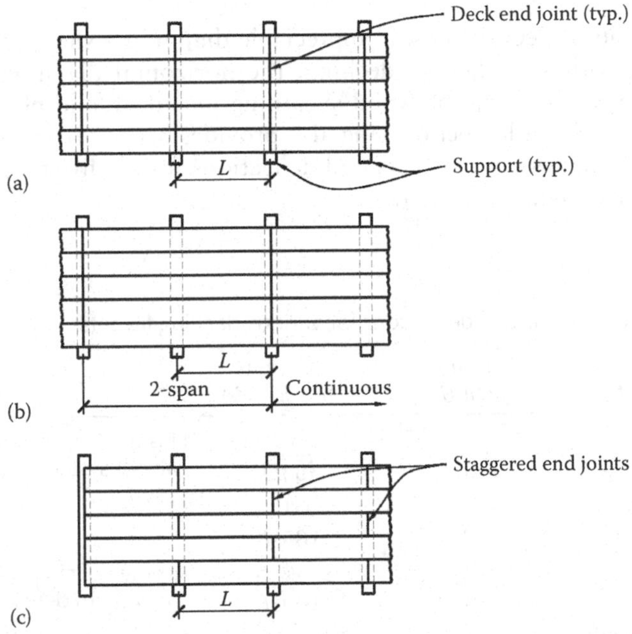

Figure 4.20 Common deck layout profiles: (a) Simple span: decking is supported at each end, (b) two-span continuous: decking is supported at each end, as well as one interior support, (c) combined simple two-span continuous: same as two-span continuous with end joints of decking alternated and alternate pieces in end span are simple span. *(Continued)*

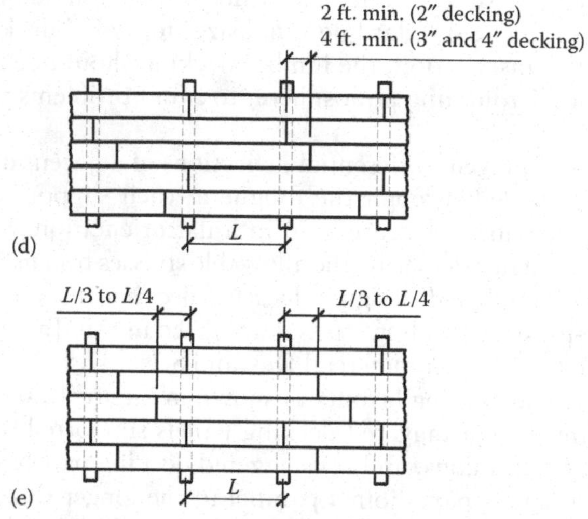

Figure 4.20 (Continued) Common deck layout profiles: (d) controlled-random layout: end joints are generally in-line, but are offset 2 ft. minimum for 2 in. decking and 4 ft. minimum for 3- and 4- in. decking, and (e) cantilevered pieces intermixed: decking is simple span in every third course: other pieces are cantilevered over the supports in alternating third or quarter points of the span.

where the panel joints are continuous. If appreciable diaphragm shear capacity is required in a floor framing with sawn-lumber decking, the horizontal diaphragm shear capacity can be enhanced by using $\frac{5}{16}$-in. plywood sheathing nailed to, and placed on top of, the sawn-lumber decking. Sawn-lumber decking also provides better fire ratings than plywood-sheathed floors. The maximum moments and deflections for the five types of sawn-lumber decking layout are shown in Table 4.14.

Table 4.14 Maximum moments and deflections for various decking layouts

Type of decking layout	*Maximum moment* (M_{max})	*Maximum deflection* (Δ)	*Remarks*
Simple span	$\frac{wL^2}{8}$	$\frac{5wL^4}{384EI}$	Has the least stiffness of all the deck layout systems
Two-span continuous	$\frac{wL^2}{8}$	$\frac{wL^4}{185EI}$	
Combined simple and two-span continuous	$\frac{wL^2}{8}$	$\frac{wL^4}{109EI}$	Average of simple-span and two-span maximum deflections
Controlled random layout	$\frac{wL^2}{10}$	$\frac{wL^4}{145EI_e}$	Maximum deflection for three-span decking used; $I_e = 0.67(I)$ for 2-in. decking and $0.80I$ for 3 and 4 in. decking
Cantilever pieces intermixed	$\frac{wL^2}{10}$	$\frac{wL^4}{105EI_e}$	

Source: Ambrose, J., *Simplified Design of Wood Structures*, 5th ed., Wiley, New York, 1994; Stalnaker, J.J. and Harris, E.C., *Structural Design in Wood*, Chapman & Hall, London, 1997.

Example 4.7: Design of sawn-lumber decking (ASD/LRFD)

Design sawn-lumber roof decking to span 10 ft 0 in. between roof trusses. The superimposed roof dead load is 16 psf and the snow load is 35 psf. Assume that dry service and normal temperature conditions apply and Douglas Fir-Larch commercial decking (DF-L Commercial Dex).

Solution:
Assume 3 × 6 in. decking (actual size = 2.5 × 5.5 in.) and simple-span decking layout in Table 4.13. The section properties for 3 × 6 in. decking are

A_{xx} = (13.75 in.^{2})(12 in./5.5 in.) = 30 in.^{2} per foot width of deck
S_{yy} = (5.729 in.^{3})(12 in./5.5 in.) = 12.5 in.^{3} per foot width of deck
I_{yy} = (7.161 in.^{4})(12 in./5.5 in.) = 15.62 in.^{4} per foot width of deck

Loads:

$$\text{Decking self-weight}\left(\frac{2.5\text{ in.}}{12}\times 32\text{ psf}\right)=6.7\text{ psf}$$

Superimposed dead loads = 16 psf

Total dead load D = 22.7 psf

Snow load S = 35 psf

The governing load combination for this roof deck will be dead load plus snow load ($D + S$):

total load = $D + S$ = 22.7 + 35 = 57.7 psf

Bending stress:

$$\text{Maximum moment } M_{\max}=\frac{wL^2}{8}=\frac{(57.7\text{ psf})(10\text{ ft})^2}{8}=722\text{ ft-lb per foot width of deck}$$

$$= 8664\text{ in.-lb per foot width of deck}$$

The tabulated stress values and appropriate adjustment factors from NDS-S Table 4E for 3 × 6 in. decking are

$F_b C_r$ = 1650 psi (repetitive member)
$F_{c\perp}$ = 625 psi
$E = 1.7 \times 10^6$
C_F = 1.04 (for 3-in. decking)
C_i = 1.0 (lumber is not incised)

The allowable stresses are calculated using the sawn-lumber adjustment factor applicability Table (Table 3.1) as follows:

$$F_b' = F_b C_r C_D C_M C_t C_L C_F C_i = (1650)(1)(1)(1)(1)(1.04)(1) = 1716\text{ psi}$$

$$E' = E C_M C_t C_i = (1.7\times 10^6)(1)(1)(1) = 1.7\times 10^6\text{ psi}$$

The applied bending stress for bending about the weak (y–y) axis is

$$f_{by}=\frac{M_{\max}}{S_{yy}}=\frac{8664\text{ lb-in.}}{12.5\text{ in}^3}=693\text{ psi} \ll \text{allowable stress } F_b'=1716\text{ psi} \qquad \text{OK}$$

Deflection:

Dead load $D = 22.7$ psf = 22.7 lb/ft per foot width of deck = 1.89 lb/in. per foot width of deck

Live load (snow) $S = 35$ psf = 35 lb/ft per foot width of deck = 2.92 lb/in. per foot width of deck. The dead-load deflection is

$$\Delta_{\text{DL}} = \frac{5wL^4}{384E'I_{yy}} = \frac{(5)(1.89\text{ lb/in.})(10\text{ ft}\times 12)^4}{(384)(1.7\times 10^6\text{ psi})(15.62\text{ in.}^4)} = 0.19\text{ in.}$$

The live-load deflection is

$$\Delta_{\text{LL}} = \frac{5wL^4}{384E'I_{yy}} = \frac{(5)(2.92\text{ lb/in.})(10\text{ ft}\times 12)^4}{(384)(1.7\times 10^6\text{ psi})(15.62\text{ in.}^4)} = 0.30\text{ in.}$$

$$< \frac{L}{360} = \frac{(10\text{ ft})(12\text{ in./ft})}{360} = 0.33\text{ in.} \qquad \text{OK}$$

Since seasoned wood in dry service conditions is assumed to be used in this building, the creep factor $k = 0.5$. The total incremental dead plus floor live-load deflection is

$$\boldsymbol{\Delta}_{\text{TL}} = k\boldsymbol{\Delta}_{\text{DL}} + \boldsymbol{\Delta}_{\text{LL}}$$

$$= (0.5)(0.19\text{ in.}) + 0.30\text{ in.} = 0.40\text{ in.} < \frac{L}{240} = \frac{(10\text{ ft})(12\text{ in./ft})}{240} = 0.5\text{ in.} \qquad \text{OK}$$

Use a 3 × 6 in. DF-L Commercial Dex with a simple-span deck layout.

LRFD Solution:

Example 4.8: Design of sawn-lumber decking

Design sawn-lumber roof decking to span 10 ft 0 in. between roof trusses. The superimposed roof dead load is 16 psf and the snow load is 35 psf. Assume that dry service and normal temperature conditions apply and Douglas Fir-Larch commercial decking (DF-L Commercial Dex).

Solution:

Assume 3 × 6 in. decking (actual size = 2.5 × 5.5 in.) and simple-span decking layout in Table 4.14. The section properties for 3 × 6 in. decking are

$A_{xx} = (13.75\text{ in}^2)(12\text{ in./}5.5\text{in.}) = 30\text{ in}^2$ per foot width of deck

$S_{yy} = (5.729\text{ in}^3)(12\text{ in./}5.5\text{in.}) = 12.5\text{in}^3$ per foot width of deck

$I_{yy} = (7.161\text{ in}^4)(12\text{ in./}5.5\text{in.}) = 15.62\text{ in}^4$ per foot width of deck

Loads:

Decking self-weight $\left(\frac{2.5\text{in.}}{12}\times 32\text{psf}\right)$	= 6.7 psf
Superimposed dead loads	= 16 psf
Total dead load *D*	= 22.7 psf
Snow load *S*	= 35 psf

The governing load combination for this roof deck will be 1.2D + 1.6S

$w_U = (1.2)(22.7)+(1.6)(35) = 83.3$ psf

Bending stress:

Maximum moment $M_U = \frac{w_U L^2}{8} = \frac{(83.3\text{ psf})(10\text{ ft})^2}{8} = 1041$ ft-lb per foot width of deck

$= 12486$ in.-lb per foot width of deck

The tabulated stress values and appropriate adjustment factors from NDS-S Table 4E for 3 × 6 in. decking are:

$F_b C_r = 1650$ psi (repetitive member)
$F_{c\perp} = 625$ psi
$E = 1.7 \times 106$
$C_F = 1.04$ (for 3-in. decking)
$C_i = 1.0$ (lumber is not incised)

The available stresses are calculated using the sawn-lumber adjustment factors applicability table (Table 3.3) as follows:

$$F'_{b,r} = F_b K_{F,b} \phi_b \lambda C_M C_t C_L C_F C_r C_{fu} C_i$$

$$= (1650)(2.54)(0.85)(0.8)(1.0)(1.0)(1.0)(1.0)(1.04) = \mathbf{2963\ psi}$$

$$E' = E C_M C_t C_i = (1.7 \times 10^6)(1)(1)(1) = 1.7 \times 10^6 \text{ psi}$$

The factored bending stress for bending about the weak *(y–y)* axis is

$$f_{by} = \frac{M_{max}}{S_{yy}} = \frac{12486\,\text{lb-in.}}{12.5\,\text{in}^3} = 999\,\text{psi} < 2963\,\text{psi} \quad \text{OK}$$

Deflection:
Dead load $D = 22.7$ psf = 22.7 lb/ft per foot width of deck = 1.89 lb/in. per foot width of deck.

Live load (snow) $S = 35$ psf = 35 lb/ft per foot width of deck = 2.92 lb/in per foot width of deck. The dead-load deflection is

$$\Delta_{DL} = \frac{5wL^4}{384E'I_{yy}} = \frac{(5)(1.89\text{ lb/in.})(10\text{ ft}\times 12)^4}{(384)(1.7\times 10^6\text{ psi})(15.62\text{ in}^4)} = 0.19\text{ in.}$$

The live-load deflection is

$$\Delta_{LL} = \frac{5wL^4}{384E'I_{yy}} = \frac{(5)(2.92\text{ lb/in.})(10\text{ ft}\times 12)^4}{(384)(1.7\times 10^6\text{ psi})(15.62\text{ in}^4)} = 0.30\text{ in.}$$

$$< \frac{L}{360} = \frac{(10\text{ ft})(12\text{ in./ft})}{360} = 0.33\text{ in.} \quad \text{OK}$$

Conservatively assuming $k = 1.0$

$$\Delta_{TL} = k\Delta_{DL} + \Delta_{LL}$$

$$= (1.0)(0.19\text{ in.}) + 0.30\text{ in.} = 0.49\text{ in.} < \frac{L}{240} = \frac{(10\text{ ft})(12\text{ in./ft})}{240} = 0.5\text{ in.} \quad \text{OK}$$

Use a 3 × 6 in. DF-L Commercial Dex with a simple-span deck layout.

4.4 MISCELLANEOUS STRESSES IN WOOD MEMBERS

In this section, several other types of stresses that can occur in wood members are discussed. Examples include amplification of the shear stress in notched beams, end-to-end or parallel-to-grain bearing stress, and bearing stress at an angle to the grain.

4.4.1 Shear stress in notched beams

It may be necessary to provide notches and holes in wood beams to allow for passage of mechanical ducts or because of headroom limitations as shown in Figure 4.21. The limitations on the size and placement of notches and holes [4,5] are shown in the figure. The sizes and location of holes in preengineered lumber is specified by the manufacturer. Preengineered lightgage metal plates that are screwed and/or glued to the sides of the joist can be used for reinforcing floor joists with larger notches and holes [6].

The notching of a beam creates stress concentrations, in addition to reducing the allowable shear stress due to a reduction in the depth of the beam at the notch. The NDS code requirements for notched sawn-lumber beams are as follows (Figure 4.21):

- No notches are permitted within the middle third of the beam span.
- No notches are permitted within the outer thirds of the beam span on the tension face (except at beam bearing supports) when the thickness of a beam b is ≥3.5 in.
- Within the outer thirds of the beam span, the depth of the notched beam, d_n, must be ≥(5/6)d.
- At the beam supports, the depth of the notched beam, d_n, must be ≥(3/4)d.
- The notch length l_n must be ≤(1/3)d.

The NDS code equation for calculating the adjusted applied shear stress when the notch is on the *tension* face is given as

$$f_v = \frac{\frac{3}{2}V}{bd_n}\left(\frac{d}{d_n}\right)^2 \le \text{allowable shear stress } F_v' \tag{4.20}$$

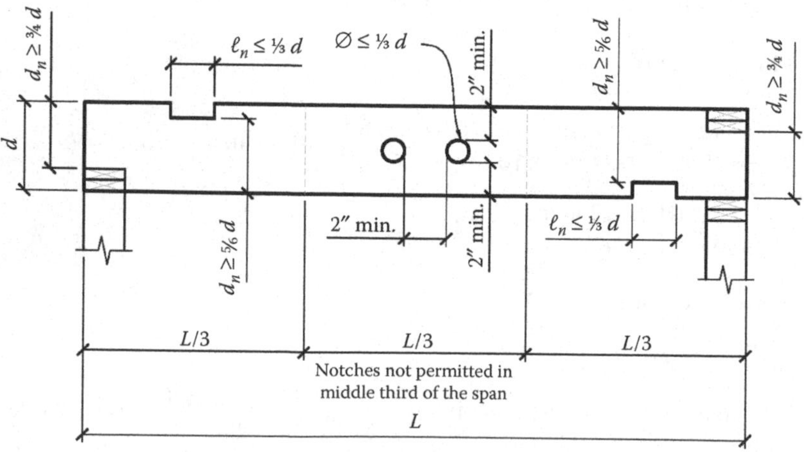

Figure 4.21 Notches and holes in sawn lumber.

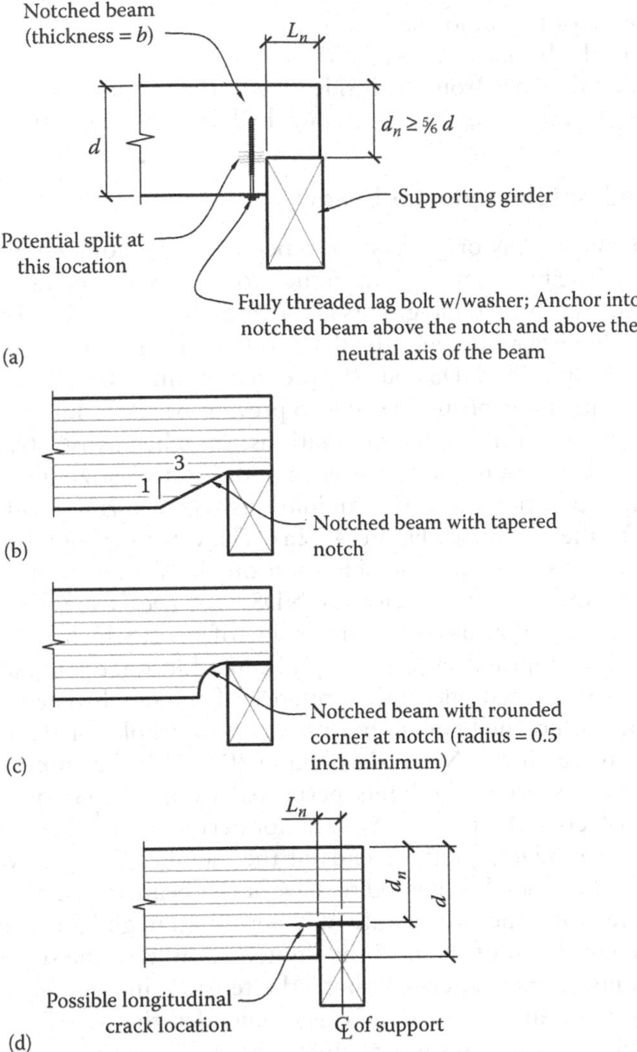

Figure 4.22 Notches in beams: (a) reinforced notch, (b) tapered notch, (c) notch with rounded corner, and (d) unreinforced square notch.

The $(d/d_n)^2$ term in Equation 4.20 accounts for the stress concentration effects at a square notch due to the abrupt change in cross section resulting from the notch. For notches with a gradual taper slope (i.e., 3 horizontal:1 vertical over a horizontal length of at least $4(d\text{-}d_n)$) or rounded notch corners with a minimum radius of 0.5 in., or a reinforced notch as discussed later in this chapter (see Figures 4.22b, 4.22c,and 4.22d), the stress concentration effects will be negligible and for those cases, the applied shear stress, f_v, becomes $\frac{3}{2}V/bd_n$, which is the shear stress of an unnotched beam with a reduced depth, d_n [7,8].

For a notch on the *compression* face, the adjusted applied shear stress is given as

$$f_v = \frac{\frac{3}{2}V}{b\{d - [(d - d_n/d_n])e\}} \qquad (4.21)$$

$$\leq \text{allowable shear stress } F_v'$$

where:

d is the total depth of the unnotched beam

d_n is the depth of the beam remaining at the notch

e is the horizontal distance from the inside face of the support to the edge of the notch (e must be $\leq d_n$; if $e > d_n$, use $e = d_n$ in calculating the shear stress)

4.4.2 Tapered and notched glulam beams

Sometimes, glulam roof beams or girders are tapered on the compression side to achieve proper roof drainage. The taper can be done in the shop when the glulam is manufactured or cut in the field [7]. Glulam beams and girders are also sometimes notched at the ends to provide bearing at the supports and reduce the floor-to-floor height of a building or to achieve lateral stability of a beam. The NDS code [9] prescribes limits on the depth and length of unreinforced notches in glulam beams in order to prevent brittle failure of the glulam beams from cracking or separation along the laminations. Notches cause stress concentrations and perpendicular to the grain tension stresses in addition to horizontal shear stresses that lead to cracking and separation along the laminations which further lead to a reduction in the effective depth of the beam (see Figure 4.22a). The effects of notches are more severe when a beam is notched on the tension side than on the compression side [8,10,11]. To account for the reduction in shear capacity, the NDS code uses a shear reduction factor for notched glulam beams, C_{vr}, that has a uniform value of 0.72, which reduces the allowable shear stress or the LRFD adjusted shear stress [7,8]. In addition, the unfactored or factored applied shear stress is magnified due to the presence of the notch. The cracking or tearing along the laminations of notched glulam beams was responsible for the failure of a pedestrian bridge under construction in North Carolina in 2015 [12]. The notches on these beams were unreinforced and exceeded the limits permitted by the NDS code for unreinforced notches [12]. The NDS code (Section 5.4.5) does not permit glulam bending members to be notched in the tension zone except at the ends of the member bearing over a support, and the maximum depth of the notch is limited to 10% of the beam depth or 3 in., whichever is smaller. In the compression zone, notches are only permitted in glulam beams at the member ends and the maximum depth of the notch is limited to 40% of the depth of the member. The code also does not permit notches in both the tension and compression zones at the same location along the beam or girder. It is recommended that notched glulam beams be avoided where possible, but in cases where glulam beams have to be notched, the notched corners should be rounded off or tapered to make the change in cross section less abrupt and more gradual, and thus reduce stress concentrations. The bearing length of the notch is chosen so as not to exceed the allowable bearing stress perpendicular-to-grain for allowable stress design or the LRFD adjusted design bearing stress perpendicular-to-grain for ultimate strength design (LRFD).

4.4.3 Reinforced notched glulam beams

The adverse effect of stress concentrations on the load capacity of notched glulam beams can be reduced by reinforcing the beam through its depth with internal or external reinforcement. Internal reinforcement may consist of vertical through-bolts or threaded lag screws near the notch corner, while external reinforcement may take the form of a U-shaped carbon fiber reinforced plastic wrap near the notch corner; the internal or external reinforcement also helps to prevent brittle failure from the tearing or cracking along the glulam laminations that typically starts at the notch corners [8,10,11]. Thus for a reinforced notch on the tension zone at the end of a beam, the shear stress is calculated using Equation 4.20 but the $(d/d_n)^2$ term or the stress concentration effect is neglected. The reinforcement will be subjected to forces parallel and perpendicular to the glulam laminations and thus should be capable of resisting a tension force perpendicular to grain of the gulam and shear forces parallel to the grain of the gulam (i.e., parallel to the laminations) [8]. The axial and shear forces acting on the internal vertical reinforcement in a reinforced notched glulam beam can be calculated approximately as follows [8] (see Figure 4.23):

$$F_{\text{vertical}} = V\,(1 - \alpha)$$

$$F_{\text{horizontal}} = 3V\,(\alpha - \alpha^2)\,(2\beta + 1)$$

where:

$\alpha = d_n/d$
$\beta = L_n/d$
d_n is the vertical depth of the beam at the notched section
d is the total overall depth of the gulam beam
L_n is the horizontal length of the notch from the centerline of the beam support to the notch corner or the beginning of the notch corner for rounded or tapered notches
V is the shear at the support
F_{vertical} is the axial force in the reinforcement
F_{vertical} is the shear force in the reinforcement

Example 4.9: Notched glulam beam (ASD)

A 13½ in. × 60 in. deep 20F-V12 glulam girder is used to support a simply supported pedestrian bridge spanning 60 ft over a ravine. A cross section of the bridge with the floor framing is shown in Figure 4.24. Assume the wood density is 36pcf.

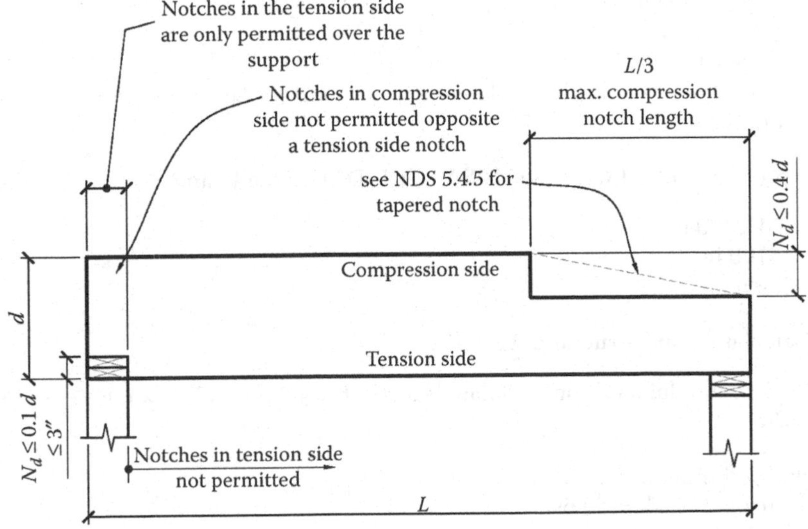

Figure 4.23 Notches and holes in glulam beams.

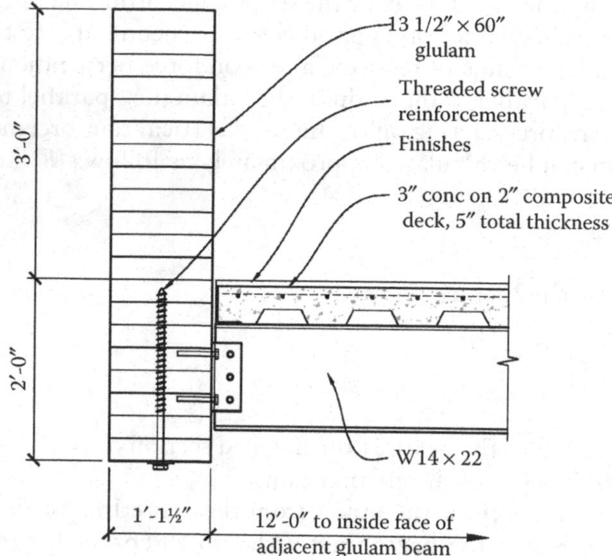

Figure 4.24 Glulam beam with reinforcement.

1. Given that the glulam beam is notched in the tension zone at both end supports, determine the allowable shear stress (ASD) and check if the beam is adequate to resist the unfactored applied shear stress. Assume that the density of the glulam is 36 pcf.
2. Determine the allowable bending stress and check if the beam is adequate in bending. The beam has lateral supports only at the end supports.

Solution:
From NDS-S (Table 5A–Expanded) tabulated, the reference design stresses are as follows:

F^{+}_{bx} = 2000 psi (tension lamination stressed in tension)
$F_{v,xx}$ = 265 psi
$F_{c\perp xx,\text{ tension lam}}$ = 560 psi
$E_x = 1.5 \times 10^6$ psi
$E_y = 1.4 \times 10^6$ psi
$E_{y\,min} = 0.74 \times 10^6$ psi ($E_{y,\,min}$, and not $E_{x,\,min}$, is used for lateral buckling of the girder about the weak axis)

Section properties of a **13.5″ × 60″ 20F-V12 DF/DF Glulam Beam:**

I_{xx} = 243,000 in.^{4}
S_{xx} = 8100 in.^{3}
A = 810 in.^{2}

Load Calculation and Structural Analysis:

Dead load on the glulam beam – Glulam beam self-weight = [(13.5″ × 60″)/144](36 pcf) = **203 lb/ft**

Floor slab and framing:
3″ concrete on 2″ deck = 50 psf
Finishes = 25 psf
Steel framing = 5 psf

Total floor slab and framing load = **80 psf**
Total dead load on the Glulam beam = 80 psf × (12 ft/2) + 203 lb/ft = 683 lb/ft
Total live load on the Glulam beam = 100 psf (12 ft/2) = 600 lb/ft
Total unfactored load ($D + L$) = 600 + 683 = **1283 lb/ft**
Total maximum shear at the supports, V = 1283 lb/ft (60 ft/2) = 38,490 lb

$$\text{Maximum moment, } M = 1283\,\frac{(60\,\text{ft})^2}{8} = 577{,}350\text{ ft-lb} = 6{,}928{,}200\text{ in.-lb}$$

Part (a)—Shear Stress:

Horizontal shear stress parallel to the grain, $F_{v,\,xx} = 265$ psi

Adjustment factors:

For $D + L$, $C_D = 1.25$
C_M (*for* F_v) = 0.875 (moisture content will be greater than 16% since the beam is exposed)
C_M (*for* F_b) = 0.8 (moisture content will be greater than 16% since the beam is exposed)
For a notched glulam beam, the shear reduction factor, $C_{vr} = 0.72$
$C_t = 1.0$
$C_{fu} = 1.0$
$C_c = 1.0$
$C_I = 1.0$

Notch Depth:

The NDS code Section 5.4.5 limits notches in the tension zone to the end supports and with a maximum depth of 10% of the unreduced depth or 3 in., whichever is smaller.

Notch depth = 10% (60″) = 6″ or 3″,
Therefore, use 3″ notch depth
Thus, $d_n = d$ – notch depth = 60″ – 3″ = 57″

(**Note**: Using a notch depth greater than 3″ for this unreinforced notched glulam beam would be in violation of the NDS code. Such was the case with the Wake Forest Pedestrian Bridge that collapsed in 2014 [12].)

Allowable horizontal shear stress (ASD) for the notched glulam beam at the bearing support:

$$F'_{v,\,xx} = F'_{v,\,xx}C_D C_M C_t C_{vr}$$
$$= (265)(1.25)(0.875)(1)(\mathbf{0.72}) = 209\text{ psi}$$

Applied stress from the unfactored loads:

$$f_v = \frac{3V}{2bd_n}\left(\frac{d}{d_n}\right)^2 = \frac{(3)(38{,}490\text{ Ib})}{(2)(13.5")(57")}\left(\frac{60"}{57"}\right)^2 = 83\,\text{psi} < F_v' = 209\,\text{psi} \qquad \text{OK}$$

Part (b)—Bending stress:

$F^+_{bx} = 2000$ psi (tension lamination stressed in tension)
$E_x = 1.5 \times 10^6$ psi
$E_y = 1.4 \times 10^6$ psi
$E_{y,\,min} = 0.74 \times 10^6$ psi ($E_{y,\,min}$, and not $E_{x,min}$, is used for lateral buckling of the girder about the weak axis)

From the adjustment factor applicability Table for glulam (Table 3.2), we obtain the allowable bending stress of the glulam girder with C_V and C_L equal to 1.0 as

$$F^{*}_{bx} = F^{+}_{bx}C_D C_M C_t C_{fu} C_c C_I = (2000)(1.25)(0.80)(1.0)(1.0)(1.0)(1.0) = \mathbf{2000\ psi}$$

The allowable pure bending modulus of elasticity E'_x and the bending stability modulus of elasticity $E'_{y,\min}$ are calculated as

$$E'_x = E_x C_M C_t = (1.5 \times 10^6)(0.833)(1.0) = 1.25 \times 10^6 \text{ psi}$$

$$E'_{y,\min} = E_{y,\min} C_M C_t = (0.74 \times 10^6)(0.833)(1.0) = 0.62 \times 10^6 \text{ psi}$$

4.4.4 Calculate the beam stability factor C_L

From Equation 3.3 the beam stability factor is calculated as follows: the unsupported length of the compression edge of the beam or the distance between points of lateral support preventing rotation and/or lateral displacement of the compression edge of the beam is

$l_u = 60$ ft $= 720$ in. (i.e., beam is unbraced for its full span between supports)

$$\frac{l_u}{d} = \frac{720 \text{ in.}}{60 \text{ in.}} = 12$$

Since the beam is uniformly loaded and using the l_u/d value of 12 above, the effective length of the beam is obtained from Table 3.10 (or NDS code Table 3.3.3) as

$$l_e = 1.63 l_u + 3d = (1.63)(720 \text{ in.}) + (3)(60 \text{ in.}) = 1354 \text{ in.}$$

$$R_B = \sqrt{\frac{l_e d}{b^2}} = \sqrt{\frac{(1354)(60 \text{ in.})}{(13.5 \text{ in.})^2}} = 21 \le 50 \qquad \text{OK}$$

$$F_{bE} = \frac{1.20 E'_{\min}}{R_B^2} = \frac{(1.20)(0.62 \times 10^6)}{(21)^2} = 1687 \text{ psi}$$

$$\frac{F_{bE}}{F_b^*} = \frac{1687 \text{ psi}}{2000 \text{ psi}} = 0.84$$

From Equation 3.3 the beam stability factor is calculated as

$$C_L = \frac{1 + F_{bE}/F_b^*}{1.9} - \sqrt{\left(\frac{1 + F_{bE}/F_b^*}{1.9}\right)^2 - \frac{F_{bE}/F_b^*}{0.95}}$$

$$= \frac{1 + 0.84}{1.9} - \sqrt{\left(\frac{1 + 0.84}{1.9}\right)^2 - \frac{0.84}{0.95}} = 0.737$$

4.4.5 Calculate the volume factor C_V

L = length of the beam in feet between the adjacent points of zero moment = 60 ft
d = depth of the beam = 60 in.
b = width of the beam, in. = 13.5 in.
x = 10 (for western species glulam)

From Equation 3.3,

$$C_V = \left(\frac{21}{L}\right)^{1/x}\left(\frac{12}{d}\right)^{1/x}\left(\frac{5.125}{b}\right)^{1/x} = \left(\frac{1291.5}{bdL}\right)^{1/x} = \left[\frac{1291.5}{(13.5\text{ in.})(60\text{ in.})(60\text{ ft})}\right]^{1/10} = 0.696 \le 1.0$$

The smaller of C_V and C_L will govern and is used in the allowable bending stress calculation. Since $C_V = 0.696 < C_L = 0.737$, use $\mathbf{C_V = 0.696}$. Using Table 3.2 (i.e., adjustment factor applicability Table for glulam), we obtain the allowable bending stress as

$$F'^{+}_{bx} = F^{+}_{bx} C_D C_M C_t C_{fu} C_c C_I \,(C_L \text{ or } C_V) = F^{*}_{b}(C_L \text{ or } C_V) = (2000)(0.696) = \mathbf{1392\ psi}$$

Using Equation 4.10, the unfactored applied bending stress is

$$f_b = \frac{M_{\max}}{S_{xx}} = \frac{6{,}928{,}200\text{ lb-in.}}{8100\text{ in}^3} = 855\text{ psi}$$

$$< \text{Allowable bending stress, } F'^{+}_{bx} = 1392\text{ psi} \qquad \textbf{OK}$$

Therefore, the notched Glulam beam is adequate in shear with a 3 in. deep notch and is also adequate in bending with the beam unbraced for its full length between supports. Lateral supports must be provided to the beam at the end supports.

4.4.6 Bearing stress parallel to the grain

Bearing stress parallel to the grain occurs at the ends of two axially loaded compression members bearing on each other end to end, as shown in Figure 4.25, and also where a column bears on a steel base plate or other surface. The applied bearing stress parallel to the grain is given as

$$f_c = \frac{P}{A} \le F_c^{*} \tag{4.22}$$

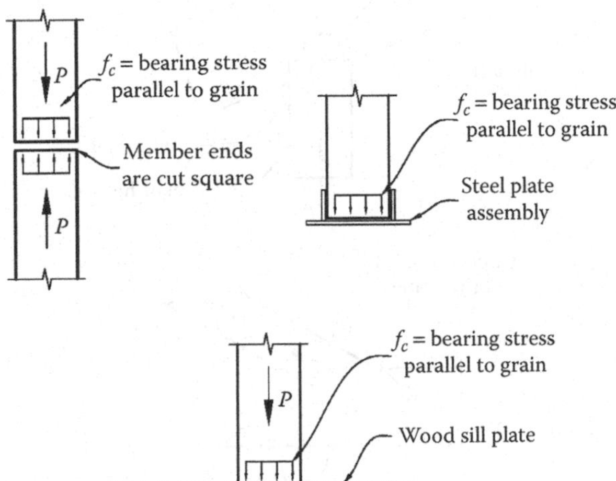

Figure 4.25 Parallel to the grain loading.

where:

A = net bearing area

$F_c^* = F_c C_D C_M C_t C_F C_i$ = allowable compression stress parallel to the grain, excluding the effects of the column stability factor, C_P

P = compression load

F_c = tabulated compression stress parallel to grain (NDS-S Tables 4A to 4D)

The adjustment or C factors are as discussed in Chapter 3. NDS code Section 3.10.1.3 requires that a steel bearing plate or other material with sufficient stiffness be provided whenever the bearing stress parallel to the grain, f_c, is greater than $0.75F_c^*$ to help distribute the load.

4.4.7 Bearing stress at an angle to the grain

This bearing situation occurs at the supports of sloped rafters or where a sloped member frames and bears on a horizontal member as shown in Figure 4.26. The rafter will be subjected to the bearing stress at an angle to the grain, f_θ, and the beam or header or horizontal member will be subjected to bearing stress perpendicular to the grain, $f_{c\perp}$. The bearing stress applied in the rafter at an angle $\boldsymbol{\theta}$ to the grain is calculated as

$$f_\theta = \frac{P}{A} \leq F_\theta' \tag{4.23}$$

where:

P is the applied load or reaction

A is the bearing area

F_θ is the allowable bearing stress at an angle $\boldsymbol{\theta}$ to the grain

$\boldsymbol{\theta}$ is the angle between the direction of the reaction or bearing stress and the direction of the grain (see Figure 4.26)

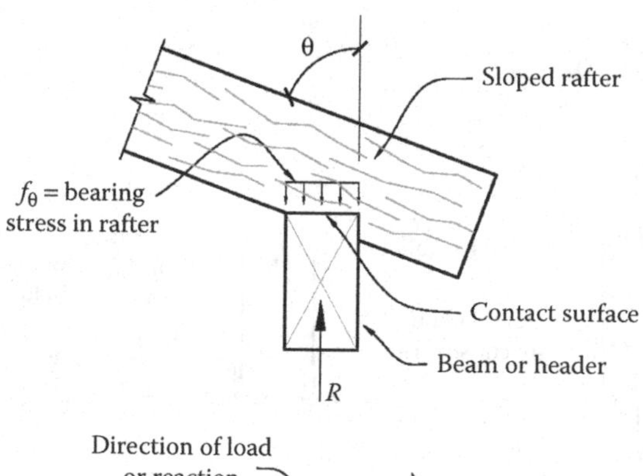

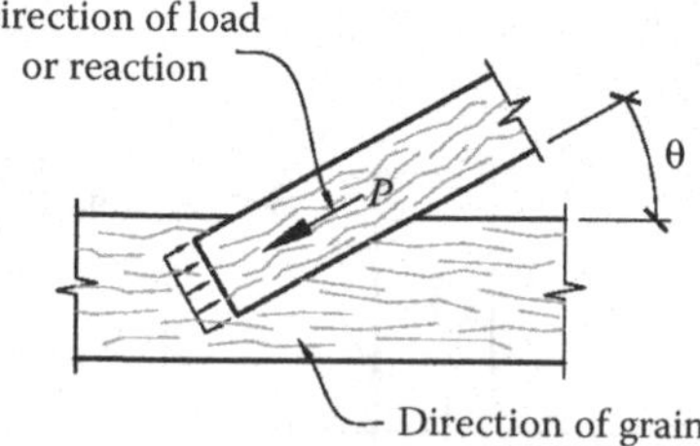

Figure 4.26 Bearing at an angle to the grain.

The allowable bearing stress at an angle to grain is given in the NDS code by the Hankinson formula as

$$F'_\theta = \frac{F_c^* F'_{c\perp}}{F_c^* \sin^2\theta + F_{c\perp}^* \cos^2\theta} \; for\, ASD \tag{4.24}$$

$$F'_{\theta,r} = \frac{F_{c,r}^* F'_{c\perp,r}}{F_{c,r}^* \sin^2\theta + F_{c\perp,r}^* \cos^2\theta} \; for\, LRFD$$

where:

$F_c^* = F_c C_D C_M C_t C_F C_i$ (ASD)

$F_{c,r}^* = F_c C_M C_t C_F C_i\, K_{F,c}\, \phi_c\, \lambda$ (LRFD)

= allowable/available bearing stress *parallel* to the grain of the rafter or sloped member excluding column stability effects

$F'_{c\perp} = F_{c\perp}\, C_M C_t C_i C_b$ (ASD)

$F'_{c\perp r} = F_{c\perp}\, C_M C_t C_b C_i\, K_{F,c\perp} \phi_c \lambda$ (LRFD)

= allowable/available bearing stress *perpendicular* to the grain of the beam or header

Note that when θ is zero, the load or reaction will be parallel to the grain, and F'_θ will be equal to F_c^*. When θ is 90°, the load or reaction will be perpendicular to the grain, and F'_θ will be equal to $F'_{c\perp}$.

4.4.8 Sloped rafter connection

The design procedure for a sloped rafter connection is as follows:

For ASD:

- If $f_\theta \le F'_\theta$ in the sloped rafter, the rafter is adequate in bearing.
- If $f_\theta > F'_\theta$, increase the bearing area A for the rafter either by increasing the thickness b of the rafter and/or the bearing length provided (i.e., the thickness b of the header).
- If $f_{c\perp} = F'_{c\perp}$. in the beam or header, the header is adequate in bearing.
- If $f_{c\perp} > F'_{c\perp}$ in the beam or header, the header is not adequate; therefore, either increase the thickness b of the header or use a steel plate to increase the bearing area $A_{bearing}$ and thus reduce the bearing stress perpendicular to the grain in the beam or header (see Figure 4.27). Note that the steel plate will only help relieve the bearing stress in the beam or header, not the sloped rafter.

For LRFD:

- If $f_{\theta u} \le F'_{\theta,r}$ in the sloped rafter, the rafter is adequate in bearing.
- If $f_{\theta u} > F'_{\theta,r}$, increase the bearing area A for the rafter either by increasing the thickness b of the rafter and/or the bearing length provided (i.e., the thickness b of the header).

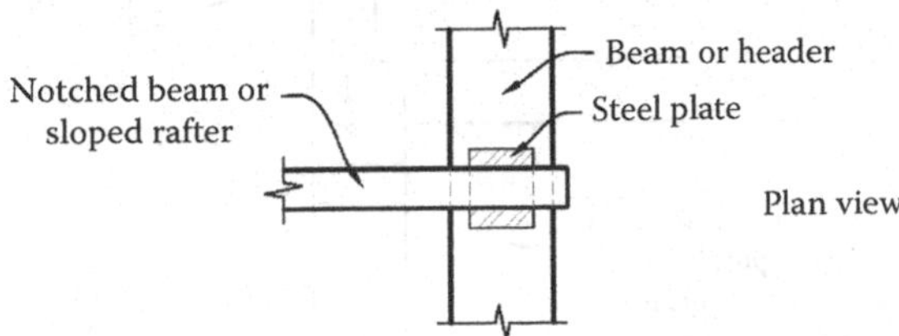

Figure 4.27 Bearing stress on a header.

- If $f_{c\perp u} = F'_{c\perp,r}$, in the beam or header, the header is adequate in bearing.
- If $f_{c\perp u} > F'_{c\perp,r}$ in the beam or header, the header is not adequate; therefore, either increase the thickness b of the header or use a steel plate to increase the bearing area A_{bearing} and thus reduce the bearing stress perpendicular to the grain in the beam or header.

The bearing location and thus critical stress can be altered as shown in Figure 4.27. If a truss has a high end bearing load, the intersection between the end web and bottom chord can be such that the bearing stress is based on loading parallel to the grain, which can have a higher capacity. However, the bearing surface would likely still be loaded perpendicular to the grain and thus might need a plate or similar connection to spread the load out.

Example 4.10: Bearing stresses at a sloped rafter-to-stud wall connection (ASD/LRFD)

The sloped rafter shown in Figure 4.28 is supported on a ridge beam at the ridge line and on the exterior stud wall. The unfactored reaction from dead load plus the snow load on the rafter at the exterior stud wall is 1800 lb (2600 lb factored). Assuming No. 1 Spruce-Pine-Fir (SPF) and normal temperature and dry service conditions, determine

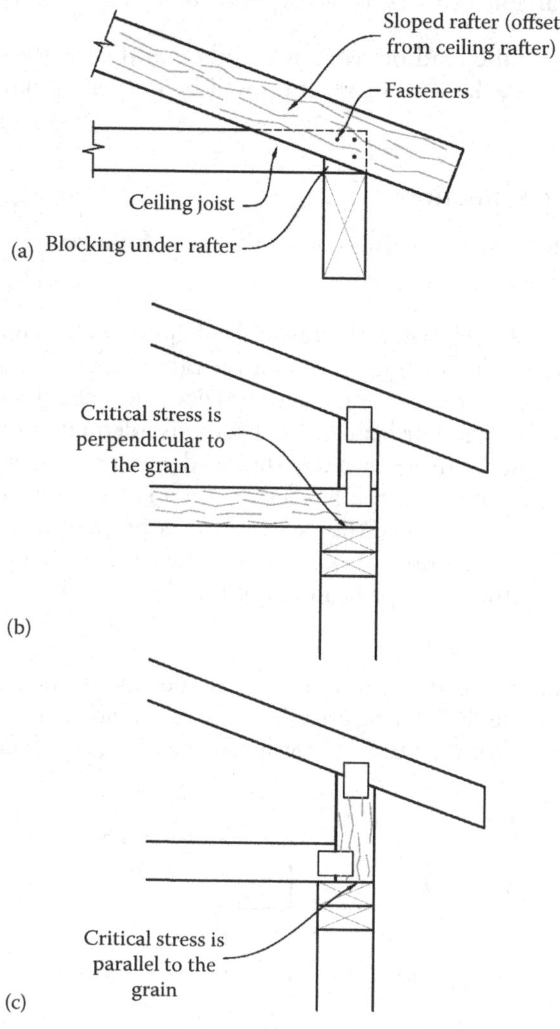

Figure 4.28 Alternate sloped rafter detail.

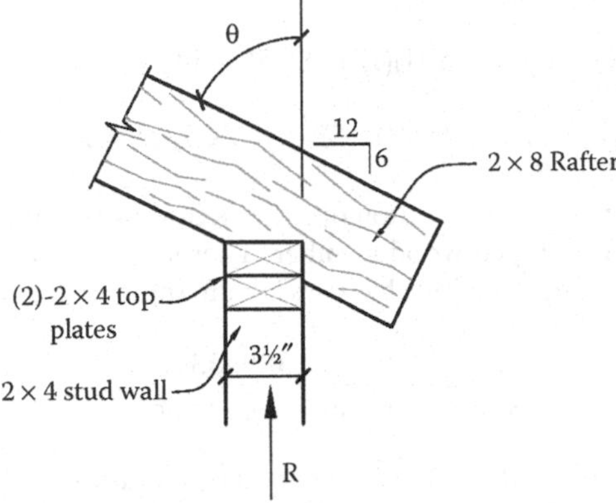

Figure 4.29 Sloped rafter-to-stud wall connection.

(a) the allowable bearing stresses at an angle to the grain F'_θ and the allowable bearing stress or allowable stress perpendicular to the grain $F'_\perp$; (b) the bearing stress at an angle to the grain f_θ in the rafter; (c) the bearing stress perpendicular to the grain $f_{c\perp}$ in the two top plates; and (d) the maximum reaction that could occur based on the above (Figure 4.29). Use the ASD and LRFD methods given below.

ASD Solution:

R = 1800 lb

2 × 8 sloped rafter: $b = 1.5$ in.; $d = 7.25$ in.

2 × 4 top plates on top of stud wall: $d = 3.5$ in.

Since the 2 × 8 rafter and 2 × 4 top plates are dimension lumber, use NDS-S Table 4A. Using the Table we obtain the tabulated stresses for SPF as follows:

bearing stress perpendicular to the grain $F_{c\perp} = 425$ psi

compression stress parallel to the grain $F_c = 1150$ psi

The stress adjustment or C factors are

$C_M = 1.0$ (dry service conditions)
$C_t = 1.0$ (normal temperature conditions)
$C_i = 1.0$ (wood is not incised)
$C_F(F_c) = 1.05$
$C_D = 1.15$ (dead load plus snow load)

The bearing length in the 2 × 4 top plate measured parallel to the grain is the same as the thickness of the sloped rafter: $l_b = b_{\text{rafter}} = 1.5$ in. Since $l_b = 1.5$ in. < 6 in. and the bearing is not nearer than 3 in. from the end of the top plate and the end of the rafter (see Chapter 3), the bearing stress factor is calculated as

$$C_b = \frac{l_b + 0.375\text{ in.}}{l_b} = \frac{1.5\text{ in.} + 0.375\text{ in.}}{1.5\text{ in.}} = 1.25$$

The bearing area at the rafter–stud wall connection is

$$A_{\text{bearing}} = (\text{thickness } b \text{ of rafter})(\text{width } d \text{ of the } 2 \times 4 \text{ top plates})$$

$$= (1.5\text{ in.})(3.5\text{ in.}) = 5.25\text{ in.}^2$$

1. *Allowable bearing stresses:*

$$F'_{c\perp} = F_{c\perp}C_M C_t C_i C_b = (425)(1)(1)(1)(1.25) = 531 \text{ psi}$$

$$F^*_c = F_c C_D C_M C_t C_F C_i = (1150)(1.15)(1)(1)(1.05)(1) = 1389 \text{ psi}$$

θ is the angle between the direction of the reaction or bearing stress and the direction of grain in the sloped wood member. Therefore, tan **θ** = 12/6 and **θ** = 63.4°. Using Equation 4.24, the allowable stress at an angle to grain is given as

$$F'_\theta = \frac{F^*_c F'_{c\perp}}{F^*_c \sin^2\theta + F'_{c\perp}\cos^2\theta} = \frac{(1389)(531)}{1389\sin^2 63.4 + 531\cos^2 63.4} = 606 \text{ psi}$$

2. *Bearing stress applied at angle* **θ** *to the grain of the rafter:*

$$f_\theta = \frac{R}{A_{\text{bearing}}} = \frac{1800 \text{ lb}}{5.25 \text{ in.}^2}$$

$$= 343 \text{ psi} < F'_\theta = 606 \text{ psi} \quad \text{OK}$$

3. *Bearing stress applied perpendicular to the grain of the top plate:*

$$f_{c\perp} = \frac{R}{A_{\text{bearing}}} = \frac{1800 \text{ lb}}{5.25 \text{ in.}^2}$$

$$= 343 \text{ psi} < F'_{c\perp} = 531 \text{ psi} \quad \text{OK}$$

4. *The critical stress is perpendicular to the grain, thus the maximum reaction is:*

$$R_{max} = (531 \text{ psi})(5.25) = 2787 \text{ lb}$$

LRFD Solution:
Use a factored load of R_u = *2600 lb*

2 × 8 sloped rafter: b = 1.5 in.; d = 7.25 in.

2 × 4 top plates on top of the stud wall: d = 3.5 in.

Since the 2 × 8 rafter and 2 × 4 top plates are dimension lumber, use NDS-S Table 4A. Using the Table we obtain the tabulated stresses for SPF as follows:

bearing stress perpendicular to the grain $F_{c\perp}$ = 425 psi

compression stress parallel to the grain F_c = 1150 psi

The stress adjustment or C factors are

C_M = 1.0 (dry service conditions)
C_t = 1.0 (normal temperature conditions)
C_i = 1.0 (wood is not incised)
$C_F(F_c)$ = 1.05

The bearing length in the 2 × 4 top plate measured parallel to the grain is the same as the thickness of the sloped rafter: $l_b = b_{\text{rafter}}$ = 1.5 in. Since l_b = 1.5 in. < 6 in. and the bearing is not nearer than 3 in. from the end of the top plate and the end of the rafter (see Chapter 3), the bearing stress factor is calculated as

$$C_b = \frac{l_b + 0.375 \text{ in.}}{l_b} = \frac{1.5 \text{ in.} + 0.375 \text{ in.}}{1.5 \text{ in.}} = 1.25$$

The bearing area at the rafter–stud wall connection is

A_{bearing} = (thickness b of rafter)(width d of the 2 × 4 top plates)

= (1.5 in.)(3.5 in.) = 5.25 in.^{2}

1. *Available bearing stresses:*

$$F'_{c\perp,r} = F_{c\perp} C_M C_t C_i C_b K_{F,c\perp} \phi_c \lambda = (425)(1)(1)(1)(1.25)(1.67)(0.9)(0.8) = 639 \text{ psi}$$

$$F^*_{c,r} = F_c C_M C_t C_F C_i K_{F,c} \phi_c \lambda = (1150)(1)(1)(1.05)(1)(2.4)(0.9)(0.8) = 2087 \text{ psi}$$

$K_{F,c\perp} = 1.67$

$K_{F,c} = 2.4$

$\phi_c = 0.9$

$\lambda = 0.8$

$\boldsymbol{\theta}$ is the angle between the direction of the reaction or bearing stress and the direction of grain in the sloped wood member. Therefore, tan $\boldsymbol{\theta}$ = 12/6 and $\boldsymbol{\theta}$ = 63.4°. Using Equation 4.24, the allowable stress at an angle to grain is given as

$$F'_{\theta,r} = \frac{F^*_{c,r} F'_{c\perp,r}}{F^*_{c,r} \sin^2\theta + F^*_{c\perp,r} \cos^2\theta} = \frac{(2087)(639)}{2087 \sin^2 63.4 + 639 \cos^2 63.4} = 742 \text{ psi}$$

2. *Factored stress applied at angle* $\boldsymbol{\theta}$ *to the grain of the rafter:*

$$f_{\theta u} = \frac{R_u}{A_{\text{bearing}}} = \frac{2600 \text{ lb}}{5.25 \text{ in.}^2}$$

$= 495 \text{ psi} < F'_{\theta,r} = 742 \text{ psi}$ OK

3. *Bearing stress applied perpendicular to the grain of the top plate:*

$$f_{c\perp} = \frac{R_u}{A_{\text{bearing}}} = \frac{2600 \text{ lb}}{5.25 \text{ in.}^2}$$

$= 495 \text{ psi} < F'_{c\perp} = 639 \text{ psi}$ OK

4. *The critical stress is perpendicular to the grain, thus the maximum factored reaction is:*

$R_{u\max}$ = *(639* psi*)(5.25) = 3354* lb

4.5 PREENGINEERED LUMBER BEAMS AND HEADERS

In this section we discuss the selection of preengineered lumber such as laminated veneer lumber (LVL), laminated strand lumber (LSL), and parallel strand lumber (PSL) using manufacturers' published load Tables. These proprietary wood members usually have fewer defects and a lower moisture content than sawn-lumber beams and columns. They are less susceptible to shrinkage and are used for spans and loadings that exceed the capacity of sawn-lumber beams. They are also more dimensionally stable. They are also sometimes

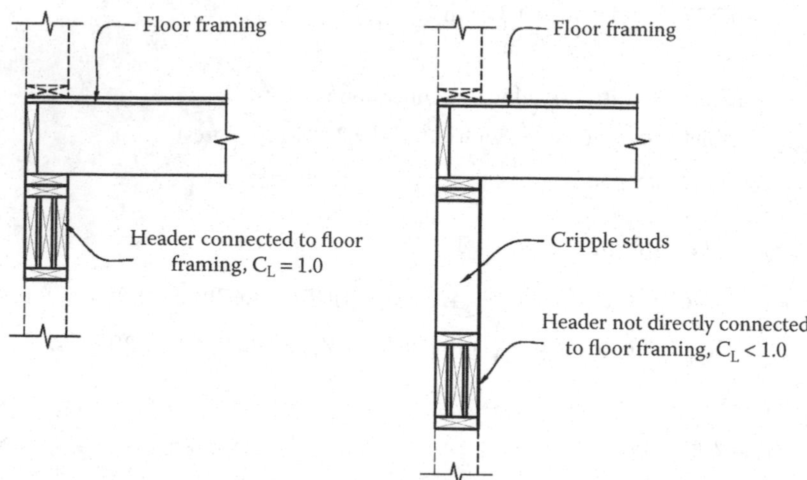

Figure 4.30 Dropped headers.

preferred over steel in cases where a steel beam would prove to be more economical in that wood products are more compatible and more readily available than steel sections. Builders experienced with wood construction often prefer to use wood as that is the material they are more accustomed to working with.

The material properties for engineered wood products are provided by the manufacturer and they are designed very similar to sawn lumber. In many cases, the manufacturer will provide load Tables giving the capacity of the available sections in a form that is easy to use for the engineer or builder, such as column load capacity or uniform load capacity for use as floor joists.

The design of beams and headers usually involves checking just bending and deflection as those are the two primary controlling design parameters. Shear and bearing stresses can control in cases with high concentrated loads. Given this typical use for engineered lumber, the material properties are often described in terms of the base allowable bending stress and modulus of elasticity. For example, the designation 2400F-1.8E would indicate that the member has baseline allowable bending stress, F_b = 2400 psi and modulus of elasticity of $E = 1.8 \times 10^6$ psi. Each manufacturer treats any applicable adjustment factors, such as load duration and size factors, in different ways, so careful attention is needed when reviewing and using the product data from various manufacturers.

For dropped beams or headers that are not directly supporting a roof or floor diaphragm and with no joists or beams framing into them, and without any diagonal braces (see Figure 4.30), there will be insufficient lateral support provided by the roof or floor diaphragm for these members. In this case the unbraced length for the dropped beam or header should be taken as equal to the span length of the beam. This will lead to a large reduction in the bending moment capacity of the dropped beam compared to the case where the beam directly supports a roof or floor diaphragm [13].

Example 4.11: Header beam design using preengineered lumber (ASD)

Refer to Figure 4.31 and the following given information:

floor dead load = 15 psf
floor live load = 40 psf
normal temperature and moisture conditions
full lateral stability is provided ($C_L = 1.0$)

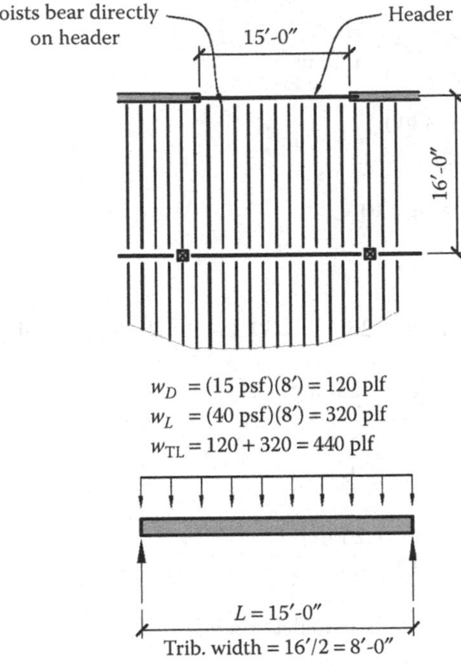

Figure 4.31 Partial floor framing plan and header loading.

Determine the required header size using prefabricated LVL with the following material properties: $F_b = 2600$ psi, $F_v = 285$ psi, and $E = 1900$ ksi.

Solution:
Calculate the maximum moment and shear in the header:

Span of header beam $L = 15$ ft
Tributary width of the header beam = 8 ft
Dead load w_{DL} = (15 psf)(8 ft) = 120 lb/ft
Live load w_{LL} = (40 psf)(8 ft) = 320 lb/ft

From Section 2.1 the governing load combination is dead plus floor live load; thus, the total load is

$$w_{TL} = w_{DL} + w_{LL} = 120 + 320 = 440 \text{ lb/ft}$$

$$M_{max} = \frac{w_{TL}L^2}{8} = \frac{(440)(15)^2}{8} = 12{,}375 \text{ ft-lb}$$

$$V_{max} = \frac{w_{TL}L}{2} = \frac{(440)(15)}{2} = 3300 \text{ lb}$$

Note: After calculating the maximum shear and moment values, the designer could select a specific LVL directly from manufacturers' literature since the manufacturer typically makes this type of information available. After selecting an appropriate product based on shear and moment capacity, the designer would then check the deflection.Since the joists bear directly on the header, the header is fully braced against lateral torsional buckling and thus $C_L = 1.0$.

Required section properties:

$$S_{\text{req'd}} = \frac{M}{F_b} = \frac{(12.375)(12)}{2600} = 57.1 \text{ in.}^3$$

$$A_{\text{req'd}} = \frac{1.5V}{F_v} = \frac{(1.5)(3300)}{285} = 17.3 \text{ in.}^2$$

Select two 1 3/4 × 11 1/4 in. LVLs.

Section properties:

$$A = 2bh = (2)(1.75 \text{ in.})(11.25 \text{ in.}) = 39.3 \text{ in.}^2 > 17.3 \text{ in.}^2 \quad \text{OK}$$

$$S_x = \frac{2bh^2}{6} = \frac{(2)(1.75 \text{ in.})(11.25 \text{ in.})^2}{6}$$

$$= 73.8 \text{ in}^3 > 57.1 \text{ in.}^3 \qquad \text{OK}$$

$$I_x = \frac{2bh^3}{12} = \frac{(2)(1.75 \text{ in.})(11.25 \text{ in.})^3}{12} = 416 \text{ in.}^4$$

Check the deflection. The live-load deflection is

$$\Delta_L = \frac{5w_{LL}L^4}{384EI} = \frac{(5)(320/12)(15\times12)^4}{(384)(1.9\times10^6)(416)} = 0.47 \text{ in.}$$

$$\Delta_{\text{allowable}} = \frac{L}{360} = \frac{(15)(12)}{360} = 0.5 \text{ in.} > 0.47 \text{ in.} \qquad \text{OK}$$

Use two 1 3/4 × 11 1/4 in. LVLs.

Note: Most manufacturers specify fastening requirements for multiple plies of LVLs or PSLs. The fastening can be either nailed or bolted, but the designer should refer to the published requirements from the specific manufacturer for this information.

Example 4.12: Use of the floor joist span tables

Using the floor joist span Tables in Appendix B, select a joist size and spacing for the following:

1. Floor dead load of 15 psf and live load of 100 psf using Hem-Fir #1, span of 13 ft
2. Floor dead load of 20 psf and live load of 40 psf using SPF #1, span of 16 ft
3. Roof dead load of 20 psf and snow load of 55 psf using Douglas Fir-Larch Select Structural, span of 20 ft

Solution:

1. To meet either an *L*/360 or *L*/480 deflection limit, use 2 × 12@12″, which has a maximum span of 14′–4″
2. To meet an *L*/360 deflection limit, use 2 × 10@12′, which has a maximum span of 16′–2″

   To meet an *L*/480 deflection limit, use 2 × 12@16″, which has a maximum span of 16′–3″
3. To meet an *L*/360 deflection limit, use 2 × 12@12″, which has a maximum span of 20′–3″

To meet an $L/240$ deflection limit, use $2 \times 12@16'$, which has a maximum span of 20′–5″. Note that it is conservative to use this Table since the actual snow load is less than the given snow load in the table.

Example 4.13: Design of sloped rafters with collar ties

A sloped rafter with a collar tie is used to create a cathedral ceiling in a residential building (see Figure 4.32). The rafters are spaced at 16 in. on centers. The total roof dead load (including the rafter and collar ties) is 20 psf on a horizontal plan area and the snow load is 35 psf. Neglect the weight of the ceiling on the collar ties. Assume Hem-Fir No. 1. Design the collar tie and the sloped rafters.

Solution:
Structural Analysis: There is no horizontal reaction at A since the top of the stud wall cannot restrain the horizontal displacement of the sloped rafter at A.
Consider the equilibrium of free body diagram ABC,

$\Sigma F_y = 0 \Rightarrow A_y = (74 \text{ lb/ft})\ (8.5 \text{ ft}) = 629 \text{ lb}$

$\Sigma F_x = 0 \Rightarrow C_x = H$

ΣM about point $A = 0 \Rightarrow (74 \text{ lb/ft})(8.5 \text{ ft})(8.5 \text{ ft}/2) + (H)(3 \text{ ft}) - (H)(6 \text{ ft}) = 0$

Therefore, $H = C_x = 891 \text{ lb}$
The maximum moment in the rafter occurs at the collar tie. At that point,

$M_B = (629 \text{ lb})(4.25 \text{ ft}) - (74 \text{ lb/ft})(4.25 \text{ ft})(4.25 \text{ ft}/2) = 2005 \text{ ft-lb}$

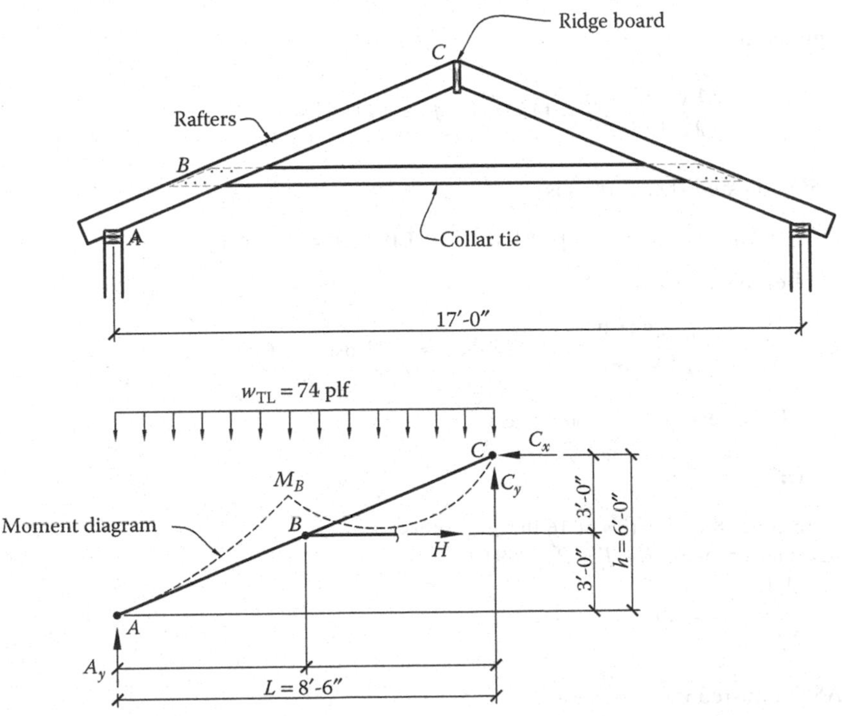

Figure 4.32 Collar tie details.

From the analysis of the free body diagram of the rafter and collar tie (see Figure 4.32), we find

$A_y = 629$ lb

$B_x = 891$ lb

$M_B = 2005$ ft-lb

Analysis Summary:

Sloped rafter: $A_y = 629$ lb; $V_{max} = 891$ lb; $M_{max} = M_B = 2005$ ft-lb = 24,060 in.-lb
Collar tie: Tension force, $B_x = 891$ lb

Design of Sloped Rafter:

Try 2 × 10 sloped rafters and 2 × 8 collar ties at 16 in. on centers
S_{xx} (2 × 10) = 21.39 in.^{3}
$C_D = 1.15$ (dead plus snow)
$C_M = C_t = 1.0$
$C_r = 1.15$ (for the rafters only)
$C_i = 1.0$
$C_F = 1.1$ (for 2 × 10)
Since rafters are braced by the roof deck, $C_L = 1.0$.
The NDS-S reference design values are
$F_b = 975$ psi
$F_v = 150$ psi

The ASD adjusted bending stress is

$$F_b' = F_b C_D C_M C_t C_L C_F C_{fu} C_i C_r = (975\text{ psi})(1.15)(1.0)(1.0)(1.0)(1.1)(1.0)(1.0)(1.15) = 1418\text{ psi}$$

The applied bending stress

$$= \frac{M_{max}}{S_{xx}} = \frac{24{,}060\text{ in.-lb}}{21.39\text{ in.}^3} = 1125\text{ psi} < F_b' = 1418\text{ psi} \qquad \text{OK}$$

The ASD adjusted shear stress is

$$F_v' = F_v C_D C_M C_t C_i = (150\text{ psi})(1.15)(1.0)(1.0)(1.0) = 173\text{ psi}$$

The applied shear stress

$$= \frac{1.5V_{max}}{A} = \frac{(1.5)(891\text{ lb})}{13.88\text{ in.}^2} = 96\text{ psi} << F_v' = 173\text{ psi} \qquad \text{OK}$$

Use 2 × 10 Rafters at 16 in. on centers

Collar tie:

Assume 2 × 8 collar ties at 16 in. on centers
Maximum tension for $T = B_x = 891$ lb
$C_D = 1.15$
$C_M = C_t = C_i = 1.0$
$C_F = 1.2$

The ASD adjusted tension stress is

$$F_t' = F_t C_D C_M C_t C_F C_i = (625)(1.15)(1.0)(1.0)(1.2)(1.0) = 863\text{ psi}$$

Assume that the connection of the collar tie to the sloped rafter is with 1 – ¾″ diameter bolt.
Bolt hole diameter = (¾″ + 1/16″) = 0.813″
Net area at bolt hole = (7.25″ – 0.813″)(1.5″) = 9.66 in.^{2}

The applied tension stress is

$$f_t = \frac{T}{A_{net}} = \frac{891\text{ lb}}{9.66\text{ in.}^2} = 92\text{ psi} < F_t' = 863\text{ psi} \qquad \text{OK}$$

Use 2 × 8 collar ties at 16″ on centers

4.6 BENDING OF TOP PLATES ABOUT ITS WEAK AXIS

Roof rafters and trusses and floor joists often times bear on a single top plate or double top plates between the vertical studs, as shown in Figure 4.33, thus causing bending of the top plates about their weaker axis. In this section, we calculate the bending moment capacity of a top plate in bending. The use of a single top plate is not preferred since a lap splice in the top plate cannot be easily achieved and the presence of at least two plates are needed for strength to transfer loads to the studs. However, in some cases it is more beneficial to reduce the overall shrinkage of the wall and use a single top plate. In practice, it is difficult to achieve a condition where the floor or roof framing is directly over a stud, so either the top plates have to be designed for this loading, or perhaps the distance from a framing member to the nearest stud is limited to about 2 in. to 3 in. If the load occurs within about 3 in. of a stud, then it can be assumed that the load transfers in shear and doesn't create bending in the top plates.

Example 4.14: Bending of top plates

Given the following

- The wood species and stress grade is Hem-Fir No. 1
- The reference design bending stress, F_b = 975 psi
- The spacing of the wall studs in Figure 4.33 is 16 in. o.c. or 1.33 ft
- Wall studs are 2 × 6 with two 2 × 6 double top plates

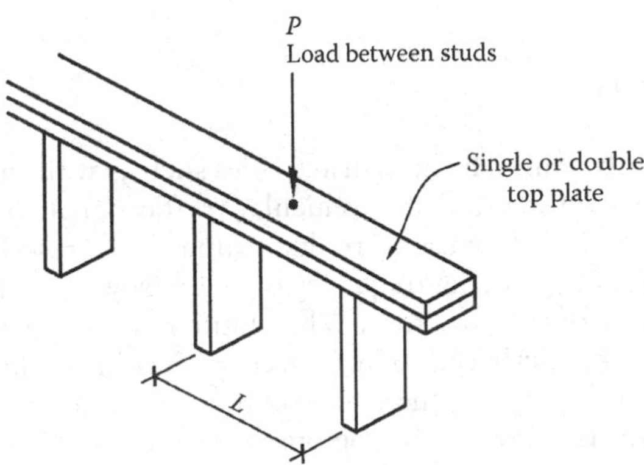

Figure 4.33 Bending of top plates.

Determine the maximum possible load to the top plates.

In practice, there are usually not enough connectors between both plates to enable them to act compositely in bending as one member. So, we are going to make the assumption that they resist the load equally.

$$S_{yy}\ (2 \times 6) = 2.063\ \text{in.}^3$$

$$C_D = C_M = C_t = 1.0$$

$$C_r = 1.0 \text{ and } C_i = 1.0$$

From NDS-S Table 4A, we obtain the following adjustment factors:

$$C_{fu}\ (2 \times 6) = 1.15$$

$$C_F = 1.3$$

Since lateral torsional buckling will be about its stronger axis and the top plates are braced at the stud locations and by the rafters or floor joists, assume $C_L = 1.0$.

The ASD adjusted bending stress is

$$F_b' = F_b C_D C_M C_t C_L C_F C_{fu} C_i C_r$$
$$= (975\ \text{psi})(1.0)(1.0)(1.0)(1.0)(1.3)(1.15)(1.0)(1.0) = 1458\ \text{psi}$$

We will make the following additional assumptions:

- The concentrated reaction from the rafters or floor joist occurs at the midspan of the top plate.
- The two top plates do not act compositely but they resist the load equally.
- The top plates are continuous over several spans.
- We will use centerline to centerline span dimensions.

Based on the above assumptions, we can calculate the applied moment as

$M_{\text{applied}} = 0.171$ PL because of the continuity = (0.171) P (16 in.) = 2.74 P in.-lb

$M_{\text{resisting}}$ (assuming double top plates) = 2 $F_b' S_{yy}$ = (2)(1458 psi)(2.063 in.^{3}) = 6016 in.-lb

$$M_{\text{resisting}} \geq M_{\text{applied}}$$

Therefore, 2.74 P = 6016.

Therefore the maximum reaction from the rafters or floor joist is **P = 2196 lb.**

For studs spaced at 24 in. on centers, **P = 1464 lb.**

4.7 FLOOR VIBRATIONS

Floor vibrations are usually caused by human activities such as walking or rhythmic activity. Machinery or other external forces, such as vehicular traffic, can also cause floor vibrations. Problems associated with floor vibration are due primarily to resonance, which is a phenomenon that occurs when the forcing frequency or frequency of the human activity is at or near the natural frequency of the structure. When human activity such as aerobics occurs, the cyclical force due to this activity produces acceleration in the floor, and as each cycle of loading adds energy to the system, the vibration increases and the acceleration reaches a maximum. This vibration is mitigated by the presence of damping elements such as partitions or ceiling elements. The NDS code does not provide any specific guidance regarding floor vibrations because it is a serviceability issue and not safety related. Traditionally, the

Table 4.15 Acceleration limits for floor vibrations (%)

Activity	*Acceleration limit*, $\frac{a_0}{g} \times 100\%$
Hospital (operating rooms)	0.25
Office, residential, church	0.50
Shopping malls	1.5
Dining, weight lifting	2
Rhythmic activity	5

Source: AISC, *Floor Vibrations Due to Human Activity*, AISC/CISC Steel Design Guide Series 11, American Institute of Steel Construction, Chicago, IL, 1997.

main serviceability criterion used to control vibrations was to limit deflection under uniform design loads to *L*/360. For shorter spans the *L*/360 limit is typically found to be adequate to control floor vibrations. With the advent of preengineered joists, combined with longer spans and more wide-open areas (and thus no partitions to provide damping), some floors have been found not to have good performance with respect to floor vibrations. In the past, higher deflection limits were used to control vibrations. This limit could vary between *L*/360 and *L*/720 and sometimes an absolute value, such as ½″, was used. The use of these limits was based on the experience and judgment of the designer. In the United States, most codes do not require that floors be designed to control vibrations, so the deflection limits to control vibrations are left up to the designer. The National Building Code of Canada does require that floors be designed to control vibrations, which is discussed later. The acceptability of floor vibrations is a function of the occupant's sensitivity to floor vibrations, which can be quite subjective and variable. In general, occupants are more sensitive to floor vibrations when they are engaged in low amounts of activity. The limits on vibration and floor acceleration are usually expressed as a percentage of the acceleration due to gravity g. Table 4.15 shows the generally accepted limits on floor acceleration.

4.7.1 Floor vibration design criteria

Rhythmic activity on light-frame floors (i.e., wood floors) is not very common. These activities usually occur on combined concrete and steel structures, and preferably should occur on the ground floor of a building, if possible. To control floor vibrations due to rhythmic activity, the natural frequency of the floor needs to be higher than the forcing frequency of the highest harmonic of the activity in question. Table 4.16 lists typical minimum floor frequencies.

The natural frequency of a floor system can estimated as follows:

$$f_n = 0.18\sqrt{\frac{g}{\Delta_T}} \tag{4.25}$$

Table 4.16 Typical minimum floor frequencies (Hertz)

Activity	*Steel/concrete floor*	*Light-frame floor*
Dancing and dining	5	10
Rhythmic activity	9	13

where:

f_n is the natural frequency of the floor system, Hz

$g = 386$ in./s^2

Δ_T = total floor deflection, in.

$= \Delta_j + \Delta_g + A_c$

Δ_j is the joist deflection, in.

Δ_g is the girder deflection, in.

Δ_c is the column deflection, in.

For framing systems without girders or columns (such as wood floor framing supported by bearing walls), the girder and column deflection terms would be ignored. A more in-depth coverage of controlling vibrations due to rhythmic activity is given in other references [14].

As stated previously, rhythmic activity on wood-framed floors is not very common. Since wood-framed floors are used primarily for residential, multifamily, and hotel occupancies, the primary design consideration is with walking vibrations. With rhythmic activity, the accumulated energy due to a cyclical load was the primary area of concern. With walking vibrations, the main source of annoyance to the occupants is floor jolts due to a person walking across the floor.

Current research by the ATC, AISC, and others [14–17] is based on response data from occupants on various floor types and is used in this book. To control walking vibrations, floor deflections with respect to beam span should be controlled in accordance with the following equation [15]:

$$\Delta_p \leq 0.024 + 0.1e^{-0.18(L-6.4)} \leq 0.08 \text{ in.} \tag{4.26}$$

where:

Δ_p is the maximum floor deflection (in.) under a concentrated load of 225 lb

L is the beam span, ft

This equation can also be shown graphically (see Figure 4.34). This is also the basis for the serviceability criteria required by the National Building Code of Canada [18], which is outlined as follows in Table 4.17.

The formula for calculating the floor deflection under a 225-lb concentrated load is given as

$$\Delta_p = \frac{C_{pd}}{N_{eff}} \frac{PL^3}{48EI_{eff}} \tag{4.27}$$

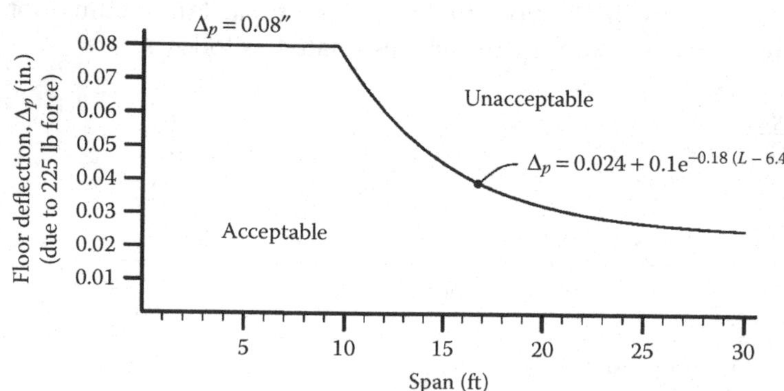

Figure 4.34 Floor deflection vs. span curve for floor vibrations.

where:

Δ_p is the maximum floor deflection, in.
C_{pd} = continuity factor for point load
= 0.7 for continuous
= 1.0 for simple span
N_{eff} is the number of effective joists, ≥1.0
$P = 225$ lb
L is the beam span, in.
EI_{eff} is the effective flexural stiffness of the floor panel, lb-in.^{2}

In calculating the maximum floor deflection due to a point load, the deflection of the supports (i.e., girders and columns) can be ignored if the deflection under total load (dead load plus live load) is less than $L/360$ or 0.5 in., whichever is smaller Table 4.17.

The effective flexural stiffness of preengineered products such as I-joists or wood trusses is usually given by the manufacturer of the specific product. For sawn lumber or where product information is not available, the effective flexural stiffness can be calculated as follows:

$$EI_{eff} = \frac{EI}{1 + \gamma EI / C_{fn} EI_m} \tag{4.28}$$

where:

EI is the stiffness of the floor panel, lb-in.^{2} (includes decking and concrete topping)
$C_{fn} = 1.0$ for simple spans (see Figure 4.33 for continuous spans)
E is the modulus of elasticity, psi
G is the modulus of rigidity, psi (for wood, use $G = 0.1 \times 10^6$ psi or $E/G = 20$)

$$\gamma = \frac{14.4}{(L/r)^2} \frac{E}{G} \quad \text{for sawn lumber} \tag{4.29}$$

$$\gamma = \frac{96 EI_m}{K_s L^2} \quad \text{for preengineered joists} \tag{4.30}$$

L is the joist span, in.
r is the radius of gyration of the joist, in.

Table 4.17 NBC Serviceability Criteria

	Deflection limit based on 225 lb (1 kN) load at midspan	
Span	*USCS*	*Metric*
Under 10 ft (3 m)	$\Delta \le 0.08$ in.	$\Delta \le 2.0$ mm
9–18 ft (3–5.5 m)	$\Delta \le \frac{37.32}{L^{1.3}}$ (L is in inches)	$\Delta \le \frac{8}{L^{1.3}}$ (L is in meters)
18–32 ft (5.5–9.9 m)	$\Delta \le \frac{25.78}{L^{0.63}}$ (L is in inches)	$\Delta \le \frac{2.55}{L^{0.63}}$ (L is in meters)
Over 32 ft (9.9 m)	$\Delta \le 0.024$ in.	$\Delta \le 0.6$ mm
All spans under uniform load of 40 psf (1.9 kN/m^2)	$\Delta \le \frac{L}{360}$	

$= \sqrt{I_m / A_m}$

I_m is the moment of inertia of the joist
K_s is the shear deflection constant
$= (0.4 \times 10^6)d$ lb (for preengineered I-joists)
$= d \times 10^6$ lb (for preengineered metal web trusses)
$= 2 \times 10^6$ lb (for preengineered metal plate-connected wood trusses)
d is the joist depth, in.

It should be noted that shear deflection is not typically critical for solid sawn members, but shear deflection is a factor to be considered for preengineered members because of the thinner webs and shear deformation at the truss joints. Shear deflection can be an issue for shorter spans, but vibrations are less of an issue at shorter spans.

With reference to Figure 4.34, the flexural stiffness of the floor panel, EI, can be determined taking into account the composite action of the decking and concrete topping. For noncomposite floor systems, the flexural stiffness of the floor panel is equal to the flexural stiffness of the joist:

$$EI = EI_m + EI_{top} + EA_m y^2 + EA_{top}(h_{top} - y)^2 \tag{4.31}$$

where:
EI_m is the flexural stiffness of the joist, lb-in.^{2}
EI_{top} is the flexural stiffness of the floor deck parallel to the joist, lb-in.^{2}

$$EI_{top} = EI_{w,par} + EI_c + \frac{EA_{w,par} EA_c h_{cw}^3}{EA_{w,par} + EA_c} \tag{4.32}$$

EA_{top} = effective axial stiffness of the floor deck parallel to the joist

$$EA_{top} = \frac{EA_{flr}}{1 + 10EA_{flr}/S_{flr}L_{flr}^2} \tag{4.33}$$

EA_m = axial stiffness of the joist
EA_{flr} = axial stiffness of the floor deck, lb (see $EA_{w,par}$ in Table 4.18 for wood panels)
EA_c = axial stiffness of the concrete deck, lb
EI_c = flexural stiffness of the concrete deck, lb-in.^{2}
S_{flr} = slip modulus for floor deck connection to the joist (see Table 4.19)
L_{flr} = width of the floor deck (distance between joints in the floor deck), in.
= 48 in. or beam span, L in inches for concrete-topped floors
s = joist spacing, in.

Table 4.18 Approximate stiffness properties of wood subflooring (OSB and plywood) per inch width

Panel thickness (in.)	*$EI_{w,par}$ (lb-in.^{2}/in. × 10^6)*	*$EI_{w,perp}$ (lb-in.^{2}/in. × 10^6)*	*$EA_{w,par}$ (lb/in. × 10^6)*	*$EA_{w,perp}$ (lb/in. × 10^6)*
5/8	0.008	0.027	0.35	0.52
3/4	0.016	0.040	0.46	0.65

Notes: $EI_{w,par}$, flexural stiffness of the wood panel parallel to the member; $EI_{w,perp}$, flexural stiffness of the wood panel perpendicular to the member; $EA_{w,par}$, axial stiffness of the wood panel parallel to the member; $EA_{w,perp}$, axial stiffness of the wood panel perpendicular to the member.

Table 4.19 Slip modulus S_{flr} (lb/in./in.) for wood deck fastening

Type of deck	*Nailed*	*Glued and nailed*
OSB or plywood	600	50,000
Tongue and groove	5800	100,000

h_{cw} = half of the total deck thickness, in.
$= 1/2(t_d + t_c)$
h_{top} = distance from the centroid of the floor deck to the centroid of the joist, in.

$$h_{top} = \frac{d}{2} + t_d + t_c - y_{deck} \tag{4.34}$$

$$y_{deck} = \frac{EA_{w,\,par}[(t_d/2) + t_c] + EA_c(t_c/2)}{EA_{w,\,par} + EA_c} \tag{4.35}$$

y_{deck} = distance from the top of the concrete to the centroid of the floor deck
t_d = thickness of the deck (plywood)
t_c = thickness of the concrete
y = distance from the centroid of the system to the centroid of the joist, in.

$$y = \frac{EA_{top}h_{top}}{EA_m + EA_{top}} \tag{4.36}$$

When calculating the axial and flexural stiffness of the concrete deck, the modulus of elasticity is multiplied by 1.2 to account for dynamic loading, thus:

$$E_c = (1.2)(33)w_c^{1.5}\sqrt{f_c'} \tag{4.37}$$

where:
w_c is the unit weight of the concrete, lb/ft^3
f_c' is the 28-day compressive strength of the concrete, psi

The base equation for the modulus of elasticity is given in ACI 318 [19].

The number of effective floor joists N_{eff} in the system is a function of the longitudinal and transverse stiffness of the floor panel. The presence of transverse blocking or bridging will increase the number of effective joists by transferring a concentrated load to adjacent joists, which reduces the point-load deflection of the floor system. However, the amount of stiffness that the blocking or bridging contributes to the system is very small when a concrete topping is used.

For wood-framed floors, the number of effective joists is determined as follows:

$$N_{eff} = \frac{1}{DF_b - DF_v} \tag{4.38}$$

where:

$$DF_b = 0.0294 + 0.536K_1^{1/4} + 0.516K_1^{1/2} - 0.31K_1^{3/4} \tag{4.39}$$

$$DF_v = -0.00253 - 0.0854K_1^{1/4} + 0.079K_1^{1/2} - 0.00327K_2 \tag{4.40}$$

$$K_1 = \frac{K_j}{K_j + \Sigma\, K_{bi}} \tag{4.41}$$

$$K_2 = \frac{\Sigma K_{vi}}{\Sigma K_{bi}} \tag{4.42}$$

$$K_j = \text{longitudinal stiffness} = EI_{eff}/L^3 \tag{4.43}$$

K_{bi} = stiffness of the transverse flexural components, lb/in.

$$= 0.585EI_{bi}L/s^3 \text{ for panel or deck components} \tag{4.44}$$

$$= (2a/L)^{1.71}EI_x/s^3 \text{ for strong backs and straps} \tag{4.45}$$

K_{vi} = stiffness of the transverse shear components, lb/in.

$$= (2a/L)^{1.71}E_vA/s \text{ for cross-bridging or solid blocking} \tag{4.46}$$

a is the distance of the element to the closest end of the joist, in.
L is the joist span, in.
s is the joist spacing, in.
E_v is the effective shear modulus, psi
= 1000 psi for 2 × 2 cross-bridging without strapping
= 2000 psi for 2 × 2 cross-bridging with strapping
= 2000 psi for 2 × blocking without strapping
= 3000 psi for 2 × blocking with strapping
A is the effective shear area, in^2 = joist depth, in. × 1.5 in.
EI_x is the effective flexural stiffness of strong backs or straps
E is the 1.35×10^6 psi for 2× strong backs or strapping (with or without strapping)
EI_{bi} is the flexural stiffness of the floor panel or deck per unit width
= $EI_{w,\,perp}$ (Table 4.18) for wood deck panels

$$= EI_{w,\,perp} + EI_c + \frac{EA_{w,\,perp}EA_c h_{cw}^3}{EA_{w,\,perp} + EA_c} \text{ for concrete on wood deck} \tag{4.47}$$

4.7.2 Remedial measures for controlling floor vibrations in wood framed floors

Walking vibrations was found not to be critical for the two example problems considered in the previous section. However, there might be situations in practice where the vibration threshold (as measured by the deflection limit) is exceeded and remedial measures have to be taken to mitigate the effects of excessive floor vibrations. The primary ways for controlling walking vibrations include increasing the natural frequency of the floor system for floors in the low frequency range; adding mass to the floor to increase damping; and adding full height partitions to the floor also to increase damping. Some available options that could be used to mitigate walking vibration problems include the following [14,15,20]:

1. Reducing the span of the floor joist by adding a new column, if it is architecturally feasible. This increases the natural frequency of the floor system.
2. Using full-height partitions, which increases the amount of damping in the floor.
3. Providing additional bridging between joists. This increases the stiffness of the floor in the direction perpendicular to the joist, thus increasing the number of joists participating in resisting the vibration effects. See Figure 4.35.

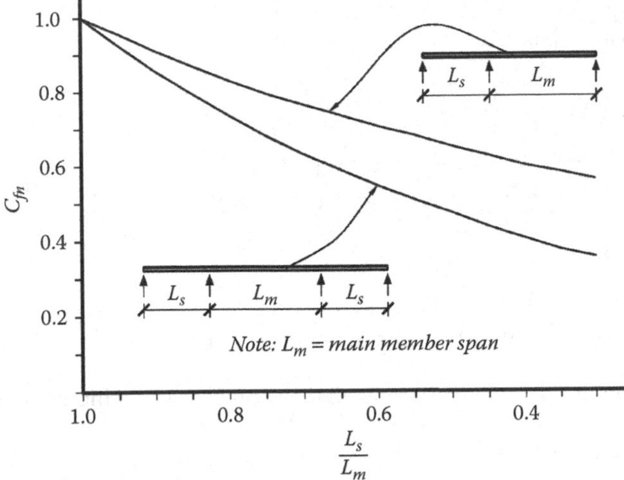

Figure 4.35 Continuity factor, C_{fn}, for multiple spans.

4. Using a thicker flooring system, preferably one that has some structural strength such as hardwood flooring. This increases the floor stiffness and hence the natural frequency of the floor system.
5. Reinforcing the existing floor joists by using flitch plates on both sides of the joist. In the case of I-joist, plywood panels could be attached to both sides of the joist to increase the floor stiffness.

Example 4.15: Floor vibrations on floor framing with sawn lumber

Determine if a floor framed with 2 × 10 sawn lumber (DF-L, Select Structural) spaced at 16 in. o.c. and a 14 ft 0 in. simple span with $\frac{3}{4}$-in. plywood nailed to the framing is adequate for walking vibration. Assume a residential occupancy. Neglect the contribution from transverse bridging.

Solution:
Floor stiffness:
From the NDS Supplement, the properties of the 2 × 10 sawn lumber are

$E = 1.9 \times 10^6$ psi

$A_m = 13.88$ in.^{2}

$I_m = 98.9$ in.^{4}

$EI_t = EI_{w,\text{par}} = 0.016 \times 10^6$ psi/in. (Table 4.18)

$(EI_{w,\text{par}})(S) = (0.016 \times 10^6)(16\text{ in.}) = 0.256 \times 10^6$ psi

$EA_{\text{flr}} = EA_{w,\text{par}} = 0.46 \times 10^6$ lb/in. (Table 4.18)

$(EA_{w,\text{par}})(S) = (0.46 \times 10^6)(16\text{ in.}) = 7.36 \times 10^6$ lb

$S_{\text{flr}} = 600$ lb/in./in. (Table 4.19)

$C_{fn} = 1.0$ (simple span, see Equation 4.28)

$L_{\text{flr}} = 48$ in. (not a concrete-topped floor)

From Equation 4.34,

$$h_{top} = \frac{9.25\text{ in.}}{2} + \frac{\frac{3}{4}\text{ in.}}{2} = 5\text{ in.}$$

From Equation 4.33,

$$EA_{top} = \frac{EA_{flr}}{1 + 10EA_{flr}/S_{flr}L_{flr}^2} = \frac{7.36\times10^6}{1+(10)(7.36\times10^6)/(600)(48)^2} = 0.135\times10^6\text{ lb}$$

From Equation 4.36,

$$\gamma = \frac{EA_{top}h_{top}}{EA_m + EA_{top}} = \frac{(0.135\times10^6)(5\text{ in.})}{(1.9\times10^6)(13.88\text{ in.}^2)+(0.135\times10^6)^2} = 0.025\text{ in.}$$

The relatively small value for y indicates that the plywood decking does not offer much stiffness.

From Equation 4.31,

$$EI = EI_m + EI_{top} + EA_my^2 + EA_{top}(h_{top} - y)^2$$

$$= (1.9\times10^6)(98.9) + (0.256\times10^6) + (1.9\times10^6)(13.88)(0.025)^2 + (0.135\times10^6)(5 - 0.025)^2$$

$$= 191\times10^6\text{ lb-in.}^2$$

From Equation 4.29,

$$\gamma = \frac{14.4}{(L/r)^2}\frac{E}{G} = \frac{14.4}{(14\times12)^2(13.88/98.9)}\left(\frac{1.9\times10^6}{0.1\times10^6}\right) = 0.069$$

where the radius of gyration is

$$r = \sqrt{\frac{I_m}{A_m}},\text{ and therefore}$$

$$\frac{1}{r^2} = \frac{A_m}{I_m} = \frac{13.88\text{ in.}^2}{98.9\text{ in.}^4}$$

From Equation 4.28, the effective flexural stiffness

$$EI_{eff} = \frac{EI}{1+\gamma EI/C_{fn}EI_m} = \frac{191\times10^6}{1+(0.069)(191\times10^6)/(1.0)(1.9\times10^6)(98.9)} = 178\times10^6\text{ lb-in.}^2$$

Number of effective joists:
From Equation 4.43, the longitudinal stiffness is

$$K_j = \frac{EI_{eff}}{L^3} = \frac{178\times10^6}{(14\times12)^3} = 37.46$$

From Equation 4.44, the stiffness of the transverse flexural deck components is

$$K_{bi} = \frac{0.585EI_{bi}L}{s^3} = \frac{(0.585)(0.040\times10^6)(14\times12)}{16^3} = 960$$

($EI_{bi} = EI_{w,\text{ perp}}$, Table 4.18)

From Equation 4.41,

$$K_1 = \frac{37.46}{37.46 + 960} = 0.0376$$

$K_2 = 0$ (blocking/bridging contribution neglected)

From Equations 4.39 and 4.40, we obtain the following parameters:

$$DF_b = 0.0294 + 0.536K_1^{1/4} + 0.516K_1^{1/2} - 0.31K_1^{3/4}$$

$$= 0.0294 + (0.536)(0.0376)^{1/4} + (0.516)(0.0376)^{1/2} - (0.31)(0.0376)^{3/4}$$

$$= \mathbf{0.339}$$

$$DF_v = -0.00253 - 0.0854K_1^{1/4} + 0.079K_2^{1/2} - 0.00327K_2$$

$$= -0.00253 - (0.0854)(0.0376)^{1/4} + 0 - 0$$

$$= \mathbf{-0.0401}$$

From Equation 4.38, the number of effective joists is

$$N_{\text{eff}} = \frac{1}{DF_b - DF_v} = \frac{1}{0.339 - (-0.0401)} = 2.64$$

From Equation 4.27, the floor deflection under a 225 lb concentrated load is

$$\Delta_p = \frac{C_{\text{pd}}}{N_{\text{eff}}} \frac{PL^3}{48EI_{\text{eff}}} = \left(\frac{1.0}{2.64}\right) \frac{(225)(14 \times 12)^3}{(48)(178 \times 10^6)} = 0.047 \text{ in.}$$

To control walking vibrations, the floor deflection calculated above must satisfy the following criteria from Equation 4.26:

$$\boldsymbol{\Delta}_p \leq 0.024 + 0.1e^{-0.18(L-6.4)} \leq 0.08 \text{ in.}$$

L = span of joist, ft = 14 ft

$$\boldsymbol{\Delta}_p \leq (0.024) + (0.1)\, e^{-0.18(14-6.4)}$$

$$= \mathbf{0.049\ in.} > 0.047 \text{ in.} \qquad \text{OK}$$

$$< 0.08 \text{ in.}$$

Using Figure 4.35, the same result is obtained. Therefore, the calculated deflection (0.047 in.) is less than the maximum allowable deflection to control walking vibration (0.049 in.) and the floor is deemed adequate for walking vibrations.

Example 4.16: Floor vibrations on floor framing with preengineered joists

With reference to the floor section shown in Figure 4.36 and the following design criteria, determine if the floor is adequate for floor vibrations (Figure 4.37):

14-in. I-joists spaced 16 in. o.c.
L = 25 ft 0 in. simple span
3/4-in. plywood glued and nailed to the framing
1.5-in. concrete topping, f_c' = 3000 psi; w_c = 110 lb/ft^3
2 × 14 blocking is provided at midspan

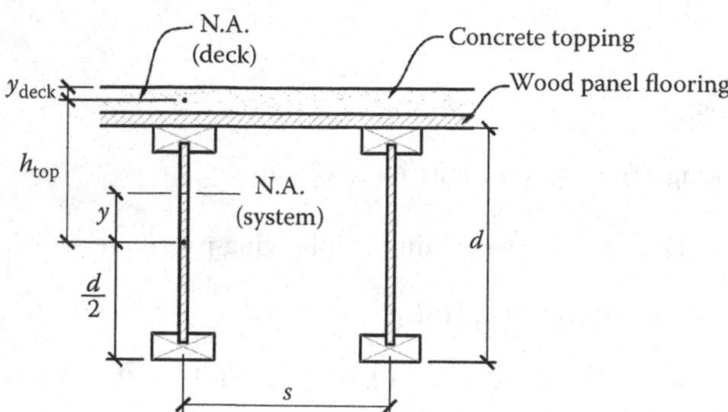

Figure 4.36 Typical floor section.

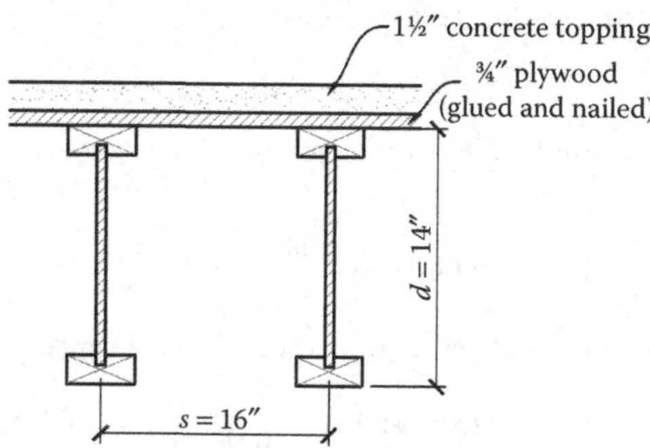

Figure 4.37 Floor section for preengineered joists.

Solution:
Floor stiffness:
The flexural and axial stiffness of the 14-in. I-joists are

$$EI_m = 600 \times 10^6 \text{ in.}^2\text{-lb}$$

$$EA_m = 20 \times 10^6 \text{ lb}$$

(*Note:* The values above should be obtained from the joist manufacturer.)

$$EI_{w,\ \text{par}} = (0.016 \times 10^6 \text{ psi/in.})(16 \text{ in.}) = 0.256 \times 10^6 \text{ psi (Table 4.18)}$$

$$EI_{w,\ \text{perp}} = 0.040 \times 10^6 \text{ psi/in. (Table 4.18)}$$

$$EA_{w,\ \text{par}} = (0.46 \times 10^6 \text{ lb/in.})(16 \text{ in.}) = 7.36 \times 10^6 \text{ lb (Table 4.18)}$$

$$EA_{w,\ \text{perp}} = 0.65 \times 10^6 \text{ lb (Table 4.18)}$$

$$S_{\text{flr}} = 50{,}000 \text{ lb/in./in. (Table 4.19)}$$

$$C_{fn} = 1.0 \text{ (simple span, see Equation 4.28)}$$

From Equation 4.37, the dynamic modulus of elasticity and stiffness of the concrete deck are

$$E_c = (1.2)(33)w_c^{1.5}\sqrt{f_c'} = (1.2)(33)(110)^{1.5}\sqrt{3000} = 2.5\times10^6 \text{ psi}$$

$$EI_c = \frac{(2.5\times10^6)(16\text{ in.})(1.5)^3}{12} = 11.25\times10^6 \text{ lb-in.}^2$$

$$EA_c = (2.5 \times 10^6)(16\text{ in.})(1.5) = 60 \times 10^6 \text{ lb}$$

The concrete deck and plywood sheathing contribute to the axial stiffness of the floor

$$EA_{flr} = EA_c + EA_{w,\,par} = (60 \times 10^6) + (7.36 \times 10^6) = 67.36 \times 10^6 \text{ lb}$$

From Equation 4.35, the distance from the top of the concrete to the centroid of the floor deck is

$$\gamma_{deck} = \frac{EA_{w,\,par}[(t_d/2)+t_c]+EA_c(t_c/2)}{EA_{w,\,par}+EA_c} = \frac{(7.36\times10^6)[(0.75/2)+1.5]+(60\times10^6)(1.5/2)}{(7.36\times10^6)+(60\times10^6)}$$

$$= 0.873 \text{ in.}$$

From Equation 4.34, the distance from the centroid of the floor deck to the centroid of the joist is

$$h_{top} = \frac{d}{2}+t_d+t_c-\gamma_{deck} = \frac{14}{2}+0.75+1.5-0.873 = 8.38 \text{ in.}$$

The effective axial stiffness of the floor deck parallel to the joist is obtained from Equation 4.33 as

$$EA_{top} = \frac{EA_{flr}}{1+10EA_{flr}/S_{flr}L_{flr}^2} = \frac{67.36\times10^6}{1+(10)(67.36\times10^6)/(50{,}000)(25\times12)^2} = 58.6\times10^6 \text{ lb}$$

The flexural stiffness of the floor deck parallel to the joist from Equation 4.32 is

$$EI_{top} = EI_{w,\,par}+EI_c+\frac{EA_{w,\,par}EA_c h_{cw}^3}{EA_{w,\,par}+EA_c}$$

$$= (0.256\times10^6)+(11.25\times10^6)+\frac{(7.36\times10^6)(60\times10^6)[(0.75+1.5/2)]^3}{(7.36\times10^6)+(60\times10^6)}$$

$$= 20.9 \times 10^6 \text{ lb-in.}^2$$

The distance from the centroid of the system to the centroid of the joist from Equation 4.36 is

$$\gamma = \frac{EA_{top}h_{top}}{EA_m+EA_{top}} = \frac{(58.6\times10^6)(8.38\text{ in.})}{(20\times10^6)+(58.6\times10^6)} = 6.25 \text{ in.}$$

From Equation 4.31, the flexural stiffness of the floor panel is

$$EI = EI_m + EI_t + EA_m y^2 + EA_t(h_t - y)^2$$

$$= (600 \times 10^6) + (20.9 \times 10^6) + (20 \times 10^6)(6.25)^2 + (58.6 \times 10^6)(8.38 - 6.25)^2$$

$$= 1668 \times 10^6 \text{ in.}^2\text{-lb}$$

From Equation 4.30,

$$\gamma = \frac{96EI_m}{K_sL^2} = \frac{(96)(600\times10^6)}{(0.4\times10^6)(14\text{ in.})(25\times12)^2} = 0.114$$

Note that $K_s = (0.4 \times 10^6)d$ for preengineered I-joists where d is the depth of the joist

$$EI_{eff} = \frac{EI}{1+\gamma EI/C_{fn}EI_m} = \frac{1668\times10^6}{1+(0.114)(1668\times10^6)/(1.0)(600\times10^6)} = 1267 \times 10^6 \text{ in.}^2\text{-lb}$$

Number of effective joists:
From Equation 4.43,

$$K_j = \frac{EI_{eff}}{L^3} = \frac{1267\times10^6}{(25\times12)^3} = 46.9$$

The flexural stiffness of the floor panel per unit width from Equation 4.47 is

$$EI_{bi} = EI_{w,\,perp} + EI_c + \frac{EA_{w,\,perp}EA_c h_{cw}^3}{EA_{w,\,perp} + EA_c}$$

$$= (0.040\times10^6) + (0.704\times10^6) + \frac{(0.65\times10^6)(3.75\times10^6)[(0.75+1.5)/2]^3}{(0.65\times10^6)+(3.75\times10^6)}$$

$$= 1.53 \times 10^6 \text{ lb-in.}^2\text{/in.}$$

From Equation 4.44, the stiffness of the transverse flexural deck components is

$$K_{bi} = \frac{0.585EI_{bi}L}{s^3} = \frac{(0.585)(1.53\times10^6)(25\times12)}{16^3} = 65{,}674 \text{ lb/in.}$$

From Equation 4.46,

$$K_{vi} = \frac{(2a/L)^{1.71}E_v A}{s}$$

$$a = \frac{25 \text{ ft} \times 12}{2} = 150 \text{ in. (blocking at midspan)}$$

A = joist depth, inches × 1.5 in.

The effective shear modulus E_v is 2000 psi for 2× blocking without strapping

$$K_{vi} = \frac{[(2)(150)/(25\times12)]^{1.71}(2000)(14)(1.5)}{16} = 2625 \text{ lb/in.}$$

From Equation 4.41,

$$K_1 = \frac{K_j}{K_j + \Sigma K_{bi}} = \frac{46.9}{46.9+65{,}674} = 0.000714$$

From Equation 4.42,

$$K_2 = \frac{\Sigma K_{vi}}{\Sigma K_{bi}} = \frac{2625}{65{,}674} = 0.03997$$

From Equations 4.39 and 4.40, we calculate the following parameters:

$$DF_b = 0.0294 + 0.536K_1^{1/4} + 0.516K_1^{1/2} - 0.31K_1^{3/4}$$

$$= 0.0294 + (0.536)(0.000714)^{1/4} + (0.516)(0.000714)^{1/2} - (0.31)(0.000714)^{3/4}$$

$$= 0.129$$

$$DF_v = -0.00253 - 0.0854K_1^{1/4} + 0.079K_2^{1/2} - 0.00327K_2$$

$$= -0.00253 - (0.0854)(0.000714)^{1/4} + (0.079)(0.03997)^{1/2} - (0.00327)(0.03997)$$

$$= -0.00083$$

The number of effective joists determined from Equation 4.38 is

$$N_{eff} = \frac{1}{DF_b - DF_v} = \frac{1}{(0.129)-(-0.00083)} = 7.68$$

From Equation 4.27, the floor deflection under a 225 lb concentrated load is

$$\Delta_p = \frac{C_{pd}}{N_{eff}} \frac{PL^3}{48EI_{eff}} = \left(\frac{1.0}{7.68}\right) \frac{(225)(25 \times 12)^3}{(48)(1267 \times 10^6)} = 0.013 \text{ in.}$$

To control walking vibrations, the floor deflection calculated above must satisfy the following criteria from Equation 4.26.

$$\Delta_p \leq 0.024 + 0.1e^{-0.18(L-6.4)} \leq 0.08 \text{ in.}$$

$$\leq (0.024) + (0.1)e^{-0.18(25-6.4)}$$

$$= \mathbf{0.028\ in.} > 0.013 \text{ in.} \qquad \text{OK}$$

$$< 0.08 \text{ in.}$$

Using Figure 4.34, the same result is obtained. Therefore, the calculated deflection (0.013 in.) is less than the maximum allowable deflection to control walking vibration (0.028 in.) and the floor is deemed adequate for walking vibrations.

PROBLEMS

4.1 Determine the maximum uniformly distributed floor live load that can be supported by a 10 × 24 girder with a simply supported span of 16 ft. Assume that the wood species and stress grade is Douglas Fir-Larch Select Structural, normal duration loading, normal temperature conditions, with continuous lateral support provided to the compression flange. The girder has a tributary width of 14 ft and a floor dead load of 35 psf in addition to the girder self-weight.

4.2 For the girder in Problem 4.1, determine the minimum bearing length required for a live load of 100 psf assuming that the girder is simply supported on columns at both ends.

4.3 Determine the joist size required to support a dead load of 12 psf (includes the self-weight of the joist) and a floor live load of 40 psf assuming a joist spacing of 24 in. and a joist span of 15 ft. Assume Southern Pine wood species, normal temperature, and dry service conditions.

4.4 Determine the joist size required to support a dead load of 10 psf (includes the self-weight of the joist) and a floor live load of 50 psf assuming a joist spacing of 16 in. and a joist span of 15 ft. In addition, the joists support an additional 10-ft-high partition wall weighing 10 psf running perpendicular to the floor joists at the midspan of the joists. Assume SPF Select Structural wood species, normal temperature, and dry service conditions.

4.5 Assuming that wood is Hem-Fir #2 and normal temperature and moisture conditions, determine the most economical member for the conditions shown based on bending and deflection ($L/480$ for live loads, $L/240$ for total loads). Assume that full lateral stability is provided and use $k = 1.0$ for deflections.

a. 2× Joists at 12″ o.c. spanning 16 ft supporting loads of $D = 15$ psf and $L = 50$ psf.
b. 2× rafters at 16″ o.c. spanning 15 ft horizontally supporting loads of $D = 20$ psf and $S = 60$ psf.
c. 2× joists at 24″ o.c. spanning 10 ft supporting loads of $D = 10$ psf and $L = 125$ psf.

d. 2× joists at 12″ o.c. spanning 14 ft supporting a uniform load of $D + L = 75$ psf and a point load at midspan of $P_{D+L} = 750$ lb.

4.6 Assuming that wood is Douglas Fir #1 and normal temperature and moisture conditions, determine the most economical member for the conditions shown based on bending and deflection ($L/360$ for live loads, $L/240$ for total loads). Assume that full lateral stability is provided unless stated otherwise and use $k = 1.0$ for deflections. Assume that all members are a Beam and Stringer.

a. Beam spanning 19 ft supporting a uniform load of $W_{D+S} = 950$ plf
b. Beam loaded and braced at one-third points with $P_{D+L} = 4000$ lb at each point and spans 24 ft
c. Beam loaded and braced at midspan with $P_{D+L} = 8000$ lb and spans 20 ft
d. Beam spanning 18 ft supporting a uniform load of $W_{D+L} = 700$ plf and loaded at midspan with $P_{D+L} = 5000$ lb

4.7 The floor plan shown in Figure 4.38 is for an office occupancy, supporting 1 level. Wood is Hem-Fir, No. 1.
Normal temperature and moisture conditions apply. All floor framing members are braced laterally by the floor sheathing. Assume the floor dead load (including the weight of the framing), $D = 16$ psf.
Assuming 2 × 10 at 16″ o.c. for the J-1:

a. bending under dead plus floor live load
b. shear under dead plus floor live load
c. live-load deflection (compare with L/360)
d. Determine the minimum bearing length

4.8 The floor plan shown in Figure 4.39 is for dining room/restaurant occupancy. All wood is Douglas Fir-Larch, Select Structural. Normal temperature and moisture conditions apply. Floor framing members are adequately braced laterally by the floor sheathing. Assume the floor dead load (including the weight of the framing), $D = 12$ psf.

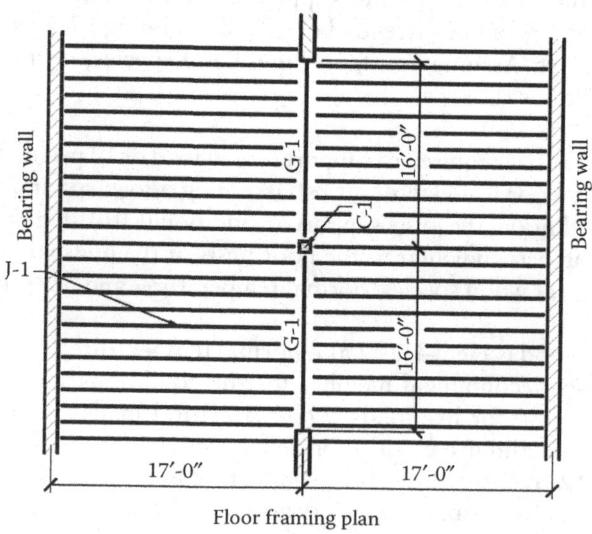

Figure 4.38 Detail for Problem 4.7.

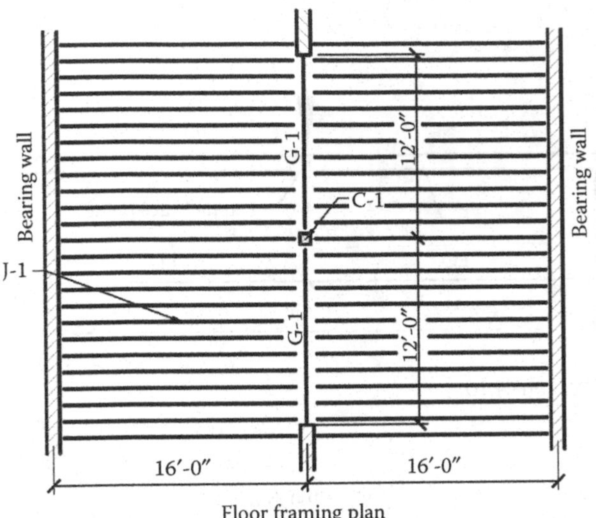

Figure 4.39 Detail for Problem 4.8.

a. Assuming sawn lumber for J-1, determine the most economical size based on bending, shear, and deflection using a joist spacing not greater than 24″.
b. Select the most economical size for G-1 assuming that it is a Beam and Stringer based on bending, shear, and deflection.
c. Select a joist hanger for J-1 from a manufacturer's catalog assuming a face-mounted connection to G-1.
d. Determine the axial load to column C-1 and select a column cap connector for G-1 from a manufacturer's catalog assuming G-1 frames over the top of C-1.
e. Provide a sketch of the floor plan design and details of the connections. Include all appropriate dimensions and fasteners.

4.9 Floor framing consists of 2 × 12 sawn-lumber joists spaced 19.2″ o.c., $L = 17$ ft. Wood is Douglas Fir-Larch, Select Structural. Normal temperature and moisture conditions apply. Loading is due to $D + L$ and floor framing members are braced laterally by the floor sheathing. Assume the floor dead load (including the weight of the framing), $D = 20$ psf.

a. Determine the following: F_b', F_v', F_t', E'.
b. Determine the maximum moment and maximum shear that a typical joist could support in ft.-lb.
c. Based on the moment calculated in (b), determine the maximum live load that could be supported by this floor, in psf; suggest an occupancy category for this floor.
d. Calculate the live-load deflection based on your response to (c) and compare with $L/360$.

4.10 Refer to the beam loading shown in Figure 4.40. Wood is Douglas Fir-Larch, Select Structural, and normal temperature and moisture conditions apply. Loads are $D + L$, and consider only the point load shown. The beam is simply supported and laterally braced at midspan as shown below

a. Find F_b', F_v', and E'.
b. Find the maximum load **P** based on bending strength.
c. Find the maximum load **P** based on shear strength.
d. Find the maximum load **P** based on deflection using a limit of $L/240$.

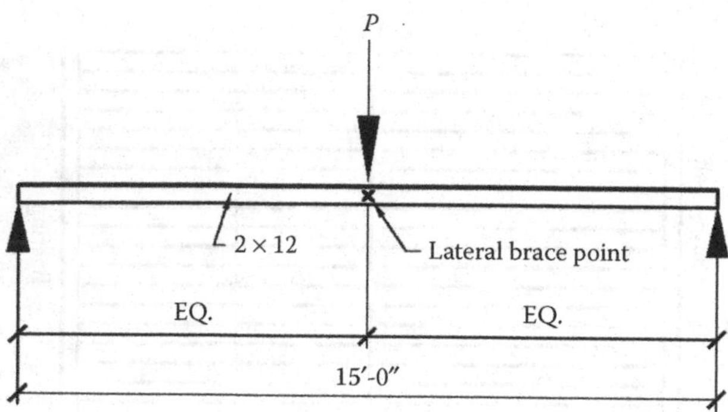

Figure 4.40 Detail for Problem 4.10.

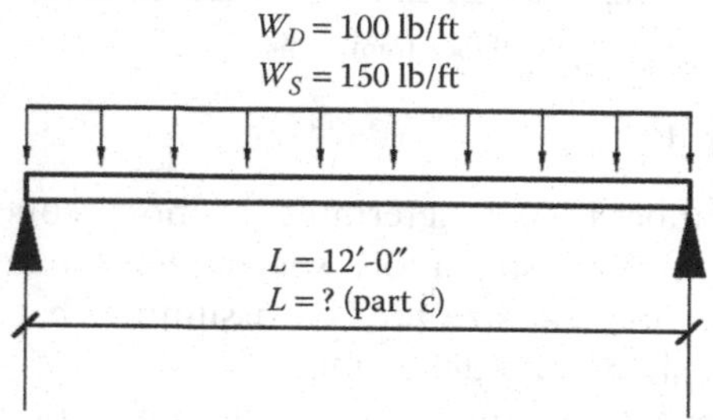

Figure 4.41 Detail for Problem 4.12.

4.11 Develop a spreadsheet to analyze a simply supported beam loaded at midspan and check Problem 4.10.

4.12 Refer to the beam loading shown in Figure 4.41. Wood is Southern Pine, No. 2, and normal temperature and moisture conditions apply. Loads are $D + S$, and consider only the uniform loads shown. The beam is simply supported and laterally braced along the full length. The beam is a 2 × 8.

a. Find E'.
b. Determine if the beam is adequate for deflection under total loads using a limit of $L/180$; use $k = 1.0$.
c. What is the maximum beam span if snow load deflection is limited to $L/360$?

4.13 Refer to the floor framing plan below showing a header supporting floor joists (Figure 4.42). Wood is Douglas Fir-Larch, Select Structural, and normal temperature and moisture conditions apply. Loads are only $D = 10$ psf and $L = 40$ psf; use $L/240$ and $L/360$ for total- and live-load deflections assuming $k = 1.0$. Assume that H-1 has full lateral stability and is supported at each end by a single 2 × 6.

a. Find the allowable bending stress, compression perpendicular to the grain, and modulus of elasticity.
b. Determine if H-1 is adequate for bending, shear, deflection, and bearing length.

4.14 Refer to the beam loading shown in Figure 4.43. Wood is Hem-Fir, #2, and normal temperature and moisture conditions apply. Loads are $D + L_r$, and consider only the

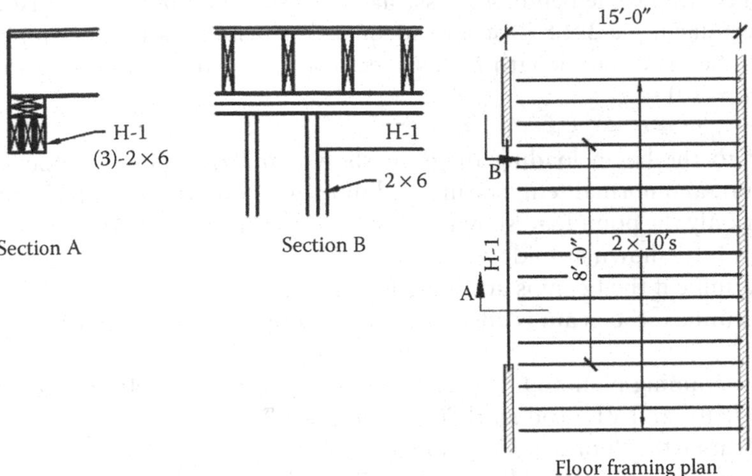

Figure 4.42 Detail for Problem 4.13.

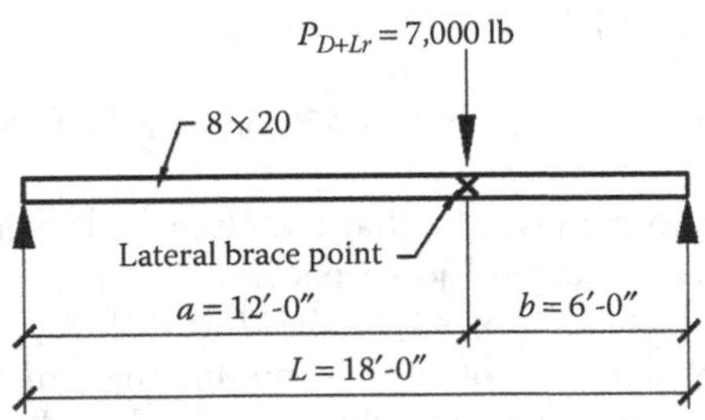

Figure 4.43 Detail for Problem 4.14.

point load shown. Loading is to the narrow (8″) face and the beam is laterally braced only as shown at the one-third point and at each end.

a. Determine if the beam is adequate for bending.
b. Determine if the beam is adequate for deflection using a limit of $L/240$.

4.15 Refer to the roof framing plan shown in Figure 4.44. Wood is Hem-Fir, #1, and normal temperature and moisture conditions apply. Total load is 60 psf, which includes self-weight and loading is $D + L_r$.

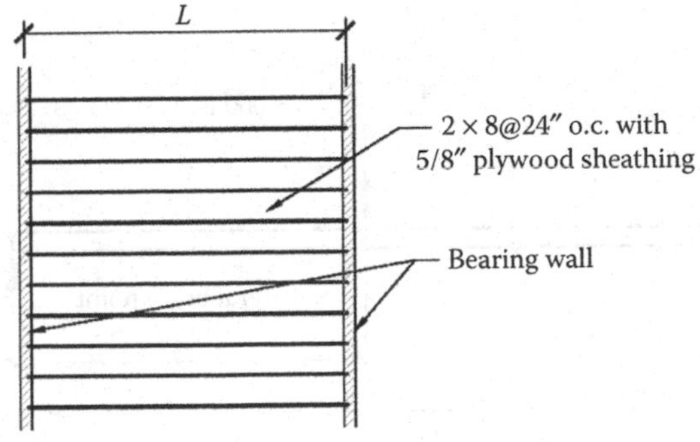

Figure 4.44 Detail for Problem 4.15.

a. Find the allowable bending stress, shear stress, and modulus of elasticity.
b. Determine if the 2 × 8 joists are adequate for bending and shear for $L = 15$ ft.
c. Find the maximum length L that meets an $L/240$ deflection limit for total loads; use $k = 1.0$.

4.16 Refer to the beam loading diagram shown in Figure 4.45. Wood is SPF, Select Structural, and normal temperature and moisture conditions apply. Loads are $D + S$; consider only the point load shown. Loading is to the narrow (6″) face and the beam is braced only as shown at midpsan.
a. Determine if the beam is adequate for bending.
b. Determine if the beam is adequate for deflection using a limit of $L/240$.

4.17 For the framing plan and floor section shown in Figure 4.46, use a manufacturer's load Table for an LVL product that is rated 2600F-1.9E.
a. Determine the floor dead load in psf.
b. Determine the dead and live loads to J-1 and G-1 in plf.
c. Determine the maximum moment and shear in J-1 and G-1.
d. Determine the maximum service load in kips to C-1.
e. Are members G-1 and J-1 adequate for bending, shear, and deflection?

4.18 For the floor framing shown in Figure 4.47, consider the following only: assume that all C-factors are 1.0.
Total dead load = 15 psf (assume that it includes self-weight), live load = 80 psf
J-1 is a 9 ½″ deep, engineered I-joist spaced at 12″ o.c.
G-1 is built up with (4)-1 ¾″ × 9 ½″ LVL (2600F-1.9E)
a. Draw a free-body diagram of each showing the uniform dead and live loads (assume that G-1 receives a uniformly distributed load).
b. Determine the maximum bending moment, shear, and deflection in J-1 and G-1.
c. Are J-1 and G-1 adequate for moment and deflection? Consider live-load deflection only against $L/480$ and use a manufacturer's data sheet for J-1 and G-1.
d. Select a joist hanger from the steel connector manufacturer and draw a detail of this connection indicating the required nails. Assume J-1 frames into the side of G-1 using a face-mounted hanger.

4.19 Develop a spreadsheet to analyze wood joists based on the following:

2 × 12 joist, Hem-Fir Select Structural, $L = 15$ ft
DL = 10 psf (includes self-weight)
LL = 40 psf

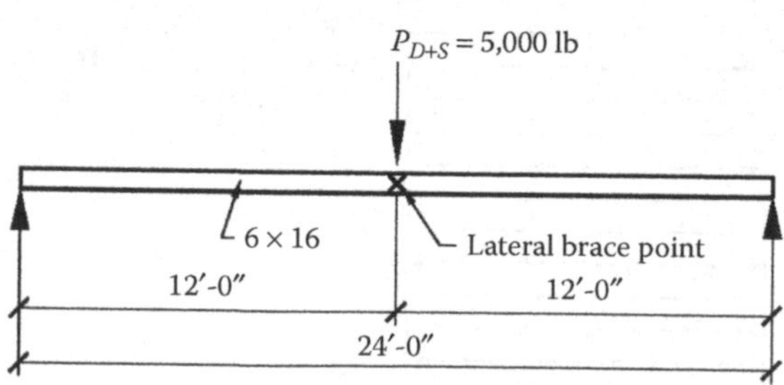

Figure 4.45 Detail for Problem 4.16.

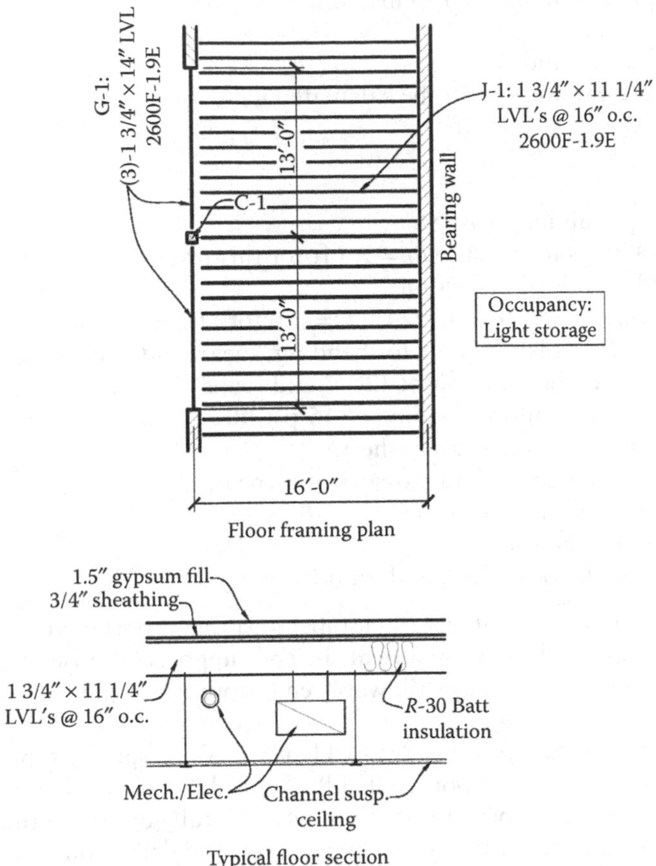

Figure 4.46 Detail for Problem 4.17.

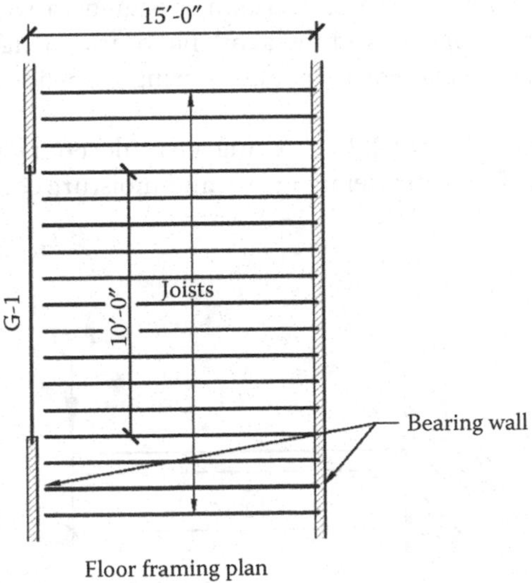

Figure 4.47 Detail for Problem 4.18.

Point load at midspan of $D = 100$ lb and $L = 300$ lb
Spacing = 16 in. o.c.
beam dimensions b and d
A_x, I_x, S_x (these variables are to be calculated using the beam dimensions)
Maximum moment and shear
f_b, f_v
F'_b, F'_v, E', E'_{min}
All C-factors pertaining to above
Include calculations for C_L (use $lu = 7.5$ ft for this problem)
Live-load and total-load deflection
Deflection limits of $L/360$ and $L/240$; $k = 1.0$ for DL deflection
Determine if the joist is adequate for bending, shear, and deflection

3.26 A wood joist measures 1 ⅞″ × 12 ½″ and spans 17 ft and is spaced 16″ on center. It is subject to a uniform loading of 15 psf (dead) and 40 psf (live). Determine the following by means of a spreadsheet:

a. Area, section modulus, and moment of inertia
b. Maximum moment and shear
c. Maximum bending stress
d. Deflection under dead, live, and total loads

4.20 A 4 × 14 Hem-Fir Select Structural lumber joist that is part of an office floor framing is notched 2.5 in. on the tension face at the end supports because of headroom limitations. Determine the maximum allowable end support reaction.

4.21 A simply supported 6 × 12 girder of Hem-Fir No. 1 species supports a uniformly distributed load (dead plus floor live load) of 600 lb/ft over a span of 24 ft. Assuming normal moisture and temperature conditions and a fully braced beam, determine if the beam is structurally adequate for bending, shear, and deflection.

4.22 A 5 ½ × 30 in. 24F–1.8E glulam beam spans 32 ft and supports a concentrated moving load of 5000 lb that can be located anywhere on the beam, in addition to its self-weight. Normal temperature and dry service conditions apply, and the beam is laterally braced at the supports and at midspan. Assuming a wood density of 36 lb/ft^3, a load duration factor C_D of 1.0, and the available bearing length l_b of 4 in., is the beam adequate for bending, shear, deflection, and bearing perpendicular to the grain?

4.23 Refer to the sketch in Figure 4.48 showing a cantilevered glulam beam that is a 16F-E6, DF/DF, 3.5″ × 12″. Normal temperature and moisture conditions apply. Bending is

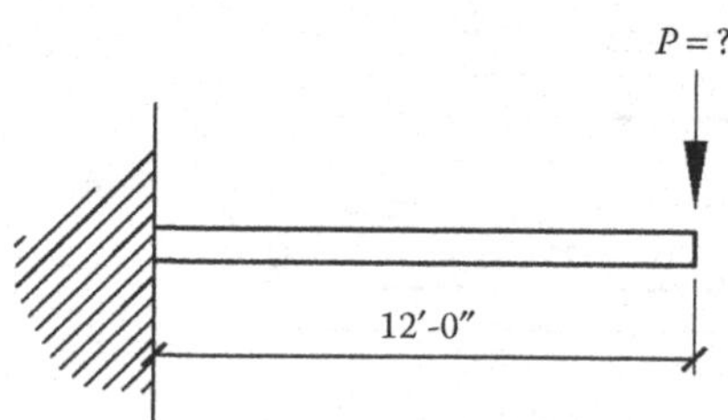

Figure 4.48 Detail for Problem 4.23.

about the strong axis, and the beam is laterally unbraced. Load **P** is based on the load combination $D + L$ only and Ignore the self-weight of the beam.

a. Find the maximum load P based on the shear strength of the beam.
b. Find the maximum load P based on the bending strength of the beam.
c. Find the maximum load P based on the deflection of the beam using $L/400$.

4.24 Refer to the beam loading diagram shown in Figure 4.49. Glulam is 20F-1.5E and normal temperature and moisture conditions apply. Bending is about the strong axis, and the beam is laterally braced at one-quarter points as shown.

a. Determine if the member is adequate in bending and shear.
b. Determine if the member is adequate for deflection, compare with $L/240$.

4.25 Refer to the beam loading diagram shown in Figure 4.50. Glulam is 24F-E4, SP/SP, and normal temperature and moisture conditions apply. Bending is about the strong axis, and the beam is fully braced laterally. Loads are due to $D + L$ only and ignore self-weight. Use $X = 12$ ft for parts a, b, and c.

a. Find F_b' and find the maximum load P based on bending strength.
b. Find E' and find the maximum load P assuming a deflection limit of $L/240$.
c. Find $F'_{c\perp}$ and find the maximum load P if the bearing length at each end is 5′. Use $C_b = 1.0$.
d. Determine if the beam is adequate for bending assuming $X = 6$ ft and $P = 7000$ lb.

4.26 Refer to the beam loading diagram shown in Figure 4.51. Glulam is 16F-E3, DF/DF, and normal temperature and moisture conditions apply. Bending is about the strong axis, and the beam is laterally braced at midspan as shown.

a. Determine if the member is adequate in bending and shear.
b. Determine if the member is adequate for deflection, compare with $L/240$.

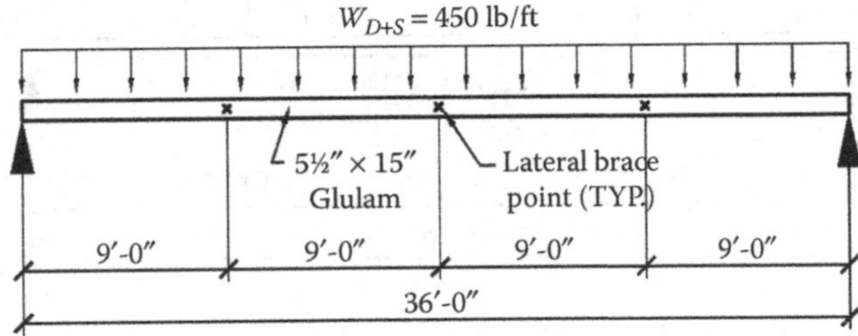

Figure 4.49 Detail for Problem 4.24.

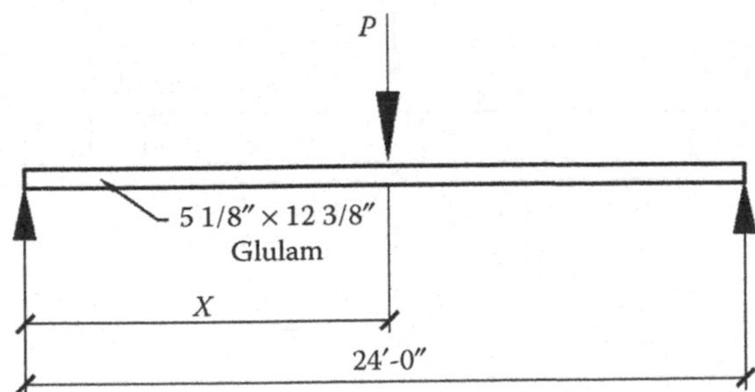

Figure 4.50 Detail for Problem 4.25.

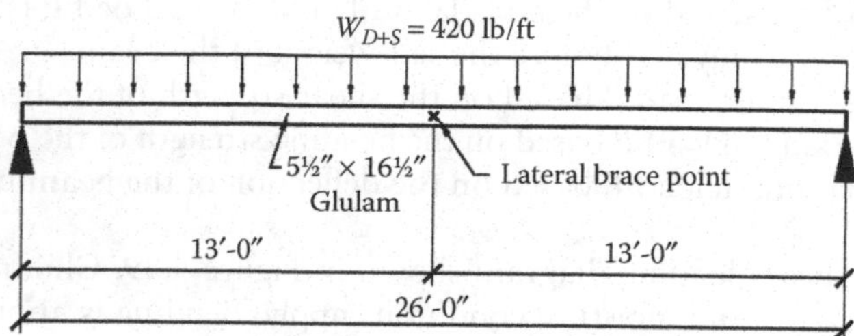

Figure 4.51 Detail for Problem 4.26.

4.27 Refer to the beam loading diagram shown in Figure 4.52. Glulam is 16F-E6, DF/DF, and normal temperature and moisture conditions apply. Bending is about the strong axis, and the beam is laterally braced at midspan as shown.
   a. Determine if the member is adequate in bending and shear.
   b. Determine if the member is adequate for deflection, compare with *L*/240.

4.28 Refer to the beam loading diagram shown in Figure 4.53. Glulam is 24F-E11, HF/HF, and normal temperature and moisture conditions apply. Bending is about the strong axis, and the beam is laterally braced at midspan as shown.
   a. Determine if the member is adequate in bending and shear.
   b. Determine the required bearing length.
   c. Determine if the member is adequate for deflection, compare with *L*/360 and *L*/240 for live and total loads; assume that the dead load is 40% of the total.

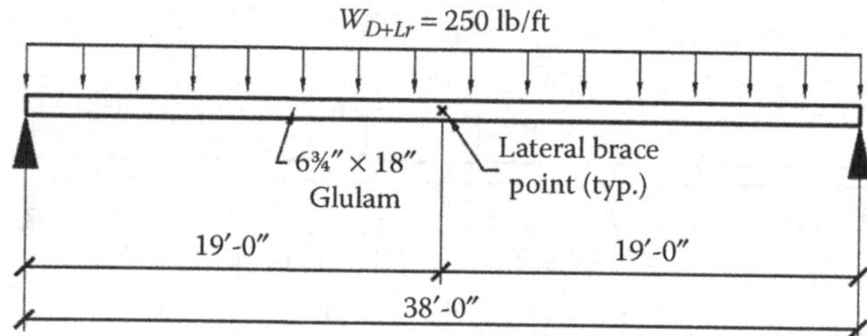

Figure 4.52 Detail for Problem 4.27.

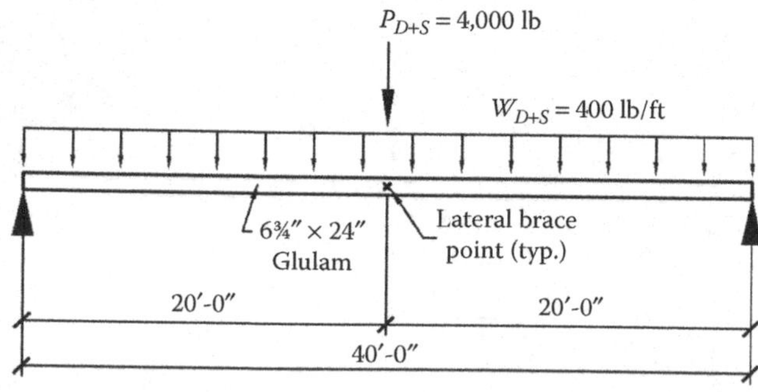

Figure 4.53 Detail for Problem 4.28.

4.29 Develop a spreadsheet to analyze a simply supported glulam beam loaded at midspan and with an applied uniform load based on Problem 4.28. Submit the output for Problem 4.28.

4.30 Assuming a glulam with properties of 2000F-1.6E and normal temperature and moisture conditions, determine the most economical member for the conditions shown based on bending and deflection (*L*/360 for live loads, *L*/240 for total loads). Assume full lateral stability is provided unless stated otherwise and use $k = 1.0$ for deflections.

a. Beam spanning 36ft supporting a uniform load of $W_{D+S} = 1200$ plf
b. Beam loaded and braced at one-third points with $P_{D+L} = 6000$ lb at each point and spans 33ft
c. Beam loaded and braced at midspan with $P_{D+L} = 8000$ lb and spans 28 ft
d. Beam spanning 24 ft supporting a uniform load of $W_{D+S} = 900$ plf and loaded at midpsan with $P_{D+S} = 6000$ lb

4.31 The floor plan shown in Figure 4.54 is for Residential occupancy. All sawn lumber is Hem-Fir, #2. Normal temperature and moisture conditions apply. Floor framing members are braced laterally by the floor sheathing. Assume the floor dead load (including the weight of the framing), $D = 20$ psf.

a. Assuming sawn lumber for J-1, determine the most economical size based on bending, shear, and deflection using a joist spacing not greater than 19.2″.
b. Select the most economical size for G-1 assuming that it is a Glulam 24F-1.6E based on bending, shear, and deflection. The width of G-1 cannot exceed the width of the wall studs.
c. Select a joist hanger for J-1 from a manufacturers catalog assuming a face-mounted connection to G-1.
d. Determine the reaction at the ends of G-1 and determine the minimum number of jack studs that are required for bearing length.
e. Provide a sketch of the floor plan design and details of the connections. Include all appropriate dimensions and fasteners.

4.32 For the detail shown in Figure 4.55, loading is $D + S$ only and wood is Douglas Fir-Larch, select structural. Normal temperature and moisture conditions apply.

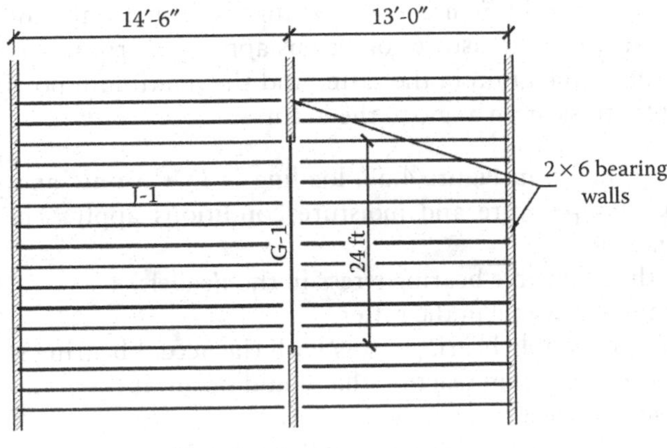

Figure 4.54 Detail for Problem 4.31.

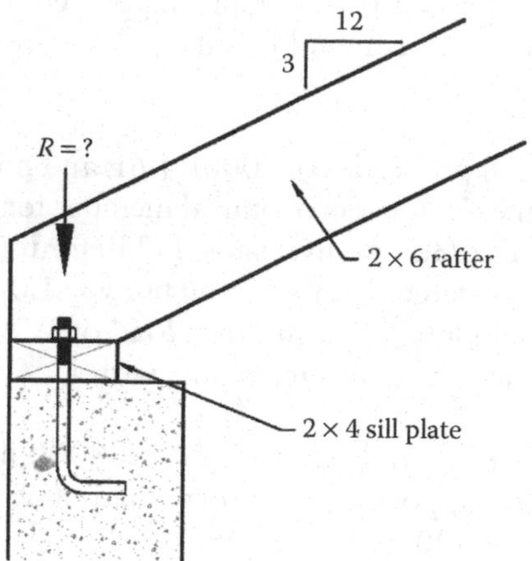

Figure 4.55 Detail for Problem 4.32.

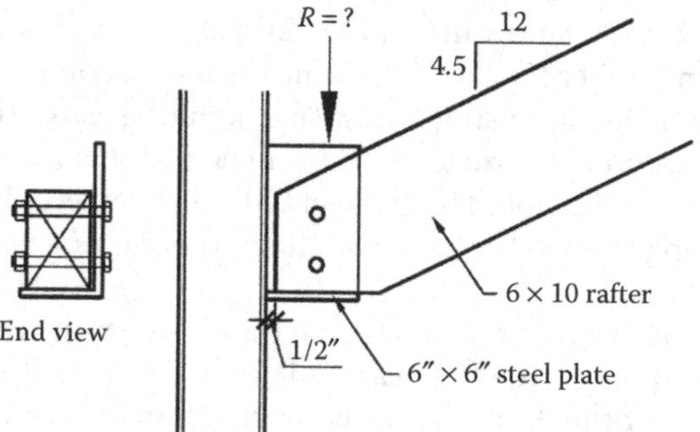

Figure 4.56 Detail for Problem 4.33.

a. Determine the allowable bearing stress at an angle to the grain in the rafter.
b. Determine the maximum possible reaction R based on (a).

4.33 For the detail shown in Figure 4.56, loading is $D + S$ only and wood is SPF, #1. Normal temperature and moisture conditions apply. Determine the allowable bearing stress at an angle to the grain in the rafter and the maximum possible reaction based on the allowable stress at an angle to the grain.

4.34 For the detail shown in Figure 4.57, loading is $D + S$ only and wood is Hem-Fir, No. 2. Normal temperature and moisture conditions apply. The rafter overhangs header a distance of 8″.

a. Determine the allowable bearing stress in the header.
b. Determine the allowable in the rafter.
c. Compare the allowable bearing stress with the actual bearing stress for both.
d. Determine the maximum reaction that could occur at the connection.
e. Develop a spreadsheet to analyze the above.

4.35 Develop a spreadsheet to analyze stress at an angle to the grain for a rafter bearing on a wood header. Submit the output for Problem 4.36.

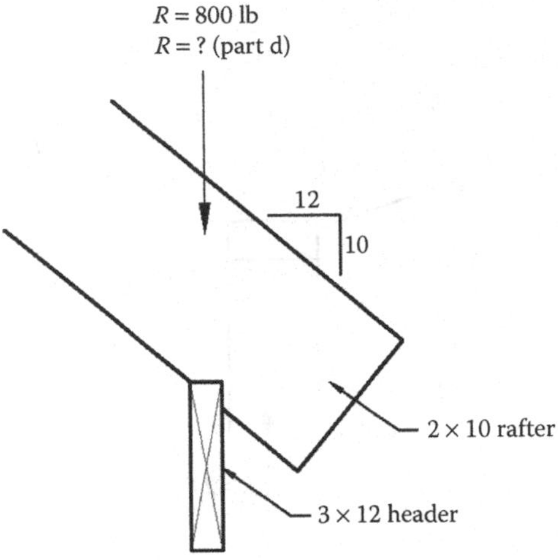

Figure 4.57 Detail for Problem 4.34.

4.36 For the detail shown in Figure 4.58, loading is $D + S$ only and wood is Hem-Fir, Select Structural. Normal temperature and moisture conditions apply.

a. Determine the allowable bearing stress in the 3 × 6 member.
b. Determine the allowable bearing stress in the rafter.
c. Compare the allowable bearing stress with the actual bearing stress for both.
d. Determine the maximum reaction that could occur at the connection based on (a) and (b).

4.37 The sloped 2 × 10 rafter shown in Figure 4.59 is supported on a ridge beam at the ridge line and on the exterior stud wall. The reaction from the dead load plus the snow load on the rafter at the exterior stud wall is 2400 lb. Assuming No. 1 Hem-Fir, normal temperature, and dry service conditions, determine the following:

a. The allowable bearing stresses at an angle to the grain F'_θ and the allowable bearing stress or allowable stress perpendicular to the grain $F'_{c\perp}$
b. The bearing stress at an angle to the grain f_θ in the rafter
c. The bearing stress perpendicular to the grain $f_{c\perp}$ in the top plate

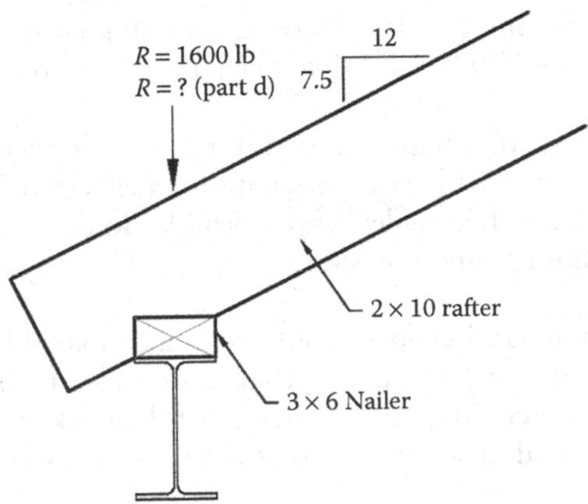

Figure 4.58 Detail for Problem 4.36.

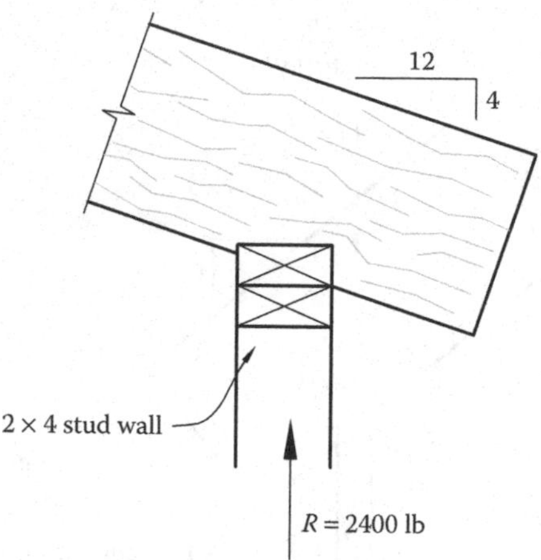

Figure 4.59 Detail for Problem 4.37.

4.38 A single wood girder spans 40 ft and is supported at each and at midspan. It is subjected to a uniformly distributed load of 500 plf. Determine the following:

a. Maximum moment and shear; draw both diagrams. Use either an analysis program or a design aid.
b. Maximum moment and shear assuming that there are two beams spanning 20 ft. Show the shear and moment diagram.
c. Explain the differences between the two framing scenarios and discuss the merits of using each one.

4.39 Develop a spreadsheet to calculate C_L, beam stability factor. Submit the results for a 2 × 10, Hem-Fir, select structural. Plot a curve showing C_L on the vertical axis and unbraced length 0 to 20 ft on the horizontal axis.

4.40 Locate data Tables for an I-joist manufacturer, select the most economical joist to span 18 ft, and cover a floor area of 32 ft. Design loading is $D = 20$ psf and $L = 40$ psf.

a. Provide complete data sheets from the manufacturer.
b. Select a joist spacing and draw a free-body diagram of the joist showing the dead and live loads in pounds per lineal foot. Common joist spacings for a floor are 12″, 16″, and sometimes 19.2″ and 24″ for heavier commercial applications. Consider total loads only.
c. Select the lightest section that will work for 2 joist depths.
d. Confirm that the selected joist is adequate for deflection for total loads only. The manufacturer's data will have deflection formulas to use.
e. Sketch the final floor framing design

4.41 Design a DF-L tongue-and-grooved roof decking to span 14 ft between roof trusses. The decking is laid out in a pattern such that each deck sits on two supports. Assume a dead load of 15 psf, including the self-weight of the deck, and a snow load of 40 psf on a horizontal projected area. Assume that dry service and normal temperature conditions apply.

4.42 Determine if a floor framed with 2 × 12 sawn lumber (Hem-Fir, No. 1) spaced at 24 in. o.c. and a 15 ft 0 in. simple span with 3/4-in. plywood glued and nailed to the framing is adequate for walking vibration. Assume a residential occupancy. Assume 2 × solid blocking at 1/3 points.

REFERENCES

1. ICC (2015), *International Building Code,* International Code Council, Washington, DC.
2. APA (1999), *Glulam Beam Camber,* Technical Note EWS S550E, April.
3. Fournier, Christopher R. (2013), *Wood-framed Stair Stringer Design and Construction, Structure Magazine,* March, pp. 44–47.
4. Hoyle, Robert J., Jr. (1978), *Wood Technology in the Design of Structures,* 4th ed., Mountain Press, Missoula, MT.
5. ICC (2003), *International Residential Code,* International Code Council, Washington, DC.
6. Kermany, Abdy (1999), *Structural Timber Design,* Blackwell Science, London.
7. AITC (2012), *Timber Construction Manual,* 6th ed., American Institute of Timber Construction, Englewood, CO.
8. Jockwer, R (2014), *Structural Behaviour of Glued Laminated Timber Beams with Unreinforced and Reinforced Notches,* PhD dissertation, ETH, Zurich, Switzerland.
9. ANSI/AWC (2015), *National Design Specification for Wood Construction,* American Wood Council, Leesburg, VA.
10. AITC (2012), Guidelines for Evaluation of Holes and Notches in Structural Glued laminated Timber Beams, Technical Note 19, July.
11. APA (2014), Field Notching and Drilling of Glued laminated Timber Beams, Report No. EWS S560H, November.
12. Ayub, M (2015), *Investigation of the November 13 and 14, 2014 Collapses of Two Pedestrian Bridges Under Construction at Wake Technical Community College Campus, Raleigh, N.C.,* U.S. Department of Labor, Occupational Safety and Health Administration, Directorate of Construction, Washington, DC.
13. Schweizer, Gary (2008), Lateral Support of Wood Beams in Residential Structures, Structure magazine, January, pp. 39–42.
14. AISC (1997), *Floor Vibrations Due to Human Activity,* AISC/CISC Steel Design Guide Series 11, American Institute of Steel Construction, Chicago, IL.
15. ATC (1997), *Minimizing Floor Vibration,* ATC Design Guide 1, Applied Technology Council, Redwood City, CA.
16. Allen, D. E. and Pernica, G. (1998), *Control of Floor Vibrations,* Construction Technology Update 22, National Research Council of Canada, Ottawa, Canada.
17. Onysko, D. M. (1988), Performance criteria for residential floors based on consumer responses, *Proceedings of the 1988 International Conference on Timber Engineering,* Vol. 1, pp. 736–745.
18. National Building Code of Canada, 2015 (NBC), NRC, Ottawa, ON.
19. ACI (2014), *Building Code Requirements for Structural Concrete and Commentary,* ACI 318-05, American Concrete Institute, Farmington Hills, MI.
20. APA (2004), *Minimizing Floor Vibration by Design and Retrofit,* Technical Note E710, September.

Chapter 5

Wood members under axial and bending loads

5.1 INTRODUCTION

Axially loaded members occur in several forms in a timber structure. Some of these include columns, wall studs, truss members, shear wall chords, diaphragm chords, and drag struts or collectors. Columns and wall studs are vertical compression members that carry gravity loads down to the foundation. In addition, they may be subjected to lateral wind loads if located on the exterior face of the building. In a typical wood-framed building, axially compressed members such as columns and wall studs occur more frequently than members subjected to axial tension. Axial tension members occur more frequently in wood structures as truss members.

The strength of axially compressed wood members is a function of the unbraced length of the member and the end support conditions, and these members fail either by buckling due to lateral instability caused by the slenderness of the member or by crushing of the wood fibers as they reach their material strength. A more critical load effect that must also be considered for axially loaded members is the case where bending occurs simultaneously with an axial load. In general, the four basic design cases for axially loaded members and the corresponding loading conditions considered in this chapter are as follows [1–5]:

1. *Tension* (Figure 5.1a): Truss members, diaphragm chords, drag struts, and shear wall chords
2. *Tension plus bending* (Figure 5.1b): Truss bottom chord members
3. *Compression* (Figure 5.1c): Columns, wall studs, truss members, diaphragm chords, drag struts, and shear wall chords
4. *Compression plus bending* (Figure 5.1d): Exterior columns, wall studs, and truss top chord members

Each of these load cases is examined in greater detail in several design examples throughout the chapter. In these examples, we distinguish between the analysis and design of axially loaded wood members. In the analysis of member strength, the cross-sectional dimensions of the member, the wood species, and the stress grade are usually known, whereas the member strength is unknown and has to be determined. In design, the applied loads are usually known, but the member size, wood species, and stress grade are unknown and have to be determined.

5.2 PURE AXIAL TENSION: CASE 1

For this design case, axial tension parallel to the grain is considered. Examples of this load case may occur in truss web members and truss bottom chord members, diaphragm chords, drag struts, and shear wall chords. The National Design Specification (NDS) code [6] does

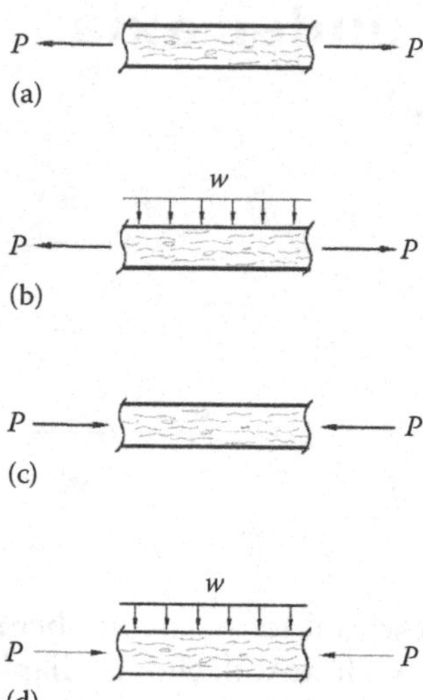

Figure 5.1 (a) Tension load case, (b) tension plus bending load case, (c) compression load case, and (d) compression plus bending load case.

not permit tension stress perpendicular to the grain (see NDS subsection 3.8.2) because of the negligible tensile strength of wood in that direction. The basic design equations for axial tension stress parallel to grain for both the allowable stress design (ASD) and load and resistance factor design (LRFD) methods are as follows:

ASD:

$$f_t = \frac{T}{A_n} \leq F_t' \tag{5.1}$$

LRFD:

$$f_{t,u} = \frac{T_u}{A_n} \leq F_{t,r}' \tag{5.2}$$

where:

f_t is the unfactored applied tension stress (ASD)
$f_{t,u}$ is the factored applied tension stress (LRFD)
T is the unfactored applied tension force (ASD)
T_u is the factored applied tension force (LRFD)
A_n is the net area at the critical section $= A_g - \Sigma A_{\text{holes}} = A_g - \Sigma\,(d_{\text{bolt}} + \frac{1}{16}\text{ in.})(\text{thickness, } b)$
A_g is the gross cross-sectional area of the member
ΣA_{holes} is the sum of the area of the bolt holes perpendicular to the load = (number of bolts perpendicular to load)($d_{\text{bolt}} + \frac{1}{16}$ in.)(member thickness, b), if bolts are present
F_t' is the allowable tension stress (ASD)
$F_{t,r}'$ is the LRFD-adjusted tension stress (LRFD)

Using the NDS code applicability table presented in Tables 3.1 through 3.4, the allowable tension stresses are calculated for both the ASD and LRFD methods as follows:

ASD:

$$F_t' = \begin{cases} F_t C_D C_M C_t C_F C_i & \text{for sawn lumber} \\ F_t C_D C_M C_t & \text{for glulams} \end{cases} \tag{5.3}$$

LRFD:

$$F'_{t,r} = \begin{cases} F_t C_M C_t C_F C_i \, K_{F,t} \, \phi_t \, \lambda & \text{for sawn lumber} \\ F_t C_M C_t \, K_{F,t} \, \phi_t \, \lambda & \text{for glulams} \end{cases} \quad (5.4)$$

where:

- F_t is the NDS-S tabulated reference design tension stress
- C is the adjustment factors, as discussed in Chapter 3
- $K_{F,t}$ is the format conversion factor = 2.70 (for tension stress)
- ϕ_t is the strength reduction factor = 0.80
- λ is the time effect factor, which is dependent on the load combination (see Table 3.11)

For bolted connections, the NDS code requires that the holes be at least $\frac{1}{32}$ to $\frac{1}{16}$ in. larger than the diameter of the bolt used (see NDS subsection 11.1.3.2). In this text, the bolt hole diameter will be assumed to be $d_{\text{bolt}} + \frac{1}{16}$ in.

5.2.1 Design of tension members

The design procedure for axially loaded tension members is given below and is illustrated in Example 5.3.

1. Perform structural analysis to determine the member force (unfactored or factored).
2. Assume a trial member size initially.
3. Analyze the trial member to determine the applied tension stress and the allowable tension stress (or tension load), as in Examples 5.1 and 5.2.
4. If the unfactored applied tension stress (ASD) in the trial member is less than the allowable tension stress, the trial member size is deemed adequate. If the applied tension stress is *much* less than the allowable stress, the member is adequate but uneconomical, and in that case, the trial member size should be reduced and the analysis should be repeated until the applied tension stress is *just* less than the allowable tension stress.
5. If the unfactored applied tension stress is greater than the allowable tension stress, the trial member size must be increased and the analysis must be repeated until the applied stress is just less than the allowable stress.

For LRFD, the factored applied tension stress is compared with the LRFD-adjusted tension stress in the steps above.

Example 5.1: (ASD/LRFD): Analysis of a sawn lumber tension member (ASD/LRFD)

A 2 × 8 wood member (Figure 5.2) is subjected to an axial tension load that consists of a dead load of 2400 lb and a snow load of 4000 lb. The lumber is Hem-Fir No. 2; normal temperature conditions apply; and the moisture content (MC) is less than 19%. Assume that the connections are made with one row of $\frac{7}{8}$-in.-diameter bolts.

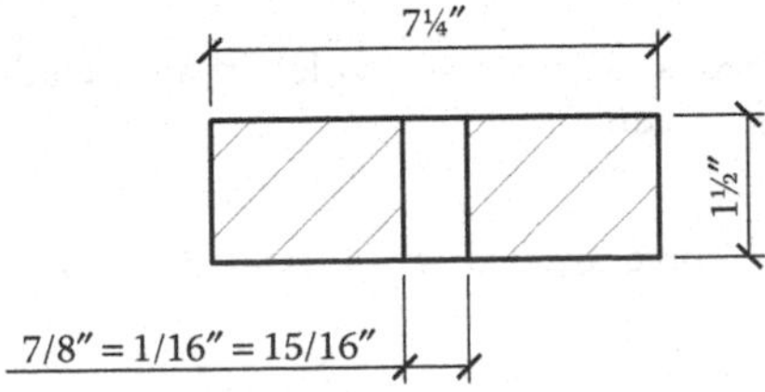

Figure 5.2 Cross section of tension member at bolt hole—Example 5.1.

a. Determine the applied tension stress f_t and the allowable tension stress F_t', and check the adequacy of this member.
b. Determine the factored applied tension stress $f_{t,u}$ and the LRFD-adjusted tension stress $F_{t,r}'$, and check the adequacy of this member.

Solution:

D = 2400 lb
S = 4000 lb

Total unfactored tension force, $T = 2400 + 4000 = 6400$ lb
Total factored tension force, $T_u = 1.2(2400) + 1.6(4000) = 9280$ lb

Adjustment Factors:

Since a 2 × 8 is dimension lumber, the applicable table is NDS-S Table 4A. The tabulated design tension stress and the adjustment factors from the table are as follows:

$F_t = 525$ psi
$C_D = 1.15$ (the C_D value for the shorter-duration load in the load combination, i.e., the snow load, is used; see subsection 3.1)
$C_M = 1.0$ (dry service conditions)
$C_t = 1.0$ (normal temperature conditions)
$C_F = 1.2$ (for tension stress)
$C_i = 1.0$ (assumed since no incision is prescribed)
$\lambda\,(1.2D + 1.6S) = 0.8$
$K_F = 2.70$ (for tension stress)
$\phi = 0.80$ (for tension stress)

Part (a) – ASD:
The gross area A_g is 10.88 in.^{2}, and the area of the bolt holes is

$$\Sigma A_{\text{holes}} = (1\text{ hole})(7/8\text{ in.} + 1/16\text{ in.})(1.5\text{ in.}) = 1.41\text{ in}^2$$

The net cross-sectional area of the member at the critical section is

$$A_n = A_g - \Sigma A_{\text{holes}} = 10.88 - 1.41 = 9.47\text{ in.}^2$$

The unfactored axial tension force applied, $T = 6400$ lb. Therefore, the unfactored applied tension stress (ASD) is

$$f_t = \frac{T}{A_n} = \frac{6400\text{ lb}}{9.47\text{ in}^2} = \mathbf{676\text{ psi}}$$

Using the NDS applicability table (Table 3.1), the ASD is given as

$$F_t' = F_t C_D C_M C_t C_F C_i$$

$$= (525)(1.15)(1.0)(1.0)(1.2)(1.0) = \mathbf{725\text{ psi}}$$

$$f_t = 676\text{ psi} < F_t' = 725\text{ psi} \qquad \textbf{OK}$$

Since the unfactored applied tension stress f_t is less than the allowable tension stress F_t', the 2 × 8 Hem-Fir No. 2 is adequate.

Part (b) – LRFD:
The factored axial tension force applied, $T_u = 9280$ lb. Therefore, the factored applied tension stress is

$$f_{t,u} = \frac{T_u}{A_n} = \frac{9280\text{ lb}}{9.47\text{ in}^2} = \mathbf{980\text{ psi}}$$

Using the NDS applicability table (Table 3.3), the LRFD-adjusted tension stress (LRFD) is given as

$$F'_{t,r} = F_t C_M C_t C_F C_i (K_F)(\phi)(\lambda) = F_t C_M C_t C_F C_i (2.70)(0.80)(0.8)$$

$$= (525)(1.0)(1.0)(1.2)(1.0)(2.70)(0.80)(0.8) = \mathbf{1089\ psi}$$

$$f_{t,u} = 980 \text{ psi} < F'_{t,r} = 1089 \text{ psi} \qquad \textbf{OK}$$

Since the factored applied tension stress $f_{t,u}$ is less than the LRFD-adjusted tension stress $F'_{t,r}$, the 2 × 8 Hem-Fir No. 2 is adequate.

Example 5.2: (ASD/LRFD): Analysis of a glulam tension member (ASD/LRFD)

For a 2½ × 6 (four laminations) 5DF glulam axial combination member (Figure 5.3) subject to a concentric axial tension dead load of 6000 lb and a floor live load of 9500 lb, assume that normal temperature conditions apply, the MC is greater than 16%, and the connection is made with one row of ¾-in.-diameter bolts.

a. Calculate the unfactored applied tension stress f_t and the allowable tension stress F'_t, and check the structural adequacy of the member.
b. Calculate the factored applied tension stress $f_{t,u}$ and the LRFD-adjusted tension stress $F'_{t,r}$, and check the structural adequacy of the member.

Solution:

Total unfactored tension force, $T = 6000 + 9500 = 15{,}500$ lb
Total factored tension force, $T_u = 1.2(6000) + 1.6(9500) = 22{,}400$ lb

Adjustment Factors:

For a 5DF glulam axial combination, use NDS-S Table 5B. The tabulated design tension stress and the applicable adjustment factors from the table are obtained as follows:

$F_t = 1650$ psi for 5DF

$C_{M(Ft)} = 0.80$ (wet service since MC > 16%)

$C_D\ (D + L) = 1.0$ (the C_D value for the shortest-duration load in the load combination, i.e., floor live load, is used; see Chapter 3)

$C_t = 1.0$ (normal temperature conditions)

$\lambda\ (1.2D + 1.6L) = 0.8$ (assuming normal occupancy)

$K_F = 2.70$ (for tension stress)

$\phi = 0.80$ (for tension stress)

Part (a) – ASD:

The gross area A for a 2½ × 6 in. glulam is 15 in.^{2} (*Note:* The glulam is specified using the actual size.)

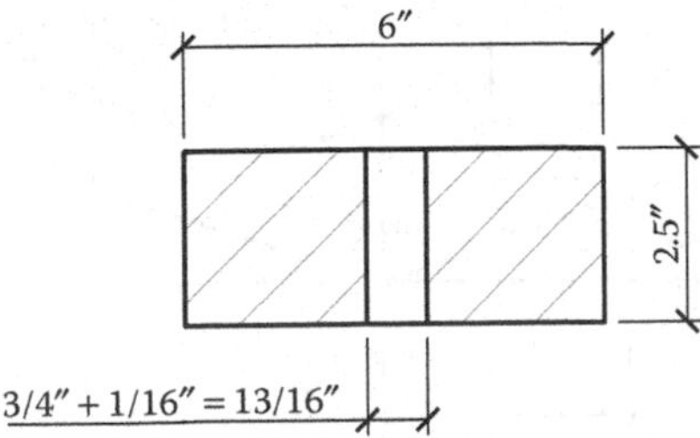

Figure 5.3 Cross section of tension member at bolt hole—Example 5.2.

The area of the bolt holes is

$$\Sigma A_{\text{holes}} = (1\text{ hole})(\tfrac{3}{4}\text{ in.} + \tfrac{1}{16}\text{ in.})(2\tfrac{1}{2}\text{ in.}) = 2.03\text{ in}^2$$

The net cross-sectional area at the critical section is

$$A_n = A_g - \Sigma A_{\text{holes}} = 15 - 2.03 = 12.97\text{ in.}^2$$

The unfactored tension stress applied f_t is

$$f_t = \frac{T}{A_n} = \frac{15{,}500\text{ lb}}{12.97\text{ in}^2} = \mathbf{1195\text{ psi}}$$

Using the NDS applicability table (Table 3.2), the allowable tension stress (ASD) is given as

$$F_t' = F_t C_D C_M C_t$$

$$= (1650)(1.0)(0.8)(1.0) = \mathbf{1320\text{ psi}} > f_t = \mathit{1195\ psi} \qquad \textbf{OK}$$

A $2\tfrac{1}{2} \times 6$ 5DF glulam tension member is adequate.

Part (b) – LRFD:
The factored applied tension stress $f_{t,u}$ is

$$f_{t,u} = \frac{T_u}{A_n} = \frac{22{,}400\text{ lb}}{12.97\text{ in}^2} = \mathbf{1727\text{ psi}}$$

Using the NDS applicability table (Table 3.4), the LRFD-adjusted tension stress is given as

$$F_{t,r}' = F_t C_M C_t\,(K_F)(\phi)(\lambda) = F_t C_M C_t\,\mathit{(2.70)(0.80)(0.8)}$$

$$= (1650)(0.8)(1.0)\mathit{(2.70)(0.80)(0.8)} = \mathbf{2281\text{ psi}}$$

$$f_{t,u} = 1727\text{ psi} < F_{t,r}' = 2281\text{ psi} \qquad \textbf{OK}$$

Since the factored applied tension stress $f_{t,u}$ is less than the LRFD-adjusted tension stress, $F_{t,r}'$, the $2\tfrac{1}{2} \times 6$ (5DF) glulam tension member, is adequate.

Example 5.3: (ASD/LRFD): Design of a wood tension member (ASD/LRFD)

Design member *AF* of the typical interior truss is shown in Figure 5.4, assuming the following design parameters:

The trusses are spaced 4′-0″.
The dead load is 22.5 psf and the snow load is 40 psf on a horizontal plan area.
The ends of the truss members are connected with one row of ½-in.-diameter bolts.
The wood species/stress grade is No. 1 Southern Pine.

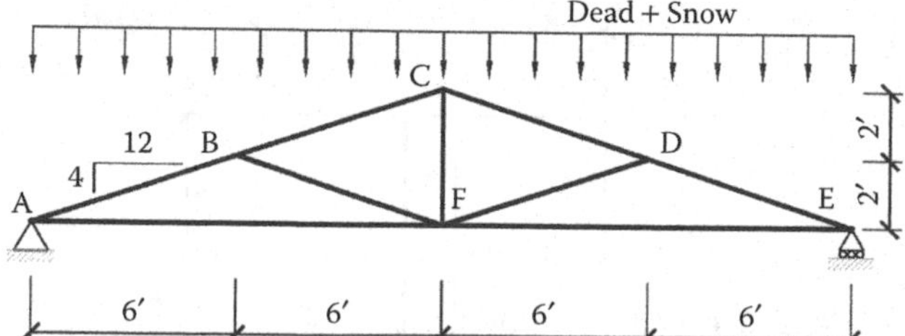

Figure 5.4 Roof truss profile—Example 5.3.

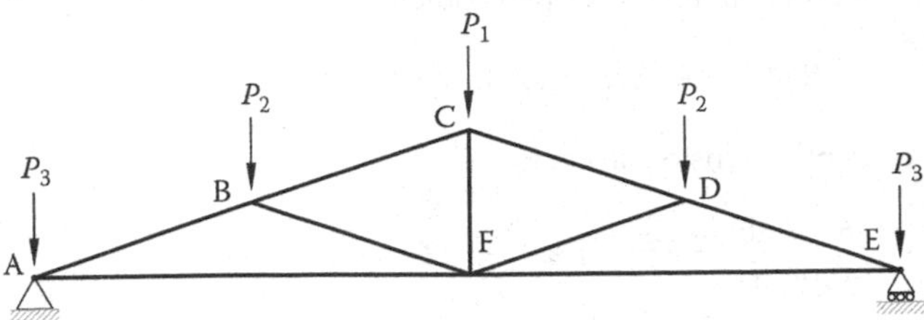

Figure 5.5 Free-body diagram of roof truss.

Solution:

1. Analyze the truss to obtain the member forces (Figure 5.5). The tributary width of the truss is 4 ft.
   Dead load $D = 22.5$ psf × 4 ft (tributary width of truss) = 90 lb/ft
   Snow load $S = 40$ psf × 4 ft = 160 lb/ft
   Total unfactored load w_{TL} = **250 lb/ft**
   Total factored load $w_{TL,u} = 1.2(90) + 1.6(160) =$ **364 lb/ft**

   Calculate the unfactored joint loads:

   $P_1 = (6\text{ ft}/2 + 6\text{ ft}/2)(250\text{ lb/ft}) = 1500\text{ lb}$
   $P_2 = (6\text{ ft}/2 + 6\text{ ft}/2)(250\text{ lb/ft}) = 1500\text{ lb}$
   $P_3 = (6\text{ ft}/2)(250\text{ lb/ft}) = 750\text{ lb}$

   By analyzing the truss using the method of joints [7] or a structural analysis software program, the unfactored force in the truss bottom chord (member *AF*) is obtained as

   $T_{AF} = 6750$ lb (tension)

   The corresponding factored tension force in member AF is

   $$= \frac{364\text{ lb/ft}}{250\text{ lb/ft}}(6750\text{ lb})$$

   $$= 9828\text{ lb}$$

   We will first design member AF using the ASD approach:
2. Assume the member size. Try a 2 × 8 (dimension lumber). Therefore, use NDS-S Table 4B.
3. From NDS-S Table 4B (Southern Pine dimension lumber), we obtain the tabulated design tension stress and the adjustment factors as follows. The reference design tension stress F_t is 800 psi and the adjustment or C factors are as follows:

   $C_D = 1.15$ (the C_D value for the shortest-duration load in the load combination, i.e., the snow load, is used; see subsection 3.1)
   $C_t = 1.0$ (normal temperature conditions)
   $C_M = 1.0$ (dry service, since the truss members are protected from weather)
   $C_F = 1.0$ (for Southern Pine dimension lumber, for $d \le 12$ in., $C_F = 1.0$, and for $d > 12$ in., see the adjustment factors section of NDS-S Table 4B)
   $C_i = 1.0$ (assumed)

   Using the NDS applicability table (Table 3.1), the allowable tension stress (ASD) is given as

   $$F_t' = F_t C_D C_M C_t C_F C_i$$
   $$= (800)(1.15)(1.0)(1.0)(1.0)(1.0) = \mathbf{920\ psi}$$

   From NDS-S Table 1B, the gross area A_g for a 2 × 8 is 10.88 in.^{2} The available net area at the critical section is

$$A_n = A_g - (\text{number of bolt holes perpendicular to force})\ \Sigma(d_{bolt} + \tfrac{1}{16}\text{ in.})(\text{thickness}, b)$$

$$= 10.88 - (1)(\tfrac{1}{2}\text{ in.} + \tfrac{1}{16}\text{ in.})(1.5\text{ in.}) = 10.04\text{ in}^2$$

4. Unfactored applied tension stress is

$$f_t = \frac{T}{A_n} = \frac{6750\text{ lb}}{10.04\text{ in}^2} = 672\text{ psi} < F_t' = 920\text{ psi} \qquad \text{OK}$$

2 × 8 No. 1 Southern Pine is adequate for the truss bottom chord.
We will now repeat the design by using the LRFD method:
The additional adjustment factors for LRFD are

$$\lambda\ (1.2D + 1.6S) = 0.8$$

$$K_F = 2.70\ (\text{for tension stress})$$

$$\phi = 0.80\ (\text{for tension stress})$$

The factored axial tension force applied, T_u = 9828 lb. Therefore, the factored applied tension stress is

$$f_{t,u} = \frac{T_u}{A_n} = \frac{9828\text{ lb}}{10.04\text{ in}^2} = 979\text{ psi}$$

Using the NDS applicability table (Table 3.3), the LRFD-adjusted tension stress (LRFD) is given as

$$F_{t,r}' = F_t C_M C_t C_F C_i (K_F)(\phi)(\lambda) = F_t C_M C_t C_F C_i (2.70)(0.80)(0.8)$$

$$= (800)(1.0)(1.0)(1.0)(1.0)(2.70)(0.80)(0.8) = \mathbf{1382\ psi}$$

$$f_{t,u} = 979\text{ psi} < F_{t,r}' = 1382\text{ psi} \qquad \text{OK}$$

Since the factored applied tension stress $f_{t,u}$ is less than the LRFD-adjusted tension stress $F_{t,r}'$, the 2 × 8 No. 1 Southern Pine is adequate.

Example 5.4: (ASD/LRFD): Design of a wood tension member (ASD/LRFD)

Design member *AF* of the truss in Example 5.3, assuming that the wood species and stress grade is Douglas Fir-Larch (DF-L) Select Structural and the total unfactored dead plus snow load is 500 lb/ft.

Repeat the design by using the LRFD method, assuming a total factored load of 728 lb/ft.

Solution:

1. Analyze the truss to obtain the member forces. The given total unfactored load (dead plus snow) w_{TL} is 500 lb/ft. Using the truss in Figure 5.5, we calculate the joint loads as follows:

$$P_1 = (6\text{ ft}/2 + 6\text{ ft}/2)(500\text{ lb/ft}) = 3000\text{ lb}$$
$$P_2 = (6\text{ ft}/2 + 6\text{ ft}/2)(500\text{ lb/ft}) = 3000\text{ lb}$$
$$P_3 = (6\text{ ft}/2)(500\text{ lb/ft}) = 1500\text{ lb}$$

By analyzing the truss using the method of joints or a structural analysis software program, the unfactored force in the truss bottom chord (member *AF*) is obtained as

$$T_{AF} = 13{,}500\text{ lb (tension)}$$

The corresponding factored tension force in member AF is

$= (728 \text{ lb/ft}/500 \text{ lb/ft})(13{,}500 \text{ lb}) = 19{,}656 \text{ lb}$

We will first design member AF by using the ASD approach:

2. Try a 2 × 8 dimension lumber. Therefore, use NDS-S Table 4A.

3. From NDS-S Table 4A (DF-L Select Structural), we obtain the tabulated design tension stress and the adjustment factors as follows. The reference design tension stress F_t is 1000 psi and the adjustment or C factors are as follows:

   $C_D = 1.15$ (the C_D value for the shortest-duration load in the load combination, i.e., the snow load, is used)
   $C_t = 1.0$ (normal temperature conditions)
   $C_M = 1.0$ (dry service, since the truss members are protected from the weather)
   $C_F(F_t) = 1.2$
   $C_i = 1.0$ (assumed)

   Using the NDS applicability table (Table 3.1), the allowable tension stress (ASD) is given as

   $$F_t' = F_t C_D C_M C_t C_F C_i$$
   $$= (1000)(1.15)(1.0)(1.0)(1.2)(1.0) = \mathbf{1380\ psi}$$

   From NDS-S Table 1B, the gross area A_g for a 2 × 8 = 10.88 in.^{2} The available net cross-sectional area at the critical section is

   $$A_n = A_g - (\text{number of bolt holes perpendicular to force}) \times \Sigma\,(d_{bolt} + \tfrac{1}{16}\text{ in.})(\text{thickness, } b)$$

   $$= 10.88 - (1)(\tfrac{1}{2}\text{ in.} + \tfrac{1}{16}\text{ in.}) \times (1.5\text{ in.}) \simeq 10.04\text{ in}^2$$

4. Unfactored applied tension stress is

   $$f_t = \frac{T}{A_n} = \frac{13{,}500\text{ lb}}{10.04\text{ in}^2} = \mathbf{1345\ psi} < F_t' = 1380\text{ psi} \qquad \textbf{OK}$$

   A 2 × 8 DF-L Select Structural is adequate for the truss bottom chord
   We will now repeat the design by using the LRFD method:
   The additional adjustment factors for LRFD are

   $\lambda\,(1.2D + 1.6S) = 0.8$
   $K_F = 2.70$ (for tension stress)
   $\phi = 0.80$ (for tension stress)

   The factored axial tension force applied, T_u = 19,656 lb. Therefore, the factored applied tension stress is

   $$f_{t,u} = \frac{T_u}{A_n} = \frac{19{,}656\text{ lb}}{10.04\text{ in}^2} = \mathbf{1958\ psi}$$

   Using the NDS applicability table (Table 3.2), the LRFD-adjusted tension stress (LRFD) is given as

   $$F_{t,r}' = F_t C_M C_t C_F C_i (K_F)(\phi)(\lambda)$$
   $$= F_t C_M C_t C_F C_i (2.70)(0.80)(0.8)$$
   $$= (1000)(1.0)(1.0)(1.2)(1.0)(2.70)(0.80)(0.8) = \mathbf{2074\ psi}$$

   $$f_{t,u} = 1958\text{ psi} < F_{t,r}' = 2074\text{ psi} \qquad \text{OK}$$

   Since the factored applied tension stress $f_{t,u}$ is less than the LRFD-adjusted tension stress $F_{t,r}'$, the 2 × 8 DF-L Select Structural is adequate for the truss bottom chord.

5.3 AXIAL TENSION PLUS BENDING: CASE 2

Some examples of members subjected to combined axial tension and bending loads include truss bottom chords and door headers subjected to bending from gravity loads and tension from acting as a chord member in a roof or floor diaphragm. For these types of structural wood members, five design checks need to be investigated. It should be noted that for each design check in the ASD method, the controlling load duration factor based on the pertinent load combination for each design check should be used (see Chapter 3). For the LRFD method, the time effect factor, λ, would depend on the load combination being considered for each design check. The required design checks are given in the following subsections.

5.3.1 Design check 1: Tension on the net area

Tension on the net area (Figure 5.6a) occurs at the ends of the member, where the gross area of the member may be reduced due to the presence of bolt holes at the end connections (e.g., at the ends of truss members).

The applied axial tension stress parallel to grain for both the ASD and LRFD methods have been calculated previously in Section 5.2 as follows:

ASD:

$$f_t = \frac{T}{A_n} \leq F_t' \tag{5.1}$$

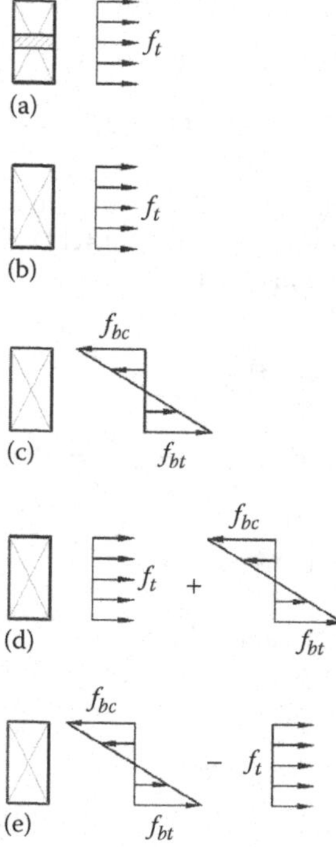

Figure 5.6 (a) Stress diagram for load case 2—design check No. 1, (b) Stress diagram for load case 2—design check No. 2, (c) Stress diagram for load case 2—design check No. 3, (d) Stress diagram for load case 2—design check No. 4, and (e) Stress diagram for load case 2—design check No. 5.

LRFD:

$$f_{t,u} = \frac{T_u}{A_n} \leq F'_{t,r} \tag{5.2}$$

where:

f_t is the unfactored applied tension stress (ASD)
T is the unfactored applied tension force (ASD)
$f_{t,u}$ is the factored applied tension stress (LRFD)
T_u is the factored applied tension force (LRFD)
A_n is the net area at the critical section $= A_g - \Sigma A_{\text{holes}} = A_g - \Sigma\,(d_{\text{bolt}} + 1/16\text{ in.})(\text{thickness}, b)$
A_g is the gross cross-sectional area of the member
ΣA_{holes} is the sum of the area of the bolt holes perpendicular to the load
$= (\text{number of bolts perpendicular to load})(d_{\text{bolt}} + 1/16\text{ in.})(\text{member thickness}, b)$
F'_t is the allowable tension stress (ASD)
$F'_{t,r}$ is the LRFD-adjusted tension stress (LRFD)

Using the NDS code applicability table (Tables 3.1 and 3.2), the allowable tension stresses (ASD) and the LRFD-adjusted design stresses are calculated as follows:

ASD:

$$F'_t = \begin{cases} F_t C_D C_M C_t C_F C_i & \text{for sawn lumber} \\ F_t C_D C_M C_t & \text{for glulams} \end{cases} \tag{5.3}$$

LRFD:

$$F'_{t,r} = \begin{cases} F_t C_M C_t C_F C_i\, K_{F,t}\, \phi_t\, \lambda & \text{for sawn lumber} \\ F_t C_M C_t\, K_{F,t}\, \phi_t\, \lambda & \text{for glulams} \end{cases} \tag{5.4}$$

where:

F_t is the NDS-S tabulated reference design tension stress
C is the adjustment factors, as discussed in Chapter 3
$K_{F,t}$ is the format conversion factor for LRFD = 2.70 (for tension stress)
ϕ_t is the strength reduction factor for LRFD = 0.80 (for tension stress)
λ is the time effect factor for LRFD and is dependent on the load combination (see Table 3.11)

5.3.2 Design check 2: Tension on the gross area

Tension on the gross area (Figure 5.6b) occurs at the midspan of the member, where there are no bolt holes, and hence, the gross area is applicable (e.g., at the midspan of truss members and wall studs). The tension stress applied on the gross area is calculated as follows:

ASD:

$$f_t = \frac{T}{A_g} \leq F'_t \tag{5.5}$$

LRFD:

$$f_{t,u} = \frac{T_u}{A_g} \leq F'_{t,r} \tag{5.6}$$

The parameters in Equations 5.5 and 5.6 are as defined in design check No. 1.

5.3.3 Design check 3: Bending only

The required design interaction equation for bending for both the ASD and LRFD methods are given below and is dependent on the maximum bending moment. Note that for prismatic and simply supported members, the maximum bending moment and maximum bending stress occur at the midspan of the member (Figure 5.6c).

ASD:

$$f_{bt} = \frac{M}{S_x} \le F_b' \tag{5.7}$$

LRFD:

$$f_{bt,u} = \frac{M_u}{S_x} \le F_{b,r}' \tag{5.8}$$

where:

f_{bt} is the *unfactored* bending stress in the tension fiber (ASD)
$f_{bt,u}$ is the *factored* bending stress in the tension fiber (LRFD)
M is the *unfactored* maximum bending moment in the member (typically, occurs at midspan)
M_u is the *factored* maximum bending moment in the member (typically, occurs at midspan)
S_x is the section modulus about the axis of bending
F_b' is the allowable bending stress (ASD)
$F_{b,r}'$ is the LRFD-adjusted design bending stress

Using the applicability table (Tables 3.1 through 3.4), the allowable bending stress (ASD) and the LRFD-adjusted stress are obtained as

ASD:

$$F_b' = \begin{cases} F_b C_D C_M C_t C_L C_F C_i C_r C_{fu} & \text{(for sawn lumber)} \\ F_b C_D C_M C_t C_L C_v C_{fu} C_c & \text{(for glulam)} \end{cases} \tag{5.9}$$

The LRFD-adjusted bending stress is obtained as

LRFD:

$$F_{b,r}' = \begin{cases} F_b C_M C_t C_L C_F C_i C_r C_{fu} K_{F,b}\, \phi_b\, \lambda & \text{(for sawn lumber)} \\ F_b C_M C_t C_L C_V C_{fu} C_c C_I K_{F,b}\, \phi_b\, \lambda & \text{(for glulam)} \end{cases} \tag{5.10}$$

where:

F_b is the NDS-S tabulated reference design bending stress
C is the adjustment factors, as discussed in Chapter 3
K_F is the format conversion factor for LRFD = 2.54 (for bending stress)
ϕ is the strength reduction factor for LRFD = 0.85 (for bending stress)
λ is the time effect factor for LRFD and is dependent on the load combination (see Table 3.11)

5.3.4 Design check 4: Bending plus tension

Bending plus tension (Figure 5.6d) occurs at the bottom of the member at the point of maximum moment, where the moment and tension force are at their maximum values, creating maximum tension stress in the tension or bottom fiber of the member. The interaction equation for bending plus tension for both the ASD and LRFD methods are given as follows:

ASD:

$$\frac{f_t}{F'_t}+\frac{f_{bt}}{F^*_b}\leq 1.0 \quad \text{(NDS code equation 3.9-1)} \tag{5.11}$$

LRFD:

$$\frac{f_{t,u}}{F'_{t,r}}+\frac{f_{bt,u}}{F^*_{b,r}}\leq 1.0 \quad \text{(NDS code equation 3.9-1)} \tag{5.12}$$

where:

f_t is the unfactored tension stress due to the axial tension force (ASD)
$f_{t,u}$ is the factored tension stress due to the axial tension force (LRFD)
f_{bt} is the unfactored flexural tension stress due to the unfactored bending moment (ASD)
$f_{bt,u}$ is the factored flexural tension stress due to the factored bending moment (LRFD)
F'_t is the unfactored allowable tension stress, as calculated in design check No. 1 (ASD)
$F'_{t,u}$ is the LRFD-adjusted tension stress, as calculated in design check No. 1 (LRFD)
F^*_b is the unfactored allowable bending stress, excluding the C_L factor (ASD)
$F^*_{b,r}$ is the LRFD-adjusted bending stress, excluding the C_L factor (LRFD)

The allowable bending stress for the ASD method and the LRFD-adjusted design bending stress are given as follows:

ASD:

$$F^*_b=\begin{cases}F_bC_DC_MC_tC_FC_iC_rC_{fu} & \text{(for sawn lumber)}\\ F_bC_DC_MC_tC_VC_{fu}C_c & \text{(for glulam)}\end{cases} \tag{5.13}$$

LRFD:

$$F^*_{b,r}=\begin{cases}F_bC_MC_tC_FC_{fu}C_iC_rK_{F,b}\,\phi_b\,\lambda & \text{(for sawn lumber)}\\ F_bC_MC_tC_VC_{fu}C_cC_IK_{F,b}\,\phi_b\,\lambda & \text{(for glulam)}\end{cases} \tag{5.14}$$

where:

F_b is the NDS-S tabulated reference design bending stress
C is the adjustment factors, as discussed in Chapter 3
K_F is the format conversion factor for LRFD = 2.54 (for bending stress)
ϕ is the strength reduction factor for LRFD = 0.85 (for bending stress)
λ is the time effect factor for LRFD and is dependent on the load combination (see Table 3.11)

Note that for this design check, the load duration factor, C_D, for the ASD method, and the time effect factor, λ, for the LRFD method, will be dependent on the load combination that causes the combined bending plus tension. It should also be noted that the allowable bending stress and the LRFD-adjusted bending stress equations above do not include the beam stability factor C_L because the fibers in the tension zone of a member subjected to combined bending plus axial tension are not susceptible to buckling instability, since this can be caused only by compressive stresses. Therefore, the allowable bending stress for ASD is denoted as F_b^* and the LRFD-adjusted design bending stress is denoted as $F_{b,r}^*$.

5.3.5 Design check 5: Bending minus tension

Bending plus tension (Figure 5.6e) occurs in the top or compression fibers of the member subjected to combined axial tension plus bending, with the compression stresses caused by the bending of the member. The compression stress f_{bc} ($f_{bc,u,}$ *for LRFD)* caused by the bending moment applied is opposed or counteracted by the tension stress resulting from the axial tension force P on the member. For this design check, the beam stability factor, C_L, has to be considered, since there is a possibility that a net compression stress could exist in the top fibers of the member. Where the unfactored axial tension stress f_t ($f_{t,u}$ *for LRFD)* exceeds the unfactored compression stress due to bending f_{bc} ($f_{bc,u,}$ *for LRFD)*, this design check will not be critical and can be ignored, since this condition will produce a net tension stress in the top fibers, which is less than the tension stress from design check No. 4.

The interaction equation for this design check is given for both the ASD and LRFD methods as follows:

ASD:

$$\frac{f_{bc} - f_t}{F_b^{**}} \leq 1.0 \quad \text{(NDS code equation 3.9-2)} \tag{5.15}$$

LRFD:

$$\frac{f_{bc,u} - f_{t,u}}{F_{b,r}^{**}} \leq 1.0 \quad \text{(NDS code equation 3.9-2)} \tag{5.16}$$

where:

$f_{bc} = M/S_x$ = unfactored bending stress in the compression fibers (ASD)
$f_{bc,u} = M_u/S_x$ = factored bending stress in the compression fibers (LRFD)

Note that for wood members with a rectangular cross section, for the ASD method,

$f_{bc} = f_{bt}$ from design check No. 3; similarly, for LRFD-method, $f_{bc,u} = f_{bt,u}$ from design check No. 3.

The allowable bending stress for the ASD method and the LRFD-adjusted design bending stress are given as follows:

ASD:

$$F_b^{**} = \begin{cases} F_b C_D C_M C_t C_L C_F C_{fu} C_i C_r & \text{(for sawn lumber)} \\ F_b C_D C_M C_t C_L C_{fu} C_c C_I & \text{(for glulam)} \end{cases} \tag{5.17}$$

LRFD:

$$F_{b,r}^{*} = \begin{cases} F_b C_M C_t C_L C_F C_{fu} C_i C_r K_{F,b}\, \phi_b\, \lambda & \text{(for sawn lumber)} \\ F_b C_M C_t C_L C_{fu} C_c C_I K_{F,b}\, \phi_b\, \lambda & \text{(for glulam)} \end{cases} \tag{5.18}$$

It is important to note that it may not be necessary to perform all of the foregoing design checks for a given problem, since some of the design checks could be eliminated by inspection.

5.3.6 Euler critical buckling stress

The Euler critical buckling load for a column is the minimum load at which the column first bows out, and this load is given as

$$P_{cr} = \frac{\pi^2 EI}{(K_e l)^2} \tag{5.19}$$

where:

E is the modulus of elasticity
I is the moment of inertia about the axis of bending
$K_e l$ is the effective length of the column = l_e
K_e is the effective length factor (= 1.0 for column with pinned end supports)
l is the unbraced length or height of column

For a rectangular column section with width b and depth d, the Euler critical buckling load becomes

$$P_{cr} = \frac{\pi^2 E(bd^3/12)}{(K_e l)^2} \tag{5.20}$$

Dividing both sides of Equation 5.20 by the cross-sectional area A (= bd) yields the Euler critical buckling stress

$$F_{cE} = \frac{\pi^2 E(bd^3/12)}{(K_e l)^2 bd} = \frac{(\pi^2/12)Ed^2}{(K_e l)^2} \tag{5.21}$$

This equation is simplified further to yield the NDS code equation for the Euler critical buckling stress, which is given as

$$F_{cE} = \frac{0.822E}{(K_e l/d)^2} \tag{5.22}$$

In the NDS code, the modulus of elasticity E in Equation 5.22 is replaced by the allowable buckling or minimum modulus of elasticity E'_{min}; hence, the equation becomes

$$F_{cE} = \frac{0.822E'_{min}}{(K_e l/d)^2} \tag{5.23}$$

5.4 PURE AXIAL COMPRESSION: CASE 3

There are two categories of columns: solid columns and built-up columns. Examples of *solid columns* include interior building columns, struts (such as trench shores), studs (such as wall studs), and truss members. *Built-up columns* are made up of individual laminations joined mechanically, with no spaces between laminations. The mechanical connection between the laminations is made with through-bolts or nails. Subsection 15.3 of the NDS code gives the mechanical connection requirements for built-up columns. Built-up columns are commonly used within interior or exterior walls of wood-framed buildings, because the component members that make up the built-up column are already available on site and can readily be fastened together on site.

The strength of a column is affected by buckling if the column length is excessive. There are two types of columns with respect to the column length: short columns and long columns. A *short column* is one in which the axial strength depends only on the crushing or material capacity of the wood. A *long column* is one in which buckling occurs before the column reaches the crushing or material capacity $F_c' A$ of the wood member. The strength of a long column is a function of the slenderness ratio $K_e l/r$ of the column, which is defined as the ratio of the effective length $K_e l$ of the column to the least radius of gyration r of the member. The radius of gyration is given as

$$r = \sqrt{\frac{I}{A}} \tag{5.24}$$

where:

I is the moment of inertia of the column member about the axis of buckling
A is the cross-sectional area of the column

The allowable stress design equation for the ASD method and the interaction equation for the LRFD method for this pure compression load case are given as follows:

ASD:

$$f_c = \frac{P}{A_g} \le F_c' \tag{5.25}$$

LRFD:

$$f_{c,u} = \frac{P_u}{A_g} \le F_{c,r}' \tag{5.26}$$

where:

f_c is the *unfactored* axial compression stress (ASD)
P is the *unfactored* axial compression force applied on the wood column
A_g is the gross cross-sectional area of the tension member
F_c' is the allowable compression stress (ASD)
$f_{c,u}$ is the *factored* axial compression stress (LRFD)
P_u is the *factored* axial compression force applied on the wood column
A_g is the gross cross-sectional area of the tension member
$F_{c,r}'$ is the LRFD-adjusted compression stress

Using the NDS adjustment factors applicability tables (Tables 3.1 through 3.4), the allowable compressive stress parallel to grain for the ASD method and the LRFD-adjusted compressive stress are given as follows:

ASD:

$$F_c' = \begin{cases} F_c C_D C_M C_t C_F C_i C_P & \text{(sawn lumber)} \\ F_c C_D C_M C_t C_P & \text{(glulam)} \end{cases} \tag{5.27}$$

LRFD:

$$F_{c,r}' = \begin{cases} F_c C_M C_t C_F C_i C_{P,r} K_{F,c} \phi_c \lambda & \text{(sawn lumber)} \\ F_c C_M C_t C_{P,r} K_{F,c} \phi_c \lambda & \text{(glulam)} \end{cases} \tag{5.28}$$

where:

F_c is the NDS-S tabulated reference design compression stress
$K_{F,c}$ is the format conversion factor for LRFD = 2.40 (for compression stress)
ϕ_c is the strength reduction factor for LRFD = 0.90 (for compression stress)
λ is the time effect factor for LRFD and is dependent on the load combination (see Table 3.11)

The adjustment or C factors are as discussed in Chapter 3, and the column stability factor C_P depends on the unbraced length of the column. The higher the column slenderness ratio $(K_e l/r)$, the smaller the column stability factor C_P. For columns that are laterally braced over their entire length, the column stability factor C_P is 1.0. For all other cases, calculate C_P *by* using the equations below:

ASD:

$$C_P = \frac{1 + F_{cE}/F_c^*}{2c} - \sqrt{\left(\frac{1 + F_{cE}/F_c^*}{2c}\right)^2 - \frac{F_{cE}/F_c^*}{c}} \quad (5.29)$$

LRFD:

$$C_{P,r} = \frac{1 + F_{cE,r}/F_{c,r}^*}{2c} - \sqrt{\left(\frac{1 + F_{cE,r}/F_{c,r}^*}{2c}\right)^2 - \frac{F_{cE,r}/F_{c,r}^*}{c}} \quad (5.30)$$

where F_{cE} = Euler critical buckling stress (ASD) $= \dfrac{0.822E'_{min}}{(l_e/d)^2}$ [see equation (5.23)]

$F_{cE,r}$ = Euler critical buckling stress (LRFD) $= \dfrac{0.822E'_{min,r}}{(l_e/d)^2}$ [see equation (5.23)]

For ASD:

$$E'_{min} = E_{min} C_M C_t C_i C_T \quad \text{(for sawn lumber)}$$
$$E'_{min} = E_{min} C_M C_t \quad \text{(for glulam)} \quad (5.31)$$

For LRFD:

$$E'_{min,r} = E_{min} C_M C_t C_i C_T K_{F,st}\, \phi_{st} \quad \text{(for sawn lumber)}$$
$$E'_{min,r} = E_{min} C_M C_t K_{F,st}\, \phi_{st} \quad \text{(for glulam)} \quad (5.32)$$

E_{min} is the tabulated minimum elastic modulus for buckling and stability calculations and is a function of the pure bending modulus of elasticity E, with a factor of safety of 1.66 applied, in addition to a 5% lower exclusion value on the pure bending modulus of elasticity).

l_e = effective length of column = $K_e l$
l = unbraced column length (about the axis of buckling)

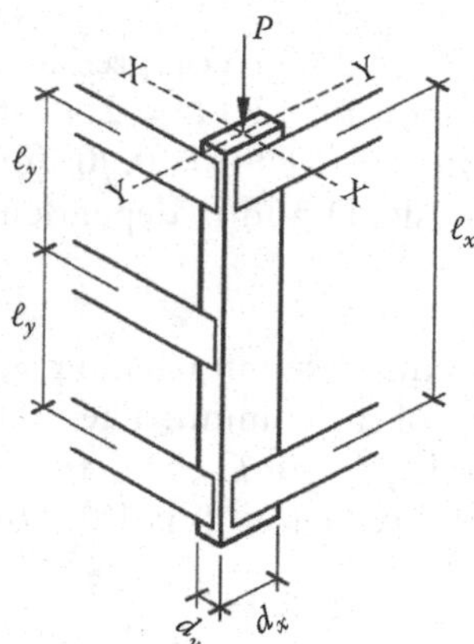

Figure 5.7 Column with unequal unbraced lengths.

l_x = unbraced length for buckling about the x–x axis (see Figure 5.7)

l_{y1} or l_{y2} = unbraced length for buckling about the y–y axis (see Figure 5.7)
(the larger of l_{y1} or l_{y2} is used in the buckling calculations, because the larger the unbraced length, the lower the allowable compressive stress)

K_e = buckling length coefficient (see Table 5.1)
(for most columns in wood buildings, pinned conditions are usually assumed at both ends of the column, even though there is some rotational restraint provided by the beam-to-column connection hardware; consequently, the effective length factor K_e is assumed to be 1.0)

Table 5.1 Buckling length coefficients, K_e

Buckling modes						
Theoretical K_e value	0.50	0.70	1.0	1.0	2.0	2.0
Recommended design K_e when ideal conditions are approximated	0.65	0.80	1.2	1.0	2.1	2.4
End condition legend	Rotation fixed, translation fixed Rotation free, translation fixed Rotation fixed, translation free Rotation free, translation free					

Source: Adapted from Faherty, K. et al., *Wood Engineering and Construction*, McGraw-Hill, New York, 1995.

d = dimension of the column section *perpendicular* to the axis of buckling. For x–x axis buckling, use $d = d_x$; for y–y axis buckling, use $d = d_y$ (the NDS code limits the slenderness ratio l_e/d to 50 to minimize the impact of slenderness on wood columns; when the l_e/d ratio is greater than 50, the designer would need to increase the column size and/or reduce the unbraced length of the column by providing lateral bracing, if possible, to bring the slenderness ratio within the limit of 50; it should be noted that the NDS code allows slenderness ratios of up to 75, but only during construction)

c = interaction factor = 0.8 (for sawn lumber)
= 0.9 (for glulam)
= 0.85 (for round timber poles and piles)

F_c^* = *unfactored* crushing strength of column at zero slenderness ratio (i.e., strength when $l_e/d = 0$), which means excluding C_P (see Equation 5.33)

$F_{c,r}^*$ = *LRFD-adjusted* crushing strength of column at zero slenderness ratio (i.e., strength when $l_e/d = 0$), which means excluding C_P (see Equation 5.34)

c = interaction factor = 0.8 (for sawn lumber)
= 0.9 (for glulam or structural composite lumber)
= 0.85 (for round timber poles and piles)

$K_{F,st}$ = format conversion factor for LRFD = 1.76 (for stability)

ϕ_{st} = strength reduction factor for LRFD = 0.85 (for stability)

For ASD:

$$F_c^* = \begin{cases} F_c C_D C_M C_t C_F C_i & \text{(sawn lumber)} \\ F_c C_D C_M C_t & \text{(glulam)} \end{cases} \tag{5.33}$$

For LRFD:

$$F_{c,r}^* = \begin{cases} F_c C_M C_t C_F C_i K_{F,c} \phi_c \lambda & \text{(sawn lumber)} \\ F_c C_M C_t K_{F,c} \phi_c \lambda & \text{(glulam)} \end{cases} \tag{5.34}$$

Example 5.5 (ASD): Axial tension plus bending—truss bottom chord

The roof section of a building shown in Figure 5.8 has 40-ft span trusses spaced at 2 ft 0 in. o.c. The roof dead load is 15 psf; the snow load is 35 psf; and the ceiling dead load is 15 psf of horizontal plan area. Design a typical bottom chord for the truss, assuming the following design parameters:

- No. 1 & Better Hem-Fir.
- Moisture content ≤19%, and normal temperature conditions apply.
- The members are connected with a single row of ¾-in.-diameter bolts.

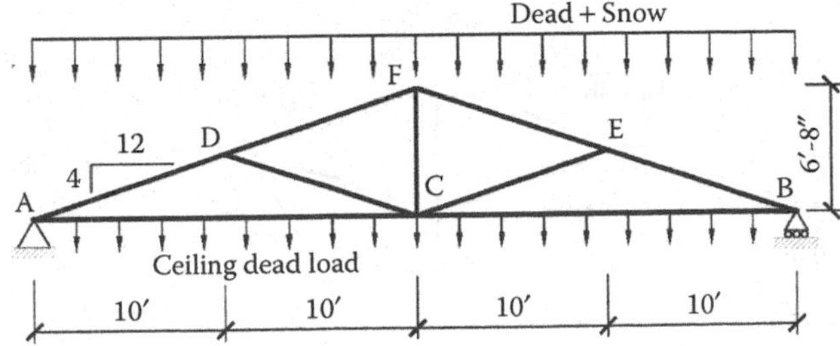

Figure 5.8 Roof truss elevation—Example 5.5.

Solution: Calculate the unfactored joint loads. The given loads in psf of horizontal plan area are as follows:

Roof dead load D (roof) = 15 psf
Ceiling dead load D (ceiling) = 15 psf
Snow load S = 35 psf

Since the tributary area for a typical roof truss is 80 ft² (i.e., 2-ft tributary width × 40-ft span), the roof live load L_r from Equation 2.6 will be 20 psf, which is less than the snow load S. Therefore, the controlling load combination from Chapter 2 will be dead load plus snow load (i.e., $D + S$).

$$\text{Total load on roof } w_{TL} = (15 + 35 \text{ psf})(2\text{-ft tributary width}) = 100 \text{ lb/ft}$$

$$\text{Total ceiling load} = (15 \text{ psf})(2 \text{ ft}) = 30 \text{ lb/ft}$$

The unfactored concentrated gravity loads at the truss joints are calculated as follows:

Roof dead load + snow load:

$$P_A \text{ (top)} = (10 \text{ ft}/2)(100 \text{ lb/ft}) = 500 \text{ lb}$$

$$P_D = (10 \text{ ft}/2 + 10 \text{ ft}/2)(100 \text{ lb/ft}) = 1000 \text{ lb}$$

$$P_F = (10 \text{ ft}/2 + 10 \text{ ft}/2)(100 \text{ lb/ft}) = 1000 \text{ lb}$$

$$P_E = (10 \text{ ft}/2 + 10 \text{ ft}/2)(100 \text{ lb/ft}) = 1000 \text{ lb}$$

$$P_B \text{ (top)} = P_A \text{ (top)} = 500 \text{ lb}$$

Ceiling dead loads:

$$P_A \text{ (bottom)} = (20 \text{ ft}/2)(30 \text{ lb/ft}) = 300 \text{ lb}$$

$$P_B \text{ (bottom)} = (20 \text{ ft}/2)(30 \text{ lb/ft}) = 300 \text{ lb}$$

$$P_C = (20 \text{ ft}/2 + 20 \text{ ft}/2)(30 \text{ lb/ft}) = 600 \text{ lb}$$

By analyzing the truss in Figure 5.9, using the method of joints *or* computer analysis software, we obtain the unfactored tension force in the bottom chord member (member AC or CB) as T_{AC} = 5400 lb. In addition to the tension force on member AC, there is a uniform ceiling load of 30 lb/ft acting on the truss bottom chord. The free-body diagram of member AC is shown in Figure 5.10.

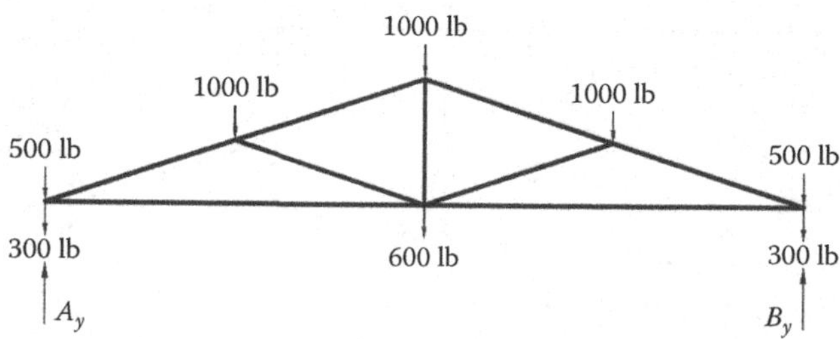

Figure 5.9 Free-body diagram of roof truss.

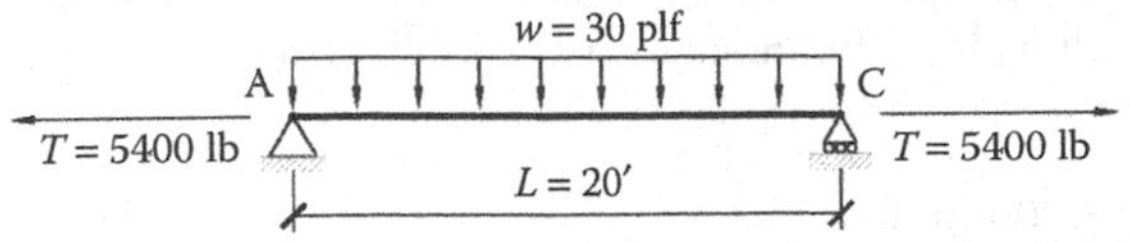

Figure 5.10 Free-body diagram of member.

Design:

We will now design the member using the ASD method. Assume a 2 × 10 sawn lumber. The gross cross-sectional area and the section modulus can be obtained from NDS-S Table 1B as follows:

$$A_g = 13.88 \text{ in.}^2$$

$$S_{xx} = 21.4 \text{ in.}^3$$

Since the trial member is dimension lumber, NDS-S Table 4A is applicable. From the table, we obtain the tabulated design stresses and stress adjustment factors.

$F_b = 1100$ psi (tabulated bending stress)
$F_t = 725$ psi (tabulated tension stress)
$C_F(F_b) = 1.1$ (size adjustment factor for bending stress)
$C_F(F_t) = 1.1$ (size adjustment factor for tension stress)
$C_D = 1.15$ (for design check with snow load)
$= 0.9$ (for design check with only dead load)
$C_L = 1.0$ (assuming that lateral buckling is prevented by the ceiling and bridging)
$C_r = 1.15$ (all three repetitive member requirements are met; see subsection 3.4.2.4)

Design Check 1:

The unfactored tension force applied, *T*, is 5400 lb (caused by dead load + snow load). The net area is

$$A_n = A_g - \Sigma A_{\text{holes}} = 13.88 - (1 \text{ hole})(3/4 \text{ in.} + 1/16 \text{ in.})(1.5 \text{ in.}) = 12.66 \text{ in}^2$$

From Equation 5.1, the unfactored tension stress applied at the supports is

$$f_t = \frac{T}{A_n} = \frac{5400 \text{ lb}}{12.66 \text{ in}^2} = 426 \text{ psi}$$

Using the NDS applicability table (Table 3.1), the allowable tension stress is given as

$$F_t' = F_t C_D C_M C_t C_F C_i$$
$$= (725)(1.15)(1.0)(1.0)(1.1)(1.0) = \mathbf{917\ psi} > f_t \qquad \text{OK}$$

Design Check 2:

From Equation 5.5, the unfactored tension stress applied at the midspan is

$$f_t = \frac{T}{A_g} = \frac{5400 \text{ lb}}{13.88 \text{ in}^2} = \mathbf{389\ psi} < F_t' = 917 \text{ psi} \qquad \text{OK}$$

Design Check 3:

The maximum unfactored moment, which for this member occurs at the midspan, is given as

$$M_{\max} = \frac{(30 \text{ lb/ft})(20 \text{ ft})^2}{8} = 1500 \text{ ft-lb} = 18{,}000 \text{ in.-lb}$$

This moment is caused by only the ceiling *dead* load (since no ceiling live load is specified); therefore, the controlling load duration factor C_D is 0.9. From Equation 5.7, the unfactored bending stress applied (i.e., tension stress due to bending) is

$$f_{bt} = \frac{M_{max}}{S_x} = \frac{18{,}000 \text{ in.-lb}}{21.4 \text{ in}^3} = 842 \text{ psi}$$

Using the NDS applicability table (Table 3.1), the allowable bending stress (ASD) is given as

$$F_b' = F_b C_D C_M C_t C_F C_i C_r C_{fu}$$

$$= (1100)(0.9)(1.0)(1.0)(1.0)(1.1)(1.0)(1.15)(1.0)$$

$$= \mathbf{1252 \text{ psi}} > f_{bt} \qquad \text{OK}$$

Design Check 4:
For this case, the applicable load is the dead load plus the snow load; therefore, the controlling load duration factor C_D is 1.15 (see Chapter 3). Using the NDS applicability table (Table 3.1), the allowable axial tension stress and the allowable bending stress are calculated as follows:

$$F_t' = F_t C_D C_M C_t C_F C_i$$

$$= (725)(1.15)(1.0)(1.0)(1.1)(1.0) = 917 \text{ psi}$$

$$F_b^* = F_b C_D C_M C_t C_F C_i C_r C_{fu}$$

$$= (1100)(1.15)(1.0)(1.0)(1.1)(1.0)(1.15)(1.0) = 1600 \text{ psi}$$

The unfactored applied axial tension stress f_t at the midspan is 389 psi, as calculated for design check No. 2, whereas the unfactored applied tension stress due to bending f_{bt} is 842 psi, as calculated for design check No. 3. From Equation 5.11, the combined unfactored axial tension plus bending interaction equation for the stresses in the tension fiber of the member is given as

$$\frac{f_t}{F_t'} + \frac{f_{bt}}{F_b^*} \le 1.0$$

Substitution into the interaction equation yields

$$\frac{389 \text{ psi}}{917 \text{ psi}} + \frac{842 \text{ psi}}{1600 \text{ psi}} = \mathbf{0.95 < 1.0} \qquad \text{OK}$$

Design Check 5:
The unfactored applied compression stress due to bending is

$$f_{bc} = \frac{M_{max}}{S_x} = \frac{18{,}000 \text{ in.-lb}}{21.4 \text{ in}^3} = 842 \text{ psi } (= f_{bt} \text{ since this is a rectangular cross section})$$

It should be noted that f_{bc} and f_{bt} are equal for this problem because the wood member is rectangular in cross section. The load duration factor for this case, C_D, is 1.15 (combined dead + snow loads). Using the NDS applicability table (Table 3.1), the allowable compression stress (for ASD) due to bending is

$$F_b^{**} = F_b C_D C_M C_t C_L C_F C_i C_r C_{fu}$$

$$= (1100)(1.15)(1.0)(1.0)(1.0)(1.1)(1.0)(1.15)(1.0) = 1600 \text{ psi}$$

From Equation 5.15, the interaction equation for this design check is given as

$$\frac{f_{bc} - f_t}{F_b^{**}} \leq 1.0$$

Substituting in the interaction equation yields

$$\frac{842 \text{ psi} - 389 \text{ psi}}{1600 \text{ psi}} = 0.283 < 1.0 \qquad \text{OK}$$

From all of the steps above, we find that all five design checks are satisfied; therefore, use 2 × 10 Hem-Fir No. 1 & Better for the bottom chord. Note that if any one of the design checks above had not been satisfied, we would have had to increase the member size and/or the stress grade until all five design checks were satisfied.

5.4.1 Built-up columns

Built-up columns (Figure 5.11) are commonly used at the end of shear walls as tension or compression chord members. They can also be used in lieu of a solid post within a stud wall to support heavy concentrated loads. Subsection 15.3 of the NDS code [6] lists the minimum basic provisions that apply to built-up columns, and these are listed here for reference:

- Each lamination is to be rectangular in cross section.
- Each lamination is to be at least 1.5 in. thick.

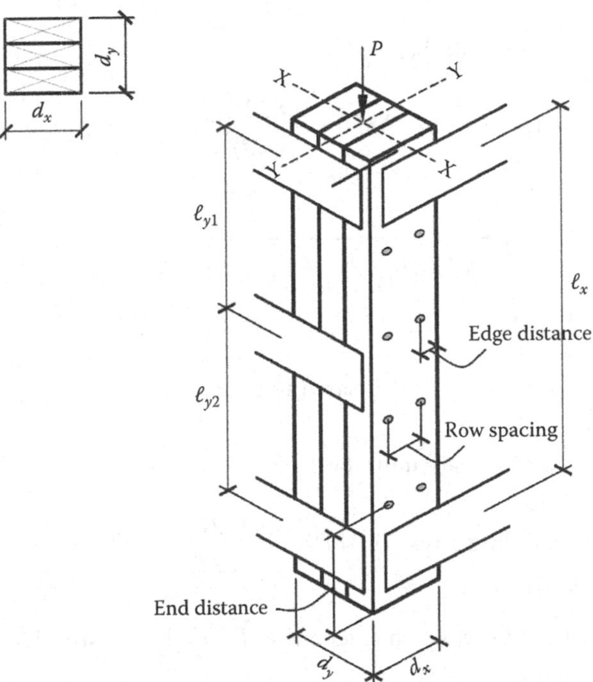

d_x = Column dimension parallel to the face of the laminations in contact with each other
d_y = Column dimension perpendicular to the face of the laminations
ℓ_x = Effective column length for buckling about the x–x axis = $K\ell_x$
ℓ_y = Effective column length for buckling about the y–y axis = $K\ell_y$

Figure 5.11 Built-up column elevation and cross section.

- All laminations are to have the same depth.
- Faces of the adjacent laminations are to be in contact.
- All laminations are full column length.
- The nailed or bolted connection requirements of subsections 15.3.3 and 15.3.4 of the NDS code are to be met.

The allowable compressive stresses parallel to grain in a built-up column for the ASD method and the LRFD-adjusted design compression stress are calculated as follows:

ASD:

$$F_c' = F_c C_D C_M C_t C_F C_i C_P \quad \text{(sawn lumber)} \tag{5.35}$$

LRFD:

$$F_{c,r}' = F_c C_M C_t C_F C_i C_{P,r} K_{F,c}\, \phi_c\, \lambda \quad \text{(sawn lumber)} \tag{5.36}$$

ASD:

$$C_P = K_f \left[\frac{1+F_{cE}/F_c^*}{2c} - \sqrt{\left(\frac{1+F_{cE}/F_c^*}{2c}\right)^2 - \frac{F_{cE}/F_c^*}{c}} \right] \tag{5.37}$$

$$E_{\min}' = E_{\min} C_M C_t C_i C_T \quad \text{(for sawn lumber)}$$

$$F_c^* = F_c C_D C_M C_t C_F C_i \quad \text{(sawn lumber)}$$

$$F_{cE} = \text{Euler critical buckling stress (ASD)} = \frac{0.822 E_{\min}'}{(l_e/d)^2} \quad \text{[see equation (5.23)]}$$

LRFD:

$$C_{P,r} = K_f \left[\frac{1+F_{cE,r}/F_{c,r}^*}{2c} - \sqrt{\left(\frac{1+F_{cE,r}/F_{c,r}^*}{2c}\right)^2 - \frac{F_{cE,r}/F_{c,r}^*}{c}} \right] \tag{5.38}$$

$$E_{\min,r}' = E_{\min} C_M C_t C_i C_T K_{F,st}\, \phi_{st} \quad \text{(for sawn lumber)}$$

$$F_{c,r}^* = F_c C_M C_t C_F C_i K_{F,c}\, \phi_c\, \lambda \quad \text{(sawn lumber)}$$

$$F_{cE,r} = \text{Euler critical buckling stress (LRFD)} = \frac{0.822 E_{\min,r}'}{(l_e/d)^2} \quad \text{[see equation (5.23)]}$$

$K_f = 0.60$ (for nailed columns when calculating F_{cE} or $F_{cE,u}$ using L_{ey}/d_y)

0.75 (for bolted columns when calculating F_{cE} or $F_{cE,u}$ using L_{ey}/d_y)

1.0 (for nailed or bolted columns when calculating F_{cE} or $F_{cE,u}$ using L_{ex}/d_x)

Note that the K_f factor in Equations 5.37 and 5.38 applies only to weak axis (y–y axis) buckling, and it is an axial load reduction factor for the built-up column relative to a solid column that accounts for the slippage or interface shear between the different laminations of the built-up column [8]. The first two K_f values above are used when the weak axis of the column controls (i.e., when l_{ey}/d_y is greater than l_{ex}/d_x). Where buckling about the strong axes controls or when l_{ex}/d_x is greater than l_{ey}/d_y, K_f is 1.0.

It should be noted that where the built-up column is embedded within a sheathed wall such that it is continuously braced for buckling about the weak axis, the column stability factor C_{pyy} will be 1.0. This assumes that the built-up column is adequately braced laterally by the wall panel, and examples of wall panels that can provide this lateral support include plywood or oriented strand board (OSB) wall panels. The use of gypsum wall board to brace a non-repetitive member such as a built-up column or a multistory built-up column is not recommended by the authors for several reasons: the gypsum wall board may be removed by in later renovations of the building and as such should not be considered a permanent bracing; the gypsum wall board can lose its bracing ability if it becomes water saturated from flooding or water backup. Built-up columns with gypsum wall board have experienced problems [8]. For the minimum nailing or bolting requirements for built-up columns, the reader is referred to subsections 15.3.3 and 15.3.4 of the NDS code [6].

Example 5.6 (ASD/LRFD): Column axial load capacity in pure compression

Using the ASD and LRFD methods, design a column for an axial compression dead load plus snow load (D = 6.8 kips and S = 10.2 kips), given the following parameters:

No. 1 DF-L wood species.
Column unbraced lengths $l_x = l_y = 12$ ft.
Dry service and normal temperature conditions apply.

Solution:

Unfactored applied load, P_s = 6.8 kips + 10.2 kips = 17 kips
Factored applied load, P_u = 1.2(6.8 kips) + 1.6(10.2 kips) = 24.5 kips

Several design aids for selecting an appropriate column size for a given load are given in Appendix B. Entering Figure B.10 with an unfactored applied load of 17 kips and an unbraced length of 12 ft, a trial size of 6 × 6 is indicated. It is important to note that the load duration factor C_D is 1.0 for this chart. Since a load duration factor C_D of 1.15 is given for this problem, it can be seen by inspection that the actual capacity of the 6 × 6 member would be slightly higher than 17 kips for this example.

A 6 × 6 member is a post and timber (P&T); therefore, NDS-S Table 4D is applicable. From the table, the stress adjustment or C factors are

C_M = 1.0 (normal moisture conditions)
C_t = 1.0 (normal temperature conditions)
C_D = 1.15 (snow load controls for the given load combination of $D + S$)
C_i = 1.0 (assumed)
C_T = 1.0

From the table, the tabulated reference design stresses are obtained as follows:

F_c = 1000 psi (compression stress parallel to grain)
$E = 1.6 \times 10^6$ psi (reference modulus of elasticity)
$E_{min} = 0.58 \times 10^6$ psi (buckling modulus of elasticity)
$C_F = (12/d)^{1/9} \leq 1.0$
$= 1.0$ for $d < 12$ in.; therefore, $C_F = 1.0$

The column dimensions are d_x = 5.5 in. and d_y = 5.5 in. The cross-sectional area A = 30.25 in.^{2} The effective length factor K_e is assumed to be 1.0, as discussed earlier in the chapter; therefore, the slenderness ratios about the x–x and y–y axes are given as

$$\frac{l_e}{d_x} = \frac{(1.0)(12\text{ ft})\times(12\text{ in./ ft})}{5.5\text{ in.}} = 26.2 < 50 \quad \text{OK}$$

$$\frac{l_e}{d_y} = \frac{(1.0)(12\text{ ft})\times(12\text{ in./ ft})}{5.5\text{ in.}} = 26.2 < 50 \quad \text{OK}$$

Part (a) – ASD Method:

Using the adjustment factors applicability table (Table 3.1), the allowable modulus of elasticity for buckling calculations is given as

$$E'_{\min} = E_{\min}\, C_M C_t C_i\, C_T$$

$$= (0.58 \times 10^6)(1.0)(1.0)(1.0)(1.0) = 0.58 \times 10^6\text{ psi}$$

$$c = 0.8\text{ (sawn lumber)}$$

The Euler critical buckling stress (ASD) is calculated as

$$F_{cE} = \frac{0.822E'_{\min}}{(l_e/d)^2} = \frac{(0.822)(0.58 \times 10^6)}{(26.2)^2} = 695\text{ psi}$$

$$F_c^* = F_c C_D C_M\, C_t C_F C_i$$

$$= (1000)(1.15)(1.0)(1.0)(1.0)(1.0) = 1150\text{ psi}$$

$$\frac{F_{cE}}{F_c^*} = \frac{695\text{ psi}}{1150\text{ psi}} = 0.604$$

From Equation 5.29, the column stability factor for the ASD method is calculated as

$$C_p = \frac{1+0.604}{(2)(0.8)} - \sqrt{\left[\frac{1+0.604}{(2)(0.8)}\right]^2 - \frac{0.604}{0.8}} = 0.502$$

The allowable compression stress is

$$F'_c = F_c^* C_P = (1150)(0.502) = 577.3\text{ psi}$$

The unfactored allowable compression load capacity of the column is

$$P_{\text{allowable}} = F'_c A_g = (577.3\text{ psi})(30.25\text{ in.}^2)$$

$$= \mathbf{17{,}463\text{ lb} = 17.5\text{ kips}} > P_s = 17\text{ kips} \quad \text{OK}$$

This value agrees with Table B.64 in the appendix.
Use a 6 × 6 No. 1 DF-L column.

Part (b) – LRFD Method:

The following additional adjustment factors are needed for the LRFD method:

$K_{F,c}$ = 2.40 (for compression stress)
$K_{F,st}$ = 1.76 (buckling modulus)
ϕ_c = 0.90 (for compression stress)
ϕ_{st} = 0.85 (for buckling modulus)
λ = 0.8 (for 1.2D + 1.6S)

Using the adjustment factors applicability table (Table 3.3), the LRFD-adjusted modulus of elasticity for buckling calculations is given as

$$E'_{\min,r} = E_{\min} C_M C_t C_i C_T K_{F,c} \phi_{st}$$

$$= (0.58 \times 10^6)(1.0)(1.0)(1.0)(1.0)(1.76)(0.85) = 0.87 \times 10^6 \text{ psi}$$

$$c = 0.8 \text{ (sawn lumber)}$$

The LRFD-adjusted Euler critical buckling stress is calculated as

$$F_{cE,r} = \frac{0.822 E'_{\min,r}}{(l_e/d)^2} = \frac{(0.822)(0.87 \times 10^6)}{(26.2)^2} = 1042 \text{ psi}$$

$$F^*_{c,r} = F_c C_M C_t C_F C_i K_{F,c} \phi_c \lambda$$

$$= (1000)(1.0)(1.0)(1.0)(1.0)(2.40)(0.90)(0.8)$$

$$= 1728 \text{ psi}$$

$$\frac{F_{cE,r}}{F^*_{c,r}} = \frac{1042 \text{ psi}}{1728 \text{ psi}} = 0.603$$

From Equation 5.30, the column stability factor for the LRFD method is calculated as

$$C_{P,r} = \frac{1+0.603}{(2)(0.8)} - \sqrt{\left[\frac{1+0.603}{(2)(0.8)}\right]^2 - \frac{0.603}{0.8}} = 0.502$$

The allowable compression stress is

$$F'_{c,r} = F^*_{c,r} C_{P,r} = (1728)(0.502) = 867 \text{ psi}$$

The LRFD-adjusted compression load capacity of the column is

$$P_r = F'_{c,r} A_g = (867 \text{ psi})(30.25 \text{ in.}^2)$$

$$= \mathbf{26{,}300 \text{ lb} = 26.3 \text{ kips}}$$

The factored load $P_u = 1.2\,(6.8) + 1.6(10.2) = 24.5$ kips $< P_r = 26.3$ kips **OK**

Use a 6 × 6 No. 1 DF-L column.

Note that for this problem, the ratio of the applied factored load to the load capacity for the LRFD method, P_u/P_r (= 24.5 kip/26.3 kip = 0.93), is lower than the ratio of unfactored applied load to the allowable load for the ASD method, $P_s/P_{\text{allowable}}$ (= 17 kip/17.5 kip = 0.97).

5.4.2 *P–δ* effects in members under combined axial and bending loads

The bending moments in a member supported at both ends and subjected to combined axial compression plus bending are amplified by the presence of that axial load. The magnitude of the moment amplification is a function of the slenderness ratio of the member. The amplification of the bending moment in the presence of axial compression load is known as the *P–δ effect*. This concept is illustrated by the free-body diagram in Figure 5.12.

Consider a pin-ended column subjected to combined axial load *P* and a lateral bending load *W* at the mid height, as shown in Figure 5.12. The lateral load causes lateral deflection of the column, which is further amplified by the presence of the axial compression load *P* until equilibrium is achieved at a deflection δ; hence, it is called the *P–δ* effect.

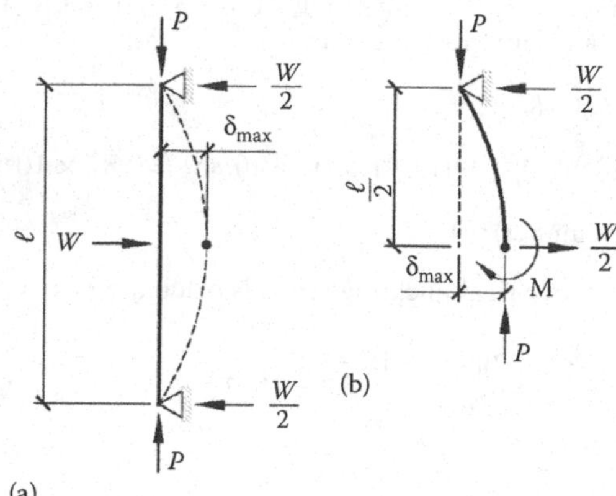

Figure 5.12 P–δ effect in axially loaded compression members. (a) Column under axial load and (b) free-body diagram.

Summing moments about point B in the free-body diagram of Figure 5.12b, we obtain

$$-P\delta - \frac{W}{2}\frac{L}{2} + M = 0$$

Therefore,

$$M = \frac{WL}{4} + P\delta \qquad (5.39)$$

where:

M is the maximum second-order or final moment

$WL/4$ is the maximum first-order or maximum moment in the absence of the axial load P

δ is the maximum lateral deflection

The maximum second-order moment M includes the moment magnification or the *P–δ moment*. Thus, the presence of the axial compression load leads to an increase in the column moment, but the magnitude of the increase depends on the slenderness ratio of the column. The moment magnification effect is accounted for in the NDS code by multiplying the bending stress term f_b/F_b' by the magnifier $1/(1 - f_c/F_{cE})$ (see Equations 5.46 through 5.49), where f_c is the applied axial compression stress in the column and F_{cE} is the Euler critical buckling stress about the axis of buckling, as calculated in Equation 5.23.

Example 5.7 (ASD): Axial load capacity of a built-up column

Calculate the axial load capacity of a 12-ft-long nailed built-up column consisting of three 2 × 6s (Figure 5.13), assuming DF-L No. 1 lumber, normal temperature and moisture conditions, and a load duration factor C_D of 1.0.

Solution: For three 2 × 6, $d_x = 5.5$ in. and $d_y = (3)(1.5\text{ in.}) = 4.5$ in. (see NDS-S Table 1B). Since a 2 × 6 is a dimension lumber, NDS-S Table 4A is applicable. From the table, the tabulated design stresses are obtained as

$$F_c = 1500\text{ psi}$$

$$E_{min} = 0.62 \times 10^6\text{ psi}$$

From NDS-S Table 4A, the size factor for axial compression stress parallel to grain is

$C_{F(Fc)} = 1.1$ (size factor for axial compression stress parallel to the grain)

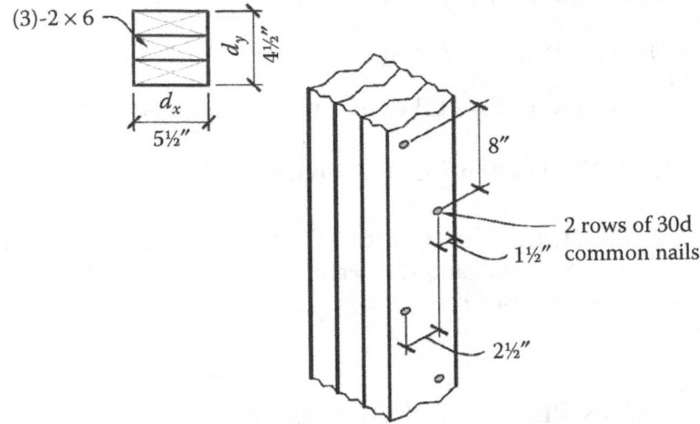

Figure 5.13 Built-up column cross section—Example 5.7.

The following stress adjustment or C factors were specified in the problem:

$C_M = 1.0$ (normal moisture conditions)
$C_t = 1.0$ (normal temperature conditions)
$C_D = 1.0$
$C_i = 1.0$ (assumed)

The unbraced length of the column $l_x = l_y = 12$ ft. $K_e = 1.0$ (building columns are typically assumed to be pinned at both ends).

$$\frac{l_{ex}}{d_x} = \frac{K_e l_x}{d_x} = \frac{(1.0)(12\text{ ft})(12\text{ in./ft})}{5.5} = 26.2 < 50 \quad \textbf{OK}$$

$$\frac{l_{ey}}{d_y} = \frac{K_e l_y}{d_y} = \frac{(1.0)(12\text{ ft})(12\text{ in./ft})}{4.5} = 32 < 50 \quad \textbf{OK}$$

Since $l_{ey}/d > l_{ex}/d_x$, the weak-axis slenderness ratio (l_{ey}/d) governs; therefore, the controlling column stability factor is about the weak axis (i.e., C_{Pyy}).

$K_f = 0.60$ (nailed built-up column)

$c = 0.8$ (visually graded lumber)

$$E'_{\min} = E_{\min} C_M C_t C_i$$

$$= (0.62 \times 10^6)(1.0)(1.0)(1.0) = 0.62 \times 10^6 \text{ psi}$$

$$F_{cEyy} = \frac{0.822 E'_{\min}}{(l_{ey}/d_y)^2} = \frac{(0.822)(0.62 \times 10^6)}{(32)^2} = 498 \text{ psi}$$

$$F_c^* = F_c C_D C_M C_t C_F C_i$$

$$= (1500)(1.0)(1.0)(1.0)(1.1)(1.0) = \mathbf{1650\ psi}$$

$$\frac{F_{cE}}{F_c^*} = \frac{498 \text{ psi}}{1650 \text{ psi}} = 0.302$$

From Equation 5.37, the controlling column stability factor (i.e., C_{Pyy}) is calculated as

$$C_P = 0.60\left[\frac{1+0.302}{(2)(0.8)} - \sqrt{\left[\frac{1+0.302}{(2)(0.8)}\right]^2 - \frac{0.302}{0.8}}\right] = 0.168$$

The allowable compression stress is given as

$$F'_c = F^*_c C_P = (1650)(0.168) = 277.0 \text{ psi}$$

The unfactored axial load capacity of the built-up column is calculated as

$$P_{\text{allowable}} = F'_c A_g = (277.0)(4.5 \text{ in.})(5.5 \text{ in.}) = \mathbf{6858 \text{ lb}}$$

From subsection 15.3.3 of the NDS code [6], the minimum nailing requirement is 30d nails spaced vertically at 8 in. o.c. and staggered 2½ in. with an edge distance of 1½ in. and an end distance of 3½ in. (see Figure 5.13).

5.5 AXIAL COMPRESSION PLUS BENDING: CASE 4

For members subjected to combined axial compression and bending loads, there are four design checks that need to be investigated. It is important to note that for each design check, the controlling load duration factor or time effect factor based on the pertinent load combination for each design check should be used (see Chapter 3). The four design checks required are as follows.

5.5.1 Design check 1: Compression on the net area

Compression on the net area (Figure 5.14a) occurs at the ends of the member (i.e., at the connections), and the column stability factor C_P is 1.0 for this case, since there can be

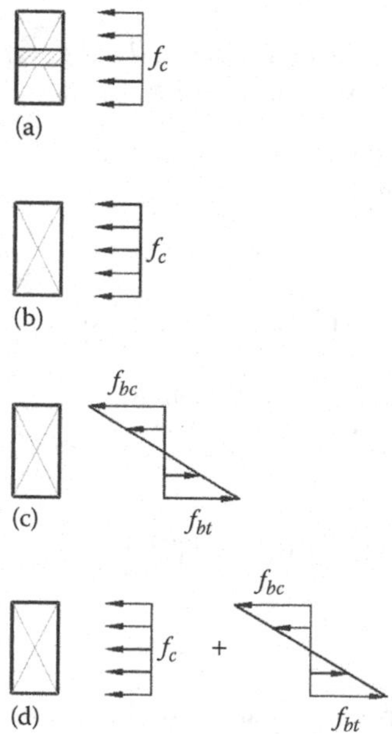

Figure 5.14 (a) Stress diagram for load case 4—design check No. 1, (b) Stress diagram for load case 4—design check No. 2, (c) Stress diagram for load case 4—design check No. 3, and (d) Stress diagram for load case 4—design check No. 4.

no buckling at the ends of the member. The allowable stress design equation for the ASD method and the interaction equation for the LRFD method for this pure compression load case based on net area are given as follows:

ASD:

$$f_c = \frac{P}{A_n} \leq F_c' \tag{5.40}$$

LRFD:

$$f_{c,u} = \frac{P_u}{A_n} \leq F_{c,r}' \tag{5.41}$$

where:

f_c is the *unfactored* axial compression stress (ASD)
P is the *unfactored* axial compression force applied on the wood column
F_c' is the allowable compression stress parallel to grain (ASD) (see Equation 5.27)
$f_{c,u}$ is the *factored* axial compression stress (LRFD)
P_u is the *factored* axial compression force applied on the wood column
$F_{c,r}'$ is the LRFD-adjusted compression stress parallel to grain (see Equation 5.28)

Note that C_P is 1.0 for this load case. A_n = net area at the critical section = $A_g - \Sigma A_{\text{holes}}$

A_g is the gross cross-sectional area of the member
ΣA_{holes} is the sum of the area of the bolt holes perpendicular to the load, if bolts are present
= (number of bolts perpendicular to load)(d_{bolt} + 1/16 in.)(member thickness, b)

Using the NDS code applicability table (Tables 3.1 through 3.4), the allowable compression stress parallel to the grain is as calculated in Equations 5.27 and 5.28.

5.5.2 Design check 2: Compression on the gross area

Compression on the gross area (Figure 5.14b) occurs at the midspan of the member (i.e., where no connections exist), and the gross cross-sectional area of the member is applicable. For this location, however, the column will be susceptible to buckling, and therefore, the column stability factor C_P may be less than 1.0 and has to be calculated. The allowable stress design equation for the ASD method and the interaction equation for the LRFD method for design check 2 (compression on *gross area*) are given as follows:

ASD:

$$f_c = \frac{P}{A_g} \leq F_c' \tag{5.42}$$

LRFD:

$$f_{c,u} = \frac{P_u}{A_g} \leq F_{c,r}' \tag{5.43}$$

The parameters in Equations 5.42 and 5.43 are as defined in design check No. 1. The allowable stress for ASD, F_c', and the LRFD-adjusted compression stress, $F_{c,r}'$, are calculated using Equations 5.27 through 5.34. In calculating the column stability factor C_P, the larger slenderness ratio is used, since this represents the most critical condition with respect to buckling.

5.5.3 Design check 3: Bending only

The required design interaction equations for bending for both the ASD and LRFD methods are given below and are dependent on the maximum bending moment. For prismatic and simply supported members, the maximum bending moment and maximum bending stress occur at the midspan of the member.

ASD:

$$f_{bc} = \frac{M}{S_x} \leq F_b' \tag{5.44}$$

LRFD:

$$f_{bc,u} = \frac{M_u}{S_x} \leq F_{b,r}' \tag{5.45}$$

where:

f_{bc} = *unfactored* bending stress in the *compression* fiber (ASD)
$f_{bc,u}$ = *factored* bending stress in the *compression* fiber (LRFD)
M = *unfactored* maximum bending moment in the member
M_u = *factored* maximum bending moment in the member
S_x = section modulus about the axis of bending
F_b' = allowable bending stress (ASD)
$F_{b,r}'$ = LRFD-adjusted design bending stress (LRFD)

Using the applicability table (Tables 3.1 through 3.4), the allowable bending stress (ASD) is obtained as

ASD:

$$F_b' = \begin{cases} F_b C_D C_M C_t C_L C_F C_i C_r C_{fu} & \text{(for sawn lumber)} \\ F_b C_D C_M C_t C_L C_V C_{fu} C_c & \text{(for glulam)} \end{cases} \tag{5.9}$$

The LRFD-adjusted bending stress is obtained as

LRFD:

$$F_{b,r}' = \begin{cases} F_b C_M C_t C_L C_F C_i C_r C_{fu} K_{F,b} \phi_b \lambda & \text{(for sawn lumber)} \\ F_b C_M C_t C_L C_V C_{fu} C_c C_I K_{F,b} \phi_b \lambda & \text{(for glulam)} \end{cases} \tag{5.10}$$

where:

F_b is the NDS-S tabulated reference design bending stress
C is the adjustment factors, as discussed in Chapter 3
$K_{F,b}$ is the format conversion factor for LRFD = 2.54 (for bending stress)
ϕ_b is the strength reduction factor for LRFD = 0.85 (for bending stress)
λ is the time effect factor for LRFD and is dependent on the load combination (see Table 3.11)

Note that for dimension-sawn lumber built-up columns with *three or more laminations* embedded within a sheathed wall and subjected to bending about its strong axis, the repetitive member factor C_r is 1.15.

5.5.4 Design check 4: Bending plus compression

Bending plus compression (Figure 5.14d) occurs at the point of maximum moment (i.e., at the midspan for prismatic members), where the moment magnification (or P–δ) factor is at its maximum value. The general design interaction equation for combined *concentric* axial compression load plus *biaxial* bending is given in the NDS code as follows:

ASD:

$$\left(\frac{f_c}{F'_c}\right)^2 + \frac{\dfrac{f_{bx}}{F'_{bx}}}{1-\dfrac{f_c}{F_{cEx}}} + \frac{\dfrac{f_{by}}{F'_{by}}}{1-\dfrac{f_c}{F_{cEy}}-\left(\dfrac{f_{bx}}{F_{bE}}\right)^2} \leq 1.0 \tag{5.46}$$

LRFD:

$$\left(\frac{f_{c,u}}{F'_{c,r}}\right)^2 + \frac{\dfrac{f_{bx,u}}{F'_{bx,r}}}{1-\dfrac{f_{c,u}}{F_{cEx,r}}} + \frac{\dfrac{f_{by,u}}{F'_{by,r}}}{1-\dfrac{f_{c,u}}{F_{cEy,r}}-\left(\dfrac{f_{bx,u}}{F_{bE,r}}\right)^2} \leq 1.0 \tag{5.47}$$

where:

For ASD:

f_c is the *unfactored* axial compression stress
F'_c = allowable compression stress parallel to the grain
f_{bx} = *unfactored* bending stress due to x–x axis bending
F'_{bx} is the allowable bending stress for bending about the x–x axis
F_{cEx} is the Euler critical stress for buckling about the x–x axis
F_{bE} is the Elastic beam buckling (lateral torsional buckling) stress
f_{by} is the bending stress due to y–y axis bending
F'_{by} is the allowable bending stress for bending about the y–y axis
F_{cEy} is the Euler critical stress for buckling about the y–y axis

$$f_c < F_{cEx} = \frac{0.822E'_{\min}}{(l_{ex}/d_x)^2}$$ (Euler critical column buckling stress about the x–x axis)

$$f_c < F_{cEy} = \frac{0.822E'_{\min}}{(l_{ey}/d_y)^2}$$ (Euler critical column buckling stress about the y–y axis)

$$f_{by} < F_{bE} = \frac{1.20E'_{y\min}}{(R_B)^2}$$ (Euler critical lateral – torsional buckling stress for y–y axis bending)

For LRFD:

$f_{c,u}$ is the *factored* axial compression stress (LRFD)
$F'_{c,u}$ is the *LRFD-adjusted* compression stress parallel to the grain (LRFD)
$f_{bx,u}$ is the *factored* bending stress due to x–x axis bending
$F'_{bx,u}$ is the *LRFD-adjusted* bending stress for bending about the x–x axis
$F_{cEx,u}$ is the *LRFD-adjusted* Euler critical stress for buckling about the x–x axis
$F_{bEx,u}$ is the *LRFD-adjusted* Elastic beam buckling (lateral torsional buckling)

f_{by} is the *factored* bending stress due to y–y axis bending
F'_{by} is the *LRFD-adjusted* bending stress for bending about the y–y axis
F_{cEy} is the *LRFD-adjusted* Euler critical stress for buckling about the y–y axis

$$f_{c,u} < F_{cEx,r} = \frac{0.822E'_{x\,\min,r}}{(l_{ex}/d_x)^2} \quad \text{(Euler critical column buckling stress about the } x\text{–}x \text{ axis)}$$

$$f_{c,u} < F_{cEy,r} = \frac{0.822E'_{y\min,r}}{(l_{ey}/d_y)^2} \quad \text{(Euler critical column buckling stress about the } y\text{–}y \text{ axis)}$$

$$f_{by,u} < F_{bE,r} = \frac{1.20E'_{y\,\min,r}}{(R_B)^2} \quad \text{(Euler critical lateral-torsional buckling stress for } y\text{–}y \text{ axis bending)}$$

R_B is the beam slenderness ratio for lateral torsional buckling $= \sqrt{(l_e d_x/b^2)} \le 50$
l_e is the effective length for bending members from Table 3.10
d_x is the column dimension perpendicular to the x–x axis (see Figure 5.15)
b is the column dimension perpendicular to the y–y axis

For combined *concentric* axial load plus *uniaxial* bending about the strong (x–x) axis of the member, the interaction equation in Equations 5.46 and 5.47 reduces to

ASD:

$$\left(\frac{f_c}{F'_c}\right)^2 + \frac{\dfrac{f_{bx}}{F'_{bx}}}{1 - \dfrac{f_c}{F_{cEx}}} \le 1.0 \tag{5.48}$$

LRFD:

$$\left(\frac{f_{c,u}}{F'_{c,r}}\right)^2 + \frac{\dfrac{f_{bx,u}}{F'_{bx,r}}}{1 - \dfrac{f_{c,u}}{F_{cEx,r}}} \le 1.0 \tag{5.49}$$

- The allowable bending stress for each axis of bending for the ASD method and the LRFD-adjusted design bending stress can be obtained from Equations 5.9 and 5.10.
- The allowable compression stress parallel to grain for the ASD method and the LRFD-adjusted design compression stress parallel to grain can be obtained from Equations 5.27 through 5.34.
- $E'_{\min}$ and $E'_{\min,r}$ can be calculated using Equations 5.31 and 5.32.

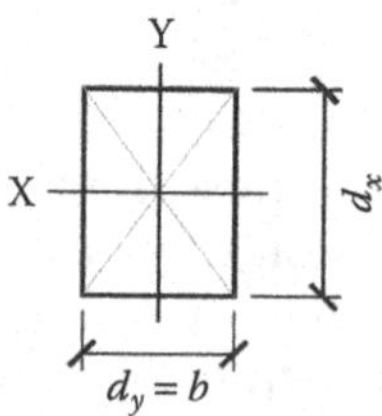

Figure 5.15 Column cross section.

In the calculation of the column stability factor C_P, the larger of the slenderness ratios about both orthogonal axes (l_{ex}/d_x or l_{ey}/d_y) should be used, since the column will buckle about the axes with the larger slenderness ratio. The higher the column slenderness ratio $K_e l/r$, the smaller the column stability factor C_P. For columns that are laterally braced over their entire length, the column stability factor C_P is 1.0. For all other cases, calculate C_P by using Equations 5.29 and 5.30.

Example 5.8 (ASD): Combined compression plus bending—truss top chord

Design the top chord member *AD* of the roof truss in Example 5.5 (Figure 5.16). Assume Hem-Fir is used.

Solution: The unfactored maximum axial compression force in member *AD* is

P_{max} (from computer analysis, method of joints, or method of sections) = 5693 lb

The unfactored maximum moment in member *AD* is

$$M_{max} = \frac{(100\text{ lb/ft})(10\text{ ft})^2}{8} = 1250\text{ ft-lb} = 15{,}000\text{ in.-lb}$$

Select a Trial Member Size

Try 2 × 10 No. 1 & Better Hem-Fir. A 2 × 10 is a dimension lumber; therefore, use NDS-S Table 4A. From NDS-S Table 1B, we obtain the section properties for the trial member size as follows:

Gross cross-sectional area $A = 13.88$ in.^{2}
Section modulus $S_x = 21.4$ in.^{3}
$d_x = 9.25$ in.
$d_y = 1.5$ in.

From NDS-S Table 4A, we obtain the tabulated design stress values and the stress adjustment or *C* factors as follows:

$F_c = 1350$ psi
$F_b = 1100$ psi
$E_{min} = 0.55 \times 10^6$ psi
$C_F(F_c) = 1.0$
$C_F(F_b) = 1.1$
$C_M = 1.0$ (normal moisture conditions)
$C_t = 1.0$ (normal temperature conditions)
$C_{fu} = 1.0$
$C_r = 1.15$ (the condition for repetitiveness is discussed in Chapter 3)

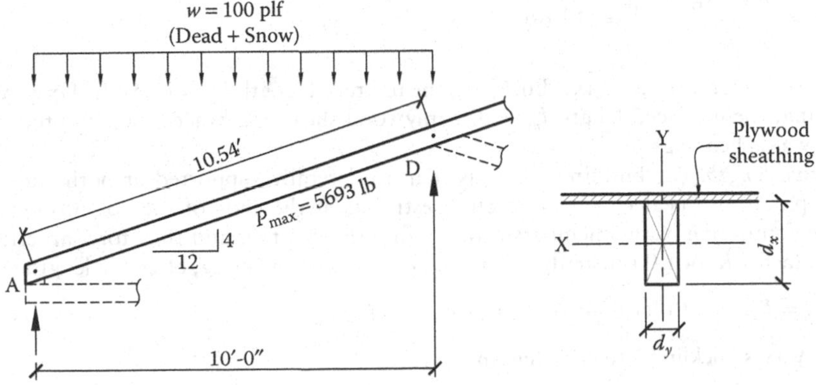

Figure 5.16 Truss top chord load diagram and cross section.

$C_D = 1.15$ (snow load controls for the load combination $D + S$)
$C_i = 1.0$ (assumed)
$C_T = 1.0$

It should be noted that for each design check, the controlling load duration factor value will correspond to the C_D value of the shortest-duration load in the load combination for that design case. Thus, the C_D value for the various design cases may vary.

***Design Check 1*: Compression on Net Area**
This condition occurs at the ends (i.e., supports) of the member. For this load case, the axial load P is caused by dead load plus snow load ($D + S$); therefore, the load duration factor C_D from Chapter 3 is 1.15. The unfactored compression stress applied is

$$f_c = \frac{P}{A_n} \le F_c'$$

Net area $A_n = A_g - \Sigma A_{\text{holes}}$

$$= 13.88 - (1 \text{ hole})(3/4 \text{ in.} + 1/16 \text{ in.})(1.5 \text{ in.})$$

Therefore,

$$A_n = 12.66 \text{ in.}^2$$

The compressive axial stress applied is

$$f_c = \frac{P}{A_n} = \frac{5693 \text{ lb}}{12.66 \text{ in}^2} = 450 \text{ psi}$$

At support locations, there can be no buckling; therefore, $C_P = 1.0$.
The allowable compression stress parallel to the grain is

$$F_c' = F_c C_D C_M C_t C_F C_i C_P$$

$$= (1350)(1.15)(1.0)(1.0)(1.0)(1.0)(1.0)$$

Therefore,

$$F_c' = 1553 \text{ psi} > f_c = 450 \text{ psi} \quad \textbf{OK}$$

***Design Check 2*: Compression on Gross Area**
This condition occurs at the *midspan* of the member. The axial load P for this load case is also caused by dead load plus the snow load ($D + S$); therefore, the load duration factor C_D from Chapter 3 is 1.15. The unfactored applied axial compressive stress is

$$f_c = \frac{P}{A_g} = \frac{5693 \text{ lb}}{13.88 \text{ in}^2} = 411 \text{ psi}$$

Unbraced length. For x–x axis buckling, the unbraced length $l_{ux} = 10.54$ ft. For y–y axis buckling, the unbraced length $l_{uy} = 0$ ft (plywood sheathing braces the truss top chord for y–y axis bending).
Effective length. For building columns that are laterally supported at both ends, it is usual practice to neglect any rotational restraints at the ends of the column provided by the connection of the members framing into the column, and therefore, an effective length factor K_e of 1.0 is usually assumed. For x–x axis buckling, effective length is

$$l_{ey} = K_e l_{ux} = (1.0)(10.54 \text{ ft}) = 10.54 \text{ ft}$$

For y–y axis buckling, effective length is

$$l_e = K_e l_u = (1.0)(0 \text{ ft}) = 0 \text{ ft}$$

Slenderness ratio: For x–x axis buckling, the effective length is

$$\frac{l_{ex}}{d_x} = \frac{(10.54\text{ ft})(12)}{9.25\text{ in}} = 13.7 < 50 \text{ (larger slenderness ratio controls)}$$

For y–y axis buckling, the effective length is

$$\frac{l_{ey}}{d_y} = \frac{0}{1.5\text{ in.}} = 0 < 50$$

Therefore,

$$(l_e/d)_{\max} = 13.7.$$

This slenderness ratio will be used in the calculation of the column stability factor C_P and the allowable compression stress parallel to the grain F_c'. The buckling modulus of elasticity is

$$E'_{\min} = E_{\min} C_M C_t C_i C_T$$

$$= (0.55 \times 10^6)(1.0)(1.0)(1.0)(1.0) = 0.55 \times 10^6 \text{ psi}$$

$$c = 0.8 \text{ (visually graded lumber)}$$

The Euler critical buckling stress about the weaker axis (i.e., the axis with the higher slenderness ratio), which for this problem happens to be the x–x axis, is

$$F_{cE} = \frac{0.822E'_{\min}}{(l_e/d)^2_{\max}} = \frac{(0.822)(0.55 \times 10^6)}{(13.7)^2} = 2409 \text{ psi}$$

$$F_c^* = F_c C_D C_M C_t C_F C_i$$

$$= (1350)(1.15)(1.0)(1.0)(1.0)(1.0) = \mathbf{1553\ psi}$$

$$\frac{F_{cE}}{F_c^*} = \frac{2409 \text{ psi}}{1553 \text{ psi}} = 1.551$$

From Equation 5.29, the column stability factor is calculated as

$$C_P = \frac{1+1.551}{(2)(0.8)} - \sqrt{\left[\frac{1+1.551}{(2)(0.8)}\right]^2 - \frac{(1.551)}{0.8}} = 0.818$$

The allowable compression stress parallel to grain is

$$F_c' = F_c^* C_P = (1553)(0.818) = \mathbf{1270\ psi} > f_c = 411 \text{ psi} \quad \text{OK}$$

***Design Check 3*: Bending Only**

This condition occurs at the point of maximum moment (i.e., at the *midspan* of the member), and the moment is caused by the uniformly distributed gravity load on the top chord (member AD), and since this load is the dead load plus the snow load, the controlling load duration factor (see Chapter 3) is 1.15. The unfactored maximum moment in member AD is

$$M_{\max} = \frac{w_{D+S} l^2}{8} = \frac{(100 \text{ lb/ft})(10 \text{ ft})^2}{8}$$

$$= 1250 \text{ ft-lb} = 15{,}000 \text{ in.-lb.}$$

The unfactored compression stress applied in member *AD* due to bending is

$$f_{bc} = \frac{M}{S_x} = \frac{15{,}000 \text{ in.-lb}}{21.4 \text{ in}^3} = 701 \text{ psi}$$

The allowable bending stress is

$$F_b' = F_b C_D C_M C_t C_L C_F C_{fu} C_i C_r$$

$$= (1100)(1.15)(1.0)(1.0)(1.0)(1.1)(1.0)(1.0)(1.15)$$

$$= \mathbf{1600 \text{ psi}} > f_b = 701 \text{ psi} \quad \textbf{OK}$$

Note that for exterior studs that provide lateral support to brick cladding, the lateral deflections should also be checked to avoid excessive deflections that could lead to cracking of the brick wall.

***Design Check 4*: Bending Plus Axial Compression Force**
This condition occurs at the *midspan* of the member. For this load case, the loads causing the combined stresses are the dead load plus the snow load ($D + S$); therefore, the controlling load duration factor C_D is 1.15. It should be noted that because the load duration factor C_D for this combined load case is the same as that for load cases 2 and 3, the parameters to be used in the combined stress interaction equation can be obtained from design checks No. 2 and 3. The reader is cautioned to be aware that the C_D values for all four design checks are not necessarily always equal.

The interaction equation for combined concentric axial load plus uniaxial bending is obtained from Equation 5.48 as

$$\left(\frac{f_c}{F_c'}\right)^2 + \frac{f_{bx}/F_{bx}'}{1 - f_c/F_{cEx}} \le 1.0$$

where:

f_c is the 411 psi (occurs at the midspan; see design check No. 2)
F_c' is the 1267 psi (occurs at the midspan; see design check No. 2)
f_{bx} is the 701 psi (occurs at the midspan; see design check No. 3)
F_{bx}' is the 1600 psi (occurs at the midspan; see design check No. 3)

The reader should note that in the interaction equation above, the Euler critical buckling stress about the *x–x axis*, F_{cEx}, is what is required, since the bending of the truss top chord member is about that axis. The effective length for *x–x* axis buckling is

$$\frac{l_{ex}}{d_x} = \frac{(10.54 \text{ ft})(12)}{9.25 \text{ in.}} = 13.7 < 50$$

The Euler critical buckling stress about the *x–x* axis is

$$F_{cEx} = \frac{0.822 E_{\min}'}{(l_{ex}/d_x)^2} = \frac{(0.822)(0.55 \times 10^6)}{(13.7)^2} = 2409 \text{ psi}$$

Substituting the parameters above into the interaction equation yields

$$\left(\frac{411}{1267}\right)^2 + \left(\frac{701/1600}{1 - 411/2409}\right) = \mathbf{0.63} < 1.0 \quad \textbf{OK}$$

2 × 10 No. 1 & Better Hem-Fir is adequate for the truss top chord.

Example 5.9 (ASD): Combined axial compression plus bending—exterior stud wall

Design the ground floor exterior stud wall for the two-story building with the roof and floor plans shown in Figure 5.17. The floor-to-floor height is 10 ft and the wall studs are spaced at 2 ft o.c. Assume DF-L wood species and the design loads shown.

Solution: First, we calculate the unfactored gravity and lateral loads acting on a typical wall stud (Figure 5.18).

Tributary width of wall stud = 2 ft
Roof dead load D = 20 psf
Roof snow load S = 40 psf
Roof live load L_r = 20 psf (see Chapter 2)

Since $L_r < S$, the snow load controls. The lateral wind load 0.6W = 15 psf (acts horizontally). The net vertical wind load is assumed to be zero for this building.

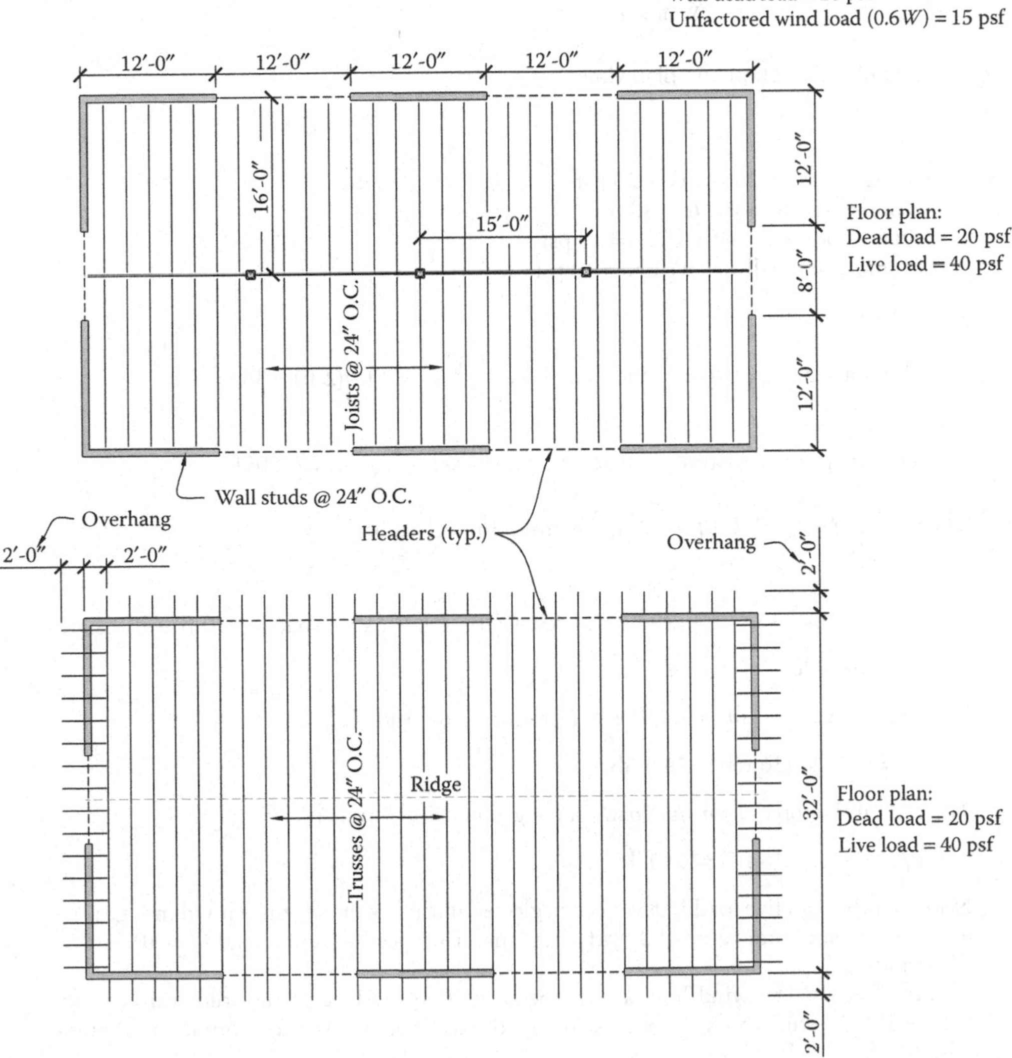

Figure 5.17 Roof and floor plans—Example 5.9.

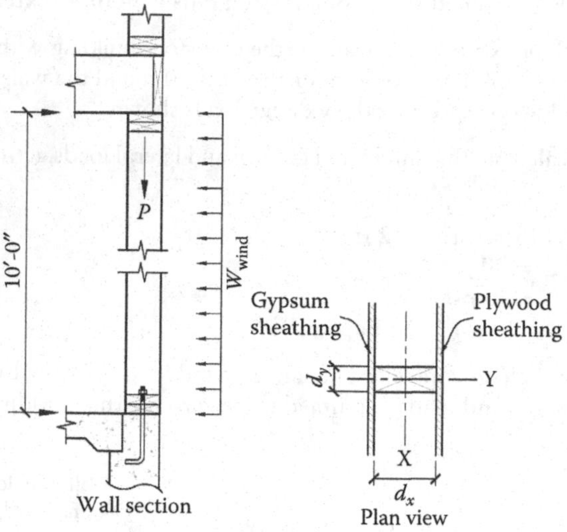

Figure 5.18 Stud wall section and plan view.

Second-floor dead load $D = 20$ psf
Second-floor live load $L = 40$ psf
Second-floor wall self-weight = 10 psf
Ground-floor wall self-weight = 10 psf

Gravity loads:

$$\text{Tributary area per stud at the } roof \text{ level} = \left(\frac{32\text{ ft}}{2} + 2\text{ ft}\right)(2\text{ ft}) = \mathbf{36\ ft^2}$$

$$\text{Tributary area per stud at the } second\text{-}floor \text{ level} = \frac{16\text{ ft}}{2}(2\text{ ft}) = \mathbf{16\ ft^2}$$

The total unfactored *dead load* on the ground-floor studs is

$$P_D = P_{D(\text{roof})} + P_{D(\text{floor})} + P_{D(\text{wall})}$$
$$= (20\text{ psf})(36\text{ ft}^2) + (20\text{ psf})(16\text{ ft}^2) + (10\text{ psf})(2\text{ ft stud spacing})(10\text{ ft wall height})$$
$$= \mathbf{1440\ lb}$$

The total unfactored *snow load* on the ground-floor studs is

$$P_S = (40\text{ psf})(36\text{ ft}^2) = \mathbf{1440\ lb}$$

The total unfactored *floor live load* on the ground-floor studs is

$$P_L = (40\text{ psf})(16\text{ ft}^2) = \mathbf{640\ lb}$$

Note that the roof live load L_r has been neglected in this example, since it will not govern, because its maximum value of 20 psf is less than the specified snow load ($S = 40$ psf) for this building.

Lateral loads. The wind load acts perpendicular to the face of the stud wall, causing bending of the stud about the x–x (strong) axis (see Figure 5.18). The unfactored lateral wind load

$$w_{\text{wind}} = (15\text{ psf})(2\text{ ft tributary width}) = 30\text{ lb/ft}$$

The unfactored maximum moment due to wind load is calculated as

$$M_w = \frac{wL^2}{8}$$

$$= \frac{(30)(10^2)}{8} = \mathbf{375\ ft\text{-}lb = 4500\ in.\text{-}lb}$$

The most critical axial load combination and the most critical combined axial and bending load combination are now determined following the normalized load procedure outlined in Chapter 3. The axial load P in the wall stud will be caused by the gravity loads D, L, and S, whereas the bending load and moment will be caused by the lateral wind load W. The unfactored load values to be used in the load combinations are as follows:

$D = 1440$ lb
$S = 1440$ lb
$L = 640$ lb
$0.6W = 4500$ in.-lb

All other loads are assumed to be zero and are therefore neglected in the load combinations. The applicable load combinations with all the zero loads neglected are shown in Table 5.2.

It can be recalled from Chapter 3 that the necessary condition to use the normalized load method is that all the loads be similar and of the same type. We can separate the load cases in Table 5.2 into two types of loads: *pure axial load cases* and *combined load cases*. The most critical load combinations are the load cases with the highest normalized load or moment. Since not all the loads on this wall stud are pure axial loads only or bending loads only, the normalized load method discussed in Chapter 3 can then be used to determine only the most critical pure axial load case, not the most critical combined load case. The most critical combined axial load plus bending load case would have to be determined by carrying out the design (or analysis) for all the combined load cases, with some of the load cases eliminated by inspection.

Pure axial load case. For the pure axial load cases, the load combination with the highest normalized load (P/C_D) from Table 5.2 is

$D + 0.75(L + S)$: $\mathbf{P = 3000\ lb}$**, with** $\mathbf{C_D = 1.15}$

Table 5.2 Applicable and governing load combinations

Load combination	Axial load, P (lb)	Moment, M (in.-lb)	Normalized load and moment		
			C_D	P/C_D	M/C_D
D	1440	0	0.9	1600	0
$D + L$	1440 + 640 = 2080	0	1.0	2080	0
$D + S$	1440 + 1440 = 2880	0	1.15	2504	0
$D + 0.75L + 0.75S$	1440 + (0.75)(640 + 1440) = **3000**	0	1.15	2609	0
$D + 0.75\ (0.6W) + 0.75L + 0.75S$	1440 + (0.75)(640 + 1440) = **3000**	0.75(4500) = 3375	1.6	1875	2109
$D + 0.6W$	1440	4500	1.6	900	2813
$0.6D + 0.6W$	(0.6)(1440) = **864**	4500	1.6	540	2813

Combined load case. For the combined load cases, the load combinations with the highest normalized load (P/C_D) and normalized moment (M/C_D) from Table 5.2 are

$$D + 0.75(L + S + 0.6W)\text{: } \mathbf{P = 3000\ lb,\ M = 3375\ in.\text{-}lb,\ with\ C_D = 1.6}$$

or

$$D + 0.6W\text{: } \mathbf{P = 1440\ lb,\ M = 4500\ in.\text{-}lb,\ with\ C_D = 1.6}$$

By inspection, it would appear as if the load combination $D + 0.75(L + S + 0.6W)$ will control for the combined load case, but this should be verified through analysis, and so, both of these combined load cases will be investigated.

Select a trial member size

Several design aids are given in Appendix B to help the reader with preliminary sizing of wall studs. Using Figure B.23, a 2 × 6 DF-L No. 2 appears to be adequate to resist the pure axial load case and the combined load cases above for the given floor-to-floor height of 10 ft.

Try a 2 × 6 DF-L No. 2.

Since 2 × 6 is a dimension lumber, use NDS-S Table 4A to determine the reference design values, and from NDS-S Table 1B, we obtain the section properties for the trial member size as follows:

Gross cross-sectional area $A_g = 8.25$ in.^{2}
Section modulus $S_x = 7.56$ in.^{3}
$d_x = 5.5$ in.
$d_y = 1.5$ in. (fully braced by sheathing)

From NDS-S Table 4A, we obtain the tabulated reference design stress values and the stress adjustment or C factors as follows:

$F_c = 1350$ psi
$F_b = 900$ psi
$F_{c\perp} = 625$ psi
$E_{min} = 0.58 \times 10^6$ psi
$C_F(F_c) = 1.1$
$C_F(F_b) = 1.3$
$C_M = 1.0$ (normal moisture conditions)
$C_t = 1.0$ (normal temperature conditions)
$C_r = 1.15$ (the condition for repetitiveness is discussed in Chapter 3)
$C_{fu} = 1.0$
$C_i = 1.0$
$C_L = 1.0$
$C_D =$ to be determined for each design check
$C_T = 1.0$

The beam stability factor C_L will be 1.0 for bending of the wall stud about the x–x (strong) axis, because the wall stud is braced against lateral torsional buckling by the plywood sheathing. The load duration factor C_D will be determined for each design check, and for each design check, the controlling load duration factor value will correspond to the C_D value of the shortest-duration load in the load combination for that design case. Thus, the C_D value for the various design cases may vary.

***Design Check 1*: Compression on Net Area**

This condition occurs at the ends (i.e., supports) of the member. Note that for this design check, which is a pure axial load case, the load combination with the highest normalized axial load P/C_D will be most critical. This controlling load combination from Table 5.2 is $D + 0.75(L + S)$, for which $\mathbf{P = 3000}$ **lb** and $\mathbf{C_D = 1.15}$. Since the ends of the wall studs are usually connected to the sill plates and top plates with nails rather than with bolts, the net area will be equal to the gross area, since there are no bolt holes to consider.

Net area $A_n = A_g - \Sigma A_{\text{holes}} = A_g = 8.25 \text{ in.}^2$

Unfactored applied compressive axial stress, $f_c = \dfrac{P}{A_n} = \dfrac{3000 \text{ lb}}{8.25 \text{ in}^2} = 364 \text{ psi}$

At support locations, there can be no buckling; therefore, $C_P = 1.0$.

The allowable compression stress parallel to the grain is

$$F_c' = F_c C_D C_M C_t C_F C_i C_P$$

$$= (1350)(1.15)(1.0)(1.0)(1.1)(1.0)(1.0)$$

$$= \mathbf{1708 \text{ psi}} > f_c = 364 \text{ psi} \qquad \textbf{OK}$$

It should be noted that design check No. 1 could have been eliminated by inspection when compared with design check No. 2, since the column stability factor C_P for design check No. 1 is 1.0, which will be greater than that for design check No. 2, and thus, the allowable stresses in design check No. 2 will be smaller.

***Design Check 2*: Compression on Gross Area**

This is also a pure axial load case that occurs at the mid height of the wall stud, and the most critical load combination will be the load combination with the highest normalized axial load P/C_D. This controlling load combination from Table 5.2 is $D + 0.75(L + S)$, for which $\mathbf{P = 3000}$ **lb** and $\mathbf{C_D = 1.15}$.

The unfactored axial compression stress is

$$f_c = \frac{P}{A_g} = \frac{3000 \text{ lb}}{8.25 \text{ in}^2} = 364 \text{ psi}$$

The unfactored compressive stress perpendicular to grain in the sill plate is

$$f_{c\perp} = \frac{P}{A_g} = \frac{3000 \text{ lb}}{8.25 \text{ in}^2} = 364 \text{ psi}$$

Unbraced length. For x–x axis buckling, the unbraced length l_{ux} is 10 ft. For y–y axis buckling, the unbraced length l_u is 0 ft (plywood sheathing braces the stud fully for y–y axis buckling).

Effective length. Wall studs are usually supported at both ends; therefore, it is usual practice to assume an effective length factor K_e of 1.0.

For x–x axis buckling, effective length $l_{ex} = K_e l_{ux} = (1.0)(10 \text{ ft}) = 10$ ft.
For y–y axis buckling, effective length $l_e = K_e l_u = (1.0)(0) = 0$ ft.

Slenderness ratio: For x–x axis buckling, the effective length is

$$\frac{l_{ex}}{d_x} = \frac{(10 \text{ ft})(12)}{5.5 \text{ in.}} = \mathbf{21.8} < 50 \text{ (larger slenderness ratio controls)}$$

For y–y axis buckling, the effective length is

$$\frac{l_{ey}}{d_y} = \frac{0}{1.5 \text{ in.}} = 0 < 50$$

Therefore,

$(l_e/d)_{max} = 21.8.$

This slenderness ratio will be used in the calculation of the column stability factor C_P and the allowable compression stress parallel to the grain F'_c:

$$\text{Buckling modulus of elasticity } E'_{min} = E_{min} C_M C_t C_i C_T$$
$$= (0.58 \times 10^6)(1.0)(1.0)(1.0) = 0.58 \times 10^6 \text{ psi}$$
$$c = 0.8 \text{ (visually graded lumber)}$$

The Euler critical buckling stress about the "weaker" axis (i.e., *the axis with the higher slenderness ratio*, which for this problem happens to be the x–x axis, because the stud is fully braced about the y–y axis by the plywood sheathing) is

$$F_{cE} = \frac{0.822E'_{min}}{(l_e/d)^2_{max}} = \frac{(0.822)(0.58 \times 10^6)}{(21.8)^2} = 1003 \text{ psi}$$

$$F^*_c = F_c C_D C_M C_t C_F C_i$$
$$= (1350)(1.15)(1.0)(1.0)(1.1)(1.0) = \mathbf{1708 \text{ psi}}$$

$$\frac{F_{cE}}{F^*_c} = \frac{1003 \text{ psi}}{1708 \text{ psi}} = 0.587$$

From Equation 5.29, the column stability factor is calculated as

$$C_P = \frac{1+0.587}{(2)(0.8)} - \sqrt{\left[\frac{1+0.587}{(2)(0.8)}\right]^2 - \frac{0.587}{0.8}} = 0.492$$

The allowable compression stress *parallel* to grain is

$$F'_c = F^*_c\, C_P = (1708)(0.492) = \mathbf{840 \text{ psi}} > f_c = 364 \text{ psi} \qquad \text{OK}$$

The allowable compression stress *perpendicular* to grain in the sill plate is

$$F'_{c\perp} = F_{c\perp}\, C_M C_t C_i = (625)(1.0)(1.0)(1.0)$$
$$= 625 \text{ psi} > f_{c\perp} = 364 \text{ psi} \qquad \text{OK}$$

***Design Check 3*: Bending Only**

This condition occurs at the *mid height* of the wall stud, and the bending moment is caused by the lateral wind load. Only the two controlling combined load cases in Table 5.2 need to be considered for this design check. The load combination with the maximum normalized moment M/C_D is $D + 0.6W$, for which the maximum moment **M = 4500 in.-lb, P = 1440 lb, and C_D = 1.6.**

The compression stress applied in the wall stud due to bending is

$$f_{bc} = \frac{M}{S_x} = \frac{4500 \text{ in.-lb}}{7.56 \text{ in}^3} = 595 \text{ psi}$$

The allowable bending stress is

$$F'_b = F_b C_D C_M C_t C_L C_F C_{fu} C_i C_r$$
$$= (900)(1.6)(1.0)(1.0)(1.0)(1.3)(1.0)(1.0)(1.15)$$
$$= 2153 \text{ psi} > f_b = 595 \text{ psi} \qquad \text{OK}$$

***Design Check 4*: Bending Plus Axial Compression Force**
This condition occurs at the *mid height* of the wall stud, and only the two controlling combined load cases in Table 5.2 need to be considered for this design check, and these are

$$D + 0.75(L + S + 0.6W): \quad \mathbf{P = 3000\ lb, M = 3375\ in.\text{-}lb}, \text{ with } \mathbf{C_D = 1.6}$$

and

$$D + 0.6W \quad \mathbf{P = 864\ lb}: \quad \mathbf{M = 4500\ in.\text{-}lb}, \text{ with } \mathbf{C_D = 1.6}$$

We investigate these load combinations separately to determine the most critical.

Load combination $D + 0.75(L + S + 0.6W)$:

$$P_{max} = 3000 \text{ lb}, \quad M_{max} = 3375 \text{ in.-lb, with } C_D = 1.6$$

The unfactored bending stress applied is

$$f_{bx} = \frac{M_{max}}{S_{xx}} = \frac{3375 \text{ in.-lb}}{7.56 \text{ in}^3} = 446 \text{ psi} \left(\begin{array}{l}\text{lateral wind load causes bending} \\ \text{about the } x\text{–}x \text{ axis}\end{array}\right)$$

The unfactored axial compression stress applied at the mid height is

$$f_c = 364 \text{ psi (from design check No. 2)}$$

The allowable bending stress is

$$F_b' = F_b C_D C_M C_t C_F C_{fu} C_i C_r$$

$$= (900)(1.6)(1.0)(1.0)(1.0)(1.3)(1.0)(1.0)(1.15)$$

$$= 2153 \text{ psi} > f_b = 446 \text{ psi} \quad \mathbf{OK}$$

Note that the beam stability factor C_L is 1.0, because the compression edge of the stud is braced laterally by the wall sheathing for bending due to loads acting perpendicular to the face of the wall. We now proceed to calculate the column stability factor C_P.

$$F_c^* = F_c C_D C_M C_t C_F C_i$$

$$= (1350)(1.6)(1.0)(1.0)(1.1)(1.0) = \mathbf{2376\ psi}$$

(Recall that C_F for F_c^* is different than C_F for F_b')

The Euler critical buckling stress about the "weaker" axis (i.e., the axis with the higher slenderness ratio), which for this problem happens to be the x–x axis, is

$$F_{cE(max)} = \frac{0.822E'_{min}}{(l_e/d)^2_{max}} = \frac{(0.822)(0.58\times 10^6)}{(21.8)^2} = 1003 \text{ psi}$$

$$\frac{F_{cE}}{F_c^*} = \frac{1003 \text{ psi}}{2376 \text{ psi}} = 0.422$$

From Equation 5.29, the column stability factor is calculated as

$$C_P = \frac{1+0.422}{(2)(0.8)} - \sqrt{\left[\frac{1+0.422}{(2)(0.8)}\right]^2 - \frac{0.422}{0.8}} = 0.377$$

The allowable compression stress *parallel* to grain is

$$F'_c = F^*_c C_P = (2376)(0.377) = \mathbf{896\ psi} > f_c = 364\ \text{psi} \qquad \textbf{OK}$$

The interaction equation for combined concentric axial load plus uniaxial bending caused by transverse loading is obtained from Equation 5.48 as:

$$\left(\frac{f_c}{F'_c}\right)^2 + \frac{f_{bx}/F'_{bx}}{1 - f_c/F_{cEx}} \le 1.0$$

The Euler critical buckling stress about the x–x axis (i.e., the axis of bending of the wall stud due to lateral wind loads) is

$$F_{cEx} = \frac{0.822E'_{min}}{(l_{ex}/d_x)^2} = \frac{(0.822)(0.58 \times 10^6)}{(21.8)^2} = 1003\ \text{psi}$$

Substituting the parameters above into the interaction equation yields

$$\left(\frac{364}{896}\right)^2 + \frac{446/2153}{1 - 364/1003} = 0.49 < 1.0 \qquad \textbf{OK}$$

Load combination D + **0.6***W:*

$$P_{max} = 864\ \text{lb}, \quad M_{max} = 4500\ \text{in.-lb, with}\ C_D = 1.6$$

The unfactored bending stress applied is

$$f_{bx} = \frac{M_{max}}{S_{xx}} = \frac{4500\ \text{in.-lb}}{7.56\ \text{in}^3} = 595\ \text{psi} \quad \left(\begin{array}{l}\text{lateral wind load causes bending}\\ \text{about the } x\text{–}x \text{ axis}\end{array}\right)$$

The unfactored axial compression stress applied is

$$f_c = \frac{P}{A_g} = \frac{864\ \text{lb}}{8.25\ \text{in}^2} = 105\ \text{psi}$$

Except for the applied bending and axial stresses, all other parameters used in the interaction equation for the preceding combined load case would also apply to this load case. Therefore, the interaction equation for this load case will be

$$\left(\frac{105}{896}\right)^2 + \frac{595/2153}{1 - 105/1003} = 0.32 < 1.0 \qquad \textbf{OK}$$

It should be noted that because the interaction values obtained are much less than 1.0, a more efficient design can be obtained by using a lower stress grade for the 2 × 6 wall stud, such as No. 3 grade.

5.5.5 Columns with eccentric axial loads plus transverse loads

Thus far, we have considered compression members subjected to combined axial load and bending due to transverse loads applied perpendicular to the longitudinal axis of the member. In some structures, wood columns may be subjected to axial compression load that is applied eccentrically about the centroid of the column in addition to transverse loads applied perpendicular to the longitudinal axis of the member (see Figure 5.19).

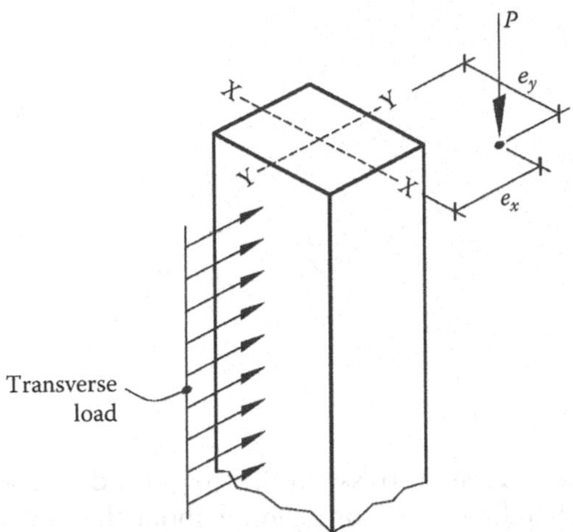

Figure 5.19 Eccentrically loaded column.

The interaction equations for combined bending and axial load that have been discussed thus far in this chapter assume that the bending moment is caused only by transverse loading on the wood member. For the special condition where the moment is caused only by eccentric axial loads or by eccentric axial loads plus transverse loads, the required interaction equation can be found in subsection 15.4 of the NDS Code. Note that there is currently no requirement in the NDS Code to design for a *minimum* column eccentricity. Where columns with eccentric axial loads are encountered in practice such as columns with brackets, the structural designer should consider the axial load eccentricity in their design. The general interaction equations for a column subjected to eccentric axial loads about both orthogonal axes plus transverse bending about both axes are given in NDS subsection 15.4.1 for the ASD method as

$$\left(\frac{f_c}{F_c'}\right)^2 + \frac{f_{bx} + f_c\left(\frac{6e_x}{d_x}\right)\left[1 + 0.234\left(\frac{f_c}{F_{cEx}}\right)\right]}{F_{bx}'\left[1 - \frac{f_c}{F_{cEx}}\right]} + \frac{f_{by} + f_c\left(\frac{6e_y}{d_y}\right)\left[1 + 0.234\left(\frac{f_c}{F_{cEy}}\right) + 0.234\left(\frac{f_{bx} + f_c\left(\frac{6e_x}{d_x}\right)}{F_{bE}}\right)^2\right]}{F_{by}'\left[1 - \frac{f_c}{F_{cEy}} - \left(\frac{f_{bx} + f_c\left(\frac{6e_x}{d_x}\right)}{F_{bE}}\right)^2\right]} \le 1.0 \tag{5.50}$$

$$\frac{f_c}{F_{cEy}} + \left(\frac{f_{bx} + f_c\left(\frac{6e_x}{d_x}\right)}{F_{bE}}\right)^2 < 1.0 \tag{5.51}$$

where:

$$f_c < F_{cEx} = \frac{0.822E'_{min}}{\left(\frac{l_{ex}}{d_x}\right)^2}$$

$$f_c < F_{cEy} = \frac{0.822E'_{min}}{\left(\frac{l_{ey}}{d_y}\right)^2}$$

$$f_{bx} < F_{bE} = \frac{1.20E'_{min}}{(R_B)^2}$$

f_c is the unfactored compression stress parallel to grain due to axial load
f_{bx} is the unfactored bending stress for bending about the x–x or strong axis of the column due to *transverse loads* acting perpendicular to the longitudinal axis of the column
f_{by} is the unfactored bending stress for bending about the y–y or weak axis of the column due to *transverse loads* acting perpendicular to the longitudinal axis of the column
R_B is the beam slenderness ratio for lateral torsional buckling $= \sqrt{l_e d_x / b^2} \leq 50$

Note: R_B must *not* be greater than 50; otherwise, the beam width, b, would need to be increased. Note that $b = d_y$

F'_c is the allowable compression stress parallel to grain (ASD)
F'_{bx} is the allowable bending stress about the x–x axis (ASD)
F'_{by} is the allowable bending stress about the y–y axis (ASD)
d_x is the depth of column for x–x axis bending (i.e., dimension of column parallel to the y axis)
d_y is the depth of column for y–y axis bending (i.e., dimension of column parallel to the x axis)
e_x is the eccentricity of column about the x–x axis, measured from the centerline of the column parallel to the y axis
e_y is the eccentricity of column about the y–y axis, measured from the centerline of the column parallel to the x axis
l_{ex} is the effective length for bending about the x–x axis = $K_x L_x$
l_{ey} is the effective length for bending about the y–y axis = $K_y L_y$

For the frequent case where there is no eccentricity of the axial load about the y–y axis (i.e., $e_y = 0$ in Equation 5.50), and where there is no transverse load perpendicular to the longitudinal axis of the column that causes bending about the y–y axis (i.e., $f_{by} = 0$ in Equation 5.50), the interaction equation 5.50 reduces to:

$$\left(\frac{f_c}{F'_c}\right)^2 + \frac{f_{bx} + f_c\left(\frac{6e_x}{d_x}\right)\left[1 + 0.234\left(\frac{f_c}{F_{cEx}}\right)\right]}{F'_{bx}\left[1 - \frac{f_c}{F_{cEx}}\right]} \leq 1.0 \qquad (5.52)$$

Note that when the eccentricity e_x is set equal to zero in Equation 5.52 (meaning concentric axial load plus transverse bending), it results in interaction Equation 5.53 which is the exact same interaction equation as previously given in Equation 5.48.

$$\left(\frac{f_c}{F_c'}\right)^2 + \frac{f_{bx}}{F_{bx}'\left[1 - \frac{f_c}{F_{cEx}}\right]} \leq 1.0 \tag{5.53}$$

Example 5.10: (ASD)—Column with eccentric axial load

A 6 in. × 6 in. structural composite lumber column with the properties given below has unbraced lengths about the x–x and y–y axes of 12 ft and is subjected to an unfactored eccentric axial load of 16 kips and a moment about the x–x axis caused by an eccentricity of $d_x/6$. There are no transverse loads on the column, and the eccentricity of the axial load about the y–y axis is zero. Determine if the column is adequate to resist the eccentric load.

$L_{ux} = 12 \text{ ft} = 144 \text{ in.}$

$L_{uy} = 12 \text{ ft} = 144 \text{ in.}$

$d_x = 5.25 \text{ in.}, d_y = 5.25 \text{ in.}$

$b = d_y = 5.25 \text{ in.}$

$A = 27.563 \text{ in.}^2\ S_x = 24.117 \text{ in.}^3, I_x = 63.308 \text{ in.}^4$

$F_b = 2400 \text{ psi}$

$F_c = 2500 \text{ psi}$

$E = 1.8 \times 10^6 \text{ psi}$

$E_{min} = 0.915 \times 10^6 \text{ psi}$

$C_D = 1.0$

$C_M = 1.0$

$C_r = 1.0$

$C_{F,c} = 1.0$

$C_{F,b} = 1.0$

$C_T = 1.0$

$C_t = 1.0$

$C_{fu} = 1.0$

$C_i = 1.0$

Solution:

Applied Loads and Eccentricities:

The unfactored applied compression stress is

$$f_c = \frac{P}{A_g} = \frac{16{,}000\ \text{lb}}{27.563\ \text{in}^2} = 581\ \text{psi}$$

$$e_x = \frac{d_x}{6} = \frac{5.25\ \text{in.}}{6} = 0.875\ \text{in.}$$

$e_y = 0$ (since there is *no* eccentricity of the axial load about the y–y axis)

$f_{bx} = 0$ (since there is *no* transverse load on the column causing moments about the x–x axis)

$f_{by} = 0$ (since there is *no* transverse load on the column causing moments about the y–y axis)

Substituting $e_y = 0$, $f_{bx} = 0$, and $f_{by} = 0$ into Equation 5.52 yields the required interaction equation as

$$\left(\frac{f_c}{F'_c}\right)^2 + \frac{f_c\left(\dfrac{6e_x}{d_x}\right)\left[1+0.234\left(\dfrac{f_c}{F_{cEx}}\right)\right]}{F'_{bx}\left[1-\dfrac{f_c}{F_{cEx}}\right]} \leq 1.0$$

$$E'_{\min} = E_{\min} C_M C_t C_i C_T$$

$$= (0.915 \times 10^6)(1.0)(1.0)(1.0)(1.0) = 0.915 \times 10^6\ \text{psi}$$

$c = 0.9$ (structural composite lumber)

$$\frac{l_{ex}}{d_x} = \frac{(12\ \text{ft})(12)}{5.25\ \text{in.}} = 27.43 < 50 \quad \text{OK}$$

$$\frac{l_{ey}}{d_y} = \frac{(12\ \text{ft})(12)}{5.25\ \text{in.}} = 27.43 < 50 \quad \text{OK}$$

The Euler critical buckling stress about the x–x axis is

$$F_{cEx} = \frac{0.822E'_{\min}}{(l_{ex}/d_x)^2} = \frac{(0.822)(0.915\times10^6)}{(27.43)^2} = 1000\ \text{psi}$$

The Euler critical buckling stress about the y–y axis is

$$F_{cEy} = \frac{0.822E'_{\min}}{(l_{ey}/d_y)^2} = \frac{(0.822)(0.915\times10^6)}{(27.43)^2} = 1000\ \text{psi}$$

To calculate the allowable compression stress parallel to grain, use the smaller of F_{cEx} and F_{cEy}

$$F_c^* = F_c C_D C_M C_t C_F C_i$$

$$= (2500)(1.0)(1.0)(1.0)(1.0)(1.0) = 2500 \text{ psi}$$

$$\frac{F_{cE}}{F_c^*} = \frac{1000 \text{ psi}}{2500 \text{ psi}} = 0.40$$

For structural composite lumber, $c = 0.9$ (see Chapter 3)
From Equation 5.29, the column stability factor is calculated as

$$C_P = \frac{1+0.40}{(2)(0.9)} - \sqrt{\left[\frac{1+0.40}{(2)(0.9)}\right]^2 - \frac{(0.40)}{0.9}} = 0.377$$

The allowable compression stress parallel to grain is

$$F_c' = F^*_c C_P = (2500)(0.377) = \mathbf{943\ psi} > f_c = 581 \text{ psi} \qquad \text{OK}$$

The allowable bending stress is

$$F_b' = F_b C_D C_M C_t C_L C_F C_{fu} C_i C_r$$

$$= (2400)(1.0)(1.0)(1.0)(1.0)(1.0)(1.0)(1.0)(1.0)$$

$$= \mathbf{2400\ psi} > f_{bx} = 0 \text{ psi} \quad \textbf{OK} \text{ (since there are no transverse loads on the column)}$$

The effective length for bending members from Table 3.9:

$$\frac{l_{ux}}{d_x} = \frac{144 \text{ in.}}{5.25 \text{ in.}} = 27.43 > 14.3$$

Therefore, the effective length, $l_e = 1.84 l_{ux} = (1.84)(144 \text{ in.}) = 265$ in.

R_B = the beam slenderness ratio for lateral torsional buckling

$$= \sqrt{\frac{l_e d_x}{b^2}} = \sqrt{\frac{(265 \text{ in.})(5.25 \text{ in.})}{(5.25 \text{ in.})^2}} = 7.1 \le 50 \qquad \text{OK}$$

$$F_{bE} = \frac{1.20 E'_{min}}{(R_B)^2} = \frac{(1.20)(0.915\times10^6)}{(7.1)^2} = 21{,}781 \text{ psi}$$

$$\left(\frac{f_c}{F_c'}\right)^2 + \frac{f_c\left(\frac{6e_x}{d_x}\right)\left[1+0.234\left(\frac{f_c}{F_{cEx}}\right)\right]}{F_{bx}'\left[1-\frac{f_c}{F_{cEx}}\right]} \le 1.0$$

$$\left(\frac{581\text{ psi}}{943\text{ psi}}\right)^2 + \frac{(581\text{ psi})\left(\dfrac{6(0.875\text{ in.})}{5.25\text{ in.}}\right)\left[1+0.234\left(\dfrac{581\text{ psi}}{1000\text{ psi}}\right)\right]}{(2400\text{ psi})\left[1-\dfrac{581\text{ psi}}{1000\text{ psi}}\right]} = 1.04 \approx 1.0 \quad \text{OK*}$$

*Note that in practice, a small amount of overstress (e.g., ≤5%) can be permissible.

We also need to check the interaction equation from Equation 5.51:

$$\frac{f_c}{F_{cEy}} + \left(\frac{f_{bx} + f_c\left(\dfrac{6e_x}{d_x}\right)}{F_{bE}}\right)^2 = \frac{581\text{ psi}}{1000\text{ psi}} + \left(\frac{0+(581\text{ psi})\left(\dfrac{6(0.875\text{ in.})}{5.25\text{ in.}}\right)}{21{,}781\text{ psi}}\right)^2 = 0.58 < 1.0 \quad \text{OK}$$

5.6 PRACTICAL CONSIDERATIONS FOR ROOF TRUSS DESIGN

The application of the four load cases discussed in this chapter to typical roof trusses, such as those shown in Figure 5.20, is discussed next. The slope or pitch of a truss is usually noted as a ratio of the vertical dimension to the horizontal dimension, with the horizontal number equal to 12 (e.g., 4:12 and 7:12). The slope or pitch of a roof truss is usually determined by the owner or the architect, not by the engineer.

The spacing of the roof truss is usually selected by the engineer and is typically 16 or 24 in. o.c. The truss spacing should be a multiple of the stud spacing in the bearing walls to ensure that the truss bears directly on a wall stud; otherwise, the top plates have to be designed to resist the truss reactions that result in bending of the plates about their weak axis. The analysis of a roof truss is usually carried out with structural analysis software, and in many cases, the design of the truss members is also performed with computer software. When performing the design of the typical roof truss in Figure 5.20 for gravity loads, the following *four* load cases have to be considered, and the compression members of the truss must be sized such that the maximum slenderness ratio l_e/d does not exceed 50.

1. Pure tension (web members *BF* and *CF*)
2. Combined tension plus bending (bottom chord members *AB*, *BC*, and *CD*)
3. Pure compression (web members *BG* and *CE*)
4. Combined compression plus bending (top chord members *AG*, *GF*, *FE*, and *ED*)

In practice, roof trusses are preengineered and built by truss manufacturers. However, the designer or engineer of record has to specify the truss spacing and profile, the support locations, and the gravity and wind loads applied, including those causing uplift of the truss.

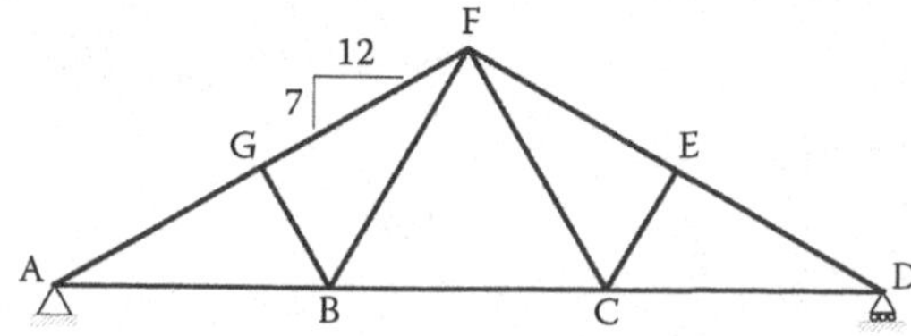

Figure 5.20 Typical roof truss.

The engineer will, however, have to check and approve the shop drawings submitted by the truss manufacturer.

Wind Uplift on Roof Trusses: If there is net wind uplift on a roof truss, stress reversals may occur in the bottom chord members (e.g., members *AB*, *BC*, and *CD*). These members, which are normally in tension, will be subjected to compression forces because of the net wind uplift. In addition, the roof truss supports will have to be tied down to the wall studs by using hurricane clips.

5.6.1 Types of roof trusses

There are two main categories of roof trusses: lightweight and heavy timber roof trusses. Lightweight roof trusses are made from dimension-sawn lumber, and heavy timber roof trusses are made from larger-sawn lumber sizes (e.g., 4 × 6 to 8 × 12). Lightweight roof trusses can be used to span up to 60–70 ft and are typically spaced at 16 or 24 in. o.c. Heavy timber trusses were used in the past to span longer distances, up to 120 ft, and are typically spaced much farther apart than the lightweight roof trusses. A maximum span-to-depth ratio of 8–10 is typically specified for flat-roof heavy timber trusses and 6–8 is specified for bow-string trusses.

Several common profiles of lightweight roof trusses were shown in Figure 1.7. The pitch or rise per foot of roof trusses varies from 3 in 12 to 12 in 12, depending on the span of the roof truss. Roof pitches shallower than 4 in 12 are adequate for shedding rainwater, but pitches higher than 4 in 12 are more efficient for proper shedding of snow and ice loads in cold climates. For a given span, the higher the roof pitch, the smaller the axial forces in the compression and tension chords of the truss. The members of lightweight roof trusses are usually connected with toothed light-gage plate connectors or nailed plate connectors designed in accordance with the Truss Plate Institute's *National Design Standard for Metal Plated Connected Wood Trusses* ANSI/TPI 1-2002 [9]. This specification is used by preengineered truss manufacturers for the computer-aided design of factory-built roof and floor trusses.

5.6.2 Bracing and bridging of roof trusses

Lateral bracing and bridging are required for lightweight roof trusses during and after erection to ensure lateral stability. During erection of roof trusses, lateral bracing and bridging perpendicular to the truss span are required to tie the top and bottom chords and web members of adjacent trusses together for lateral stability and to prevent the trusses from tipping or rolling over on their sides before attachment of the roof diaphragms. Lateral braces and bridging should be provided per BCSI 1-06 [10]. Typically, the top chord members are temporarily braced to maintain stability until the roof sheathing is installed, whereas the web and bottom chord members are permanently braced. The temporary top chord bracing consists of horizontal 2× continuous wood members placed in the same plane as the top chord and running perpendicular to the truss span. They are usually spaced 10 ft on centers for truss spans up to 30 ft, 8 ft on centers for truss spans between 30 ft and 45 ft, and 6 ft on centers for truss spans between 45 ft and 60 ft. Permanent braces for the truss bottom chord members consist of continuous horizontal 1× or 2× wood members placed perpendicular to the truss span and in the plane of the bottom chord, spaced at 8 to 10 ft on centers, and preferably located at a truss web member. These continuous members are located on top of the bottom chord to avoid interference and should be lapped over at least two trusses. In addition, 2× diagonal braces should be provided in the plane of the bottom chord between all the horizontal braces, and these should occur every 10 truss spaces or at 20 ft on centers maximum. The permanent bracing for the truss web members includes diagonal braces in the vertical plane

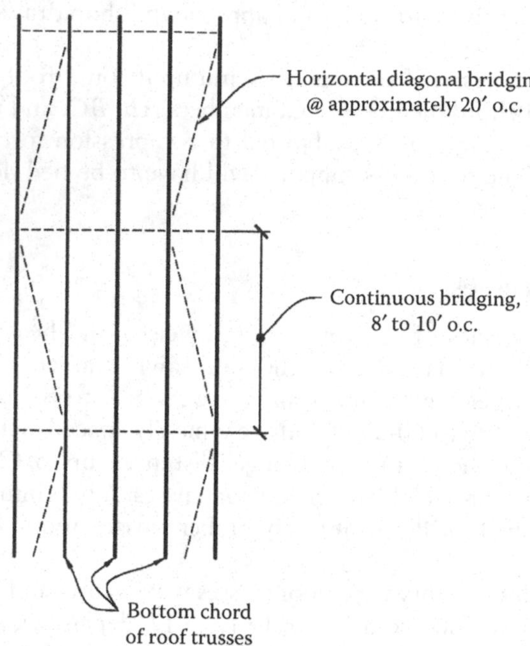

Figure 5.21 Typical truss bottom chord bridging.

(perpendicular to the truss span) that is connected to the webs of several trusses, and the location of these diagonal braces should coincide with the location of the bottom-chord horizontal braces. The diagonal braces should occur at every 10 truss spaces or at 20 ft on centers maximum. Several collapses of roof trusses during erection have been reported, so it is imperative that roof trusses be braced adequately to prevent collapse. In some cases, roof trusses may be erected in modules of three or more adjacent trusses laterally braced together on the ground for stability, with or without the roof sheathing attached. Without adequate lateral bracing or bridging, it is highly possible for the compression chord members to buckle out of the plane of the truss, even under the self-weight of the truss, and buckling failures are usually catastrophic and sudden [10]. The force in the lateral bracing for compression is usually assumed to be at least 2% of the compression force in the braced member.

During construction, temporary braces should be provided to brace the first truss to the ground. In addition, diagonal bridging in the vertical plane has to be provided to tie the compression chord to the tension chord of adjacent trusses. The temporary continuous horizontal bridging is located at the bottom of the top chord to avoid interference with installation of the roof sheathing. The slenderness ratio of the 2× bracing and bridging elements must not be greater than 50. A minimum of two 16d nails is used to connect the bridging to the truss member. Figure 5.21 shows typical bridging for a roof truss bottom chord. For more information, the reader should refer to the Truss Plate Institute (TPI) publication, *Guide to Good Practice for Handling, Installing and Bracing Metal Connected Wood Trusses* (BCSI 1-03) [10].

PROBLEMS

5.1 A 4 × 8 wood member is subjected to an axial tension load of 8000 lb caused by the dead load plus the snow load plus the wind load. Determine the applied tension stress f_t and the allowable tension stress F_t', and check the adequacy of this member. The lumber is DF-L No. 1; normal temperature conditions apply; and the MC is greater than 19%. Assume that the end connections are made with a single row of ¾-in.-diameter bolts.

5.2 A 2 × 10 Hem-Fir No. 3 wood member is subjected to an axial tension load of 6000 lb caused by the dead load plus the snow load. Determine the applied tension stress f_t and the allowable tension stress F_t', and check the adequacy of this member. Normal temperature and dry service conditions apply, and the end connections are made with a single row of ¾-in.-diameter bolts.

5.3 A 2½ × 9 (six laminations) 5DF glulam axial combination member is subjected to an axial tension load of 20,000 lb caused by the dead load plus the snow load. Determine the applied tension stress f_t and the allowable tension stress F_t', and check the adequacy of this member. Normal temperature and dry service conditions apply, and the end connections are made with a single row of ⅞-in.-diameter bolts.

5.4 For the roof truss elevation shown in Figure 5.22, determine if a 2 × 10 Spruce-Pine-Fir No. 1 member is adequate for the bottom chord of the typical truss, assuming that the roof dead load is 20 psf, the snow load is 40 psf, and the ceiling dead load is 15 psf of the horizontal plan area. Assume that normal temperature and dry service conditions apply and that the members are connected with a single row of ¾-in.-diameter bolts. The trusses span 36 ft and are spaced at 2 ft o.c. (Figure 5.22).

5.5 For the roof truss in Problem 5.4, determine if a 2 × 10 Spruce-Pine-Fir No. 1 member is adequate for the top chord.

5.6 Determine if 10-ft-high 2 × 6 Hem-Fir No. 1 interior wall studs spaced at 16 in. o.c. are adequate to support a dead load of 300 lb/ft and a snow load of 600 lb/ft. Assume that the wall is sheathed on both sides and that normal temperature and dry service conditions apply. Neglect interior wind loads.

5.7 Determine the axial load capacity of a nailed built-up column of four 2 × 6s with a 12-ft unbraced height. Assume Hem-Fir No. 2, normal temperature and dry service conditions, and a load duration factor C_D of 1.0.

5.8 Determine if 2 × 6 DF-L Select Structural studs spaced at 2 ft o.c. are adequate at the ground-floor level to support the following loads acting on the exterior stud wall of a three-story building.

Roof: Dead load = 300 lb/ft; snow load = 800 lb/ft; roof live load = 400 lb/ft. Third floor: Dead load = 400 lb/ft; live load = 500 lb/ft. Second floor: Dead load = 400 lb/ft; live load = 500 lb/ft.

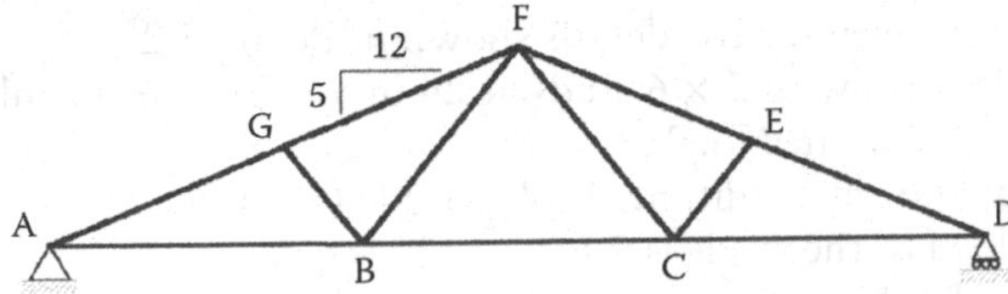

Figure 5.22 Roof truss elevation—Problem 5.4.

The unfactored lateral Wind load acting perpendicular to the face of the wall 0.6W = 15 psf. The floor-to-floor height is 10 ft; the wall studs are spaced at 2 ft o.c.; and assume that the studs are sheathed on both sides.

5.9 Find F_b', F_c', E', and E_{min}' for the following:
   a. 2 × 6 wall studs at 12 in. o.c. in an interior condition braced at mid height only and 12 ft tall. Wood is Spruce-Pine-Fir, Select Structural. Loading is $D + S + W$.
   b. 6 × 6 wood column supporting a residential deck in an exterior condition. Height is 10 ft, and wood is Southern Pine, Select Structural.
   c. 6 × 6 wood column used for shoring construction loading (gravity only). Height is 8 ft, and wood is Spruce-Pine-Fir, No. 2.
   d. 2 × 4 members in the top chord of a roof truss, with trusses spaced 24 in. on center in an interior condition under D-W loading, with continuous roof sheathing. Wood is MSR 2000F-1.6E.
   e. 12 × 12 wood column used for a heavy-storage occupancy, with high humidity and moisture conditions. Height is 14 ft, and wood is Southern Pine, Select Structural.

5.10 Assume the following:
- A bearing wall with 2 × 6 studs at 24″ o.c., L = 12 ft
- All lumber is Hem-Fir, No. 2
- Normal temperature conditions apply, and MC < 19%
- Wall sheathing provides full lateral stability to the weak axis
- Uniform loads on the wall are $w_D = 300$ plf; $w_S = 300$ plf
- Unfactored wind pressure on the wall is $0.6W = 10$ psf
- Consider the load combination of $D + 0.75(S + 0.6W)$ only
   a. Calculate the axial load, P, and the moment in a typical stud. Sketch a typical wall stud showing the loads.
   b. Determine the applied axial stress, f_c, and applied bending stress, f_b.
   c. Determine the allowable axial stress, F_c'.
   d. Determine the allowable bending stress, F_b'.
   e. Determine if the member is adequate for combined bending/compression.

5.11 For the framing plan shown in Figure 5.23:
   a. Find the allowable axial load for C-1.
   b. If the dead load is 15 psf, what is the allowable snow load in psf for the roof based on the axial strength of C-1?

   Assume the following:
   - Normal temperature and moisture conditions apply
   - Loads are $D + S$ only
   - Column is pinned at each end and is 13 ft tall
   - Column C-1 is a glulam, 5 ½ × 9 in. and *15-HF* series

5.12 The following is given for the details shown in Figure 5.24:
- A storage platform with 2 × 6 studs at 24 in. o.c., $L = 6$ ft (unbraced length)
- All lumber is Hem-Fir, No. 2
- Normal temperature conditions apply, and MC < 19%
- Total dead load to the studs is 7 psf
- Consider the load combination of $D + L$ only; loads are uniform across the top surface

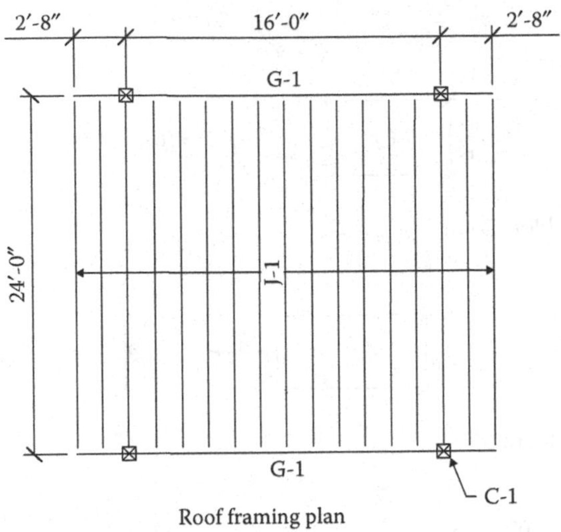

Figure 5.23 Detail for Problem 5.11.

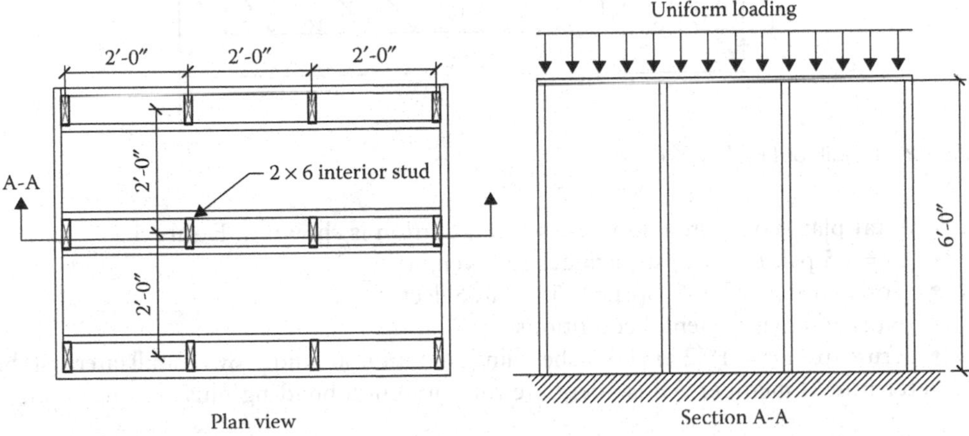

Figure 5.24 Detail for Problem 5.12.

a. Find the design axial strength, F'_c, in psi, and design load, in pounds, to the typical interior stud.
b. Determine the live load capacity in psf for the platform, based on the axial strength of the studs. Is the platform adequate for "Light Warehouse Storage"?

5.13 Built-up 2 × 8's loaded in tension are shown in Figure 5.25. Normal temperature and moisture conditions apply. Loads are based on the combination $D + L_r$ only, and wood is Hem-Fir, Select Structural.

a. Find F'_t.
b. Determine the maximum bolt diameter that could be used, based on tension on the net area of the 2 × 8's.

5.14 Find the most economical 2× member to support a tension load of 7500 lb (see Figure 5.26). Lumber is Hem-Fir, Select Structural, and normal temperature and moisture conditions apply.

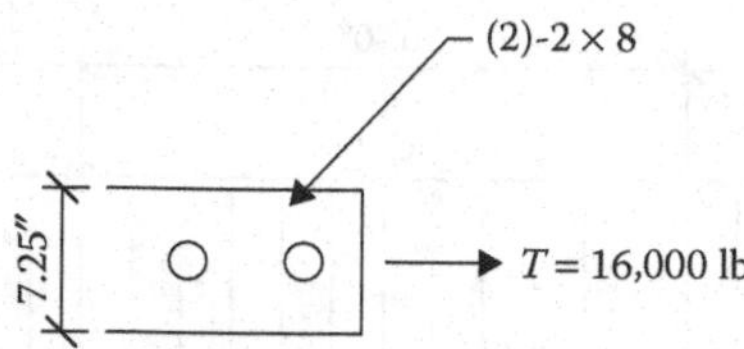

Figure 5.25 Detail for Problem 5.13.

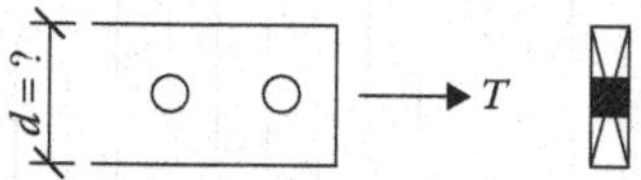

Figure 5.26 Detail for Problem 5.14.

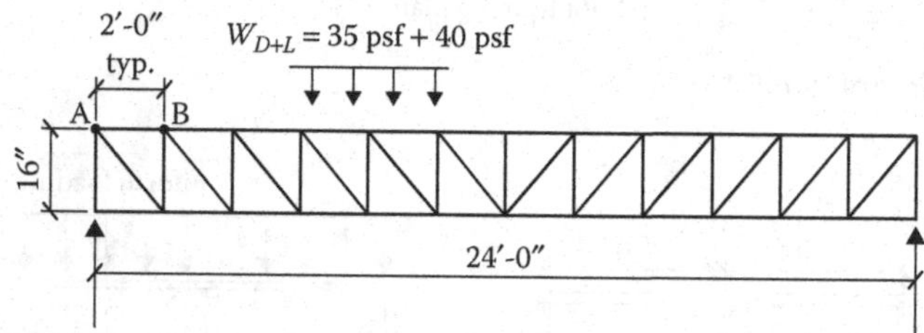

Figure 5.27 Detail for Problem 5.15.

5.15 Metal-plate-connected floor truss configuration is shown in Figure 5.27

- $D = 35$ psf, $L = 40$ psf (includes self-weight)
- Top chord is a 2 × 6, Spruce-Pine-Fir, Select
- Normal environmental conditions
- Truss spacing = 19.2 in.; floor sheathing is continuous and provides full lateral stability

Determine if member AB is adequate for combined bending plus compression.

5.16 Trusses shown in Figure 5.28 have a 2 × 6 bottom chord, with the weak axis fully braced. Member AB has an applied axial tension load of 3500 lb and a vertical load of 750 lb in the middle of AB.

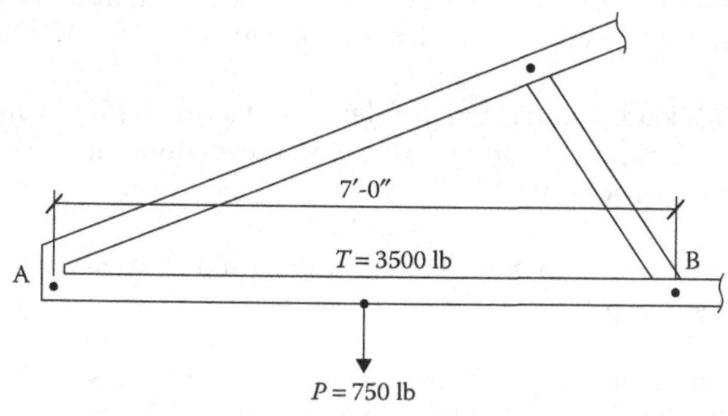

Figure 5.28 Detail for Problem 5.16.

Trusses are spaced 24″ O.C. and are continuously sheathed. Wood is Hem-Fir, Select Structural, and normal temperature and moisture conditions apply. Loading is based on $D + 0.75(L + L_r)$ only.

a. Find F_b' and F_t' for member AB.
b. Determine if member AB is adequate for combined bending plus tension loading.

5.17 Details given in Figure 5.29 are for a metal-plate-connected truss with a 2 × 4 bottom chord, with the weak axis fully braced. Member BC has an applied axial tension load and moment as shown; consider stresses at midspan of member BC only. Axial tension is constant, and moment varies, as shown across the member.

Trusses are spaced 16 in. o.c. and are continuously sheathed. All wood is Hem-Fir, #1. Normal temperature and moisture conditions apply. Loading is based on $D + S$ only, and ignore the self-weight.

a. Find F_b' and F_t' for member BC.
b. Determine if member AB is adequate for combined loading (bending plus tension) at midspan.
c. Sketch applied loads on the truss that could correspond to the member loads in BC.

5.18 Develop a spreadsheet based on Problem 5.6 to calculate the allowable axial loads for the following:

L_u = 8 ft and stud spacing = 24 in.
L_u = 12 ft and stud spacing = 12 in.

Submit the output for each case, showing the allowable axial load, in pounds, for each case.

5.19 a. Develop a spreadsheet to check Problem 5.8 for combined bending plus compression.
b. Using the variables in the spreadsheet, determine the maximum axial load that could be applied at the ground level, keeping all other variables constant.

5.20 Develop a spreadsheet and submit the following output to determine if the truss bottom chord shown below is adequate for combined tension plus bending. Loads are $D + S$ only, and normal environmental conditions apply (Figure 5.30).

5.21 The details shown in Figure 5.31 are for a metal-plate-connected wood truss, with trusses spaced 2′-0″ on center. Bottom chord load is 10 psf, and the loading is to the

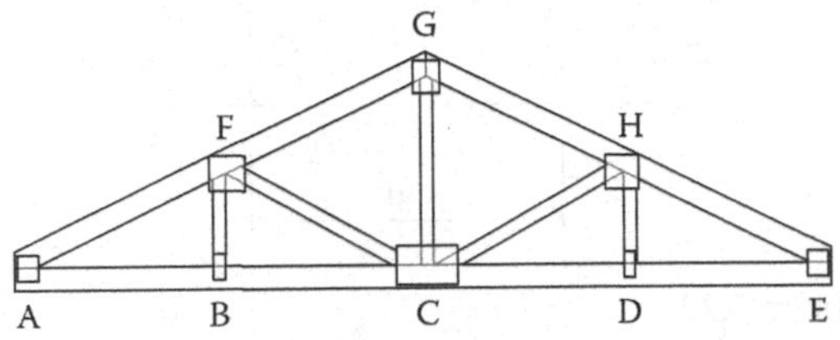

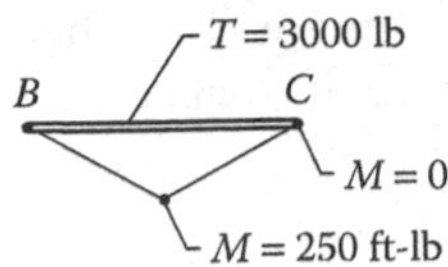

Figure 5.29 Detail for Problem 5.17.

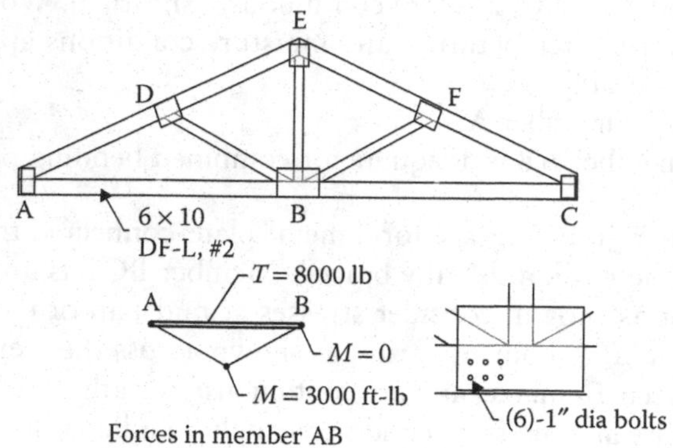

Figure 5.30 Detail for Problem 5.20.

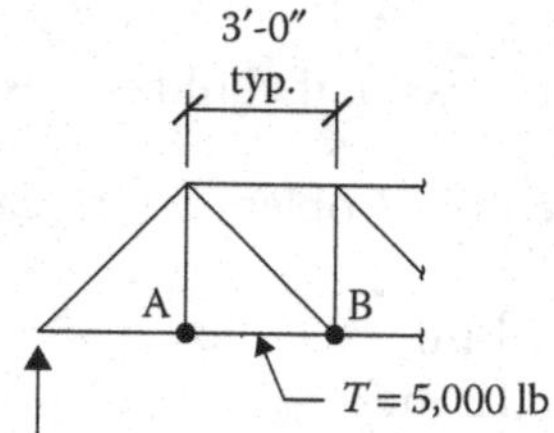

Figure 5.31 Detail for Problem 5.21.

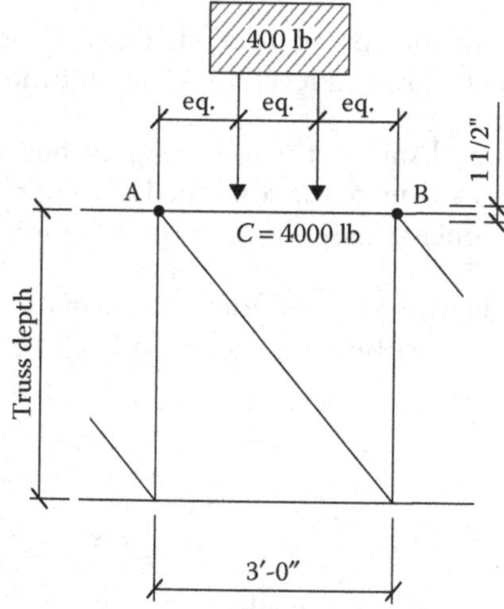

Figure 5.32 Detail for Problem 5.22.

wide face of a 2 × 6 between A and B (i.e., 1.5 in. vertical and 5.5 in. horizontal). Determine if member A–B is adequate for combined tension and bending. Consider the load combination $D + S$ only. Lumber is Spruce-Pine-Fir, #1, and normal temperature and moisture conditions apply.

5.22 Details in Figure 5.32 show a metal-plate-connected truss, with a 2 × 6 top chord, oriented flatwise. Load from mechanical unit is equally distributed to the

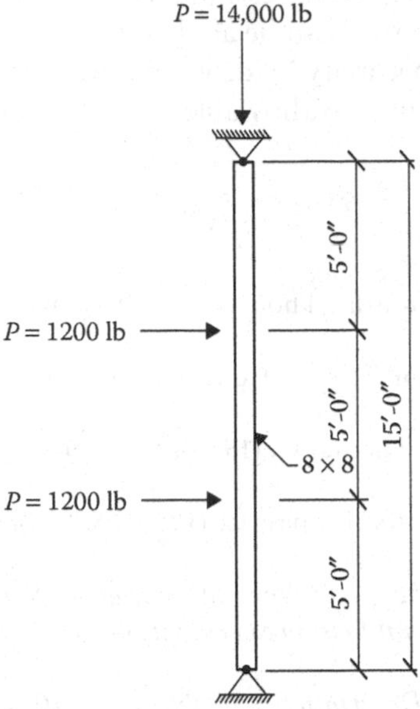

Figure 5.33 Detail for Problem 5.23.

points shown. Determine if member A–B is adequate for combined compression plus bending. Lumber is Spruce-Pine-Fir, Select Structural, and normal temperature and moisture conditions apply. Consider $D + S$ loading only.

5.23 Determine if the 8 × 8 beam-column member in Figure 5.33 is adequate for combined compression and bending loading at midspan. The member is laterally unbraced for the 15 ft length.

Wood is DF-L, Select Structural. Normal temperature and moisture conditions apply, and loads are based on the combination $D + 0.75(S + 0.6\ W)$ only.

5.24 The details in Figure 5.34 show wall studs loaded in compression only. Normal temperature and moisture conditions apply, and MC < 19%. Wood is Spruce-Pine-Fir, Select Structural. Loads are based on the combination $D + L_r$ only.

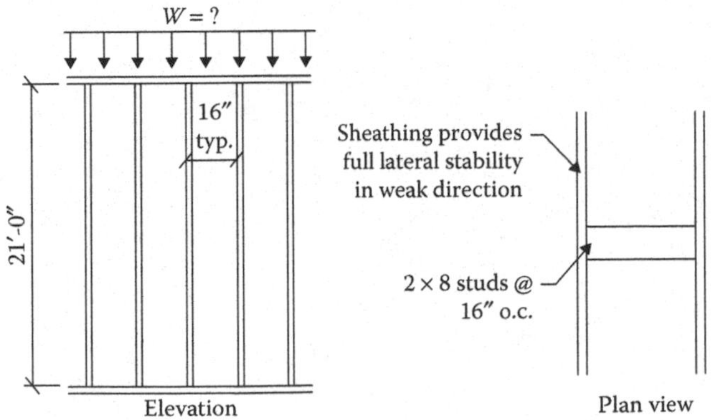

Figure 5.34 Detail for Problem 5.24.

a. Determine F_c' (allowable compression stress parallel to the grain), and the maximum load, in pounds, that can be applied to each stud.
b. Calculate the uniform gravity load, *W,* in lb/ft that can be applied to the top of the wall, as shown based on the allowable stress from part (a).

REFERENCES

1. Faherty, Keith F., and Williamson, Thomas G. (1995), *Wood Engineering and Construction,* McGraw-Hill, New York.
2. AITC (American Institute for Timber Construction) (2012), *Timber Construction Manual,* 6th ed., Wiley, New York.
3. Halperin, Don A., and Bible, Thomas G. (1994), *Principles of Timber Design for Architects and Builders,* Wiley, New York.
4. Stalnaker, Judith J., and Harris, Earnest C. (1997), *Structural Design in Wood,* Chapman & Hall, London.
5. Breyer, Donald et al. (2003), *Design of Wood Structures—ASD,* 5th ed., McGraw-Hill, New York.
6. ANSI/AF&PA (2015), *National Design Specification for Wood Construction,* American Wood Council, Leesburg, VA.
7. French, Samuel E. (1996), *Determinate Structures: Statics, Strength, Analysis and Design,* Delmar Publishers, Albany, NY.
8. Chancy, Ryan, Verhulst, Stewart M., Fowler, David W., and Ahuja Deepak (2009), Built-up column stability in wood construction, *ASCE Forensic Engineering Congress 2009: Pathology of the Built Environment,* Fifth Forensic Engineering Congress, Washington, DC, November 11–14.
9. ANSI/TPI 1-2002, *National Design Standard for Metal Plated Connected Wood Truss Construction,* Truss Plate Institute, Alexandria, VA.
10. BCSI 1-06, *Guide to Good Practice for Handling, Installing, and Bracing Metal Connected Wood Trusses,* Truss Plate Institute, Alexandria, VA.

Chapter 6

Roof and floor sheathing under vertical and lateral loads (horizontal diaphragms)

6.1 INTRODUCTION

There are numerous uses for plywood in wood structures as well as in other structure types. In a wood structure, plywood is used primarily as roof and floor sheathing and wall sheathing. The floor and roof sheathing supports gravity loads and also acts as a horizontal diaphragm to support lateral wind and seismic loads. Horizontal diaphragms behave like deep horizontal beams spanning between the vertical lateral force resisting systems, (i.e., shear walls). The wall sheathing supports transverse wind loads and acts as a vertical diaphragm to support lateral wind and seismic loads. Plywood is also used in composite structural components, such as preengineered I-joists or as sheathing in a preengineered insulated roof panel. *Plywood* is manufactured by peeling a log of wood into thin plies or veneers that are approximately $\frac{1}{16}$ to $\frac{5}{16}$ in. thick. Each ply is bonded together with adhesive, heat, and pressure, and is typically oriented 90° to adjacent plies. The thickness of the final product typically varies between $\frac{1}{4}$ and $1\frac{1}{8}$ in.

The use of *oriented strand board* (OSB) has also gained wide acceptance. OSB is composed of thin strands of wood that are pressed and bonded into mats oriented 90° to adjacent mats. The strength characteristics of OSB are similar to those of plywood and the two are often used interchangeably.

The standard size of plywood or OSB is in 4 × 8 ft sheets. Most plywood conforms to Product Standard PS 1-09, *Construction and Industrial Plywood,* published by the American Plywood Association (APA)—The Engineered Wood Association [1]. OSB typically conforms to Product Standard PS 2-10, *Performance Standard for Wood-Based Structural-Use Panels,* also published by the APA [2]. PS-1 and PS-2 are voluntary product standards that contain information pertaining to the use and design properties of plywood and OSB panels.

6.1.1 Plywood grain orientation

The inner plies of plywood are typically oriented such that the grain of each adjacent ply is at an angle of 90° with adjacent plies. The grain in the outer plies is oriented parallel to the 8-ft edge, which is the strongest configuration when the sheet of plywood is oriented perpendicular to the supporting structure (see Figure 6.1). In three-ply construction, there is one inner band oriented 90° with the outer plies or face grain. In four-ply construction, the two inner plies are oriented in the same direction. In five-ply construction, the bands typically alternate in grain orientation (see Figure 6.2).

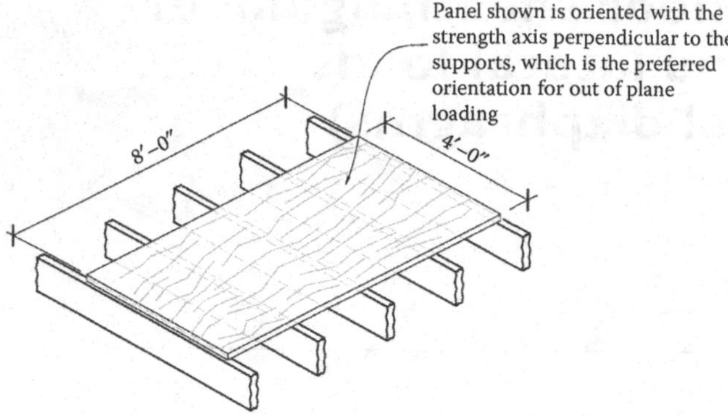

Figure 6.1 Plywood orientation on framing members.

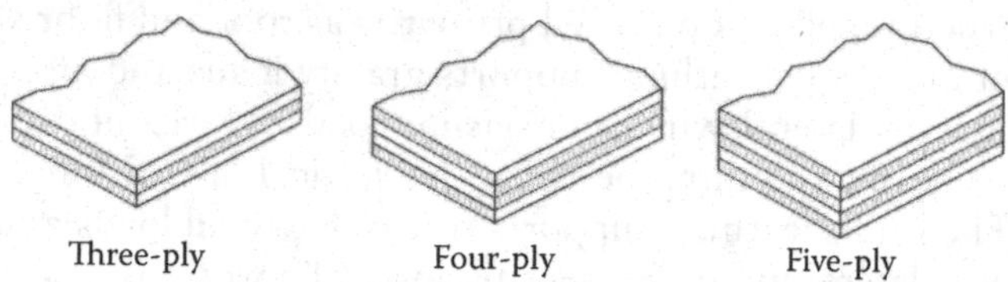

Figure 6.2 Plywood grain orientation in cross section.

6.1.2 Plywood species and grades

Plywood can be manufactured from a variety of wood species. PS-1 defines the various species and classifies them into five groups, with group 1 representing the strongest and group 5 representing the weakest. The term *Structural I* refers to plywood made from a group 1 species, and these panels are quite frequently specified by designers because of the certainty of the quality of wood species from which the component plies of the plywood are derived. The grade of the plywood is a function of the quality of the veneer. In PS-1, there are several veneer grades, which are summarized in Table 6.1 in order of decreasing quality. Grades C and D are the most commonly used for structural applications. Grade C is the minimum grade required for exterior use, and grade D is not permitted for exterior use. A plywood panel could have different veneer grades on each face. For example, the most common

Table 6.1 Plywood veneer grades

Grade	*Description*
N	Intended for natural finish; free from knots and other defects; synthetic fillers permitted for small defects
A	Suitable for painting; free from knots and other defects; synthetic fillers permitted for small defects
B	Slightly rough grain permitted; minor sanding and patching for up to 5% of panel area; small open defects permitted
C-plugged	Improved C grade; knotholes limited to $\frac{1}{4} \times \frac{1}{2}$ in.
C	Knotholes less than 1 in. in diameter permitted; knotholes of 1.5 in. permitted in limited cases
D	Knotholes less than 2 in. in diameter permitted; knotholes of 2.5 in. permitted in limited cases

structural types of plywood are C–C (grade C on both outer plies) and C–D (grade C on the face and grade D on the back).

There are four main exposure categories for plywood: exterior, exposure 1, exposure 2 [intermediate glue (IMG)], and interior. This classification is based on the ability of the bond in the glue to resist moisture. The plies of exterior plywood are glued with fully waterproof glue and are grade C or better. Exterior plywood should be used when the moisture content in service exceeds 18%, or when the panel is exposed permanently to the weather. The plies of exposure 1 plywood are glued with fully waterproof glue, but the veneer may be grade D. Exposure 1 plywood is recommended for use when the moisture content in service is relatively high or when the panels experience prolonged exposure to the weather during construction. The plies of exposure 2 or IMG plywood is glued with an adhesive that has intermediate resistance to moisture. It is recommended for protected applications where the panel is not exposed continuously to high humidity. Interior plywood is recommended for permanently protected interior applications. The glue used for interior plywood has a moderate resistance to moisture and can sustain only short periods of high humidity in service.

Section 2304.8.2 of the International Building Code (IBC) [3] requires that roof sheathing be constructed with exterior glue. For wall panels, exterior glue is to be used when the panel is exposed to the weather (see IBC Section 2304.6.1). When the wall panels are on an exterior wall but not exposed, the glue can be exposure 1. Interior structural panels use exposure 2 glue. Table 6.2 shows examples of various designations that could be used to identify plywood. Each sheet of plywood is typically stamped for identification. Sample grade stamps are shown in Figure 6.3.

6.1.3 Span rating

The span rating for floor and roof sheathing appears as two numbers separated by a sloped line (see Figure 6.3b and d). The left-hand number or numerator indicates the maximum recommended spacing of supports in inches when the panel is used for *roof sheathing* spanning over three or more supports with the grains perpendicular to the supports. The right-hand number represents the maximum recommended spacing of supports in inches when the panel is used as *floor sheathing* (subfloor).

The span rating for APA-rated STURD-I-FLOOR and APA-rated siding appears as a single number (see Figure 6.3a and c). APA-rated STURD-I-FLOOR is intended for single-floor applications that do not have floor finishes that require an underlayment (such as carpet

Table 6.2 Plywood sheathing grades

Type	*Description*	*Uses*
C–C EXT	Exterior plywood with C-grade veneer on the face, back, and cross-band plies; uses *Exterior* glue	Used for exterior exposed applications (exposed permanently to the weather)
C–C EXT STRUCT I	Same as the above but uses group 1 wood species	Same as the above
C–D INT	Interior plywood with C-grade veneer on the face plies; D-grade veneer on the back and cross-band plies; uses *Interior* glue	Interior applications only
C–DX	Same as C–D INT but uses *Exterior* glue	Used for interior applications and exterior protected applications (e.g., roof of covered building)
C–D INT STRUCT I	Same as C–D INT but uses only group 1 wood species	Interior applications only

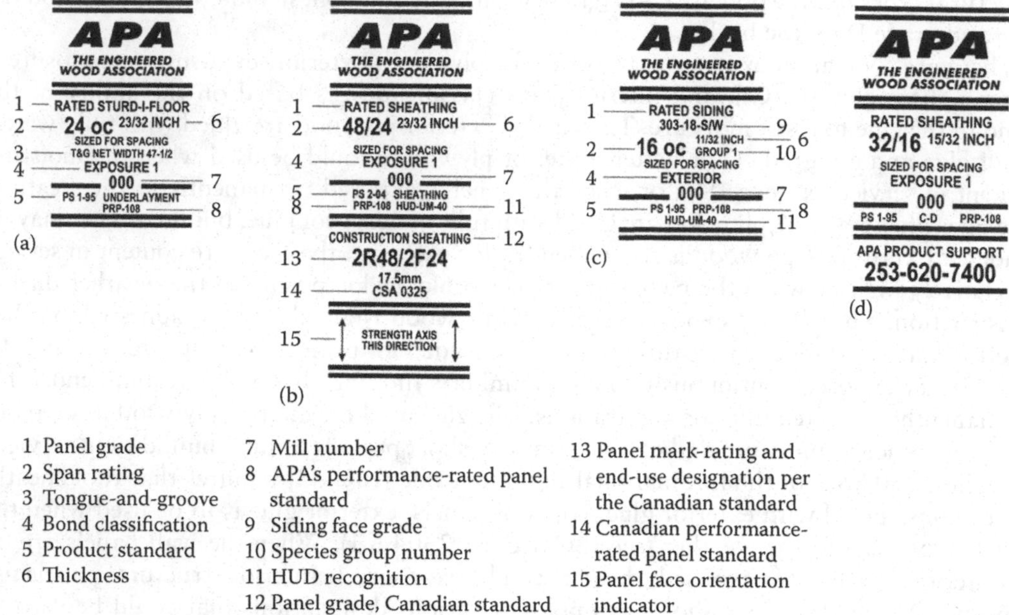

Figure 6.3 Plywood grain stamps. (Courtesy of APA–The Engineered Wood Association, Tacoma, WA.)

and pad) and are typically manufactured with span ratings of 16, 20, 24, 32, and 48 in. The span ratings for APA-rated STURD-I-FLOOR are valid when the panel spans over three or more supports with the grains perpendicular to the supports. APA-rated siding is typically manufactured with span ratings of 16 and 24 in. The span rating for APA-rated siding is valid when the panel is oriented parallel or perpendicular to the supports, provided that blocking is used on the unsupported edges.

6.2 ROOF SHEATHING: ANALYSIS AND DESIGN

To design roof sheathing for gravity loads, there are several options to choose from. References 1–8 all have various forms of load versus span tables, and some have section properties. In some limited applications, the section and material properties are used directly to determine the stresses and deflections, similar to the design of beams. In the use of some of these tables, the edge support of the panels needs to be considered. The edges of the panels can be supported by one of the following: tongue-and-grooved edges, panel edge clips, or wood blocking (see Figure 6.4). When panel edge clips are used, they are to be placed halfway between the supports, except that two panel clips are required at third points when the supports are 48 in. apart [see IBC Table 2304.8(3)].

The primary approach taken in this book for the design of wood panels for gravity and lateral loading will be the use of the tables given in the IBC, but other methods will be examined for comparison. For roof sheathing continuous over two or more spans perpendicular to the supports, Table 2304.8(3) should be used. This table correlates the span rating and panel thickness with uniform total and live loads, depending on the edge support condition (see Example 6.1). The load capacity shown is based on deflections of $L/180$ for total loads and $L/240$ for live loads.

For roof sheathing continuous over two or more spans parallel to the supports, Table 2304.8(5) should be used. The deflection criteria are based on $L/180$ for total loads and

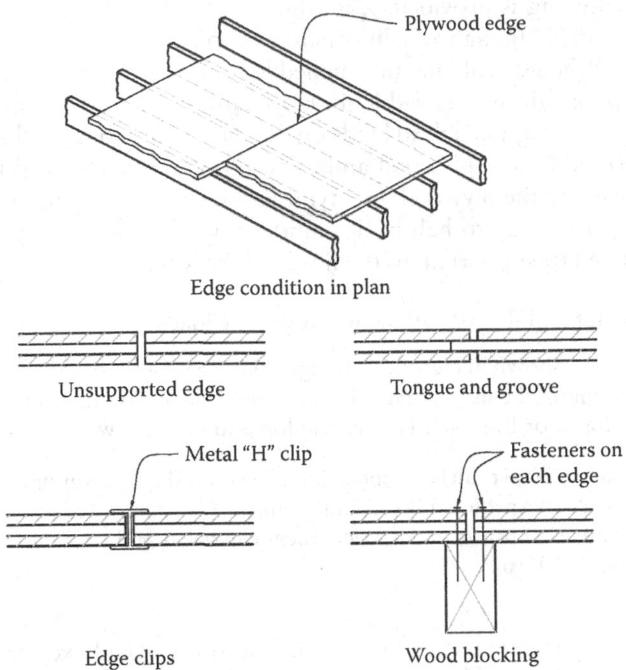

Figure 6.4 Plywood edge supports.

L/240 for live loads, and the design must include edge supports. This table is also based on five-ply panels, with the exception that four-ply panels can be used with a 15% reduction in some cases, as noted in the table.

6.3 FLOOR SHEATHING: ANALYSIS AND DESIGN

Two types of systems are used for floor sheathing. The first is a single-layer system, which is also referred to as a *single-floor* or *combination subfloor–underlayment*. In this system, the panel serves as the structural support and the underlayment for the floor finish. The span rating for panels intended for single-floor use is designated by a single number (see Figure 6.3a). The APA-rated STURD-I-FLOOR panels are intended for single-floor use. Single-floor systems are typically used when floor finishes have some structural strength and can therefore bridge between adjacent plywood panels (e.g., hardwood floor, concrete, or gypcrete topping).

The second type is a *two-layer system* in which the bottom layer, or *subfloor*, acts as the primary structural support. The upper layer, referred to as the *underlayment*, provides a surface for the floor finish and is typically ¼ to ½ in. thick. *Two-layer systems* are typically used where floor finishes have little structural strength (e.g., terrazzo).

As discussed previously, a variety of resources are available to design floor sheathing for gravity loads for various loading conditions and parameters. IBC Tables 2304.8(3) and 2304.8(4) will be used directly in the some examples to follow. These tables use the single-floor and two-layer span rating designations and are based on panels oriented perpendicular to the supports continuous over two or more spans. The total uniform load capacity is 100 psf, which is typically adequate for any wood floor structure, and the deflection is based on *L*/360. There is a unique case when the support spacing is 48 in., where the total uniform load capacity is reduced to 65 psf. When using these tables, the panels are required to have edge support (see Figure 6.4) unless one of the following is provided: ¼-in.-thick underlayment with the panel edges staggered over the subfloor, 1½ in. lightweight concrete topping over the subfloor, or a floor finish of ¾-in. wood strips. When calculating the applied load to the wood panel, note that the panel does not usually support all of the dead load. For example, in a typical floor system, the plywood might support direct applied dead loads on the top surface, but the floor joists and ceiling would not be supported by the plywood and the ceiling and mechanical fixtures would also not likely be supported by the plywood. In a typical floor system, the total dead load might be around 15 psf, but perhaps up to half of this amount would not be supported by the panels. The panels are assumed to support all of the live load, however.

Example 6.1: Roof and floor sheathing under gravity loads

For the framing plan shown in Figure 6.5, design (a) the roof and (b) the floor sheathing for gravity loads assuming the use of OSB. Use a live load deflection limit of *L*/240 for snow load and *L*/360 for floor live load. The vertical loads are as follows:

Roof dead load = 20 psf (includes mechanical/electrical and framing)
Roof snow load = 40 psf (on a horizontal plane)
Floor dead load = 20 psf (includes mechanical/electrical and framing)
Floor live load = 40 psf

Solution:
(a) *Roof sheathing.* The total dead load on the sheathing would exclude the weight of the mechanical and electrical fixtures and the framing. For this problem, that weight will be assumed to be 10 psf. Therefore, the dead load on the plywood sheathing is

$$DL_{sheathing} = 20 \text{ psf} - 10 \text{ psf} = \mathbf{10\ psf}$$

Snow load = **40 psf**

Total load = **50 psf**

The span of the plywood sheathing is 24 in. (spacing of framing).

Enter IBC Table 2304.8(3) under the "Roof" section and look for plywood with a maximum span greater than or equal to the span required (i.e., 24 in.). From this table, the following is selected:

Span rating = 24⁄16
C–DX plywood = 7⁄16-in.-thick (recall that roof panels have an exterior glue)
Edges without support

Allowable total load = **50 psf** = 50 psf applied **OK**

Allowable live load = **40 psf** = 40 psf applied **OK**

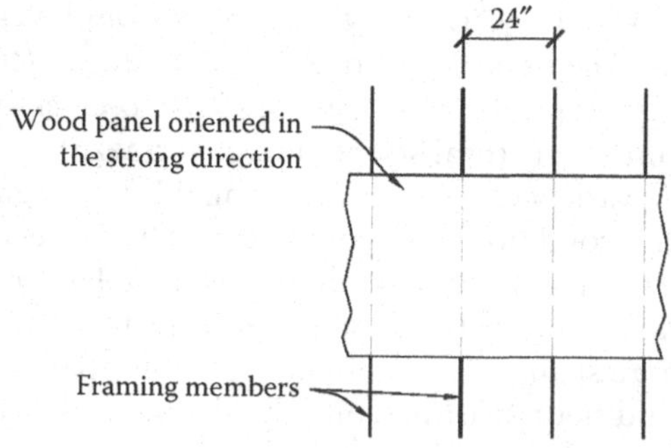

Figure 6.5 Partial roof and floor framing plans.

Use 7⁄16-in.-thick C–DX plywood with a span rating of 24⁄16 and unsupported edges. Note that a ½-in.-thick panel could also have been selected, but it is more economical to select the smaller panel for the same span rating. Because the span rating is the same with or without edge support, it is more economical to select a system without edge supports. Note also that the selection of a 24/0-rated panel would not be adequate because the total and live load capacities are 40 and 30 psf, respectively, which are less than the loads applied.

An alternate solution will now be examined based on Reference 8. From Table 2a, an allowable load of 51 psf is found for an OSB panel with a 24⁄16 span rating to meet the *L*/240 deflection limit. For bending and shear, the allowable load capacities are 80 and 133 psf, respectively, for a load duration factor of 1.0. These strength capacities could be increased by 1.15 because the loading is based on snow, but the deflection-based capacity of 51 psf controls is not adjusted for the duration of load. There is also no adjustment for the span condition because a three-span continuous is assumed for 24 in. spacing (or any spacing less than 32 in. o.c., according to this standard).

The panel thickness associated with a 24⁄16 span rating is found in Table 11 of Reference 5. The acceptable thicknesses are 7⁄16, 15⁄32, and ½ in. Section 4.5.6 of this standard discusses edge support requirements, but the use of edge support is only recommended and not required.

Alternate solution summary: Select 7⁄16 in. CD–X, with a 24⁄16 span rating.

(b) *Floor sheathing*. The floor joists are spaced at 24 in. o.c. Similar to the roof sheathing solution, the total dead load on the sheathing would exclude the weight of the mechanical and electrical fixtures and the framing. Assuming that weight to be 10 psf, the dead load on the plywood sheathing is

$$DL_{sheathing} = 20 \text{ psf} - 10 \text{ psf} = \mathbf{10 \text{ psf}}$$

Live load = **40 psf**

Total load = **50 psf**

Two types of floor systems are available.

1. *Two-layer floor system* [use IBC Table 2304.8(3)]. To design a two-layer system, enter the last column of IBC Table 2304.8(3) and look for a row with a plywood span greater than or equal to 24 in. From the table, the following is selected:

   Span rating = 48⁄24.
   C–D plywood = 23⁄32 in. thick

   Allowable total load (see footnote in Table 2304.8(3)) = 100 psf > 50 psf **OK**

   According to footnote (d) in IBC Table 2304.8(3), the edge support is required unless one of the following is provided: ¼-in.-thick underlayment with the panel edges staggered over the subfloor, 1½-in. lightweight concrete topping over the subfloor, or a floor finish of ¾-in. wood strips.

2. *Single-layer floor system* [either IBC Table 2304.8(3) or 2304.8(4)]. Using IBC Table 2304.8(3) to find a panel with a maximum span greater than or equal to the span required (i.e., 24 in.), the following is selected:

   Span rating = 24
   C–D plywood = 23⁄32 in. thick

   Allowable total load (see footnote in Table 2304.8(3)) = 100 psf > 50 psf **OK**

   Alternatively, the following is selected from IBC Table 2304.8(4):
   Assuming group 1 species plywood ⇒ ¾-in.-thick C–D plywood
   Assuming group 2 or 3 species plywood ⇒ 7⁄8-in.-thick C–D plywood
   Assuming group 4 species plywood ⇒ 1-in.-thick C–D plywood

Allowable total load (see footnote in Table 2304.8(4)) = **100 psf** > 50 psf

Similar to the two-layer system, the edge supports are required unless one of the following is provided: ¼-in.-thick underlayment with the panel edges staggered over the subfloor, 1½ in. of lightweight concrete topping over the subfloor, or a floor finish of ¾-in. wood strips.

An alternate solution will now be examined based on Reference 8. From Table 2b, an allowable load of 50 psf is found for an OSB panel with a 32/16 span rating to meet the *L*/360 deflection limit. For bending and shear, the allowable load capacities are 93 and 147 psf, respectively, for a load duration factor of 1.0. Therefore, the deflection-based capacity is 50 psf. There is no adjustment for the span condition because a three-span continuous is assumed for 24 in. spacing.

The panel thickness associated with a 32/16 span rating is found in Table 11 of Reference 5. The acceptable thicknesses are 15/32, ½, 19/32, and ⅝ in. Because this is floor loading, this standard recommends to use the panel edge supports, which is a good practice to always include fpr floor loading.

Using a single floor grade, a span rating of 16 in. o.c. is selected from Table 2c of Reference 8. This has an allowable load of 65 psf, which is found for an OSB STURD-I-FLOOR panel where the design is controlled by the *L*/360 deflection limit. For bending and shear, the allowable load capacities 104 and 182 psf, respectively, for a load duration factor of 1.0. Therefore, the deflection-based capacity is 65 psf. There is no adjustment for the span condition because a three-span continuous is assumed for 24 in. spacing. The panel thickness associated with a 16 in. o.c. span rating is found in Table 11 of Reference 5. The acceptable thicknesses are 19/32 and ⅝ in.

Alternate solution summary: Select either a 15/32 in. C–D panel with a 32/16 span rating or a 19/32 in. C–D panel with a 16 in. span rating.

6.3.1 Extended use of the IBC tables for gravity loads on sheathing

To extend the use of IBC Tables 2304.8(3) and 2304.8(4) beyond the loads and span combinations given in the table, the following allowable total and live load equations could be used assuming moment equivalence:

$$W_{a\mathrm{TL}} = \left(\frac{L_{\max}}{L_a}\right)^2 W_{\mathrm{TL}}$$

$$W_{a\mathrm{LL}} = \left(\frac{L_{\max}}{L_a}\right)^2 W_{\mathrm{LL}}$$

where:

$W_{a\mathrm{TL}}$ is the adjusted total load capacity (psf)
$W_{a\mathrm{LL}}$ is the adjusted live load capacity (psf)
W_{TL} is the total load capacity corresponding to the span rating (psf)
W_{LL} is the live load capacity corresponding to the span rating (psf)
$L_{\max}$ is the maximum allowable span (span rating) (in.)
L_a is the actual span (joist spacing) (in.)

6.4 PANEL ATTACHMENT

The *Wood Structural Panels Supplement* [9] gives the minimum nailing requirements for roof and floor sheathing. Note that greater nailing requirements may be required based on the panel strength as a diaphragm (see Section 6.5). In addition to nailing, the panels can

Table 6.3 Minimum nailing requirements for roof and floor sheathing

		Nail spacing (in.)	
Panel description (span rating, panel thickness)	*Nail size*	*Panel edges*	*Intermediate supports*
Single floor			
16, 20, 24, ¾ in. thick or less	6d	6	12
24, ⅞, or 1 in. thick	8d	6	12
32-, 48-, and 32-in. c-c span or less	8d	6	12
48-in. and 48-in. c-c span or less	8d	6	6
Subfloor			
7⁄16 to ⅓ in. thick	6d	6	12
⅞ in. thick or less	8d	6	12
Greater than ⅞ in. thick	10d	6	6
Underlayment			
¼ in.	3d	3	6
11⁄32	3d	6	8
Roof			
5⁄16 to 1 in. thick	8d	6	12
Greater than 1 in. thick	8d	6	12
Spacing 48 in. and greater	8d	6	6

Source: Adapted from AWC, *National Design Specification Supplement: Design Values for Wood Construction*, American Wood Council, Leesburg, VA, 2015, Table 6.1.

be glued to the supports, which reduce squeaking due to the plywood slipping on the supports, and it also helps in the control of floor vibrations (see Section 4.7). A summary of the nailing requirements is shown in Table 6.3. Nails at the intermediate supports are typically referred to as *field nailing*. Nails at the edges could be along a main framing member and at the blocking, if blocking is provided. Fasteners should be spaced at least ⅜ in. from the panel edges (see Section 4.2.7 of SDPWS of [6]).

Example 6.2: Extended use of the IBC tables for gravity loads on sheathing

(a) Given a 7⁄16-in.-thick C–DX OSB sheathing (edges unsupported) with a span rating of 24⁄16, determine the allowable total load and allowable live load if it is used as a roof sheathing with a span of 19.2 in.

(b) Repeat part (a) assuming that the plywood is used as floor sheathing (two-layer system) with a span of 12 in.

Solution:

(a) *Roof sheathing.* From IBC Table 2304.7(3), the following is obtained for 24⁄16 roof sheathing (edges unsupported):

$$L_{max} = 24 \text{ in. (maximum span of the plywood when used as a roof sheathing)}$$

$$W_{TL} = 50 \text{ psf}$$

$$W_{LL} = 40 \text{ psf}$$

The actual span of the plywood sheathing is L_a = 19.2 in. The adjusted allowable total and live loads are calculated as follows:

$$W_{aTL} = \left(\frac{L_{\max}}{L_a}\right)^2 W_{TL} = \left(\frac{24}{19.2}\right)^2 (50) = \mathbf{78\ psf} \text{ (adjusted total load capacity)}$$

$$W_{aLL} = \left(\frac{L_{\max}}{L_a}\right)^2 + W_{LL} = \left(\frac{24}{19.2}\right)^2 (40) = \mathbf{62\ psf} \text{ (adjusted live load capacity)}$$

As an alternate solution, Reference 8 will be used. From Table 2a of Reference 8, the following is obtained (Table 6.4):

Total load capacity = 105 psf

(b) *Floor sheathing*. From IBC Table 2304.8(3), the following is obtained for 24⁄16 floor sheathing:

L_{max} = 16 in. (maximum span of the plywood when used as a floor sheathing)

L_a = 12 in.

W_{TL} = 100 psf (see footnote for tabulated total load capacity)

The actual span of the plywood sheathing is L_a = 16 in. The adjusted allowable total load is calculated as follows:

$$W_{aTL} = \left(\frac{L_{\max}}{L_a}\right)^2 W_{TL} = \left(\frac{16}{12}\right)^2 (100) = \mathbf{177\ psf} \text{ (adjusted total load capacity)}$$

As an alternate solution, Reference 8 will be used. From Table 2a of Reference 8, the following is obtained (Table 6.5):

Total load capacity = 286 psf

Table 6.4 Plywood design summary for Example 6.2a

Parameter	*Total load capacity*	*Notes*
L/240 deflection	105 psf	Controls
Bending	125 psf	Could be increased for duration of load
Shear	169 psf	Could be increased for duration of load

Source: APA (2011), *Load-Span Tables for APA Structural-Use Panels (Q225G)*, APA—The Engineered Wood Association, Tacoma, WA.

Table 6.5 Plywood design summary for Example 6.2b

Parameter	*Total load capacity*	*Notes*
L/360 deflection	339 psf	
Bending	321 psf	
Shear	286 psf	Controls

Source: APA (2011), *Load-Span Tables for APA Structural-Use Panels (Q225G)*, APA—The Engineered Wood Association, Tacoma, WA.

6.5 HORIZONTAL DIAPHRAGMS

In a general structural sense, a diaphragm is a plate like structural element that is subjected to in-plane loading. In concrete building or a steel-framed building with a concrete slab, the concrete slab acts as the diaphragm. In a wood structure, the diaphragm is mainly the roof and floor sheathing, but drag struts and chords are also critical members in the diaphragm and load path (Figure 6.6). The sequence of the load path can be summarized as follows:

- Lateral wind or seismic loads are imposed on the diaphragm.
  - For wind loads, the wind pressure is applied to the wall sheathing and wall studs, which is then transferred to the diaphragm and the foundation at the base.
  - For seismic loads, the lateral force is assumed to transfer directly to the diaphragm.
- Loads are transferred from the diaphragm sheathing to the floor framing through the diaphragm fasteners.
- Loads are transferred from the framing to the drag struts (where drag struts are not present, the loads are transferred directly to the shear walls).
- Loads are transferred from the drag struts to the shear walls (vertical diaphragms).
- Loads are transferred from the wall sheathing to the wall framing through the shear wall fasteners.
- Loads are transferred from the wall framing to the sill plate.
- Loads are transferred from the sill plate to the anchor bolts.
- Loads are transferred from the anchor bolts to the foundation.

The design of each of these elements and connections is necessary to maintain a continuous load path to the foundation. Each of these items is covered later in the book.

The diaphragm can be modeled as a beam with simple supports at the shear wall locations. This beam consists of a web, which resists the shear (sheathing), flanges that resist the tension and compression forces resulting from a moment couple (chords), and supports that transfer the load to the shear walls (drag struts or collectors). Figure 6.7 illustrates the beam action of the diaphragm.

In a typical wood-framed structure, the shear force is resolved into a unit shear across the diaphragm. The unit shear can be defined as follows:

$$v_d = \frac{V}{b}$$

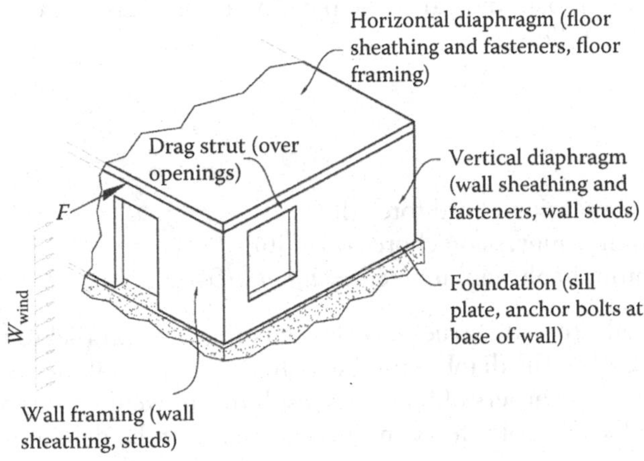

Figure 6.6 Load path in a building.

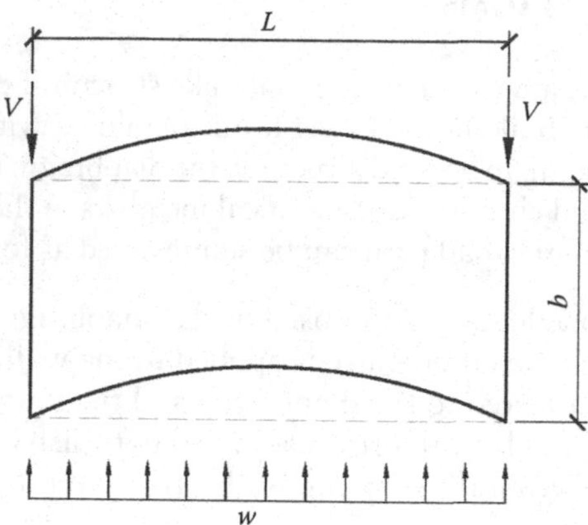

Figure 6.7 Beam action of a diaphragm.

where:

v_d is the unit shear in the diaphragm (lb/ft)
V is the shear force (= $wL/2$ at supports) (lb)
b is the width of the diaphragm parallel to the lateral force (distance between the chords) (ft)
L is the length of the diaphragm perpendicular to the lateral force (distance between the shear walls) (ft)

The shear force V transferred to the shear walls is distributed uniformly in a similar manner. The unit shear in the shear walls can be defined as follows:

$$v_w = \frac{V}{\Sigma L_w}$$

where:

v_w is the unit shear in the shear walls along the diaphragm support line (lb/ft)
V is the total shear force along the diaphragm support line (lb)
ΣL_w is the summation of the shear wall lengths along the diaphragm support line (ft)

The top plates in the walls act as the chords and drag struts. The chords are oriented perpendicular to the lateral load and carry the bending moment in the diaphragm. The chord force can be defined as follows:

$$T = C = \frac{M}{b}$$

where:

T is the maximum tension chord force (lb)
C is the maximum compression chord force (lb)
M is the maximum diaphragm moment = $wL^2/8$ (ft-lb)

The drag struts or collectors are structural elements that are parallel to the lateral load and they help to drag or collect the diaphragm shears to the shear walls across door and window openings and at reentrant corners of buildings, and thus prevent stress concentrations at the ends of the shear walls. The total force in the drag strut can be defined as

$$F_{ds} = v_d L_{ds}$$

where:

F_{ds} is the total force in the drag strut (lb)
v_d is the unit shear in the diaphragm (lb/ft)
L_{ds} is the length of the drag strut (ft)

The total force F_{ds} can be entirely a tension force or a compression force, or it can be distributed as a tension force at one end of the drag strut and a compression force at the other end. The total force is distributed in this latter manner when the drag strut is located between two shear walls and is due to the combined *pushing* and *pulling* of each shear wall. See Example 6.3 for further explanation of the calculation of drag strut forces.

6.5.1 Horizontal diaphragm strength

In the course of a building design, the floor and roof sheathing thicknesses are usually sized for gravity loads prior to calculating lateral loads. For most cases, this is typically economical because the panel thickness is not the most critical factor in the strength of the diaphragm. The limiting factor is usually the capacity of the nails or staples that fasten the sheathing to the framing members. In general, there is a direct relationship between the amount of fasteners and the shear strength of the diaphragm.

There are two basic types of diaphragms: blocked and unblocked (see Figure 6.8). A *blocked diaphragm* is one in which the sheathing is fastened to a wood framing member under all four panel edges as well as at the intermediate supports. An *unblocked diaphragm* is fastened only at two panel edges and at the intermediate supports. It is usually more economical to design an unblocked diaphragm because of the added field labor required to provide blocking.

With reference to Table 6.3, the minimum nailing requirements are the same for blocked and unblocked diaphragms. However, decreasing the fastener spacing at the panel edges can increase the diaphragm capacity. SDPWS Table 4.2A, 4.2B, or 4.2C [6] lists the nominal shear capacities for various diaphragm types. Several diaphragm configurations and loading conditions are also identified in this table (load cases 1 through 6). The loading direction and sheathing orientation will define which load case to use (see Figure 6.9).

The nominal diaphragm shear strength values shown in SDPWS Table 4.2A, 4.2B, or 4.2C show an increase of 40% for wind design. The reason for this increase is due to the historically good performance of wood diaphragms during high-wind events. Wind and seismic loads are listed separately in this table, and because they are nominal values, they have to be adjusted for ASD or LRFD loads. For ASD, the allowable unit shear capacity is determined by dividing the value in the table by $\Omega = 2.0$. For LRFD, the design unit shear strength is found by multiplying the value in the table by $\phi_D = 0.8$ plus any other adjustment factors in the footnotes.

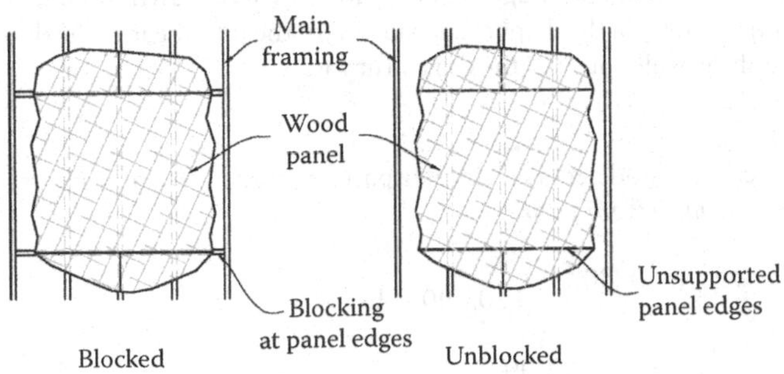

Figure 6.8 Blocked and unblocked diaphragms.

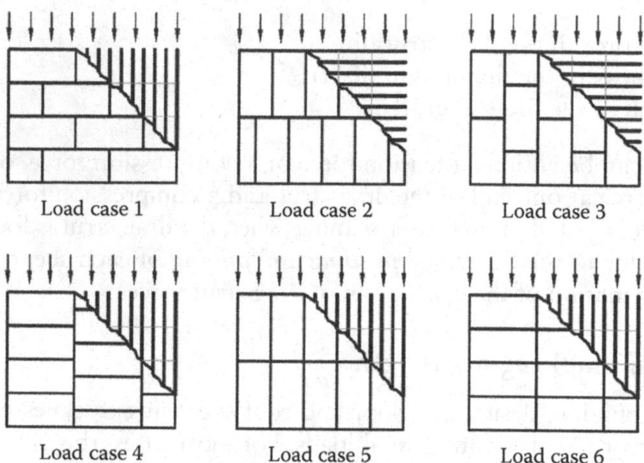

Figure 6.9 Diaphragm load cases.

The nominal diaphragm shear strength values shown in the table are based on framing members of Douglas Fir-Larch or Southern Pine when using nails. For nails in framing members of other species, multiply the following specific gravity adjustment factor (SGAF) by the tabulated unit shear for the *actual* grade of plywood specified.

$$\text{SGAF} = 1 - (0.5 - G) \leq 1.0 \text{ for nails}$$

where:

SGAF is the specific gravity adjustment factor
G is the specific gravity of wood (NDS Table 12.3.2A) [10]

For wood panels used with staples, use IBC Table 2306.2(1) to determine the allowable diaphragm shear capacity. For staples in framing members of other species, multiply the following specific gravity adjustment factors by the tabulated allowable unit shear for *structural* I grade, regardless of the plywood grade specified.

$$\text{SGAF} = \begin{cases} 0.82 & \text{for staples when } G \geq 0.42 \\ 0.65 & \text{for staples when } G < 0.42 \end{cases}$$

Example 6.3: Horizontal diaphragm forces

Based on the floor plan and diaphragm loading due to wind shown in Figures 6.10 through 6.12, calculate (a) the force in the diaphragm chords, (b) the unit shear in the diaphragm, (c) the unit shear in the shear walls, and (d) the drag strut forces.

Solution:

The diaphragm can be modeled as a simple span beam, as shown in Figure 6.9.

(a) *Force in the diaphragm chords*

$$M = \frac{wL^2}{8} = \frac{(150)(80)^2}{8} = 120{,}000 \text{ ft-lb}$$

$$T_{CG} = C_{AD} = \frac{M}{b} = \frac{120{,}000}{30} = \mathbf{4000\ lb}$$

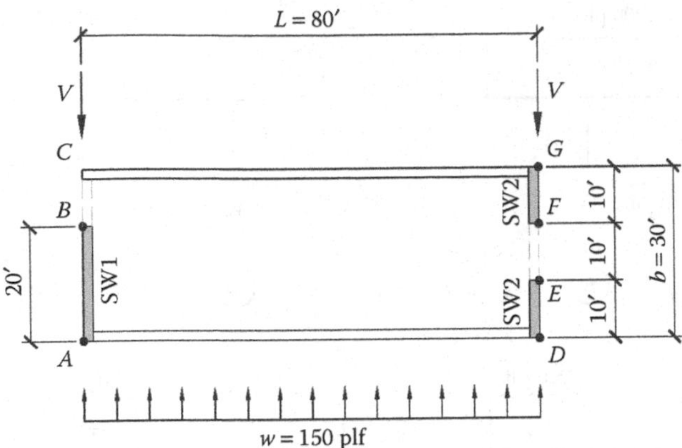

Figure 6.10 Example floor plan.

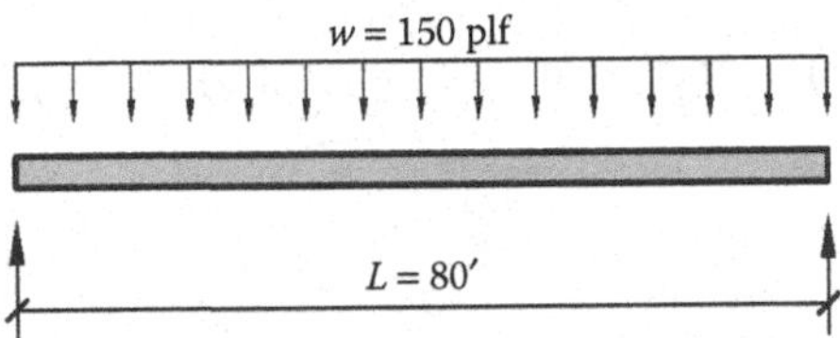

Figure 6.11 Example diaphragm loading.

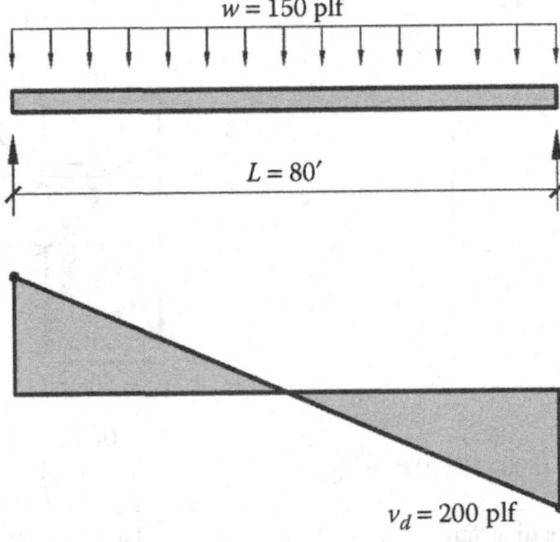

Figure 6.12 Variation in the unit shear across a diaphragm.

Note: Since the lateral wind load can act in the reverse direction, chords *AD* and *CG* would have to be designed for both tension and compression loads.

(b) *Unit shear in the diaphragm*

$$V_{max} = \frac{wL}{2} = \frac{(150)(80)}{2} = 6000 \text{ lb}$$

$$v_d = \frac{V}{b} = \frac{6000}{30} = \mathbf{200\ lb/ft} \text{ (maximum unit shear)}$$

The variation in unit shear can be shown as in Figure 6.13.

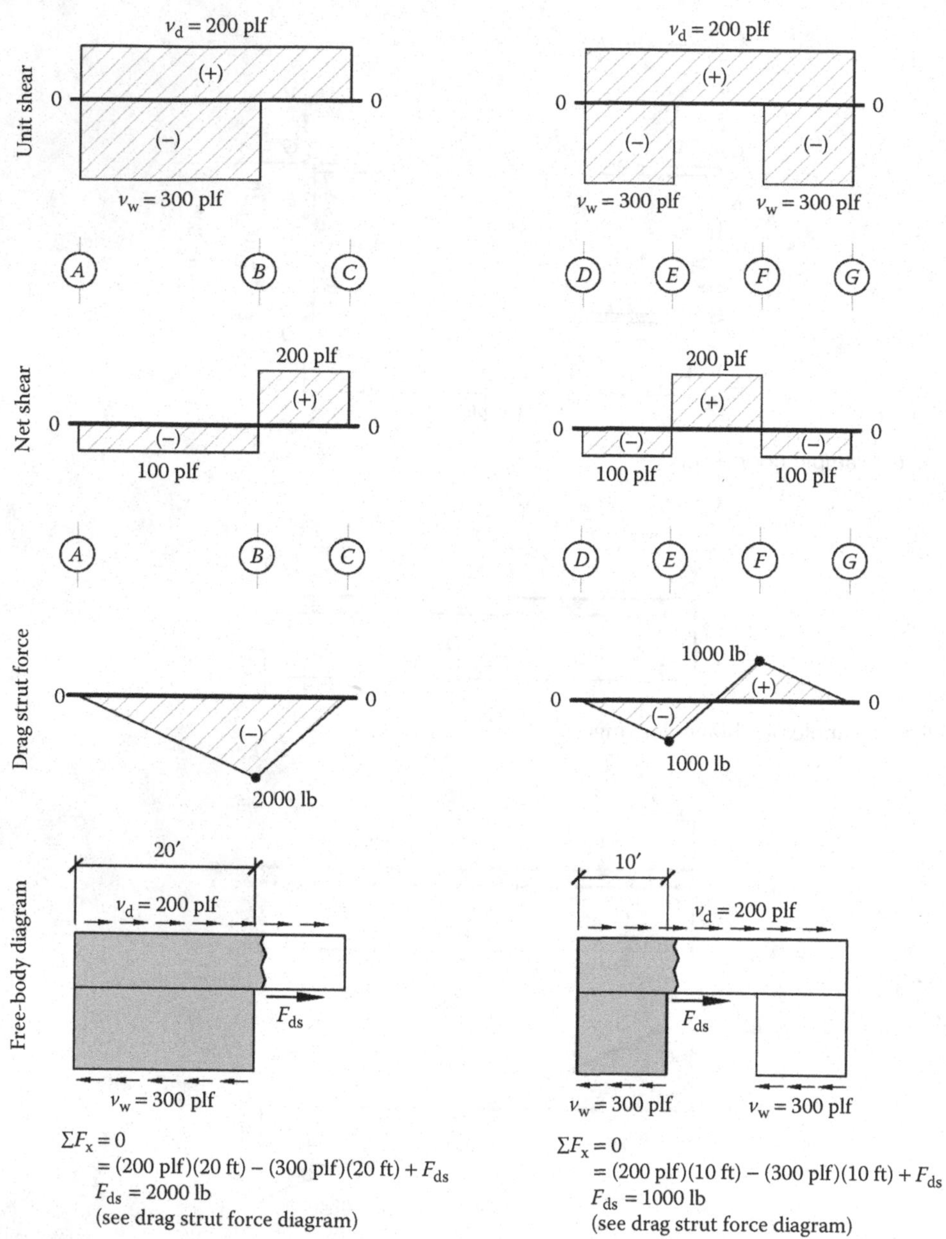

Figure 6.13 Unit shear, net shear, drag strut force, and free-body diagrams.

(c) *Unit shear in the shear walls*
Along line *ABC:*

$$v_w = \frac{V}{\sum L_w} = \frac{6000}{20} = \mathbf{300\ lb/ft}$$

Along line *DEFG:*

$$v_w = \frac{V}{\sum L_w} = \frac{6000}{10+10} = 300 \text{ lb/ft}$$

(d) *Drag strut forces*
The total force in each drag strut is

$$F_{CB} = F_{EF} = v_d L_{ds} = (200)(10) = \mathbf{2000\ lb}$$

To determine the maximum drag strut force, the variation in diaphragm and shear wall unit shears has to be plotted as shown in Figure 6.11. Therefore, the maximum drag strut forces are as follows:

$F_{CB} = \mathbf{2000\ lb}$ (tension or compression)
$F_{EF} = \mathbf{1000\ lb}$ (tension or compression)

The maximum drag strut force can also be determined by using the free body diagrams in Figure 6.11. Equilibrium of forces in the free body diagrams yields the same results as was obtained using the unit shear diagram. Note that the total force in the drag strut *CB* is 2000 lb and can be either tension or compression.

The code limits the aspect ratio of horizontal diaphragms. When all of the panel edges are supported and fastened, the maximum ratio of the diaphragm length to the width (L/b) (see Figure 6.14) is limited to 4. For unblocked diaphragms, the ratio is limited to 3 (see SDPWS Table 4.2.4). The distribution of lateral forces to the shear walls in a building is dependent on the rigidity of the horizontal diaphragm, and the diaphragm rigidity can be classified as flexible or rigid based on the ratio of diaphragm lateral deflection to the shear wall lateral deflection. For the design of wood-framed buildings, it is common in practice to assume a flexible diaphragm, and this assumption has been confirmed by others [3]. ASCE 7-10 [11] permits wood structural panels to be classified as flexible diaphragms, provided they have no more than 1½ in. of nonstructural topping. Furthermore, the simplified analysis procedure for seismic loads presented in Chapter 2 assumes a flexible diaphragm (ASCE 7, Section 12.3). This procedure is permitted for wood-framed structures that are three stories and less in height, which is the category that most wood structures fall into.

For a flexible diaphragm, the distribution of lateral seismic forces is based on the plan tributary widths or areas of each shear wall. For wind loads, the tributary area of the shear walls associated with the vertical exterior wall surfaces perpendicular to the wind direction is used. With reference to the flexible diaphragm shown in Figure 6.15, the lateral force is distributed equally to each shear wall, even though one wall is longer than the other. Because the lateral force to each shear wall is equal, a shorter wall will have a larger lateral deflection. In this case, the diaphragm is assumed to be flexible enough to absorb this unbalanced shear wall deflection.

For a rigid diaphragm, a torsional moment is developed and must be considered in the design of shear walls. This moment develops when the centroid of the lateral load is eccentric

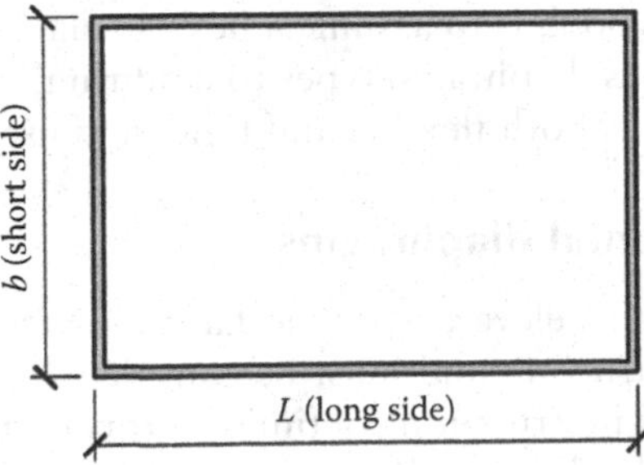

Figure 6.14 Diaphragm aspect ratios.

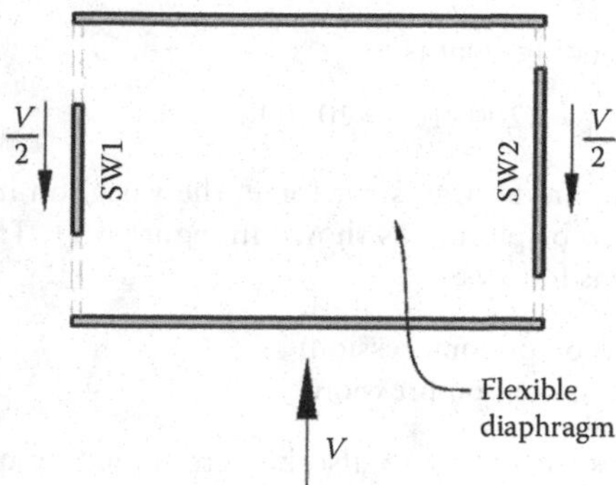

Figure 6.15 Flexible diaphragm vs. shear wall force distribution.

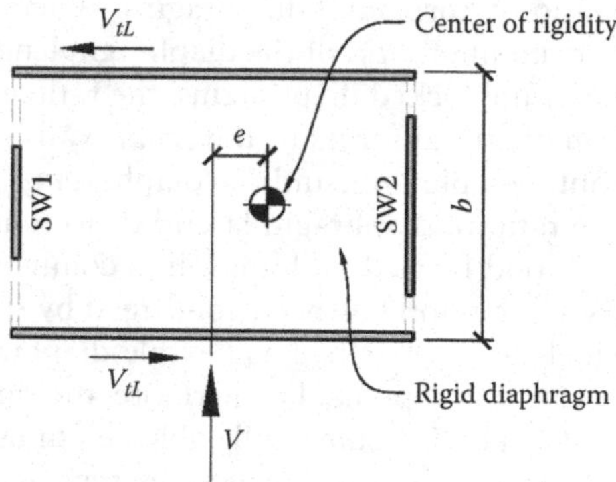

Figure 6.16 Rigid diaphragm with an eccentric lateral load.

to the center of rigidity of the shear walls. One approximate method for resisting this torsional moment is a couple in the shear walls oriented perpendicular to the lateral load. This is shown in Figure 6.16, in which the center of rigidity of the transverse shear walls does not coincide with the centroid of the lateral load. This results in a torsional moment that is resolved by forces in the longitudinal walls as shown in the figure.

In some cases, a rigid diaphragm is necessary, such as an open-front structure. SDPWS Section 4.2.5 limits the aspect ratios of rigid diaphragms in open-front structures. The length of the diaphragm normal to the opening cannot exceed 25 ft, and the length-to-width ratio cannot exceed 1.0 for single-story structures and 0.67 for structures taller than one story (see Figure 6.17).

The approach taken in this book is to assume a flexible diaphragm. However, the reader is encouraged to examine various diaphragm types to confirm this assumption. One approach is to analyze the diaphragm as both flexible and rigid, and use the larger forces for design.

6.5.2 Openings in horizontal diaphragms

Floor openings such as for stairs, elevators, or mechanical chases, and roof openings for skylights may occur in a wood structure and must be considered in the design of the horizontal diaphragm. As with any break in structural continuity, proper attention should be given to the design and detailing of these diaphragms due to the localized stress increases that occur, and it is recommended that drag struts be provided along the edges of the opening. The ASCE 7 load

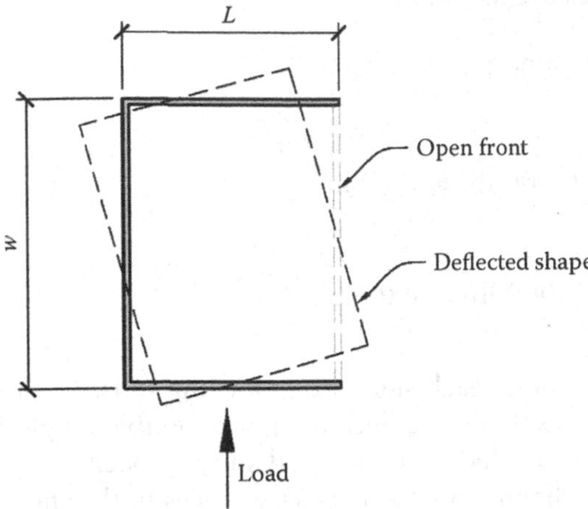

Figure 6.17 Rigid diaphragm in an open-front structure.

specification classifies a building with the area of diaphragm opening greater than 50% of the total diaphragm area as having plan structural irregularity for the purposes of seismic load calculations (ASCE 7, Section 12.3). Most wood buildings have diaphragm openings that are much smaller than 50%, and thus can be classified as regular. Consider the diaphragm loading and shear diagram as shown in Figure 6.18.

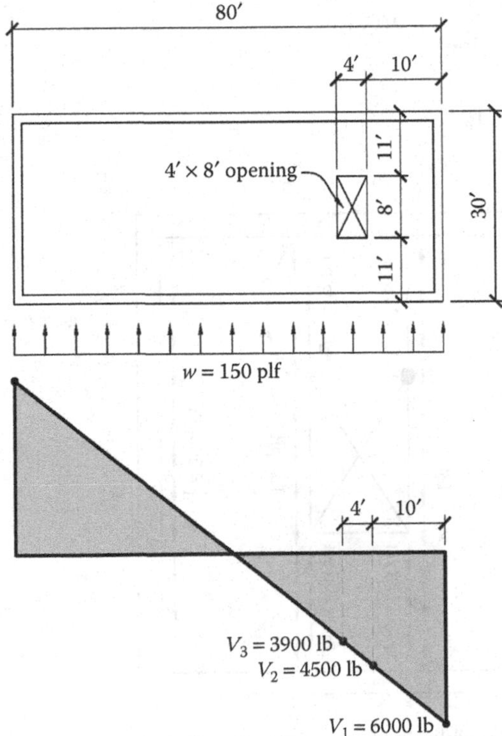

Figure 6.18 Opening in a diaphragm.

Where the shear forces are as follows:

$$v_1 = \frac{(150\text{ plf})(80\text{ ft})}{2} = 6000\text{ lb}$$

$$v_2 = \left(\frac{40\text{ ft} - 10\text{ ft}}{40\text{ ft}}\right)(6000\text{ lb}) = 4500\text{ lb}$$

$$v_3 = \left(\frac{40\text{ ft} - 14\text{ ft}}{40\text{ ft}}\right)(6000\text{ lb}) = 3900\text{ lb}$$

Note that the total shear along each side of the opening must be distributed over a reduced length of diaphragm that excludes the opening. For a flexible diaphragm, the fastener layout may have to be modified to achieve the required shear capacity on each side of the opening (Figure 6.19). For a rigid diaphragm, the increased forces in the chords due to the diaphragm opening have to be transferred into the diaphragm, and this will lead to an increase in the number of fasteners used to connect the diaphragm to the chords. A unit shear diagram can be constructed as shown in Figure 6.20.

The unit shears are as follows:

$$v_1 = \frac{6000\text{ lb}}{30\text{ ft}} = 200\text{ plf}$$

$$v_{2a} = \frac{4500\text{ lb}}{30\text{ ft}} = 150\text{ plf}$$

$$v_{2b} = \frac{4500\text{ lb}}{(30\text{ ft} - 8\text{ ft opening})} = 205\text{ plf}$$

$$v_{3a} = \frac{3900\text{ lb}}{30\text{ ft}} = 130\text{ plf}$$

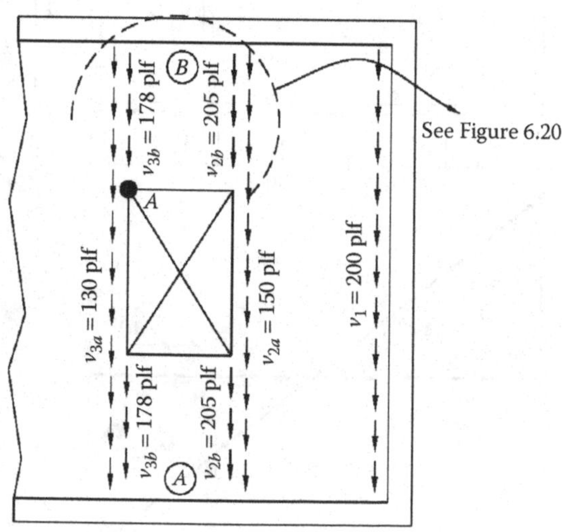

Figure 6.19 Unit shear in a flexible diaphragm with an opening.

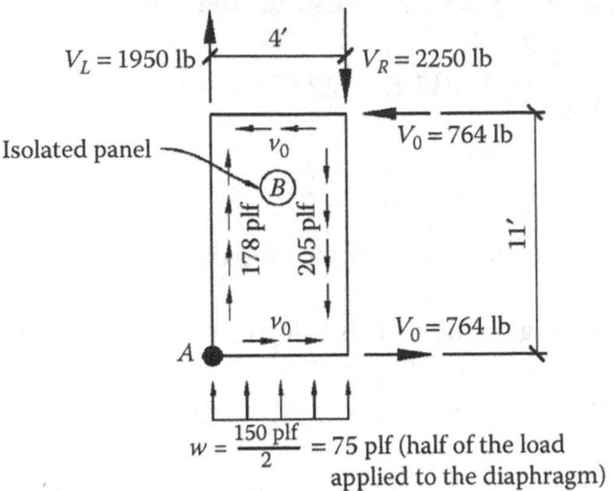

Figure 6.20 Unit shear in a rigid diaphragm with an opening.

$$v_{3b} = \frac{3900 \text{ lb}}{(30 \text{ ft} - 8 \text{ ft opening})} = 178 \text{ plf}$$

These unit shear values would now be compared to IBC Table 2306.3.1 to select an appropriate fastener layout for the flexible diaphragm case.

For a rigid diaphragm, the unit shear diagram in the transverse direction remains as shown in Figure 6.20. However, there is a torsional moment that is developed in each diaphragm panel on each side of the opening. This torsional moment can be resolved into a couple of forces along the panel edges that are perpendicular to the lateral load (see Figure 6.20). In the figure,

V_L = total force along the left side of the panel

$$= \frac{11 \text{ ft}}{11 \text{ ft} + 11 \text{ ft}}(3900 \text{ lb}) = 1950 \text{ lb}$$

V_R = total force along the right side of the panel

$$= \frac{11 \text{ ft}}{11 \text{ ft} + 11 \text{ ft}}(4500 \text{ lb}) = 2250 \text{ lb}$$

The total force on the diaphragm along each side of the diaphragm opening is proportional to the length of the diaphragm beyond the opening and parallel to the lateral load.

v_L = unit shear along the left side of the panel

$$= \frac{1950 \text{ lb}}{11 \text{ ft}} = 178 \text{ plf}$$

v_R = unit shear along the right side of the panel

$$= \frac{2250 \text{ lb}}{11 \text{ ft}} = 205 \text{ plf}$$

Considering the equilibrium of panel B and summing moments about point A yields

$$\Sigma M_A = (75\text{ plf})(4\text{ ft})\left(\frac{4\text{ ft}}{2}\right) + (V_0)(11\text{ ft}) - (2250\text{ lb})(4\text{ ft}) = 0$$

Therefore,

$V_0 = 764$ lb (total force to be added to the chords)

The unit shear in the diaphragm due to this additional force is as follows:

$$v_0 = \frac{764\text{ lb}}{4'} = 191\text{ plf}$$

The value of 191 plf for the unit shear values along the panel sides would now be compared to Table 4.2A, 4.2B, or 4.2C of the SDPWS to select an appropriate fastener layout.

6.5.3 Chords and drag struts

The chords and drag struts are the top plates in horizontal diaphragms. The chords are perpendicular to the direction of the lateral loads, whereas the drag struts are parallel to the lateral load. The drag struts are used to ensure continuity of the lateral load path in a diaphragm, and the chords act as the tension and compression flanges of the horizontal diaphragm as it bends in-plane due to the lateral loads. These top plates are usually not continuous, that is, they have to be spliced because of length limitations. Staggered splices are normally used (Figure 6.21) and the plates are commonly nailed together in the field.

For nailed or bolted double top plates, only one member is effective to resist the axial tension loads because of the splice. However, for double top plates connected with splice plates, both members will be effective in resisting the axial tension forces. For large chord or drag strut forces, more than two plates may be required. For a triple top plate system, only two members are effective in resisting the axial tension loads.

The chords and drag struts must be designed for the worst-case tension and compression loads. Typically, the tension load controls because the tension strength is generally less than the compression strength. Furthermore, the tension force cannot be transferred across a splice, whereas the compression force can be transferred by end bearing. The ASCE 7 load standard requires that certain structural elements, such as drag struts, be designed for special seismic forces E_m, but light-framed construction, which includes wood buildings, is exempt from this requirement.

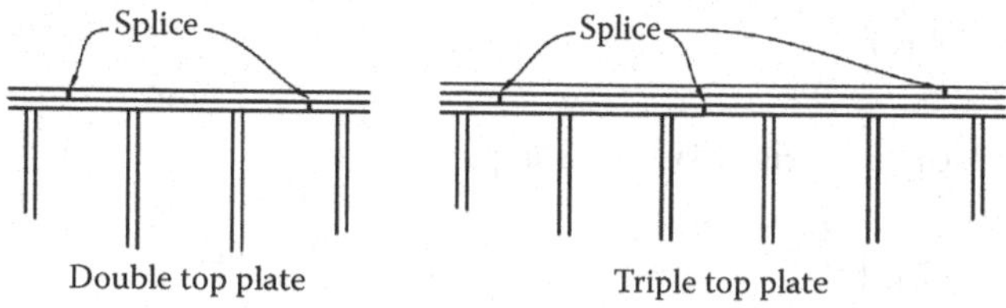

Figure 6.21 Double and triple top plates.

Example 6.4: Design of horizontal diaphragm elements

Design the roof diaphragm of the building shown in Figure 6.22 for a lateral wind load of 20 psf. The floor-to-floor height of the building is 10 ft, and the height from the roof datum or level to the peak of the roof truss is 6 ft. Assume Hem-Fir for all wood framing and that normal temperature and moisture conditions apply.

Roof dead load = 20 psf

Roof snow load = 40 psf (on a horizontal plane)

Floor dead load = 20 psf

Floor live load = 40 psf

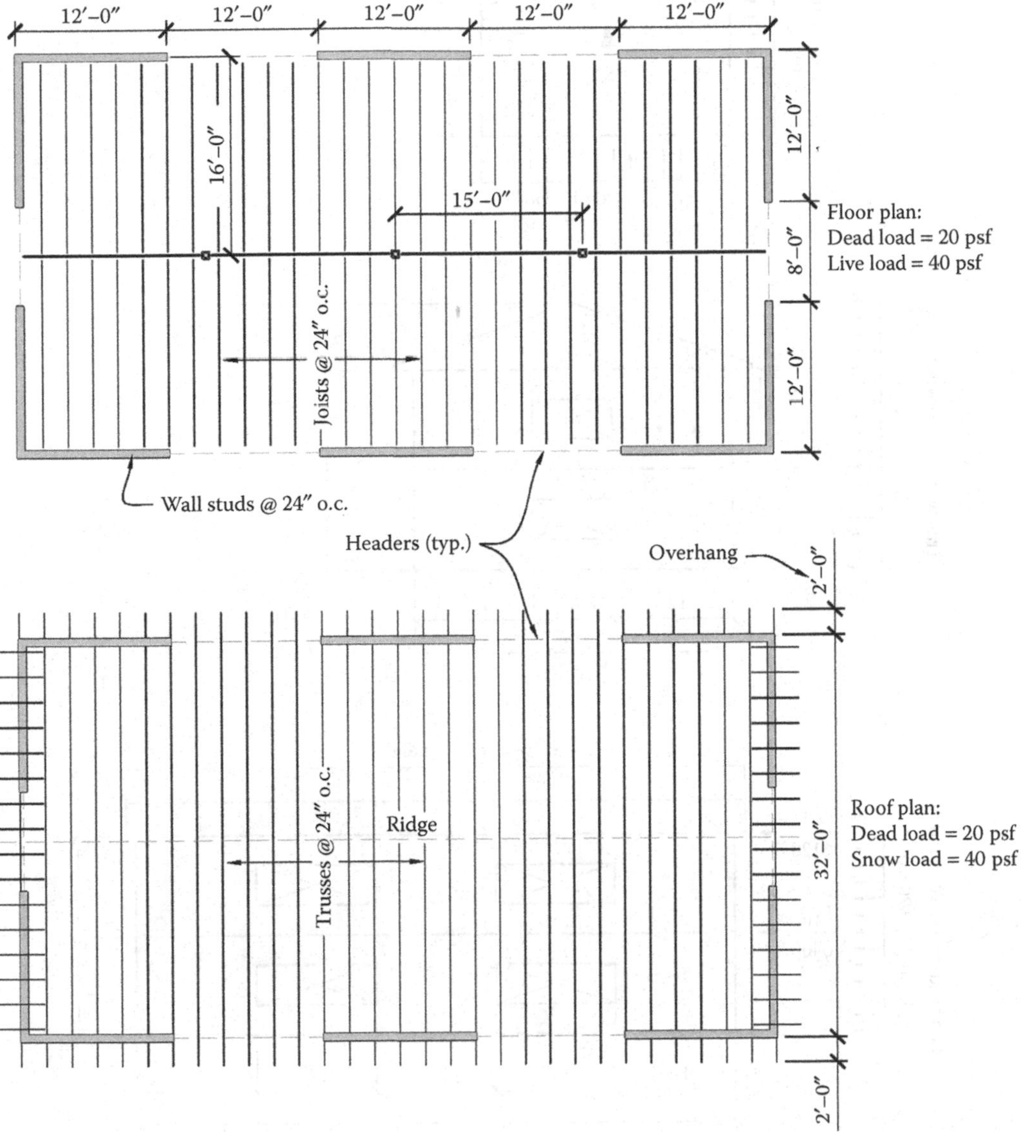

Figure 6.22 Roof and floor plans for Example 4.

Solution:

Note: Only the roof diaphragm has been designed in this example problem; the reader should follow the same procedure to design the floor diaphragm.

First calculate the wind loads (see Figure 6.23). In the N–S direction, the uniformly distributed lateral loads in the horizontal diaphragm are as follows:

$$\text{Roof lateral load } w_r = (20\text{ psf})\left(6\text{ ft} + \frac{10\text{ ft}}{2}\right) = \mathbf{220\ lb/ft}$$

$$\text{Second floor lateral load } w_2 = (20\text{ psf})\left(\frac{10\text{ ft}}{2} + \frac{10\text{ ft}}{2}\right) = \mathbf{220\ lb/ft}$$

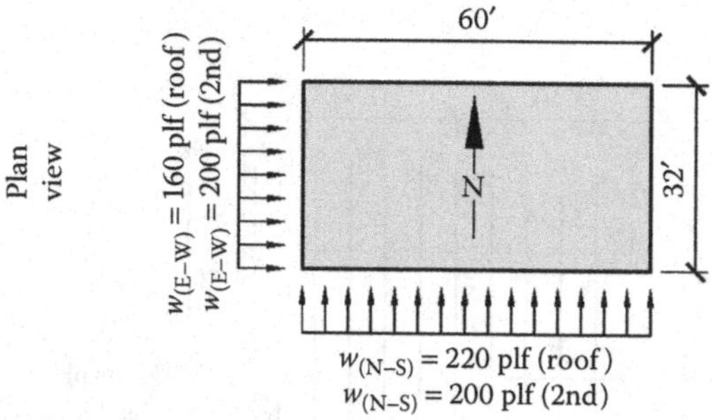

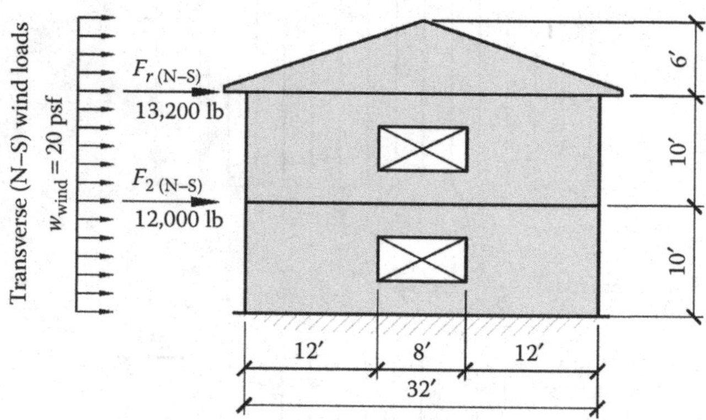

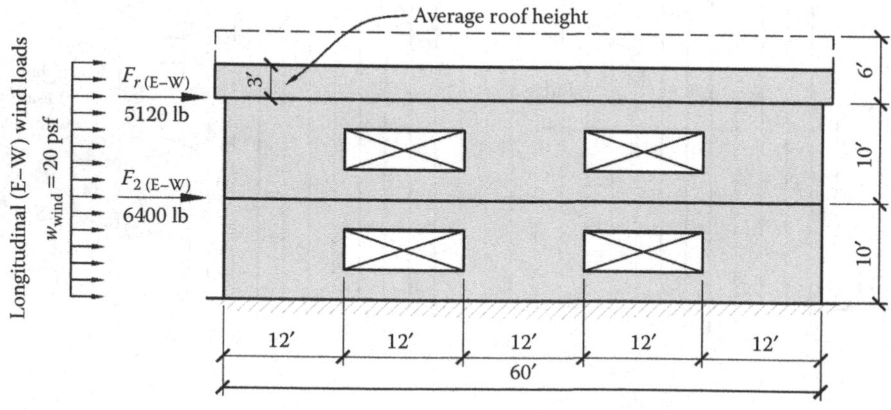

Figure 6.23 Wind load summary.

Total lateral force at the roof level F_{roof} = (220 plf)(60 ft) = **13,200 lb**

Total lateral force at the second floor level F_{2nd} = (200 plf) (60 ft) = **12,000 lb**

In the E–W direction, the uniformly distributed lateral loads in the horizontal diaphragm are as follows:

$$\text{Roof lateral load } w_r = (20 \text{ psf})\left(\frac{6 \text{ ft}}{2} + \frac{10 \text{ ft}}{2}\right) = \mathbf{160 \text{ lb/ft}}$$

$$\text{Second floor lateral load } w_2 = (20 \text{ psf})\left(\frac{10 \text{ ft}}{2} + \frac{10 \text{ ft}}{2}\right) = \mathbf{220 \text{ lb/ft}}$$

Total lateral force at the roof level F_{roof} = (160 plf)(32 ft) = **5120 lb**

Total lateral force at the second floor level F_{2nd} = (200 plf)(32 ft) = **6400 lb**

Roof diaphragm design

In designing the roof diaphragm for lateral loads, we consider the horizontal projection of the roof; thus, the roof slope or pitch is not a factor in designing for lateral loads (Figure 6.24). Check the diaphragm aspect ratio:

$$\frac{60 \text{ ft}}{32 \text{ ft}} = 1.88 < 3 \quad \text{OK}$$

Note: SDPWS Table 4.2.4 requires that the diaphragm aspect ratio be less than 4 if blocking is provided and 3 if blocking is omitted. Because the aspect ratio is less than 3 for this example, we can proceed with the design with or without blocking.

N–S direction. The lateral load w_r is perpendicular to continuous panel joints and perpendicular to unblocked panel edges; therefore, use load case 1 from Figure 6.9 (or SDPWS Table 4.2C):

$$\text{Maximum horizontal shear (reaction) } V_{N\text{-}S} = \frac{w_{N\text{-}S}L}{2} = \frac{(220)(60)}{2} = 6600 \text{ lb}$$

$$\text{Maximum moment } M_{N\text{-}S} = \frac{w_{N\text{-}S}L^2}{8} = \frac{(220)(60)^2}{8} = 99,000 \text{ ft-lb}$$

The maximum unit shear in the diaphragm is as follows:

$$v_d = \frac{V}{b} \text{(note that overhangs are excluded)}$$

$$= \frac{6600}{32 \text{ ft}} = \mathbf{206.3 \text{ lb/ft}} \approx 207 \text{ lb/ft}$$

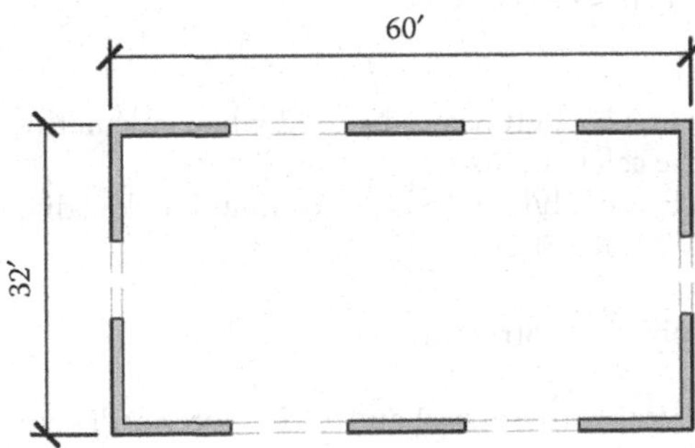

Figure 6.24 Free body diagram of diaphragm.

Note: The exact value of 206.3 lb/ft has to be used when calculating the drag strut forces to retain accuracy in the calculations.

For load case 1 with $v_d \approx 207$ lb/ft and assuming an unblocked diaphragm, the following is obtained from Table 4.2C:

⅜-in.-thick C–DX plywood, Structural I
8d common nails
Nails spaced at 6 in. (supported edges) and 12 in. (intermediate framing; see footnote b in Table 4.2C)

$$v_w = 607 \text{ plf}$$

$$v_{all} = \frac{v_w}{\Omega} = \frac{670}{2.0} = 335 \text{ plf}$$

The values above are valid only if the framing is Douglas Fir-Larch or Southern Pine. Because Hem-Fir wood species is specified, the value obtained from the table must be adjusted as follows:

$$\text{SGAF} = 1 - (0.5 - G) \leq 1.0, \text{ for nails} \qquad G = 0.43 \text{ (NDS Table 12.3.2A)}$$

$$= 1 - (0.5 - 0.43) = 0.93$$

$$v_{adjust} = (335)(0.93) = \mathbf{312\ lb/ft} > v_d = 207 \text{ lb/ft} \quad \textbf{OK}$$

By inspecting IBC Table 2304.8(3), the minimum panel thickness is at least ⅜ in. for roof framing members spaced 24 in. Therefore, it would be necessary to select a ⅜-in. panel for gravity loads.

E–W direction. The lateral load w_r is parallel to continuous panel joints and parallel to unblocked panel edges; therefore, use load case 3 from Figure 6.9 (or SDPWS Table 4C):

$$\text{Maximum horizontal shear (reaction) } V_{E\text{-}W} = \frac{w_{E\text{-}W}L}{2} = \frac{(160)(32)}{2} = 2560 \text{ lb}$$

$$\text{Maximum moment } M_{E\text{-}W} = \frac{w_{E\text{-}W}L^2}{8} = \frac{(160)(32)^2}{8} = 20,480 \text{ ft-lb}$$

The maximum unit shear in the diaphragm is as follows:

$$v_d = \frac{V}{b}$$

$$= \frac{2560}{60 \text{ ft}} = \mathbf{42.7\ lb/ft} \approx 43 \text{ lb/ft}$$

Note: The exact value of 42.7 lb/ft has to be used when calculating the drag strut forces to retain accuracy in the calculations.

For load case 3 with $v_d \approx 43$ lb/ft and assuming an unblocked diaphragm, the following is obtained from SDPWS Table 4.2C:

⅜-in.-thick C–DX plywood, Structural I
8d common nails
Nails spaced at 6 in. (supported edges) and 12 in. (intermediate framing; see footnote b in Table 4.2C)

$$v_w = 505 \text{ plf}$$

$$v_{all} = \frac{v_w}{\Omega} = \frac{505}{2.0} = 252 \text{ plf}$$

Adjusted allowable shear (using SGAF = 0.93 from the previous part and using the 40% increase allowed by the code):

$$v_{adjust} = (0.93)(252 \text{ lb/ft}) = \mathbf{234 \text{ lb/ft}} > v_d = 43 \text{ lb/ft} \quad \text{OK}$$

The allowable shear is much greater than the actual shear, but the ⅜-in. panel was the minimum required for the N–S direction.

Chord forces (Figure 6.25). In the N–S direction, *A* and *B* are the chord members. The axial forces in chord members *A* and *B* are as follows:

$$T = C = \frac{M_{N-S}}{b_{N-S}} = \frac{(99{,}000 \text{ ft-lb})}{32 \text{ ft}} = \mathbf{3100 \text{ lb tension or compression}}$$

where b_{N-S} is the distance between chords *A* and *B*.

In the E–W direction, *C* and *D* are the chord members. The axial forces in chord members *C* and *D* are as follows:

$$T = C = \frac{M_{E-W}}{b_{E-W}} = \frac{(20{,}480 \text{ ft-lb})}{60 \text{ ft}} = \mathbf{342 \text{ lb tension or compression}}$$

where b_{E-W} is the distance between chords *C* and *D*.

Drag struts. In the N–S direction, *C* and *D* are the drag struts. The unit shear applied in the horizontal diaphragm in the N–S direction is $v_d = 206.3$ lb/ft. The unit shear in the N–S shear walls is as follows:

$$v_w = \frac{V_{N-S}}{\sum L_w} = \frac{6{,}600 \text{ lb}}{12 \text{ ft} + 12 \text{ ft}} = \mathbf{275 \text{ lb/ft}}$$

Note: In drawing the unit shears diagram (Figure 6.26), the horizontal diaphragm unit shears are plotted as positive and the shear wall unit shears are plotted as negative. The combined unit shears are a summation of the unit shear in the diaphragm and the unit shear in the walls. The drag strut force diagram is obtained by summing the area under the combined unit shear diagram.

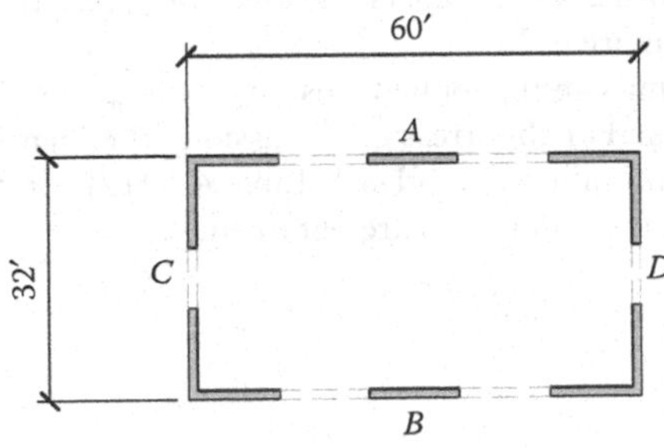

Figure 6.25 Diaphragm chords.

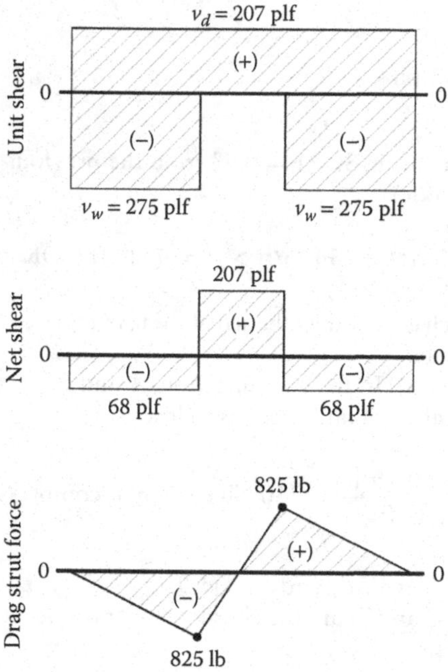

Figure 6.26 Unit shear, net shear, and drag strut force diagrams (N–S direction).

For N–S wind, the maximum drag strut force in *C* and *D* is the largest force in the drag strut force diagram which is 825 lb (tension or compression). In the E–W direction, *A* and *B* are the drag struts. The unit shear applied in the horizontal diaphragm in the E–W direction is $v_d = 42.7$ lb/ft. The unit shear in the E–W shear walls is as follows:

$$v_w = \frac{V_{E-W}}{\Sigma L_w} = \frac{2560 \text{ lb}}{12 \text{ ft} + 12 \text{ ft} + 12 \text{ ft}} = \mathbf{71.1\ lb/ft}$$

Following the same procedure as that used in the construction of N–S unit shear, the combined unit shear and drag strut force diagrams are shown in Figure 6.27.

For E–W wind, the maximum drag strut force in *A* and *B* is the largest force in the drag strut force diagram which is 342 lb (tension or compression).

The chord and drag strut forces are summarized in Table 6.6. The maximum force (tension and compression) for which the top plates should be designed is 3100 lb.

Design of top plates. Try double 2 × 6 top plates (same size as the wall studs given). As a result of the splicing of the double plates, only one 2 × 6 plate is effective in resisting the tension loads, whereas both 2 × 6 members are effective in resisting the axial compression force in the top plates (Figure 6.28).

Design for tension. The maximum applied tension force $T_{max} = \mathbf{3100\ lb}$ (see the summary of chord/drag strut forces). For the stress grade, assume the Hem-Fir stud grade. Because a 2 × 6 is the dimension lumber, use NDS-S Table 4A [12] for design values. From the table, the following design stress values are obtained:

$$F_t = 400 \text{ psi}$$

$$C_D(\text{wind}) = 1.6$$

$$C_M = 1.0$$

$$C_t = 1.0$$

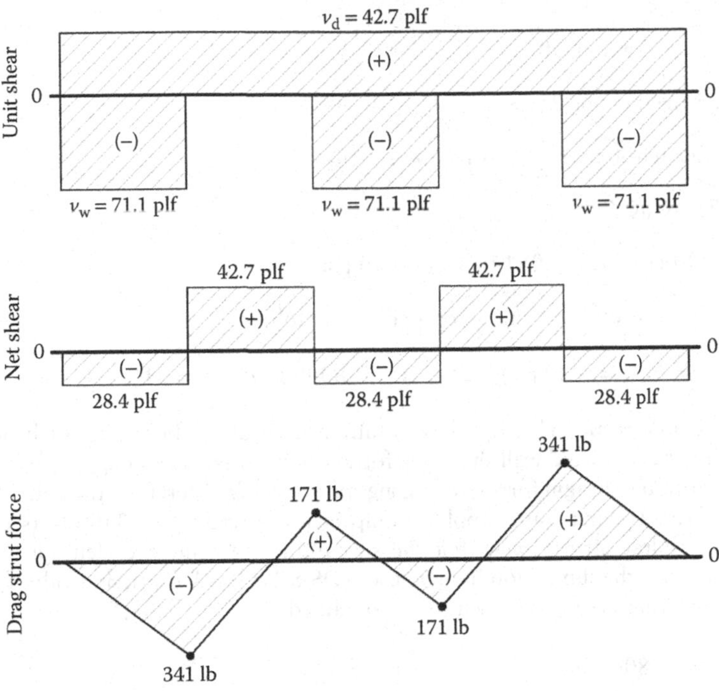

Figure 6.27 Unit shear, net shear, and drag strut force diagrams (E–W direction).

Table 6.6 Summary of chord and drag strut forces (lb), Example 6.4

	N–S wind		*E–W wind*	
Side	*Chord*	*Drag strut*	*Chord*	*Drag strut*
A	3100	–	–	342
B	3100	–	–	342
C	–	825	342	–
D	–	825	342	–

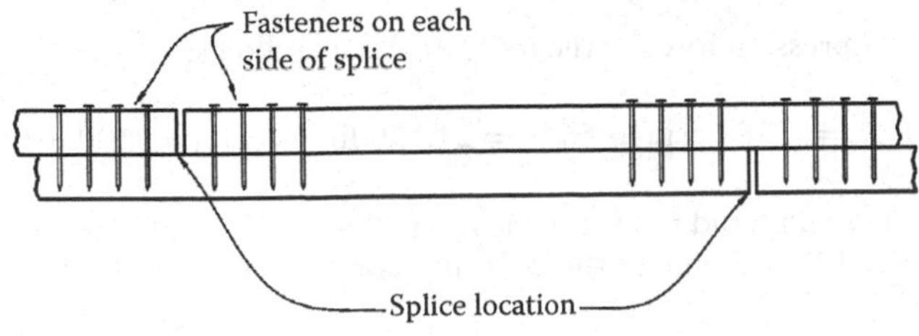

Figure 6.28 Top plate splice details.

$$C_F(F_t) = 1.0$$

$$C_i = 1.0 \text{ (assumed)}$$

$$A_g \; 8.25 \text{ in}^2 \text{ (for one } 2 \times 6\text{)}$$

The allowable tension stress is as follows:

$$F'_t = F_t C_D C_M C_t C_F C_i$$

$$= (400)(1.6)(1.0)(1.0)(1.0)(1.0) = 640 \text{ psi}$$

The allowable tension force in the top plates is as follows:

$$T_{\text{allowable}} = F'_t A_g = (640 \text{ psi})(8.25 \text{ in}^2) = \mathbf{5280\ lb} > T_{\text{applied}} = 3100 \text{ lb} \quad \textbf{OK}$$

Design for compression. The top plate is fully braced about both axes of bending. It is braced by the stud wall and wall sheathing for *y–y* or weak-axis bending, and it is braced by the roof or floor diaphragm for *x–x* or strong-axis bending, Therefore, the column stability factor $C_P = 1.0$. The maximum applied compression force, $P_{\text{max}} = $ **3100 lb** (see the summary of chord/drag strut forces). For the stress grade, assume the Hem-Fir stud grade. Because a 2 × 6 is the dimension lumber, use NDS-S Table 4A for design values. From the table, the following design stress values are obtained:

$$F_c = 800 \text{ psi}$$

$$C_D(\text{wind}) = 1.6$$

$$C_M = 1.0$$

$$C_t = 1.0$$

$$C_F(F_c) = 1.0$$

$$C_i = 1.0 \text{ (assumed)}$$

$$C_P = 1.0$$

$$A_g = 16.5 \text{ in}^2 \text{ (for two } 2 \times 6\text{'s)}$$

The allowable compression stress is as follows:

$$F'_c = F_c C_D C_M C_t C_F C_i C_P$$

$$= (800)(1.6)(1.0)(1.0)(1.0)(1.0)(1.0) = 1280 \text{ psi}$$

The allowable compression force in the top plates is as follows:

$$P_{\text{allowable}} = F'_c A_g = (1280 \text{ psi})(16.5 \text{ in}^2) = \mathbf{21{,}120\ lb} > P_{\text{applied}} = 3100 \text{ lb} \quad \textbf{OK}$$

Use two 2 × 6's Hem-Fir stud grade for the top plates.
Note: See Chapter 10 for design of the lap splice connection.

Example 6.5: Design of a horizontal diaphragm with partial wood blocking

Considering the diaphragm loading shown in Figure 6.29, determine the extent for which wood blocking is required assuming the floor framing is similar to that shown for Example 6.4.

Solution:
For load case 1 and assuming a blocked diaphragm, the following is obtained from SDPWS Table 4.2A:

$\frac{3}{8}$-in.-thick C–DX plywood, Structural I
8d common nails
2× blocking required
Nails spaced at 6 in. (all edges) and 12 in. (intermediate framing, see footnote b in Table 4.2A)

$$v_w = 755 \text{ plf}$$

$$v_{all} = \frac{v_w}{\Omega} = \frac{755}{2.0} = 377 \text{ plf}$$

SGAF = 0.93 (from Example 6.4). The adjusted allowable shear is as follows:

$$v_{allowable} = (0.93)(377 \text{ lb/ft}) = \mathbf{351\ lb/ft} > v_d = 338 \text{ lb/ft} \quad \textbf{OK}$$

Extent of wood blocking. Wood blocking is provided only in the zones where the unblocked diaphragm allowable unit shear is less than the unit shear applied. The zone on the shear diagram in which the unit shear is less than the unblocked diaphragm allowable unit shear (i.e., 312 lb/ft calculated in Example 6.4) is determined using similar triangles (Figure 6.29).

$$\frac{338 \text{ lb/ft}}{30 \text{ ft}} = \frac{312 \text{ lb/ft}}{30 \text{ ft} - x} \quad x = 2.31 \text{ ft, say 4 ft (because the joist spacing is 24 in.)}$$

Therefore, the horizontal diaphragm needs to be blocked in the N–S direction only in the zones within the first 2 ft from the shear walls (hatched area in Figure 6.29).

Thus far, we have considered only buildings with shear walls along the exterior faces, which results in single-span simply supported horizontal diaphragms. When a building becomes too long with respect to its width, the diaphragm aspect ratio may exceed the 4:1 maximum allowed by the IBC, in which case interior shear walls will be required. With interior shear walls, the horizontal diaphragms are considered as separate simply supported diaphragms that span between the shear walls. We now consider an example building with both interior and exterior shear walls.

Example 6.6: Horizontal diaphragms with interior and exterior shear walls

Determine the forces in the chords and drag struts, and the diaphragm unit shears at the roof level for a building with the roof plan shown in Figure 6.30. Consider only the N–S wind loads given.

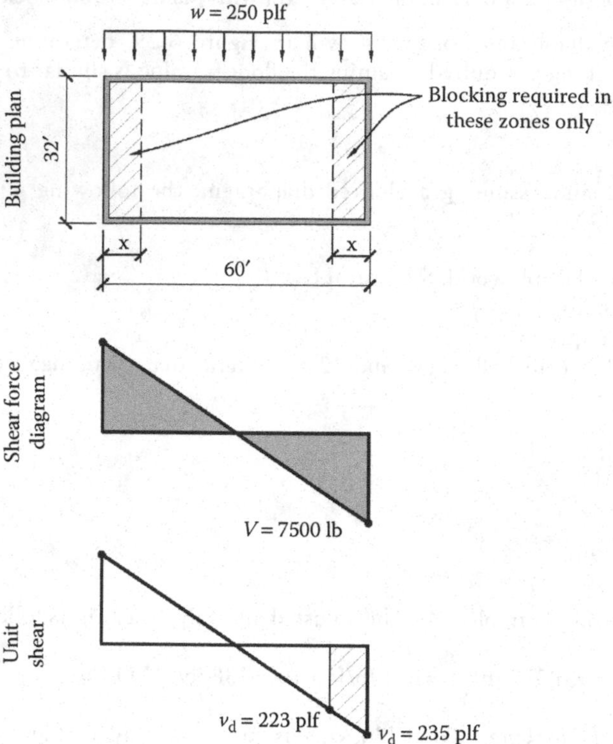

Figure 6.29 Building plan, shear force, and unit shear diagrams.

Solution:

Diaphragm D1. For the N–S wind, *A* and *B* are the chords; *E* and *F* are the drag struts.

$$\text{Maximum moment } M_{D1} = \frac{w_{D1}L^2}{8} = \frac{(200)(50)^2}{8} = 62{,}500 \text{ ft-lb}$$

The axial tension and compression force in chords *A* and *B* is as follows:

$$T = C = \frac{M_{D1}}{b_{D1}} = \frac{62{,}500 \text{ ft-lb}}{75 \text{ ft}} = \mathbf{833\ lb\ tension\ or\ compression}$$

$$v_{d1} = \frac{V_1}{b} = \frac{5000 \text{ lb}}{75 \text{ ft}} = \mathbf{66.7\ lb/ft} \text{ (unit shear in the diaphragm along line 1)}$$

$$v_{d2(\text{left})} = \frac{V_2}{b} = \frac{5000 \text{ lb}}{75 \text{ ft}} = \mathbf{66.7\ lb/ft} \text{ (unit shear in the diaphragm along line 2)}$$

Diaphragm D2. For the N–S wind, *C* and *D* are the chords; *F* and *G* are the drag struts.

$$\text{Maximum moment } M_{D2} = \frac{w_{D2}L^2}{8} = \frac{(200)(100)^2}{8} = 250{,}000 \text{ ft-lb}$$

The axial tension and compression force in chords *A* and *B* is as follows:

$$T = C = \frac{M_{D2}}{b_{D2}} = \frac{250{,}000 \text{ ft-lb}}{75 \text{ ft}} = \mathbf{3334\ lb\ tension\ or\ compression}$$

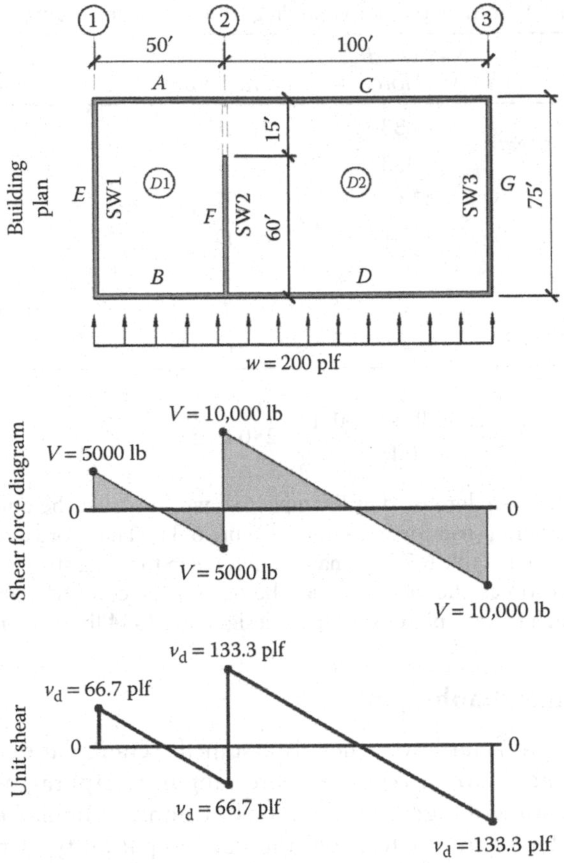

Figure 6.30 Building plan, shear force, and unit shear diagrams for Example 6.

$$v_{d2(\text{right})} = \frac{V_2}{b} = \frac{10{,}000 \text{ lb}}{75 \text{ ft}} = \mathbf{133.3\ lb/ft} \text{ (unit shear in the diaphragm along line 2)}$$

$$v_{d3} = \frac{V_3}{b} = \frac{10{,}000 \text{ lb}}{75 \text{ ft}} = \mathbf{133.3\ lb/ft} \text{ (unit shear in the diaphragm along line 3)}$$

Drag strut forces (N–S wind):

1. The force in the drag strut on line 1 = 0 because there are no wall openings along line 1, and therefore, there is no force to be "dragged."
2. The force in the drag strut on line 3 = 0 because there are no wall openings along line 3, and therefore, there is no force to be dragged
3. To calculate the force in the drag strut on line 2 (i.e., at the interior wall), the unit shear, net shear, and drag strut force diagrams must be constructed. The total diaphragm unit shear along line 2 is as follows:

$$v_{d(\text{total})} = v_{d(\text{left})} + v_{d(\text{right})}$$

$$= 66.7 \text{ lb/ft} + 133.3 \text{ lb/ft} = \mathbf{200\ lb/ft}$$

The unit shear in the shear wall along line 2 is

Table 6.7 Summary of chord and drag strut forces (lb), Example 6.6

Chords	*N–S wind chord force*	*Drag strut*	*N–S wind drag strut force*
A	833	*A*	–
B	833	*B*	–
C	3334	*C*	–
D	3334	*D*	–
E	–	*E*	0
F	–	*F*	3000
G	–	*G*	0

$$v_{w2} = \frac{V_2}{\sum L_w} = \frac{10{,}000 \text{ lb} + 5000 \text{ lb}}{60 \text{ ft}} = 250 \text{ lb/ft}$$

Following the procedure used in Example 6.4, we construct the unit shear, combined unit shear, and drag strut force diagrams (Figure 6.31). The chord and drag strut forces are summarized in Table 6.7. The maximum force in the drag strut on line 2 = 3000 lb.

The axial forces shown above can be tension or compression. The maximum force for which the top plates should be designed is 3334 lb, tension or compression.

6.5.4 Nonrectangular diaphragms

The examples presented thus far have generally assumed rectangular diaphragms. In practice, it is common to encounter several types of nonrectangular diaphragms. Although there are countless possibilities, some design considerations of nonrectangular diaphragms are summarized in Figure 6.32. The aim is to avoid the noncompatibility of the lateral deflections in nonrectangular diaphragms by using drag struts and collectors. The drag struts effectively convert a nonrectangular diaphragm into several rectangular diaphragm segments, thus ensuring the compatibility of the lateral deflections of the various diaphragm segments.

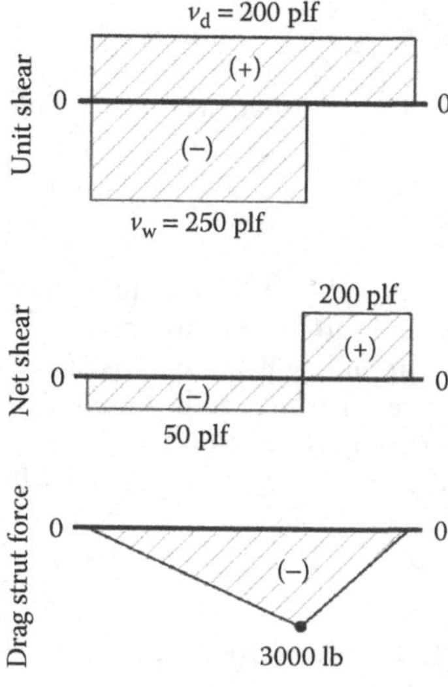

Figure 6.31 Unit shear, net shear, and drag strut force diagrams for Example 6.

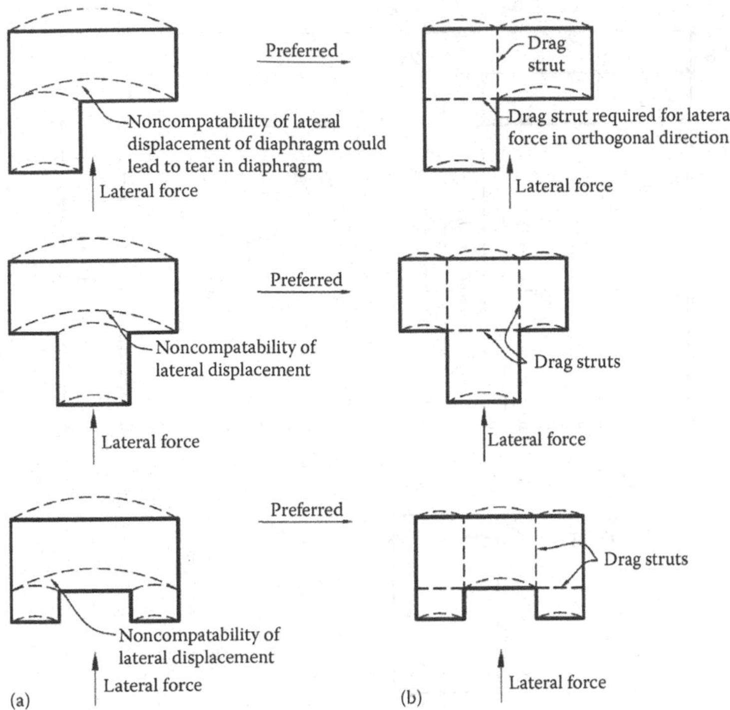

Figure 6.32 Lateral bending action of nonrectangular diaphragms.

PROBLEMS

6.1 For a total roof load of 60 psf on 2× framing members at 24 in. o.c., determine an appropriate plywood thickness and span rating assuming panels with edge support and panels oriented perpendicular to the framing. Specify the minimum fasteners and spacing.

6.2 Repeat Problem 6.1 for panels oriented parallel to the supports.

6.3 For a total floor load of 75 psf on 2× floor framing at 24 in. o.c., determine an appropriate plywood thickness and span rating assuming panels with edge support and panels oriented perpendicular to the framing. Specify the minimum fasteners and spacing.

6.4 A one-story building is shown in Figure 6.33. The roof panels are $^{15}\!/_{32}$-in.-thick CD–X with a span rating of $^{32}\!/_{16}$. The wind load is 25 psf. Assume that the diaphragm is unblocked and that the framing members are Spruce-Pine-Fir. Determine the following for the transverse direction:

a. Unit shear in the diaphragms
b. Required fasteners and spacing
c. Maximum chord forces
d. Maximum drag strut forces (construct unit shear, net shear, and drag strut force diagrams)

6.5 Repeat Problem 6.4 for the longitudinal direction.

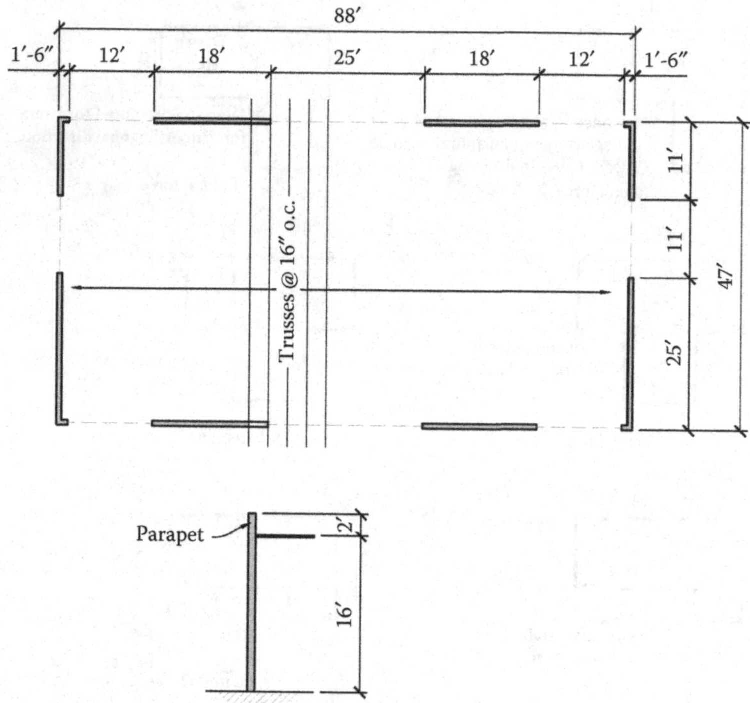

Figure 6.33 Building plan for Problems 6.4 and 6.5.

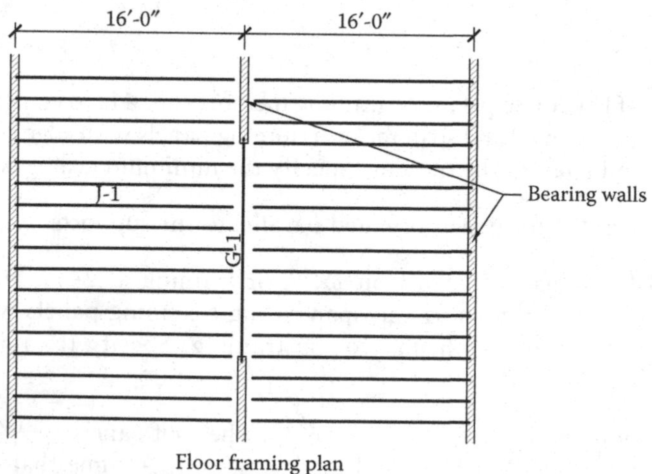

Figure 6.34 Details for Problem 6.6.

6.6 For the following floor framing plan shown in Figure 6.34:

- Total loads to the floor sheathing are $D = 10$ psf and $L = 80$ psf.
- Spacing = 19.2 in. o.c.
- Plywood sheathing is oriented with the strong axis perpendicular to the supports. Do the following:

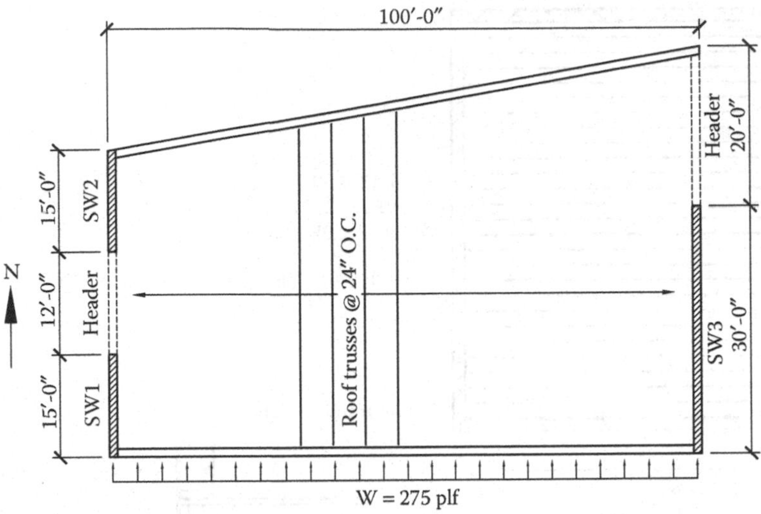

Figure 6.35 Details for Problem 6.7.

a. Select a plywood thickness and span rating for gravity loads and also a nailing pattern. Provide the full specification for the plywood and nailing.
b. Sketch a typical 4 ft. × 8 ft. plywood panel showing the actual mailing pattern assuming unblocked edges.

6.7 For the following roof framing plan shown in Figure 6.35:

- Total loads to the roof sheathing are $D = 10$ psf and $S = 50$ psf
- Plywood sheathing is oriented with the strong axis perpendicular to the supports.
- Lateral load is due to wind.

Do the following:

a. Select a plywood thickness and span rating and also a nailing pattern. Provide the full specification for the plywood and nailing.
b. Draw the unit shear diagram in the roof diaphragm.
c. For the maximum diaphragm unit shear (in plf) from (b), determine a plywood thickness and nailing pattern assuming an unblocked diaphragm. Assume Douglas Fir-Larch lumber.
d. Determine the unit shear (in plf) in each shear wall.
e. Determine the maximum drag strut force (in pounds) at each header; provide diagrams showing the calculations.
f. Describe the load path for the wind load.

6.8 For the following floor framing plan shown in Figure 6.36:

- The diaphragm receives a lateral load from the soil pressure.
- Plywood sheathing is oriented with the strong axis perpendicular to the supports; blocking is provided.

Do the following:

a. Determine the reaction at the top of the wall and determine the maximum unit shear in the diaphragm.
b. For the maximum diaphragm shear from (a), select a nailing pattern assuming $^{15}/_{32}$ in. Struct I sheathing.
c. Describe the load path for the lateral load.

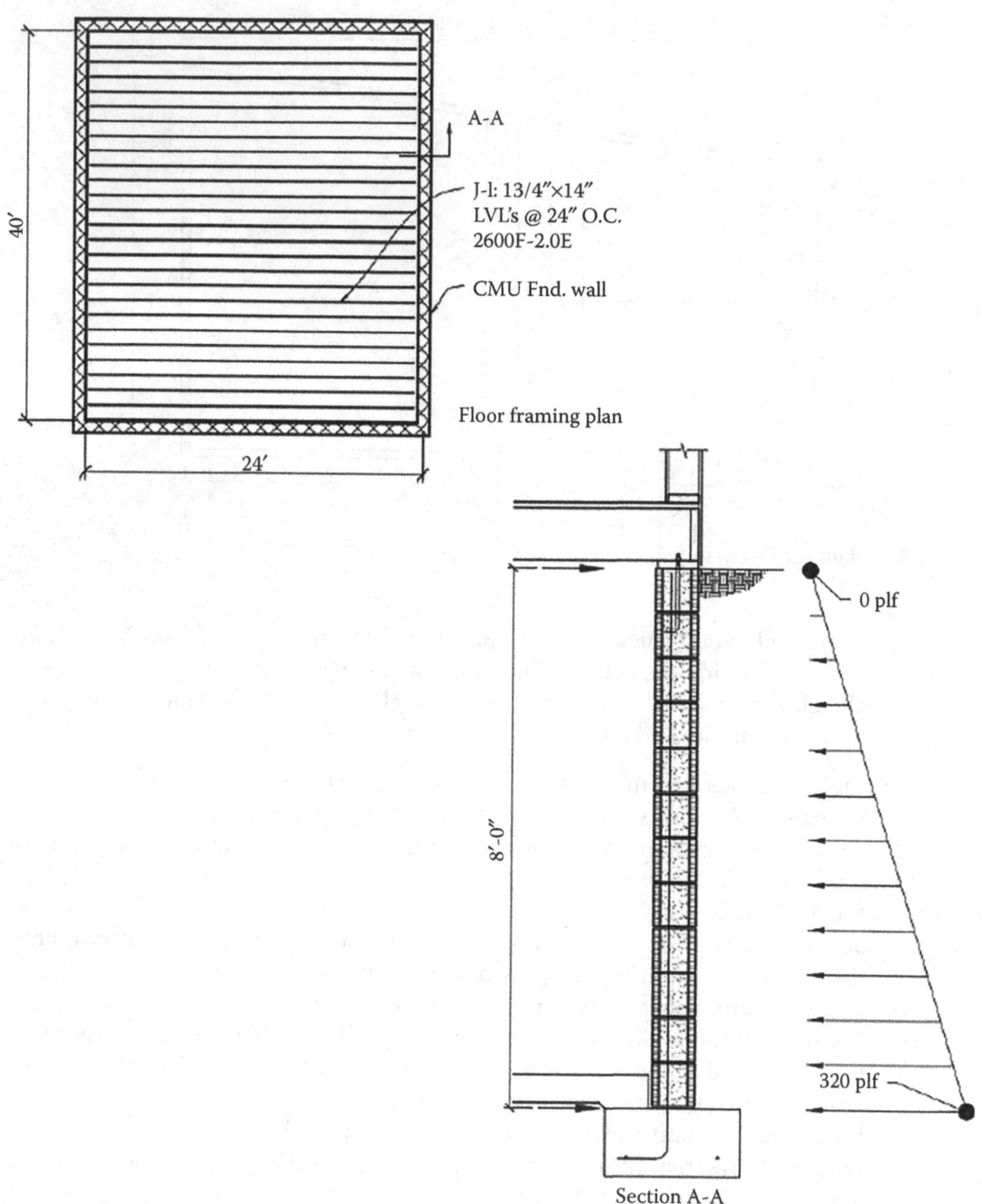

Figure 6.36 Details for Problem 6.8.

6.9 Refer to the following sketch showing loads on a roof diaphragm shown in Figure 6.37:

- Normal temperature and moisture conditions apply.
- The diaphragm shear capacities are as follows: blocked = **400 plf**, unblocked = **240 plf.**
  a. Determine the extent of blocking required for loads in the N–S direction and sketch it on the plan view shown.

6.10 Refer to the following sketch showing the building plan for a one-story building. Lateral load is wind and framing is Douglas Fir-Larch.

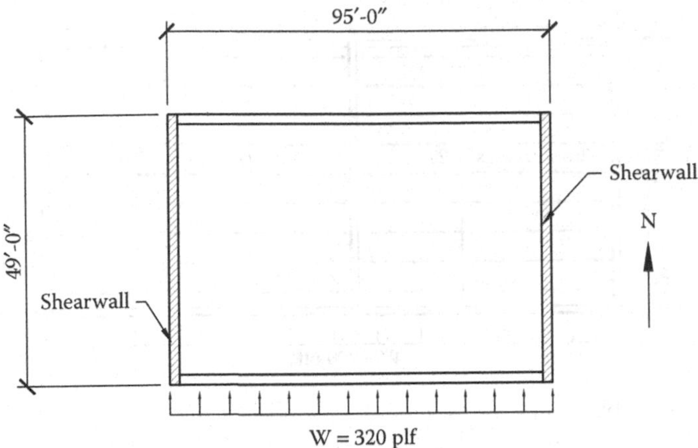

Figure 6.37 Details for Problem 6.9.

a. Draw the unit shear, net shear, and axial force diagrams for the roof diaphragm drag struts, and determine the maximum axial force in the drag struts in the three shear wall lines.
b. Determine the unit shear in the diaphragm and select the fastening and edge support requirements assuming Structural 1 sheathing and blocking provided. Give the full specification (Figure 6.38).

6.11 Refer to the following drawing showing the building plan for a one-story building.

- Lateral load is wind and framing is Douglas Fir-Larch.
  a. Draw the unit shear, net shear, and axial force diagrams for the roof diaphragm drag struts, and determine the maximum axial force in the drag struts for the SW1 and SW2 lines.

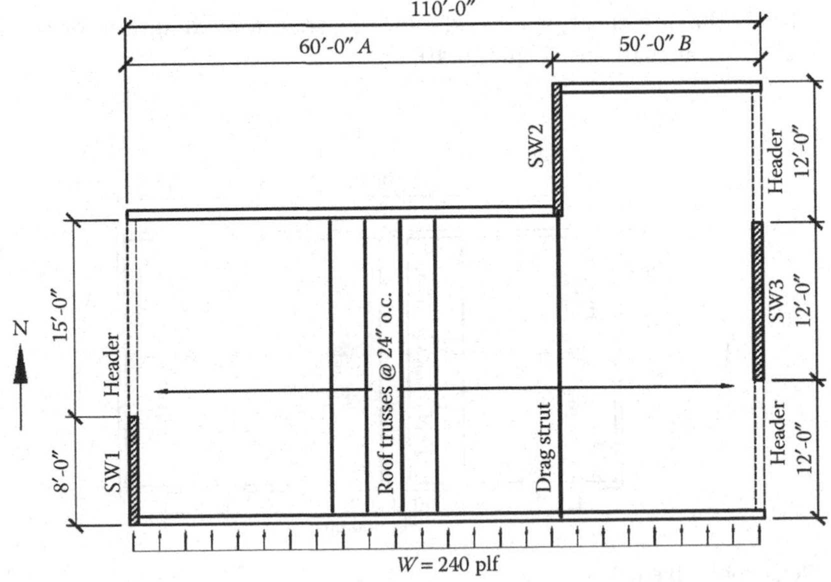

Figure 6.38 Detail for Problem 6.10.

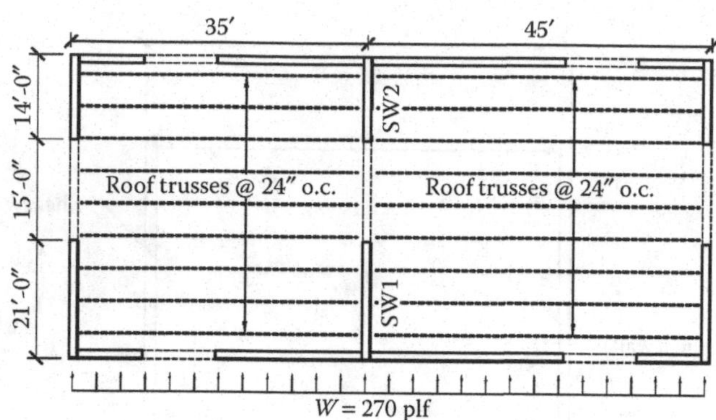

Figure 6.39 Detail for Problem 6.11.

b. Determine the lateral force to each shear wall (SW1, SW2) in pounds.
c. For SW1, select the fastening and edge support requirements assuming Structural 1 sheathing. Assume the plywood is applied directly to the framing and is on one side only. Give the full specification.
d. Determine the unit shear in the diaphragm along the SW3 and SW4 lines and select the fastening and edge support requirements assuming ⅜ ft. Structural 1 sheathing and an unblocked diaphragm (Figure 6.39).

6.12 Refer to the following drawing showing loads on a roof diaphragm.

- Normal temperature and moisture conditions apply.
- The diaphragm shear capacities are as follows: blocked = **400 plf**, unblocked = **240 plf.**

a. Determine the extent of blocking required for loads in the N–S direction and sketch it on the plan view shown.
b. Draw the unit shear, net shear, and drag strut force diagram for the 18 in. and 22 ft shear wall line (Figure 6.40).

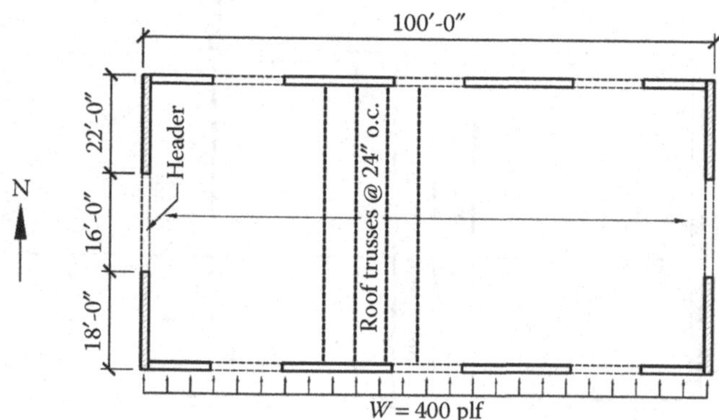

Figure 6.40 Detail for Problem 6.12.

6.13 Refer to the following drawing showing loads on a roof diaphragm and the following given information.

- Roof framing is 2 × 14 at 24 in. o.c., Hem-Fir Select Structural.
- Normal temperature and moisture conditions apply.
- Assume load case 1 type, blocked layout.
- Consider loads in the N–S direction only.
- Gravity loads supported by the plywood are as follows: $D = 10$ psf, $S = 35$ psf.
  a. Select the required plywood thickness and span rating to support gravity loads.
  b. For the thickness specified in part (a), determine the fastening and edge support requirements for the roof diaphragm. Give the full specification (Figure 6.41).

6.14 Refer to the following sketch showing loads on a roof diaphragm.

- Normal temperature and moisture conditions apply.
- The diaphragm shear capacities are as follows: blocked = 375 plf, unblocked = 225 plf.

Determine if the extent of blocking shown is adequate for loads in the N–S direction (Figure 6.42).

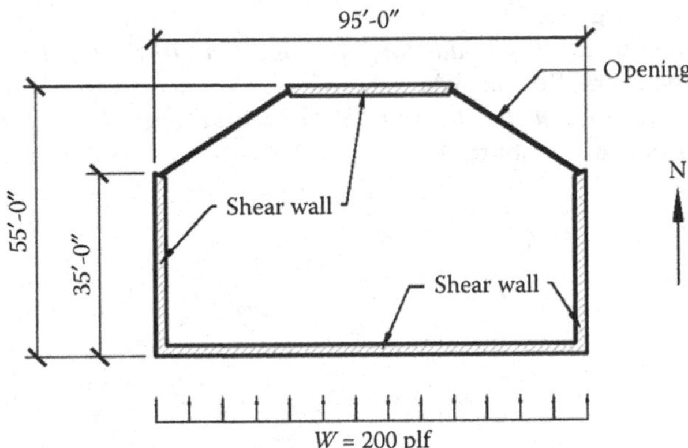

Figure 6.41 Detail for Problem 6.13.

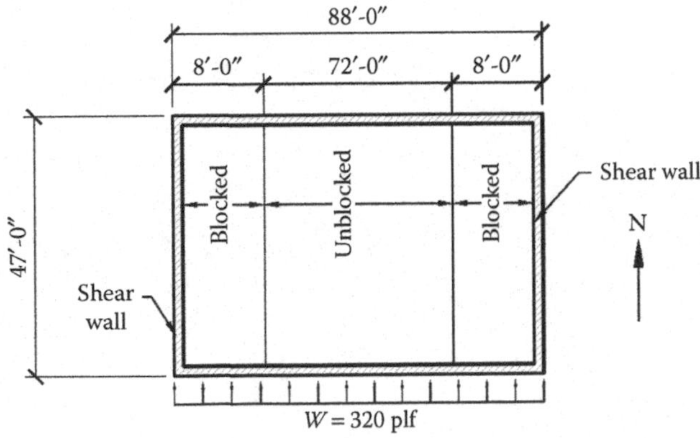

Figure 6.42 Detail for Problem 6.14.

REFERENCES

1. APA (2010), *Construction and Industrial Plywood,* PS 1-09, APA—The Engineered Wood Association, Tacoma, WA.
2. APA (2014), *Performance Standard for Wood-Based Structural-Use Panels,* PS 2-10, APA—The Engineered Wood Association, Tacoma, WA.
3. ICC (2006, 2015), *International Building Code*, International Code Council, Washington, DC.
4. APA (2016), *Engineered Wood Construction Guide (E30W),* APA—The Engineered Wood Association, Tacoma, WA.
5. APA (2012), *Panel Design Specification (D510C),* APA—The Engineered Wood Association, Tacoma, WA.
6. AWC (2015), *Special Design Provisions for Wind and Seismic*, American Wood Council, Washington, DC.
7. APA (2002), *Proper Selection and Installation of APA Plywood Underlayment (R350C),* APA—The Engineered Wood Association, Tacoma, WA.
8. APA (2011), *Load-Span Tables for APA Structural-Use Panels (Q225G),* APA—The Engineered Wood Association, Tacoma, WA.
9. ANSI/American Forest & Paper Association (2001), *Wood Structural Panels Supplement*, AF&PA, Washington, DC.
10. AWC (2015), *Allowable Stress Design Manual for Engineered Wood Construction*, American Wood Council, Leesburg, VA.
11. ASCE (2010), *Minimum Design Loads for Buildings and Other Structures,* ASCE-7, American Society of Civil Engineers, Reston, VA.
12. AWC (2015), *National Design Specification Supplement: Design Values for Wood Construction*, American Wood Council, Leesburg, VA.

Chapter 7

Vertical diaphragms under lateral loads (shear walls)

7.1 INTRODUCTION

The design of vertical diaphragms or shear walls is one of the most critical elements in wood-framed structures. The lateral forces due to wind or seismic loads are transmitted from the horizontal diaphragm to the shear walls, which then transfer these loads to the foundation. The straps, ties, and hold-down anchors must be detailed to ensure a continuous load path from the roof through the floors to the foundation of the building. The most common sheathing materials used are either plywood or gypsum wall board (GWB).

The lateral load distribution to vertical diaphragms or shear walls is a function of the relative rigidity of the horizontal roof and floor diaphragms. Horizontal diaphragms can be classified as *flexible* or *rigid*, and most wood buildings can be assumed to have flexible diaphragms. Therefore, lateral wind loads are distributed to the shear walls based on the tributary area of the shear wall relative to the vertical wall surface on which the wind acts, whereas seismic loads are distributed based on the plan tributary area of the shear walls.

The capacity of a plywood shear wall is greater than that of a shear wall with GWB, especially when comparing their resistance to seismic loads. A typical wood-framed structure has plywood sheathing on the exterior walls only, with the remaining interior walls sheathed with GWB. If a building design requires the use of plywood sheathing on the interior walls, the plywood is typically covered with GWB for esthetic purposes, so the designer must consider the economic impact of using interior plywood shear walls versus strengthening the exterior plywood shear walls.

Wall sheathing must also act to transfer the horizontal wind load to the wall studs on the exterior walls. International Building Code (IBC) Section 1607.14[1] requires that interior walls be adequate to resist a lateral load of 5 psf for walls greater than 6 ft in height, but this lateral pressure is typically small enough not to affect the design of the interior wall sheathing or studs.

7.1.1 Wall sheathing types

Although several types of wall sheathing are available for use in shear walls, in this book we focus on plywood sheathing only. There are two common types of plywood wall sheathing: the first is *exposed plywood panel siding* (e.g., APA 303-rated siding). These panels are manufactured with various surface textures and perform a dual purpose as a structural panel as well as being the exterior surface of the building. Several proprietary products are available that are recognized by the IBC for use as shear walls. Exposed plywood siding can be fastened directly to wall studs or over a layer of ½- or ⅝-in.-thick GWB to meet fire rating requirements. IBC Table 2308.6.3(2) specifies the minimum thickness for exposed plywood as ⅜ in. for studs spaced 16 in. o.c. and ½ in. for studs spaced 24 in. o.c.

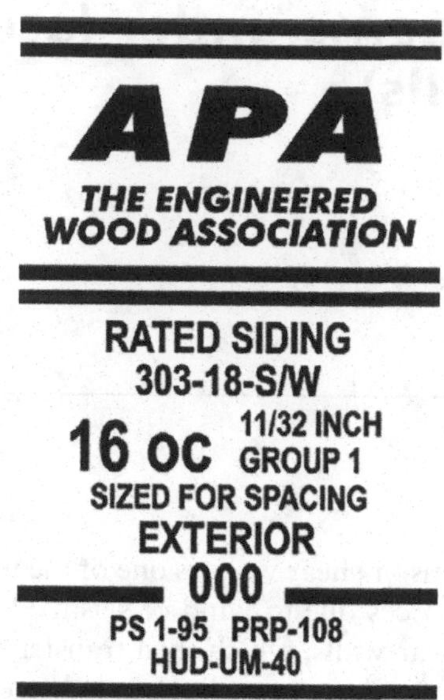

Figure 7.1 Wall sheathing grade stamp. (Reprinted with the permission of APA–The Engineered Wood Association.)

The second type is *wood structural panel wall sheathing*. This is commonly used on exterior walls applied directly to the wall framing with a finish material applied to the plywood. IBC Table 2308.6.3(3) specifies the minimum thickness for plywood not exposed to the weather.

The span rating for wall panels is usually indicated by one number, either 16 or 24 in. (see Figure 7.1). This number refers to the maximum wall stud spacing. Panels rated for wall sheathing are performance tested, with the short direction spanning across the supports (i.e., the weaker direction); therefore, the wall panels can be placed either horizontally or vertically. Loads normal to the wall panels typically do not control the thickness of the sheathing, rather the unit shear in the wall panels due to lateral loads will usually determine the panel thickness. The out-of-plane wind load capacity of wood panels and panels of other materials is given in SDPWS Table 3.2.1, and the out-of-plane wind load capacity of wood roof panels is in SDPWS Table 3.2.2 [2].

7.1.2 Plywood as a shear wall

When lateral loads act parallel to the face of the wall, the plywood wall sheathing acts as a shear wall. Shear walls are the main elements used to resist lateral loads due to wind and earthquake in wood-framed buildings. They act as vertical cantilevers that are considered fixed at the ground level (see Figure 7.2).

The shear wall diaphragm (i.e., the plywood) resists the shear, whereas the shear wall chords (i.e., the vertical members at the ends of the shear wall) resist the moments caused by the lateral forces; this moment creates a tension-and-compression force couple in the shear wall chords. The SDPWS specifies that the plywood be attached to the framing members with fasteners spaced not less than ⅜ in. from a panel edge, a minimum of 6 in. o.c. along the panel edges and 12 in. o.c. in the field—that is, along the intermediate framing members (Figure 7.3). Shear walls can be blocked or unblocked; however, reduction factors are applied to the shear capacity for unblocked walls along with other detailing requirements in the SDPWS. It is

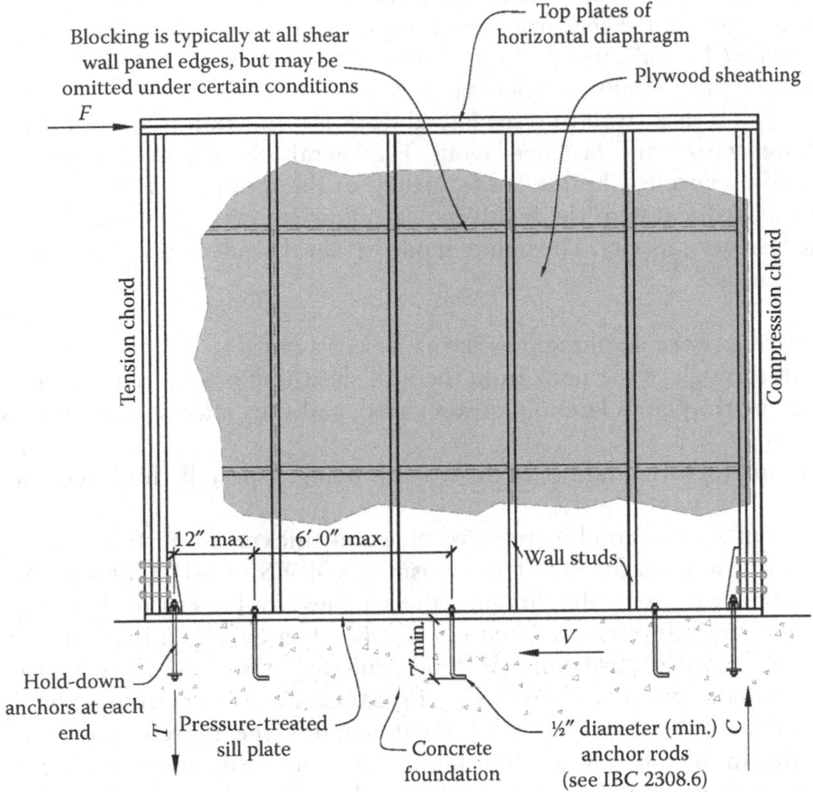

Figure 7.2 Shear wall as a vertical cantilever.

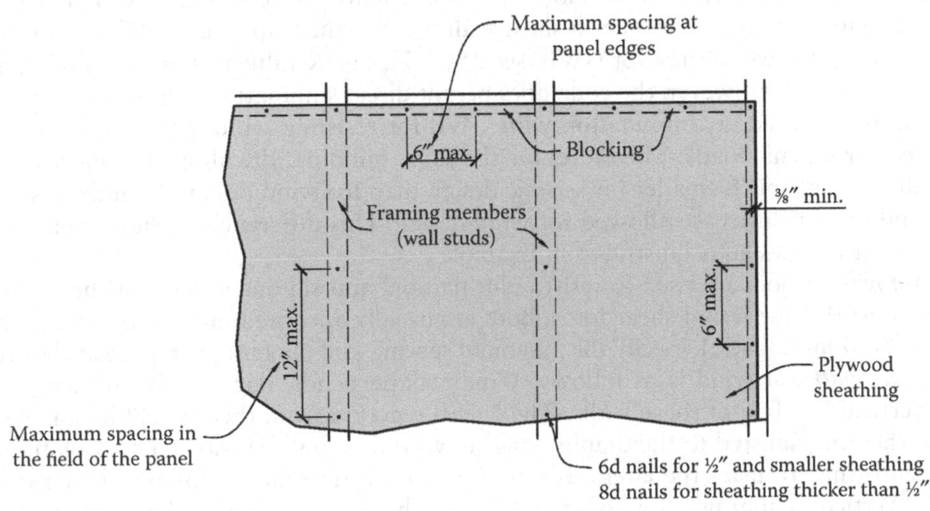

Figure 7.3 Typical minimum fastening for wall panels.

recommended that the location of the nails from the edge of the stud framing be about the same as the distance of the nail from the panel edge. Table 6.1 from the *Wood Structural Panels Supplement* [3] indicates that the minimum nail size should be 6d for wood panels ½ in. and thinner and 8d nails for thicker panels.

The capacity of a shear wall to resist lateral loads is a function of several items: plywood thickness, fastener size, and fastener layout. The lateral load capacity of wood shear wall panels is largely controlled by the load capacity of the fasteners. The plywood itself will rarely fail in shear because of the relatively high shear strength of the wall sheathing compared to the fastener capacity. The failure mode for shear walls includes one or more of the following scenarios:

1. Edge tearing of the diaphragm as the nails bear against it
2. The pull-through of the nails from the wall sheathing or failure of the nails in bending, leaving the panels laterally unsupported, and thus making the panels susceptible to buckling
3. Longitudinal splitting failure of the wood framing, especially for larger nail sizes

The capacities of several combinations of plywood thickness and fasteners are listed in Special Design Provisions for Wind and Seismic (SDPWS of [4]) Tables 4.3A, 4.3B, and 4.3C. As was the case with diaphragms, these tables are broken up between wind and seismic loads as the capacity for wind loading is taken as 40% higher than for seismic loads. The capacities listed are nominal values, and thus, they have to be adjusted for ASD or LRFD loads. For ASD, the allowable unit shear capacity is determined by dividing the value in the table by $\Omega = 2.0$. For LRFD, the design unit shear strength is found by multiplying the value in the table by $\phi = 0.8$ plus any other adjustment factors in the footnotes. These shear capacities apply to plywood fastened to one side of the studs. For plywood fastened to both sides of the studs, the shear capacity is doubled if the sheathing material and fasteners are the same on each side. When the shear capacity is not equal on both sides, the total capacity is the larger of either twice the smaller capacity, or the capacity of the stronger side.

In some cases, such as in exterior walls, plywood is applied on the exterior face and gypsum GWB is applied on the interior face. Although the GWB and fasteners have some capacity to act as a shear wall to resist lateral loads [5], it is usually not recommended that GWB and plywood be used in combination as a shear wall because the response modification factor R is much greater for wood than for GWB (see ASCE 7). The R value is indirectly proportional to the seismic load. However, the code does permit direct summation of the shear capacities for wood panels used in combination with GWB for resisting wind loads only. This is not permitted for seismic loads. Therefore, for the same building, the allowable shear capacity of a wall could be much smaller for seismic design than for wind design given the discussion above and the 40% increase allowed for wind design. This difference in shear capacities for wind and seismic design is illustrated in Example 7.3.

Hold-down anchors are used to resist uplift tension, and sill anchors (or anchor rods) are used to resist the horizontal shear force. Both are usually anchored into the concrete foundation. IBC Sections 2308.3.1 specify the minimum spacing and dimensions for the anchor rods.

The *load path* for wind is as follows: Wind acts perpendicular to wall surfaces, resulting in vertical bending of these walls. The lateral reaction from the vertical bending of the walls is then transmitted to the diaphragms above and below the walls. The roof and floor diaphragms then transfer the lateral reactions to the vertical lateral force-resisting systems (i.e., the vertical diaphragms or shear walls), and the shear walls transfer the lateral loads safely to the ground or foundation.

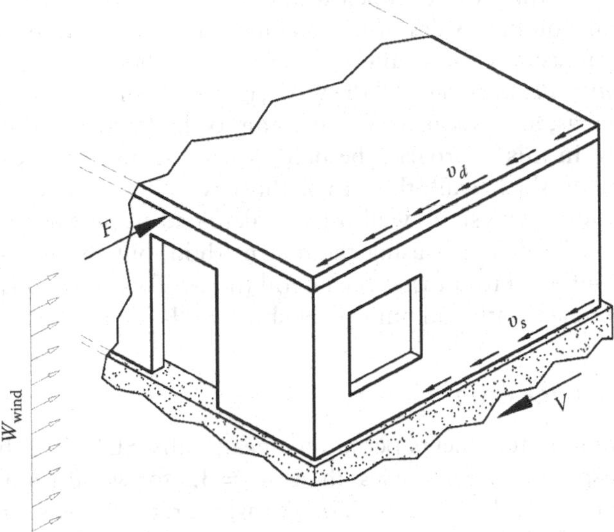

Figure 7.4 Typical load path for lateral loads.

For seismic loads, the ground motion results in the back-and-forth lateral displacement of the base of the building, causing a shear force at the base of the building. This base displacement causes lateral displacements of the diaphragms in the direction opposite to the base displacement. These lateral displacements result in equivalent lateral forces at the various diaphragm levels of the building. The equivalent lateral forces are then transmitted through the roof and floor diaphragms to the vertical lateral force-resisting systems (i.e., the vertical diaphragms or shear walls), and the shear walls transfer the lateral loads safely to the ground or foundation (Figure 7.4).

From Figure 7.4,

F is the lateral force at the horizontal diaphragm for wind loads, F is the resultant from the tributary wind load at the diaphragm for seismic loads, F is calculated (e.g., equivalent lateral force)

V is the base shear

v_d is the unit shear in the horizontal diaphragm

v_s is the unit shear in the vertical diaphragm (shear wall)

7.2 SHEAR WALL ANALYSIS

There are three methods given in Section 4.3.5 of the SDPWS for shear wall analysis:

1. *Segmented approach:* The full height segments between window and door openings are assumed to be the only walls resisting lateral loads; thus, the stiffness contribution of the walls above and below the window or door openings is neglected. In this approach, the hold-down anchors are located at the ends of each wall segment. The shear wall forces calculated using this approach are usually the largest of the three approaches. This is the method adopted in this book and is the most commonly used method in practice.
2. *Force transfer method:* The walls are assumed to behave as a coupled shear wall system. In this case, the hold-down anchors are also located at the extreme ends of the entire shear wall; thus, each shear wall line will typically have only two hold-down anchors. To ensure coupled shear wall action, horizontal straps or ties would have to

be added to tie the wall segments below and above the window and door openings to the wall segments on both sides of the opening. The shear wall forces calculated using this approach are usually the smallest of the three methods.

3. *Perforated shear wall approach:* All the walls, including those above and below window and door openings, are assumed to contribute to the lateral load-carrying capacity of the shear walls. In this approach, the hold-down anchors are located at the extreme ends of the *overall* shear wall; thus, each shear wall line will typically have only two hold-down anchors. The shear wall forces calculated using this approach are usually smaller than those calculated using the first method, but the edges of the wall openings must be reinforced to transfer the lateral forces. This method is sometimes used in buildings where the length and number of shear walls are limited.

7.2.1 Shear wall aspect ratios

Several IBC aspect ratio requirements pertain to shear walls. SDPWS Table 4.3.4 states that the height-to-width aspect ratio must be less than $h/b_s = 3.5$ for wood-framed walls with wood shear wall panels that are blocked (Figure 7.5). The aspect ratio decreases to $h/b_s = 2.0$ for the unblocked case. When the height-to-aspect ratio is greater than 2.0, the following reduction factor is applied to the unit shear capacity of wood structural panel shear walls:

$$R_{WSP} = 1.25 - 0.125\,(h/b_s)$$

where:

R_{WSP} is the reduction factor for wood structural panels
h is the height of the shear wall segment
b_s is the length of the shear wall segment

Example 7.1: Definition of a shear wall

Determine which wall panels can be used as a shear wall in the building elevation shown in Figure 7.6.

Solution:

$$\text{Wall 1:}\quad \frac{h}{b_s} = \frac{12\text{ ft}}{2\text{ ft}} = 6 > 3.5$$

Therefore, wall 1 does not meet the SDPWS and will be neglected for shear action.

$$\text{Wall 2:}\quad \frac{h}{b_s} = \frac{12\text{ ft}}{5\text{ ft}} = 2.4 < 3.5$$

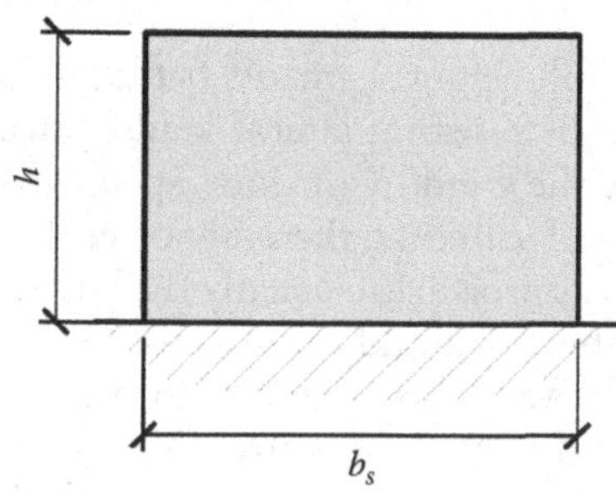

Figure 7.5 Shear wall aspect ratios.

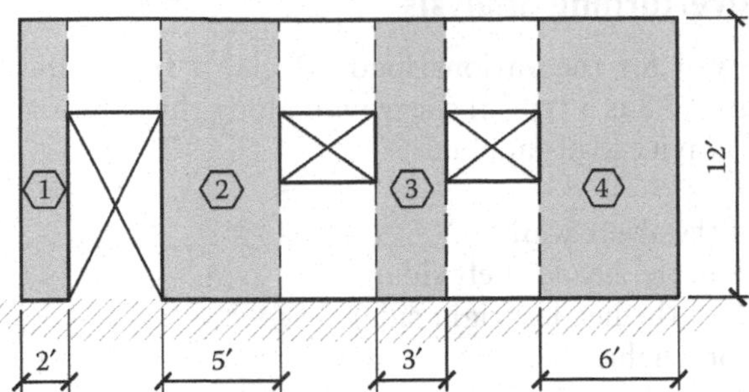

Figure 7.6 Shear wall aspect ratios for Example 7.1.

Therefore, wall 2 meets the SDPWS, but the shear capacity will have to be adjusted by the following factor:

$$WSP_2 = 1.25 - 0.125(h/b_s) = 1.25 - 0.125(12/5) = 0.95$$

Wall 3: $\frac{h}{w} = \frac{12\text{ ft}}{3\text{ ft}} = 4 > 3.5$

Therefore, wall 3 does not meet the SDPWS and will be neglected for shear action.

Wall 4: $\frac{h}{w} = \frac{12\text{ ft}}{6\text{ ft}} = 2 < 3.5$

Therefore, wall 4 meets the SDPWS and does not require a reduction in capacity.
Use wall 2 and wall 4 only to resist lateral loads.

7.2.1.1 Multistory shear wall aspect ratio

When calculating the shear wall aspect ratio in accordance with IBC requirements, the value used for h is the height from the individual wall in question. For proper continuity in the load path for the shear walls, all of the walls in a structure should align from top to bottom (Figure 7.7).

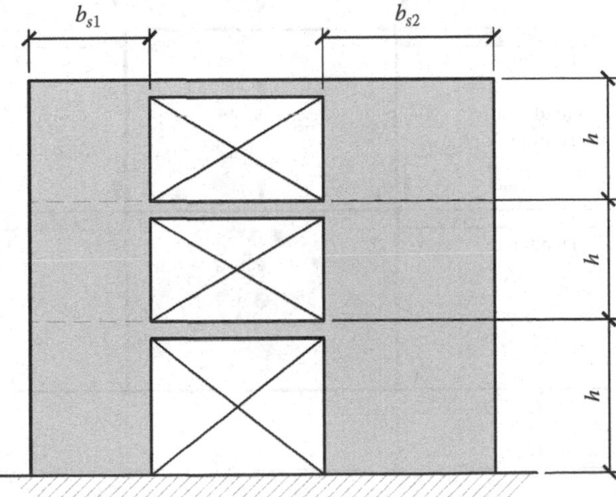

Figure 7.7 Multistory shear wall aspect ratios.

7.2.2 Shear wall overturning analysis

To analyze a shear wall for the various load effects, a free-body diagram must first be drawn. Shown in Figure 7.8 is a typical shear wall along the exterior wall of a building with tributary lateral and gravity loads applied.

From Figure 7.8,

w is the length of the shear wall
$L_{H(1)}$ is the length of the header (left side)
$L_{H(2)}$ is the length of the header (right side)
H_1 is the first-floor height
H_2 is the second-floor height
F_{roof} is the force applied to the shear wall at the roof level (either wind or seismic)
F_{2nd} is the force applied to the shear wall at the second floor (either wind or seismic)
W_{2nd} is the self-weight of the shear wall at the second floor
W_{1st} is the self-weight of the shear wall at the first floor
R_{roof} is the force due to uniform load on the shear wall at the roof = $w_{roof}\, w$
R_{2nd} is the force due to uniform load on the shear wall at the second floor = $w_{2nd} w$

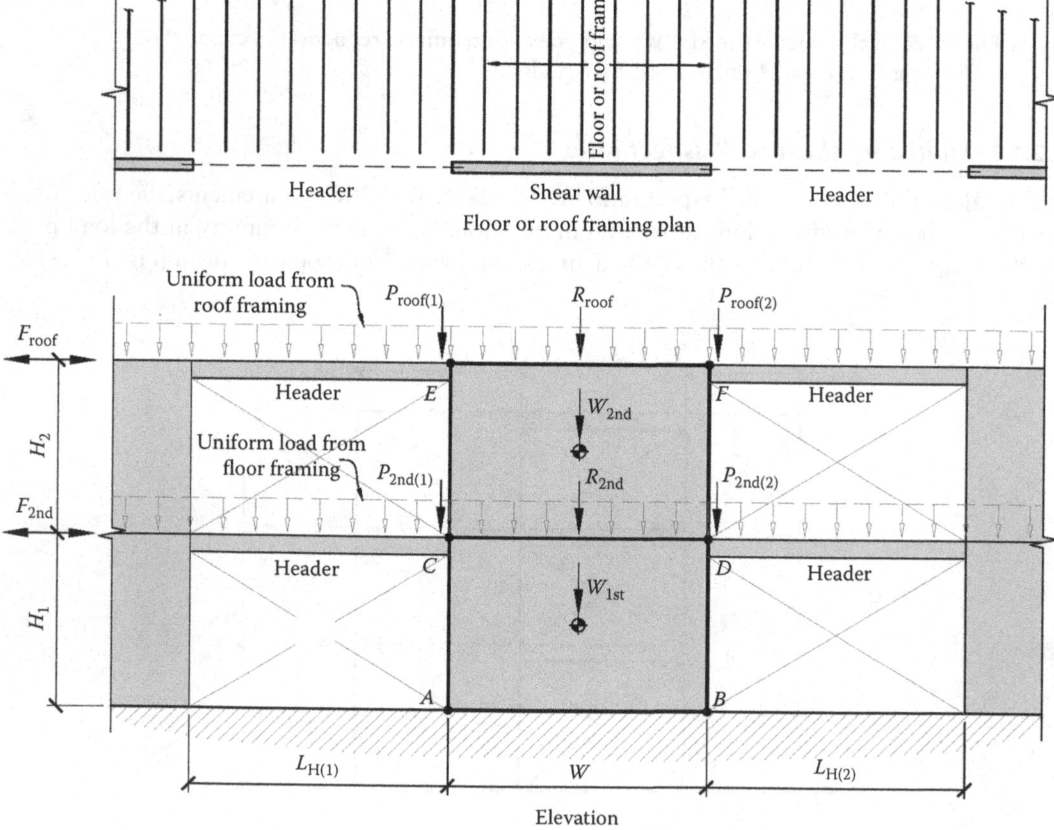

Figure 7.8 Loads on a typical shear wall: (a) Floor or roof framing plan and (b) elevation.

P_{roof} is the force due to uniform load on the header at the roof

$$= \frac{w_{roof}L_{H1}}{2} \text{ or } \frac{w_{roof}L_{H2}}{2}$$

P_{2nd} is the force due to uniform load on the header at the second floor

$$= \frac{w_{2nd}L_{H1}}{2} \text{ or } \frac{w_{2nd}L_{H2}}{2}$$

The uniform loads are resolved into resultant forces to simplify the overturning analysis. The uniform loads along the wall (i.e., the load from the roof or floor framing and the self-weight of the wall) act at the centroid of the wall, or at a distance of $w/2$ from the end of the shear wall.

The lateral wind or seismic loads acting on the shear wall can act in either direction. If the reactions from the left and right headers are equal, the overturning analysis has to be performed with the lateral load in only one direction. If the header reactions are not equal, the overturning analysis must be performed with the lateral applied in both directions, as will be shown later.

The loads that act on the roof level that result in the applied forces R_{roof} and P_{roof} can be some combination of dead, roof live, snow, or wind loads, depending on the governing load combination. Similarly, the loads that act on the second floor that result in the applied forces R_{2nd} and P_{2nd} can be some combination of dead and floor live loads. Figure 7.8 is somewhat generic but is typical for most shear walls. It is important to note that it is possible to have other gravity loads acting on the shear wall (such as a header from another beam framing perpendicular into the shear wall). These loads must also be accounted for in the overturning analysis.

The free-body diagram shown is specific for a two-story structure; however, it can be seen by inspection that this analysis can be used for a building with any number of stories. For simplicity, a two-story example is used here. The overturning analysis must also be performed at each level for multistory structures. This will require another isolated free-body diagram for each level, as shown in Figure 7.9 for the two-story structure. From Figure 7.9,

T_1 and C_1 = shear wall *tension chord* and *compression chord* forces at the ground floor level

T_2 and C_2 = shear wall *tension chord* and *compression chord* forces at the second floor

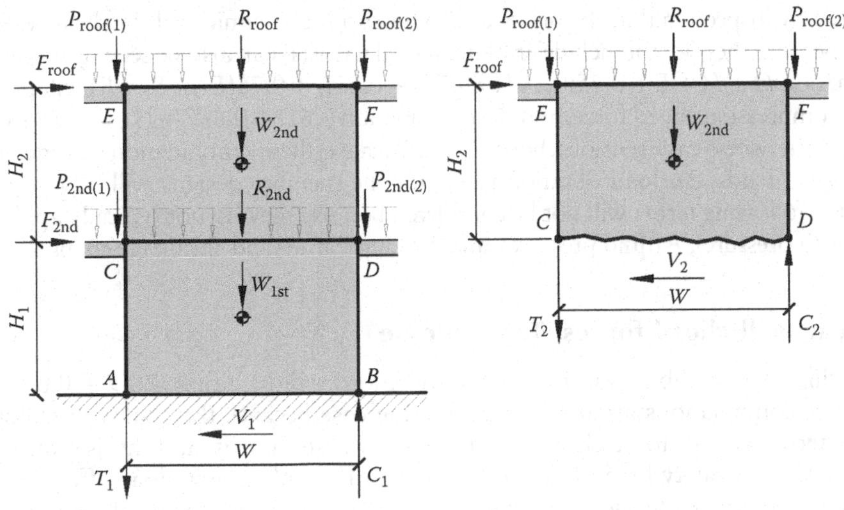

Figure 7.9 Loads and reactions on a typical shear wall.

Note: For calculating the T_1 and T_2 chord forces, the *full* dead loads on the shear wall are used, whereas for C_1 and C_2 chord forces, only the *tributary* gravity loads apply; therefore, only the gravity loads in the compression chords are used.

V_1 is the base shear at the ground level ($= F_{roof} + F_{2nd}$)

V_2 is the base shear at the second floor ($= F_{roof}$)

The lateral forces in this example are shown acting in one direction to show the resulting tension and compression chord forces.

The overturning moment can be calculated at each level as follows: At the second floor summing the moments about point C or D yields

$$OM_2 = F_{roof} H_2 \tag{7.1}$$

At the ground level, summing the moments about point A or B yields

$$OM_1 = [F_{roof}(H_2 + H_1)] + (F_{2nd} H_1) \tag{7.2}$$

The gravity loads prevent the shear wall from overturning or toppling over by creating a resisting moment that opposes the overturning moment. The magnitude of this *resisting moment* is the summation of moments about the same point that the overturning occurs. The applied loads in the overturning analysis are to be in accordance with the code-required load combinations for the ASD method, which are shown below for reference. Equations 1 through 7 are for the LRFD method and are not considered here.

8. D
9. $D + H + L$
10. $D + H + (L_r$ or S or $R)$
11. $D + H + 0.75\ (L) + 0.75\ (L_r$ or S or $R)$
12. $D + H + (0.6W$ or $0.7E)$
13. $D + H + 0.75\ (0.6W) + 0.75L + 0.75\ (L_r$ or S or $R)$
14. $D + H + 0.75\ (0.7E) + 0.75L + 0.75\ (S)$
15. $0.6D + 0.6W + H$ (D always opposes W and H)
16. $0.6D + 0.7E + H$ (D always opposes E and H)

It can be seen by inspection that the first four combinations are eliminated for shear wall analysis and design because they do not include any lateral load terms. It can also be seen by inspection that the combination $D + H + F + 0.75(0.6W$ or $0.7E) + 0.75L + 0.75(L_r$ or S or $R)$ will produce the worst-case compression chord force, and the combination $0.6D + 0.6W + H$ or $0.6D + 0.7E + H$ will produce the worst-case tension chord forces. Because these combinations all include either wind or seismic loads, the load duration factor for the member designs will be $C_D = 1.6$. For simplicity, the following terms will not be considered because they do not typically apply to shear walls: H (earth pressure), F (fluid pressure), and T (temperature and shrinkage change).

7.2.3 Shear wall chord forces: Tension case

The controlling load combination for the tension chord will be either $0.6D + 0.6W$ or $0.6D + 0.7E$. These combinations are used for the tension case because the goal is to maximize the value in the tension chord to determine what the worst-case value is, and this is done by having the least amount of gravity load present in the model. The only gravity load effect present that resists overturning for the tension case is the dead load. Given that the dead load is multiplied by a 0.6 term, it is implied that there is a factor of safety of 1.67 (or $1/0.6$) against overturning.

When calculating the maximum tension chord force (either T_1 or T_2 in Figures 7.8 and 7.9), the point at which to sum moments must first be determined. For the tension case, the following rules apply:

1. When the header reactions are equal, the summation of moments can be taken about any point (e.g., *A* or *B* at the base, or point *C* or *D* at the second level in Figure 7.8 or 7.9).
2. When the header reactions are not equal, the summation of moments should be taken about a point that lies on the same vertical line as the *larger* header reaction.

When the header reactions are not equal, it should be observed that the gravity load is minimized (and thus the tension chord force is maximized) when the summation of moments is taken about a point under the larger header reaction.

7.2.3.1 Wind load combination

Knowing that the governing load combination for wind loads is $0.6D + 0.6W$, the value T_1 or T_2 can be calculated from the free-body diagram in accordance with the following equilibrium equations by summing the moments about the compression chord:

$$\mathrm{OM}_2 - 0.6\mathrm{RM}_{D2} - (T_2 w) = 0$$

Solving for T_2 yields

$$T_2 = \frac{\mathrm{OM}_2 - 0.6\mathrm{RM}_{D2}}{w} \tag{7.3}$$

where:

OM_2 is the **unfactored** overturning moment at the second level caused by the unfactored wind load $0.6W$ (see Equation 7.1)
RM_{D2} is the summation of the dead load resisting moments at the second level
w is the shearwall length

Similarly, an equation for T_1 can also be developed:

$$T_1 = \frac{\mathrm{OM}_1 - 0.6\mathrm{RM}_{D1}}{w} \tag{7.4}$$

where:

OM_1 is the unfactored overturning moment caused by the unfactored wind load $0.6W$
RM_{D1} is the summation of dead load resiting moments at the ground level

7.2.3.2 Seismic load combination

The governing load combination for seismic loads is $0.6D + 0.7E$. Section 12.4.2 of ASCE 7 [6] defines the term E as follows:

$$E = \rho Q_E + 0.2S_{DS}D \quad \text{or} \quad E = \rho Q_E - 0.2S_{DS}D$$

Q_E is the horizontal effect of the seismic loads
S_{DS} is the design spectral response acceleration (see Section 11.4.4 of ASCE 7)
D is the dead load
ρ is the redundancy coefficient, see Section 12.3.4 of ASCE 7

Note: When using the simplified procedure to calculate seismic loads, $\rho = 1.0$ for seismic design categories A, B, and C. For seismic design categories D, E, and F, the reader is referred to Section 12.3.4 of ASCE 7.

By inspection, it can be seen that the equation $E = \rho Q_E - 0.2S_{DS}D$ will govern for the design of the shear wall tension chord because the dead load effect is minimized, whereas $E = \rho Q_E + 0.2S_{DS}D$ will govern for the design of the compression chord.

The value of ρ can either be 1.0 or 1.3, and the value of S_{DS} can vary from 0.1 in low-seismic hazard areas to over 3.0 in a high-seismic hazard areas.

For the tension chord, $E = \rho QE - 0.2S_{DS}D$

Substituting this in the governing load combination ($0.6D + 0.7E$) yields

$0.6D + 0.7(\rho Q_E - 0.2S_{DS}D) = (0.6 - 0.14S_{DS})D + 0.7\rho Q_E$ governing load combination

The value T_1 or T_2 can now be determined for the seismic load case as follows:

$$T_2 = \frac{(0.7\rho)\text{OM}_2 - (0.6 - 0.14S_{DS})\text{RM}_{D2}}{w} \tag{7.5}$$

$$T_1 = \frac{(0.7\rho)\text{OM}_1 - (0.6 - 0.14S_{DS})\text{RM}_{D1}}{w} \tag{7.6}$$

Note: If T_1 or $T_2 \leq 0$, this implies no uplift on tension chord
If T_1 or $T_2 \geq 0$, this implies uplift on tension chord

7.2.4 Shear wall chord forces: Compression case

The controlling load combination for the compression chord is $D + 0.75(0.6W$ or $0.7E) + 0.75L + 0.75(L_r$ or S or $R)$. The code allows the effects from two or more transient loads to be reduced by a factor of 0.75. Therefore, the governing load combination can be rewritten as follows:

$D + 0.75[0.6W + L + (L_r \text{ or } S \text{ or } R)]$ for the wind load case

$D + 0.75[0.7E + L + (L_r \text{ or } S \text{ or } R)]$ for the seismic load case

The term that accounts for the rain load effect R is ignored in this example because it typically does not control for most wood buildings. Also, the *larger* of the roof live load L_r and the snow load S should be used in the overturning analysis.

The general equations for determining the worst-case compression chord force can be developed similar to the equations developed earlier for the tension chord. The equations developed for the compression case will have to include an expanded form of the resisting moment RM term to include both a dead load and a transient load term because different numerical coefficients will apply to each term. Recall that the tension case involved only dead load effects, whereas the compression case will include other gravity load effects because the goal is to maximize the compression chord force. It should be recalled that only the *tributary* gravity loads on the compression chord are used to calculate the compression chord forces.

When calculating the maximum compression chord force (either C_1 or C_2 in Figures 7.8 and 7.9), the point about which to sum moments must first be determined. For the compression case, the following rules apply:

1. When the header reactions are equal, the summation of moments can be taken about any point (e.g., *A* or *B* at the base, or point *C* or *D* at the second level in Figure 7.8 or 7.9).
2. When the header reactions are not equal, the summation of moments should be taken about a point that lies on the same vertical line as the *smaller* header reaction.

When the header reactions are not equal, it should be observed that the gravity load is maximized (and thus the compression chord force is maximized) when the summation of moments is about a point under the smaller header reaction.

7.2.4.1 Wind load combination

The value of C_1 or C_2 can be calculated from the free-body diagram in accordance with the following equilibrium equations by summing the moments about the tension chord at each level:

$$0.75\text{OM}_2 + \text{RM}_{D2} + 0.75\text{RM}_{T2} - C_2 w = 0$$

Solving for C_2 yields

$$C_2 = \frac{0.75\text{OM}_2 + \text{RM}_{D2} + 0.75\text{RM}_{T2}}{w} \tag{7.7}$$

where:

OM_2 is the unfactored overturning moment at the second level caused by $0.6W$ (see Equation 7.1)
RM_{D2} is the summation of the tributary dead load resisting moments at the second level
RM_{T2} is the summation of the tributary transient load resisting moments at the second level

Similarly, an equation for C_1 can also be developed by summing the moments about point *A*:

$$C_1 = \frac{0.75\text{OM}_1 + \text{RM}_{D1} + 0.75\text{RM}_{T1}}{w} \tag{7.8}$$

where:

OM_1 is the unfactored overturning moment caused by the unfactored wind load $0.6W$
RM_{D1} is the summation of dead load resiting moments at the ground level
RM_{T1} is the summation of transient (live) load resisting moments at the ground level

7.2.4.2 Seismic load combination

From the seismic load effect for the tension case, we can write

$$E = \rho Q_E + 0.2S_{DS}D$$

Combining this with the governing load combination yields

$$(1.0 + 0.105S_{DS})D + 0.75[0.7\rho Q_E + L + (L_r \text{ or } S \text{ or } R)]$$

or

$$(1.0 + 0.105S_{DS})D + 0.525\rho Q_E + 0.75L + 0.75\,(L_r \text{ or } S \text{ or } R) \quad \text{governing load combination}$$

The value of C_1 or C_2 for the seismic case can now be calculated as follows:

$$C_2 = \frac{0.525\rho OM_2 + (1.0 + 0.105S_{DS})RM_{D2} + 0.75RM_{T2}}{w} \tag{7.9}$$

$$C_1 = \frac{0.525\rho OM_1 + (1.0 + 0.105S_{DS})RM_{D1} + 0.75RM_{T1}}{w} \tag{7.10}$$

where OM_1 and OM_2 are the overturning moments caused by the factored horizontal seismic force, Q_E. Note that the load factors that converts Q_E to unfactored loads is already embedded in Equations 7.9 and 7.10.

In calculating the resisting moments (RM_{D1}, RM_{D2}, RM_{T1}, RM_{T2}) in Equations 7.7 through 7.10, it should be noted that the full *P* loads or header reactions (see Figure 7.10) and the tributary *R* and *W* (i.e., distributed loads) loads should be used because the compression chords only support distributed gravity loads that are tributary to it, in addition to the concentrated header reactions or *P* loads. The tributary width of the compression chord is one-half of the stud spacing. Thus,

$$R_{trib} = \frac{R(\frac{1}{2} \times \text{stud spacing})}{w}$$

$$W_{trib} = \frac{W(\frac{1}{2} \times \text{stud spacing})}{w}$$

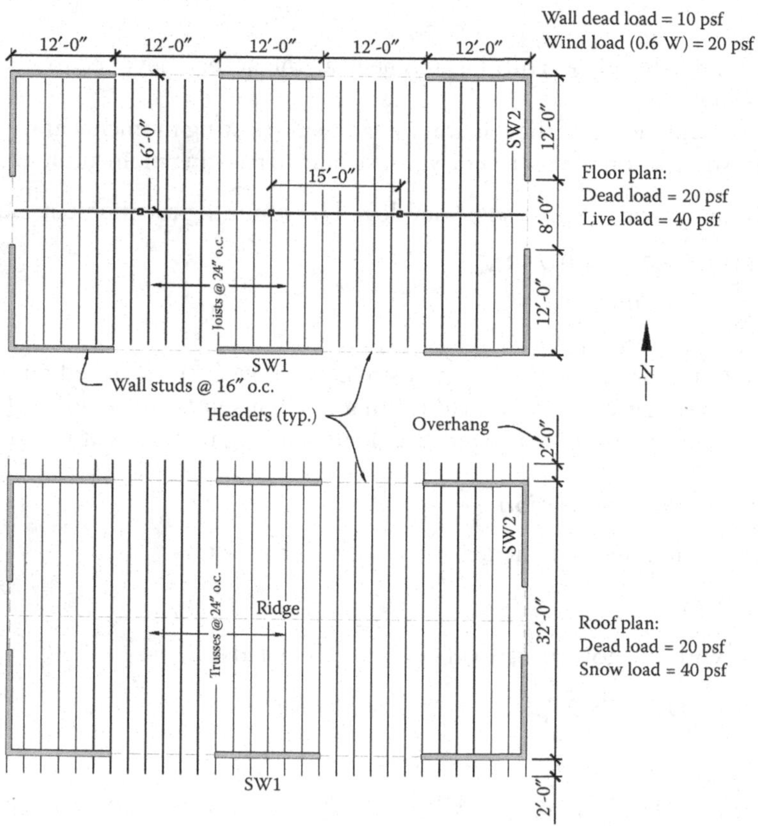

Figure 7.10 Building plan for Example 7.2.

Recall that R and W are the total loads on the whole shear wall, whereas the terms R_{trib} and W_{trib} represent the amounts of gravity load that is tributary to the compression chords. Because these loads are located at the compression chord (i.e., at the end of the wall), the moment arm relative to the tension chord for the terms R_{trib} and W_{trib} is equal to the length of the wall w and should be used accordingly in the resisting moment (RM_{D1}, RM_{D2}, RM_{T1}, RM_{T2}) calculations. The resisting moment about the tension chord for the R_{trib} and W_{trib} terms can be written as follows:

$$\text{RM}(R_{\text{trib}}) = \frac{R \times (\frac{1}{2} \times \text{stud spacing})}{w} \times w = R(\tfrac{1}{2} \times \text{stud spacing})$$

$$\text{RM}(W_{\text{trib}}) = \frac{W \times (\frac{1}{2} \times \text{stud spacing})}{w} \times w = W \times (\tfrac{1}{2} \times \text{stud spacing})$$

7.3 SHEAR WALL DESIGN PROCEDURE

The procedure for designing shear walls is as follows:

1. Calculate the lateral forces at each level (wind and seismic loads) and the gravity loads that act on the shear wall. Draw free-body diagrams similar to those shown in Figures 7.8 and 7.9.
2. Design the shear wall for the unit shear in the shear wall at each level. Give the full specification for the shear wall as follows:
   a. Plywood grade and thickness
   b. Nail size and penetration
   c. Nail spacing (at edges and field). Recall that blocking is required at all panel edges
3. Calculate the shear wall chord forces (T_1, T_2, C_1, C_2, etc.). Design the chords for these forces. Recall that the weak axis of the wall studs are braced by the wall sheathing, and the unbraced length of the chord for bending about the strong axis is equivalent to the floor-to-floor height.
4. Design the connections:
   a. Hold-down anchors to resist T_1 or T_2
   b. Sill anchors to resist the base shear V_1
   c. Between the horizontal and vertical diaphragms to ensure a continuous load path

Example 7.2: Design of shear walls
Design shear walls SW1 and SW2 for the two-story building shown in Figure 7.10 considering wind loads only. Assume a floor-to-floor height of 10 ft. Hem-Fir lumber framing is used and assume an unfactored lateral wind load of 20 psf, which has been normalized for service level (i.e., 20 psf = 0.6W). See Figures 7.11 and 7.12 for additional information.

Solution:
East–West wind:

$$F_{r(\text{E-W})} = (20\text{ psf})\left(\frac{6\text{ ft}}{2} + \frac{10\text{ ft}}{2}\right)(32\text{ ft}) = \mathbf{5120\text{ lb}}$$

$$F_{2(\text{E-W})} = (20\text{ psf})\left(\frac{10\text{ ft}}{2} + \frac{10\text{ ft}}{2}\right)(32\text{ ft}) = \mathbf{6400\text{ lb}}$$

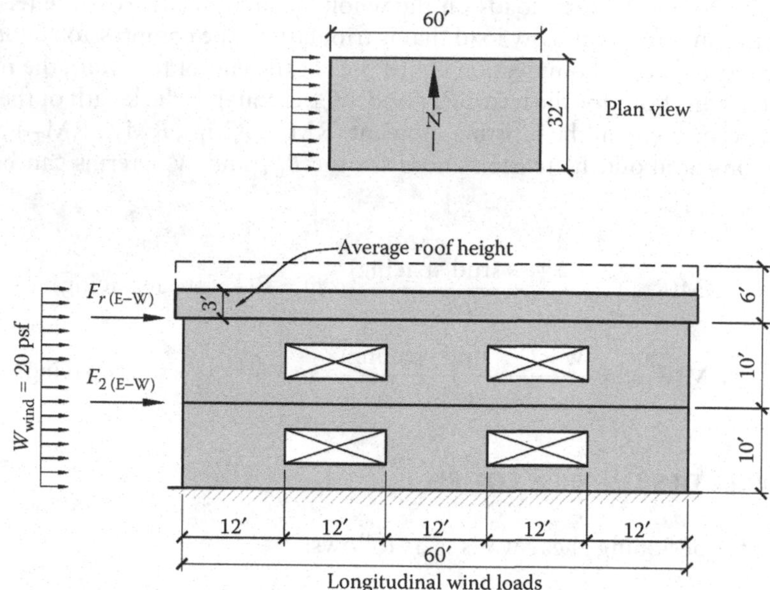

Figure 7.11 Longitudinal elevation for Example 7.2.

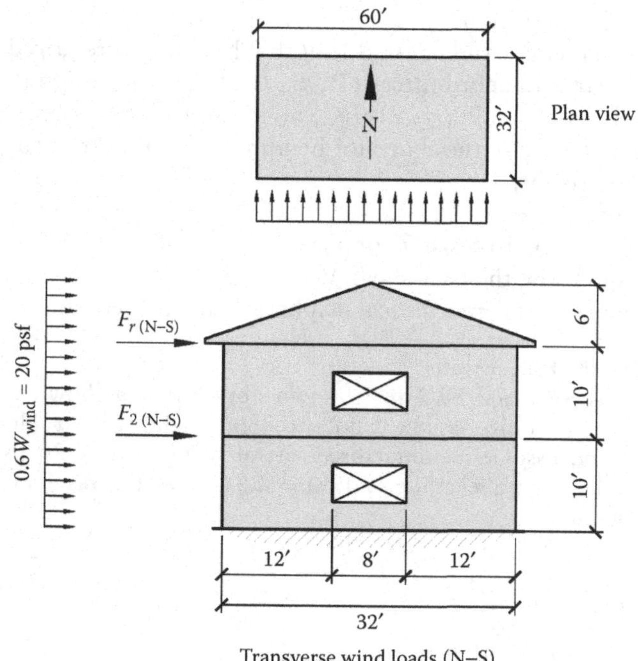

Figure 7.12 Transverse elevation for Example 7.2.

These values represent the total force at each level. They will have to be distributed to each shear wall in the E–W direction.

North–South wind:

$$F_{r(\text{N-S})} = (20 \text{ psf})\left(6 \text{ ft} + \frac{10 \text{ ft}}{2}\right)(60 \text{ ft}) = \mathbf{13,200 \text{ lb}}$$

$$F_{2(\text{N-S})} = (20 \text{ psf})\left(\frac{10 \text{ ft}}{2} + \frac{10 \text{ ft}}{2}\right)(60 \text{ ft}) = \mathbf{12,000 \text{ lb}}$$

These values represent the total force at each level. They will have to be distributed to each shear wall in the N–S direction.

Lateral loads on the E–W shear wall (SW1). Research has shown that wood diaphragms are typically classified *as flexible diaphragms* (see further discussion in Chapter 6); thus, the amount of lateral load on any individual shear wall is proportional to the tributary area of the shear wall in question. For this example, the lateral load on SW1 is proportional to the length of the wall relative to the other east–west shear walls. We must first check that SW1 meets the aspect ratio requirements in IBC Section 2305.3.3:

$$\frac{h}{w} \le 2.0 \text{ (assumes no reduction in strength)}$$

$$\frac{10 \text{ ft} + 10 \text{ ft}}{12 \text{ ft}} = \mathbf{1.67 \le 2.0}$$

Therefore, SW1 meets the ideal aspect ratio requirements. There are six shear walls, each 12 ft long, in the E–W direction, so the lateral load to SW1 at each level is

$$F_r = \frac{\text{Length of SW1}}{\sum \text{All shear wall lengths}} F_{r(\text{E-W})}$$

$$= \frac{12 \text{ ft}}{(6)(12 \text{ ft})}(5120) = \mathbf{854 \text{ lb}}$$

$$F_2 = \frac{\text{Length of SW1}}{\sum \text{All shear wall lengths}} F_{2(\text{E-W})}$$

$$= \frac{12 \text{ ft}}{(6)(12 \text{ ft})}(6400)$$

$$= \mathbf{1067 \text{ lb}}$$

Gravity loads on the E–W shear wall (SW1). At the roof level:

Direct load on the wall:

$$R_D = (20 \text{ psf})\underbrace{\left(\frac{32 \text{ ft}}{2} + 2 \text{ ft}\right)}_{\text{(TW+overhang)}}\underbrace{(12 \text{ ft})}_{\text{(wall length)}} = \mathbf{4320 \text{ lb}}$$

$$R_S = (40 \text{ psf})\left(\frac{32 \text{ ft}}{2} + 2 \text{ ft}\right)(12 \text{ ft}) = \mathbf{8640 \text{ lb}}$$

$$W_D = (10 \text{ psf})\underbrace{(10 \text{ ft})}_{\text{(wall height)}}\underbrace{(12 \text{ ft})}_{\text{(wall length)}} = \mathbf{1200 \text{ lb}}$$

Reaction from headers (both ends equal):

$$P_D = (20 \text{ psf})\left(\frac{32 \text{ ft}}{2} + 2 \text{ ft}\right)\underbrace{\left(\frac{12 \text{ ft}}{2}\right)}_{\text{(half of header length)}} = \mathbf{2160 \text{ lb}}$$

$$P_S = (40 \text{ psf})\left(\frac{32 \text{ ft}}{2} + 2 \text{ ft}\right)\left(\frac{12 \text{ ft}}{2}\right) = \mathbf{4320 \text{ lb}}$$

At the second floor:
Direct load on the wall:

$$R_D = (20 \text{ psf})\underbrace{\left(\frac{16 \text{ ft}}{2}\right)}_{\text{(TW)}}\underbrace{(12 \text{ ft})}_{\text{(wall length)}} = \mathbf{1920 \text{ lb}}$$

$$R_L = (40 \text{ psf})\left(\frac{16 \text{ ft}}{2}\right)(12 \text{ ft}) = \mathbf{3840 \text{ lb}}$$

$$W_D = (10 \text{ psf})\underbrace{(10 \text{ ft})}_{\text{(wall height)}}\underbrace{(12 \text{ ft})}_{\text{(wall length)}} = \mathbf{1200 \text{ lb}}$$

Reaction from headers (both ends equal):

$$P_D = (20 \text{ psf})\left(\frac{16 \text{ ft}}{2}\right)\underbrace{\left(\frac{12 \text{ ft}}{2}\right)}_{\text{(half of header length)}} = \mathbf{960 \text{ lb}}$$

$$P_L = (40 \text{ psf})\left(\frac{16 \text{ ft}}{2}\right)\left(\frac{12 \text{ ft}}{2}\right) = \mathbf{1920 \text{ lb}}$$

See Figure 7.13 for the loads on shear wall SW1.

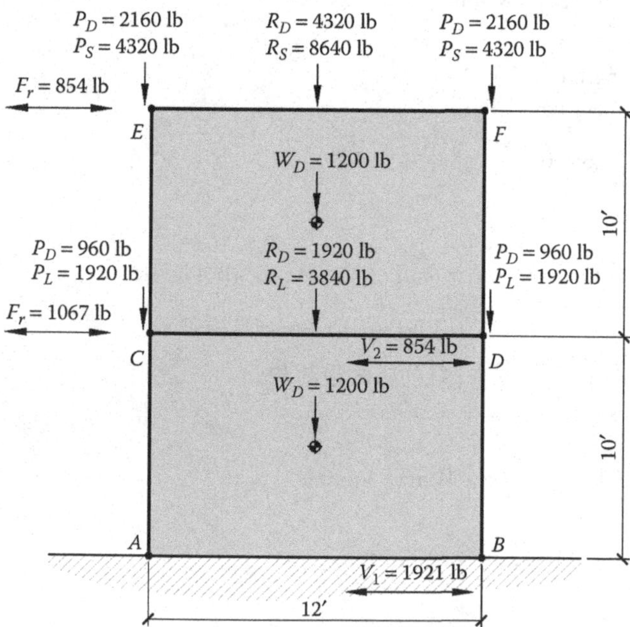

Figure 7.13 Free-body diagram of SW1 in the E–W direction.

Lateral loads on the N–S shear wall (SW2). The lateral load on SW2 is proportional to the length of the wall relative to the other N–S shear walls. We must first check that SW2 meets the aspect ratio requirements in the SDPWS:

$$\frac{h}{w} \leq 2.0 \text{ (assumes no reduction in strength)}$$

$$\frac{10\text{ ft} + 10\text{ ft}}{12\text{ ft}} = \mathbf{1.67 \leq 2.0}$$

Therefore, SW2 meets the ideal aspect ratio requirements. There are four shear walls, each 12 ft long, in the N–S direction, so the lateral load to SW2 at each level is

$$F_r = \frac{\text{Length of SW2}}{\Sigma \text{ All shear wall lengths}} F_{r(\text{N-S})} = \frac{12\text{ ft}}{(4)(12\text{ ft})}(13{,}200) = \mathbf{3300\ lb}$$

$$F_2 = \frac{\text{Length of SW2}}{\Sigma \text{ All shear wall lengths}} F_{2(\text{N-S})} = \frac{12\text{ ft}}{(4)(12\text{ ft})}(12{,}000) = \mathbf{3000\ lb}$$

Gravity loads on the E–W shear wall (SW2). At the roof level:
Direct load on the wall:

$$R_D = (20\text{ psf})\underbrace{\left(\frac{2\text{ ft}}{2}\right)}_{\text{(truss spacing)}}\underbrace{(12\text{ ft})}_{\text{(wall length)}} = \mathbf{240\ lb}$$

$$R_S = (40\text{ psf})\left(\frac{2\text{ ft}}{2}\right)(12\text{ ft}) = \mathbf{480\ lb}$$

$$W_D = (10\text{ psf})\underbrace{(10\text{ ft})}_{\text{(wall height)}}\underbrace{(12\text{ ft})}_{\text{(wall length)}} = \mathbf{1200\ lb}$$

Reaction from the left-side header ($P_{(1)}$):

$$P_D = \left(20\text{ psf} \times \underbrace{\frac{2\text{ ft}}{2}}_{\text{truss spacing}}\right) \times \frac{8\text{ ft}}{2} = 80\text{ lb}$$

$$P_S = \left(40\text{ psf} \times \underbrace{\frac{2\text{ ft}}{2}}_{\text{truss spacing}}\right) \times \frac{8\text{ ft}}{2} = 160\text{ lb}$$

Reaction from the right-side header ($P_{(2)}$):
Because there is no header on the right-hand side of SW2,

$$P_D = 0$$

$$P_S = 0$$

At the second floor:
Direct load on the wall:

$$R_D = (20\text{ psf})\underbrace{\left(\frac{2\text{ ft}}{2}\right)}_{\text{(joist spacing)}}\underbrace{(12\text{ ft})}_{\text{(wall length)}} = \mathbf{240\ lb}$$

$$R_L = (40\text{ psf})\left(\frac{2\text{ ft}}{2}\right)(12\text{ ft}) = \mathbf{480\ lb}$$

$$W_D = (10 \text{ psf})(10 \text{ ft})(12 \text{ ft}) = \mathbf{1200 \text{ lb}}$$
(wall height) (wall length)

Reaction from the left-side header:

$$\text{Tributary area of header reaction} = \frac{\overset{\text{(trib. width)}}{(16 \text{ ft}/2 + 16 \text{ ft}/2)} \; \overset{\text{(half of girder length)}}{(15 \text{ ft}/2)}}{2} = 60 \text{ ft}^2$$

$$P_{D1} = (20 \text{ psf})(60 \text{ ft}^2) = \mathbf{1200 \text{ lb}}$$

$$P_{L1} = (40 \text{ psf})(60 \text{ ft}^2) = \mathbf{2400 \text{ lb}}$$

Reaction from the right-side header:
There is no header on the right end; therefore,

$$P_{D2} = 0$$
$$P_{L2} = 0$$

See Figure 7.14 for the loads on shear wall SW2.
Design of wall sheathing for SW1.
Unit shears:

$$v_2 = \frac{F_r}{\text{Wall length}} = \frac{854 \text{ lb}}{12 \text{ ft}} = \mathbf{72 \text{ lb/ft}}$$

$$v_1 = \frac{F_r + F_2}{\text{Wall length}} = \frac{(854 \text{ lb} + 1067 \text{ lb})}{12 \text{ ft}} = \mathbf{160 \text{ lb/ft}}$$

The shear wall–fastening pattern could be different for each level, but both values are relatively low enough such that a minimum plywood thickness and fastening pattern will

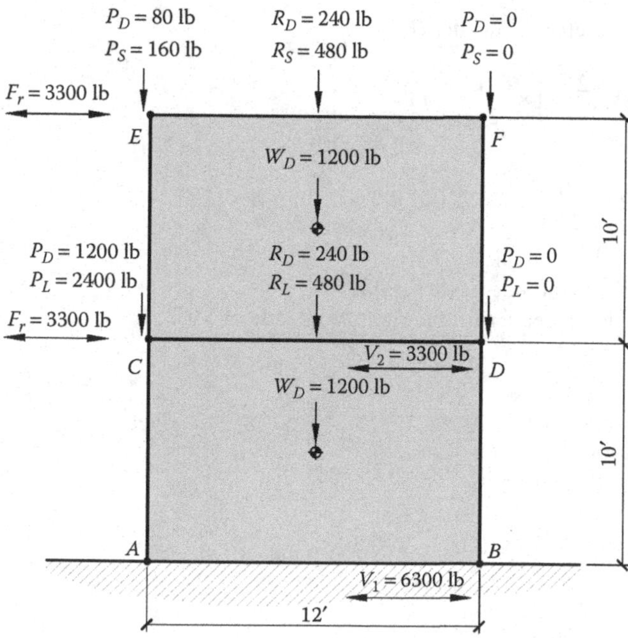

Figure 7.14 Free-body diagram of SW2 in the N–S.

be required. Plywood of ⅜ in. thickness is selected as a minimum practical size to allow flexibility in architectural finishes (cf. SDPWS Table 3.2.1). From SDPWS Table 4.3A, select ⅜-in.-thick Structural I CD–X, 8d nails at 6 in. o.c. (panel edges) and 12 in. o.c. (panel field), minimum penetration = 1⅜ in. Note that blocking is required for all shear walls with an aspect ratio greater than 1.5 (see SDPWS Table 4.3.4). $v_w = 645$ plf.

This value is valid only for wall framing with Douglas Fir-Larch or Southern Pine. For wall framing with other wood species, the values must be reduced in accordance with SDPWS Table 4.3A. From NDS Table 12.3.2A [7], the specific gravity of Hem-Fir wood species is

$$\gamma_{\text{Hem-Fir}} = 0.43$$

Adjusted shear capacity. As Applying the specific gravity adjustment factor and the strength reduction factor:

$$\text{SGAF} = 1 - (0.5 - G)$$

$$\text{SGAF}_{\text{Hem-Fir}} = 1 - (0.5 - 0.43) = 0.93$$

$$v_{\text{cap}} = \frac{v_w}{\Omega}\text{SGAF} = \frac{645}{2.0}(0.93) = \mathbf{299\ plf}$$

$$v_{\text{cap}} = (230\ \text{plf})\ (0.93) = 213\ \text{plf}$$

Comparing the actual with the allowable values, we have

$$\mathbf{v_1 = 160\ plf < v_{cap} = 299\ plf \quad OK}$$

Tension chord force (SW1). Isolating the upper level of the shear wall and solving for T_2 from the free-body diagram by summing the moments about point D (Figure 7.15) yields

$$\text{OM}_2 = (854\ \text{lb})(10\ \text{ft}) = 8540\ \text{ft-lb}$$

$$\text{RM}_{D2} = [(2160)(12)] + \left[(4320 + 1200)\left(\frac{12\ \text{ft}}{2}\right)\right]$$

$$= 59{,}040\ \text{ft-lb}$$

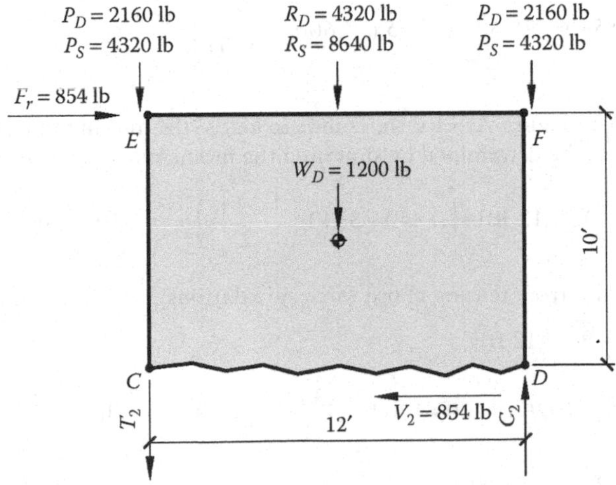

Figure 7.15 Tension chord force, SW1.

$$T_2 = \frac{OM_2 - 0.6RM_{D2}}{w} \text{ (from Equation 7.3)}$$

$$= \frac{(8540) - (0.6)(59{,}040)}{12 \text{ ft}} = -2240 \text{ lb} \quad \text{(the minus sign indicates no net uplift)}$$

By referring back to Figure 7.13 for the complete loads, the maximum chord force at the ground level can be determined by summing the moments about point B:

$$OM_1 = [(854)(10 \text{ ft} + 10 \text{ ft})] + [(1067)(10 \text{ ft})] = 27{,}750 \text{ ft-lb}$$

$$RM_{D1} = [(2160 + 960)(12 \text{ ft})] + \left[(4320 + 1920 + 1200 + 1200)\left(\frac{12 \text{ ft}}{2}\right)\right] = 89{,}280 \text{ ft-lb}$$

$$T_1 = \frac{OM_1 - 0.6RM_{D1}}{w} \text{ (from Equation 7.4)}$$

$$T_1 = \frac{(27{,}750) - (0.6)(89{,}280)}{12 \text{ ft}} = -2152 \text{ lb (the minus sign indicates no net uplift)}$$

Compression chord force (SW1). Referring back to Figure 7.15, and solving for C_2 from the free-body diagram by summing the moments about point C, and accounting for the R and W loads that are *tributary* to the compression chord (stud spacing is 16 in. or 1.33 ft; see Figure 7.10):

$$RM_{T2} = [(4320)(12 \text{ ft})] + \left[(8640)\left(\frac{1.33 \text{ ft}}{2}\right)\right] = 57{,}600 \text{ ft-lb}$$

$OM_2 = 8540$ ft-lb (from tension chord force calculations)

$$RM_{D2} = [(2160)(12 \text{ ft})] + \left[(4320 + 1200)\left(\frac{1.33 \text{ ft}}{2}\right)\right] = 29{,}600 \text{ ft-lb}$$

$$C_2 = \frac{0.75OM_2 + RM_{D2} + 0.75RM_{T2}}{w} \text{ (from Equation 7.7)}$$

$$= \frac{(0.75)(8540) + 29{,}600 + (0.75)(57{,}600)}{12 \text{ ft}} = 6601 \text{ lb}$$

By referring back to Figure 7.13 for the complete loads, the maximum chord force at the ground level C_1 can be determined by summing the moments about point A:

$$RM_{T1} = [(4320 + 1920)(12 \text{ ft})] + \left[(8640 + 3840)\left(\frac{1.33 \text{ ft}}{2}\right)\right] = 83{,}200 \text{ ft-lb}$$

$OM_1 = 27{,}750$ ft-lb (from tension chord force calculations)

$$RM_{D1} = [(2160 + 960)(12 \text{ ft})]$$

$$+ \left[(4320 + 1920 + 1200 + 1200)\left(\frac{1.33 \text{ ft}}{2}\right)\right] = 43{,}200 \text{ ft-lb}$$

$$C_1 = \frac{0.75OM_1 + RM_{D1} + 0.75RM_{T1}}{w} \text{ (from Equation 7.8)}$$

$$= \frac{(0.75)(27,750) + 43,200 + (0.75)(83,200)}{12 \text{ ft}} = \mathbf{10,535 \ lb}$$

Design of wall sheathing (*SW*2).

Unit shears:

$$v_2 = \frac{F_r}{\text{Wall length}} = \frac{3300 \text{ lb}}{12 \text{ ft}} = \mathbf{275 \ lb/ft}$$

$$v_1 = \frac{F_r + F_2}{\text{Wall length}} = \frac{(3300 \text{ lb} + 3000 \text{ lb})}{12 \text{ ft}} = \mathbf{525 \ lb/ft}$$

The shear wall–fastening pattern could be different for each level, but for simplicity only, the unit shear in the ground floor is considered.

From SDPWS Table 4.3A, select 15/32-in.-thick Structural I CD–X, 8d nails at 4 in. o.c. (panel edges) and 12 in. o.c. (panel field), minimum penetration = 1⅜ in., blocking required, $v_w = 1205$ plf.

This value is valid only for wall framing with Douglas Fir-Larch or Southern Pine. From the previous SW1 wall sheathing design, we obtained the specific gravity adjustment factor, SGAF, for Hem-Fir wood species as 0.93. Therefore,

$$v_{cap} = \frac{v_w}{\Omega} \text{SGAF} = \frac{1205}{2.0}(0.93) = 560 \text{ plf}$$

Comparing the actual with the allowable values, we have

$$v_1 = \mathbf{525 \ plf} < v_{cap} = \mathbf{560 \ plf} \quad \mathbf{OK}$$

Tension chord force (*SW*2). Isolating the upper level of the shear wall and solving for T_2 from the free-body diagram by summing the moments about point *C* (Figure 7.16) yields

$$\text{OM}_2 = (3300)(10 \text{ ft}) = 33,000 \text{ ft-lb}$$

$$\text{RM}_{D2} = \left[(240 + 1200)\left(\frac{12 \text{ ft}}{2}\right)\right] = 8640 \text{ ft-lb}$$

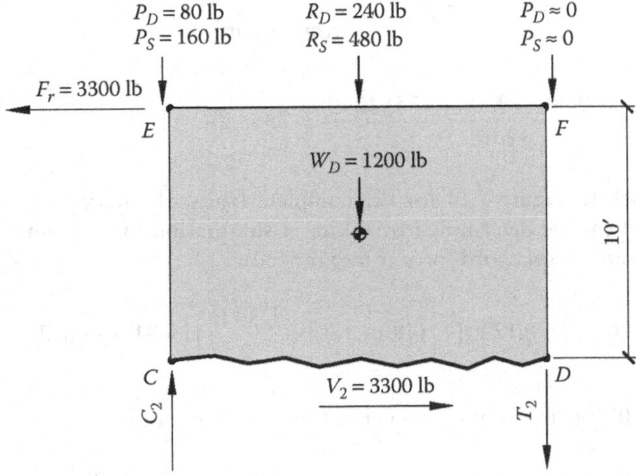

Figure 7.16 Partial free-body diagram.

$$T_2 = \frac{\text{OM}_2 - 0.6\text{RM}_{D2}}{w} \text{ (from Equation 7.3)}$$

$$= \frac{(33{,}000) - (0.6)(8640)}{12 \text{ ft}} = \textbf{2318 lb} \text{ (net uplift)}$$

By referring back to Figure 7.14 for the complete loads, the maximum chord force in wall SW2 at the ground level can be determined by taking a summation of moments about point *A* (because gravity loads are minimized; see Section 7.2):

$$\text{OM} = [(3300)(10 \text{ ft} + 10 \text{ ft})] + [(3000)(10 \text{ ft})] = 96{,}000 \text{ ft-lb}$$

$$\text{RM}_{D1} = \left[(240 + 240 + 1200 + 1200)\left(\frac{12 \text{ ft}}{2}\right)\right] = 17{,}280 \text{ ft-lb}$$

$$T_1 = \frac{\text{OM}_1 - 0.6\text{RM}_{D1}}{w} \text{ (from Equation 7.4)}$$

$$= \frac{(96{,}000) - (0.6)(17{,}280)}{12 \text{ ft}} = \textbf{7136 lb} \text{ (net uplift)}$$

Compression chord force (*SW2*). Referring back to Figure 7.16, and solving for C_2 from the free-body diagram by summing the moments about point *D* (yields higher forces than summing the moments about *C*) and accounting for the *R* and *W* loads that are tributary to the compression chord (stud spacing is 16 in. or 1.33 ft; see Figure 7.10); we obtain

$$\text{RM}_{T2} = \left[(480)\left(\frac{1.33 \text{ ft}}{2}\right)\right] + 160(12 \text{ ft}) = 2240 \text{ ft-lb}$$

$\text{OM}_2 = 33{,}000$ ft-lb (from tension chord force calculations)

$$\text{RM}_{D2} = \left[(240 + 1200)\left(\frac{1.33 \text{ ft}}{2}\right)\right] + 80(12 \text{ ft}) = 1920 \text{ ft-lb}$$

$$C_2 = \frac{0.75\text{OM}_2 + \text{RM}_{D2} + 0.75\text{RM}_{T2}}{w} \text{ (from Equation 7.7)}$$

$$= \frac{(0.75)(33{,}000) + 1920 + (0.75)(2240)}{12 \text{ ft}} = \textbf{2363 lb}$$

By referring back to Figure 7.14 for the complete loads, the maximum chord force at the ground level can be determined by taking a summation of moments about point *B* (because the compression chord force is maximized):

$$\text{RM}_{T1} = [(2400 + 160)(12 \text{ ft})] + \left[(480 + 480)\left(\frac{1.33 \text{ ft}}{2}\right)\right] = 31{,}360 \text{ ft-lb}$$

$\text{OM}_1 = 96{,}000$ ft-lb (from tension chord force calculations)

$$\text{RM}_{D1} = [(1200 + 80)(12 \text{ ft})] + \left[(240 + 240 + 1200 + 1200)\left(\frac{1.33 \text{ ft}}{2}\right)\right] = 17{,}280 \text{ ft-lb}$$

Table 7.1 Summary of shear wall chord design (lb), Example 7.2

	Level 2			*Ground level*		
	V_2	T_2	C_2	V_1	T_1	C_1
SW1	854	−2240[a]	6601	1921	−2152[a]	10,535
SW2	3300	2318	2363	6300	7136	9400

[a] A negative value actually indicates net compressive force; theoretically, hold-down anchors would not be required.

$$C_1 = \frac{0.75\text{OM}_1 + \text{RM}_{D1} + 0.75\text{RM}_{T1}}{w} \text{(from Equation 7.8)}$$

$$= \frac{(0.75)(96{,}000) + 17{,}280 + (0.75)(31{,}360)}{12 \text{ ft}} = \mathbf{9400 \text{ lb}}$$

The shear wall chord design is summarized in Table 7.1. The chord members can now be designed for the worst-case tension and compression forces (see examples in Chapter 8).

Connection design. A connection between the upper shear wall and the lower shear wall to resist the net uplift at the second floor is required. From the chord design,

$$T_{2\max} = 2318 \text{ lb}$$

A steel strap is selected to span across the floor framing and tie the upper shear wall chord to the lower shear wall chord (Figure 7.17). A preengineered connector could be selected from a manufacturer's catalog (using a load duration factor of 1.6% or 160%), which is common in practice. A design example using a steel strap is provided in Chapter 9.

A connection between the lower shear wall and the foundation to resist the net uplift at the ground floor is required. From the chord design, $T_{1\max} = 7136$ lb. A preengineered connector (Figure 7.18) and the required embedment into the concrete foundation could be selected from a manufacturer's catalog (using a load duration factor of 1.6% or 160%), which is common in practice. The actual design of the hold-down anchor and embedment involves principles of steel and concrete design, which is beyond the scope of this book. It should be noted that steel anchors loaded in tension should be deformed or hooked (i.e., not smooth) in order to engage the concrete properly.

The uplift connectors are typically provided at both ends of the shear wall because the lateral loads can act in the reverse direction. The tension chord members need to be

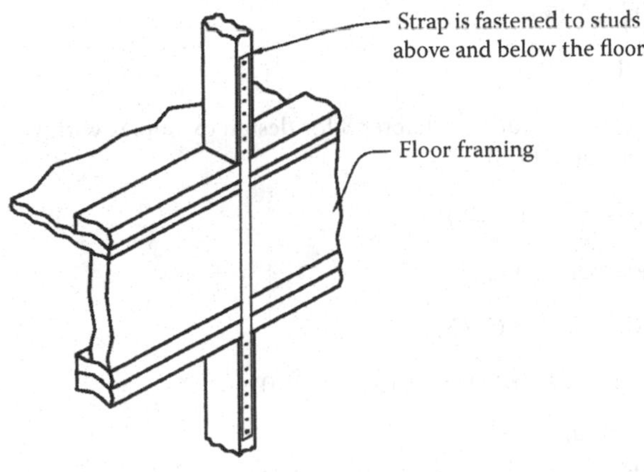
Strap is fastened to studs
above and below the floor
Floor framing

Figure 7.17 Hold-down strap.

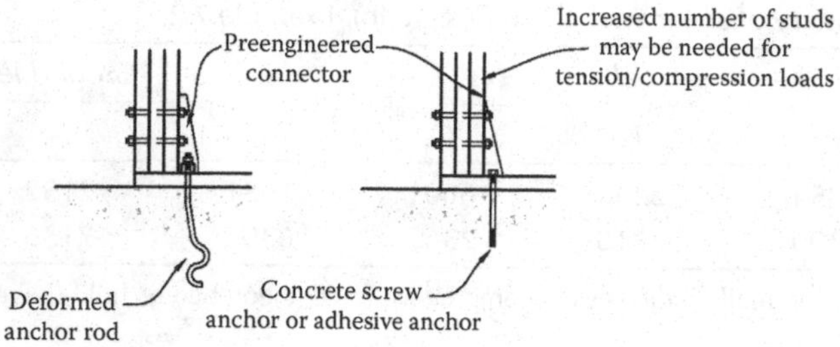

Figure 7.18 Hold-down anchor types.

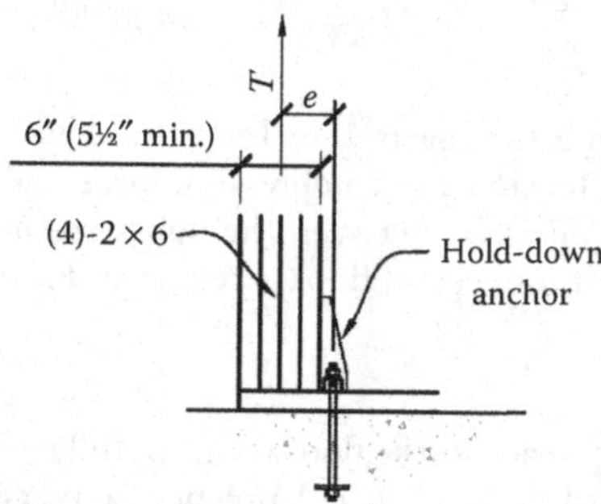

Figure 7.19 Hold-down anchor with eccentricity.

checked for stresses due to the tension chord force, T_1 and T_2, in addition to the bending moment (T_1e and T_2e) resulting from the eccentricity e of the hold-down anchors (see Figure 7.19). Use the ***net area*** of the chord member to account for the bolt holes required for the hold-down or tie-down anchors.

The sill anchors should also be designed to resist the base shear. Considering the worst-case base shear $V_1 = 6300$ lb, the allowable shear parallel to the grain in the sill plate is

$$Z' = ZC_DC_MC_tC_gC_\Delta C_{eg}C_{di}C_{tn}$$

$C_D = 1.6$ (wind)

$C_M = 1.0$ (MC $\leq 19\%$)

$C_t = 1.0$ ($T \leq 100°$F)

C_g, C_Δ, C_e, C_{di}, $C_{tn} = 1.0$ (see Chapter 8 for design examples with C_g, C_Δ, C_e, C_{di}, C_{tn}, or NDS code Section 11.3)

$Z_\parallel = 590$ lb (NDS Table 12E)

The allowable shear per bolt is

$$Z' = ZC_DC_MC_tC_g\ C_AC_{eg}\ C_{di}C_{tn}$$

$$= (590)(1.6)(1.0)(1.0)(1.0)(1.0)(1.0)(1.0)(1.0)$$

$$= 944 \text{ lb per bolt}$$

Therefore, the number of sill anchor bolts required = (6300 lb)/(944 lb/bolt) ≃ 7 bolts. The maximum bolt spacing is

$$\frac{(12\text{-ft wall length} - 1\text{-ft edge distance} - 1\text{-ft edge distance})}{7 \text{ bolts} - 1}$$

$$= 1.67 \text{ ft} \approx 1 \text{ ft } 8 \text{ in.}$$

Use ½-in.-diameter anchor bolts at 1 ft 8 in. o.c. (see Figure 7.2 for a typical layout).

For ease of wall placement, a contractor may request that the sill anchor bolt holes be oversized. One way to do this, and still ensure a snug tight connection between the anchors and sill plate so that the sill anchors are able to transfer the lateral base shear, is to sleeve the bolt holes with metal tubing and fill the areas between the bolt and the inside face of the sleeve with expansive cement [8].

The compression chord members also need to be checked for stresses due to the compression chord forces C_1 and C_2. When checking the stresses in the chord members at mid-height, include buckling (i.e., include the C_p factor), and when checking the stresses at the base of the chords, use the ***net area*** of the chord member to account for the bolt holes required for the hold-down or tie-down anchors. Design the sill plate for compression perpendicular to the grain due to the compression chord forces C_1 and C_2.

Example 7.3: Difference in shear wall capacity for wind and seismic designs

A 3-ft-long × 8-ft-tall shear wall panel has the following design parameters:

¾-in. plywood sheathing with 8d nails at 4 in. o.c. [edge nailing (EN)] and 12 in. o.c. [field nailing (FN)] on the outside face.

½ in. gypsum wall board (GWB) on the inside face of the wall. The GWB is unblocked and fastened per IBC Table 2306.3.(3) with an allowable unit shear capacity of 100 plf.

Determine the allowable unit shear capacity of the shear wall panel for both wind and seismic designs.

Solution: Wall aspect ratio = 8 ft/3 ft = 2.67.

Per section 4.3.4 of the SDPWS, the shear capacities are to be adjusted as follows:

$$WSP_2 = 1.25 - 0.125(h/b_s)$$
$$= 1.25 - 0.125(8/3) = 0.916$$

The allowable unit shear for the wood panels is

$$v_w = 1105 \text{ plf (SDPWS Table 4.3A)}$$

$$v_{cap} = \frac{v_w}{\Omega} = \frac{1010}{2.0} = 505 \text{ plf}$$

Wind design:

- The addition of the unit shear capacity of walls of dissimilar materials when designing for wind loads is permitted.

Total allowable unit shear of wall adjusting for the aspect ratio:

$$v_{adj} == (0.916)(505 + 100) = 554 \text{ plf}$$

Seismic design:

- Addition of unit shear capacities of dissimilar materials is not allowed for seismic design. Consequently, only the plywood sheathing will be considered because it has the higher shear capacity.

The allowable unit shear for the wood panels is

$v_w = 720$ plf (SDPWS Table 4.3A)

$$v_{cap} = \frac{v_w}{\Omega} = \frac{720}{2.0} = 360 \text{ plf}$$

Total allowable unit shear of wall adjusting for the aspect ratio:

$$v_{adj} == (0.916)(360) = 330 \text{ plf}$$

Note: From the results above, the allowable unit shear for the same shear wall panel is quite a bit higher for wind design as it is for seismic design. Although the wind load may have been the controlling load for this building, the shear capacity under seismic loads is quite low compared to the capacity under wind loads, which indicates that checking this building and shear wall panel for seismic loads is also necessary. This example points to the need to compare not only the wind and seismic loads but also the respective shear capacities of the shear wall under both loading conditions. The importance of this difference in shear wall capacities has been highlighted in Reference 9.

7.4 COMBINED SHEAR AND UPLIFT IN WALL SHEATHING

In high wind areas of the country, several hold-down anchors may be required between floors to resist the uplift due to wind loads, and this can often lead to increased cost and there may also be nailing interference because of the relatively large number of fasteners that may be required for these anchor devices. In lieu of providing hold-down straps between floors (see Figure 7.17) to resist uplift loads, the exterior sheathing can be fastened in such a way to create a vertical plywood panel splice across the floor to transfer the uplift force. Figure 7.20 illustrates two possible details that can be used for this condition. In both cases, nails are added to resist the uplift force in addition to the nails required for shear.

The detail shown in Figure 7.20a should be used where the rim joist is sawn lumber or any lumber that is susceptible to shrinkage. The exterior sheathing is preengineered and is generally dry when it is installed, whereas sawn lumber will have some water content at the time of installation and will shrink over time. The exterior sheathing and plywood shim should have a clear distance of about ½ in. at the ends to allow for shrinkage. The shim should also

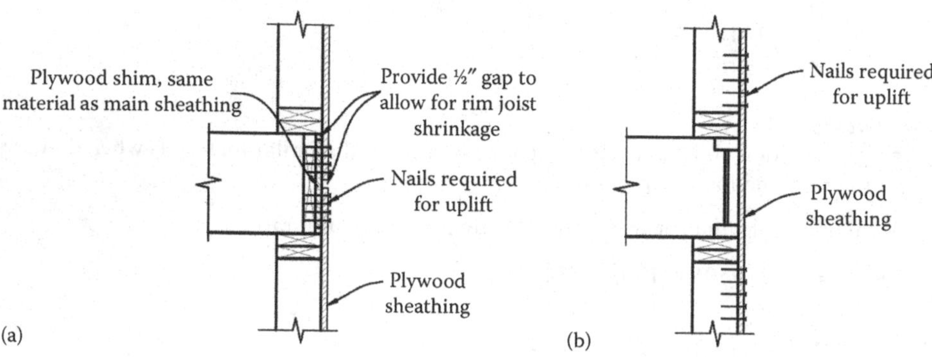

Figure 7.20 Uplift details using exterior sheathing: (a) sawn lumber rim joist and (b) preengineered rim joist.

be made from the same material and have the same thickness and orientation as the exterior sheathing. Furthermore, the plywood shim acts to transfer the tension force because sawn lumber has very little capacity when loaded perpendicular to the grain. When an OSB or plywood rim joist is used, a shim is not required because these products are not susceptible to shrinkage and have sufficient tension capacity when loaded in this manner.

The detail shown in Figure 7.20b should be used when the rim joist is a preengineered I-joist or laminated veneer lumber (LVL) because shrinkage in sawn lumber would likely cause the exterior sheathing to buckle in this case.

Using the exterior sheathing to support both uplift and in-plane shear creates a situation where the sheathing and fasteners are subjected to combined loading. However, the shear capacity of a shear wall is controlled by the fasteners such that it is generally difficult to add enough fasteners so that the sheathing fails. Therefore, the design approach taken here will be twofold: Limit the tension in the sheathing due to uplift to a reasonably low value and add fasteners to resist the uplift in addition to the fasteners required for lateral shear. The reader is referred to Section 4.4 of the SDPWS for more information on the use and applicability of this method.

PROBLEMS

7.1 For the building plan shown in Figure 7.21 and based on the following given information:

- Wall framing is 2 × 6 at 16 in. o.c., Hem-Fir, Select Structural.
- Normal temperature and moisture conditions apply.
- Roof dead load = **15 psf**, wall dead load = **10 psf**, snow load P_f = **40 psf.**
- Floor-to-roof height = 18 ft.
- Consider loads in the N–S direction only.

a. For the 18-ft and 20-ft shear walls on the east and west faces only, select the required plywood thickness, fastening, and edge support requirements. Assume that the plywood is applied directly to the framing and is on one side only. Give the full specification.

b. Determine the maximum tension force in the shear wall chords of the 20-ft-long shear wall.

c. Determine the maximum compression force in the shear wall chords of the 20-ft-long shear wall.

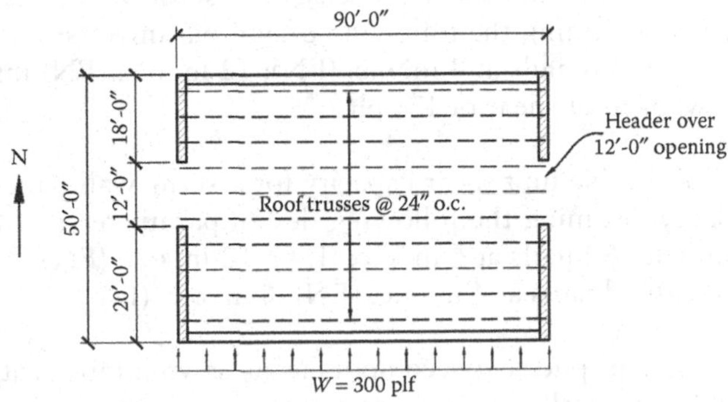

Figure 7.21 Building plan for Problem 7.1.

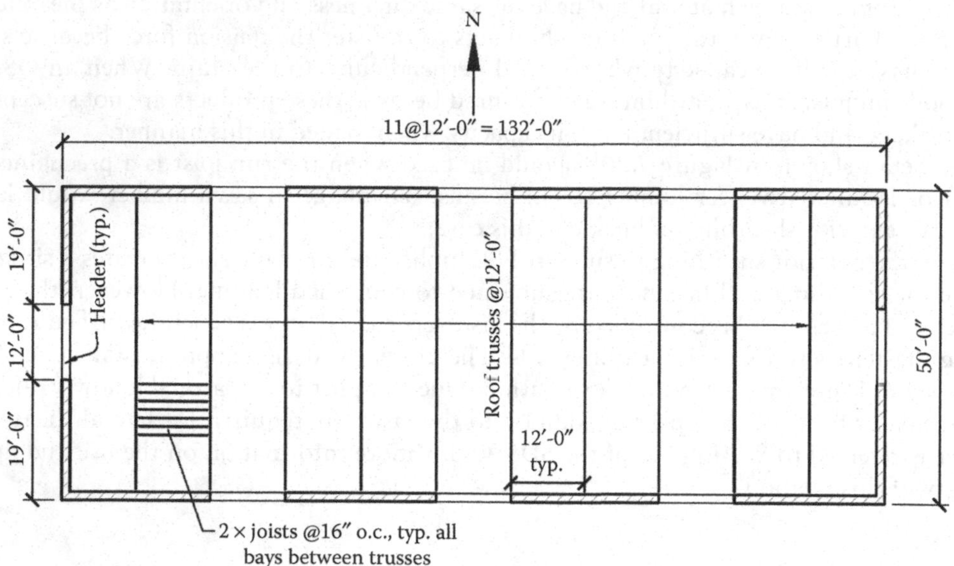

Figure 7.22 Building plan for Problem 7.2.

7.2 For the building plan shown in Figure 7.22 for a one-story building and based on the following given information:

- Wall framing is 2 × 6 at 16 in. o.c., Douglas Fir-Larch, No. 1.
- Normal temperature and moisture conditions apply.
- Roof dead load = 15 psf, wall dead load = 10 psf, roof live load = 20 psf.
- Floor-to-roof height = 18 ft.
- Applied wind load = 250 plf to the diaphragm.

a. For the typical 19-ft shear wall, select the required plywood thickness, fastening, and edge support requirements. Assume that the plywood is applied directly to the framing and is on two sides. Give the full specification.

b. Determine the maximum tension force in the shear wall chords for the interior shear wall.

c. Determine the maximum compression force in the shear wall chords for the interior shear wall.

d. Select an appropriate anchor rod layout to resist the lateral loads applied.

7.3 Determine the allowable unit shear capacity for a shear wall panel under both wind and seismic loads, assuming the following design parameters: exterior face of wall: $^{15}/_{32}$-in. plywood with 8d nails at 4 in. o.c. (EN); 12 in. o.c. (FN) interior face of wall; GWB with allowable unit shear of 175 plf.

7.4 Determine the allowable unit shear capacity for a shear wall panel under both wind and seismic loads, assuming the following design parameters: exterior face of wall: $^{15}/_{32}$-in. plywood with 8d nails at 4 in. o.c. (EN); 12 in. o.c. (FN) interior face of wall: $^{3}/_{8}$-in. plywood with 8d nails at 4 in. o.c. (EN); 6 in. o.c. (FN).

7.5 Design a plywood lap splice between floors using sawn lumber with 2 × 6 wall studs at 16 in. o.c. and $^{7}/_{16}$ in. wall sheathing for a net uplift of 600 plf. Stud spacing is 16 in., and lumber is Douglas Fir-Larch No. 1. Sketch the design.

7.6 Refer to the drawing below Figure 7.23 showing the building plan for a one-story building:

- Wall framing is 2 × 6 at 16 ft o.c., Douglas Fir-Larch, No. 1.
- Normal temperature and moisture conditions apply.
- Roof dead load = 18 psf, wall dead load = 10 psf, snow load, $P_f = 40$ psf.
- Floor-to-roof height = 18 ft.
- Consider loads in the N–S direction only.

a. Determine the lateral force to each shear wall (SW1, SW2).
b. For SW1, draw the unit shear, net shear, and axial force diagrams for the roof diaphragm drag struts, and determine the maximum axial force in the drag struts.
c. For SW1, select the required plywood thickness, fastening, and edge support requirements. Assume the plywood is applied directly to the framing and is on one side only. Give the full specification.
d. For SW1, determine the maximum tension force in the shear wall chords (Figure 7.23).

7.7 Refer to Figure 7.24 showing the building plan for a one-story building. Lateral load is wind and framing is Douglas Fir-Larch.

a. Draw the unit shear, net shear, and axial force diagrams for the roof diaphragm drag struts, and determine the maximum axial force in the drag struts in the three shear wall lines.
b. Determine the lateral force in pounds and select the fastening and edge support requirements assuming Structural I sheathing for each of the shear walls. Assume the plywood is applied directly to the framing and is on two sides. Give the full specification.

7.8 Figure 7.25 shows wind loads on a two-story building in the N–S direction. Framing is Spruce-Pine-Fir and the diaphragms are flexible. Do the following:

a. Draw the shear force diagram (in pounds) and the unit shear diagram (in pounds per foot) for the two-story shearwalls SW2, SW3, and SW4.
b. Assuming Struct I wall sheathing on one side only, select a plywood thickness and fastening pattern for the lower level of SW2.
c. Explain any differences that might exist between the loads to SW1 and SW5.

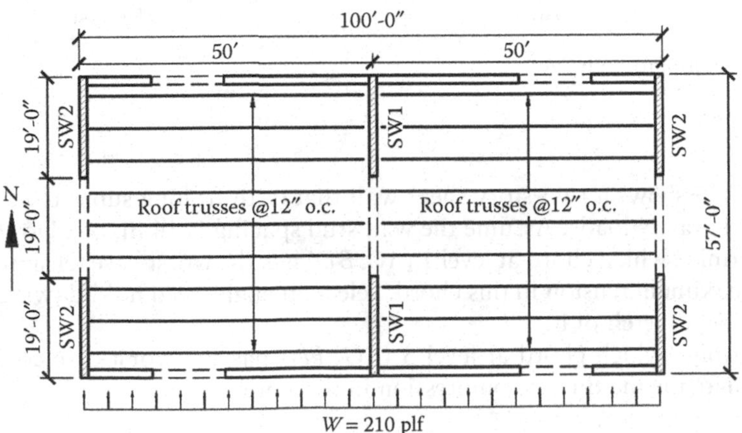

Figure 7.23 Detail for Problem 7.6.

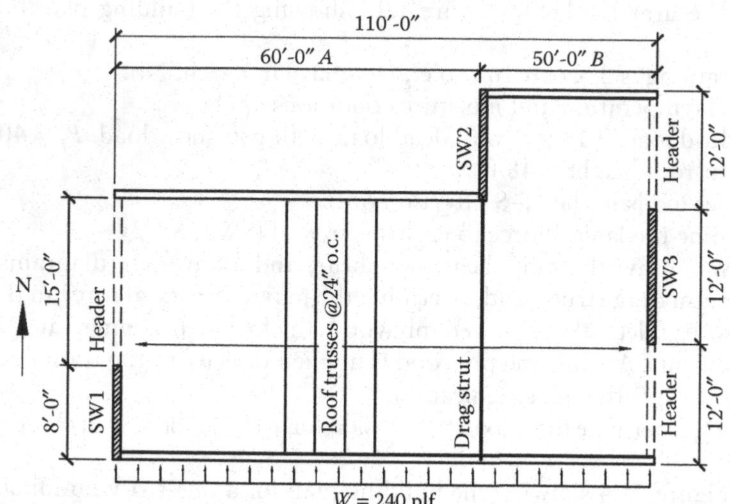

Figure 7.24 Detail for Problem 7.7.

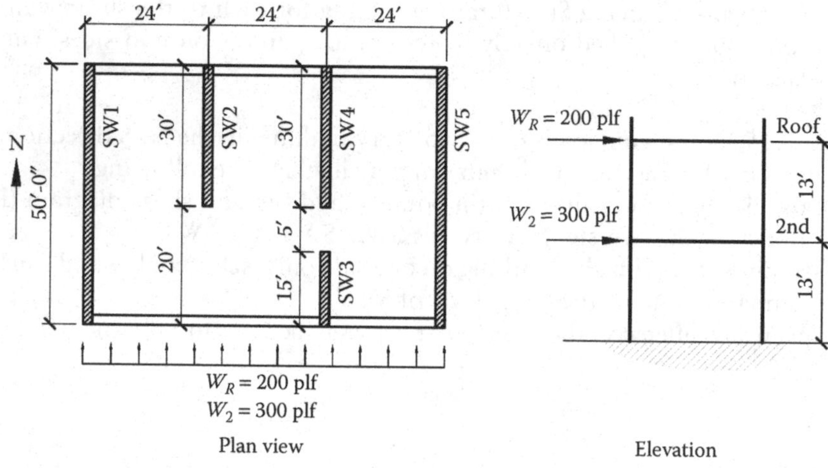

Figure 7.25 Detail for Problem 7.8.

7.9 Figure 7.26 shows a two-story shear wall that is 16 ft. long subjected to lateral wind loads and gravity loads. Assume the wall stud spacing is 16 in. o.c. Do the following:

a. Determine which chord at level 1 (*AC*, *BD*) has the worst-case tension and calculate the maximum tension in this chord. Select an engineered hold-down connector and provide a sketch of it.

b. Determine which chord at level 3 (*EG*, *FH*) has the worst-case compression and calculate the maximum compression in this chord.

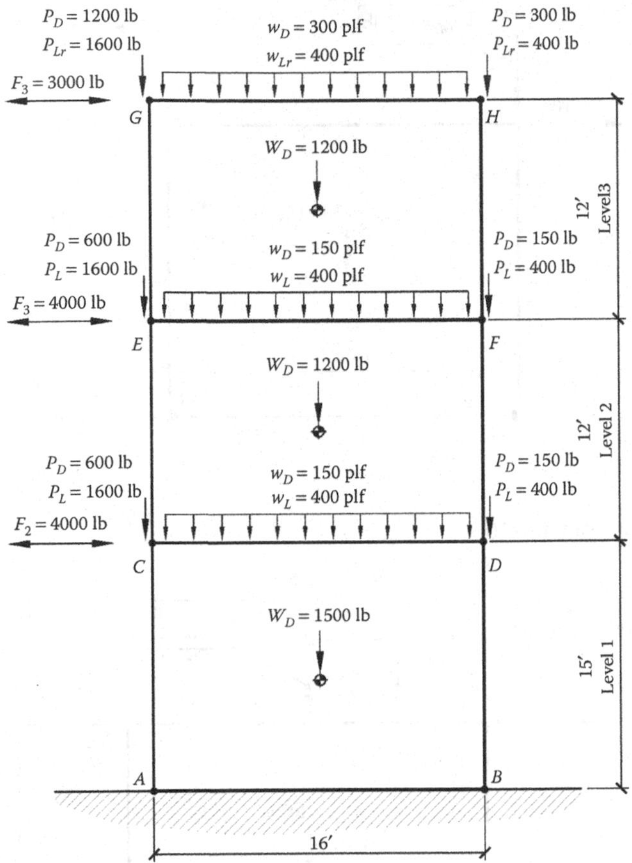

Figure 7.26 Detail for Problem 7.9.

7.10 For the shear wall shown in Figure 7.27, loads are due to $D + W$ only. Determine the dead required (P_{DL}) to prevent uplift at the tension chord.

7.11 Refer to Figure 7.28 showing a shear wall elevation with applied loads:

- Lateral load is due to wind and can act in either direction.
- Wall framing is 2 × 6 at 16 ft. o.c, Douglas Fir-Larch, Select Structural.
- Normal temperature and moisture conditions apply.
- Plywood sheathing is on both sides of the wall.
  a. Select the required plywood thickness, fastening, and edge support requirements for the shear wall considering the lateral wind load only. Give the full specification.
  b. Without doing any calculations, determine which chord will have the worst-case tension (uplift) and worst-case compression force based on the given loading and explain your response.
  c. Determine the worst-case tension and worst-case compression chord forces.
  d. Select a hold-down connector from a manufacturer's catalog and sketch the design.

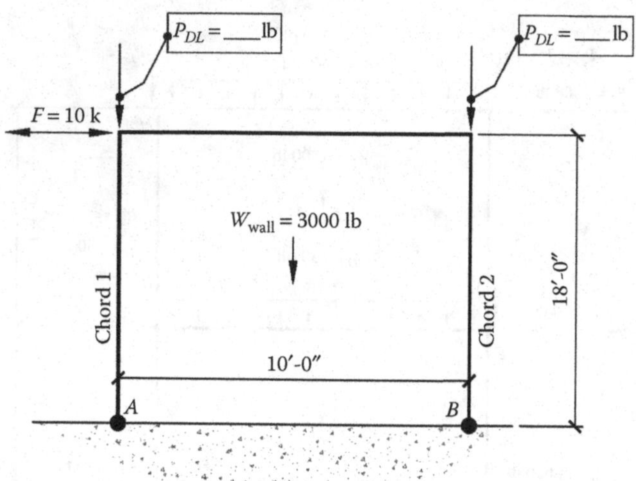

Figure 7.27 Detail for Problem 7.10.

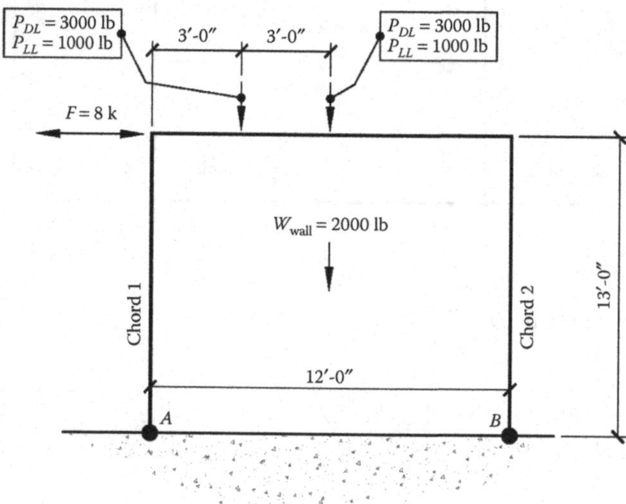

Figure 7.28 Detail for Problem 7.11.

7.12 Refer to Figure 7.29 showing a shear wall elevation with applied loads:

- Lateral load is due to wind and can act in either direction.
- Wall framing is 2 × 6 at 16 ft o.c., Hem-Fir, Select Structural.
- Normal temperature and moisture conditions apply.
- Plywood sheathing is on one of the wall.
  a. Select the required plywood thickness, fastening, and edge support requirements for the shear wall considering the lateral wind load only. Give the full specification.
  b. Determine the worst-case tension and worst-case compression chord forces.
  c. Select a hold-down connector from a manufacturer's catalog and sketch the design.

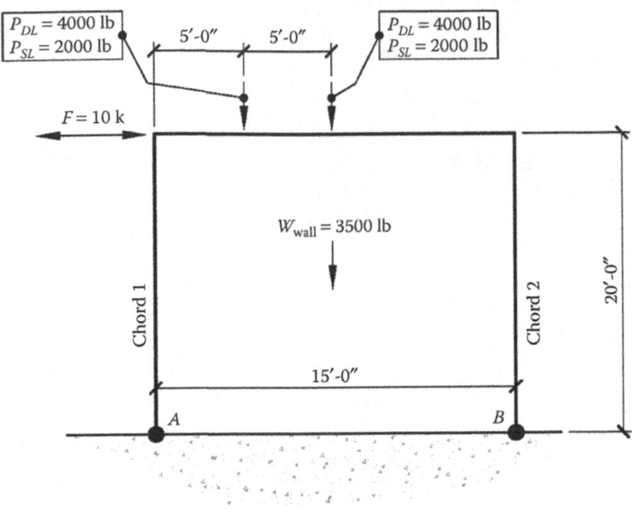

Figure 7.29 Detail for Problem 7.12.

REFERENCES

1. ICC (2015), *International Building Code,* International Codes Council, Falls Church, VA.
2. AWC (2015), *Special Design Provisions for Wind and Seismic,* American Wood Council, Washington, DC.
3. ANSI/American Forest & Paper Association (2001), *Wood Structural Panels Supplement,* AF&PA, Washington, DC.
4. Tarpy, Thomas S., Thomas, David J., and Soltis, Lawrence A. (1985), Continuous timber diaphragms, *J. Struct. Div. ASCE* 111(5): 992–1002.
5. GA (2008), *Shear Values for Screw Application of Gypsum Board on Walls,* GA-229-08, Gypsum Association, Washington, DC.
6. ASCE (2010), *Minimum Design Loads for Buildings and Other Structures,* ASCE-7, American Society of Civil Engineers, Reston, VA.
7. AWC (2015), *Allowable Stress Design Manual for Engineered Wood Construction,* American Wood Council, Leesburg, VA.
8. Knight, Brian (2006), High Rise Wood Frame Construction—Cornerstone Condominiums, STRUCTURE, pp. 68–70, June.
9. Beebe, Karyn (2006), Enhanced lateral design with wood structural panels, *Struct. Eng.,* March.

Chapter 8

Connections

8.1 INTRODUCTION

Connections within any structure are usually very small relative to the supported members, but they are often limited by the connection. The capacity of the beams and other major structural members are only as strong as their connectors. In practice, the design of connections is oftentimes overlooked and is thus delegated by default to either someone preparing shop drawings or to the builder in the field. Although this process typically results in a structure with adequate connections, proper attention to the design and detailing of connections can help limit the amount of field-related problems and design issues.

In practice, the designer has three basic options for the design of connections:

1. *Use a preengineered connector.* Several proprietary products are available, and each manufacturer usually makes available a catalog detailing each connector with a corresponding load capacity. The manufacturer usually also indicates the required placement and quantity of nails, screws, or bolts required for each connector. This is usually the least costly option from a design standpoint and is often the preferred method from the builder's standpoint
2. *Use the design equations in the National Design Specification (NDS).* This can be a tedious process that is best carried out with the aid of a computer software.
3. *Use the connector capacity tables in the NDS code.* Several tables are provided in the NDS code for various connectors and conditions and are based on the equations in the NDS code. Occasionally, conservative assumptions must be made for loading conditions not contained in the tables, but they are not as tedious as using the equations directly

In all cases, it is important that complete and accurate details are provided to ensure that the fasteners are placed properly in the field.

Several types of connectors are available, but the following connector types will be covered in this book because they are the most commonly used: (1) bolts (Table L1), (2) lag screws (Table L2), (3) wood screws (Table L3), and (4) nails (Table L4). The tables indicated are found in the appendix of the NDS code [1]. These tables list the various dimensional characteristics of each connector type. Bolts have the highest strength but require the most labor. Conversely, nails generally have the lowest strength but require the least labor and equipment. Figure 8.1 illustrates each of these connector types.

Bolts are greater than $\frac{1}{4}$ in. in diameter and are installed completely through the connected members through predilled holes that are at least $\frac{1}{32}$ in. greater in diameter than bolts, but less than $\frac{1}{16}$ in. (NDS code Section 12.1.3). Bolts are not permitted to be forcibly driven.

Lag screws are greater than $\frac{1}{4}$ in. in diameter and are installed into a wood member. Lag screws typically have a threaded and a smooth portion (see Figures 8.1 and 8.2). The NDS

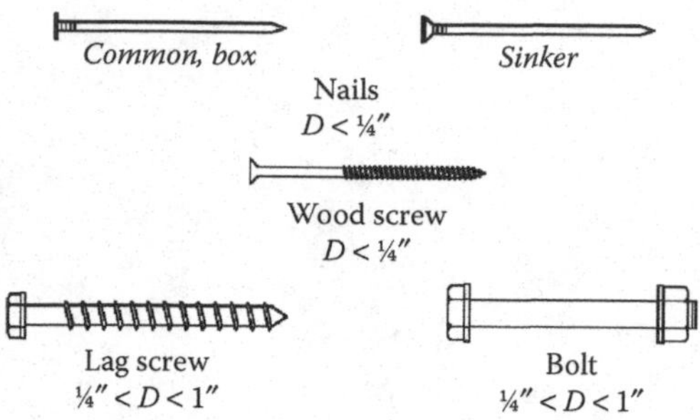

Figure 8.1 Fastener types.

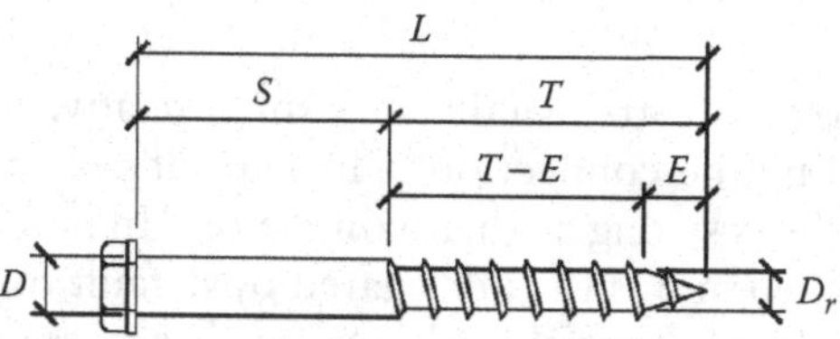

Figure 8.2 Elements of a lag screw. *D*, diameter; D_r, root diameter; *L*, overall length; *S*, unthreaded shrank length; *T*, thread length; *E*, tapered tip length.

code gives the required diameter of clearance holes and lead holes for installing lag screws to prevent splitting of the wood member during construction (NDS code Section 12.1.4). The *clearance hole* is the area around the unthreaded shank and should have the same diameter and length as the unthreaded shank. The *lead hole* is the area of the wood around the threaded portion of the shank. The diameter of the lead hole is generally smaller than the diameter of the shank. The lead hole diameter is a function of the specific gravity of the wood, see Section 12.1.14 of the NDS.

Wood screws are similar to lag screws but are generally less than ¼ in. in diameter. The NDS code (Section 12.1.4.2) gives the required diameter of lead holes to prevent splitting.

Nails are generally less than ¼ in. in diameter and are forcibly driven by a hammer or other mechanical means. Nails can have numerous head and shank types, but the discussion of these is beyond the scope of this book. We limit our discussion to *common nails,* which are representative of most structural nails. Lead holes are not specifically required in the NDS code, but are allowed (see NDS code Section 12.1.6.3).

The sizes of common screws and nails are given in Tables 8.1 and 8.2.

Table 8.1 Screw sizes

	No. 6	*No. 7*	*No. 8*	*No. 9*	*No. 10*	*No. 12*	*No. 14*
D	0.138″	0.151″	0.164″	0.177″	0.19″	0.216″	0.242″

Table 8.2 Common nail sizes

	6d	*7d*	*8d*	*10d*	*12d*	*16d*	*20d*
L	2″	2¼″	2½″	3″	3¼″	3½″	4″
D	0.113″	0.113″	0.131″	0.148″	0.148″	0.162″	0.192″

8.2 DESIGN STRENGTH

There are two basic types of loads that connections are designed to resist: lateral loads and withdrawal loads. As with the other member designs in this book, each connector or group of connectors has a base design value that has to be modified to determine the allowable or adjusted design values. Several factors have an effect on the load-carrying capacity of a connection. These factors can be seen by inspection of the adjustment factors indicated in Table 10.3.1 of the NDS code. These adjusted design values are

$$Z' = ZC_DC_MC_tC_g\, C_\Delta C_{eg}C_{di}C_{tn} \text{ for ASD} \tag{8.1}$$

$$Z'_r = ZC_MC_tC_g\, C_\Delta C_{eg}C_{di}C_{tn}\, K_F\, \phi_z\, \lambda \text{ for LRFD}$$

$$W' = WC_DC_MC_tC_g\, C_{tn} \text{ for ASD} \tag{8.2}$$

$$W'_r = WC_MC_tC_g\, C_{tn}\, K_F\, \phi_z\, \lambda \text{ for LRFD}$$

where:

Z is the nominal lateral design value for a single fastener
Z' is the adjusted lateral design value for a single fastener (ASD)
Z'_r is the adjusted lateral design value for a single fastener (LRFD)
W is the nominal withdrawal design value for a single fastener
W' is the adjusted withdrawal design value for a single fastener (ASD)
W'_r is the adjusted withdrawal design value for a single fastener (LRFD)
C_D is the load duration factor, ASD only (≤ 1.6, see Section 11.3.2 of the NDS code)
C_M is the wet service factor (Table 11.3.3 of the NDS code)
C_t is the temperature factor (Table 11.3.4 of the NDS code)
C_g is the group action factor (Section 11.3.6 of the NDS code)
C_Δ is the geometry factor (Section 12.5.1 of the NDS code)
C_{eg} is the end grain factor (Section 12.5.2 of the NDS code)
C_{di} is the diaphragm factor (Section 12.5.3 of the NDS code)
C_{tn} is the toenail factor (Section 12.5.4 of the NDS code)
K_F is 3.32 for all connections
ϕ_z is 0.65 for all connections
λ is the time effect factor (see NDS Table N3)

8.3 ADJUSTMENT FACTORS FOR CONNECTORS

There are several geometric factors that affect the capacity of a connection such as the direction of the load relative to the grain, the geometry of the connectors, and the location of the connector within the wood member. The density of the wood is also a fairly significant factor. In general, the density of the wood is proportional to the strength of the connection. The adjustment factors indicated in Section 8.2 are intended to account for all of these variables. Each of these adjustment factors is now discussed in greater detail.

8.3.1 Load duration factor C_D

The load duration factor is found in Table 2.3.2 of the NDS code and is usually the same factor used for the connected member. Section 11.3.2 of the NDS code does limit the value of C_D to 1.6 because the laboratory testing of connections at short durations is limited (i.e., impact loads). The load duration factor also does not apply when the capacity of the connection is controlled by metal strength or concrete or masonry. The load duration factor does not apply to LRFD.

8.3.2 Wet service factor C_M

Because wood has a tendency to lose water and shrink over time, the capacity of a connector depends on the moisture content of the wood at the time of fabrication and while in service. Wood that is dry, or wood with a moisture content below 19%, does not result in a reduced connection capacity when moisture is considered and $C_M = 1.0$. For wood with a moisture content greater than 19% either at the time of fabrication or while in service, refer to Table 11.3.3 of the NDS code for the C_M factor.

8.3.3 Temperature factor C_t

Although the temperature range for most wood-framed buildings is such that the connection capacity does not have to be reduced, Table 11.3.4 of the NDS code gives the values for C_t for various temperature ranges.

8.3.4 Group action factor C_g

When a series of fasteners aligned in a row are loaded, the load distribution to each fastener is not equal. The group action factor accounts for this condition. It does not apply to smaller diameter fasteners ($D < ¼$ in.). Note that a row is defined as a group of fasteners aligned and parallel to the load. The value of C_g is defined in Section 11.3.6 of the NDS code as follows (NDS code Equation 11.3–1):

$$C_g = \frac{m(1 - m^{2n})}{n[(1 + R_{EA}m^n)(1 + m) - 1 + m^{2n}]} \frac{1 + R_{EA}}{1 - m} \tag{8.3}$$

where:

n is the number of fasteners in a row.

$$R_{EA} = \text{the lesser of } \frac{E_s A_s}{E_m A_m} \text{ or } \frac{E_m A_m}{E_s A_s} \tag{8.4}$$

See Figure 8.3 for the main and side member description

where:

E_m is the modulus of elasticity of the main member (psi)
E_s is the modulus of elasticity of the side member (psi)

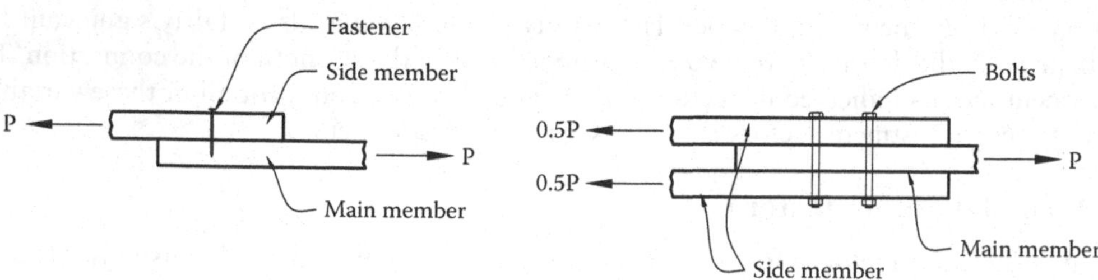

Figure 8.3 Main member/side member.

A_m is the gross cross-sectional area of the main member (in.^{2})
A_s is the sum of the gross cross-sectional areas of the side member (in.^{2})

$$m = u - \sqrt{u^2 - 1} \tag{8.5}$$

$$u = 1 + \gamma \frac{s}{2}\left(\frac{1}{E_m A_m} + \frac{1}{E_s A_s}\right) \tag{8.6}$$

where:

s is the center-to-center spacing between adjacent fasteners in a row
γ = load/slip modulus for a connection (lb/in.) (8.7)
= $180{,}000D^{1.5}$ for dowel-type fasteners in wood-to-wood connections
= $270{,}000D^{1.5}$ for dowel-type fasteners in wood-to-metal connections
D is the diameter of fastener, in.

It can be seen by inspection that the group action factor is a function of the number of fasteners, the stiffness of the main and side members, the spacing of the fasteners, the slip modulus, and the diameter of the fasteners. The "main member" is closest to the threaded end of the bolt or the pointed end of the nail, whereas the "side member" is closest to the head of the bolt or nail. It can also be seen that the equation is rather tedious. The NDS code does give the value of C_g for various connection geometries in Tables 11.3.6A through 11.3.6D. Example 8.1 illustrates that use of the tables is a reasonable alternative to using the equation for most cases.

8.3.5 Geometry factor C_Δ

Connectors with inadequate spacing have a tendency to split (see Figure 8.4); thus, to maximize the load-carrying capacity of a connection, the designer must consider the geometry of the connection. The geometry factor applies to all connectors greater than ¼ in. in diameter and is a function of the following: edge distance, end distance, center-to-center spacing, and row spacing. For connectors with $D <$ ¼ in., $C_\Delta = 1.0$, but minimum spacing requirements must be met. For connectors with $D <$ ¼ in., the NDS code indicates that the spacing "shall be sufficient to prevent splitting of the wood" (see Sections 12.1.5.7 and 12.1.6.6 of the NDS code). Although this is too vague for any direct application, there is sufficient guidance given in the NDS commentary (C12.1.4.7 and C12.1.5.6) relative to the spacing requirements for small-diameter connectors, as summarized in Table 8.3. It can be seen from the table that having prebored holes reduces the spacing required, because the prebored hole reduces the possibility of splitting when the fastener is installed.

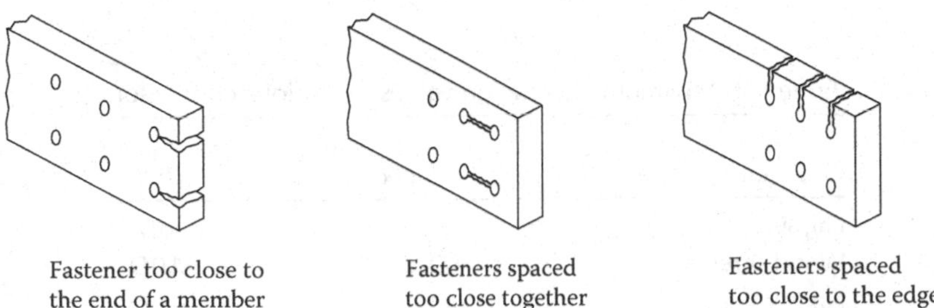

Figure 8.4 Splitting due to inadequate fastener spacing.

There is also a required minimum penetration into the main member for screws and nails that must be met (NDS code Section 12.1), as well as a required minimum penetration to achieve the full design value. The design value for a connector that has a penetration depth less than the full design penetration is multiplied by $p/8D$ for lag screws and $p/10D$ for wood screws or nails. The minimum penetration required by the NDS code is $4D$ for lag screws and $6D$ for wood screws and nails (see Table 8.4).

For larger-diameter connectors ($D > \frac{1}{4}$ in.), there are two options for spacing from Table 8.3:

1. Meet the *base* spacing requirements for the full design value of the connector from Table 8.3; for this case, $C_\Delta = 1.0$.
2. Provide a reduced spacing that is less than the base spacing required for the maximum design value but greater than the code *minimum*. For this case,

$$C_\Delta = \frac{\text{Spacing provided}}{\text{Base spacing for maximum design value from Table 8.3}}$$

Table 8.3 Recommended minimum spacing requirements for connectors with $D < \frac{1}{4}$ in. ($C_\Delta = 1.0$)

	Type of loading	*Minimum spacing for connection configuration:* *Wood side member without prebored holes*	*Wood side member with prebored holes or steel side plates without prebored holes*	*Steel side plates with prebored holes*
End distance	|| to grain, tension	$15D$	$10D$	$5D$
	|| to grain, compression	$10D$	$5D$	$3D$
Edge distance	Any	$2.5D$	$2.5D$	$2.5D$
Center-to-center spacing (pitch)	|| to grain	$15D$	$10D$	$5D$
	⊥ to grain	$10D$	$5D$	$2.5D$
	–	–	–	–
	–	–	–	–
Row spacing (gage)	Staggered	$2.5D$	$2.5D$	$2.5D$
	Inline	$5D$	$3D$	$2.5D$

Note: $C_\Delta = S_{actual}/S_{full}$; ||, parallel to the grain; ⊥, perpendicular to the grain.

Table 8.4 Minimum penetration values for various connectors

Connector	P_{min} *(reduced)*	P_{min} *(full)*
Lag screws	$4D$	$8D$
Wood screws	$6D$	$10D$
Nails	$6D$	$10D$

Note: $C_\Delta = p_{actual}/p_{full}$.

The NDS code provides values for both the base spacing and the minimum spacing for all of the following parameters: edge distance, end distance, center-to-center spacing, and row spacing. There is a unique value for C_Δ for each of these four spacing requirements. The C_Δ used for design is the smallest value obtained from these unique values. Table 8.5 summarizes the base spacing and minimum spacing requirements for larger-diameter connectors.

8.3.6 End grain factor C_{eg}

When fasteners are oriented parallel to the grain of a wood member, the connection capacity is reduced. This is the case when fasteners are inserted into the end grain of a member (see Figure 8.4). For withdrawal loads W, lag screws have an end grain factor of $C_{eg} = 0.75$. Wood screws and nails have virtually no capacity in withdrawal when loaded in the end grain of a member and thus are not permitted. For lateral loads Z, both nails and screws are adjusted by $C_{eg} = 0.67$.

8.3.7 Diaphragm factor C_{di}

The capacity of a wood diaphragm is a function of the connector type and spacing, and is covered in Chapter 6. The capacity of a diaphragm can be found in other references (such as [1]); however, there may be conditions where the diaphragm capacity has to be calculated directly. For this case, the capacity of the connectors is multiplied by $C_{di} = 1.1$. This factor applies only to wood panels attached to framing members. It does not apply to other connectors in the load path, such as drag struts or chords.

Table 8.5 Minimum spacing requirements for connectors with $D > \frac{1}{4}$ in.

	Type of loading	*Minimum spacing* Reduced $C = 0.5$	*Minimum spacing* Full (base) $C = 1.0$
End distance	⊥ to grain	$2D$	$4D$
	|| to grain, compression	$2D$	$4D$
	|| to grain, tension (softwood)	$3.5D$	$7D$
	|| to grain, tension (hardwood)	$2.5D$	$5D$
Edge distance	|| to grain, $l/D \leq 6$	N/A	$1.5D$
	|| to grain, $l/D > 6$	N/A	$1.5D$ or ½ row span (whichever is greater)
	⊥ to grain, loaded edge	N/A	$4D$
	⊥ to grain, unloaded edge	N/A	$1.5D$
Center-to-center spacing	|| to grain	$3D$	$4D$
	⊥ to grain	N/A	$3D$
Row spacing	|| to grain	N/A	$1.5D$
	⊥ to grain, $l/D \leq 2$	N/A	$2.5D$
	⊥ to grain, $2 < l/D \leq 6$	N/A	$(5l + 10D)/8$
	⊥ to grain, $l/D > 6$	N/A	$5D$

Note: l/D (bolt slenderness) is the smaller of l_m/D or l_s/D; l_m, length of the fastener in wood main member; l_s, total length of fastener in wood side member(s); D, diameter of the fastener; l, length of the fastener; $C_\Delta = S_{actual}/S_{full}$.

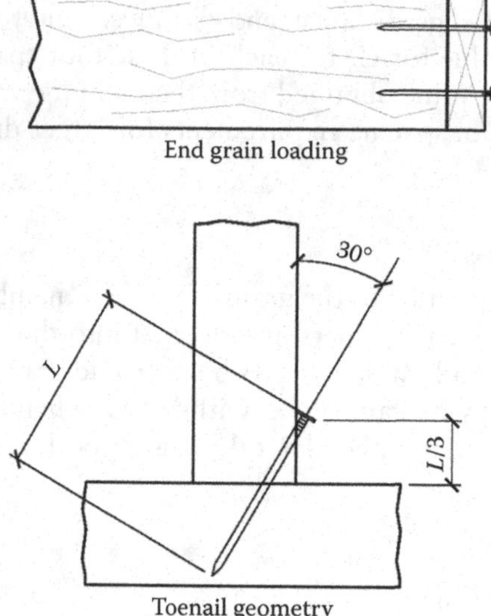

Figure 8.5 End grain and toenail connection geometry.

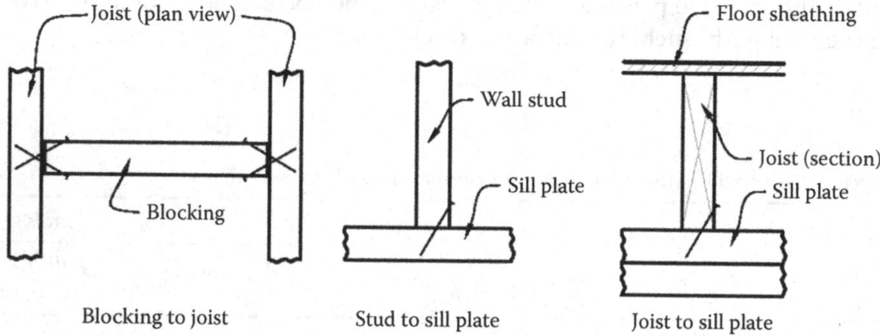

Figure 8.6 Typical toenail connections.

8.3.8 Toenail factor C_{tn}

Section 12.1.6.4 of the NDS code gives the geometric configuration of a toenail connection (see Figure 8.5). This is the optimal configuration to prevent the side member from splitting vertically. Toenail connections are typically used where the loads are minimal, such as beam to sill plate, stud to sill plate, or blocking to a beam (see Figure 8.6). For toenailed connections loaded in withdrawal (W), $C_{tn} = 0.67$. Note that the wet service factor C_M does not apply to toenailed connections loaded in withdrawal; therefore, $C_M = 1.0$. For lateral loads (Z), $C_{tn} = 0.83$.

Example 8.1: Group action factor—wood-to-wood connection

With reference to Figure 8.7, determine the group action factor C_g. Verify the results from the appropriate NDS table.

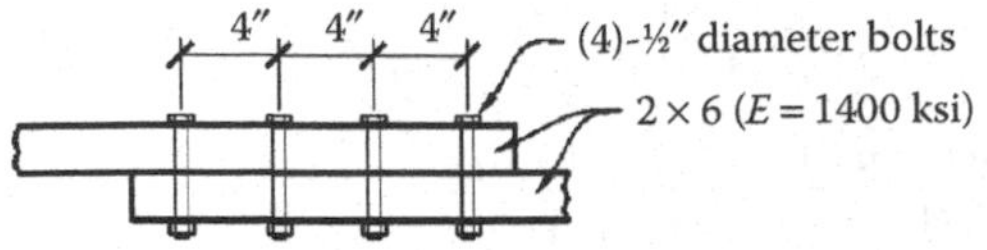

4″ 4″ 4″

(4)-½″ diameter bolts

2 × 6 (E = 1400 ksi)

Figure 8.7 Group action factor: wood-to-wood.

Solution:
From Equation 8.3:

$$C_g = \frac{m(1-m^{2n})}{n[(1+R_{EA}m^n)(1+m)-1+m^{2n})]}\,\frac{1+R_{EA}}{1-m}$$

For this case,

$$R_{EA} = \text{the lesser of } \frac{E_s A_s}{E_m A_m} \text{ or } \frac{E_m A_m}{E_s A_s} = 1.0$$

$$\gamma = 180{,}000D^{1.5} = (180{,}000)(0.5^{1.5}) = 63{,}640 \text{ (see Section 8.3)}$$

$$u = 1 + \gamma\frac{s}{2}\left(\frac{1}{E_m A_m} + \frac{1}{E_s A_s}\right)$$

$$= 1 + (63{,}640)\left(\frac{4\text{ in.}}{2}\right)\left[\frac{1}{(1.4\times 10^6)(8.25)} + \frac{1}{(1.4\times 10^6)(8.25)}\right] = 1.022$$

$$m = u - \sqrt{u^2-1}$$

$$= (1.022) - \sqrt{(1.022)^2-1} = 0.811$$

$n = 4$ (number of fasteners in a row)

Then

$$C_g = \left[\frac{(0.811)(1-0.811^{(2)(4)})}{(4)[(1+(1.0)(0.811)^4)(1+0.811)-1+(0.811)^{(2)(4)}]}\right]\left(\frac{1+1.0}{1-0.811}\right) = 0.979$$

From NDS Table 10.3.6A with $A_s/A_m = 1.0$ and $A_s = 8.25$ in^2, $C_g = 0.94$. This table is conservative for values of D less than 1 in., so the calculated value is consistent with the value obtained from the table. It is also within 5%, so the use of the NDS table for design is appropriate.

Example 8.2: Group action factor—wood-to-steel connection

With reference to Figure 8.8, determine the group action factor C_g. Verify the results from the appropriate NDS table.

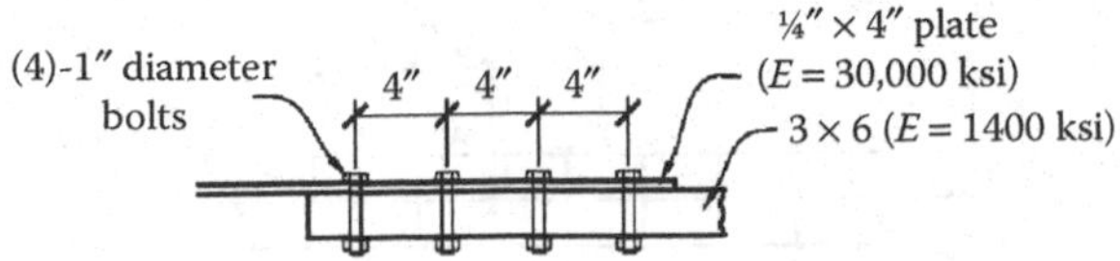

Figure 8.8 Group action factor: wood-to-steel.

Solution:
From Equation 8.3:

$$C_g = \frac{m(1-m^{2n})}{n[(1+R_{EA}m^n)(1+m)-1+m^{2n})]}\,\frac{1+R_{EA}}{1-m}$$

For the case at hand,

$$R_{EA} = \text{ the lesser of } \frac{E_sA_s}{E_mA_m} \text{ or } \frac{E_mA_m}{E_sA_s}$$

$$= \frac{(30\times10^6)(0.25\times4)}{(1.4\times10^6)(13.75)} = 1.56 \text{ or } \frac{(1.4\times10^6)(13.75)}{(30\times10^6)(0.25\times4)} = 0.642$$

$\gamma = 270{,}000D^{1.5} = (270{,}000)(1^{15}) = 270{,}000$ (see Section 8.3)

$$u = 1+\gamma\frac{s}{2}\left(\frac{1}{E_mA_m}+\frac{1}{E_sA_s}\right)$$

$$= 1+(270{,}000)\left(\frac{4\text{ in.}}{2}\right)\left[\frac{1}{(1.4\times10^6)(13.75)}+\frac{1}{(30\times10^6)(0.25)}\right] = 1.046$$

$$m = u-\sqrt{u^2-1}$$

$$= (1.046)-\sqrt{(1.046)^2-1} = 0.739$$

$n = 4$ (number of fasteners in a row)

Then,

$$C_g = \frac{(0.739)(1-0.739^{(2)(4)})}{(4)[(1+(0.642)(0.739)^4)(1+0.739)-1+(0.739)^{(2)(4)}]}\,\frac{1+0.642}{1-0.739} = 0.912$$

From NDS Table 10.3.6C with $A_m/A_s = 13.75$ and $A_m = 13.75$ in.², $C_g = 0.91$ (by interpolation), thus verifying the value from the equation.

Example 8.3: Geometry factor for a nailed connection

For the lap splice connections shown in Figure 8.9, it has been found that 12–10d nails are required. Determine if the fastener layout is adequate for the full design value. Assume that the holes are not prebored.

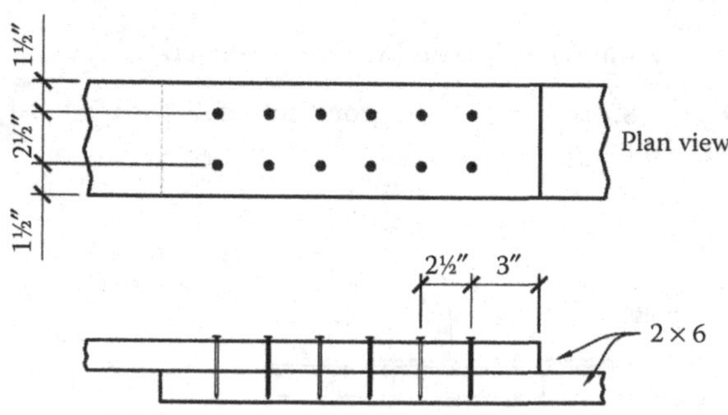

Figure 8.9 Nailed lap splice connection.

Solution:
Recall that $C_\Delta = 1.0$ for nailed connections; thus, to obtain the full design value for the fasteners, the following minimum spacings are required (see Table 8.3):

End distance: $15D$
Edge distance: $2.5D$
Center-to-center spacing: $15D$
Row spacing: $5D$

For 10d nails, $D = 0.148$ in., $L = 3$ in. (Table L4 NDS code). Thus,

End distance: (15)(0.148 in.) = 2.2 in. < 3 in. OK
Edge distance: (2.5)(0.148 in.) = 0.37 in. < 1.5 in. OK
Center-to-center spacing: (15)(0.148 in.) = 2.2 in. < 2.5 in. OK
Row spacing: (5)(0.148 in.) = 0.75 in. < 2.5 in. OK
Minimum penetration: $p = 10D = (10)(0.148 \text{ in.}) = 1.48 \text{ in.} < 1.5 \text{ in.}$ OK

The fastener layout is adequate.

Example 8.4: Geometry factor for a bolted connection

For the lap splice connections shown in Figure 8.10, determine the geometry factor for the connection shown. The lumber is Douglas Fir-Larch.

Solution:
l/D = smaller of:

$$\frac{l_m}{D} = \frac{1.5}{0.75} = 2$$

$$\frac{l_s}{D} = \frac{1.5}{0.75} = 2 \text{ (same result; use } l/D = 2)$$

The following distances are provided:

End distance: 4 in
Edge distance: 1.5 in
Center-to-center spacing: 2.5 in
Row spacing: 2.5 in

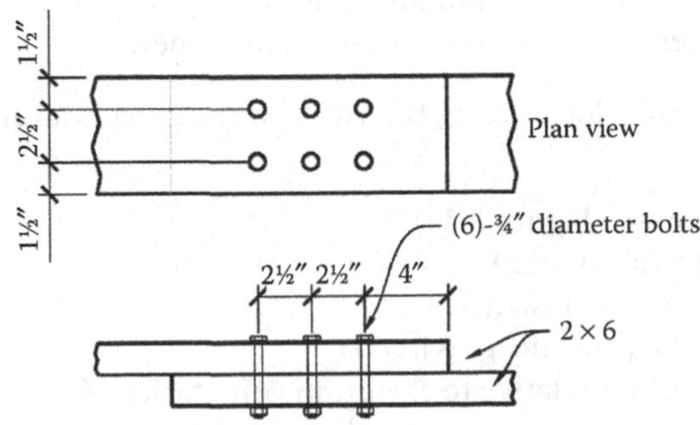

Figure 8.10 Bolted lap slice connection.

The following minimum spacings are required (see Table 8.5):

	Minimum	*Full*
End distance	3.5*D*	7*D*
Edge distance	N/A	1.5*D*
Center-to-center spacing	3*D*	4*D*
Row spacing	N/A	1.5*D*
thus:		

	Minimum		*Full*
End distance	(3.5)(0.75 in.) = 2.63 in.	to	(7)(0.75 in.) = 5.25 in.
Edge distance	N/A		(1.5)(0.75 in.) = 1.13 in.
Center-to-center spacing	(3)(0.75 in.) = 2.25 in.	to	(4)(0.75 in.) = 3 in.
Row spacing	N/A		(1.5)(0.75 in.) = 1.13 in.

By inspection, the spacings provided are all greater than the minimum value required. For the parameters that do not have a minimum value, the full or base spacing must be provided. The geometry factor must now be calculated for the other distance parameters are as follows:

$$C_\Delta = \frac{\text{Spacing provided}}{\text{Base spacing}} \tag{8.8}$$

End distance: $C_\Delta = 4$ in./5.25 in. = 0.76
Edge distance: $C_\Delta = 1.0$ (full or base spacing must be provided)
Center-to-center spacing: $C_\Delta = 2.5$ in./3 in. = 0.83
Row spacing: $C_\Delta = 1.0$ (full or base spacing must be provided)

The lowest value for C_Δ controls; therefore, C_Δ **= 0.76.**

8.4 BASE DESIGN VALUES: LATERALLY LOADED CONNECTORS

In Section 12.3 of the NDS code, a *yield limit model* is presented to determine the nominal lateral design value for a single fastener, *Z*, based on the following assumptions:

1. The faces of the connected members are in contact.
2. The load acts perpendicular to the axis of the dowel or fastener.
3. Edge distance, end distance, center-to-center spacing, and row spacing requirements are met (either base values or minimum values).
4. The minimum fastener penetration requirement is met.

The nominal lateral design value *Z* using the yield limit model is a function of the following factors:

- The mode of failure (see Figure 8.11)
- The diameter of the fastener, *D*
- The bearing length of the fastener, *l*
- The wood species (i.e., specific gravity *G*)
- The direction of loading relative to the grain orientation, θ
- The yield strength of the fastener in bending, F_{yb}
- The dowel bearing strength, F_e

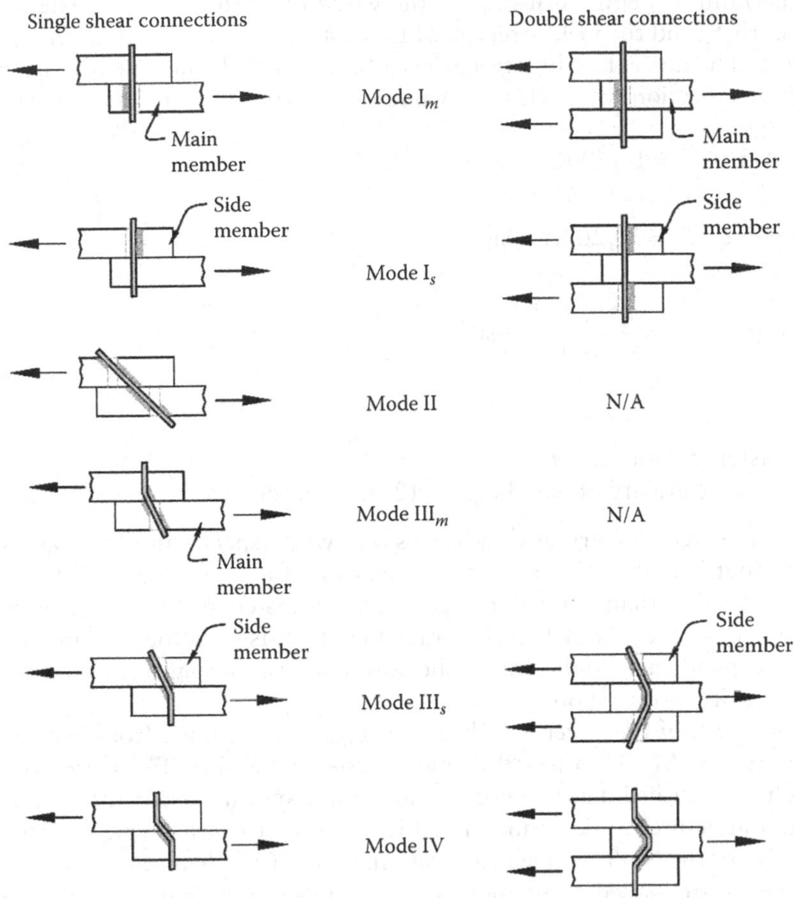

Figure 8.11 Connection failure modes.

There are four basic modes of failure for connectors and the analysis of each mode leads to a corresponding set of equations for determining the design load capacity. Figure 8.11 illustrates each of these modes in single and double shear.

1. *Mode I* is a bearing failure such that the connector crushes the wood member it is in contact with. This failure can occur in either the main member (mode I_m) or the side member (mode I_s).
2. *Mode II* is a failure such that the connector pivots within the wood members (i.e., the fastener does not yield). There is localized crushing of the wood fibers at the faces of the wood members. This type of failure occurs only in single shear connections.
3. *Mode III* is a combined failure of the connector such that a plastic hinge forms in the connector in conjunction with a bearing failure in either the main member (mode III_m) or the side member (mode III_s).
4. *Mode IV* is a failure of the connector such that two plastic hinges form with limited localized crushing of the wood fibers near the shear plane(s).

Note: In two-member connections with nails or screws, the main member is usually the member that receives the "pointed" end; in three-member connections, the main member is the middle member (see Figure 8.11).

Because the failure is either crushing of the wood or yielding of the fastener, the dowel bearing strength F_e and the yield strength of the fastener F_{yb} are the most critical factors in determining Z. The dowel bearing strength is a function of the fastener diameter, the wood species, and the direction of the loading. Values for F_e are calculated as follows:

$$\text{For } D < \tfrac{1}{4}\text{ in.:} \quad F_e = 16{,}600G^{1.84}\text{(psi)} \tag{8.9}$$

$$\text{For } D \geq \tfrac{1}{4}\text{ in.:} \quad F_{e\parallel} = 11{,}200G \text{ (psi)} \tag{8.10}$$

$$\text{For } D \geq \tfrac{1}{4}\text{ in.:} \quad F_{e\perp} = \frac{6100G^{1.45}}{\sqrt{D}}\text{(psi)} \tag{8.11}$$

where:

D is the fastener diameter (in.)
G is the specific gravity of wood (Table 12.3.3A NDS code)

Values for F_e are given for various dowel sizes and wood species in the NDS Table 12.3.3. It can be seen that load direction relative to the grain of the wood is not a factor for small-diameter fasteners (less than ¼ in.). For larger diameter fasteners, the dowel bearing strength is constant when loaded parallel to the grain but decreases as the fastener size increases when loaded perpendicular to the grain. The dowel bearing strength is also directly proportional to the density of the wood.

The yield strength of the fastener in bending, F_{yb}, is determined from tests performed in accordance with ASTM F1575 (small-diameter fasteners) and ASTM F606 (large-diameter fasteners). When published data are not available for a specific fastener, the designer should contact the manufacturer to determine the fastener yield strength. Typical values for F_{yb} are given in Table I1 of the NDS code, which is summarized in Table 8.6.

When fasteners are loaded at an angle, θ, to the grain (see Figure 8.12), the Hankinson formula shall be used to determine the dowel bearing strength:

$$F_{e\theta} = \frac{F_{e\parallel}F_{e\perp}}{F_{e\parallel}\sin^2\theta + F_{e\perp}\cos^2\theta} \tag{8.12}$$

When connections are made with steel side plates, Section I.2 of the NDS code explains that the nominal bearing strength of steel is $2.4F_u$ for hot-rolled steel and $2.2F_u$ for

Table 8.6 Yield strength in bending of various fasteners

Fastener type	*Diameter (in.)*	*F_{yb} (psi)*
Bolt, lag screw	$D \geq 0.375$	45,000
SAE J429 Grade 1, $F_y = 36$ ksi		
$F_u = 60$ ksi		
Common, box, or sinker nail; lag screw; wood screw (low to medium carbon steel)	$0.344 < D \leq 0.375$	45,000
	$0.273 < D \leq 0.344$	60,000
	$0.236 < D \leq 0.273$	70,000
	$0.177 < D \leq 0.236$	80,000
	$0.142 < D \leq 0.177$	90,000
	$0.099 \leq D \leq 0.142$	100,000

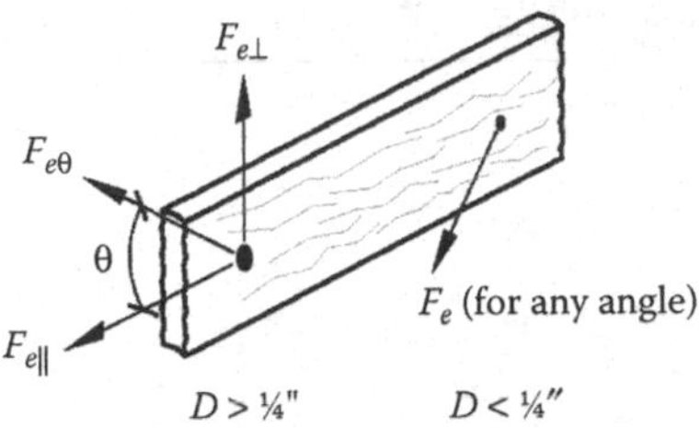

Figure 8.12 Fasteners loaded at various angles to the grain.

cold-formed steel. Applying a load duration factor of 1.6, the dowel bearing strength on the steel is as follows:

$$F_e = 1.5F_u \text{ (hot-rolled steel)}$$

$$F_e = 1.375F_u \text{ (cold-formed steel)}$$

where:

F_u = ultimate tensile strength
= 58,000 psi for ASTM A36 steel, ¼ in. and thicker
= 45,000 psi, three gage and thinner ASTM A653, grade 33

When fasteners are embedded in concrete, such as anchor bolts for a sill plate (see Chapter 9), the dowel bearing strength can be assumed to be 7500 psi for concrete with a compressive strength of at least $f'_c = 2500$ psi. Table 12E of the NDS code gives lateral design values Z for various wood species and bolt sizes. Note that this table assumes a minimum embedment of 6 in. into the concrete. Having discussed the controlling factors in determining the lateral design value Z' for a single fastener, we can now look at the equations to calculate Z'. The direction of loading with respect to the wood grain is a factor for fasteners greater than 1/4 inch in diameter, but is not a factor for fasteners that are less than 1/4 inch in diameter (see Figure 8.12). The NDS code gives a yield limit equation to determine the nominal lateral design value Z, each corresponding to a specific mode of failure. The smallest value determined from these equations is the controlling value to use for design. The equations are as follows:

F_{em} or $F_{es} = F_e$ for $D \leq$ ¼ in.

= $F_{e\parallel}$ or $F_{e\perp}$ for applied loads either parallel or perpendicular to the grain

= $F_{e\theta}$ for applied loads at an angle to the grain

= 87,000 psi for ASTM A36 steel side plates (1.5 Fu)

= 61,850 psi ASTM A653, grade 33 steel side plates (1.375 Fu)

= 7500 psi (fastener embedded into concrete)

For lag screws, the diameter D is typically taken as the root diameter D_r or the reduced body diameter at the threads (see Figure 8.2). Section 12.3.5 of the NDS code stipulates that the main member dowel length l_m does not include the length of the tapered tip when the penetration depth into the main member is less than 10D. When the penetration is greater than 10D, l_m would include the length of the tapered tip.

Yield Mode	Single Shear	Double Shear	
I_m	$Z = \frac{Dl_m F_{em}}{R_d}$	$Z = \frac{Dl_m F_{em}}{R_d}$	(8.13)
I_s	$Z = \frac{Dl_s F_{es}}{R_d}$	$Z = \frac{2Dl_s F_{es}}{R_d}$	(8.14)
II	$Z = \frac{k_1 Dl_s F_{es}}{R_d}$		(8.15)
III_m	$Z = \frac{k_2 Dl_m F_{em}}{(1+2R_e)R_d}$		
III_s	$Z = \frac{k_3 Dl_s F_{em}}{(2+R_e)R_d}$	$Z = \frac{2k_3 Dl_s F_{em}}{(2+R_e)R_d}$	(8.16)
IV	$Z = \frac{D^2}{R_d}\sqrt{\frac{2F_{em}F_{yb}}{3(1+R_e)}}$	$Z = \frac{2D^2}{R_d}\sqrt{\frac{2F_{em}F_{yb}}{3(1+R_e)}}$	(8.17)

where:

$$k_1 = \frac{\sqrt{R_e + 2R_e^2(1+R_t+R_t^2) + R_t^2 R_e^3} - R_e(1+R_t)}{(1+R_e)} \quad (8.18)$$

$$k_2 = -1 + \sqrt{2(1+R_e) + \frac{2F_{yb}(1+2R_e)D^2}{3F_{em}l_m^2}} \quad (8.19)$$

$$k_3 = -1 + \sqrt{\frac{2(1+R_e)}{R_e} + \frac{2F_{yb}(2+R_e)D^2}{3F_{em}l_s^2}} \quad (8.20)$$

D is the diameter of connector (in.)
$F\gamma_b$ is the dowel bending strength (psi) (see NDS code Table I1)
R_d is the reduction term (NDS code, Table 12.3.1B or see Table 8.7)
$K_\theta = 1 + (0.25)(\theta/90)$

$$R_e = \frac{F_{em}}{F_{es}}$$

$$R_t = \frac{l_{em}}{l_s}$$

l_m is the connector length in the *main* member (in.)
l_s is the connector length in the *side* member (in.)
F_{em} is the main member dowel bearing strength (psi)
F_{es} is the side member dowel bearing strength (psi)
θ is the maximum angle in degrees between the applied load and the grain of the *main* member in the connection

Table 8.7 Reduction factor, R_d

Fastener size (in.)	*Yield mode*	R_d
$0.25 \le D < 1$	I_m, I_s	$4K_\theta$
	II	$3.6K_\theta$
	III_m, III_s, IV	$3.2K_\theta$
$0.17 < D < 0.25$	I_m, I_s, II, III_m, III_s, IV	$10D + 0.5$
$D \le 0.17$	I_m, I_s, II, III_m, III_s, IV	2.2

As an alternative to solving for Z using the yield limit equations for lag screws, Tables 12J and 12K are provided in the NDS code. These tables are for lag screws in single shear (wood-to-wood or wood-to-steel) because lag screws are not typically loaded in double shear. The values in this table are valid when the penetration into the main member is at least $8D$. When the penetration is such that $4D < p < 8D$, the value for Z in Table 12J or 12K is adjusted by $p/8D$. Note that NDS Tables 12A through 12J use the following notations:

Z_{II} = lateral design shear strength of a fastener in single shear with the *side* member loaded *parallel* to the grain and the *main* member loaded *parallel* to the grain
$Z_{\perp}$ = lateral design shear strength of a fastener in single shear with the *side* member loaded *perpendicular* to the grain and the *main* member loaded *perpendicular* to the grain
$Z_{m\perp}$ = lateral design shear strength of a fastener in single shear with the *side* member loaded *parallel* to the grain and the *main* member loaded *perpendicular* to the grain
$Z_{s\perp}$ = lateral design shear strength of a fastener in single shear with the *side* member loaded *perpendicular* to grain and the *main* member loaded *parallel* to the grain

The following examples illustrate the calculation of Z for various loading conditions.

Example 8.5: Laterally loaded nail, single shear

For the connection shown in Figure 8.13, determine the nominal lateral design value Z and compare the results from the appropriate table in the NDS code. Lumber is Spruce-Pine-Fir for the main and side members.

Solution:

$G = 0.42$ (NDS code Table 12.3.3A)
$D = 0.131$ in. (NDS code Table L4)
$l = 2.5$ in. (NDS code Table L4)
$F_{yb} = 100{,}000$ psi (Table 8.6)
$F_{em} = F_{es} = 16{,}600G^{184}$

$= (16{,}600)(0.42)^{184} = 3364$ psi (agrees with Table 12.3.3 of the NDS code)

$$R_e = \frac{F_{em}}{F_{es}} = \frac{3364}{3364} = 1.0$$

$$R_t = \frac{l_m}{l_s} = \frac{2.5\text{ in.} - 1.5\text{ in.}}{1.5\text{ in.}} = 0.667$$

$R_d = 2.2$

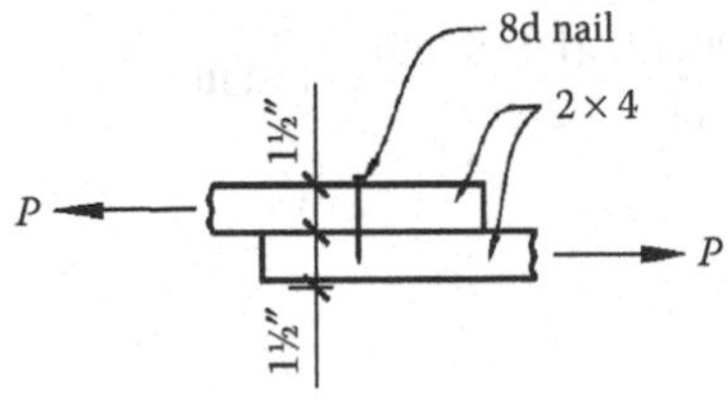

Figure 8.13 Laterally loaded nail.

$$k_1 = \frac{\sqrt{R_e + 2R_e^2(1+R_t+R_t^2)+R_t^2R_e^3} - R_e(1+R_t)}{(1+R_e)}$$

$$= \frac{\left[\begin{array}{l}\sqrt{1+(2)(1)^2\,[1+0.667+(0.667)^2]+(0.667)^2(1)^3} \\ -\ (1)(1+0.667)\end{array}\right]}{1+1} = \mathbf{0.357}$$

$$k_2 = -1 + \sqrt{2(1+R_e) + \frac{2F_{yb}(1+2R_e)D^2}{3F_{em}l_m^2}}$$

$$= -1 + \sqrt{(2)(1+1) + \frac{(2)(100{,}000)[1+(2)(1)](0.131)^2}{(3)(3364)(1)^2}} = \mathbf{1.24}$$

$$k_3 = -1 + \sqrt{\frac{2(1+R_e)}{R_e} + \frac{2F_{yb}(2+R_e)D^2}{3F_{em}l_s^2}}$$

$$= -1 + \sqrt{\frac{2(1+1)}{(1)} + \frac{(2)(100{,}000)(2+1)(0.131)^2}{(3)(3364)(1.5)^2}} = \mathbf{1.11}$$

Calculate Z for each mode of failure.
Mode I_m:

$$Z = \frac{Dl_mF_{em}}{R_d} = \frac{(0.131)(1)(3364)}{2.2} = \mathbf{200\ lb}$$

Mode I_s:

$$Z = \frac{Dl_sF_{es}}{R_d} = \frac{(0.131)(1.5)(3364)}{2.2} = \mathbf{300\ lb}$$

Mode II:

$$Z = \frac{k_1l_sF_{es}}{R_d} = \frac{(0.357)(0.131)(1.5)(3364)}{2.2} = \mathbf{107\ lb}$$

Mode III_m:

$$Z = \frac{k_2Dl_mF_{em}}{(1+2R_e)R_d} = \frac{(1.24)(0.131)(1)(3364)}{[1+(2)(1)](2.2)} = \mathbf{82\ lb}$$

Mode III_s:

$$Z = \frac{k_3Dl_sF_{em}}{(2+R_e)R_d} = \frac{(1.11)(0.131)(1.5)(3364)}{(2+1)(2.2)} = \mathbf{111\ lb}$$

Mode IV:

$$Z = \frac{D^2}{R_d}\sqrt{\frac{2F_{em}F_{yb}}{3(1+R_e)}}$$

$$= \frac{(0.131)^2}{2.2}\sqrt{\frac{(2)(3364)(100{,}000)}{3(1+1)}} = \mathbf{82\ lb}$$

The lowest value was 82 lb (modes III_m and IV); therefore, $Z = 82$ lb, which agrees with Table 12N of the NDS code.

Example 8.6: Laterally loaded bolt, double shear

For the connection shown in Figure 8.14, determine the nominal lateral design value $Z_\perp$ and compare the results from the appropriate table in the NDS code. The lumber is Hem-Fir for the main member, and the steel plates are ¼ in. A36.

Solution:

$G = 0.43$ (NDS code, Table 12.3.2A)
$D = 0.625$ in.
$l_m = 2.5$ in. (3 × main member)
$F_{yb} = 45{,}000$ psi (Table 8.8)

$$F_{em} = F_{e\perp} = \frac{6100G^{1.45}}{\sqrt{D}} = \frac{(6100)(0.43)^{1.45}}{\sqrt{0.625}}$$

$= 2269$ psi (agrees with Table 12.3.3 of the NDS code)

$F_{es} = 87{,}000$ psi (A36 side plates)

$$R_e = \frac{F_{em}}{F_{es}} = \frac{2269}{87{,}000} = 0.026$$

$$R_t = \frac{l_m}{l_s} = \frac{2.5\text{ in.}}{0.25\text{ in.}} = 10$$

$\theta = 90°$

$K_\theta = 1 + (0.25)(\theta/90) = 1 + (0.25)(90/90) = 1.25$

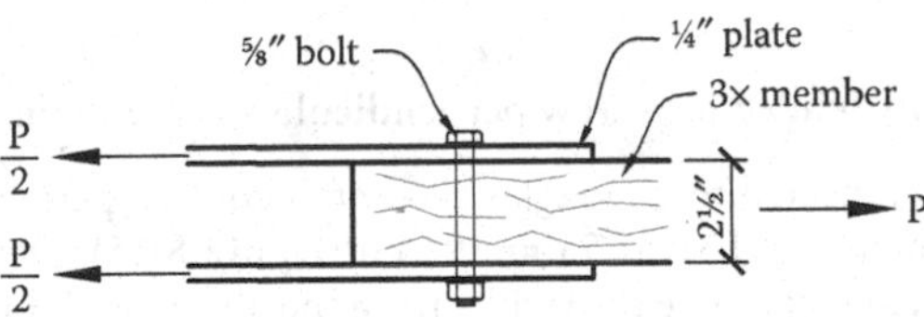

Figure 8.14 Laterally loaded bolt.

Table 8.8 Spacing requirements for lag screws not loaded in withdrawal

Distance	*Minimum value*
Edge distance	1.5*D*
End distance	4*D*
Spacing	4*D*

$$k_3 = -1 + \sqrt{\frac{2(1+R_e)}{R_e} + \frac{2F_{yb}(2+R_e)D^2}{3F_{em}l_s^2}}$$

$$= -1 + \sqrt{\frac{(2)(1+0.026)}{0.026} + \frac{(2)(45{,}000)(2+0.026)(0.625)^2}{(3)(2269)(0.25)^2}} = 14.7$$

Calculate Z for each mode of failure.
Mode I_m ($R_d = 4K_\theta$):

$$Z = \frac{Dl_mF_{em}}{R_d} = \frac{(0.625)(2.5)(2269)}{(4)(1.25)} = \mathbf{709\ lb}$$

Mode I_s ($R_d = 4K_\theta$):

$$Z = \frac{2Dl_sF_{es}}{R_d} = \frac{(2)(0.625)(0.25)(87{,}000)}{(4)(1.25)} = \mathbf{5437\ lb}$$

Mode II: Not applicable
Mode III_m: Not applicable
Mode III_s ($R_d = 3.2K_\theta$):

$$Z = \frac{2k_3Dl_sF_{em}}{(2+R_e)R_d} = \frac{(2)(14.7)(0.625)(0.25)(2269)}{(2+0.026)(3.2)(1.25)} = \mathbf{1285\ lb}$$

Mode IV:

$$Z = \frac{2D^2}{R_d}\sqrt{\frac{2F_{em}F_{yb}}{3(1+R_e)}} = \frac{(2)(0.625)^2}{(3.2)(1.25)}\sqrt{\frac{(2)(2269)(45{,}000)}{3(1+0.026)}} = \mathbf{1591\ lb}$$

The lowest value was 709 lb (mode I_m); therefore, $Z_\perp = 709$ lb, which agrees with Table 12G of the NDS code.

Example 8.7: Laterally loaded lag screw perpendicular to the grain

A ⅜-in. lag screw is connected to a Douglas Fir-Larch column perpendicular to the grain with a 10-gage side plate (ASTM A653 grade 33) (Figure 8.15). Determine the nominal lateral design value Z, and compare the results from the appropriate table in the NDS code. Assume that the penetration into the main member is at least $10D$.

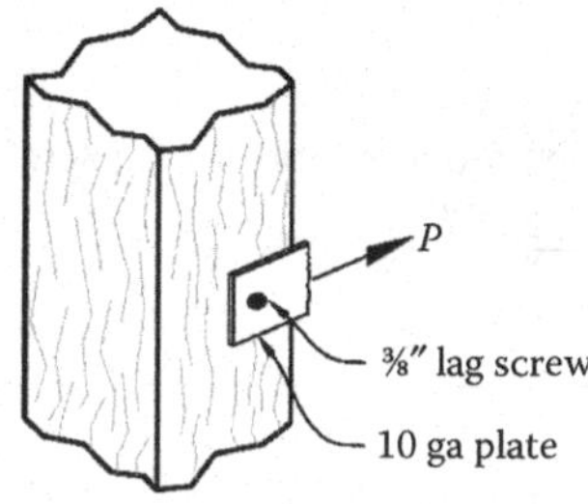

Figure 8.15 Laterally loaded lag screw.

Solution:

$G = 0.50$ (NDS code Table 12.3.3A)
$D = 0.375$ in.
$D_r = 0.265$ in. (Table L2)
$l_m = 10D = (10)(0.265) = 2.65$ in.
$l_s = 0.134$ in. (10 gage)
$F_{yb} = 45{,}000$ psi (Table 8.8)

$$F_{em} = F_{e\perp} = \frac{6100G^{1.45}}{\sqrt{D}} = \frac{(6100)(0.50)^{1.45}}{\sqrt{0.375}} = 3646 \text{ psi}$$

$F_{es} = 61{,}850$ psi (ASTM A653, grade 33)

$$R_e = \frac{F_{em}}{F_{es}} = \frac{3646}{61{,}850} = 0.059$$

$$R_t = \frac{l_m}{l_s} = \frac{2.65 \text{ in.}}{0.134 \text{ in.}} = 19.78$$

$\theta = 90°$

$K_\theta = 1 + 0.25(\theta/90) = 1 + 0.25(90/90) = 1.25$

$$k_1 = \frac{\sqrt{R_e + 2R_e^2(1 + R_t + R_t^2) + R_t^2 R_e^3} - R_e(1 + R_t)}{(1 + R_e)}$$

$$= \frac{\left\{\sqrt{\begin{matrix}0.059 + (2)(0.059)^2\ [1 + 19.78 \\ +(19.78)^2] + (19.78)^2(0.059)^3\end{matrix}}\right\} - (0.059)(1 + 19.78)}{1 + 0.059}$$

$$= 0.480$$

$$k_2 = -1 + \sqrt{2(1 + R_e) + \frac{2F_{yb}(1 + 2R_e)D^2}{3F_{em}l_m^2}}$$

$$= -1 + \sqrt{2(1 + 0.059) + \frac{(2)(45{,}000)[1 + (2)(0.059)](0.625)^2}{(3)(3646)(2.65)^2}} = 0.487$$

$$k_3 = -1 + \sqrt{\frac{2(1 + R_e)}{R_e} + \frac{2F_{yb}(2 + R_e)D^2}{3F_{em}l_s^2}}$$

$$= -1 + \sqrt{\frac{2(1 + 0.059)}{0.059} + \frac{(2)(45{,}000)2 + 0.059(0.265)^2}{(3)(3646)(0.134)^2}} = 9.11$$

Calculate Z for each mode of failure.

Mode I_m ($R_d = 4K_\theta$):

$$Z = \frac{Dl_m F_{em}}{R_d} = \frac{(0.265)(2.65)(3646)}{(4)(1.25)} = \mathbf{512\ lb}$$

Mode I_s ($R_d = 4K_\theta$):

$$Z = \frac{2Dl_s F_{es}}{R_d} = \frac{(0.265)(0.134)(61{,}850)}{(4)(1.25)} = \mathbf{439\ lb}$$

Mode II ($R_d = 3.6K_\theta$):

$$Z = \frac{k_1 Dl_s F_{es}}{R_d} = \frac{(0.480)(0.265)(0.134)(61{,}850)}{(3.6)(1.25)} = \mathbf{234\ lb}$$

Mode III_m ($R_d = 3.2K_\theta$):

$$Z = \frac{k_2 Dl_m F_{em}}{(1 + 2R_e)R_d} = \frac{(0.487)(0.265)(2.65)(3646)}{(1+(2)(0.059)(3.2)(1.25)} = \mathbf{279\ lb}$$

Mode III_s ($R_d = 3.2K_\theta$):

$$Z = \frac{2k_3 Dl_s F_{em}}{(2 + R_e)R_d} = \frac{(9.11)(0.265)(0.134)(3646)}{(2+0.059)(3.2)(1.25)} = \mathbf{143\ lb}$$

Mode IV ($R_d = 3.2K_\theta$):

$$Z = \frac{D^2}{R_d}\sqrt{\frac{2F_{em}F_{yb}}{3(1+R_e)}} = \frac{(0.265)^2}{(3.2)(1.25)}\sqrt{\frac{(2)(3646)(45{,}000)}{3(1+0.059)}} = \mathbf{178\ lb}$$

The lowest value was 143 lb (mode III_s); therefore, $Z_\perp = 143$ lb, which agrees with Table 12K of the NDS code ($Z_\perp = 140$ lb).

Example 8.8: Knee brace connection

For the knee brace connection shown in Figure 8.16, design the connectors assuming 16d common nails. The wood species is Spruce-Pine-Fir and is exposed to the weather. Loads are due to lateral wind.

Solution:
From NDS Table L-4: $D = 0.162$ in., $L = 3½$ in.
Minimum distances (from Table 8.3):

End distance: 15D
Edge distance: 2.5D
Center-to-center spacing: 15D
Row spacing: 5D

Thus,

End distance: (15)(0.162 in.) = 2.43 in.
Edge distance: (2.5)(0.162 in.) = 0.405 in.
Center-to-center spacing: (15)(0.162 in.) = 2.43 in.
Row spacing: (5)(0.162 in.) = 0.81 in.
Minimum penetration: $p = 10D = (10)(0.162\text{ in.}) = 1.62$ in.

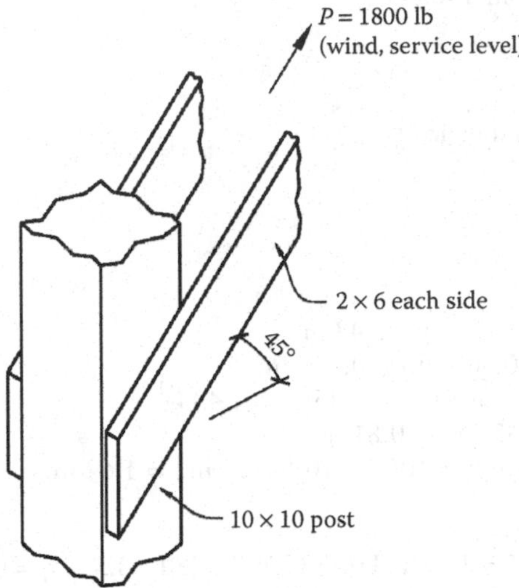

Figure 8.16 Knee brace connection.

From NDS Table 12N, Z = 120 lb. From NDS Table 11.3.3, C_M = 0.7 and C_D = 1.6 (wind). All other C factors = 1.0.

$$Z' = ZC_DC_MC_tC\ C_\Delta C_{eg}C_{di}C_{tn}$$

$$= (120)(1.6)(0.7)(1.0)(1.0)(1.0)(1.0)(1.0)(1.0) = 134 \text{ lb}$$

The number of nails required

$$N_{\text{req'd}} = \frac{(0.5)(1800)}{134}$$

$$= 6.7 \text{ nails} \simeq 7 \text{ nails, use 8 nails for symmetry}$$

Use 4–16d nails on each side of the splice. See Figure 8.17.

LRFD solution:
From NDS Table L-4: D = 0.162 in., L = 3½ in.

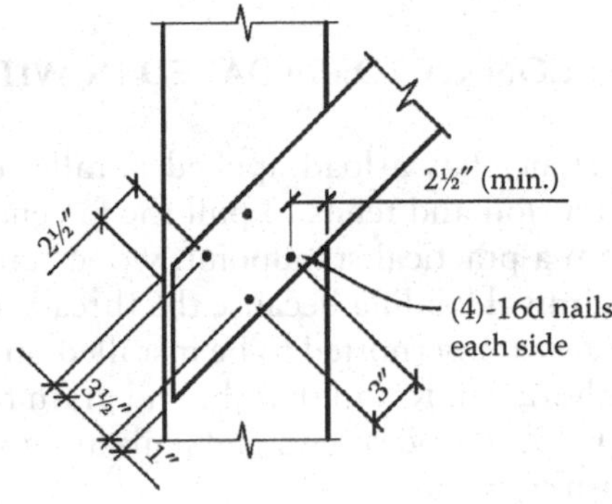

Figure 8.17 Nailed knee brace connection detail.

Minimum distances (from Table 8.3):

End distance: 15*D*
Edge distance: 2.5*D*
Center-to-center spacing: 15*D*
Row spacing: 5*D*

Thus,

End distance: (15)(0.162 in.) = 2.43 in.
Edge distance: (2.5)(0.162 in.) = 0.405 in.
Center-to-center spacing: (15)(0.162 in.) = 2.43 in.
Row spacing: (5)(0.162 in.) = 0.81 in.
Minimum penetration: $p = 10D = (10)(0.162 \text{ in.}) = 1.62 \text{ in.}$

From NDS Table 12N, $Z = 120$ lb. From NDS Table 11.3.3, $C_M = 0.7$. All other C factors = 1.0.

$$Z'_r = ZC_DC_MC_tC\ C_\Delta C_{eg}C_{di}C_{tn}\ K_F\ \phi_z\ \lambda$$

$$= (120)(1.0)(0.7)(1.0)(1.0)(1.0)(1.0)(1.0)(1.0)(3.32)(0.65)(1.0) = 181 \text{ lb}$$

$$K_F = 3.32$$

$$\phi_z = 0.65$$

$\lambda = 1.0$, see Table N3

The number of nails required using a load factor of 1.6 for wind loads:

$$N_{\text{req'd}} = \frac{(1.6)(0.5)(1800)}{181} = 7.95 \text{ nails} \approx 8 \text{ nails}$$

Use 4–16d nails on each side of the splice. See Figure 8.17.

8.5 BASE DESIGN VALUES: CONNECTORS LOADED IN WITHDRAWAL

In *withdrawal loading,* a fastener has a load applied parallel to its length such that the load stresses the fastener in tension and tends to pull the fastener out of the main member in which it is embedded. From a practical standpoint, wood screws and lag screws are the preferred connector for withdrawal loading because the threads create a more positive connection to the wood. Lag screws are permitted to be installed on either the side grain or the end grain when loaded in withdrawal. Recall that the end grain factor applies for lag screws loaded in withdrawal ($C_{eg} = 0.75$). Wood screws and nails are not permitted to be loaded in withdrawal from the end grain of wood.

Section 12.2 of the NDS code gives the withdrawal design values W for various connectors as follows:

$$W = \begin{cases} 1800G^{3/2}D^{3/4} & \text{(lag screws)} \quad (8.21) \\ 2850G^2D & \text{(wood screws)} \quad (8.22) \\ 1380G^{5/2}D & \text{(nails)} \quad (8.23) \end{cases}$$

where:

W is the withdrawal design value (lb) (per inch of penetration)
G is the specific gravity (see NDS code Table 12.3.2A)
D is the fastener diameter (in.)

The value for W calculated from Equations 8.21 through 8.23 represents the withdrawal design value per inch of penetration into the main member, per fastener. NDS code Tables 12.2A, 12.2B, and 12.2C give tabulated values for W. The fastener could also fail by yielding in tension and should be checked for this condition, but wood connections are not usually controlled by the fastener strength in pure tension.

Example 8.9: Fasteners loaded in withdrawal

With reference to Figure 8.18, determine the adjusted withdrawal values W' (= P) for the following connector types: (a) ⅜-in. lag screw; (b) no. 12 wood screw; (c) 16d common nail. Normal temperature and moisture conditions apply and loading is due to wind. Lumber is Hem-Fir.

Solution:
From NDS code Table 12.3.3A, G = 0.43 (Hem-Fir).

a. *⅜-in. lag screw*:

$$W = 1800G^{3/2}D^{3/4}$$

$$= (1800)(0.43)^{3/2}(0.375)^{3/4} = 243 \text{ lb/in. (agrees with NDS code Table 12.2A)}$$

Adjusted for penetration:

$$W = (3 \text{ in.})(243 \text{ lb/in.}) = 729 \text{ lb}$$

$$P_{max} = W' = WC_DC_MC_tC_{eg}C_{tn}$$

$$= (729)(1.6)(1.0)(1.0)(1.0)(1.0) = \mathbf{1166\ lb}$$

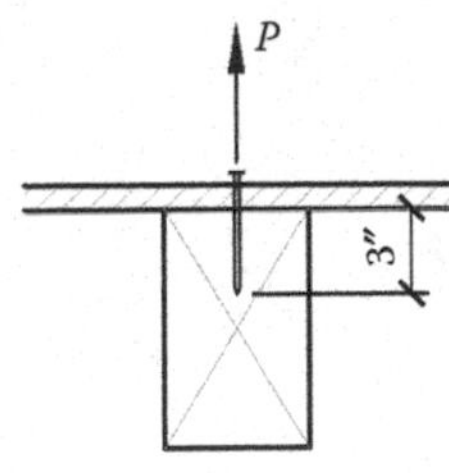

Figure 8.18 Fasteners loaded in withdrawal.

b. *No. 12 wood screw.* From Table L3 of the NDS code:

$D = 0.216$ in.

$W = 2850G^2D$

$= (2850)(0.43)^2(0.216) = 114$ lb/in. (agrees with Table 12.2B of the NDS code)

Adjusted for penetration:

$W = (3\text{ in.})(114\text{ lb/in.}) = 342$ lb

$P_{max} = W' = WC_DC_MC_tC_{eg}C_{tn}$

$= (342)(1.6)(1.0)(1.0)(1.0)(1.0) = \mathbf{546\ lb}$

c. *16d common nail.* From Table L3 of the NDS code:

$D = 0.162$ in.

$W = 1380G^{5/2}D$

$= (1380)(0.43)^{5/2}(0.162) = 27$ lb/in. (agrees with Table 12.2C of the NDS code)

Adjusted for penetration:

$W = (3\text{ in.})(27\text{ lb/in.}) = 81$ lb

$P_{max} = W' = WC_DC_MC_tC_{eg}C_{tn}$

$= (81)(1.6)(1.0)(1.0)(1.0)(1.0) = \mathbf{130\ lb}$

LRFD solution:
With reference to Figure 8.18, determine the adjusted withdrawal values W'_r (= P_u) for the following connector types: (a) ⅜-in. lag screw; (b) no. 12 wood screw; (c) 16d common nail. Normal temperature and moisture conditions apply and loading is due to wind. Lumber is Hem-Fir.

Solution:
$G = 0.43$, Hem-Fir, NDS Table 12.3.3A

$K_F = 3.32$

$\phi_z = 0.65$

$\lambda = 1.0$, see Table N3

$C_D = 1.0$ for LRFD

a. ⅜-in. lag screw:

$W = 1800G^{3/2}D^{3/4}$

$= (1800)(0.43)^{3/2}(0.375)^{3/4} = 243$ lb/in. (agrees with NDS code Table 12.2A)

Adjusted for penetration:

$W = (3 \text{ in.})(243 \text{ lb/in.}) = 729 \text{ lb}$

$P_{U\max} = W'_r = WC_DC_MC_tC\ C_{tn}\ K_F\ \phi_z\ \lambda$

$= (729)(1.0)(1.0)(1.0)(1.0)(1.0)(3.32)(0.65)(1.0) = \mathbf{1573\ lb}$

b. No. 12 wood screw:

From Table L3 of the NDS code,

$D = 0.216 \text{ in.}$

$W = 2850G^2D$

$= (2850)(0.43)^2(0.216) = 114 \text{ lb/in.}$ (agrees with Table 12.2B of the NDS code)

Adjusted for penetration:

$W = (3 \text{ in.})(114 \text{ lb/in.}) = 342 \text{ lb}$

$P_{U\max} = W'_r = WC_DC_MC_tC\ C_{tn}\ K_F\ \phi_z\ \lambda$

$= (342)(1.0)(1.0)(1.0)(1.0)(1.0)(3.32)(0.65)(1.0) = \mathbf{738\ lb}$

c. 16d common nail:

From Table L4of the NDS code,

$D = 0.162 \text{ in.}$

$W = 1380G^{5/2}D$

$= (1380)(0.43)^{5/2}(0.162) = 27 \text{ lb/in.}$ (agrees with Table 12.2C of the NDS code)

Adjusted for penetration:

$W = (3 \text{ in.})(27 \text{ lb/in.}) = 81 \text{ lb}$

$P_{U\max} = W'_r = WC_DC_MC_tC\ C_{tn}\ K_F\ \phi_z\ \lambda$

$= (81)(1.0)(1.0)(1.0)(1.0)(1.0)(3.32)(0.65)(1.0) = \mathbf{174\ lb}$

8.6 COMBINED LATERAL AND WITHDRAWAL LOADS

With reference to Figure 8.19, there are conditions in which connectors are subjected to lateral and withdrawal load simultaneously. For this case, the NDS code specifies the following interaction equations for combined loading:

$$Z'_\alpha = \begin{cases} \dfrac{W'pZ'}{W'p\cos^2\alpha + Z'\sin^2\alpha} & \text{for lag screws or wood screws} \quad (8.24) \\ \dfrac{W'pZ'}{W'p\cos\alpha + Z'\sin\alpha} & \text{for nails} \quad (8.25) \end{cases}$$

where:

Z'_a is the adjusted resulting design value under combined loading
W' is the adjusted withdrawal design value
p is the length of penetration into the main member (in.)
Z' is the adjusted lateral design value
α is the angle between the wood surface and the direction of load applied

The value Z'_a would then be compared to the resultant load P shown in Figure 8.19.

Example 8.10: Fastener loaded by lateral and withdrawal loads

A connection similar to that shown in Figure 8.19 requires a no. 14 wood screws under combined lateral and withdrawal loads. Determine P_{max} when $\alpha = 30°$ and a penetration, $p = 1.5$ in. The lumber is Hem-Fir. The side plate is 10 gage (ASTM A653 grade 33) and the loads are $D + S$. Assume that normal temperature and moisture conditions apply.

Solution:

$D = 0.242$ in. (NDS code Table L3)
$W = 127$ lb/in. (NDS code Table 12.2B)
$Z = 160$ lb (NDS code Table 12N)
$p = 1.5$ in. (greater than $6D$ but less than $10D$)

Therefore,

$$Z = (160 \text{ lb})(p/10D) = (160)[1.5/(10)(0.242)] = \mathbf{99\ lb}$$

$$W' = WC_DC_MC_tC_{eg}C_{tn}$$

$$= (127)(1.15)(1.0)(1.0)(1.0)(1.0) = \mathbf{146\ lb}$$

$C_g = 1.0$ ($D < ¼$ in.)

$C_\Delta = 1.0$, but minimum spacing must be provided for $D < ¼$ in.

From Table 8.3, required spacings (assume no prebored holes):

End distance: $10D$
Edge distance: $2.5D$

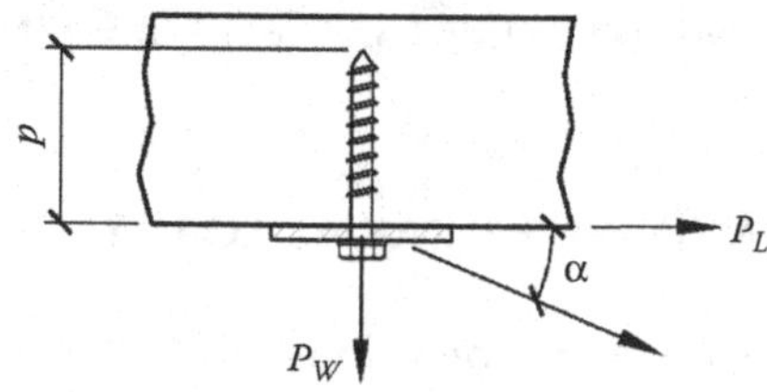

Figure 8.19 Combined lateral and withdrawal loading.

Center-to-center spacing: 10D
Row spacing: 3D

Therefore,

End distance: (10)(0.242 in.) = 2.42 in
Edge distance: (2.5)(0.242 in.) = 0.605 in
Center-to-center spacing: (10)(0.242 in.) = 2.42 in
Row spacing: (3)(0.242 in.) = 0.726 in

$$Z' = ZC_D C_M C_t C_g C_A C_{eg} C_{di} C_{tn}$$

$$= (99)(1.15)(1.0)(1.0)(1.0)(1.0)(1.0)(1.0)(1.0) = \mathbf{113\ lb}$$

$$Z'_\alpha = \frac{W'pZ'}{W'p\cos^2\alpha + Z'\sin^2\alpha}$$

$$= \frac{(146)(1.5)(113)}{(146)(1.5)\cos^2 30 + (113)\sin^2 30} = \mathbf{128\ lb}$$

$$P_{max} = 128 \text{ lb}$$

Example 8.11: Diaphragm fastener loaded by lateral and withdrawal loads

For the diaphragm connection shown in Figure 8.20, determine the capacity of the 10d toenail for (a) lateral loads and (b) withdrawal loads. Loads are due to wind and normal temperature and moisture conditions apply. The lumber is Douglas Fir-Larch.

Solution:

D = 0.148 in. (NDS code Table L3)

L = 3 in. (NDS code Table L4)

$$p = 3 \text{ in.} - \frac{(L/3)}{\cos 30} = 1.84 \text{ in.} > 10D = 1.48 \text{ in.} \quad \textbf{OK}$$

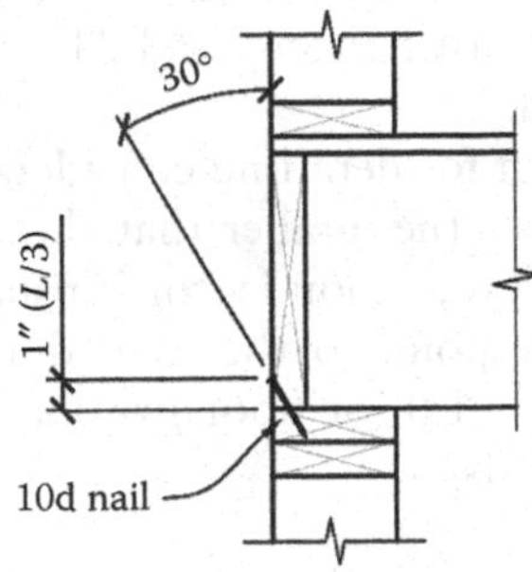

Figure 8.20 Diaphragm connection detail.

a. From NDS Table 12N, $Z = 118$ lb (side member thickness = 1.5 in.):

$Z' = ZC_D C_M C_t C_g C_\Delta C_{eg} C_{di} C_{tn}$

$C_D = 1.6$ (wind)

$C_{di} = 1.1$ (NDS code Section 11.5.3)

$C_{tn} = 0.83$ (NDS code Section 11.5.4.2)

All other C factors = 1.0.

$Z' = (118)(1.6)(1.0)(1.0)(1.0)(1.0)(1.0)(1.1)(0.83) =$ **172 lb**

b. From NDS Table 12.2C, $W = 36$ lb/in. ($G = 0.50$, see NDS Table 12.3.3A): $p = 1.84$ in. from part (a); therefore,

$W = (36\text{ lb/in.})(1.84\text{ in.}) = 66$ lb

$W = WC_D C_M C_t C_{eg} C_{tn}$

$C_D = 1.6$ (wind)

$C_{tn} = 0.67$ (NDS Section 11.5.4.1)

All other C factors = 1.0 ($C_M = 1.0$ for toenail connections; see NDS code Section 12.5.4).

$W' = (66)(1.6)(1.0)(1.0)(1.0)(0.67) =$ **71 lb**

8.7 PREENGINEERED CONNECTORS

There are a wide variety of preengineered connectors available for wood construction. Most manufacturers will readily supply a catalog of their connectors with corresponding load-carrying capacity as well as installation requirements. The selection of a manufacturer usually depends on the regional availability or the builder's preference. For most cases, the load tables and connector details from the different manufacturers are very similar. A generic load table is shown in Figure 8.21. A brief explanation of some of the elements of the table is as follows.

1. The connector designation is usually based on the geometric properties of the connector. JH could stand for "joist hanger" and 214 would represent the joist that the connector is intended to hold.
2. The dimensions are important for detailing considerations.
3. The manufacturer usually lists the lumber that should be used with each connector. For other species of lumber, a reduction factor is usually applied.
4. The percentages noted correspond to the load duration factor: 100% or full load duration corresponds to $C_D = 1.0$ (or floor live load), whereas 160% corresponds to $C_D = 1.6$ (wind or seismic loads).

Figure 8.22 illustrates other preengineered connector types and where they are typically used in a wood structure.

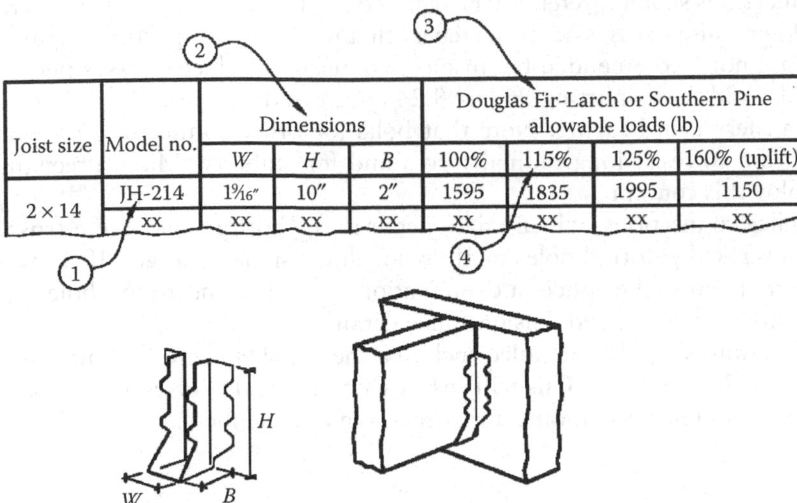

Joist size	Model no.	Dimensions			Douglas Fir-Larch or Southern Pine allowable loads (lb)			
		W	*H*	*B*	100%	115%	125%	160% (uplift)
2 × 14	JH-214	1 9/16″	10″	2″	1595	1835	1995	1150
	xx	xx	xx	xx	xx	xx	xx	xx

Figure 8.21 Preengineered connector load table.

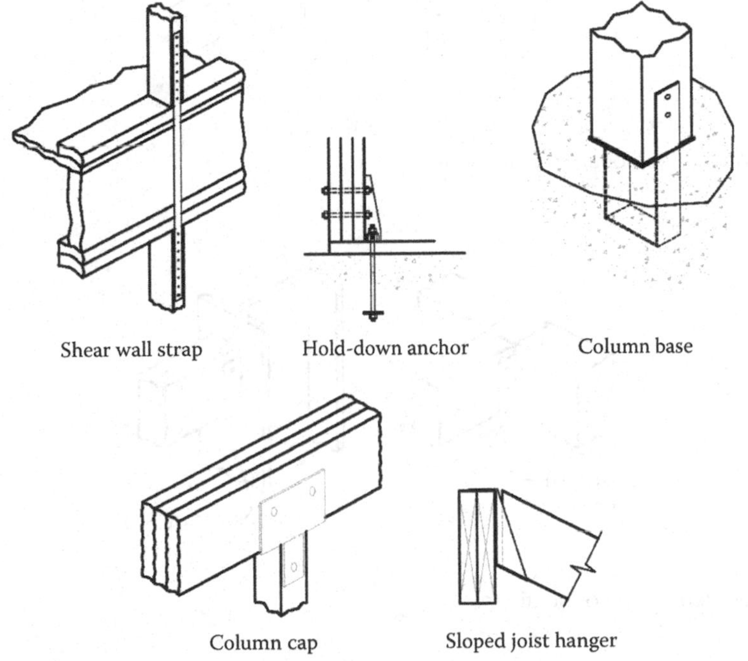

Figure 8.22 Preengineered connector types.

8.8 PRACTICAL CONSIDERATIONS

Wood members within a structure will experience dimensional changes over time due mainly to the drying of the wood. The ambient humidity, which can change with the seasons, can also contribute to the shrinking and swelling of wood members. Because of this, wood connections should be detailed to account for dimensional changes. In general, wood connections should be detailed such that the stresses that can cause splitting are avoided.

Beam connections should preferably be detailed such that the load is transferred in direct bearing and not cause stresses at the fasteners that might cause splitting. Figure 8.23 show details that are not recommended because the connectors induce stresses perpendicular to the grain and could cause splitting. Figure 8.24 indicates the preferred connections in which loads are transferred in bearing. Note that bolts (or other connectors) are still provided within the beam to resist longitudinal forces and for stability. Many preengineered joist hangers employ this concept.

Tension splice connections, which might occur in the bottom chord of a truss, should be detailed with vertically slotted holes to allow for dimensional changes. If shear forces have to be transferred across the splice, a combination of slotted and round holes can be used. Figure 8.25 shows the preferred tension splice details.

Truss connections should be detailed such that the member centerlines intersect at a common work point (Figure 8.26). This will ensure that the axial loads in the truss members do not induce shear and bending moment or stresses in the members.

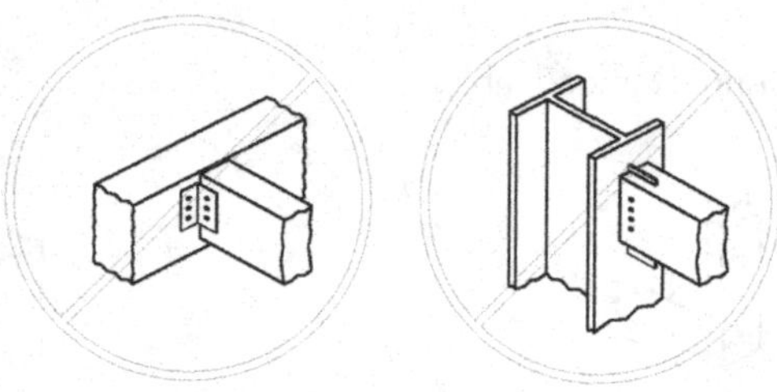

Figure 8.23 Nonpreferred beam connections.

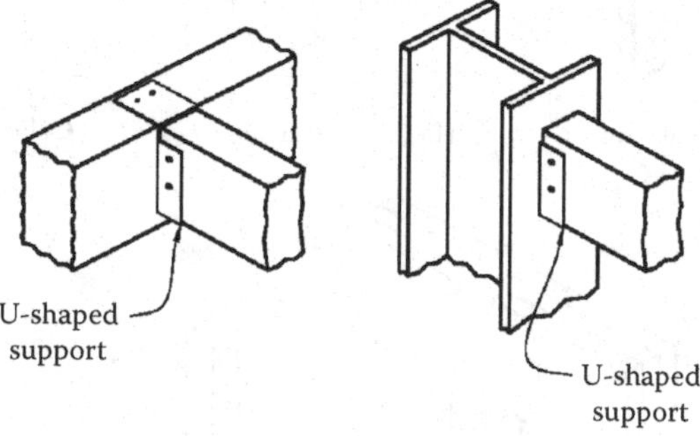

Figure 8.24 Preferred beam connections.

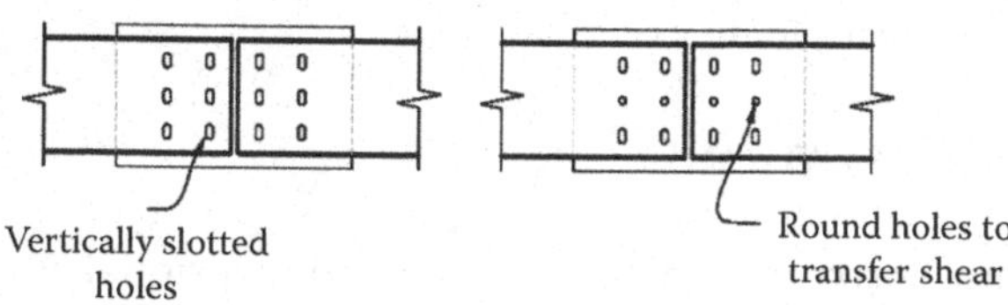

Figure 8.25 Preferred tension splice details.

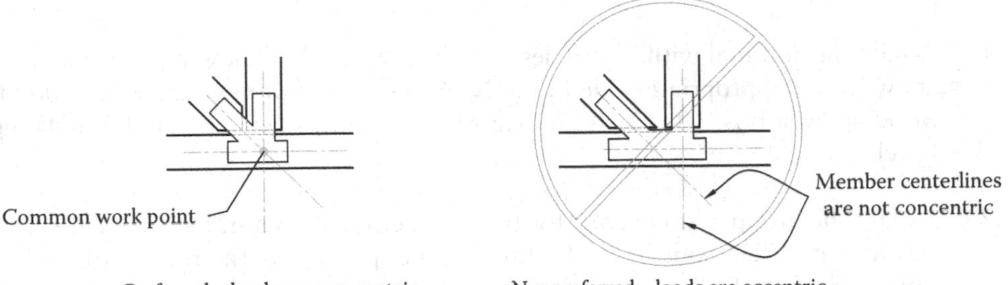

Figure 8.26 Preferred and nonpreferred truss details.

Base connections for wall studs and posts should be kept sufficiently high above the concrete foundation to avoid transfer to the wood member moisture that can cause decay. Figure 8.27 indicates the base connections that are not recommended because they allow the wood and connectors to be exposed to direct moisture. The preferred details for these conditions are shown in Figure 8.28. It is advisable to avoid members with notched connections.

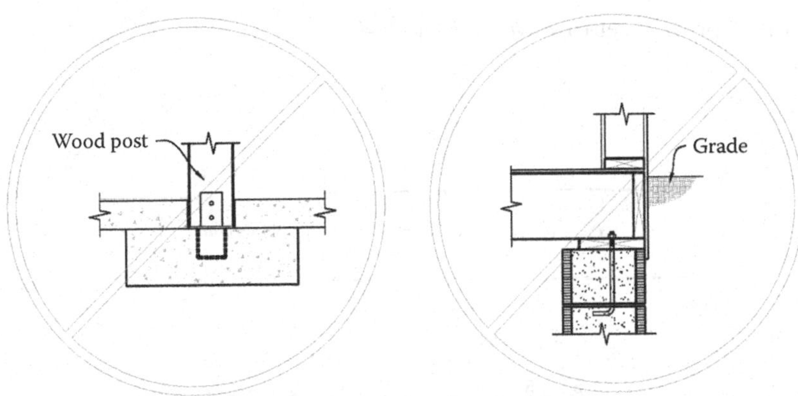

Figure 8.27 Nonpreferred base details.

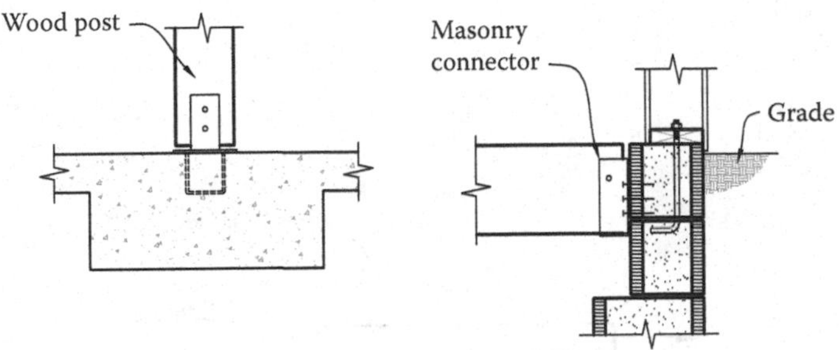

Figure 8.28 Preferred base details.

PROBLEMS

8.1 Calculate the nominal withdrawal design values W for the following and compare the results with the appropriate table in the NDS code: (a) common nails: 8d, 10d, 12d; (b) wood screws: nos. 10, 12, 14; (c) lag screws: ½ in., ¾ in. The lumber is Douglas Fir-Larch.

8.2 Calculate the group action factor for the connection shown in Figure 8.29, compare the results with NDS code Table 10.3.6A, and explain why the results are different. The lumber is Hem-Fir Select Structural.

8.3 For the following connectors, calculate the dowel bearing strength parallel to the grain ($F_{e\|}$), perpendicular to the grain ($F_{e\perp}$), and at an angle of 30° to the grain ($F_{e\theta}$). The lumber is Spruce-Pine-Fir: (a) 16d common nail; (b) wood screws: nos. 14 and 18; (c) bolts: ⅝ in., 1 in. Compare the results with Table 11.3.2 of the NDS code.

8.4 Using the yield limit equations, determine the nominal lateral design value Z for the connection shown in Figure 8.30. Compare the results with the appropriate table in the NDS code. The lumber is Douglas Fir-Larch.

8.5 Develop a spreadsheet to solve Problem 8.4.

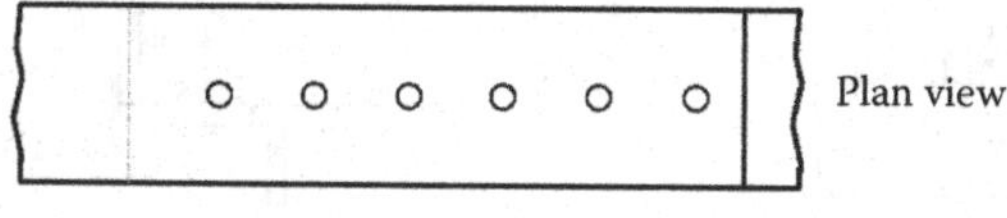

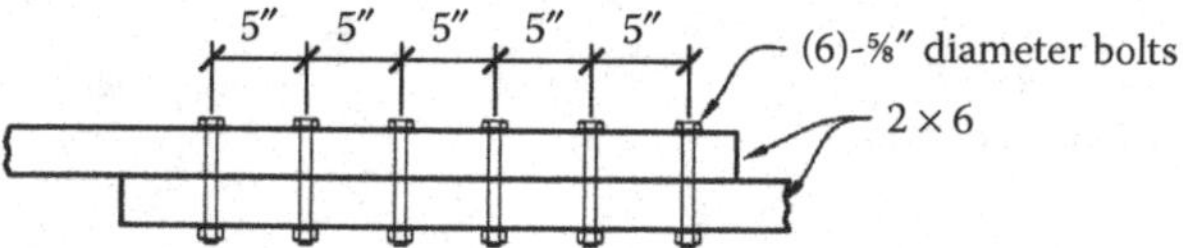

Figure 8.29 Problem 8.2.

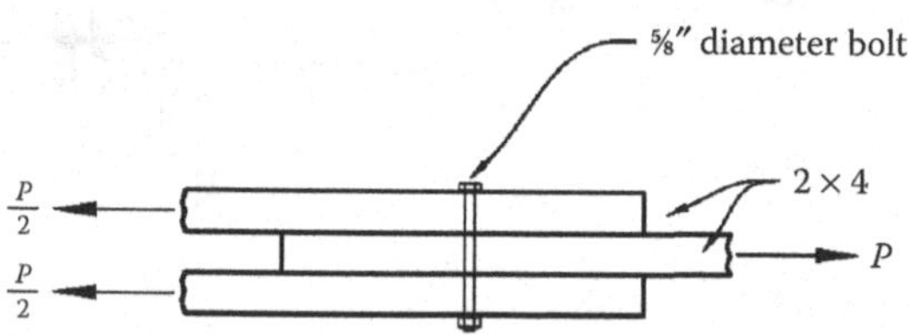

Figure 8.30 Problem 8.4.

8.6 For the connection detail shown in Figure 8.31, determine the maximum reaction that could occur based on the strength of the connection only. The lumber is Hem-Fir and the normal temperature and moisture conditions apply. Consider $D + L$ loading only.

8.7 For the stair framing shown below for an office building:

a. Determine the maximum tread length, L, based on shear and bending.
b. Calculate the maximum deflection based on the critical loading determined from (a).
c. Determine if the (2)-1/2″ bolts connecting the angle to the stringer are adequate to support the tread for gravity loading. Use NDS Table 11B directly (Figure 8.32).

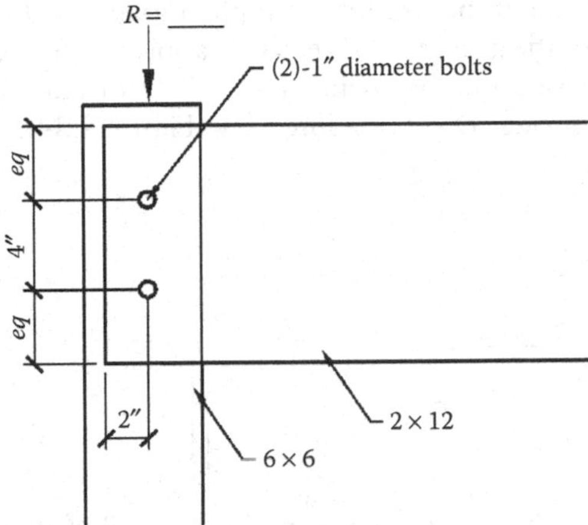

Figure 8.31 Detail for Problem 8.6.

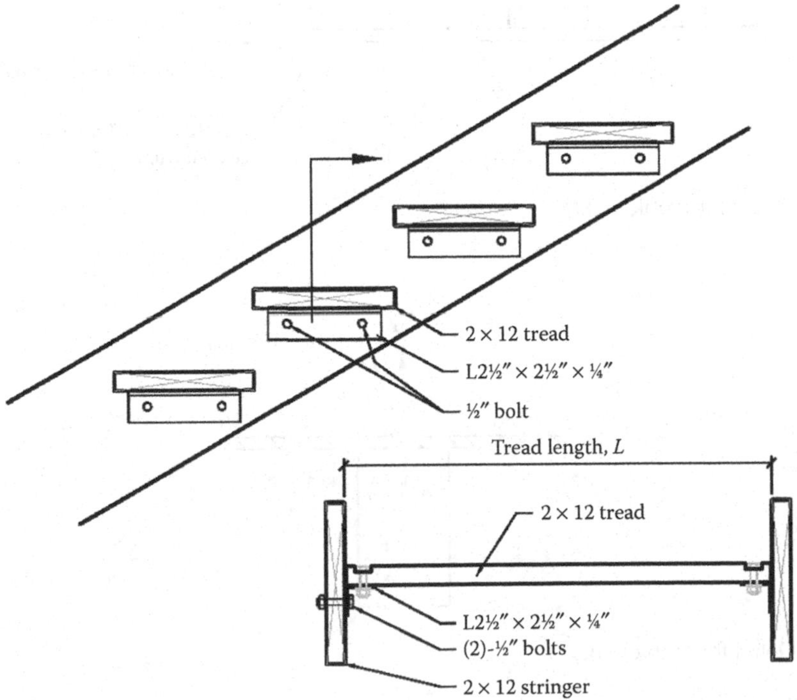

Figure 8.32 Detail for Problem 8.7.

Assume all C-factors are 1.0. The total dead load is just the self-weight of the tread and the live load is as specified by the code (two cases—uniform and concentrated; they are not concurrent) All wood is Hem-Fir, no.2.

8.8 For the connection detail shown in Figure 8.33:

a. Find the design withdrawal strength in pounds for one screw.
b. Find the maximum possible load, *P*, that could occur.

All wood is Hem-Fir and loads are due to dead only. Loading is based on a single hanging load located at any point.

8.9 Based on the given detail below, for an applied load of 800 lb., determine the minimum value of *G* (specific gravity) and select an appropriate species of wood that would satisfy this detail. Use an embedment of $e = 5''$. Normal temperature and moisture conditions apply. Consider $D + L$ loading only (Figure 8.34).

a. Use a wood screw.
b. Use a nail.

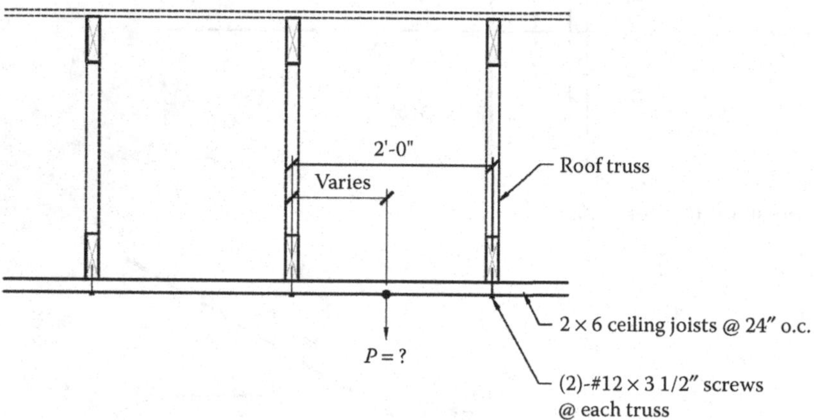

Figure 8.33 Detail for Problem 8.8.

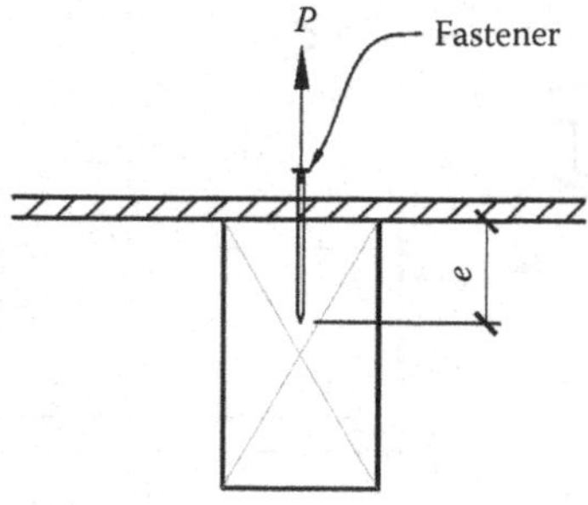

Figure 8.34 Detail for Problem 8.9.

8.10 Based on the given in Figure 8.35, (6)-20d common nails and Southern Pine lumbers were specified by the engineer for the tension splice. The contractor would prefer to use a #10 wood screws and Hem-Fir lumber instead. Assuming the full capacity of the nails is required, determine the number of wood screws required. Check spacing requirements and use the NDS tables.

8.11 For the shear wall detail given below showing the loading on an 8 and 10ft shear wall separated by a 4 ft opening:

a. Draw the unit shear, net shear, and drag strut force diagram.
b. Design a lap splice using 1/4″ diameter lag screws in a single line centered in 2-2 × 6 top plates using the worst-case force found in (a) and sketch your design showing the required dimensions. Use the NDS Tables.

All wood is Southern Pine and the load is due to wind only. Normal temperature and moisture conditions apply, and moisture content < 16% (Figure 8.36).

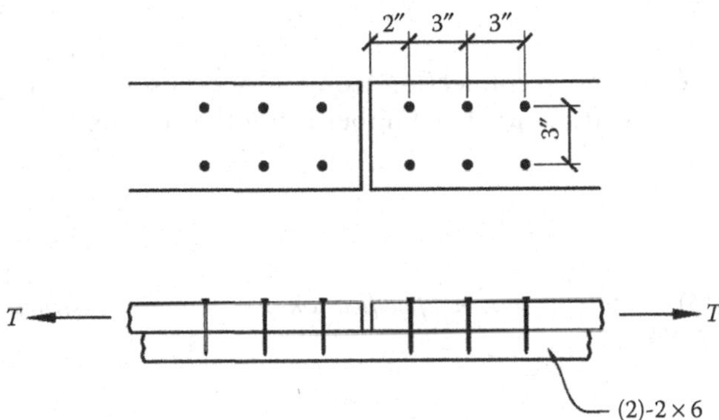

Figure 8.35 Detail for Problem 8.10.

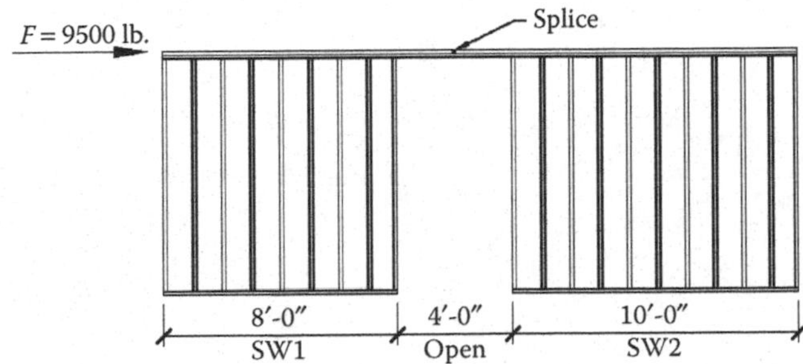

Figure 8.36 Detail for Problem 8.11.

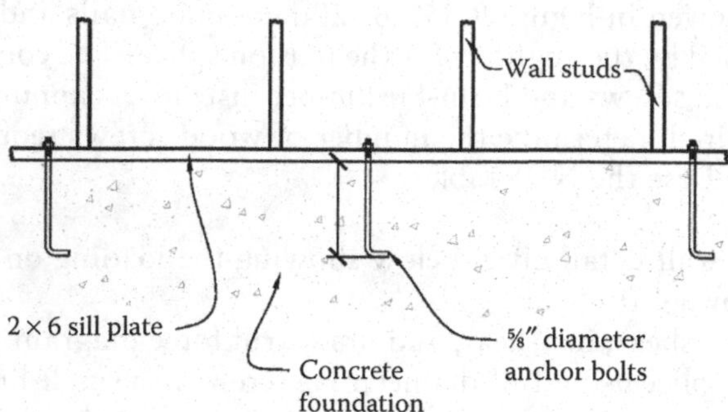

Figure 8.37 Detail for Problem 8.12.

8.12 For the sill plate connection shown in Figure 8.37, calculate the design shear capacity Z' parallel to the grain. The concrete strength is $f'_c = 3000$ psi and the lumber is Spruce-Pine-Fir. Loads are dues to seismic (E), and normal temperature and moisture conditions apply.

8.13 Design a 2 × 6 lap splice connection using (a) wood screws and (b) bolts. The load applied is 1800 lb due to wind. The lumber is Southern Pine.

REFERENCE

1. ANSI/AWC (2015), *National Design Specification for Wood Construction,* American Wood Council, Leesburg, VA.

Chapter 9

Practical considerations in the design of wood buildings

In this section, we cover the practical issues that need to be considered in the design of wood buildings. Several topics are covered including cross-laminated timber (CLT) used in modern wood multistory buildings, rules of thumb for wood design, and fire design of heavy timber members.

9.1 MULTISTORY WOOD BUILDINGS

Wood buildings are undergoing a renaissance as taller buildings are now constructed of wood than have been done in the past. Currently, there are a number of multistory buildings of up to 14 stories that have been built in Europe using CLT wall and floor panels [1–5]. In these highrise wood buildings, lateral stability is provided by the concrete corewalls at the elevators and stairs or by the in-plane shear resistance of the CLT walls. It has been suggested that wood buildings—using CLT—of up to 50 stories are feasible if a concrete core is used for lateral stability [6,7]. Some of the advantages of using wood in multistory buildings include reduced total weight of the structure, a relatively shorter construction time compared to concrete or steel highrise buildings, and a smaller carbon footprint-approximately 25 to 40 percent of the carbon footprint of steel or concrete structures [1,5,7,8].

Examples of existing multistory wood buildings include the nine-story Stadthaus building in the United Kingdom and the ten-story CLT building in Australia [3]. The Stadthaus building is a hybrid structure with a podium level or first story built out of reinforced concrete and the upper eight stories built out of wood using a platform framing system. CLT panels were used for the walls, the elevator cores, and the floor slabs as well as the roof slabs. The connections between the CLT wall panels and the CLT floor panels are made with steel angle brackets that are secured with screws to the walls and floor panels. To prevent progressive collapse, structural redundancies were built into the design of this building by designing the floor panels to be able to span over two spans or cantilever over one span to simulate the accidental loss of an interior wall panel [6].

Another example of a multistory wood building is the seven-story student housing in the West Campus of the University of Washington, Seattle, WA. The building has a two-story concrete podium structure, on top of which was built a five-story light frame wood structure of Type V-A construction. Currently, the tallest wood building is a 14-story apartment building in Bergen, Norway, and the University of British Columbia in Canada is currently building an 18-story wood dormitory highrise that is set to open in 2017. Other multistory wood buildings currently in the works include a 21-story building in Amsterdam and a proposed 40-story wood skyscraper in Stockholm, Sweden [5].

In multistory structures, the load path needs careful attention as bearing walls and continuity is often interrupted. It is ideal to lay out the space such that walls and openings align from top to bottom, which is often the case with hotels and multi-unit residential construction, but the wall layout can vary for several reasons. In hotel construction, the ground floor often has more wide-open areas for public use, whereas the upper floors often have a more regular layout with a majority of the walls positioned to stack vertically. When large open areas exist with relatively large bearing wall loads, transfer beams are needed and these often have to be made of structural steel in multistory construction.

When the bearing walls align, the load path is more easily resolved in a global sense, but there exists localized areas where load continuity needs to be maintained. The wall elevation in Figure 9.1 shows a typical framing configuration for a multistory bearing wall with an

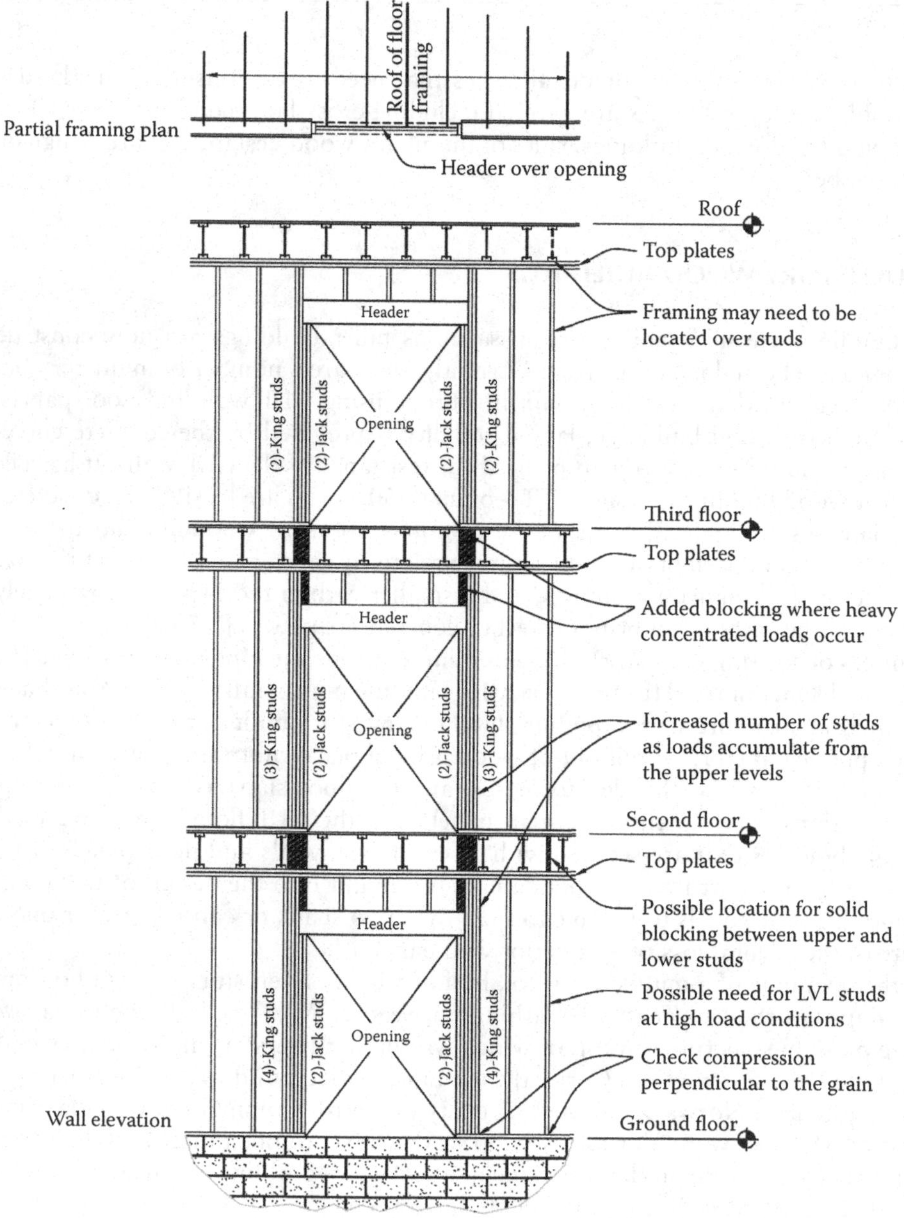

Figure 9.1 Multistory wall framing details.

opening in the same location throughout the wall height. Headers and jack studs are provided at the opening, but as the reactions from each header and stud assembly accumulate down to the foundation, it may be necessary to increase the number of studs to support the increased load as it makes its way to the foundation. Furthermore, as the load is transferred from one level to the next, solid blocking may be needed within the floor framing to maintain the load path.

Blocking between wood studs at the underside of the top plates may not be required if the top plates have adequate capacity in bending about the weak axis to support the concentrated load from the roof frusses or floor joints and spread it to the wall studs. In the design of the top plates for such a concentrated load, the section properties should conservatively be taken as the sum of the individual sections used as multiple top plates because the plates would not typically have enough friction between the plies to act as a composite section. It is also reasonable to take advantage of a continuous span condition in the design of the top plates instead of a simple span analysis, provided that attention is given to the splice locations for the top plates. It might seem ideal to require the placement of floor framing to be aligned directly over studs, but it is very difficult to achieve this during construction, so either designing the top plates for applicable concentrated loads or detailing the framing to have solid blocking within the load path is recommended.

At the lower levels of multistory buildings, consideration needs to be given to the stresses that accumulate perpendicular to the grain in the bottom plates, which can often control the design. The use of a higher grade lumber, or even engineered lumber, may be necessary. In hotels and other similar structures, the first level can often be taller for functional and esthetic purposes. In this case, the stresses in the wall studs are magnified by the increased height and the accumulated gravity load, and thus, the wall stud design for the lower levels can be quite different than that for the upper levels. The design of the lower level wall studs can often include using smaller stud spacing, a higher grade of lumber, and possible use of engineered lumber. In some cases where the wall height is relatively high or even closer to 20 ft, the use of engineered lumber is preferred because the studs are often much straighter and more dimensionally stable than sawn lumber.

9.2 SHRINKAGE CONSIDERATION IN MULTISTORY WOOD BUILDINGS

One important consideration in multistory wood buildings is the total cumulative axial shortening as well as differential movement between dissimilar materials (e.g., shrinkage of wood vs. swelling or expansion of the brick veneer or shrinkage of wood vs. steel framing that does not shrink). The more stories there are in a building, the greater the axial shrinkage in multistory wood buildings [9]. The calculation of the cumulative shrinkage in a multistory light frame wood building was discussed in Chapter 1. From Example 1.2, the total vertical shrinkage per floor for the framed section shown in Figure 1.23 was calculated as 0.3 in. In a typical multistory building, if green lumber instead of kiln dried lumber is used, the total vertical shortening per floor level due to shrinkage alone can be as high as 0.5–0.75 in., which would yield a total vertical shortening due to shrinkage of between 2.5 and 3.75 in. for a five-story platform light frame wood building [10,11]. This does not yet take into account the additional axial shortening due to axial load compounded by the relatively lower elastic modulus of wood compared to steel or concrete. Uniform and differential shortening should be carefully addressed in the design and detailing of multistory wood buildings to avoid unintended stresses in the wood members, their connections, and any supported dissimilar materials such as brick veneer. The adverse effects of differential vertical movement in wood buildings include the following [11]:

- Damage to interior finishes resulting from unintentional loading on the finishes.
- Damage to exterior brick cladding resulting from unintentional compression loading on the brick veneer that may cause outward bulging of the cladding.
- Back-slope of exterior waterproof decks or cantilevers toward the exterior wall of the building (i.e., negative slope that drains rainwater toward the building exterior wall instead of away from it) due to differential shrinkage can render the drainage of the deck or cantilever ineffective and lead to failure of the waterproofing systems due to infiltration of rainwater into the covered wood structure at the exterior wall–deck interface. This effect is more severe in multistory buildings where the effect of shrinkage and axial shortening is cumulative. This can eventually cause the decay of the wood members and exterior wall supporting the deck or cantilever, leading to subsequent failure of the exterior deck or cantilever.

The use of CLT wall and floor panels can reduce the vertical shortening due to shrinkage considerably, and in the nine-story Stadthaus building, the total vertical deformation due to shrinkage was "less than 1 in." [6]. The vertical shrinkage can be minimized by requiring in the project specifications and contract drawings that the measured *in situ* moisture content of the wood members prior to installation of the gypsum wall board finishes be less than 12% [9]. The negative effects of axial shortening or differential vertical movement can also be mitigated by adequate detailing, especially at the connections [12].

For brick veneer in wood-framed building construction, one way to mitigate the differential movement effects between the expanding brick veneer and the shrinking vertical wood frame is to support the brick on shelf angles independently of the wood framing or support the brick at each story from the wood framing with proper detailing to take care of the differential movement. This later approach was adopted for the University of Washington West Campus Housing Project—a seven-story wood building (a five-story light frame wood structure built on top of a two-story reinforced concrete podium structure) [13,14].

9.3 USE OF CLT IN MULTISTORY BUILDINGS

CLT which "consists of orthogonally stacked layers of typically low grade wood glued together on the wide faces and sometimes also on the narrow faces" was developed in Germany and Austria in the 1990s and has been used in Europe for some time for the construction of multistory wood buildings [1,13]. The CLT panels (see Figure 9.2) are used as walls, floor plates, and beams, and the connections between the wall and the floor or roof panel elements are made with steel angle brackets screws drilled into the panels. In CLT wall panels, the lumber in the outer layers run parallel to the direction of the applied axial load on the wall, whereas for the CLT floor or roof panels, the outer laminations are oriented in a direction parallel to the span of the floor and roof slab [8]. The panels consist of three, five, seven, or more layers or plies, and in North America, they can come in sizes as large as 10 ft × 64 ft though panel sizes varies depending on the CLT manufacturer [14,15]. The National Design Specification (NDS) code limits the total thickness of CLT panels to 20 in., and the thickness of the laminations varies between 5/8 and 2 in. The minimum total thickness of the panel is 2½ in. Finger-jointed CLT panels are currently not covered by the ANSI PRG-320 Standard though they are available in Europe [14].

Some of the factors that have slowed the use of wood in multistory buildings higher than five to six stories in height include the limited fire resistance of sawn lumber as well as their limited load-carrying capabilities. In addition to having a relatively higher load-carrying

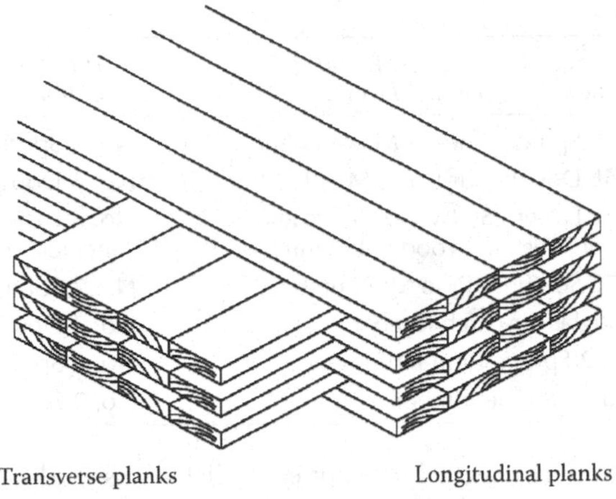

Figure 9.2 CLT plan configuration.

capacity compared to sawn lumber, CLT offers a solution to the fire resistance issue because it "self-protects" in a fire by developing a charring layer on the surfaces exposed to the fire; hence, it is classified for fire resistance purposes as heavy timber. For the nine-story Stadthaus Building in the United Kingdom, fire resistance ratings of 60 and 90 min were achieved with use of the charring layer in the CLT panels combined with encapsulation or covering of the CLT panels with gypsum board (dry wall) [6]. Because the CLT panels are prefabricated in the factory with all the door, window, and floor openings, the erection of the wall and floor panels on site is relatively quick, and thus, the construction time for a typical CLT building is shorter than for a comparable concrete building or wood building constructed with light frame wood members. In the proposal for the Stadthaus building, the architect, Andrew Waugh, found in a comparative study that the cost of the CLT high-rise building was 15% less than a comparable concrete building and weighed about 25% of a comparable concrete building; in addition, the CLT building has a much lower environmental footprint [6]. The advent of CLT is making possible the use of wood for high-rise buildings with multiple levels that far exceed the maximum five to six stories in height that has typically been allowed in the International Building Code (IBC) code for light frame wood construction [6,10,15,16]. However, it should be noted that "CLT is more prone to time-dependent deformations under load (i.e., creep) than other engineered wood products such as glued-laminated timber" [8].

9.3.1 Design of CLT members

In Europe, the manufacturers of CLT provide the mechanical properties for their proprietary CLT product, and there is no unifying standard for the design of these members. However, in North America, the design, manufacture, and qualifications of CLT members is governed by ANSI/APA PRG-320, the Standard for Performance-Rated Cross-Laminated Timber [16,17]. Determination of the capacity of CLT members in bending, shear, and axial compression and tension follows a similar conceptual approach to that used for glulam and sawn wood member in that the reference design values for CLT are multiplied by the adjustment factors to obtain the adjusted design values. For the design of CLT members in the United States and Canada, the composite bending and shear stiffness properties for CLT members are obtained from the Standard for Performance-Rated Cross-Laminated Timber—ANSI/APA PRG-320 [17,18]. In the Standard, CLT is classified into seven stress grades: E1, E2, E3, E4, V1, V2, and V3. The "E" classes use mechanically graded or machine stress-rated (MSR) lumber in the CLT laminations in the major strength direction, whereas the "V" classes use visually graded lumber.

Table 9.1 CLT stress grades

CLT stress grade	*Lumber used in* major *strength direction (i.e.,* parallel *layers)*	*Lumber used in* minor *strength direction (i.e.,* perpendicular *layers)*
E1	1950f-1.7E Spruce Pine-Fir MSR lumber	No. 3 Spruce-Pine-Fir lumber
E2	1650f-1.5E Douglas Fir-Larch MSR lumber	No. 3 Douglas Fir-Larch lumber
E3	1200f-1.2E Eastern Softwoods, Northern Species, or Western Woods MSR lumber	No. 3 Eastern Softwoods, Northern Species, or Western Woods lumber
E4	1950f-1.7E Southern Pine MSR lumber	No. 3 Southern Pine lumber
V1	No. 2 Douglas Fir-Larch lumber	No. 3 Douglas Fir-Larch lumber
V2	No. 1/No. 2 Spruce-Pine-Fir lumber	No. 3 Spruce-Pine-Fir lumber
V3	No. 2 Southern Pine lumber	No. 3 Southern Pine lumber

The description of the lumber material used for each CLT stress grades in the major and minor strength directions is shown in Table 9.1. Custom CLT products can also be used, provided they conform to the requirements of the PRG-320 Standard. The applicable adjustment factors for CLT are obtained from Table 9.2.

The design of CLT panels for out-of-plane bending, shear, deflection, and axial tension and compression is carried out using the reference design values and the properties for the seven grades of CLT given in Tables A2 and A1, respectively, of ANSI PRG-320 (see Tables 9.3 and 9.4). The structural properties were derived using the Shear anology method [8,18]. Because CLT is a proprietary material, the manufacturers of CLT also provide their properties and load tables or charts. The following are two examples to illustrate the use of PRG-320 Tables A2 and A1.

Example 9.1: Out-of-plane bending and shear capacity of a CLT panel

Design a CLT floor panel to span 20 ft assuming a superimposed dead load of 25 psf and a floor live load of 50 psf. Assume normal moisture and temperature conditions apply; the panel is fully braced and the load duration factor is 1.0. The density of the CLT panel is assumed to be 32 pcf.

Solution:

The adjustment factors (see Chapter 3) are

$C_D = 1.0$ (dead + floor live load)

$C_M = 1.0$

$C_t = 1.0$

$C_L = 1.0$

Assume a five-ply CLT panel with a layer thickness of 1.375 in. for a total thickness of 6.875 in.

$$\text{Self-weight of CLT panel} = \left(\frac{6.875\,\text{in}}{12}\right)(32\,\text{pcf}) = 20\,\text{psf}$$

$$\text{Total load} = D + L = (20\text{ psf} + 25\text{ psf}) + 50\text{ psf} = 95\text{ psf} = 95\text{ lb/ft per ft width of panel}$$

Maximum applied moment,

$$M_{\text{applied}} = \frac{wL^2}{8} = \frac{(95)(20\text{ ft})^2}{8} = 4750\text{ lb-ft/ft width of CLT panel}$$

$$\text{Maximum applied shear, } V_{\text{applied}} = \frac{wL}{2} = \frac{(95)(20\text{ ft})}{2} = 950\text{ lb/ft width of CLT panel}$$

Table 9.2 Applicability of adjustment factors for CLT

	ASD only	ASD and LRFD					LRFD only		
	Load duration factor	Wet service factor	Temperature factor	Beam stability factor[a]	Column stability factor	Column stability factor	Format conversion factor K_F	Resistance factor ϕ	Time effect factor
$F'_b S_{eff} = F_b\, S_{eff}\ \times$	C_D	C_M	C_t	C_L	—	—	2.54	0.85	λ
$F'_t A_{parallel} = F_t A_{parallel}\ \times$	C_D	C_M	C_t	—	—	—	2.70	0.80	λ
$F'_v(t_v) = F_v(t_v)\ \times$	C_D	C_M	C_t	—	—	—	2.88	0.75	λ
$F'_s(Ib/Q)_{eff} = F_s(Ib/Q)_{eff}\ \times$		C_M	C_t	—	—	—	2.88	0.75	—
$F'_c A_{parallel} = F_c A_{parallel}\ \times$	C_D	C_M	C_t	—	C_P	—	2.40	0.90	λ
$F'_{c\perp} A = F_{c\perp} A\ \times$	—	C_M	C_t	—	—	C_b	1.67	0.90	—
$EI'_{app} = EI_{app}\ \times$	—	C_M	C_t	—	—	—	—	—	—
$EI'_{app\text{-}min} = EI_{app\text{-}min}\ \times$	—	C_M	C_t	—	—	—	1.76	0.85	—

Source: ANSI/AWC, *National Design Specification for Wood Construction*, American Wood Council, Leesburg, VA. 2015.

[a] For out-of-plane bending of CLT roof or floor panels, the beam stability factor, $C_L = 1.0$.

S_{eff} = effective section modulus.
$A_{parallel}$ = cross-sectional area of the CLT layers with fibers that run parallel to the axial load.
A = bearing area.
$F'_b S_{eff}$ = adjusted bending moment strength or capacity.
$F'_t A_{parallel}$ = adjusted axial tensile strength or capacity.
$F'_v(Ib/Q)_{eff}$ = adjusted shear strength or capacity.
$F'_c A_{parallel}$ = adjusted axial compression strength or capacity.
$F'_{c\perp} A$ = adjusted compression strength or capacity perpendicular to grain.
EI'_{app} = adjusted apparent bending stiffness used for deflection calculations.
$EI'_{app\text{-}min}$ = adjusted apparent minimum bending stiffness used for buckling and stability calculations.
$EI'_{app\text{-}min} = 0.5184 EI'_{app}$.
GA = shear rigidity.

Table 9.3 The allowable design properties for PRG 320 CLT (for use in the United States)

	Major strength direction						*Minor strength direction*					
CLT Grades	$F_{b,0}$ *(psi)*	E_0 *(10^6 psi)*	$T_{t,0}$ *(psi)*	$F_{c,0}$ *(psi)*	$F_{v,0}$ *(psi)*	$F_{s,0}$ *(psi)*	$F_{b,90}$ *(psi)*	E_{90} *(10^6 psi)*	$F_{t,90}$ *(psi)*	$F_{c,90}$ *(psi)*	$F_{v,90}$ *(psi)*	$F_{s,90}$ *(psi)*
E1	1950	1.7	1375	1800	135	45	500	1.2	250	650	135	45
E2	1650	1.5	1020	1700	180	60	525	1.4	325	775	180	60
E3	1200	1.2	600	1400	110	35	350	0.9	150	475	110	35
E4	1950	1.7	1375	1800	175	55	575	1.4	325	825	175	55
V1	900	1.6	575	1350	180	60	525	1.4	325	775	180	60
V2	875	1.4	450	1150	135	45	500	1.2	250	650	135	45
V3	975	1.6	550	1450	175	55	575	1.4	325	825	175	55

For SI: 1 psi = 0.006895 MPa.
Note: See Section 4 of ANSI/APA PRG 320-2012 [18] for symbols.
Tabulated values are allowable design values and not permitted to be increased for the lumber size adjustment factor in accordance with the NDS. The design values shall be used in conjunction with the section properties provided by the CLT manufacturer based on the actual layup used in manufacturing the CLT panel (see Table 9.4). Custom CLT grades that are not listed in this table shall be permitted in accordance with Section 7.2.1 of [18].

Bending moment capacity:

From Table A2 of ANSI PRG-320 (see Table 9.4), select a five-ply CLT of Stress Grade E4 with 1.375 in. layer thickness and a total panel thickness of 6.875 in.

The tabulated reference design values from Table 9.4 is given as follows:

Resisting moment, $M_{\text{resisting}} = 10{,}425$ lb-ft/ft width of panel
Panel bending stiffness, $EI_{\text{eff},o} = 441 \times 10^6$ lb-in.²/ft width of panel
Panel shear rigidity, $GA_{\text{eff},o} = 1.1 \times 10^6$ lb/ft width of panel

The shear deformation adjustment factor, K_s, is obtained from NDS Table 10.4.1.1 as 11.5 (for a simply supported beam with uniform load).

Therefore, the apparent bending stiffness for deflection calculations is determined from Reference 13 (similar to NDS Equation 10.4-1) as follows:

$$EI_{\text{app}} = \frac{EI_{\text{eff}}}{1+\left(K_s EI_{\text{eff}}/GA_{\text{eff}}L^2\right)} = \frac{441\times10^6\ \text{lb-in}^2/\text{ft}}{1+\dfrac{(11.5)(441\times10^6\ \text{lb-in}^2/\text{ft})}{(1.1\times10^6\ \text{lb-in}^2/\text{ft})(20\,\text{ft}\times12\,\text{in./ft})^2}} = 408\times10^6\ \text{lb-in}^2/\text{ft}$$

The adjusted resisting moment is

$$M'_{\text{resisting}} = (10{,}425\ \text{lb-ft/ft})C_D C_M C_t C_L = 10{,}425(1.0)(1.0)(1.0)(1.0)$$
$$= 10{,}425\ \text{lb-ft/ft width of panel}$$

Therefore, $M'_{\text{resisting}} > M_{\text{applied}} = 4750$ lb-ft/ft width of panel   OK

Shear capacity:

For five-ply CLT Grade E4, we obtain the following from Table 9.4:

Panel bending stiffness, $EI_{\text{eff},o} = 441 \times 10^6$ lb-in.²/ft width of panel
Panel shear rigidity, $GA_{\text{eff},o} = 1.1 \times 10^6$ lb/ft width of panel

The following properties are obtained from Table 9.3:

$F_v = 175$ psi
$F_s = 55$ psi
$E_0 = 1.7 \times 10^6$ psi
$E_{90} = 1.4 \times 10^6$ psi

Table 9.4 The allowable bending capacities for CLT listed in Table 9.3 (for use in the United States)

	Lamination thickness (in.) in CLT layup								Major strength direction			Minor strength direction		
CLT grade	*CLT t (in.)*	=	⊥	=	⊥	=	⊥	=	$F_bS_{eff,0}$ *(lbf-ft/ft)*	$EI_{eff,0}$ *(10^6 lbf-in.^{2}/ft)*	$GA_{eff,0}$ *(10^6 lbf/ft)*	$F_bS_{eff,90}$ *(lbf-ft/ft)*	$EI_{eff,90}$ *(10^6 lbf-in.^{2}/ft)*	$GA_{eff,90}$ *(10^6 lbf/ft)*
	4 1/8	1 3/8	1 3/8	1 3/8					4,525	115	0.46	160	3.1	0.61
E1	6 7/8	1 3/8	1 3/8	1 3/8	1 3/8	1 3/8			10,400	440	0.92	1,370	81	1.2
	9 5/8	1 3/8	1 3/8	1 3/8	1 3/8	1 3/8	1 3/8	1 3/8	18,375	1,089	1.4	3,125	309	1.8
	4 1/8	1 3/8	1 3/8	1 3/8					3,825	102	0.53	165	3.6	0.56
E2	6 7/8	1 3/8	1 3/8	1 3/8	1 3/8	1 3/8			8,825	389	1.1	1,430	95	1.1
	9 5/8	1 3/8	1 3/8	1 3/8	1 3/8	1 3/8	1 3/8	1 3/8	15,600	963	1.6	3,275	360	1.7
	4 1/8	1 3/8	1 3/8	1 3/8					2,800	81	0.35	110	2 3	0.44
E3	6 7/8	1 3/8	1 3/8	1 3/8	1 3/8	1 3/8			6,400	311	0.69	955	61	0.87
	9 5/8	1 3/8	1 3/8	1 3/8	1 3/8	1 3/8	1 3/8	1 3/8	11,325	769	1.0	2,180	232	1.3
	4 1/8	1 3/8	1 3/8	1 3/8					4,525	115	0.53	180	3.6	0.63
E4	6 7/8	1 3/8	1 3/8	1 3/8	1 3/8	1 3/8			10,425	441	1.1	1,570	95	1.3
	9 5/8	1 3/8	1 3/8	1 3/8	1 3/8	1 3/8	1 3/8	1 3/8	18,400	1,090	1.6	3,575	360	1.9
	4 1/8	1 3/8	1 3/8	1 3/8					2,090	108	0.53	165	3.6	0.59
V1	6 7/8	1 3/8	1 3/8	1 3/8	1 3/8	1 3/8			4,800	415	1.1	1,430	95	1.2
	9 5/8	1 3/8	1 3/8	1 3/8	1 3/8	1 3/8	1 3/8	1 3/8	8,500	1,027	1.6	3,275	360	1.8
	4 1/8	1 3/8	1 3/8	1 3/8					2,030	95	0.46	160	3.1	0.52
V2	6 7/8	1 3/8	1 3/8	1 3/8	1 3/8	1 3/8			4,675	363	0.91	1,370	81	1.0
	9 5/8	1 3/8	1 3/8	1 3/8	1 3/8	1 3/8	1 3/8	1 3/8	8,275	898	1.4	3,125	309	1.6
	4 1/8	1 3/8	1 3/8	1 3/8					2,270	108	0.53	180	3.6	0.59
V3	6 7/8	1 3/8	1 3/8	1 3/8	1 3/8	1 3/8			5,200	415	1.1	1,570	95	1.2
	9 5/8	1 3/8	1 3/8	1 3/8	1 3/8	1 3/8	1 3/8	1 3/8	9,200	1,027	1.6	3,575	360	1.8

For SI: 1 in. = 25.4 mm; 1 ft = 304.8 mm; 1lbf = 4.448N.
See Section 4 of ANSI/APA PRG 320-2012 [18] for symbols.
This Table represents one of many possibilities that the CLT could be manufactured by varying lamination grades, thicknesses, orientations, and layer arrangements in the layup. Custom CLT grades that are not listed in this Table shall be permitted in accordance with Section 7.2.1 of ANSI/APA PRG 320-2012 [18].

Shear Stress in CLT Panels

There are two types of shear in CLT panel members due to the laminations:

1. The first type of shear is *through-the-thickness shear* in the panel and is caused by *loads that are parallel to the plane of the laminations* (i.e., in-plane loads), for example, in-plane lateral loads acting on the panel when it functions as a shear wall or as a horizontal floor or roof diaphragm, when the CLT panel is used as a beam or header to support gravity loads. The ASD-adjusted allowable design shear stress for this in-plane shear is

$$F_v(t_v)' = F_v(t_v)C_D C_M Ct$$

2. The second type of shear is *rolling shear* or shear in the plane of the laminations (in-plane shear) where the lamination shear relative to each other due to *loads applied perpendicular* to the plane of the laminations. This type of shear is called rolling shear because the laminations that are perpendicular to the direction of the applied in-plane shear stress have the tendency to roll past each other when subjected to the in-plane shear. This type of shear occurs in CLT panels when they are used as a floor or roof slab and subjected to gravity loading (i.e., loads perpendicular to the plane of the laminations). The ASD-adjusted allowable design rolling shear stress is

$$F_s\left(\frac{Ib}{Q_{\text{eff}}}\right)' = F_s\left(\frac{Ib}{Q_{\text{eff}}}\right)C_M C_t$$

For this problem, the shear stress in play is rolling shear and the procedure for calculating the allowable rolling shear strength is as follows:

$$\text{Calculate } F_s\left(\frac{Ib}{Q_{\text{eff}}}\right) = F_s\frac{EI_{\text{eff}}}{\sum_{I=1}^{n/2} E_i h_i z_i}$$

where:

z is the distance from the centroid of each layer to the neutral axis, except for the middle layer, use the distance from the neutral axis to the centroid of the top half of the middle layer or laminations

h is the thickness of each layer, except for middle layer that lies on the neutral axis use one-half the thickness

n = number of the CLT layers or laminations

Adjusting for bending perpendicular to the strong axis per ANSI PRG-320 [18]: In a five-ply layup (see Figure 9.3), the equivalent modulus of elasticity for layers 2 and 4,

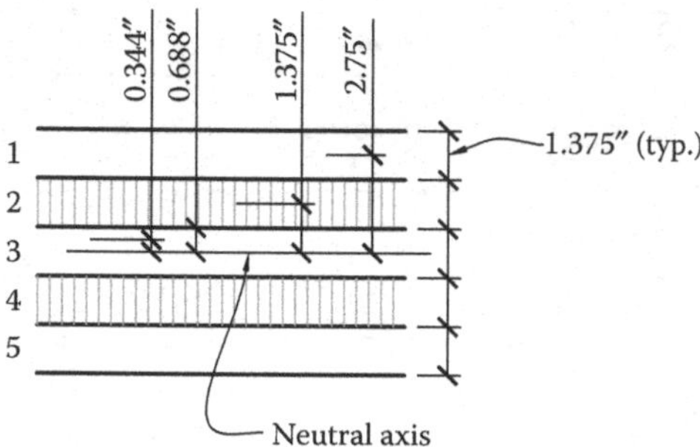

Figure 9.3 Five-ply layup of CLT.

which have laminations that are perpendicular to the direction of the laminations in the outer layers and the direction of the rolling shear, will be:

$$\text{Equivalent } E_0 \text{ for layer 2} = \frac{1}{30} \times (E_{90} \text{ for layer 2}) = (1.4 \times 10^6)/30 = 0.047 \times 10^6 \text{ psi}$$

$$\text{Equivalent } E_0 \text{ for layer 4} = \frac{1}{30} \times (E_{90} \text{ for layer 4}) = (1.4 \times 10^6)/30 = 0.047 \times 10^6 \text{ psi}$$

From Table 9.5,

$$\sum_{i=1}^{n/2} E_i h_i z_i = 6.92 \times 10^6 \text{ lb}$$

Note that only 2½ layers are considered (i.e., n = 2½)

$$\left(\frac{Ib}{Q_{\text{eff}}}\right) = \frac{EI_{\text{eff}}}{\sum_{i=1}^{n/2} E_i h_i z_i} = \frac{441 \times 10^6}{6.92 \times 10^6} = 63.7 \text{ in}^2$$

$$F_s\left(\frac{Ib}{Q_{\text{eff}}}\right) = (55 \text{ psi})(63.7 \text{ in}^2) = 3504 \text{ lb/ft width}$$

$$F_s\left(\frac{Ib}{Q_{\text{eff}}}\right)' = F_s\left(\frac{Ib}{Q_{\text{eff}}}\right) C_M C_t = (3504 \text{ lb/ft width})(1.0)(1.0) = 3504 \text{ lb/ft width}$$

$$V'_{\text{resisting}} = F_s\left(\frac{Ib}{Q_{\text{eff}}}\right)' = 3504 \text{ lb/ft width}$$

Therefore, $V'_{\text{resisting}} >> V_{\text{applied}} = 950$ lb/ft width of panel OK

Through-Thickness Shear of CLT Panel

If the five-ply Grade E4 CLT panel that was used for the slab will also be used as a shear wall, we would need to calculate the *through-thickness* shear capacity.

From Table A1 of PRG-320 (see Table 9.3), the reference design shear stress is $F_{vo} = 175$ psi

The ASD-adjusted allowable through-thickness shear strength is calculated as

Table 9.5 Calculation of (Ib/Q_{eff})

CLT layer	Direction of laminations	Modulus of elasticity of CLT, E (psi)	Effective modulus of elasticity, E_i (psi)	Distance from centroid of lamination to the neutral axis of the section, z_i (in.)	h_i (in.)	$E_i h_i z_i$
1	Parallel (i.e., "0" values)	E_0, 1.7 × 10⁶	1.7 × 10⁶	2.75	1.375	6.43 × 10⁶
2	Perpendicular (i.e., "90" values)	E_{90}, 1.4 × 10⁶	0.047 × 10⁶	1.375	1.375	0.089 × 10⁶
3 (½ Layer)	Parallel (i.e., "0" values)	E_0, 1.7 × 10⁶	1.7 × 10⁶	0.344	0.688	0.402 × 10⁶
					Σ	6.92 × 10⁶

$$F_{v,0}(t_v)' = F_{v,0}(t_v)C_D C_M C_t = (175\text{ psi})(6.875\text{ in.})(1.0)(1.0)(1.0) = 1203\text{ lb/in}$$

From Chapter 4, we can write the through-thickness shear capacity, V' as

$$V'_{\text{resisting}} = \left(\frac{2}{3}\right)(F_{v,0}t_v)'(12'') = \left(\frac{2}{3}\right)(1203\text{ lb/in})(12'') = 9{,}624\text{ lb/ft width of CLT panel}$$

Relationships between CLT Structural Material Properties and the Direction of the Laminations

The modulus of elasticity of a lamination when the direction of stress is *parallel* to the lamination (i.e., major strength direction) is denoted as E_0.

The modulus of elasticity of a lamination when the direction of stress is *perpendicular* to the lamination (i.e., minor strength direction) is denoted as E_{90}.

The relationships between the structural material properties of CLT panels with regard to the direction of the laminations vis-a-vis the direction of the stress are given in ANSI PRG-320 [18] as follows:

$$E_{90} = E_0/30$$
$$G_0 = E_0/16$$
$$G_{90} = G_0/10$$

Deflection check:

For deflection calculations, unfactored loads are used.

Total unfactored dead plus live load, $D + L = (95\text{ psf})(1\text{ ft}) = 95\text{ lb/ft}$ per ft. width of panel $= 7.917\text{ lb/in.}$ per ft. width of panel

The unfactored live load, $L = (50\text{ psf})(1\text{ ft}) = 50\text{lb/ft}$ per ft. width of panel $= 4.167\text{ lb/in.}$ per ft. width of panel

We will now check the total load deflection, Δ_{DL+LL}, and the live load deflection, Δ_{LL}, against the $L/240$ and $L/360$ limits, respectively.

The adjusted bending stiffness is

$$EI' = EI_{\text{app}}\, C_M C_t = (408 \times 10^6\text{ lb-in.}^2/\text{ft})(1.0)(1.0) = 408 \times 10^6\text{lb-in.}^2/\text{ft width of panel}$$

$$\Delta_{DL+LL} = \frac{5wL^4}{384EI'_{\text{app}}} = \frac{(5)(7.917\text{ lb/in})(20\text{ ft} \times 12)^4}{(384)(408 \times 10^6\text{ lb-in.}^2/\text{ft})} = 0.84\text{ in.} = \frac{L}{286} < \frac{L}{240} \quad \text{OK}$$

The live load deflection is

$$\Delta_{LL} = \frac{5wL^4}{384EI'_{\text{app}}} = \frac{(5)(4.167\text{ lb/in})(20\text{ ft} \times 12)^4}{(384)(408 \times 10^6\text{ lb-in.}^2/\text{ft})} = 0.44\text{ in.} = \frac{L}{545} < \frac{L}{360} \quad \text{OK}$$

Example 9.2: Axial compression capacity of a CLT wall panel

Calculate the axial load capacity of a 10-ft-high five-ply CLT wall panel (Grade E4). Assume the normal moisture and temperature conditions apply; the panel is braced at the top and bottom of the wall and the load duration factor is 1.0. The density of the CLT panel is assumed to be 32 pcf.

Solution:

The pertinent adjustment factors are

$$C_D = 1.0$$
$$C_M = 1.0$$

$C_t = 1.0$
$C_L = 1.0$

For a five-ply CLT panel with a lamination thickness of 1.375 in. and a total thickness of 6.875 in.

From Tables A2 and A1 of ANSI/APA PRG-320 (i.e., Tables 9.4 and 9.3), the tabulated reference design values for five-ply Grade E4 are obtained as follows:

$F_{co} = 1800$ psi (compression stress parallel to grain)
$F_b S_{\text{eff},o} = 10{,}425$ lb-ft/ft

Panel bending stiffness, $EI_{\text{eff},o} = 441 \times 10^6$ lb-in.^{2}/ft width of panel
Panel shear rigidity, $GA_{\text{eff},o} = 1.1 \times 10^6$ lb/ft width of panel
Though there are five plies, only three plies are oriented parallel to the length of the column. Therefore, the cross-sectional area of the parallel plies is

$$A_{\text{parallel}} = 1.375 \text{ in.} \times 3 \text{ plies} \times 12 \text{ in. width} = 49.5 \text{ in}^2$$

The axial load capacity of the wall if buckling is ignored is

$$P_c = F_{c,o} A_{\text{parallel}} = (1800 \text{ psi})(49.5 \text{ in.}^2) = 89{,}100 \text{ lb/ft width of wall}$$

The unbraced height of the wall is 10 ft; therefore, the slenderness ratio of the wall for out-of-plane buckling is

$$\frac{l_e}{h} = \frac{(1.0)(10 \text{ ft}) \times (12 \text{ in./ft})}{6.875 \text{ in.}} = 17.5 < 50 \qquad \textbf{OK}$$

The shear deformation adjustment factor, K_s, is obtained from NDS Table 10.4.1.1 as 11.8 (for a member pinned at both ends and with constant moment)

Therefore, the apparent bending stiffness for deflection calculations is determined from Reference 14 (similar to NDS Equation 10.4-1) as

$$EI_{\text{app}} = \frac{EI_{\text{eff}}}{1 + \left(K_s EI_{\text{eff}} / GA_{\text{eff}} L^2\right)}$$

$$= \frac{441 \times 10^6 \text{ lb-in}^2\text{/ft}}{1 + \left[\left((11.8)(441 \times 10^6 \text{ lb-in.}^2\text{/ft})\right) / \left((1.1 \times 10^6 \text{ lb-in.}^2\text{/ft})(10 \text{ ft} \times 12 \text{ in./ft})^2\right)\right]} = 332 \times 10^6 \text{ lb-in.}^2\text{/ft}$$

The minimum apparent bending stiffness for buckling and stability calculations (see Table 9.2 footnote) is calculated as

$$EI_{\text{app,min}} = 0.5184\ EI_{\text{app}} = (0.5184)(332 \times 10^6 \text{ lb-in.}^2\text{/ft}) = 172 \times 10^6 \text{ lb-in.}^2\text{/ft}$$

The adjusted minimum apparent bending stiffness is

$$EI'_{\text{app,min}} = EI_{\text{app,min}}\, C_M C_t = \left(172 \times 10^6 \text{ lb-in.}^2\text{/ft}\right)(1.0)(1.0) = 172 \times 10^6 \text{lb-in.}^2\text{/ft width of panel}$$

The wall buckling stability factor, C_P, is calculated using the following equation (adapted from a modified form of Equation 5.29):

$$C_P = \frac{1 + \left(P_{cE}/P_c^*\right)}{2c} - \sqrt{\left[\frac{1 + \left(P_{cE}/P_c^*\right)}{2c}\right]^2 - \frac{P_{cE}/P_c^*}{c}}$$

where:

$$P_{cE} = \frac{\pi^2 EI'_{\text{app,min}}}{L^2} = \frac{(\pi^2)(172\times10^6)}{(10\,\text{ft}\times 12\,\text{in./ft})^2} = 117{,}887\ \text{lb}$$

Because P_c (the axial load capacity of the wall with buckling neglected) was previously calculated as 89,900 lb/ft width of wall; therefore,

$$P_c^* = P_c C_D C_M C_t = 89{,}100(1.0)(1.0)(1.0) = 89{,}100\ \text{lb/ft}$$

$$\frac{P_{cE}}{P_c^*} = \frac{117{,}887}{89{,}100} = 1.323$$

$c = 0.9$ for laminated timber (e.g., CLT or glulam)
Substituting the above values into the C_P equation yields,

$$C_P = \frac{1+P_{cE}/P_c^*}{2c} - \sqrt{\left(\frac{1+P_{cE}/P_c^*}{2c}\right)^2 - \frac{P_{cE}/P_c^*}{c}} = \frac{1+1.323}{2(0.9)} - \sqrt{\left(\frac{1+1.323}{2(0.9)}\right)^2 - \frac{1.323}{0.9}} = 0.848$$

The unfactored allowable compression load capacity of the CLT wall panel is

$$P_{\text{allowable}} = P_c^* C_P = (89{,}100\ \text{lb/ft width})(0.848) = 75{,}557\ \text{lb/ft width of wall panel}$$

9.3.2 CLT floor span limits for floor vibrations

To control floor vibrations in CLT floors, the prescriptive span limit is given as [8,19]:

$$\ell \le \frac{1}{12.05}\frac{(EI_{\text{app}})^{0.293}}{(\rho A)^{0.122}}$$

where:

l is the floor span in ft
EI_{app} is the apparent stiffness of 1 foot width of a simply supported CLT floor panel under uniform load
ρ is the specific gravity of the CLT panel
A is the cross-sectional area of a 1 foot wide CLT panel

In addition to the prescribed floor span limit above, [8] also recommends that the fundamental natural frequency of the CLT floor panel be no less than 9 Hz. That is,

$$f = \frac{2.188}{2\ell^2}\sqrt{\frac{EI_{\text{app}}}{\rho A}} \ge 9\ \text{Hz}$$

Note that the above equations are iterative because EI_{app} is a function of the span length. The equations apply to both simply supported and continuous floor panels.

9.3.3 CLT horizontal and vertical diaphragms

In addition to supporting gravity loads, CLT roof and floor panels also act as horizontal diaphragms in resisting the lateral forces on a building from wind or seismic loads. The applied horizontal in-plane shear on the CLT roof/floor panels acting as horizontal diaphragms can be calculated using engineering mechanics principles similar to what was presented in Chapter 6. The in-plane shear capacity of the CLT panels will be the *through-thickness* shear capacity (see Example 9.1) which must be greater than or equal to the applied diaphragm shear. The diaphragm capacity will also depend on the connection between the CLT floor panels and the CLT vertical wall panels. Connection design for CLT members is covered in NDS code Section 12 [20], similarly, for CLT wall panels acting as vertical shear walls, the applied shear is calculated using the same principles discussed previously in Chapter 7. This applied shear is then compared with the *through-thickness* shear capacity [19].

Seismic design coefficients for CLT structures are under development in the United States and are expected to be included in the next ASCE 7 standard. Based on recent tests and nonlinear numerical analyses of CLT walls and buildings, a seismic force modification factor, R, of approximately 2.0 was found to yield a low probability of failure of the CLT structural system under the maximum considered earthquake (MCE) [8].

9.4 RULES OF THUMB FOR PRELIMINARY DESIGN

Some rules of thumb for the preliminary design of wood structural members are given in Tables 9.6 and 9.7 [17,19]. In laying out floor and roof framing in wood buildings, the following should be borne in mind [21,22]:

- Avoid framing nonload bearing wall partitions to the underside of the structural framing above. Provide a gap detail between the nonload bearing wall and the structural framing above to avoid unintended load transfer to the nonload bearing wall (see Figure 9.4). A gap may be difficult to achieve if a dry wall finish is desired at the interface between the underside of the framing and the nonloading bearing wall.
- Minimize drastic differences between the spans of adjacent beams or joists to minimize differential displacement issues [21]. Minimize the mixing of sawn lumber framing and engineered wood framing in adjacent framed areas as the shrinkage characteristics are different and could cause cracking in the dry wall.
- For roof and floor framing, consider using open web trusses to allow for the passage of mechanical, electrical, and plumbing (MEP) systems. If an open web system is not used, then the designer should consider how the MEP systems are configured with respect to the structure to avoid unnecessary field cutting of the framing members.
- In laying out the framing for exterior cantilevers from the face of a building, the designer should be aware of situations where there is an upward deflection of the tip of the cantilever due to the length of the back-span and the load on the back-span. This upward

Table 9.6 Rules of thumb for wood design

Structural member/system	*Load*	*Depth (inches) L is in inches*	*Economical spans*
Floor joists	50 psf	*L*/15	
Floor joists	40 psf	*L*/20	
Residential floor joists	–	*L*/24 + 2 in.	
Roof joists	Roof load or snow load	*L*/24	
Pitched roof truss (triangular profile)	Roof load or snow load	*L*/6	
Parallel chord truss (rectangular profile)	Floor live load	*L*/6 to *L*/10	
Glulam roof beams/girders			Up to 100 ft
Glulam floor beams			Up to 100 ft
Glulam trusses—triangular or rectangular profiles			Up to 150 ft

Table 9.7 Roof truss rules of thumb

Configuration (Truss details)	*Pitch*	*Maximum span range*
Common	2:12	Up to 35 ft
	2.5:12	Up to 40 ft
	3:12	Up to 50 ft
	3.5:12	Up to 60 ft
	4:12	Up to 65 ft
	5:12	Up to 70 ft
	6:12	Up to 75 ft
	7:12	Up to 80 ft
Monolope	2:12	Up to 35 ft
	2.5:12	Up to 40 ft
	3:12	Up to 45 ft
	3.5:12	Up to 50 ft
	4:12	Up to 55 ft
	5:12	Up to 60 ft
Scissor	6:12/2:12	Up to 55 ft
	6:12/2.5:12	Up to 50 ft
	6:12/3:12	Up to 45 ft
	6:12/3.5:12	Up to 40 ft
	4:12	Up to 35 ft
Flat	Depth:	
	16″	Up to 25 ft
	18″	Up to 28 ft
	20″	Up to 31 ft
	24″	Up to 34 ft
	28″	Up to 37 ft
	32″	Up to 40 ft
	36″	Up to 43 ft

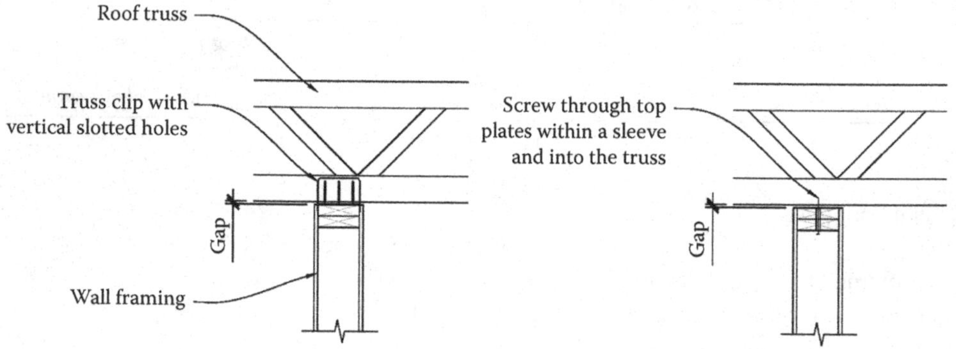

Figure 9.4 Gap detail at nonloading bearing wall.

deflection of the cantilever may negate the positive slope intended by the architect on the exterior cantilever to drain the rainwater away from the building exterior wall.

- For floors with solid sawn lumber joists with a maximum span of 20 ft, use a joist deflection limit of *L*/480 under a live load of 40 psf to minimize walking vibrations.
- The maximum practical length for sawn lumber joists is around 16 ft. Contact local suppliers to verify the maximum available length, but consider using engineered lumber for spans greater than 16 ft (e.g., I-joists, LVL's). Engineered lumber can be made in much longer lengths based on the maximum length of a delivery truck (e.g., 40 ft), so consider using continuous construction where possible with this in mind.
- Wall studs are also limited to about 16 ft for sawn lumber. Lengths above this are also not practical in that the wood may have a tendency to warp and make it so that the gypsum wall board (GWB) will not lay out flat across the studs and thus create an aesthetic problem. Consider using engineered lumber (e.g., LVL's) for studs of this length.
- For residential floors, use girder deflection limit of *L*/600 under a live load of 40 psf to minimize walking vibrations.
- For long-term vibration control, glue and screw the floor sheathing to the floor framing.

9.5 STRENGTHENING OF EXISTING WOOD STRUCTURES

Existing wood structural members such as beams or columns or trusses may need to be strengthened in order to resist additional loads or to take care of a deteriorated member. Some of the methods available for retrofitting wood structures include the following (see Figure 9.5):

1. Adding steel plates or channels to both sides of the existing wood beam; this is often called a flitch beam. The design of flitch beams is covered in Section 9.7.
2. Adding carbon fiber-reinforced plastic (CFRP) wrap around the existing wood member to increase the load capacity.
3. Adding a new column to support a beam.
4. Replacing the deteriorated or overloaded member. The structure would need to be shored and jacked up while replacing the deteriorated member.
5. Adding external post-tensioning on both sides of the existing wood beam.
6. For trusses, adding wood or plywood reinforcement is the most effective reinforcement method to increase the load capacity of a truss or to repair damaged truss members.

In retrofitting an existing beam that is loaded, it is advisable to jack up the beam to relieve it of the existing load and deflection before adding the strengthening material such as steel plates on both sides of the beam to form a flitch beam, or before adding CFRP wrap to the beam. This would ensure that the total load on the new composite system is shared between the existing wood beam and the added reinforcement in proportion to their relative stiffnesses. If the existing beam or structural member is not unloaded before adding the reinforcement, then only the new additional load above and beyond the existing load on the structure will be distributed between the existing wood beam and the new reinforcement in proportion to their relative stiffnesses. The designer must ensure that the load capacities of the existing wood beam and the added reinforcement are not exceeded.

9.5.1 Historical design values for wood

The allowable stress in existing wood members is often unknown, especially older structures. When examining existing structures for available capacity, the presence of visible grade

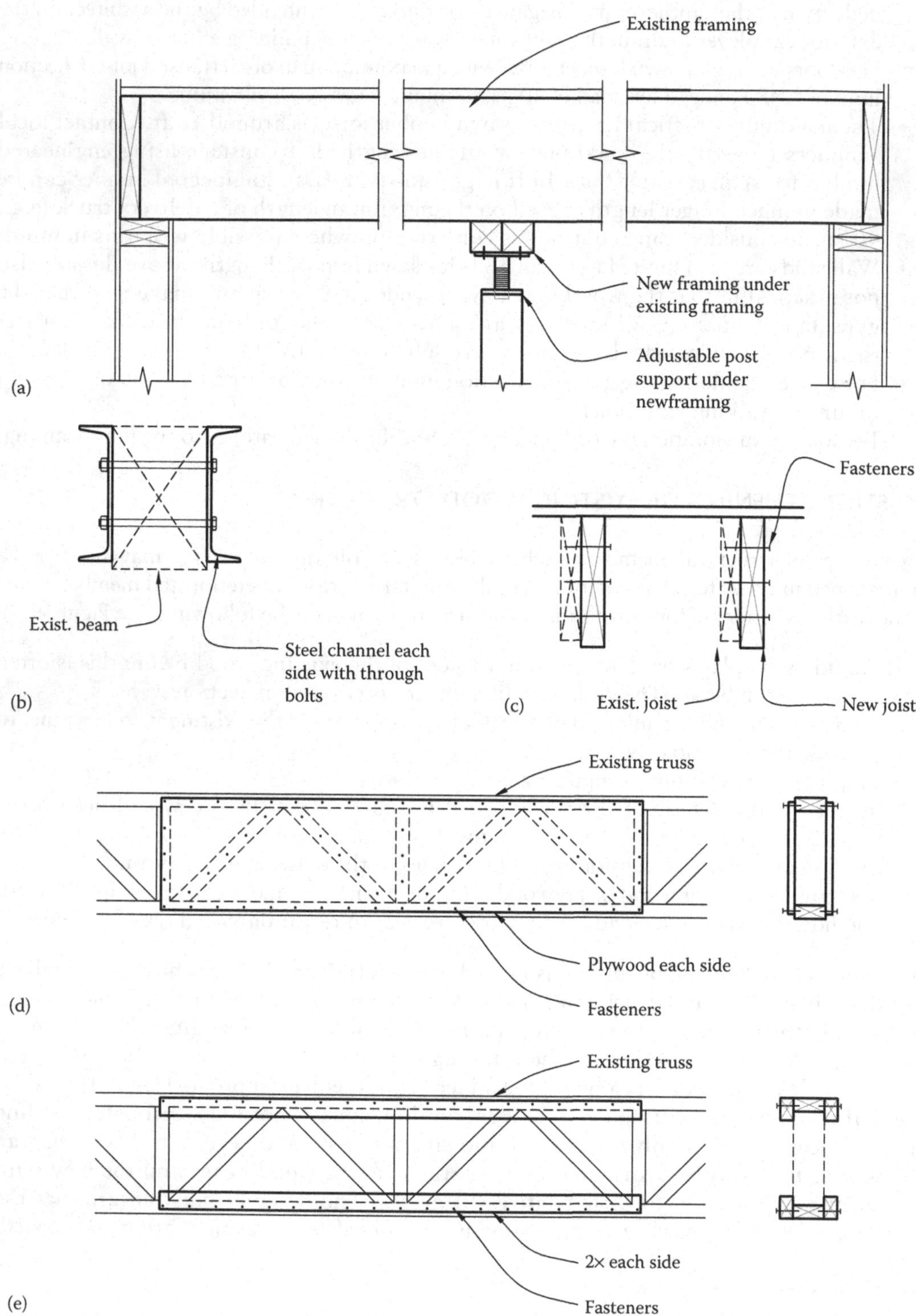

Figure 9.5 Wood reinforcing details: (a) beam reinforcing by adding new framing, (b) beam reinforcing by adding new steel framing, (c) joist reinforcing by adding new wood framing, (d) truss reinforcing with plywood, and (e) truss reinforcing with wood.

Table 9.8 Historical design values for wood, circa 1930

Stress	*Species*	*Range of values (psi)*
F'_b	Yellow pine, long leaf	1300–1600
	Yellow pine, short leaf	1000–1300
	Douglas fir	1100–1300
	Oak	1200–1400
	Spruce	700–1200
	White pine	700–1200
F'_c	Yellow pine, long leaf	1100–1600
	Yellow pine, short leaf	800–1000
	Douglas fir	1000–1600
	Oak	900–1400
	Spruce	700–1200
	White Pine	700–1000
$F'_{c\perp}$	Yellow pine, long leaf	250–350
	Yellow pine, short leaf	250–325
	Douglas fir	200–350
	Oak	500
	Spruce	200–250
	White pine	200
F'_v	Yellow pine, long leaf	110–150
	Yellow pine, short leaf	88–120
	Douglas fir	90–150
	Oak	125–200
	Spruce	80–100
	White pine	80–100

Source: Kidder, Frank E. and Parker, Harry, *Kidder-Parker Architects' and Builders' Handbook*, 18th edition, John Wiley & Sons, New York, 1956.

stamps or existing drawings is helpful, but when the wood properties are not known, assumptions have to be made. When the load capacity of a structure has to be increased or when the structure undergoes an occupancy change, it is difficult to justify such a change without reinforcing unless it is clear that there is significant reserve capacity using low assumed values.

Table 9.8 summarizes some historical design values for wood [23].

9.6 FIRE PROTECTION IN WOOD BUILDINGS

Fire safety is one of the most important considerations in the design of building structures. The resistance of a building system or component to fire is usually specified in terms of its fire rating or the number of hours of fire protection required before the structural member loses its strength and collapses. The primary goal of the building codes in relation to fire protection is to ensure that structural stability is not compromised during a fire event to allow the building occupants to safely exit the building and for firefighting operations to be carried out safely.

The IBC classifies buildings into five categories or types of construction (Types I through V) in relation to their fire resistance and the materials permitted for their construction:

Type V: Wood frame with exterior walls, partition walls, floors and roofs of sawn wood framing and joists of 2× construction. Both *noncombustible* and *combustible* materials can be used for the exterior and interior walls, and for the floor and roof framing.

Type IV: Heavy timber construction with exterior walls of *noncombustible* or *fire-retardant treated wood (FRTW)*; and heavy timber (i.e., laminated timber such as glulam or CLT) is used on the interior walls and framing.
Type III: Uses *noncombustible* materials or *FRTW* in the exterior walls; both *noncombustible* and *combustible* materials can be used for the interior walls and partitions, and floor and roof framing.
Types I and II: Building must use *noncombustible* construction materials (e.g., concrete, steel and masonry buildings).

Types I–V can be further classified into the subcategories A and B, where the "A" subcategory has more stringent requirements. For example, in Type V-A, the exterior and interior walls as well as the roof and floor framing are required to have a 1-h fire resistance rating. In Type V-B, the exterior and interior walls as well as the roof and floor framing are not required to have any resistance rating [24].

Most light framed wood buildings fall under categories III and V, while buildings with heavy timber construction (e.g., CLT highrise buildings) fall under category IV. For the different construction types listed above, the IBC prescribes limits on the building height and floor area based on the fire resistance, the materials of construction used, the occupancy or use of the building, and whether or not automatic fire suppression systems (e.g., sprinklers) are used. For example, the maximum allowable height for a Type IIIA building that has automatic sprinklers and uses 1 hr fire rated light wood framing in the interior of the building and a 2 hr fire rated exterior wall is 85 ft. Fully sprinklered business occupancy buildings are limited to six stories and multistory residential buildings are limited to five stories [4,25].

It is increasingly common to see multistory wood buildings built on top of one- or two-story concrete or steel "podium" buildings where the podium is considered as a separate structure from the wood building above. Architects use this type of hybrid construction to achieve an increased height for the building as a whole as well as being able to achieve a higher fire rating—and therefore able to accommodate a different occupancy—at the podium level(s). The connection of the wood building above and the podium level below is critical. Pre- or post-installed anchors can be used but proper coordination between the different trades is paramount. The advantages of multistory hybrid buildings include the following [8]:

- The podium level(s) can be sued for commercial occupancies.
- The podium story of steel or concrete can be used to protect the upper level wood building against vehicular impacts or flood damage in flood-prone regions.

A number of "five over one" (i.e., Five stories of light framed wood building built on top of a one story concrete or steel structure) or "six over one" buildings have been constructed in recent times [4,13].

One of the disadvantages of wood as a structural material is that wood can burn, but heavy timber members have been found to perform well in fires [17]. Light wood frame is very susceptible to fire damage because of its small member size. Consequently, fire protection for light frame wood members is accomplished by covering all sides of the framing with a fire-resistant layer of finish materials such as ceiling, gypsum wall board, and stud framing [26].

When heavy timber members or glulams are exposed to fire, a self-protective insulating outer layer known as the char layer is formed; therefore, these members could be oversized so that an adequate cross section remains after a fire event to resist the design

Table 9.9 Minimum size of heavy timber members

	Floors	*Roof*
Columns	8 × 8	6 × 8
Beams and girders	6 × 10	4 × 6

Source: ANSI/AWC, *National Design Specification for Wood Construction*, American Wood Council, Leesburg, VA. 2015.

loads [27]. The thickness of the charring layer depends on the fire rating or the number of hours of fire protection required. The fire rating is the number of hours a structural member can support its tributary load in a fire event before losing strength and collapsing. The minimum sizes required for wood members to be classified as "heavy timber" are given in Table 9.9.

9.6.1 Char rate

Wood burns at an approximate rate of 1.5 in./h, so the NDS code assumes a nominal charring rate of 1.5 in./h, and in addition, it accounts for damaged wood beyond the charred zone to determine the effective char layer thickness. The char rate is applicable to sawn lumber, timber, glulam, structural composite lumber (SCL), and CLT. The effective char rate β_{eff} in in./h is calculated per the NDS code [21] as

$$\beta_{eff} = \frac{1.2\beta_n}{t^{0.187}}$$

where:

β_n is the nominal char rate, in./h, based on 1-h exposure = 1.5 in./h for sawn timbers and glulam

t is the number of hours of exposure of the wood member to the fire

For 1-, 1½-, and 2-h fire rating, the effective char rate and the effective char layer thickness, that is, the thickness of the outer protective zone of wood that will be rendered ineffective with regard to load-carrying capacity, are obtained from in Table 9.10 [17,26]. This is similar to NDS code Table 16.2.1. The effective char thicknesses for CLT members are adopted from Reference 23 and given in Table 9.11. The values in Table 9.10 were derived based on a nominal char rate of 1.5 in./h.

The charring layout of a heavy timber member (see Figure 9.6) will depend on the surfaces of the wood member exposed to fire, and even though there is increased charring at the corners of a burning wood member that results in rounded corners, the NDS code allows a residual

Table 9.10 Effective char rates and thicknesses

Required fire rating (h)	*Effective char rate (in./h)*	*Effective char layer thickness, a_{char} (in.)*
1 h	1.8	1.8
1½ h	1.67	2.5
2 h	1.58	3.2

Source: ANSI/AWC, *National Design Specification for Wood Construction*, American Wood Council, Leesburg, VA. 2015.

Table 9.11 Effective char thicknesses for CLT

Required fire rating (h)	Effective char depth, in.						
	CLT lamination thickness, in.						
	5/8	3/4	7/8	1	1.25	1.375	1.5
1 h	2.2	2.2	2.1	2.0	2.0	1.9	1.8
1½ h	3.4	3.2	3.1	3.0	2.9	2.8	2.8
2 h	4.4	4.3	4.1	4.0	3.9	3.8	3.6

Source: ANSI/AWC, *Technical Report No. 10—Calculating the Fire Resistance of Exposed Wood Members,* American Wood Council, Leesburg, VA, 2015.

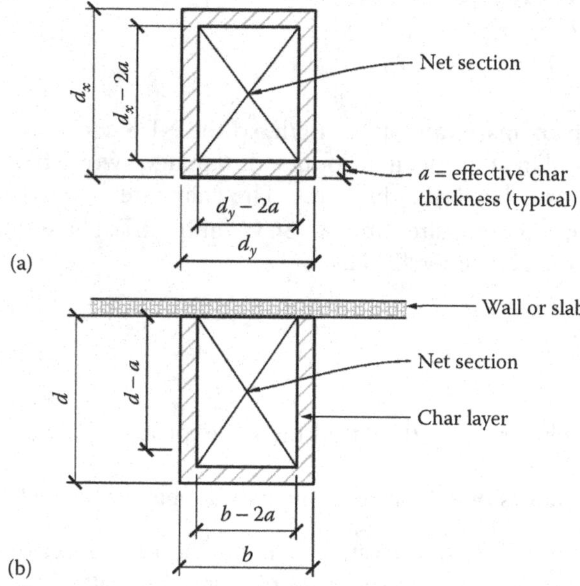

Figure 9.6 Effective char layer and net dimension: (a) Member exposed to fire on all sides and (b) Member exposed to fire on three surfaces.

rectangular section to be assumed for the net section [28]. Exposed large timber beams and girders supporting roofs or floors will typically be exposed to fire on three surfaces (i.e., three-sided exposure), whereas columns may be exposed to fire on all their four sides (i.e., four-sided exposure) unless the column is covered by a fire-resistant wall on one or more of their sides. The cross-sectional properties of the net section remaining after charring due to fire are calculated and used to determine the average ultimate strength or capacity of the fire-damaged member.

With regard to other methods for resisting fire in wood structures, it is not usual to pressure treat large timber members or glulam with fire-retardant treatment (FRT). Only smaller members such as wood studs or plywood sheathing may be pressure treated. Some of the negative effects of pressure treating large timber members with fire retardant are as follows [17]:

1. Reduction in strength
2. Corrosion effect on metal wood connectors of the fire retardant
3. Increased char rate
4. Limited to interior use only

9.6.2 Adjustment factors for fire design (NDS Code Table 16.2.2)

For wood members subject to fire, the applied stresses are calculated using the net cross-sectional dimensions, and these are compared to the average ultimate strength values calculated using Table 9.12. The factors in parentheses (e.g., "2.85") in Table 9.12 are used to convert the nominal allowable design capacity (ASD) to the average ultimate strength of the fire-damaged wood member [29].

Example 9.3: Capacity of a fire-damaged wood beam

An exposed 8 × 20 No. 1 Douglas Fir-Larch wood beam spans 24 ft and supports a 1-h rated floor assembly. Assuming a tributary width of 10 ft, determine the maximum unfactored uniform load that can be supported by the beam in a fire event. Assume the beam is laterally braced such that C_L is 1.0.

Solution:

Because 8 × 20 is a Beam & Stringer (i.e., timber), use NDS-S Table 4D to obtain the reference design bending stress (see Chapter 3):

$$F_{bx} = 1350 \text{ psi}$$

Actual size of 8 × 20:

$$b = 7.5 \text{ in.}$$

$$d = 19.5 \text{ in.}$$

The girder supports a floor, so it is exposed on three sides to the fire. For a 1-h fire rating, the thickness of the charred layer, $a = 1.8$ in. (see Table 9.10).

The *net* dimensions of the beam after charring for the three-sided exposed girder are calculated as follows:

$$b_{\text{net}} = b - 2a = 7.5'' - 2(1.8'') = 3.9''$$

$$d_{\text{net}} = d - a = 19.5'' - 1.8'' = 17.7''$$

Table 9.12 Average ultimate strength of fire-damaged wood members

Ultimate design strengths of fire-damaged wood member	*Applicable section properties for calculating the C-factors*
$F'_{bx} = F_{bx}(2.85)\, C_F C_{fu} C_V$ or $F'_{bx} = F_{bx}(2.85)\, C_F C_{fu} C_L$	C_F, C_{fu}, and C_V are calculated using the *initial* dimensions of the member prior to the fire (i.e., b, d or d_x, d_y) C_L is calculated using the *net* dimensions of the member after the fire (i.e., b_{net}, d_{net} or $d_{x,net}$, $d_{y,net}$)
$F'_t = F_t(2.85)\, C_F$	C_F is calculated using the *initial* dimensions of the member prior to the fire (i.e., b, d or d_x, d_y)
$F'_c = F_c(2.58)\, C_F C_P$ $F^*_c = F_c(2.58)\, C_F$	C_F is calculated using the *initial* dimensions of the member prior to the fire (i.e., b, d or d_x, d_y) C_P is calculated using the *net* dimensions of the member after the fire (i.e., b_{net}, d_{net} or $d_{x,net}$, $d_{y,net}$)
$F'_{bE} = F_{bE}(2.03)$ $F'_{cE} = F_{cE}(2.03)$	F_{bE} is calculated using the net dimensions after the fire F_{cE} is calculated using the *net* dimensions after the fire

$$A_{net} = (3.9'')(17.7'') = 69.03\text{ in}^2$$

$$S_{net} = \frac{(3.9'')(17.7'')^2}{6} = 203.64\text{ in}$$

$$I_{net} = \frac{(3.9'')(17.7'')^3}{12} = 1802.2\text{ in}^4$$

Recall that C_L and C_P are calculated using the *net* dimensions of the member after the fire (i.e., b_{net}, d_{net}), and C_F, C_{fu}, and C_V are calculated using the *initial* dimensions of the member prior to the fire (i.e., b, d)

$$C_L = 1.0$$

$$C_{fu} = 1.0$$

$$C_F = \left(\frac{12}{d}\right)^{1/9} \leq 1.0$$

$$= \left(\frac{12}{19.5}\right)^{1/9} = 0.95 < 1.0$$

The average ultimate bending stress of the fire-damaged beam is calculated as

$$F'_{bx} = F_{bx}(2.85)C_F C_{fu} C_L = 1350\text{ psi}(2.85)(0.95)(1.0)(1.0) = 3655\text{ psi}$$

The average ultimate moment strength of the fire-damaged beam is

$$M_{ultimate} = F'_b S_{net}$$

$$= (3655\text{ psi})(203.64\text{ in}^3) = 744{,}304\text{ in.-lb} = 62{,}025\text{ ft-lb}$$

This average ultimate moment capacity under a fire event will now be compared with the applied moment due to the unfactored loads:

$$M_{applied} = \frac{wL^2}{8} = \frac{w(24\text{ ft})^2}{8} \leq M_{ultimate}$$

Therefore, the maximum load *in lb/ft* that can be supported by the beam during the fire event is

$$w(\text{lb/ft}) = \frac{8M_{ultimate}}{(24\text{ ft})^2}$$

$$= \frac{8(62{,}025\text{ ft-lb})}{(24\text{ ft})^2} = 861.5\text{ lb/ft}$$

Because the beam has a tributary width of 10 ft, the maximum uniformly distributed load *in psf* that can be supported by the beam during the fire event is

$$w(\text{psf}) = \frac{861.5\text{ lb/ft}}{10\text{ ft}} = 86.2\text{ psf}$$

Example 9.4: (ASD) Capacity of a fire-damaged wood beam timber column

An exposed 10 × 10 No. 1 Douglas Fir-Larch wood column is supporting a floor framing and designed for a 1-h fire resistance rating. If the effective length of the column is 12 ft, determine the maximum axial load that the column can support in a fire event.

Solution:

Because 10 × 10 is a Post & Timber (i.e., timber), use NDS-S Table 4D (see Chapter 3) to obtain the reference design values:

$$F_c = 1000 \text{ psi}$$

$$E_{min} = 0.58 \times 10^6 \text{psi}$$

Actual size of 10 × 10:

$$b = 9.5 \text{ in.}$$

$$d = 9.5 \text{ in.}$$

The column is exposed on all *four* sides to the fire. For a 1-h fire rating, the thickness of the charred layer, a = 1.8 in. (see Table 9.10).

Net dimensions of the column after charring for the four-sided exposed column are calculated as follows:

$$d_{x,\text{net}} = d_x - 2a = 9.5'' - 2(1.8'') = 5.9''$$

$$d_{y,\text{net}} = d_y - 2a = 9.5'' - 2(1.8'') = 5.9''$$

Recall that C_L and C_P are calculated using the *net* dimensions of the member after the fire (i.e., $d_{x,\text{net}}$, $d_{y,\text{net}}$), and C_F , C_{fu}, and C_V are calculated using the *initial* dimensions of the member prior to the fire (i.e., d_x, d_y)

$$C_F = \left(\frac{12}{d}\right)^{1/9} \leq 1.0$$

$$= \left(\frac{12}{9.5}\right)^{1/9} = 1.0$$

$$\frac{\ell_e}{d_{x,\text{ net}}} = \frac{12 \text{ ft}(12 \text{ in./ft})}{5.9''} = 24.4 < 50 \qquad \text{OK}$$

$$\frac{\ell_e}{d_{y,\text{net}}} = \frac{12 \text{ ft}(12 \text{ in./ft})}{5.9''} = 24.4 < 50 \qquad \text{OK}$$

The Euler Buckling stress is calculated using the column *net* dimensions as follows:

$$F'_{cE,\text{net}} = (2.03)\frac{(0.822E_{\text{min}})}{(\ell_e/d_{\text{min, net}})^2}$$

$$= (2.03)\frac{(0.822)(0.58 \times 10^6)}{(24.4)^2} = 1626 \text{psi}$$

$$F^*_{c,\text{net}} = 2.58F_cC_F = (2.58)(1000 \text{psi})\,(1.0) = 2580 \text{ psi}$$

$$\frac{F'_{cE,\text{net}}}{F^*_{c,\text{net}}} = \frac{1626 \text{psi}}{2580 \text{psi}} = 0.63$$

c = 0.8 (sawn lumber)

The column stability factor is calculated using the *net* dimensions as follows:

$$C_{P,\text{net}} = \frac{1+F'_{cE,\text{net}}/F^*_{c,\text{net}}}{2c} - \sqrt{\left(\frac{1+F'_{cE,\text{net}}/F^*_{c,\text{net}}}{2c}\right)^2 - \frac{F'_{cE,\text{net}}/F^*_{c,\text{net}}}{c}}$$

$$= \frac{1+0.63}{2(0.8)} - \sqrt{\left(\frac{1+0.63}{2(0.8)}\right)^2 - \frac{0.63}{0.8}} = 0.518$$

The average ultimate compression stress parallel to grain during the fire event is

$$F'_{c,\text{net}} = F^*_{c,\text{net}}\, C_{P,\text{net}} = (2580\,\text{psi})(0.518) = 1336\,\text{psi}$$

The average ultimate axial compression capacity for the fire-damaged column is calculated as

$$P_{\text{ultimate}} = F'_{c,\text{net}}\left(d_{x,\text{net}}\right)\left(d_{y,\text{net}}\right)$$

$$= (1336\text{ psi})(5.9'')(5.9'') = 46{,}506\text{ lb} = \mathbf{46.5\ kips}$$

9.6.3 Post-fire damage assessment of heavy timber members

After a fire event, it is usually necessary to evaluate the remaining load-carrying capacity of the fire-damaged wood structure to ascertain if the structure is safe or if demolition of the structure is warranted. Also, the species and stress grade of the heavy timber member may need to be determined, and for older buildings, the wood members may not have been grade stamped. Note that the charring rate of 1.5 in./h assumed in the fire design of heavy timber members in Section 9.6.1 is not to be used for the assessment of fire-damaged wood members [30]. Instead, the actual depth of the charred layer is measured, and an additional depth of a damaged wood layer beyond the base of the char layer is added to the measured char depth to determine the total depth of the damaged and ineffective wood layer. The net cross section of the remaining undamaged section of the wood member is then determined.

Note that wood decay in the heavy timber member has not been considered in this section, but wood decay, if present and depending on its extent, will cause a drastic reduction in the load-carrying capacity of the wood member. Decayed wood members, as discussed in Chapter 1, should be seriously investigated because of the drastic loss of section and strength and the sudden nature of failure that can occur in a decayed wood structure [31,32]. The method for fire design discussed in this section applies also to glued laminated timber (glulam). The adhesives used to glue the glulam laminations together are not detrimental to the fire resistance of glulam members during a fire event. However, in full-scale fire tests of CLT members, the laminations have been found to fall off. Therefore, in the fire design of CLT members, the residual net cross section is determined by calculating the number of CLT layers that will potentially fall off due to the fire based on the char rate and the required fire resistance rating; those layers that will fall off are neglected when calculating the net dimensions of the CLT panel that is used to calculate the residual strength [28]. The procedure for assessing the residual load-carrying capacity of a fire-damaged sawn heavy timber member, which assumes no prior wood decay, is as follows [30,33]:

1. Remove and measure the actual char depth of the fire-damaged wood member, a_{actual}. Note that the actual char layer thicknesses on each exposed surface may be different depending on how much and how long these surfaces were exposed to the fire.

2. For calculating the effective net cross section of the remaining wood member, assume an additional thickness, a_{add}, of damaged wood beyond the base of the charred layer as shown in Table 9.13. It is assumed that this additional *zero-strength* layer is not removed from the fire-damaged wood member. The depth of the zero-strength layer depends on the duration of the fire and the amount of exposure a surface was exposed to the fire, but using the conservative upper limits in Table 9.13 is recommended.
3. The total depth, a_{total}, of the charred layer is taken as $a_{actual} + a_{add}$, and this is used to determine the effective remaining cross section of the fire-damaged wood member.
4. Use the net cross section to determine the effective section properties of the wood member.
5. Determine the species and stress grade of the wood member. For older buildings where the wood members have not been grade stamped, Table 9.8 could be used or a grading agency may need to be consulted to help determine the species, stress grade, and hence the reference design values to be used in the residual strength calculations. Note also that existing graded wood members that are fire damaged may need to be regraded because of the reduction in size of the wood member as a result of the fire.
6. Determine the ultimate capacity of the wood member using the equations in Table 9.12 as demonstrated in Examples 9.3 and 9.4 in Section 9.6.2.
7. Use the average ultimate capacity from step 6, determine the allowable load capacity of the fire-damaged wood member.
8. The above steps have only considered the residual strength of the heavy timber member, but the post-fire capacity of the connections of the wood members should also be evaluated because the residual connection capacity may result in a lower load capacity for the structure.

Example 9.5: Capacity of a fire-damaged heavy timber beam

A fire-damaged 8 × 20 No. 1 Douglas Fir-Larch wood beam was exposed to fire at the bottom and on the sides of the beam. A site investigation has revealed a char layer thickness of approximately 0.6 in. on all three exposed surfaces. Determine the effective cross section of the wood member after the fire to be used for calculating the residual bending moment capacity. Assume there is no wood decay.

Solution:

The actual sizes of 8 × 20 are as follows:

$b = 7.5$ in.

$d = 19.5$ in.

The measured char layer thickness, $a_{actual} = 0.6$ in. (on all three exposed surfaces).

Table 9.13 Additional thickness of damaged wood beyond the base of the charred layer

Type of stress	*Additional thickness of damaged wood beyond the base of the charred layer, a_{add} (in.)*
Compression stress	0.1–0.3
Tension stress	0.3–0.5
Bending stress	0.3–0.5

Source: White, Robert H., Kukay, Brian, Wacker, James P., and Sannon, Jenson S., Options for Evaluating Fire Damaged Components of Historic Covered Bridges, Second National Covered Bridge Conference, Dayton, Ohio, June 5–8, 2013; White, Robert H. and Woeste, Frank E., *STRUCTURE Magazine*, November, pp. 38–40, 2013.

The thickness of the additional zero strength layer, a_{add}, for *bending stress* calculations is as follows:

$a_{add} = 0.5''$ (for bending stress; see Table 9.13)

The total depth of the damaged wood on each exposed surface to be used for the residual bending moment capacity calculations is as follows:

$a_{total} = 0.6'' + 0.5'' = 1.1$ in.

The effective or net cross sections of the fire-damaged beam to be used for calculating the residual bending moment capacity are as follows:

$b_{net} = b - 2a_{total} = 7.5'' - 2(1.1'') = 5.3$ in.

$d_{net} = d - a_{total} = 19.5'' - 1.1'' = 18.4$ in.

Example 9.6: Residual capacity of a fire-damaged heavy timber column

A fire-damaged 10 × 10 No. 1 Douglas Fir-Larch wood column was exposed to fire on all four sides of the column. A site investigation has revealed a char layer thickness of approximately 0.6 in. on all four exposed surfaces. Determine the effective cross section of the wood column after the fire to be used for calculating the residual axial load capacity. Assume there is no wood decay.

Solution:

The actual sizes of 10 × 10 are follows:

$b = 9.5$ in.

$d = 9.5$ in.

The measured char layer thickness, $a_{actual} = 0.6$ in. (on all four exposed surfaces).

The thickness of the additional zero strength layer, a_{add}, for *compression stress* calculation is as follows:

$a_{add} = 0.3''$ (for compression stress; see Table 9.13)

The total depth of the damaged wood on each exposed surface to be used for the residual axial load capacity calculations is as follows:

$a_{total} = 0.6'' + 0.3'' = 0.9$ in.

The effective or net cross sections of the fire-damaged column to be used for calculating the residual axial load capacity of the column are as follows:

$$b_{net} = b - 2a_{total} = 9.5'' - 2(0.9'') = 7.7 \text{ in.}$$

$$d_{net} = d - 2a_{total} = 9.5'' - 2(0.9'') = 7.7 \text{ in.}$$

9.7 FLITCH BEAMS

A *flitch beam* is a composite section composed of a steel plate placed between wood members or steel plates or channels attached to opposite sides of an existing wood beam. These are typically used where a solid wood member is not practical, such as depth limitations or heavy loads, or where an existing wood beam needs to be strengthened to resist higher loads. Historically, flitch beams have been used as an economical alternative to higher strength preengineered wood members. Currently, preengineered wood members such as laminated veneer lumber and parallel strand lumber are typically readily available, so the use of flitch beams is not very common.

A flitch beam consists of dissimilar materials, so the section properties will have to be transformed in the design process so that the two bonded materials will experience the same strain and deformation. In a transformed section, one material is transformed into an equivalent quantity of the other material. In this case, we transform the steel plate into an equivalent amount of wood material. This equivalence is based on the ratio between the modulus of elasticity of steel and the modulus of elasticity of wood:

$$n = \frac{E_s}{E_w}$$

where:

n is the modular ratio
E_s is the modulus of elasticity of steel (= 29,000 ksi)
E_w is the modulus of elasticity of wood (varies with the wood species)

Note that connection design is covered in Chapter 8. The example here is intended to clarify the full scope of what is required for flitch beam design.

Example 9.7: Flitch beam design

For this example, the following will be assumed (see Figure 9.7):

Flitch beam supports 2 × 8′s at 16 in. spanning 11 ft on each side.
Maximum depth of the flitch beam is 7¼ in. (depth of a 2 × 8).
Total load = 55 psf (dead plus live).
Lumber is Douglas Fir-Larch, Select Structural.

A⅜ × 7¼ in. steel plate will be assumed. The approach taken here will be to transform the ⅜-in. width of steel section to an equivalent wood section while maintaining the same depth.

Solution:

From NDS Table 4A,

$F_b = 1500$ psi

$F_v = 180$ psi

$E_w = 1900$ ksi

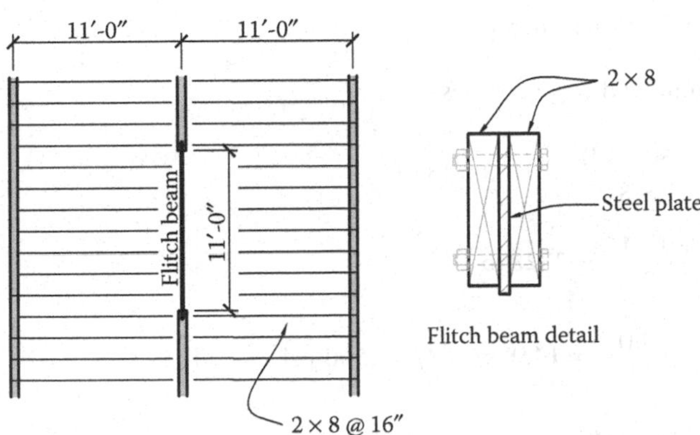

Figure 9.7 Framing plan for flitch beam design.

The modular ratio is

$$n = \frac{E_s}{E_w} = \frac{29{,}000}{1900} = 15.2$$

The section properties of the steel plates and wood side members are as follows:

$$A_p = nbh = (15.2)(0.375 \text{ in.})(7.25 \text{ in.}) = 41.4 \text{ in.}^2$$

$$S_p = \frac{nbh^2}{6} = \frac{(15.2)(0.375)(7.25)^2}{6} = 50.1 \text{ in.}^3$$

$$I_p = \frac{nbh^3}{12} = \frac{(15.2)(0.375)(7.25)^3}{12} = 181 \text{ in.}^4$$

$$A_w = (2)(10.88 \text{ in.}^2) = 21.75 \text{ in.}^2$$

$$S_w = (2)(13.14 \text{ in.}^2) = 26.28 \text{ in.}^3$$

$$I_w = (2)(47.6 \text{ in.}^2) = 95.3 \text{ in.}^4$$

Allowable stress values: (all C-factors are 1.0 except $C_f = 1.2$):

$$F_b' = 1500 \text{ psi}$$

$$(1.2) = 1800 \text{ psi}$$

$$F_v' = 180 \text{ psi}$$

$$E_w' = 1900 \text{ ksi}$$

Combining the composite section properties yields

$$A_c = 41.4 + 21.75 = 63.1 \text{ in.}^2$$

$$S_c = 50.1 + 26.28 = 76.3 \text{ in.}^3$$

$$I_c = 181 + 95.3 = 276 \text{ in.}^4$$

The total load on the beam is

$$W_{TL} = (55 \text{ psf})(11 \text{ ft}) = 605 \text{ plf}$$

Check the bending and shear stresses:

$$M = \frac{wL^2}{8} = \frac{(605)(11)^2}{8} = 9150 \text{ ft-lb}$$

$$V = \frac{wL}{2} = \frac{(605)(11)}{2} = 3328 \text{ lb}$$

$$f_b = \frac{M}{S} = \frac{(9150)(12)}{76.3} = \mathbf{1439\ psi} < \mathbf{F_b = 1500\ psi} \qquad \text{OK}$$

$$f_v = \frac{1.5V}{A} = \frac{(1.5)(3328)}{63.1} = \mathbf{79.1\ psi} < \mathbf{F_v = 150\ psi} \qquad \text{OK}$$

Check the deflection: (ignore creep effects because of the presence of the steel plates)

$$\Delta_{DL+LL} = \frac{5wL^4}{385EI} = \frac{(5)(605/12)(11\text{ ft}\times 12)^4}{(384)(1.9\times 10^6)(276)}$$

$$= \mathbf{0.38\ in.} < \frac{L}{240} = \frac{(11\text{ ft})(12)}{240} = \mathbf{0.55\ in.} \qquad \text{OK}$$

The weight of this steel plate is just over 100 pounds and it could be reasonably assumed that this is a manageable weight to be handled in the field. When the design of a flitch beam is such that the steel plate beam is difficult to handle, consider using two smaller plate beams with a total thickness equal to the required design thickness.

The composite section is adequate for bending, shear, and deflection. One other item that may need to be designed is the connectors between the wood side members and the plate. If the joists frame on top of the flitch beam, only nominal connectors need to be provided. If the joists frame into the side of the wood members, the size and spacing of the connectors become critical. One possible solution is to provide a top flange type of connector (see Figure 9.8).

For this example, it will be assumed that the joists frame into the side member with a standard joist hanger connection. The total load to one wood side member is

$$W_{side} = (55\text{ psf})(5.5\text{ ft}) = 303\text{ plf}$$

For practical reasons, the connectors should be spaced in increments consistent with the joist spacing. In this example, the joist spacing is 16 in. Assuming ½-in. connectors, the allowable shear perpendicular to the grain of the wood side member for one connector is

$Z_{\perp} = 310$ lb (NDS Table 12B)

The required spacing is then

$$S_{req'd} = \frac{303\text{ lb}}{310\text{ plf}} = 0.97\text{ ft or } 11.7\text{ in.}$$

Based on the joist spacing, provide ½-in. bolts at 8 in. o.c. The bolts should also be staggered for stability (see Figure 9.9). The minimum spacing for the connectors (from Table 8.3) is as follows:

Edge distance = $4D$ = (4)(½ in.) = **2.0 in.** (2 in. provided)
Center-to-center = $3D$ = (3)(½ in.) = **1.5 in.** (16 in. provided)
Row spacing = $5D$ = (5)(½ in.) = **2.5 in.** (3.25 in. provided)

Note that connection design is covered in Chapter 8. The example here is intended to clarify the full scope of what is required for flitch beam design.

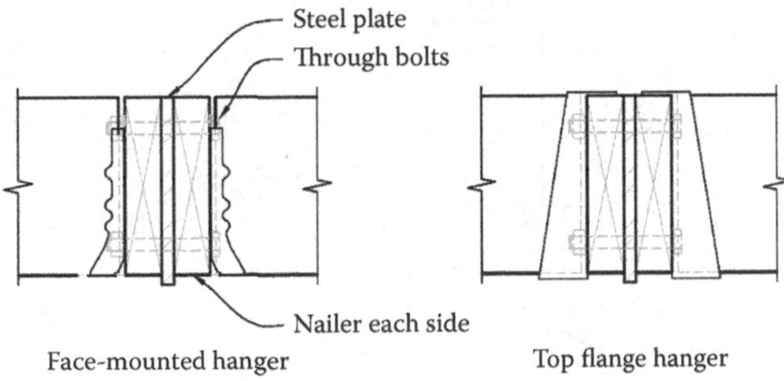

Figure 9.8 Various connections to flitch beam.

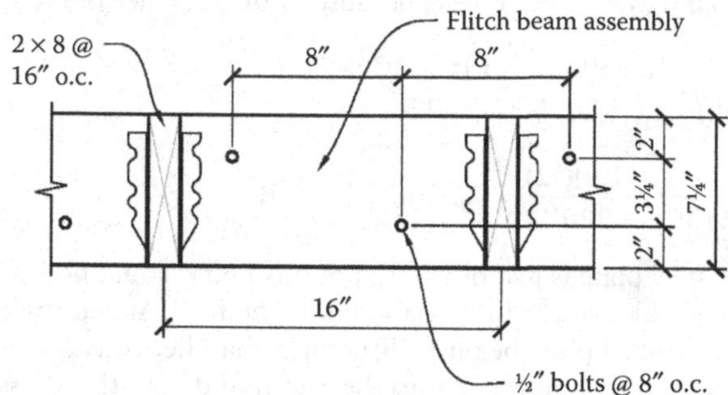

Figure 9.9 Flitch beam connection detail.

9.8 STAIRS AND RAILINGS

The design of stairs and railings in wood structures is often overlooked and is thus often constructed without many details for the builder to use as a guide. Some experienced builders are familiar with code requirements for stairs and railings, and are able to build stairs that meet the requirements of the code. The design and detailing of stairs should be given much more consideration, especially at the beginning of a project. In the authors' experience, details related to stairs and railings that are left up to the builder to resolve can lead to problems during the construction phase of the project. For example, floor framing and decking might be installed for an exterior deck prior to addressing the railing attachment, and this creates a situation where it is very difficult without significant labor and expense to install a properly anchored handrail post within the framing.

The IBC [25] prescribes several loading requirements that need to be considered in the design of stairs and railings, which are summarized as follows (see Figures 9.10 and 9.11).

9.8.1 Stairs

Live load

Uniform loading: 40 psf for residential, 100 psf for all others
Concentrated load: 300 lb on a 2″ × 2″ area of the tread and not concurrent with the uniform loading

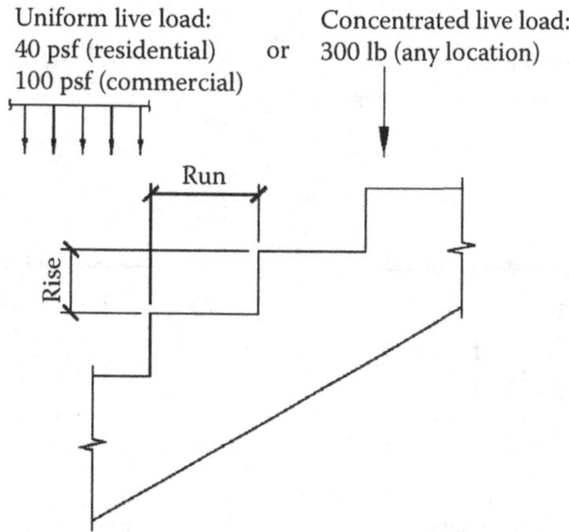

Figure 9.10 Stair dimensions and loading.

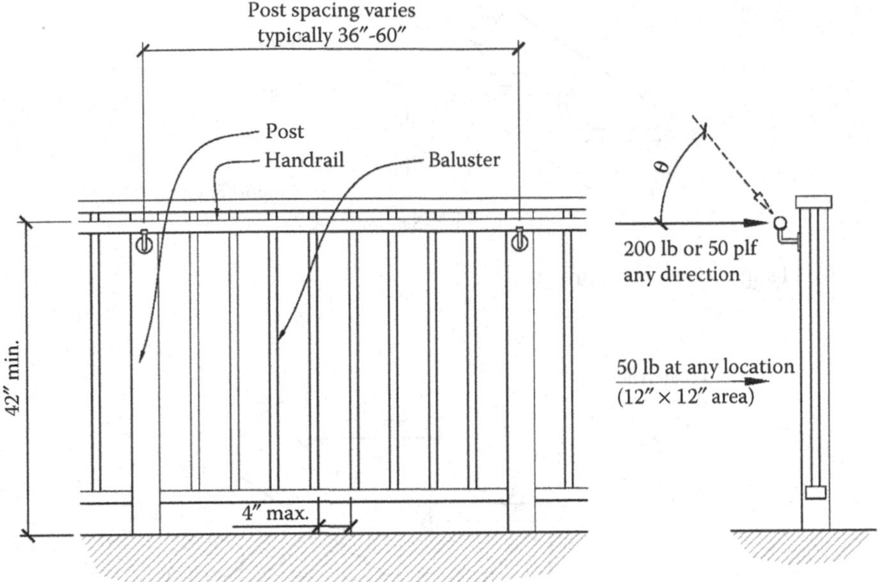

Figure 9.11 Handrail loading and dimension.

Dimensions:

	IBC [25]	*IRC [20] or (IBC R-2)*
Riser (rise)	7″ maximum 4″ minimum	7¾″ maximum
Tread (run)	11″ minimum	10″ minimum

9.8.2 Railings

In the design of railings, a horizontal line load of 50 pounds per linear foot (plf) at the top is required with some exceptions for other occupancies per IBC Section 1607.8.1. This line load is not required for residential occupancies

A concentrated load 200 lb is also required to be applied at the top, but not concurrently with the line load.

The intermediate rails and balusters must be designed for a 50 lb concentrated load at any point, and it is not concurrent with any other load.

The detailing of the supports for stairs and railings is critical in the design process. Stair stringers should have positive connections at the top and bottom that can adequately transfer the gravity and horizontal loads. The treads sometimes bear on notched stringers (see Figure 9.13), but one alternate connection that can eliminate the notching of stringers is a side-mounted connection for the treads (see Figure 9.12). The notching of stringers, while not preferred from a strength standpoint, can be made easier by drilling a small hole to avoid the overcutting and thus strength reduction from notching [34]. As discussed in Chapter 4, the drilling of the hole also helps to round the notch corners, which reduces stress concentrations.

For handrails and posts, the most critical connection is the post attachment at the base. This is a commonly overlooked detail, yet it is one with significant life safety implications. The moment that is created by the horizontal loading is such that basic nail and screw fasteners are not going to be sufficient to transfer the loading. Lag bolts and the use of engineered connectors are common

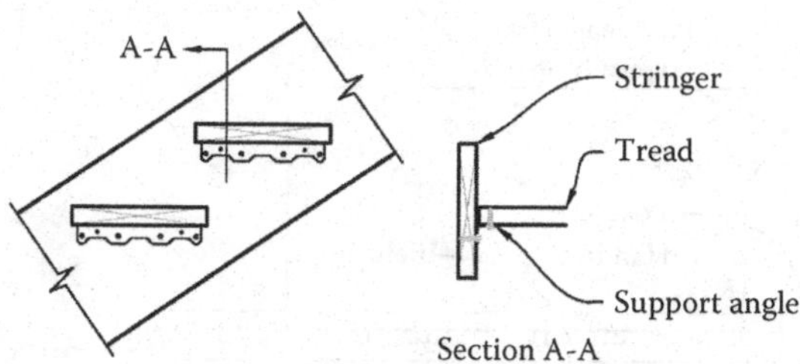

Figure 9.12 Tread supports (side mounted).

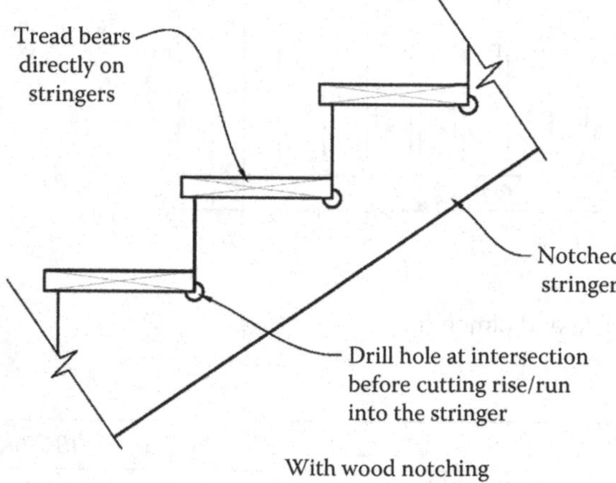

Figure 9.13 Tread supports (notched stringers).

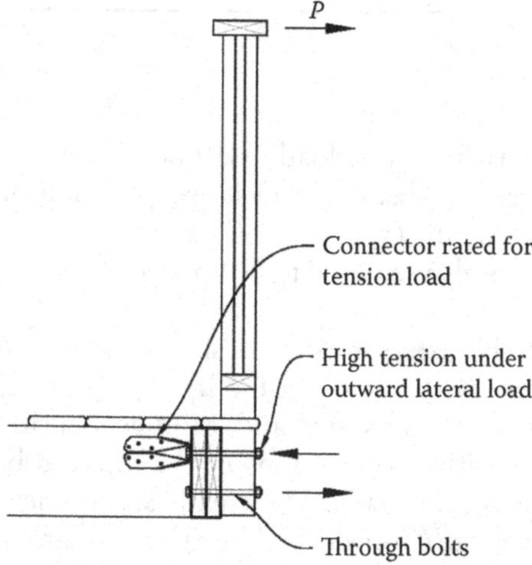

Figure 9.14 Handrail post supports.

ways to properly resist this loading (see Figure 9.14). Note that the code requires that the handrail loading be considered in all directions, but it is common practice, especially with residential structures to design and detail for outward loading with the top connector designed to resist the highest tension load. Two rows of connectors are needed at the bottom of the post to resist the moment due to the horizontal load at the top of the post.

There are several other dimensional requirements that need to be considered for a complete design of stairs and railings (see IBC Section 1009 and IRC Section 311), but the most common dimensions and loadings are given here.

9.9 OPEN WOOD FRAMES

Wood frames without shear walls are used for various structures such as gazebos, trellises, park shelters, storage structures, and structures commonly referred to as *pole barns* or *pole structures*. The lateral stability of these structures is often overlooked given the relatively minor importance of these structures. The design for gravity loading follows normal procedures, except in the area of connections where heavy beam reactions often need to be resisted.

With open frames, there are three common systems used to resist the lateral loading: knee braces, X-braces, and cantilevered posts. Knee braces (see Figure 9.15) can be effective for smaller structures, but the location of the knee brace is critical and it is ideal structurally to place the knee brace as low as possible. This is not ideal architecturally as the knee brace creates concerns with headroom. Knee braces are not usually an effective means to provide lateral stability for an entire building, and thus, they are best used for smaller open frame structures. The joints in this structure are usually modeled as *pinned*, including the base supports of the column as it is difficult to achieve connections in wood structures that can transfer moments.

An open structure with greater resistance to lateral loads would be one where X-bracing could be incorporated. The presence of X-bracing reduces the amount of potential openings in walls, but they can usually be strategically located to allow for an efficient balance between the bracing needed for stability and the open space needed for the proper functioning of the structure. Steel straps are often used as bracing material because the connections can be made with screws or nails at the ends. Wood bracing could also be used, but the design of the connections becomes more difficult, given that the loading is relatively higher and also the length of the wood members is much more limited compared to the length of steel straps. Steel straps could vary from 16 ga. up to ¼ in. thick and widths varying between

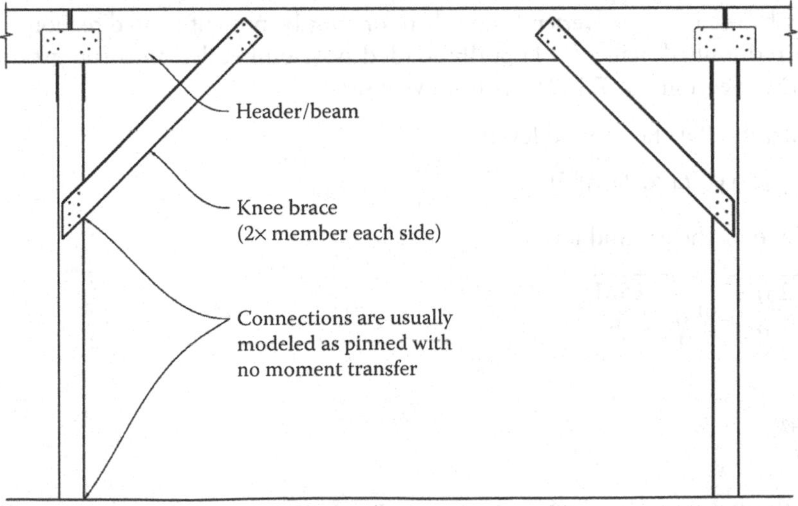

Figure 9.15 Knee-braced structure.

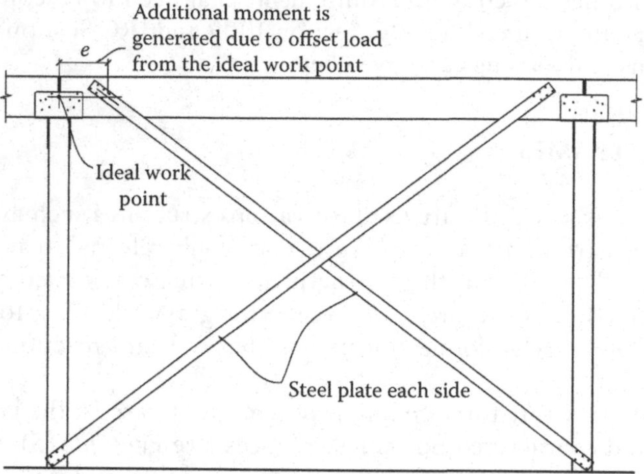

Figure 9.16 X-braced structure.

2 in. and 4 in. The joints of an X-braced structure should be detailed such that the work point or intersection of the member centerlines goes through a common point (Figure 9.16). It may be difficult to achieve this in practice, and thus, the resulting localized moments need to be considered due to the offset load (distance *e*). Note that moments in wood connections may result in cross-grain tension or tension perpendicular to grain which is undesirable.

Open structures with posts embedded into the soil or into a concrete footing are commonly used for storage or similar buildings that are not very sensitive to lateral deflections. The lateral stability of these structures is achieved by the embedment of the columns into the ground; thus, the column acts as a vertical cantilever that relies on the lateral resistance provided by the soil to develop the applied moment. Chapter 18 of the IBC provides empirical equations to determine the embedment depth required based on certain parameters. One important parameter is the presence of adequate horizontal restraint at the ground level. When the restraint is provided at the ground level, such as by the presence of a rigid pavement or a rigid concrete slab, the required embedment of the column into the soil can be reduced. It is conservative to assume that such horizontal restraint at the ground surface is not present because it is possible that during the life of the structure, the restraint may be removed inadvertently and therefore not be present in all directions.

The required embedment of a laterally loaded post embedded into the ground is calculated from IBC Section 1807.3 [25] as follows:

Nonconstrained at the ground level:

$$d = 0.5A[1 + [1 + (4.36h/A)]^{1/2}]$$

Constrained at the ground level:

$$d = \sqrt{\frac{4.25Ph}{S_3 b}} \text{ or } \sqrt{\frac{4.25M_g}{S_3 b}}$$

where:

$$A = \frac{2.34P}{S_1 b}$$

b is the diameter of round post or footing or diagonal dimension of square post or footing (in ft)

d is the depth of embedment into the earth but not greater than 12 ft for the purposes of calculating the lateral soil pressure (in ft)
h is the distance from the ground surface to force P (in ft)
P is the applied force (in pounds)
M_g is the applied moment at the ground surface = Ph
S_1 is the allowable lateral soil bearing pressure at a depth of one-third of the embedment as set forth in IBC Table 1806.2 (in psf)
S_3 is the allowable lateral soil bearing pressure at a depth equal to the depth of embedment as set forth in IBC Table 1806.2 (in psf)

In post-type structures, the top connection can be critical because of the large reactions that could be present at the connection to the column (Figures 9.17 and 9.18). One effective solution is the use of an engineered column cap connector as shown in Figure 9.19a. Where this is not possible, bolted connections can be used with a vertical reinforcing member as shown in Figure 9.19b.

Example 9.8: Embedded post design

Determine the required embedment depth of a concrete footing subject to a lateral load of 1500 lbs located 10 ft above the ground surface. The contractor would like to use an 18″ diameter concrete footing due to equipment availability and the soil type is sandy gravel. Compare the results assuming nonconstrained and constrained conditions at the ground surface.

Solution:

Nonconstrained solution:

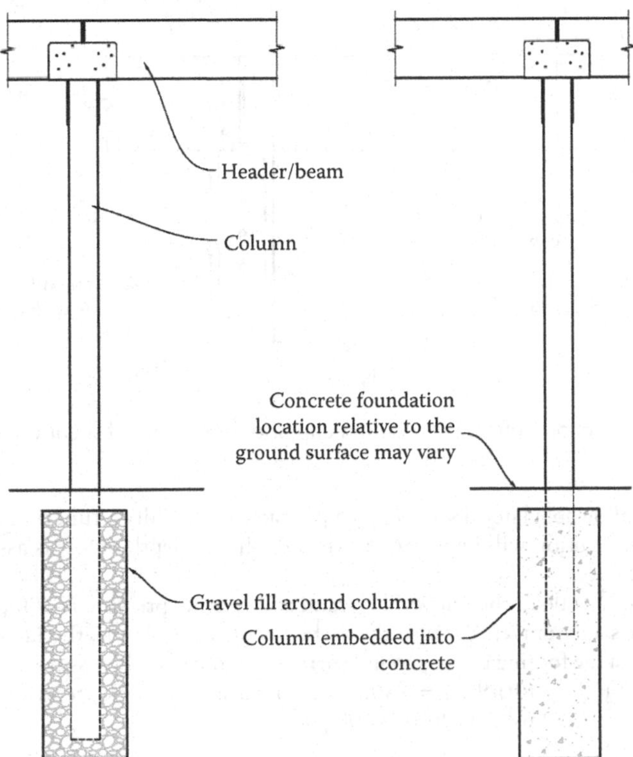

Figure 9.17 Pole structure foundation types.

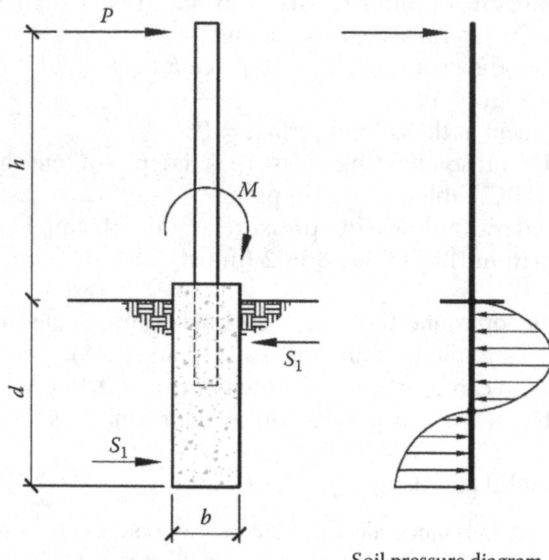

Figure 9.18 Embedded post design parameters.

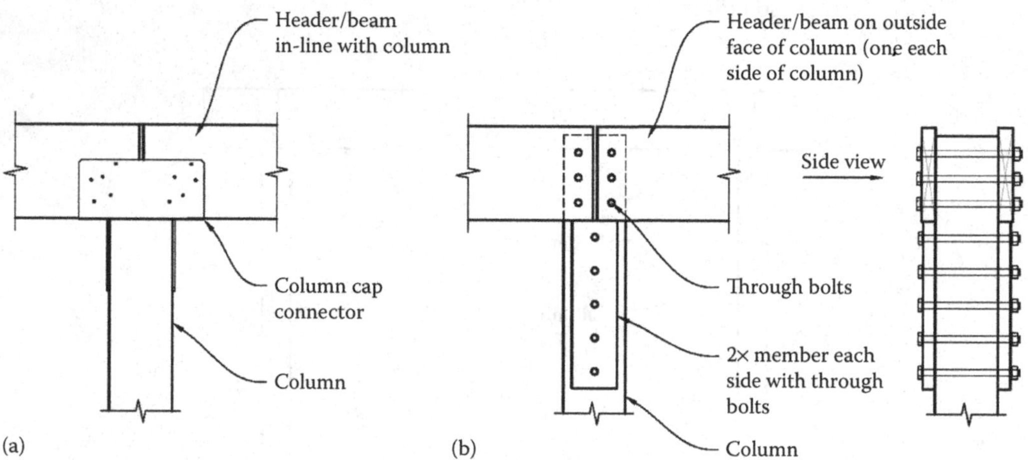

Figure 9.19 Beam connection options: (a) column cap and (b) reinforced shear connection.

Because the allowable lateral soil bearing pressure is partially a function of the embedment depth, the solution will be iterative. An embedment depth of 9 ft is assumed, which will be verified later.

From IBC Table 1806.2, the allowable lateral soil bearing pressure is 200 psf per foot of embedment for sandy gravel. Because the embedment depth is 9 ft, the lateral pressure is 0 psf at the top of the footing at the ground surface and the lateral pressure, S_3 at the bottom of the footing is (200 psf/ft)(9 ft) = 1800 psf. Therefore, the pressure at one-third of the embedment is S_1 = (200 psf/ft)(9 ft/3) = 600 psf.

$$A = \frac{2.34P}{S_1 b} = \frac{2.34(1500)}{(600)(1.5)} = 3.9$$

where:

$b = 18$ in. or 1.5 ft

$P = 1500$ lb

$$d = 0.5A[1+[1+(4.36h/A)]^{1/2}]$$

$$= 0.5(3.9)[1+[1+(4.36(10)/(3.9)]^{1/2}] = 8.75 \text{ ft}$$

Because the required embedment depth was calculated to be less than the assumed embedment depth of 9 ft, the design is adequate at a depth of 9 ft. Further iterations would yield an exact solution of $d = 8.85$ ft, but from a practical standpoint, a depth of 9 ft should be selected for construction.

Constrained solution:
Because the allowable lateral soil bearing pressure is partially a function of the embedment depth, the solution will be iterative. An embedment depth of 6 ft is assumed, which will be verified later.

The lateral soil pressure at the bottom of the footing is

$$S_3 = (200 \text{ psf/ft})(6) = 1200 \text{ psf}$$

$$d = \sqrt{\frac{4.25Ph}{S_3 b}} = \sqrt{\frac{4.25(1500)(10)}{(1200)(1.5)}} = 5.95 \text{ ft}$$

where:

$P = 1500$ lb

$h = 10$ ft

$b = 18'' = 1.5$ ft

The required embedment depth is very close to the assumed embedment depth, so further iterations are not necessary.

Comparing the nonconstrained and constrained cases, the amount of concrete and excavation is 50% [i.e., (9 ft–6 ft)/6 ft] more for the nonconstrained case, which could be a fairly significant cost on a large project. On a small project, the difference in material and excavation costs is minimal because the mobilization and concrete costs are not going to be much different. The designer would then have to weigh all of these factors in the decision between assuming constrained or nonconstrained parameters.

9.10 DRAWINGS AND SPECIFICATIONS

Construction drawings are essential in the execution of a project as they represent the legal documents that are the basis for contracts related to the construction of any structure. The drawings and any written specifications are often referred to as *contract documents* or *construction documents*. The proper representation of framing plans, details, and notes cannot be overstated. Drawings that are complete and accurate can allow a project to run more smoothly during the construction phase and can lead to fewer errors and questions in the field. Conversely, drawings that are poorly organized and incomplete can lead to problems, delays, and even litigation.

There are some basic items that are required on the contract documents from a code standpoint. Section 1603 of the IBC [27] states that "Construction documents shall show the size, section, and relative location of structural members with floor levels, column centers and offsets dimensioned." This same section goes on to state that design loads and other similar information also be provided on the documents. It is left up to the judgment of the designer as to what information to show in the plans and details, but there needs to be enough information for the builder to be able to carry out the construction of the project with as few questions as possible for the designer. It would be difficult to provide an exact list of the information that needs to be on the set of drawings that could be used for any project, but Figures 10.35 through 10.39 provide a sampling of the kinds of information that should be on the drawings used in wood construction.

One area that is often overlooked, misunderstood, and even misused is the subject of delegated design as it relates to the design and construction process. There are several factors that complicate the delegated design process, such as varying laws in each state governing delegated design, industry practices of individual companies involved (design firms, contractors, specialty engineers), and basic misunderstanding of the roles and responsibilities of those involved. The basic idea of delegated design is that a design professional for a project delegates the design of certain elements to the contractor, and the contractor then engages a specialty engineer to design the elements in question. The design professional is usually a licensed architect or structural engineer. In wood construction, the most common use of delegated design is in the design of wood trusses for roofs and also floors, so the focus of this section will be on wood trusses. In this process, the design professional would indicate what elements are delegated and also what loading or other design parameters need to be considered. The specialty engineer then designs and certifies compliance with the applicable codes or standards and contract documents, and submits the design to the engineer of record who then reviews the calculations and details, and typically must provide an approval for the delegated design to be used in the project. There can be confusion on a project if the roles and responsibilities of all those involved is not understood or well defined early on in the project life cycle. In general, the drawings and specifications for the main structure contain enough information so that the builder could then engage a wood truss supplier to provide the wood trusses and the engineering design calculations for the trusses. These drawings and specifications are often referred to as the contract documents. According to ANSI/TPI 1-2014 [35], the following information should be provided on the drawings by an architect or engineer:

1. All trusses and framing members
2. Information to fully determine the truss profiles
3. All support conditions and allowable bearing stresses
4. All applicable loads showing locations and magnitudes
5. Connections to the trusses and framing members needed for permanent bracing or other non-delegated elements
6. Serviceablity criteria, such as
    a. Allowable deflection (horizontal, vertical)
    b. Creep deflection for ponding loads
    c. Camber
    d. Differential deflection between trusses or framing members
    e. Vibration criteria, including required strong-back or bridging/blocking locations
7. Environmental conditions impacting temperature, moisture, or corrosion

Assuming that the above is adequately provided in the contract documents, the wood truss supplier would then engage a specialty engineer to assist in the design of the wood trusses. The wood truss supplier would then generate a package of documents for review by the architect or engineer to confirm that the truss layout and design can be adequately incorporated into the project. These documents would typically contain a truss placement drawing, individual truss drawings and calculations, and permanent truss bracing required for individual trusses.

According to ANSI/TPI 1-2014 [35] and the IBC [25], the following information should be provided in these documents:

1. Slope, depth, span, and spacing of all trusses
2. Location of all joints and support locations
3. Number of truss plies
4. Required bearing width
5. Design loads used at each chord, including special loads and environmental criteria
6. Other lateral loads, including drag strut loads
7. Adjustment values used in the design of the wood truss members and connection plates, including the fabrication tolerance (see ANSI/TPI 1-2014 Section 6.4.10)
8. Maximum reactions showing magnitude and direction at all supports
9. Metal connection plate sizes and locations
10. Size, species, and grade for all wood members
11. Truss-to-truss connections and field assembly requirements
12. Calculated span-to-deflection ratios and horizontal and vertical deflections
13. Maximum tension and compression forces in each wood member
14. Required permanent bracing for individual truss members and the details for this restraint (see IBC 2015 Section 2303.4.1.2)
15. Certification by a licensed engineer

A sample set of drawings and details which the truss designer would have at their disposal are shown in Figures 9.20 through 9.22. A sample truss placement drawing is indicated in Figure 9.23 and a sample truss design calculation is shown in Figure 9.24. The architect or

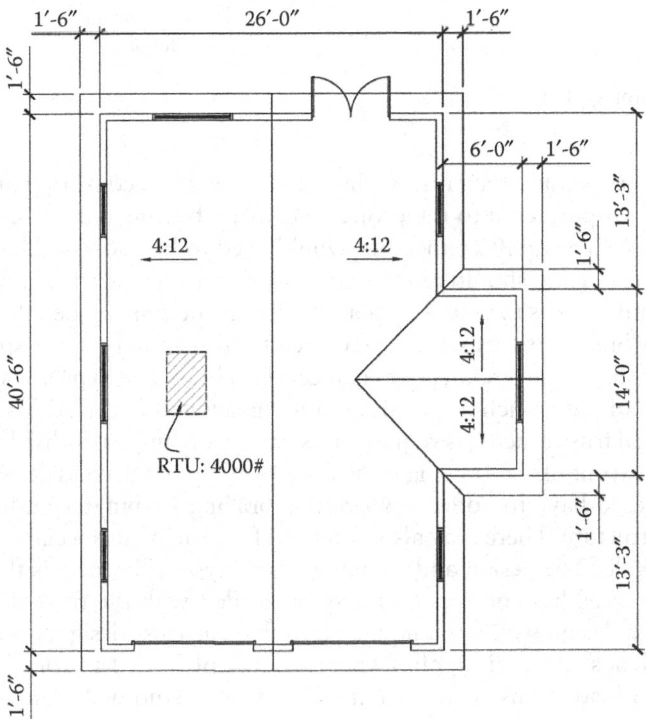

Figure 9.20 Roof layout plan.

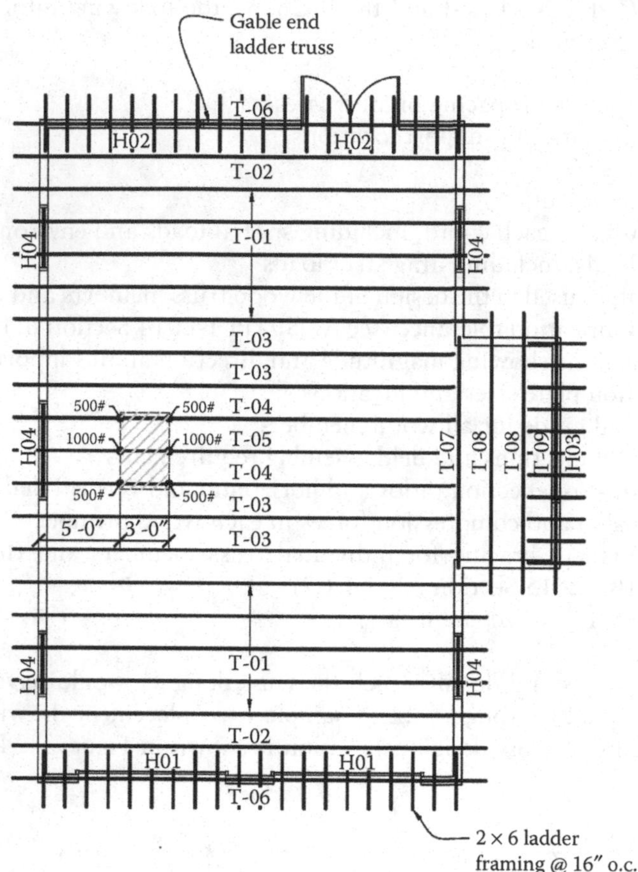

Figure 9.21 Roof framing plan.

engineer for the project should then review these documents and confirm that the design is adequate and can be incorporated into the project. The truss bearing conditions and the need for special connectors, such as uplift connectors, would need to be addressed by the project architect or engineer. The bracing should also be addressed, but there are various types of bracing, and there can be confusion as to who is responsible for the design and detailing of each bracing. Permanent stability bracing is that which is required for the stability of the structure as a whole and is the responsibility of the architect or engineer. This bracing is typically needed to ensure a complete lateral load path, such as diaphragm to shear wall connections (see Figures 9.25 and 9.26). Individual truss bracing is required to stabilize or support individual truss members, and the design and layout of this bracing is the responsibility of the truss designer, but the architect or engineer would have to address where this bracing terminates and transfers the load back to the main structure. There may also be a need for bracing in special cases, such as in the case of gable end walls. The design and detailing of this type of bracing is the responsibility of the architect or engineer, but coordination may be needed with the truss designer in order to implement it. The gable end wall is a unique case in that the truss designer often designs a truss to support gravity loads only and supplies what is commonly called a *gable end truss* or "*ladder truss*" as the gravity loads transfer through it as if it were a stud wall. The truss designer does not usually design this truss for wind loading, so it is important for the architect or engineer to ensure that the contract drawings clearly indicate whether or not wind loading is a design

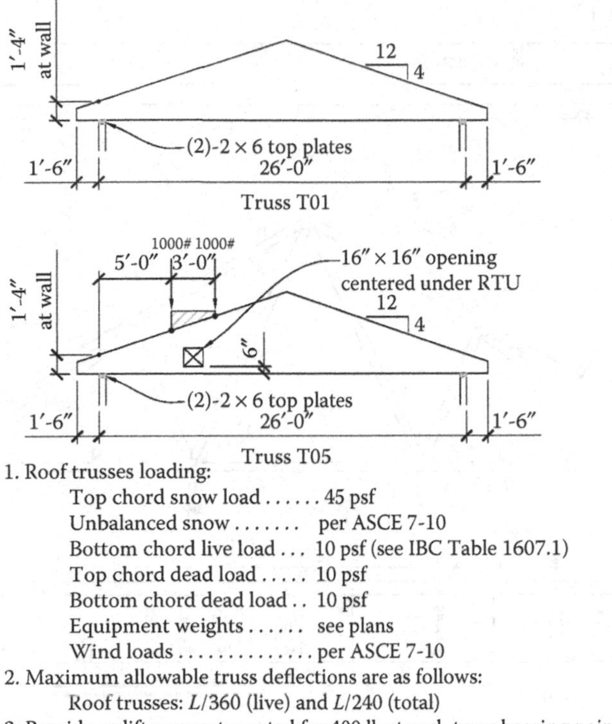

Figure 9.22 Truss loading.

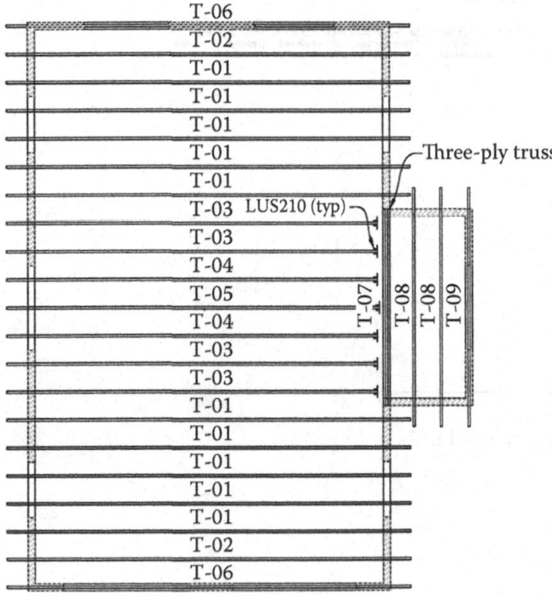

Figure 9.23 Sample truss layout plan.

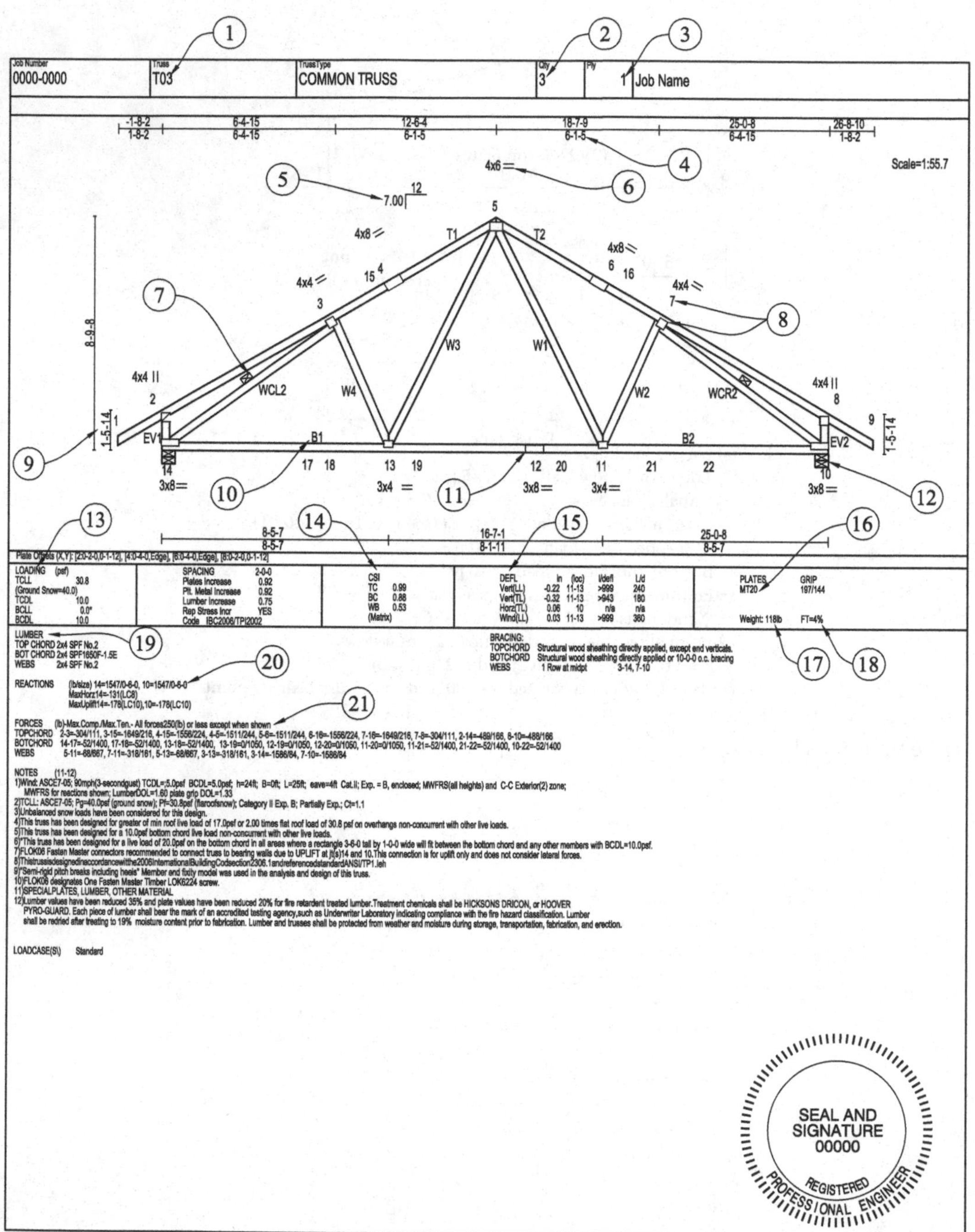

Figure 9.24 Sample preengineered truss output.

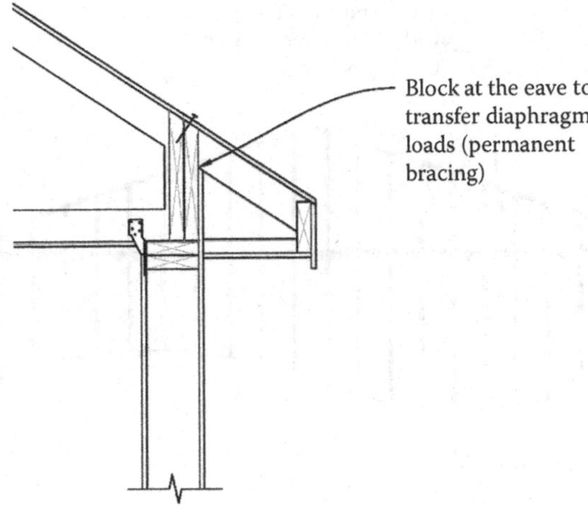

Figure 9.25 Eave blocking detail.

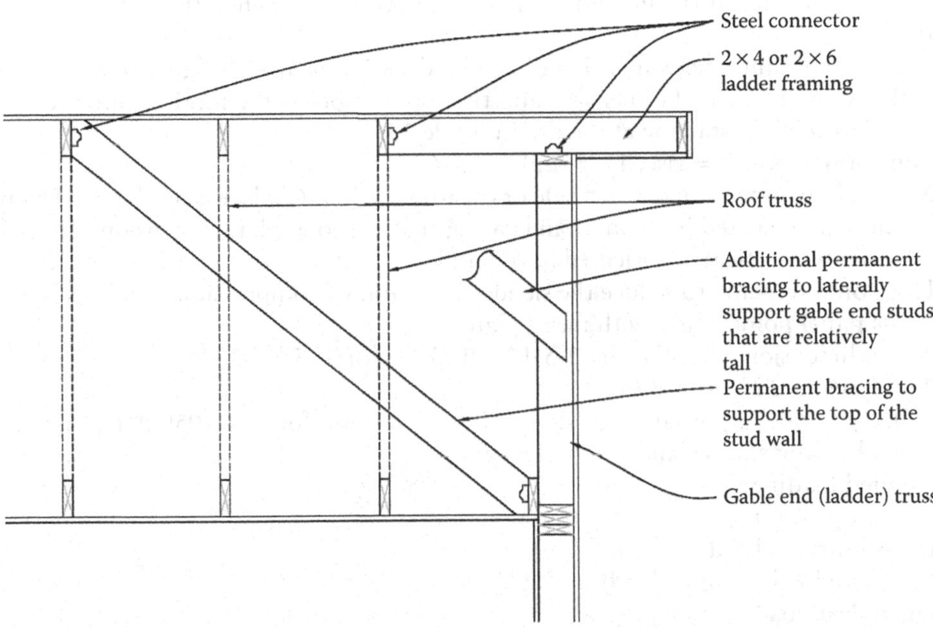

Figure 9.26 Gable end truss detail.

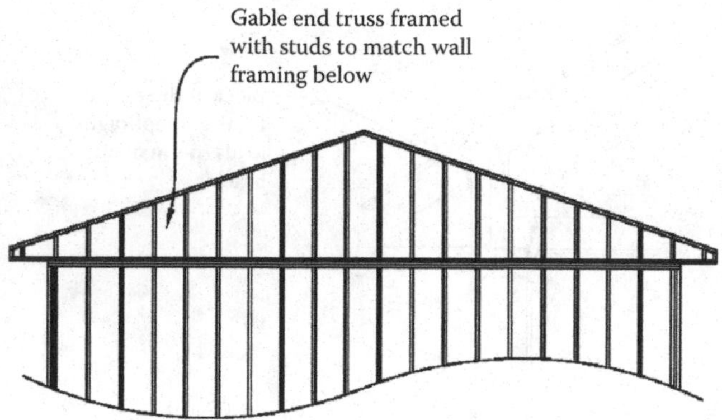

Figure 9.27 Ladder truss over stud wall.

criteria to consider, or additional bracing may be provided to support the members in the gable end (see Figure 9.27).

Legend for preengineered truss output (see Figure 9.24):

1. Truss mark; to be used on the plan sheets showing the layout of all trusses
2. The total quantity of trusses to be used on the project
3. Number of plies in the individual truss (one-ply = 1.5 inched thick, two-ply = 3 in. thick
4. Truss panel dimensions (6-1-5 = 6′-1⁵⁄₁₆″and 18-7-9 = 18′-7⁹⁄₁₆″); the bottom number is the individual panel dimension and the top number is the total running dimension from left to right starting at the origin (node 14)
5. Top chord slope (7 = rise, 12 = run)
6. Metal plate dimension on each side of the joint; a 4″ × 6″ plate is used here. The parallel lines are oriented horizontal, indicating that the long plate dimension is horizontal (4″ oriented vertical, 6″ oriented horizontal)
7. Horizontal bracing to reduce the slenderness ratio of compression members
8. Truss panel point (joint) with metal plate
9. Truss dimension – heel height (1-5-14 = 1′-5 ¹⁴⁄₁₆″or 1′-5 ⅞″)
10. Truss member identification
11. Splice (at node 12); needs to be designed for a tension force of 1050 lb (see note 21)
12. Truss bearing surface and reaction location
13. Applied loading:
    TC = top chord
    BC = bottom chord
    LL = live load (or snow load)
    DL = dead load
14. CSI = combined stress interaction or unity check; must be less than 1.0 for any member
15. Deflection checks; actual total load (TL) deflection for the bottom chord member from nodes 11–13 is 0.32 in. (negative sign indicates downward deflection); this deflection is equivalent to $L/943$ and is less than the code limit of $L/180$
16. Metal plates are product-type MT20 (20 gage thickness); metal plates are provided in pairs (one each side)
17. Total truss weight (needed for shipping purposes)

18. FT = fabrication tolerance and allowable stresses reduced by 4% for this particular manufacturer (number can vary between 0% and 20%)
19. Wood species and grade:
    Bottom chord is Spruce-Pine-Fir species and is a 2 × 4 for $F_b = 1650$ psi and $E = 1.5 \times 10^6$ psi (i.e., machine rated)
20. Maximum downward reaction occurring at nodes 14 and 10 and is 1547 lb
    Maximum uplift is 178 lb (negative sign indicates upward direction); truss will need a connection to resist both loads; load combination (LC) 10 controls; the load combinations are not stated here and are internal to the software used.
21. Truss member forces:
    Bottom chord between nodes 13–14, maximum tension force = 1400 lb (positive value indicates a tension load)
    Top chord between nodes 5–7, maximum compression force = 1649 lb (negative value indicates a compression load); compression load varies along the length of this member.

REFERENCES

1. Karsh, Eric and Holt, Rebecca. (2015), Overview of the Survey of International Tall Wood Buildings, *STRUCTURE Magazine*, April.
2. Layne, Evans (2013), *Cross Laminated Timber—Taking Wood Buildings to the Next Level,* Engineering News Record (ENR), October 14.
3. Taban, Nabih (2013), LCT ONE—Case Study of an Eight-Story Timber Office Building, *STRUCTURE magazine*, September, pp. 20–23.
4. Cover, Jennifer. (2013), University of Washington Invests in Student Housing, *STRUCTURE Magazine*, July.
5. American Society for Engineering Education (2015), TallTimber, PRISM *magazine*, ASEE, p. 15.
6. Ward, Roxane (2009), Building the World's Tallest Mixed-use Wood Structure, *STRUCTURE Magazine*, August.
7. Johnson, P E. Horos, D., and Baker, W. (2014), Timber Tower Research Projea, STRUCTURE *magazine*, March, pp. 22–23.
8. Karacabeyii, Erol and Douglas Brad (2013), *CLT Handbook, U.S. Edition,* FPInnovations and Binational Softwood Lumber Council, Pointe-Claire, Quebec.
9. APEGBC (2009), *Structural, Fire Protection and Building Envelope Professional Engineering Services for 5- and 6- Story Wood Frame Residential Building Projects,* Technical and Practice Bulletin, Association of Professional Engineers and Geologists of British Columbia, Belmont, CA.
10. Knight Brian. (2006), High Rise Wood Frame Construction, *STRUCTURE Magazine*, June.
11. Martin, Zeno and Anderson, Eric. (2012), Multistory Wood Frame Shrinkage Effects on Exterior Deck Drainage, *STRUCTURE Magazine*, April.
12. Howe, Richard W. (2011), Accommodating Movement in High Rise Wood Frame Building Construction, *STRUCTURE Magazine*, June.
13. Wood Works. (2013), *Student Housing Gets Extra College Credit from Wood*, Wood Works. www.woodworks.org/wp-content/uploads/UW_CaseStudy.pdf, Accessed April 18, 2016.
14. Sanders, Stevens L. (2011), *Behavior of Interlocking Cross-Laminated Timber*, MS Thesis, Brigham Young University, Provo, UT.
15. Podesto Lisa. (2012), Is North America Ready for Wood High-Rises?, *STRUCTURE Magazine*, June.
16. Showalter, John "Buddy." (2013), New Wood Materials—Cross Laminated Timber (CLT), *STRUCTURE Magazine*, September.

17. AITC (2012), *Timber Construction Manual,* 6th Edition, American Institute of Timber Construction, Hoboken, NJ.
18. ANSI/APAPRG 320-2012 (2012), *Standard for Performance-Rated Cross-Laminated Timber,* APA—The Engineered Wood Association, Tacoma, WA.
19. Breneman, Scott. (2016), Cross-Laminated Timber Structural Floor and Roof Design, STRUCTURE *magazine*, June, pp. 2–14.
20. ANSI/AWC (2015), *National Design Specification for Wood Construction,* American Wood Council, Leesburg, VA.
21. Martin, Zeno and Anderson, Jim. (2013), Differential Deflection in Wood Floor Framing, *STRUCTURE Magazine*, November.
22. Woeste, Frank and Dolan, Daniel. (2012), *Design to Minimize Annoying Wood-Floor Vibrations*, SEAoO Conference, September 13.
23. Kidder, Frank E. and Parker, Harry (1956), *Kidder-Parker Architects' and Builders' Handbook*, 18th edition, John Wiley & Sons, New York.
24. Stone, Jeffrey B. (2013), *Fire Protection in Wood Buildings—Expanding the Possibilities of Wood Design*, American Wood Council. http://www.awc.org/pdf/education/bcd/ReThinkMag-BCD200A1-DesigningForFireProtection-150801.pdf, accessed April 17, 2016.
25. ICC (2015), *International Building Code,* International Code Council, Washington, DC.
26. ANSI/AWC (2015), *National Design Specification for Wood Construction,* American Wood Council, Leesburg, VA.
27. ANSI/AWC (2015), *Manual for Engineered Wood Construction (ASD/LRFD),* American Wood Council, Leesburg, VA.
28. ANSI/AWC (2015), *Technical Report No. 10—Calculating the Fire Resistance of Exposed Wood Members,* American Wood Council, Leesburg, VA.
29. Douglas, Bradford K. (2004), Calculating the Fire Resistance of Exposed Wood Members, *STRUCTURE Magazine*, February, pp. 11–14.
30. White, Robert H. and Woeste, Frank E. (2013), Post-Fire Analysis of Solid Sawn Heavy Timber Beams, *STRUCTURE Magazine*, November, pp. 38–40.
31. Dunham, Lee. (2013), Decayed Wood Structures, *STRUCTURE Magazine*, October, pp. 21–24.
32. Bloom, William R. (2003), Dry Rot Problem and Solutions, *STRUCTURE Magazine*, December.
33. White, Robert H., Kukay, Brian, Wacker, James P., and Sannon, Jenson S. (2013), Options for Evaluating Fire Damaged Components of Historic Covered Bridges, Second National Covered Bridge Conference, Dayton, Ohio, June 5–8.
34. Fournier, Christopher R. (2013), Wood-framed Stair Stringer Design and Construction, *STRUCTURE Magazine*, March, pp. 44–47.
35. ANSI/TPI 1 (2014), *National Design Standard for Metal Plate Connected Wood Truss Construction*, Truss Plate Institute, Alexandria, VA.

Chapter 10

Building design case study

10.1 INTRODUCTION

The design of a simple wood building structure (see Figure 10.1) is presented in this chapter to help the reader tie together the various structural concepts presented in earlier chapters. This design case study building is a wood-framed office building and is carried out using the ASD method. It consists of roof trusses, plywood roof sheathing, stud walls with plywood sheathing, plywood floor sheathing, floor joists, a glulam floor girder, wood columns supporting the floor beam, and stairs leading from the ground floor to the second floor. The plan dimensions of the building are 30 ft × 48 ft and the floor-to-floor height is 12 ft. The exterior walls have ⅝-in. gypsum wall board on the inside and ⅝-in. wood shingle on plywood sheathing on the outside. The ceiling is ⅝-in. gypsum wall board and the second floor is finished with ⅞-in.-thick hardwood flooring. The roofing is asphalt shingles. Spruce-Pine-Fir wood species is specified for all sawn lumber members and Hem-Fir for the glulam girder. The design aids presented in Appendix B are used for selection of the ground-floor column, exterior wall stud, and the floor joists. Note that for any given project, the designer should verify what species and grade of lumber and sheathing are available where the building is to be constructed. The designer should also verify which codes and standards govern the project. For example, the International Building Code (IBC) 2015, ASCE 7-10, and NDS 2015 will primarily be used [1–4].

10.1.1 Checklist of items to be designed

The following is a checklist of all the items to be designed for this case study building.

Sheathing for floor, roof, and walls: Initially, select the roof and floor sheathing grade, minimum thickness, span rating, and edge support requirements for *gravity* loads based on the IBC tables. The plywood grade and thickness obtained for the gravity load condition should then be checked to determine if it is adequate to resist the *lateral* loads when the roof and floor sheathing act as horizontal diaphragms.

Roof truss: The truss is designed for the case of dead load plus snow load, and a structural analysis software is used to analyze the roof truss for the member forces. The computer results can be verified using a hand calculation method such as the method of sections or the method of joints.

Stud walls: All the IBC load cases have to be considered and 2 × 6 studs are assumed.

Floor joist, glulam floor girder, and columns: The critical load case is dead load and floor live load.

Roof diaphragm: Select the nailing and edge support requirements. Calculate the chord and drag strut forces and design the roof diaphragm chord and drag strut (i.e., the top plates).

Floor diaphragm: Select the nailing and edge support requirements. Calculate the chord and drag strut forces and design the floor diaphragm chord and drag strut (i.e., the top plates).

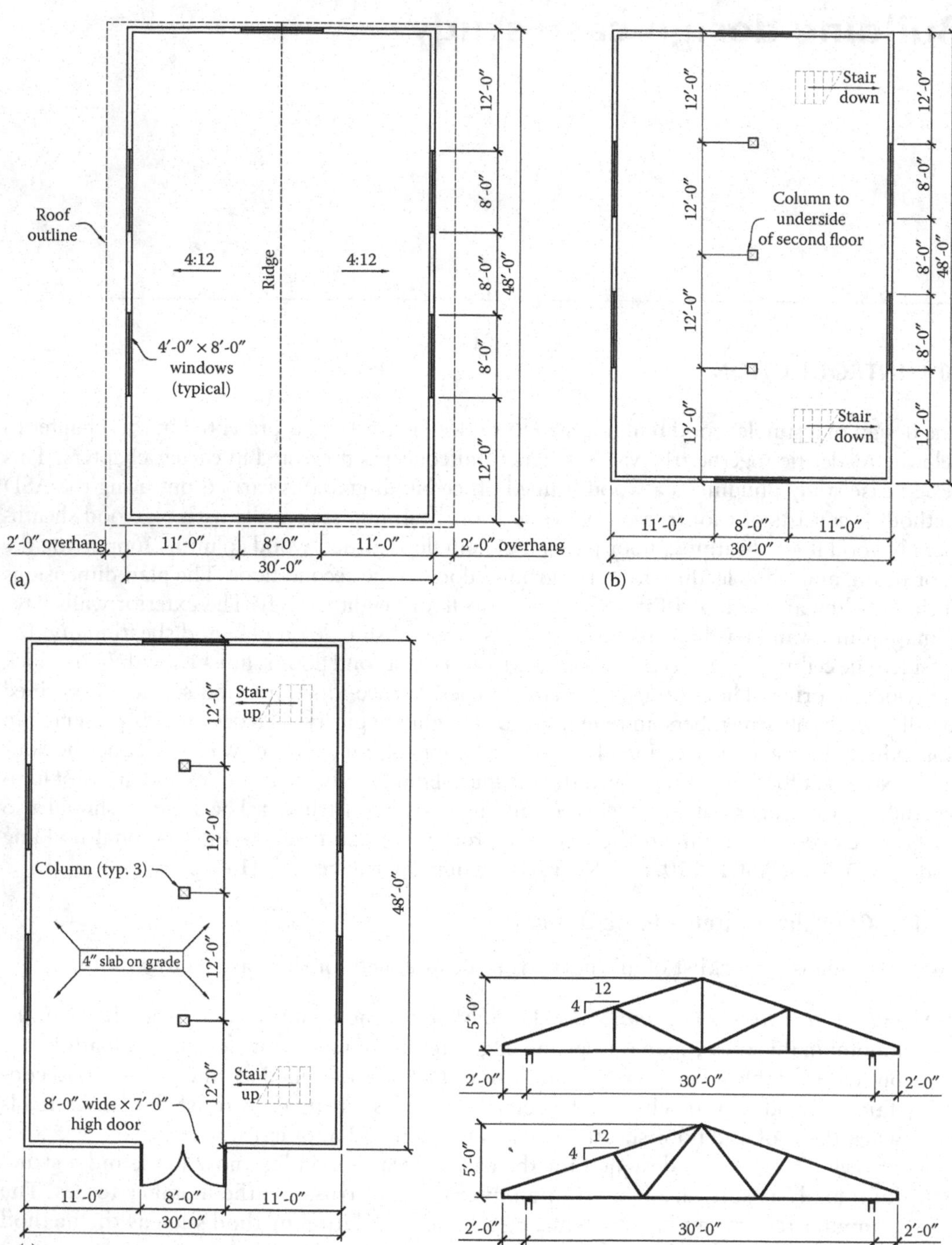

Figure 10.1 Case study building: (a) roof plan, (b) second-floor plan, (c) ground-floor plan, and (d) possible roof truss profiles.

Shear walls: Analyze and design the most critical panel in *each direction*, checking the tension and compression chords; indicate whether hold-down devices are needed; and specify the vertical diaphragm nailing requirements.

Lintel/headers: Design lintels and headers at door and window openings for gravity loads from above (i.e., dead and snow loads for lintels/headers below the roof, and dead and floor live loads for lintels below the second floor).

10.2 GRAVITY LOADS

In this section, we calculate the roof dead and live loads, the floor dead and live loads, and the wall dead load (i.e., the self-weight of the wall).

10.2.1 Roof dead load

To calculate the roof dead load, we assume roof trusses spaced at 2 ft o.c., plywood roof sheathing, asphalt shingle roofing, and ⅝-in. gypsum wall board ceiling. Other dead loads that will be included are the weight of mechanical and electrical fixtures, insulation, and possible reroofing, assuming that the existing roof will be left in place during the reroofing operation. The dead loads are obtained as follows using the dead load tables in Appendix A:

Roofing (5/16 in. asphalt shingles)	= 2.5 psf
Reroofing (i.e., future added roof)	= 2.5 psf (assumed)
⅜-in. plywood sheathing (= 0.4 psf/⅛ in. × .3)	= 1.2 psf
Framing (truss span = 30 ft; see Appendix A)	= 5.0 psf
1 in. fiber glass insulation (supported by the bottom chord)	= 1.1 psf
⅝-in. gypsum ceiling attached to the truss bottom chord (5 psf/in. × ⅝ in.)	= 3.2 psf
Mechanical and electrical (supported off the bottom chord)	= 5.0 psf
Total roof dead load D_{roof}	= 20.5 psf of sloped roof area

The dead load supported by the roof truss bottom chord includes the insulation, the gypsum ceiling, and the mechanical and electrical loads. Therefore, the total dead load on the bottom chord is 9.3 psf of horizontal plan area, whereas the dead load supported by the truss top chord is 11.2 psf. Note that the load on the top chord is in psf of sloped roof area and must be converted to psf of horizontal plan area before combining it with the loads on the truss.

The total roof dead load in psf of the *horizontal plan area* is given as

$$w_{DL}(\text{top chord}) = (11.2 \text{ psf})\left(\frac{15.8 \text{ ft}}{15 \text{ ft}}\right) \approx 12 \text{ psf}$$

$$w_{DL}(\text{bottom chord}) \approx 10 \text{ psf}$$

When calculating the roof wind uplift load, the dead load must not be overestimated because it is not conservative to do so. For example, the 2.5 psf reroofing dead load should not be considered for this case because reroofing is a future event and may not be there for quite a number of years. Therefore, for uplift load calculations, only the dead loads that are likely to be on the structure during a wind event should be used to calculate the net uplift wind pressure. For this building, the applicable roof dead load is calculated as

$$w_{DL}(\text{for uplift calculations only}) = (11.2\text{ psf} - 2.5\text{ psf})\left(\frac{15.8\text{ ft}}{15\text{ ft}}\right) + 9.3\text{ psf} = 18.5\text{ psf}$$

10.2.2 Roof live load

The snow load S and the roof live load L_r acting on the roof of this building will now be calculated.

10.2.2.1 Roof live load L_r

From Section 2.4, because $F = 4$, the roof slope factor is obtained as $R_2 = 1.0$.

The tributary area (TA) of the roof truss = 2 ft × 30 ft = 60 ft² < 200 ft²; therefore, $R_1 = 1.0$.
The roof live load on the truss $L_r = 20R_1R_2 = (20)(1.0)(1.0) = 20$ psf. 12 psf $\leq L_r \leq$ 20 psf OK
The tributary roof area for a typical exterior stud = (2 ft)(30 ft/2) = 30 ft² < 200 ft²; therefore, $R_1 = 1.0$.
The roof live load on the exterior wall stud $L_r = 20R_1R_2 = (20)(1.0)(1.0) = 20$ psf. 12 psf $\leq L_r \leq$ 20 psf OK

10.2.2.2 Snow load

The roof slope (4 in 12) implies that tan θ = 4/12; therefore, θ = 18.43°. The design snow load on the roof would be calculated based on the procedures in Chapter 2. For simplicity, a flat roof snow load of pf = 35 psf is assumed here.

The design total load in psf of the *horizontal plan area* is obtained using the IBC load combinations (in Section 2.1). Where downward wind loads on the roof are not critical (this will be checked later), as is often the case, the controlling load combination, with the deal load modified to account for the roof slope, is given by Equation 2.1 as

$$w_{TL} = D\frac{L_1}{L_2} + (L_r \text{ or } S \text{ or } R) \quad \text{psf of the horizontal plan area}$$

Because the roof live load L_r (20 psf) is less than the design snow load S (35 psf), the snow load is more critical and will be used in calculating the total roof load.

L_1 = sloped length of roof truss top chord = 15.8 ft
L_2 = horizontal projected length of roof truss top chord = 15 ft

The total load on the top chord and bottom chords of the roof truss will be calculated separately.

$$w_{TL}(\textbf{top chord}) = 12\text{ psf} + 35\text{ psf}$$

$$= \mathbf{47\text{ psf}} \text{ of the horizontal plan area}$$

$$w_{TL}(\textbf{bottom chord}) \approx \mathbf{10\text{ psf}} \text{ of the horizontal plan area (this is from the ceiling load only)}$$

The total load on the roof truss $w_{TL} = 47 + 10 =$ **57 psf** of the horizontal plan area.
The tributary width (TW) for a typical interior truss = 2 ft.
The total load in pounds per horizontal linear foot (lb/ft) on the roof truss is given as

$$w_{TL}(\text{lb/ft}) = w_{TL}(\text{psf})(\text{TW})$$

$$= (57 \text{ psf})(2 \text{ ft})$$

$$= \mathbf{114 \text{ lb/ft}}$$

The maximum shear force in the roof truss is

$$V_{max} = w_{TL}\frac{L_2}{2} = (114)\left(\frac{30 \text{ ft}}{2}\right) = 1710 \text{ lb}$$

The total load in pounds per horizontal linear foot (lb/ft) on the top chord and bottom chords are calculated as

$$w_{TL}(\text{top chord}) = 47 \text{ psf} \times 2 \text{ ft} = 94 \text{ lb/ft}$$

$$w_{TL}(\text{bottom chord}) = 10 \text{ psf} \times 2 \text{ ft} = 20 \text{ lb/ft}$$

10.2.3 Floor dead load

Floor covering/finishes (assume ⅞-in. hardwood)	4.0 psf
1-in. plywood sheathing (0.4 psf/⅛ in. × 8)	3.2 psf
Framing (assume 2 × 14 at 12 in. o.c.)	4.4 psf
⅝-in. gypsum ceiling (= 5 psf/in. × ⅝ in.)	3.2 psf
Mechanical and electrical	5.0 psf
Partitions (assumed)	~20.0 psf
Total floor dead load D_{floor}	~40 psf

Note that the minimum value allowed for partitions per ASCE 7 [2], Section 4.3.2 is 15 psf.

10.2.4 Floor live load

We neglect floor live load reduction for this design example. For light-framed wood structures, the effect of floor live load reduction is usually minimal. For office buildings, the floor live load $L = 50$ psf (load is higher in corridor areas). The live load reduction factor for the various structural elements of this building will be determined in the design sections for these structural elements.

Without live load reduction, the total floor load (in psf) obtained using the IBC load combinations (Section 2.1) is

$$w_{TL} = D + L = 40 + 50 = 90 \text{ psf}$$

10.2.5 Wall dead load

½ in. plywood (exterior face of wall)	= 1.6 psf
Stud wall framing (assume 2 × 6 studs at 2 ft o.c.)	= 0.9 psf
Wood shingles	= 3.0 psf
⅝ gypsum wall board (interior face of wall)	= 3.2 psf
1 in. fiberglass insulation	= 1.1 psf
Total wall dead load D_{roof}	≈ 10 psf of vertical surface area

10.3 SEISMIC LATERAL LOADS

The lateral seismic loads on the building would need to be calculated based on the procedures covered in Chapter 2. In order to determine the seismic loads, it is necessary to tabulate the dead loads because the seismic load is a function of the dead load

Given:

Floor dead load = 40 psf (partitions included)
Roof dead load ≈ 22.0 psf
Wall dead load = 10.0 psf (exterior walls)
Snow load $P_s = 35$ psf
$S_{DS} = 0.267$
$\rho = 1.0$

Calculate W for each level (see Table 10.1). W is the seismic weight tributary to each level; it includes the weight of the floor structure, the weight of the walls, and the weight of a portion of the snow load. Where the flat roof snow load P_f is greater than 30 psf, 20% of the flat roof snow load is included in the tributary seismic weight at the roof level (ASCE 7, Section 12.7.2).

There are a few methods available that can be used to determine the factored seismic base shear and force at each level, and each would require calculating the weight tributary to each level (see Chapter 2). For simplicity, the following unfactored seismic forces (0.7E) are assumed to be acting on this structure:

Force at the roof, $F_R = 2.38$ kips
Force at the roof, $F_2 = 3.46$ kips
Base shear, $V = 5.84$ kips

These values will be used later for the analysis and design of the diaphragm and shear walls. The seismic loads are summarized in Figure 10.2.

10.4 WIND LOADS

The wind loads on the building would be calculated using the methods given in Chapter 2. For simplicity, the overhangs are neglected for this example (see Figure 10.3 and Table 10.2).
Given:

$V = 90$ mph
$I_w = 1.0$
Mean roof height = 12 ft + 12 ft + (5 ft/2) = 26.5 ft
Exposure = C (assumed)

Table 10.1 Self-weight of each level

			Weight		
Level	*Area (ft²)*	*Tributary height (ft)*	*Floor*	*Exterior walls*	*W total*
Roof	30 ft × 48 ft **= 1440 ft²**	(12 ft/2) + (5 ft/2) **= 8.5 ft (short)** (12 ft/2) = **6 ft (long)**	(1440 ft²)[22 psf + (0.2 × 35 psf)] **= 41.8 kips**	[(8.5 ft)(10 psf)(2)(30 ft)] + [(6 ft)(10 psf)(2)(48 ft)] = **10.9 kips**	41.8 kips + 10.9 kips **= 52.7 kips**
Second floor	30 ft × 48 ft **= 1440 ft²**	(12 ft/2) + (12 ft/2) **= 12 ft**	(1440 ft²)(40 psf) **= 57.6 kips**	(12 ft)(10 psf)(2)(30 ft + 48 ft) = **18.8 kips**	57.6 kips + 18.8 kips **= 76.4 kips**
			Total weight = 52.7 kips + 76.4 kips = **129.1 kips**		

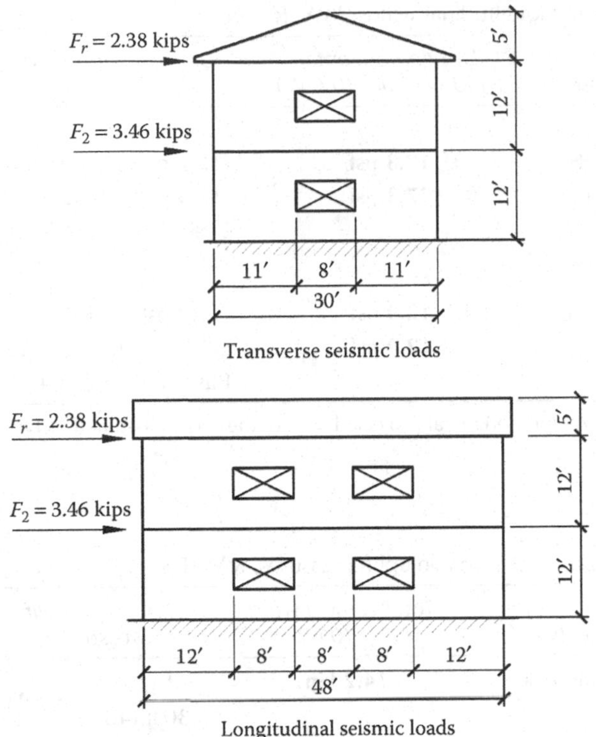

Figure 10.2 Summary of unfactored seismic loads.

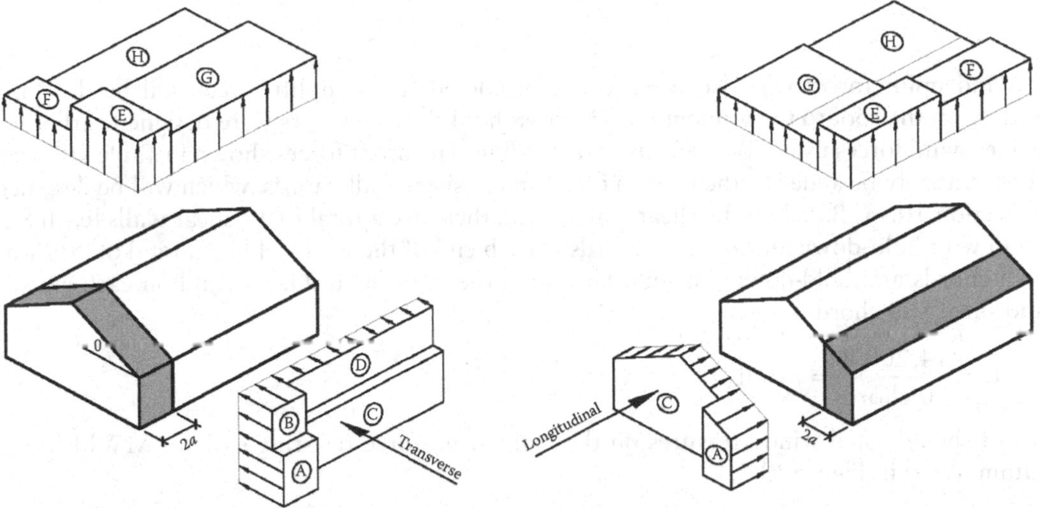

Figure 10.3 Wind loads on MWFRS.

$$\theta = \tan^{-1}\frac{4}{12} = 18.4$$

A wind analysis yields the pressures and forces using the average pressure as given in Table 10.2. All wind loads are assumed to be equal to 0.6 *W*, that is, the ASCE 7 factored wind load, *W*, have been multiplied by the ASD wind load factor of 0.6.

Table 10.2 Transverse and longitudinal winds, MWFRS

Level	*Elevation*	*Design horizontal wind pressure (0.6 W)*	*Total unfactored wind load on the building at each level*
Transverse wind			
Roof	24 ft	**17.3 psf**	(17.3 psf)(12 ft/2)(48 ft) = **5.0 kips**
Second floor	12 ft	**17.3 psf**	(17.3 psf)(12 ft/2 + 12 ft/2)(48 ft) = **10.0 kips**
			Base shear = 5.0 + 10.0 = **15.0 kips**
Longitudinal wind			
Roof	26.5 ft	**12.9 psf**	(12.9 psf)(12 ft/2 + 5 ft/2)(30 ft) = **3.3 kips**
Second floor	12 ft	**12.9 psf**	(12.9 psf)(12 ft/2 + 12 ft/2)(30 ft) = **4.7 kips**
			Base shear = 3.3 + 4.7 = **8.0 kips**

Note: The wind pressure on the horizontal projected area of the roof is assumed to be zero since this pressure is negative in ASCE 7.

Table 10.3 Average uplift pressures, MWFRS

Direction	*Total vertical wind force (0.6 W)*	*Average uplift pressure, q_u*
Transverse	**14.2 kips**	$\frac{14{,}200}{(30\text{ ft})(48\text{ ft})} = \mathbf{9.9\ psf}$
Longitudinal	**13.6 kips**	$\frac{13{,}600}{(30\text{ ft})(48\text{ ft})} = \mathbf{9.5\ psf}$

Wind uplift forces will also need to be considered. These uplift forces will need a load path from the roof to the foundation. The truss hold-down anchors were designed earlier for higher wind forces using components and cladding. The uplift forces shown in Table 10.3 can conservatively be added to the tension forces in the shear wall chords, which will be designed in Section 10.11. Based on the shear wall layout, there are a total of 10 shear walls (each 8 ft long) with hold-down anchors and chords at each end of the wall yielding a total of 20 shear wall chords and hold-down anchors. Therefore, the uplift load due to wind on each anchor and shearwall chord is

$$U = \frac{14{,}200\text{ lb}}{20\text{ chords}} = 710\text{ lb}$$

All of the adjusted wind pressures on the main wind force resisting system (MWFRS) are summarized in Figure 10.4.

10.5 COMPONENTS AND CLADDING WIND PRESSURES

The tabulated components and cladding (C&C) wind pressures on projected vertical and horizontal surface areas of the building are obtained from ASCE 7 and the procedures in Chapter 2. The following would need to be considered in the determination of the components and cladding wind loads:

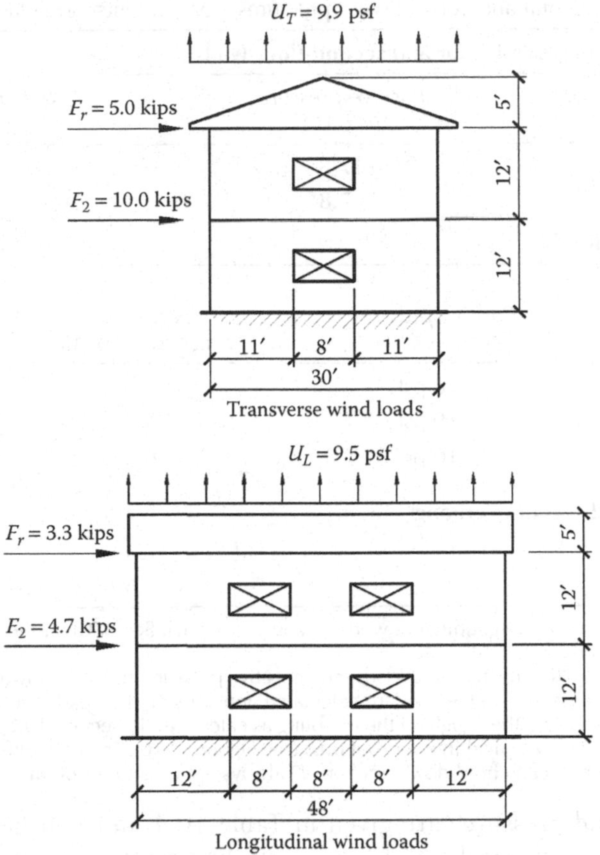

Figure 10.4 Unfactored wind loads on MWFRS.

Typical exterior wall stud:

Floor-to-floor height = 12 ft
Spacing of stud = 2 ft
Effective width per ASCE 7, Section 26-2 = ⅓ span = (⅓)(12 ft) = 4 ft
Effective wind area A_e = (4 ft)(12 ft) = 48 ft^2

Typical roof truss:

Main span = 30 ft
Spacing of trusses = 2 ft
Effective width of truss per ASCE 7, Section 26-2 = ⅓ span = (⅓)(30 ft) = 10 ft
Effective wind area A_e = (10 ft)(30 ft) = 300 ft^2
Roof overhang length = 2 ft
Effective width of roof overhang per ASCE 7, Section 26-2 = ⅓ span = (⅓)(2 ft) = 0.67 ft
Effective wind area A_e = (0.67 ft)(2 ft) = 1.34 ft^2

Table 10.4 Design horizontal and vertical wind pressures, components, and cladding (psf)

Horizontal pressures on ground-floor and second-floor walls[a]		
Zone	*Positive pressure (0.6 W)*	*Negative pressure or suction (0.6 W)*
Wall interior zone 4	17.8	**–19.6**
Wall end zone 5	17.8	**–22.6**

Vertical Pressures on Roof [b]			
	Positive pressure (0.6 W)	*Negative pressure (0.6 W)*	*Net uplift pressure (psf) (0.6D + 0.6W)*
Roof interior zone 1	10 psf	–16.6	–5.5 psf
Roof end zone 2	10 psf	**–23.3**	–12.2 psf
Roof corner zone 3	10 psf	–36.9	–25.8 psf
Vertical pressures on 2-ft roof overhang[b]			
Roof end zone 2^{ov}		–37.3	–26.2 psf
Roof corner zone 3^{ov}		**–62.6**	–51.5 psf

[a] Horizontal wind pressures for longitudinal as well as transverse winds. See ASCE 7, Table 30.7-2 for definition of wall and roof zones 1 through 5.

[b] Positive pressure indicates downward wind loads, and negative pressure indicates upward wind load that causes uplift. Net uplift wind pressures on the roof are calculated above using the dead load of 18.5 psf, which is the total roof dead load, excluding the weight of the reroofing as calculated in Section 10.2, the net uplift wind pressures, and the uplift load on a typical interior roof truss (see Figure 10.10). The net uplift load at the truss support is usually resisted with hurricane hold-down anchors. *D* always opposes *W* for uplift.

Assumed C&C wind pressures are given in Table 10.4 and will be used later for the design of the exterior wall stud and the design of the roof truss hold-down strap anchors (Figure 10.5).

10.6 ROOF FRAMING DESIGN

The roof framing for this building consists of trusses spaced at 2 ft o.c. The truss configuration is shown in Figure 10.6. We will assume that the ends of the truss members are connected by a toothed plate or nailed plywood gusset plate; therefore, there is no reduction in the gross area. The span of the roof truss center to center of the exterior wall supports is 30 ft less the stud wall thickness. If we assume 2 × 6 studs, the actual center-to-center span of the roof truss will be 29.54 ft (i.e., 30 ft–5.5 in./12). However, for simplicity, we will assume a span of 30 ft in the analysis and design of the roof truss, and this will not significantly affect or alter the truss design because of the very small difference (less than 2%) between the actual span and the 30-ft span used in our design.

10.6.1 Analysis of a roof truss

The total roof load of 57 psf was calculated in Section 10.2 without consideration of the downward wind load acting on the roof. We will now determine if this downward wind load in combination with other loads is critical for this building.

The calculated roof dead and snow loads from Section 10.2 and the calculated downward wind load from Section 10.5 are as follows:

$$D = 22 \text{ psf}$$

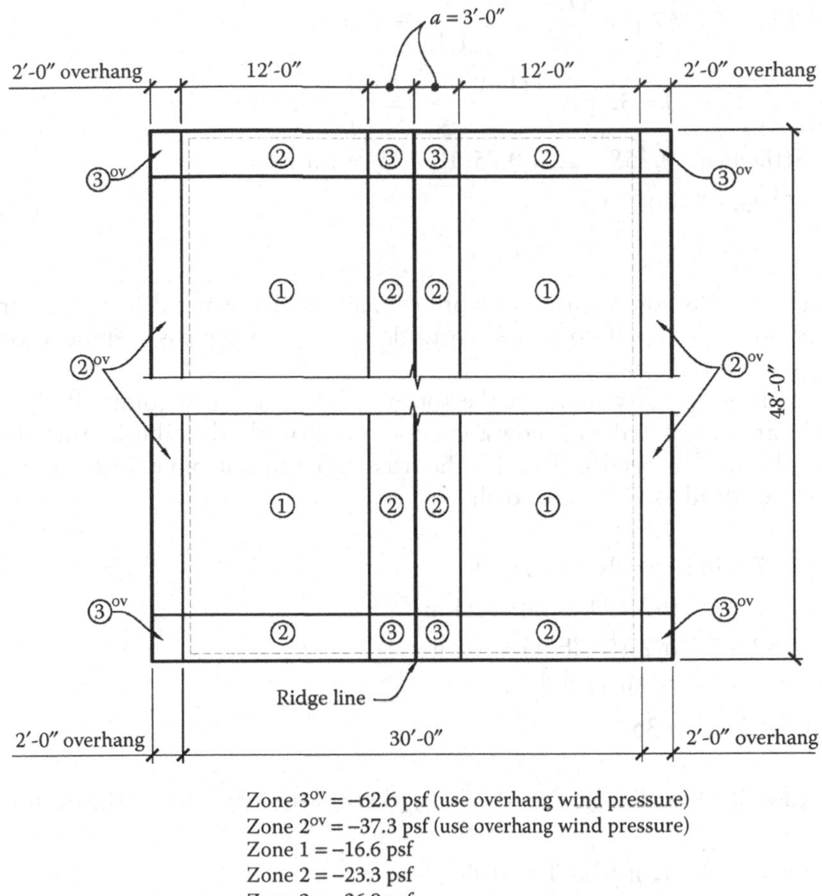

Figure 10.5 Uplift pressure on roof: components and cladding.

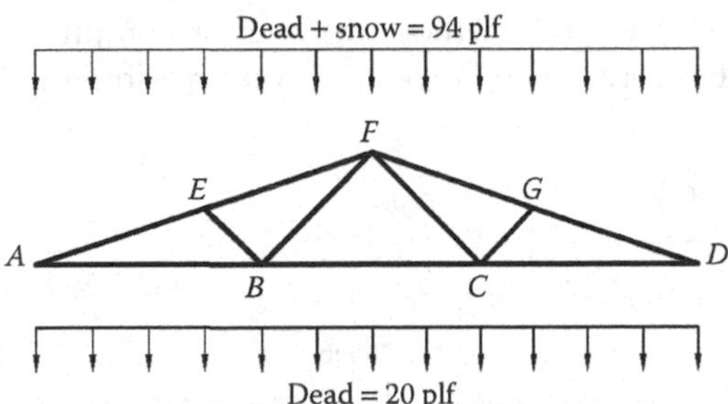

Figure 10.6 Roof truss profile and loading.

$$S = 35 \text{ psf}$$

$$W = 10 \text{ psf (downward)}$$

Using the normalized load method presented in Chapter 3 and using the applicable load combinations from Section 2.1 yield

$$D + S = 22 + 35 = 57 \text{ psf}; \ \frac{(D+S)}{C_D} = \frac{57}{1.15} = 49.6 \text{ psf (governs)}$$

$$D + 0.6W = 22 + 10 = 32 \text{ psf}, \ \frac{(D+0.6W)}{C_D} = \frac{32}{1.6} = 20 \text{ psf}$$

$$D + 0.75(0.6)W + 0.75S = 22 + 0.75(10 + 35) = 56 \text{ psf}$$

$$\frac{[D+0.75(0.6)W+0.75S]}{C_D} = \frac{56}{1.6} = 35 \text{ psf}$$

Therefore, the $D + S$ combination is more critical than the wind load combinations as originally assumed. We will now proceed to design the roof truss members for the total load of 57 psf with a C_D of 1.15.

The concentrated gravity loads at the top and bottom chord joints of the roof truss (Figure 10.7) are calculated as follows, using the uniformly distributed total loads of 94 and 20 lb/ft obtained in Section 10.2 for the truss top and bottom chords, respectively: Top chord—roof dead load + snow load:

$$P_A(\text{top}) = (7.5 \text{ ft/2})(94 \text{ lb/ft}) = 353 \text{ lb}$$
$$P_E = (7.5 \text{ ft/2} + 7.5 \text{ ft/2})(94 \text{ lb/ft}) = 705 \text{ lb}$$
$$P_F = (7.5 \text{ ft/2} + 7.5 \text{ ft/2})(94 \text{ lb/ft}) = 705 \text{ lb}$$
$$P_G = (7.5 \text{ ft/2} + 7.5 \text{ ft/2})(94 \text{ lb/ft}) = 705 \text{ lb}$$
$$P_D(\text{top}) = P_A(\text{top}) = 353 \text{ lb}$$

Bottom chord—dead load only, due to ceiling (there is no live load on the bottom chord of this truss):

$$P_A(\text{bottom}) = (10 \text{ ft/2})(20 \text{ lb/ft}) = 100 \text{ lb}$$
$$P_C = (10 \text{ ft/2} + 10 \text{ ft/2})(20 \text{ lb/ft}) = 200 \text{ lb}$$
$$P_C = (10 \text{ ft/2} + 10 \text{ ft/2})(20 \text{ lb/ft}) = 200 \text{ lb}$$
$$P_D(\text{bottom}) = (10 \text{ ft/2})(20 \text{ lb/ft}) = 100 \text{ lb}$$

The truss is then analyzed using the method of joints *or* computer structural analysis software to obtain the member forces as follows (C = compression and T = tension): Top chord members:

$$AE = GD = 3980 \text{ lb } (C)$$
$$EF = FG = 3420 \text{ lb } (C)$$

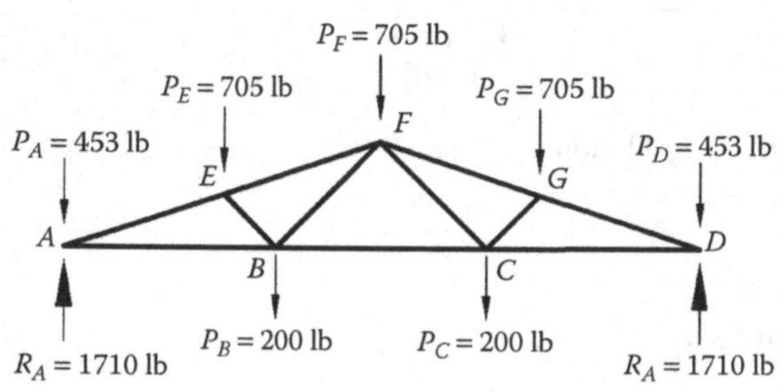

Figure 10.7 Free-body diagram of roof truss.

Bottom chord members:

$$AB = CD = 3770 \text{ lb } (T)$$
$$BC = 2520 \text{ lb } (T)$$

Web members:

$$EB = GC = 750 \text{ lb } (C)$$
$$FC = BF = 1030 \text{ lb } (T)$$

In addition to the axial forces on the top chord and bottom chord members, there are bending loads of 94 and 20 lb/ft on the top and bottom chords, respectively. These loads will be considered in the design of the top and bottom chord members of the roof truss.

10.6.2 Design of truss web tension members

1. The maximum tension force in the web tension members from the structural analysis discussed previously is 1030 lb.
2. Assume member size: Try a 2 × 4, which is the dimension lumber. Use NDS-S Table 4A. From NDS-S Table 1B, the gross area A for 2 × 4 = 5.25 in.^{2}. The wood species (Spruce-Pine-Fir) to be used is specified in the design brief; we assume a no. 3 stress grade.
3. From NDS-S Table 4A (SPF No. 3), we obtain the tabulated design tension stress and the adjustment factors as follows: Design tension stress F_t = 250 psi. Adjustment or C factors are

   $C_D = 1.15$ (the C_D value for the shortest duration load in the load combinationis used, i.e., snow load; see Section 3.4.2.1)

   $C_t = 1.0$ (normal temperature conditions)

   $C_M = 1.0$ (dry service, because the truss members are protected from weather)

   $C_F(F_t) = 1.5$

   Using the NDS applicability table (Table 3.1), the allowable tension stress is given as

   $$F_t' = F_t C_D C_M C_t C_F C_i$$
   $$= (250)(1.15)(1.0)(1.0)(1.5)(1.0)$$
   $$= \mathbf{431.3\ psi}$$

   Assuming toothed plate or nailed plywood gusset plate connections, the net area will be equal to the gross area because there are no bolt holes that will lead to a reduction in the gross area.

   $$A_n = A_g = 5.25 \text{ in.}^2$$

4. Applied tension stress

   $$f_r = \frac{T}{A_n} = \frac{1030 \text{ lb}}{5.25 \text{ in.}^2} = 197 \text{ psi} < F_t' = \mathbf{431.3\ psi} \quad \textbf{OK}$$

   A 2 × 4 SPF No. 3 is adequate for the truss web tension member.

10.6.3 Design of truss web compression members

Because a 2 × 4 member was selected for the truss web tension member, we assume a 2 × 4 trial size for the web compression member. As determined for the web tension member, the 2 × 4 is the dimension lumber, for which NDS-S Table 4A is applicable. From the table, the stress adjustment or C factors are

$$C_M = 1.0 \text{ (normal moisture conditions)}$$

$$C_t = 1.0 \text{ (normal temperature conditions)}$$

$C_D = 1.15$ (the C_D value for the shortest duration load in the load combination is used, i.e., snow load; see Section 3.4.2.1)

$$C_F(F_c) = 1.15$$

$$C_i = 1.0 \text{ (no incisions are specified in the wood for pressure treatment)}$$

From NDS-S Table 4C, the tabulated design stresses are obtained as follows:

$$F_c = 650 \text{ psi (compression stress parallel to the grain)}$$

$$E = 1.2 \times 10^6 \text{ psi (reference modulus of elasticity)}$$

$$E_{min} = 0.44 \times 10^6 \text{ psi (buckling modulus of elasticity)}$$

The member dimensions are $d_x = 3.5$ in. and $d_y = 1.5$ in.
The member length $= \sqrt{(2.5 \text{ ft})^2 + (2.5 \text{ ft})^2} = 3.54$ ft.
The cross-sectional area is $A_g = 5.25$ in.^{2}

The effective length factor K_e is assumed to be 1.0 (see Section 5.4); therefore, the slenderness ratios about the x–x and y–y axes are given as

$$\frac{l_e}{d_x} = \frac{(1.0)(3.54 \text{ ft} \times 12)}{3.5 \text{ in.}} = 12.1 < 50 \quad \textbf{OK}$$

$$\frac{l_e}{d_y} = \frac{(1.0)(3.54 \text{ ft} \times 12)}{1.5 \text{ in.}} = 28.3 \text{ (governs)} < 50 \quad \textbf{OK}$$

Using the applicability table (Table 3.1), the allowable modulus of elasticity for buckling calculations is given as

$$E'_{min} = E_{min} C_M C_t C_i$$

$$= (0.44 \times 10^6)(1.0)(1.0)(1.0)$$

$$= 0.44 \times 10^6 \text{ psi}$$

$$c = 0.8 \text{ (sawn lumber)}$$

The Euler critical buckling stress is calculated as

$$F_{cE} = \frac{0.822E'_{min}}{(l_e/d)^2} = \frac{(0.822)(0.44\times10^6)}{(28.3)^2} = 452 \text{ psi}$$

$$\begin{aligned}F_c^* &= F_c C_D C_M C_t C_F C_i \\ &= (650)(1.15)(1.0)(1.0)(1.15)(1.0) \\ &= 860 \text{ psi}\end{aligned}$$

$$\frac{F_{cE}}{F_c^*} = \frac{452 \text{ psi}}{860 \text{ psi}} = 0.526$$

From Equation 5.29, the column stability factor is calculated as

$$C_P = \frac{1+0.526}{(2)(0.8)} - \sqrt{\left[\frac{1+0.526}{(2)(0.8)}\right]^2 - \frac{0.526}{0.8}} = 0.452$$

The allowable compression stress is

$$F'_c = F_c^* C_P = (860)(0.452) = \mathbf{389 \text{ psi}}$$

The applied compression stress is

$$f_t = \frac{F}{A_g} = \frac{750 \text{ lb}}{5.25 \text{ in.}^2} = 143 \text{ psi} < F'_c = \mathbf{389 \text{ psi}}$$

A 2 × 4 SPF No. 3 is adequate for the truss web compression member.

10.6.4 Design of truss bottom chord members

The bottom chord members are subjected to combined axial tension and bending caused by the uniformly distributed load of 20 lb/ft from the insulation, ceiling, and mechanical and electrical equipment loads. The most critical bottom chord member is member *AB* or *CD*, and the forces in this member include an axial tension force of 3770 lb (caused by dead load plus snow load) combined with a uniformly distributed dead load of 20 lb/ft, as shown in Figure 10.8.

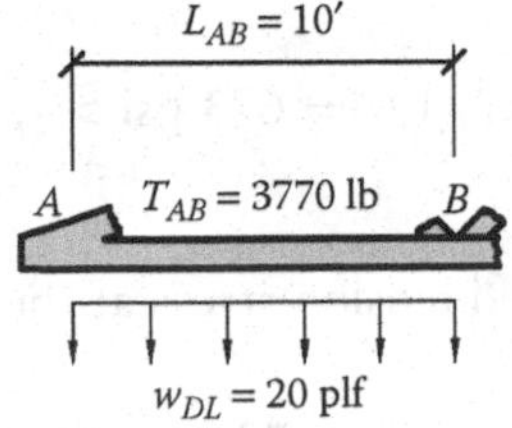

Figure 10.8 Free-body diagram of member AB.

10.6.4.1 Selection of a trial member size

Assume a 2 × 6 SPF No. 2 sawn lumber. The gross cross-sectional area and the section modulus can be obtained from NDS-S Table 1B as follows:

$$A_g = 8.25 \text{ in.}^2$$

$$S_{xx} = 7.563 \text{ in.}^3$$

The length of this member = 10 ft. Because the trial member is the dimension lumber, NDS-S Table 4A is applicable. From the table, we obtain the tabulated design stresses and stress adjustment factors.

$$F_b = 875 \text{ psi (tabulated bending stress)}$$

$$F_t = 450 \text{ psi (tabulated tension stress)}$$

$$C_F(F_b) = 1.3 \text{ (size adjustment factor for bending stress)}$$

$$C_F(F_t) = 1.3 \text{ (size adjustment factor for tension stress)}$$

$$C_D = 1.15 \text{ (for design check with snow load)}$$

$$C_D = 0.9 \text{ (for design check with dead load only)}$$

$$C_L = 1.0 \text{ (assuming lateral buckling is prevented by the ceiling and bridging)}$$

$$C_r = 1.15 \text{ (all three repetitive member requirements are met; see Section 3.4.2.4)}$$

$$C_i = 1.0 \text{ (there are no incisions specified in the wood for pressure treatment)}$$

Design check 1:

The applied tension force $T = 3770$ lb (caused by dead load + snow load).

The net area $A_n = A_g$ (toothed plated or nailed plywood plate connection assumed) = 8.25 in.^{2}

From Equation 5.1, the applied tension stress at the supports,

$$f_t = \frac{T}{A_n} = \frac{3770 \text{ lb}}{8.25 \text{ in.}^2} = 457 \text{ psi}$$

Using the NDS applicability table (Table 3.1), the allowable tension stress is given as

$$F'_t = F_t C_D C_M C_t C_F C_i$$

$$= (450)(1.15)(1.0)(1.0)(1.3)(1.0) = \mathbf{673 \text{ psi}} > f_t \quad \text{OK}$$

Design check 2:

From Equation 5.5, the applied tension stress at the midspan is

$$f_t = \frac{T}{A_g} = \frac{3770 \text{ lb}}{8.25 \text{ in.}^2} = 457 \text{ psi} < F'_t = \mathbf{673 \text{ psi}} \quad \text{OK}$$

Design check 3:

The maximum moment which for this member occurs at its midlength is given as

$$M_{max} = \frac{(20 \text{ lb/ft})(10 \text{ ft})^2}{8} = 250 \text{ ft-lb} = 3000 \text{ in.-lb}$$

This moment is caused by the ceiling *dead* load only (because no ceiling live load is specified); therefore, the controlling load duration factor C_D for this case is 0.9. Using the NDS applicability table (Table 3.1), the allowable bending stress is given as

$$F_b' = F_b C_D C_M C_t C_L C_F C_i C_r C_{fu}$$

$$= (875)(0.9)(1.0)(1.0)(1.0)(1.3)(1.0)(1.15)(1.0)$$

$$= \mathbf{1177 \text{ psi}}$$

From Equation 5.7, the applied bending stress (i.e., tension stress due to bending) is

$$f_{bt} = \frac{M_{max}}{S_x} = \frac{3000 \text{ in.-lb}}{7.563 \text{ in.}^3} = 397 \text{ psi} < F_b'$$

Design check 4:

For this case, the applicable load is dead load and snow load; therefore, the controlling load duration factor C_D is 1.15 (see Section 3.4.2.1). Using the NDS applicability table (Table 3.1), the allowable axial tension stress and the allowable bending stress are calculated as follows:

$$F_t' = F_t C_D C_M C_t C_F C_i$$

$$= (450)(1.15)(1.0)(1.0)(1.3)(1.0)$$

$$= \mathbf{673 \text{ psi}}$$

$$F^*_b = F_b C_D C_M C_t C_F C_i C_r C_{fu}$$

$$= (875)(1.15)(1.0)(1.0)(1.3)(1.0)(1.15)(1.0)$$

$$= 1504 \text{ psi}$$

The applied axial tension stress f_t at the midspan is 457 psi, as calculated for design check 2, whereas the applied tension stress due to bending f_{bt} is 397 psi, as calculated in design check 3. From Equation 5.11, the combined axial tension and bending interaction equation for the stresses in the tension fiber of the member is given as

$$\frac{f_t}{F_t'} + \frac{f_{bt}}{F_b^*} \leq 1.0$$

Substitution into the interaction equation yields

$$\frac{457 \text{ psi}}{673 \text{ psi}} + \frac{397 \text{ psi}}{1504 \text{ psi}} = \mathbf{0.94 < 1.0 \quad OK}$$

Design check 5:

The applied compression stress due to bending is

$$f_{bc} = \frac{M_{max}}{S_x} = \frac{3000 \text{ in.-lb}}{7.563 \text{ in.}^3} = 397 \text{ psi} = (f_{bt} \text{ because this is a rectangular cross section})$$

It should be noted that f_{bc} and f_{bt} are equal for this problem because the wood member is rectangular in cross section. The load duration factor for this case, $C_D = 1.15$ (combined dead + snow loads). Using the NDS applicability table (Table 3.1), the allowable compression stress due to bending is

$$F'_b = F_b C_D C_M C_t C_L C_F C_i C_r$$

$$= (875)(1.15)(1.0)(1.0)(1.0)(1.3)(1.0)(1.15)$$

$$= 1504 \text{ psi}$$

From Equation 5.15, the interaction equation for bending minus tension is given as

$$\frac{f_{bc} - f_t}{F'_b} \leq 1.0$$

Substituting in the interaction equation yields

$$\frac{397 \text{ psi} - 457 \text{ psi}}{1504 \text{ psi}} = -0.04 < 1.0 \quad \text{OK}$$

From all of the steps above, we find that all the five design checks are satisfied; therefore, use a 2 × 6 SPF No. 2 for the bottom chord.

10.6.5 Design of truss top chord members

The top chord members are subjected to combined axial compression and bending caused by the uniformly distributed dead load and snow load of 94 lb/ft. The most critical top chord member is member *AE* or *GD*, and the forces in this member include an axial compression force of 3980 lb combined with a uniformly load of 94 lb/ft, as shown in Figure 10.9.

10.6.5.1 Selection of a trial member size

Assume a 2 × 6 SPF sawn lumber. The design aids in Appendix B are utilized to design this member. The length of this member $= \sqrt{(7.5 \text{ ft})^2 + (2.5 \text{ ft})^2} = 7.91$ ft. It should be noted that for each design check, the controlling load duration factor value will correspond to the C_D value of the

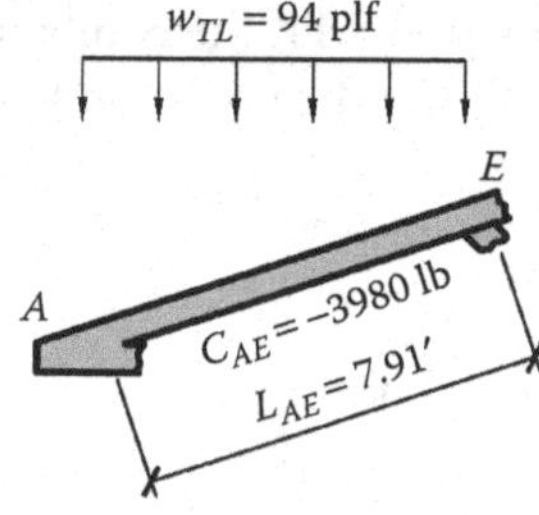

Figure 10.9 Load on truss top chord.

shortest duration load in the load combination for that design case. Thus, the C_D value for the various design cases may vary. The reader will find that design check 4 will govern, and therefore the design aids in Appendix B can conservatively be used to select an appropriate member.

Design check 1: Compression on net area

This condition occurs at the *ends* (i.e., *supports*) of the member. For this load case, the axial load P is caused by dead load and snow load ($D + S$); therefore, the load duration factor C_D from Chapter 3 is 1.15. The worst-case axial load for this case is $P = 3980$ lb.

Design check 2: Compression on gross area

This condition occurs at the *midspan* of the member. The axial load P for this load case is also caused by dead load and snow load ($D + S$); therefore, the load duration factor C_D from Chapter 3 is 1.15. The worst-case axial load for this case is $P = 3980$ lb.

Design check 3: Bending only

This condition occurs at the point of maximum moment (i.e., at the *midspan* of the member), and the moment is caused by the uniformly distributed gravity load on the top chord (member AE), and because this load is dead load and snow load, the controlling load duration factor (see Chapter 3) is 1.15.

The horizontal span of the sloped member, $AE = 7.5$ ft

The maximum moment in member AD is

$$M_{max} = \frac{w_{D+S}L^2}{8}$$

$$= \frac{(94 \text{ lb/ft})(7.5 \text{ ft})^2}{8} = 661 \text{ ft-lb}$$

Design check 4: Bending and axial compression force

This condition occurs at the *midspan* of the member. For this load case, the loads causing the combined stresses are dead load plus snow load ($D + S$); therefore, the controlling load duration factor C_D is 1.15. It should be noted that because the load duration factor C_D for this combined load case is the same as those for load cases 2 and 3, the parameters to be used in the combined stress interaction equation can be obtained from design checks 2 and 3. The reader is cautioned to be aware that the C_D value for all four design checks are not necessarily always equal and has to be determined for each case.

The loads for this case are

$$P = 3980 \text{ lb}$$

$$M = 661 \text{ ft-lb}$$

By inspection, the loads from this case will control. From Figure B.43, the axial load capacity at a moment of 661 ft-lb is ~4000 lb. Note that the load duration factor used in Figure B.43 is $C_D = 1.0$, so the use of this design aid is somewhat conservative, but appropriate. Therefore, use a 2 × 6 SPF Select Structural for the top chord.

The truss member sizes and stress grades are summarized in Table 10.5.

Gable and hip roofs must not only be designed for the balanced snow loads discussed in Chapter 2 and used to design the roof truss in this section, but the unbalanced snow loads

Table 10.5 Summary of truss member sizes and stress grades

Truss member	*Member size and stress grade*
Top chord	2 × 6 SPF Select Structural
Bottom chord	2 × 6 SPF No. 2[a]
Web members	2 × 4 SPF No. 3

[a] For simplicity and to avoid construction errors, it may be prudent to use the same size and stress grade for the top and bottom chords. In that case, 2 × 6 SPF Select Structural would be used for the bottom chord as well as the top chord.

prescribed in the ASCE 7 load standard must also be considered in design, because oftentimes, these unbalanced loads might lead to higher stresses in the truss members compared to the stresses due to the balanced snow load. In our building design case study, the applied stresses due to unbalanced snow load are smaller than those due to the balanced snow load except in the truss web members where there was approximately a 20% increase in stress due to unbalanced snow loads. However, because the applied stress in the actual web members selected was only about 50% of the allowable stress, the roof truss members in this building design case study would still be adequate to support the unbalanced snow loads. The reader should refer to Sections 7.6 through 7.9 of ASCE 7 for a full discussion of unbalanced snow loads, snow drift loads, and sliding snow loads.

10.6.6 Net uplift load on a roof truss

The net uplift load on a typical interior zone roof truss is calculated using the design uplift wind pressures calculated in Table 10.4. This uplift load will be used to determine the size of the roof truss hold-down straps or hurricane clips. Using Table 10.4, Figures 10.5 and 10.10a, the total net uplift load at the typical interior roof truss supports is 310 lb. The roof truss hold-down strap will be designed to resist this net uplift force of 310 lb. A hurricane tie-down anchor HT-1 is adequate to resist this force and has an allowable uplift resistance of 390 lb with a load duration factor of 1.6 as indicated in the generic connector selection table shown in Table 10.10b.

10.7 SECOND-FLOOR FRAMING DESIGN

In this section the typical sawn lumber floor joist and typical glulam girder are designed (Figure 10.11). It is assumed that the floor joist will be simply supported and supported off the face of the glulam girder with joist hangers. To illustrate the use of the design aids presented in Appendix B, the floor joist will be designed using the design aids.

10.7.1 Design of a typical floor joist

Joist span = 15 ft (the actual center-to-center span of the joist will be slightly less than this). From Section 10.2, the dead load was calculated as 40 psf and the floor live load as 50 psf. The total floor load (in psf) using the IBC load combinations (see Section 2.1) is

$$w_{TL} = D + L = 40 + 50 = 90 \text{ psf}$$

TA of typical floor joist = (2 ft)(15 ft) = 30 ft^2 < 150 ft^2

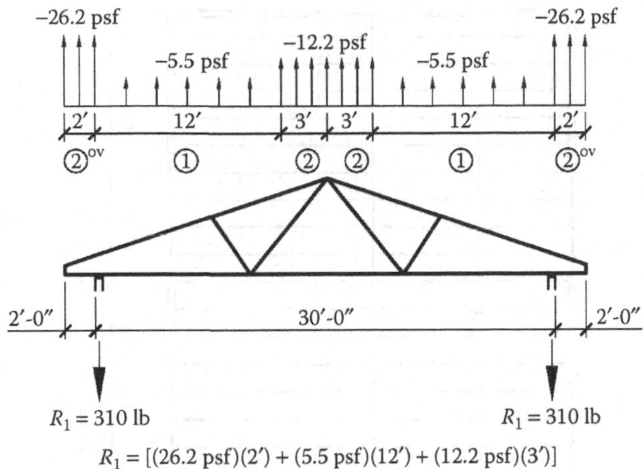

(a)

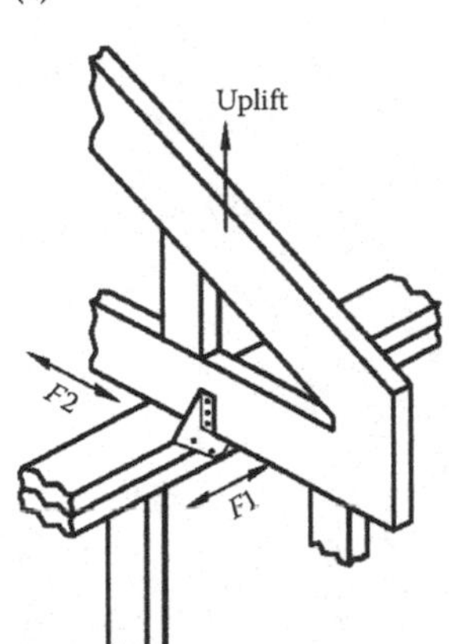

Model #	Fasteners		DF-L or S.P. Allowable loads (lb)			S.P.F. or H.F. Allowable loads (lb)		
	To truss	To plates	Uplift (160%)	Lateral (160%)		Uplift (160%)	Lateral (160%)	
				F1	F2		F1	F2
HT1	(6)-8d	(4)-8d	585	450	160	390	405	120
xx	xx	xx	xx	xx	xx	xx	xx	xx

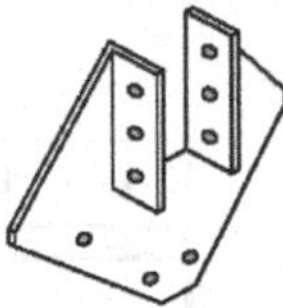

(b)

Figure 10.10 (a) Net uplift pressures on a typical interior roof truss. (b) Roof truss uplift connector.

From Section 2.4.3.1,

$$A_T = \text{TA} = 30 \text{ ft}^2 \text{ and } K_{LL} = 2(\text{interior beam})$$

$$K_{LL}A_T = (2)(30 \text{ ft}^2) = 60 \text{ ft}^2 < 400 \text{ ft}^2$$

Therefore, live load reduction is *not* permitted.

The tributary width of a typical floor joist = 2 ft. The total uniform load on the joist that will be used to design the joist for bending, shear, and bearing is

$$w_{TL} = (D + L)(\text{TW}) = (40 \text{ psf} + 50 \text{ psf})(2 \text{ ft}) = 180 \text{ lb/ft}$$

Using the free-body diagram of the typical floor joist (Figure 10.12), the load effects are calculated as follows:

$$\text{Maximum shear } V_{\max} = \frac{w_{TL}L}{2} = \frac{(180 \text{ lb/ft})(15 \text{ ft})}{2} = 1350 \text{ lb}$$

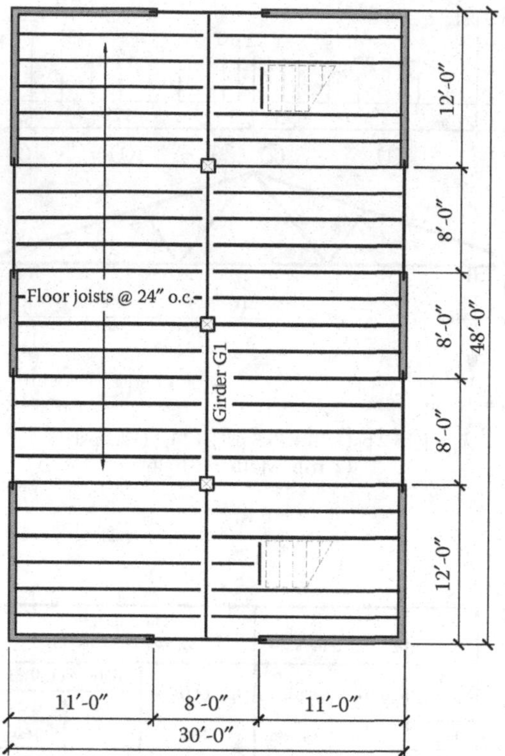

Figure 10.11 Second-floor framing plan.

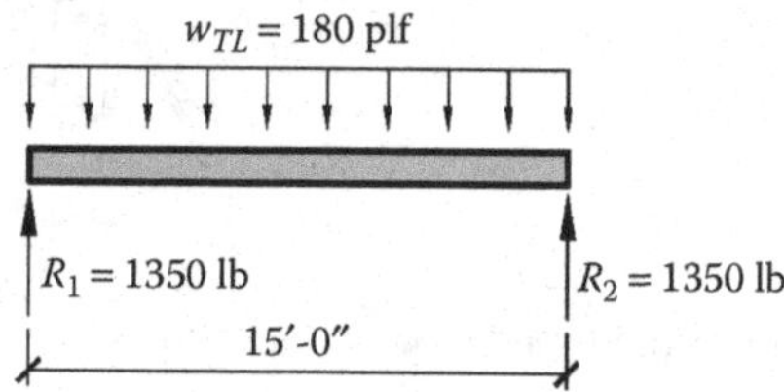

Figure 10.12 Free-body diagram of a typical floor joist.

$$\text{Maximum reaction } R_{\max} = 1350 \text{ lb}$$

$$\text{Maximum moment } M_{\max} = \frac{w_{TL}L^2}{8} = \frac{(180 \text{ lb/ft})(15 \text{ ft})^2}{8} = 5063 \text{ ft-lb} = 60{,}756 \text{ in.-lb}$$

Using Figure B.8, the allowable uniform load for a 2 × 14 SPF No. 1/No. 2 is approximately 120 plf. Therefore, use two 2 × 14′s at 24 in. o.c. because the total load-carrying capacity is (2)(120 plf) = 240 plf, which is greater than the applied load of 180 plf. Alternatively, Table B.11 could be used. Assuming a deflection limit of *L*/360, the maximum span is 16′-8″ for 2 × 14 joists spaced at 12″ o.c. Note that this table is for a total load of 95 psf, with 80 psf live load, which is then conservative to use.

The required adjustment factors for the floor joists are as follows:

Load duration C_D = 1.0 (dead load plus live load).
Moisture factor C_M =1.0 (dry service conditions assumed).

Temperature factor $C_t = 1.0$ (normal temperature conditions apply).
Bearing factor $C_b = 1.0$.

10.7.1.1 Check of bearing stress (compression stress perpendicular to the grain)

Maximum reaction at the support $R_1 = 1350$ lb.
Thickness of two 2 × 14 sawn lumber joists $b = (2) \times (1.5 \text{ in}) = 3$ in.
The allowable bearing stress or compression stress parallel to grain is

$$F'_{c\perp} = F_{c\perp} C_M C_t C_b = (425)(1)(1)(1) = 425 \text{ psi}$$

From Equation 4.9, the minimum required bearing length l_b is

$$l_{b,\text{ req'd}} \geq \frac{R_1}{bF'_{c\perp}} = \frac{1350 \text{ lb}}{(3.0 \text{ in.})(425 \text{ psi})} = 1.1 \text{ in.}$$

The floor joists will be connected to the floor girder using face-mounted joist hangers with the top of the joists at the same level as the top of the girders (see Figure 4.6b). The reader should select from one of the proprietary wood connector catalogs a joist hanger that meets the following requirements:

- The bearing length provided by the joist hanger should be at least $1\frac{1}{8}$ in.
- The joist hanger should have enough width to accommodate the two 2 × 14 floor joists.
- The joist hanger must have a capacity of at least 1350 lb using a load duration factor of 100% (i.e., $C_D = 1$ for dead load + floor live load).

Use two 2 × 14's at 24 in. o.c. SPF No. 1/No. 2 for the floor joists.

10.7.2 Design of a glulam floor girder

This glulam girder could be designed as continuous over several supports, which will result in a smaller member compared to simply supported girders. However, hauling a 48-ft-long girder to the building site and maneuvering a beam that long on-site may present problems for the contractor. Alternatively, the glulam girder could be delivered to the site in two 24-ft-long pieces, thus indicating two two-span continuous girders. However, to give the contractor wide flexibility in choosing the best way to install this girder, we design the girders as four simply supported members with 12-ft spans. This assumption will result in a larger member than if a single 48-ft-long girder continuous over several supports was used. It is assumed that the floor joists are supported on the face of the girder with face-mounted joist hangers; thus, the top of the glulam girder will be at the same elevation as the top of the floor joists. Assume that the girder self-weight = 15 lb/ft (this will be checked later). As calculated in Section 10.2, the dead and live loads are as follows:

Dead load $D = 40$ psf
Floor live load $L = 50$ psf

$$\text{TW of girder} = \frac{15 \text{ ft}}{2} + \frac{15 \text{ ft}}{2} = 15 \text{ ft}$$

$$\text{TA of girder G1} = (15 \text{ ft})(12 \text{ ft}) = 180 \text{ ft}^2$$

From Section 2.4.3.1,

$$A_T = \text{TA} = 180 \text{ ft}^2 \quad \text{and} \quad K_{LL} = 2 \text{ (interior girder)}$$

$$K_{LL}A_T = (2)(180 \text{ ft}^2) = 360 \text{ ft}^2 < 400 \text{ ft}^2$$

Therefore, live load reduction is *not* permitted.

The total uniform load on the girder that will be used to design for bending, shear, and bearing is obtained below using the dead load, the live load, and the assumed self-weight of the girder, which will be checked later:

$$w_{TL} = (D+L)(\text{TW}) + \text{girder self-weight}$$

$$= (40 \text{ psf} + 50 \text{ psf})(15 \text{ ft}) + 15 \text{ lb/ft} = \mathbf{1365 \text{ lb/ft}}$$

Using the free-body diagram of the girder (Figure 10.13), the load effects are calculated as follows:

$$\text{Maximum shear } V_{\max} = \frac{w_{TL}L}{2} = \frac{(1365 \text{ lb/ft})(12 \text{ ft})}{2} = 8190 \text{ lb}$$

$$\text{Maximum reaction } R_{\max} = 8190 \text{ lb}$$

$$\text{Maximum moment } M_{\max} = \frac{w_{TL}L^2}{8} = \frac{(1365 \text{ lb/ft})(12 \text{ ft})^2}{8} = 24{,}570 \text{ ft-lb} = 294{,}840 \text{ in.-lb}$$

The following loads will be used for calculating the joist deflections:

Uniform dead load w_{DL} = (40 psf)(15 ft) + 15 lb/ft = 615 lb/ft = 51.3 lb/in.
Uniform live load w_{LL} = (50 psf)(15 ft) = 750 lb/ft = 62.5 lb/in.

Note that if a four-span continuous girder had been assumed for this design, the maximum moment would have been 21,052 ft-lb [i.e., (0.1071)(1365 lb/ft)(12 ft)(12 ft)], a 17% reduction. The maximum shear would have been 9943 lb [i.e., (0.607)(1365 lb/ft)(12 ft)]. With a two-span continuous girder, the maximum moment would have been 24,570 ft-lb [i.e., (0.125)(1365 lb/ft)(12 ft)(12 ft)], which is the same maximum moment calculated for a simply supported girder. The maximum shear would have been 10,238 lb [i.e., (0.625)(1365 lb/ft)(12 ft)].

We will now proceed with the design following the steps described in Section 4.1.

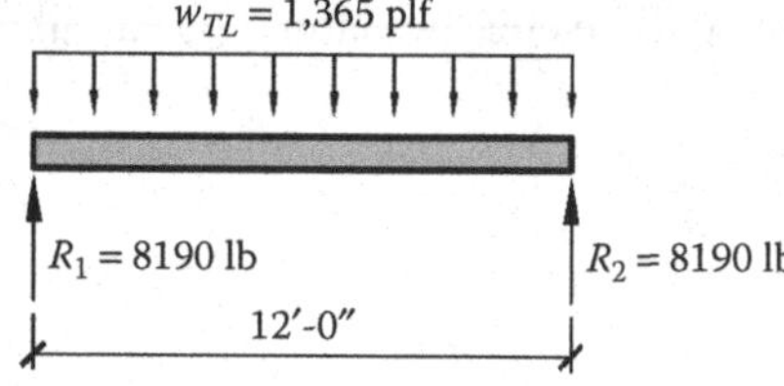

Figure 10.13 Free-body diagram of girder GI.

10.7.2.1 Check of bending stress (girders)

1. Summary of load effects (the self-weight of the girder was assumed, but this will be checked later in the design process):

   Maximum shear $V_{max} = 8190$ lb
   Maximum reaction $R_{max} = 8190$ lb
   Maximum moment $M_{max} = 294{,}840$ in.-lb
   Uniform dead load $w_{DL} = 51.3$ lb/in.
   Uniform live load $w_{LL} = 62.5$ lb/in.

2. For a glulam used primarily in bending, use NDS-S Table 5A-Expanded.
3. Using NDS-S Table 5A-Expanded, assume a 24F–E11 HF/HF glulam, therefore, $F_{bx}^{+} =$ 2400 psi. Assume initially that $F_{bx}^{\prime +} = F_{bx}^{+} = 2400$ psi. From Equation 4.1 the required approximate section modulus of the member is given as

$$S_{xx,\,\text{req'd}} \geq \frac{M_{max}}{F_{b,\,\text{NDS-S}}} = \frac{294{,}840\ \text{in.-lb}}{2400\ \text{psi}} = 122.9\ \text{in.}^3$$

4. From NDS-S Table 1C (for Western species glulam), the possible trial sizes that satisfy the section modulus requirement of step 3 are

SIZE (in.)	b (in.)	d (in.)	S_{xx} (in³)	A (in²)	I_x (in⁴)
2½ × 18	2.5	18	135	45	1215
3⅛ × 16½	3.125	16.5	141.8	51.56	1170
3½ × 15	3.5	15	131.3	52.5	984.4

   Although the 2½ × 18 in. member has the least area, it is prudent in this case to limit the depth of the girder because we will be using the bottom edge of the girder to provide lateral support to the columns because these girders are supported on top of the columns. To ensure this lateral support, the difference between the joist depth and the girder depth should be kept to a minimum.
   Try 3⅛ × 16½ in.

   $b = 3.125$ in. and $d = 16.5$ in.
   S_{xx} provided $= 141.8\ \text{in}^3 > 122.9\ \text{in}^3$ OK
   A provided $= 51.56\ \text{in}^2$
   I_{xx} provided $= 1170\ \text{in}^4$

5. The NDS-S Table 5A-Expanded tabulated stresses are

$$F_{bx}^{+} = 2400\ \text{psi}$$

$$F_{vxx} = 215\text{psi}$$

$$F_{c\perp xx,\text{tension lam}} = 500\ \text{psi}$$

$$E_x = 1.8 \times 10^6\ \text{psi}$$

$$E = 1.5 \times 10^6\ \text{psi}$$

$$E_{y,\min} = 0.78 \times 10^6\ \text{psi}$$

Table 10.6 Stress adjustment or *C* factors for glulam floor girders

Adjustment factor	*Symbol*	*Value*	*Rationale for the value chosen*
Beam stability factor	C_L	0.986	See calculation
Volume factor	C_V	1.0	See calculation
Curvature factor	C_c	1.0	Glulam girder is straight
Flat use factor	C_{fu}	1.0	Glulam is bending about its strong *x–x* axis
Moisture or wet service factor	C_M	1.0	The equilibrium moisture content is <16%
Load duration factor	C_D	1.0	The largest C_D value in the load combination of dead load plus snow load (i.e., *D* + *L*)
Temperature factor	C_t	1.0	Insulated building; therefore, normal temperature conditions apply
Bearing stress factor	C_b	1.0	$C_b = \dfrac{l_b + 0.375}{l_b}$ for $l_b \le 6$ in. = 1.0 for $l_b > 6$ in. = 1.0 for bearings at the ends of a member (see Section 3.4.3.3)

($E_{y,\min}$, not $E_{x,\min}$, is required for lateral buckling of the girder about the weak axis.)
The adjustment or *C* factors are given in Table 10.6.

From the adjustment factor applicability table for glulams (Table 3.2), we obtain the allowable bending stress of the glulam girder with C_V and C_L equal to 1.0 as

$$F_{bx}^{*+} = F_{bx}^{+} C_D C_M C_t C_{fu} C_c = (2400)(1.0)(1.0)(1.0)(1.0)(1.0) = \mathbf{2400\ psi}$$

The allowable pure bending modulus of elasticity E'_x and the bending stability modulus of elasticity $E'_{y,\min}$ are calculated as

$$E'_x = E_x C_M C_t = (1.8 \times 10^6)(1.0)(1.0) = 1.8 \times 10^6 \text{ psi}$$

$$E'_{y,\min} = E_{y\min} C_M C_t = (0.78 \times 10^6)(1.0)(1.0) = 0.78 \times 10^6 \text{ psi}$$

Calculating the beam stability factor C_L

From Equation 3.3, the beam stability factor is now calculated: The unsupported length of the compression edge of the beam *or* distance between points of lateral support preventing rotation and/or lateral displacement of the compression edge of the beam is

$$l_u = 2 \text{ ft} = 24 \text{ in. (i.e., the distance between lateral supports provided by the joists)}$$

$$\frac{l_u}{d} = \frac{24 \text{ in.}}{16.5 \text{ in.}} = 1.45$$

We have previously assumed a uniformly loaded girder; therefore, using the l_u/d value, the effective length of the beam of the beam is obtained from Table 3.10 (or NDS Code Table 3.3.3) as

$$l_e = 2.06 l_u = (2.06)(24 \text{ in.}) = 50 \text{ in.}$$

$$R_B = \sqrt{\frac{l_e d}{b^2}} = \sqrt{\frac{(50)(16.5\text{ in.})}{(3.125\text{ in.})^2}} = 9.2 \le 50 \quad \textbf{OK}$$

$$F_{bE} = \frac{1.20E'_{\min}}{R_B^2} = \frac{(1.20)(0.78\times 10^6)}{(9.2)^2} = 11{,}058\text{ psi}$$

$$\frac{F_{bE}}{F_b^*} = \frac{11.058\text{ psi}}{2400\text{ psi}} = 4.61$$

From Equation 3.3, the beam stability factor is calculated as

$$C_L = \frac{1+F_{bE}/F_b^*}{1.9} - \sqrt{\left(\frac{1+F_{bE}/F_b^*}{1.9}\right)^2 - \frac{F_{bE}/F_b^*}{0.95}}$$

$$= \frac{1+4.61}{1.9} - \sqrt{\left(\frac{1+4.61}{1.9}\right)^2 - \frac{4.61}{0.95}} = 0.986$$

Calculating the volume factor C_V

L = length of beam in feet between points of zero moment = 12 ft (conservatively, assume that L = span of beam)

d = depth of beam = 16.5 in.

b = width of beam, in. = 3.125 in.

x = 10 (for Western species glulam)

From Equation 3.3,

$$C_V = \left(\frac{21}{L}\right)^{1/x}\left(\frac{12}{d}\right)^{1/x}\left(\frac{5.125}{b}\right)^{1/x} = \left(\frac{1291.5}{bdL}\right)^{1/x}$$

$$= \left[\frac{1291.5}{(3.125\text{ in.})(16.5\text{ in.})(12\text{ ft})}\right]^{1.10} = 1.08 \le 1.0$$

Therefore, C_V is 1.0.

The smaller of C_V and C_L will govern and is used in the allowable bending stress calculation. Because $C_L = 0.986 < C_V = 1.0$, use $C_L = 0.0.986$.

Using Table 3.2 (i.e., adjustment factor applicability table for glulam), we obtain the allowable bending stress as (see Table 10.6)

$$F'_b = F_b C_D C_M C_t C_F C_r (C_L \text{ or } C_V) = F_b^*(C_L \text{ or } C_V) = (2400)(0.986) = \mathbf{2366.4\text{ psi}}$$

6. Using Equation 4.10, the actual applied bending stress is

$$f_b = \frac{M_{\max}}{S_{xx}} = \frac{294{,}840\text{ lb-in.}}{141.8\text{ in.}^3} = 2079.3\text{ psi} < F'_b = 2366.4\text{ psi} \quad \textbf{OK}$$

10.7.2.2 Check of shear stress

V_{max} = 8190 lb. The beam cross-sectional area A = 51.56 in.^{2}. The applied shear stress in the beam at the centerline of the girder support is

$$f_v = \frac{1.5V}{A} = \frac{(1.5)(8190\text{ lb})}{51.56\text{ in.}^2} = 238.3\text{ psi}$$

Using the adjustment factor applicability table for glulam (Table 3.2), we obtain the allowable shear stress as

$$F_v' = F_v C_D C_M C_t = (215)(1)(1)(1) = 215 \text{ psi} < f_v \quad \text{N.G.}$$

Recalculate the shear stress at a distance d from the face of the column support, assuming an 8 × 8 column at this stage (Figure 10.14). Therefore,

$$V' = \frac{(6 \text{ ft}) - [(7.5 \text{ in.}/2)/12] - (16.5 \text{ in.}/12)}{6 \text{ ft}}(8190 \text{ lb}) = 5887 \text{ lb}$$

Recalculating the shear stress at a distance of d from the face of support yields

$$f_v' = \frac{1.5V'}{A} = \frac{(1.5)(5887 \text{ lb})}{51.56 \text{ in.}^2} = 171.3 \text{ psi} < F_v' \quad \text{OK}$$

10.7.2.3 Deflection

The allowable pure bending modulus of elasticity for strong (i.e., x–x)-axis bending of the girder was calculated previously (see Table 10.7):

$$E_x' = E_x C_M C_t = (1.8 \times 10^6)(1.0)(1.0) = 1.8 \times 10^6 \text{ psi}$$

The moment of inertia about the strong axis $I_{xx} = 1170$ in.^{4}

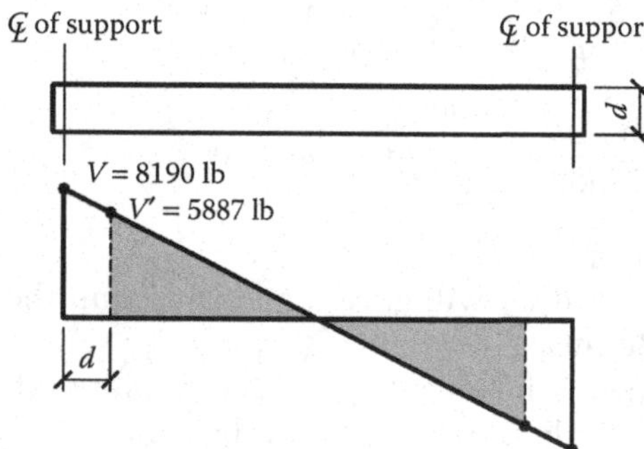

Figure 10.14 Shear force diagram of glulam girder.

Table 10.7 Floor girder deflection limits

Deflection	*Deflection limit*
Live load deflection Δ_{LL}	$\frac{L}{360} = \frac{(12 \text{ ft})(12)}{360} = 0.4$ in.
Incremental long-term deflection due to dead load plus live load (including creep effects), $\Delta_{TL} = k\Delta_{DL} + \Delta_{LL}$	$\frac{L}{240} = \frac{(12 \text{ ft})(12)}{240} = 0.6$ in.

The uniform dead load is w_{DL} = 51.3 lb/in.
The uniform live load is w_{LL} = 62.5 lb/in.

The dead load deflection is

$$\Delta_{DL} = \frac{5wL^4}{384EI} = \frac{(5)(51.3 \text{ lb/in.})(12 \text{ ft} \times 12)^4}{(384)(1.8 \times 10^6 \text{ psi})(1170 \text{ in.}^4)} = 0.14 \text{ in.}$$

The live load deflection is

$$\Delta_{LL} = \frac{5wL^4}{384EI} = \frac{(5)(62.5 \text{ lb/in.})(12 \text{ ft} \times 12)^4}{(384)(1.8 \times 10^6 \text{ psi})(1170 \text{ in}^4)} = 0.17 \text{ in.} < \frac{L}{360} = 0.4 \text{ in.} \quad \textbf{OK}$$

Because seasoned wood in dry service conditions is assumed to be used in this building, the creep factor k = 1.0. The total incremental dead plus floor live load deflection is

$$\Delta_{TL} = k\Delta_{DL} + \Delta_{LL}$$

$$= (1.0)(0.14 \text{ in.}) + 0.17 \text{ in.} = 0.31 \text{ in.} < \frac{L}{240} = 0.6 \text{ in.} \quad \textbf{OK}$$

10.7.2.4 Check of bearing stress (compression stress perpendicular to the grain)

Maximum reaction at the support R_1 = 8190 lb.
Width of $3\frac{1}{8} \times 16\frac{1}{2}$ in. glulam girder, b = 3.125 in.
The allowable bearing stress or compression stress parallel to the grain is

$$F'_{c\perp} = F_{c\perp} C_m C_t C_b = (500)(1)(1)(1) = 500 \text{ psi}$$

From Equation 4.9, the minimum required bearing length L_b is

$$l_{b,\text{req'd}} \geq \frac{R_1}{bF'_{c\perp}} = \frac{8190 \text{ lb}}{(3.125 \text{ in.})(500 \text{ psi})} = 5.24 \text{ in., say, } 5.25 \text{ in.}$$

The girders will be connected to the column using a U-shaped column cap detail as shown in Figure 4.6d, and the minimum length of the column cap required is

$$(5.25 \text{ in.})(2) + \tfrac{1}{2}\text{-in. clearance between ends of girders} = 11.0 \text{ in.}$$

10.7.2.5 Check of assumed self-weight of girder

The density of Hem-Fir is 27 lb/ft³ and is used to calculate the actual weight of the girder selected.

- Assumed a girder self-weight = 15 lb/ft.
- Actual weight of the $3\frac{1}{8} \times 16\frac{1}{2}$ in. glulam girder selected (from NDS-S Table 1C) is

$$\frac{(51.56 \text{ in.}^2)(27 \text{ lb/ft}^3)}{(12 \text{ in.})(12 \text{ in.})} = 9.7 \text{ lb/ft, which is less than the assumed girder selfweight}$$

of 15 lb/ft OK

Use a $3\frac{1}{8} \times 16\frac{1}{2}$ in. 24F–E11 HF/HF glulam girder.

10.7.3 Design of header beams

With reference to Figure 10.15, there are several header beams with different loading and span conditions. However, they are all very similar, and therefore, only two header beams will be considered (H-1 and H-2 at the second floor).

10.7.3.1 Design of header beam H-1

Dead load on header:

(40 psf)(15 ft/2)	=	300 lb/ft (second-floor dead load)
(10 psf)(6 ft height above opening)	=	60 lb/ft (weight of wall above opening)
Self-weight (assume three 2 × 12′s)	=	12 lb/ft
w_{DL}	=	**372 lb/ft**

Live load on header:

$$w_{LL} = (50\text{ psf})\left(\frac{15\text{ ft}}{2}\right) = \mathbf{375\ lb/ft} \quad \text{(second-floor live load)}$$

Total load on header:

$$w_{TL} = 372 + 375 = 747\text{ lb/ft}$$

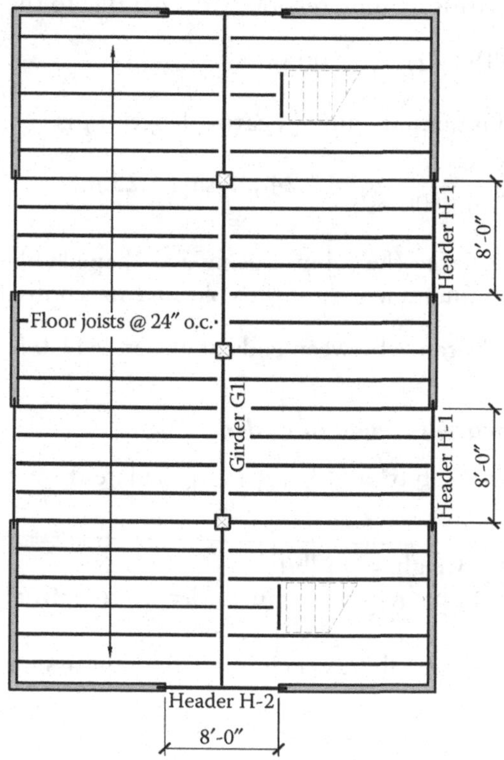

Figure 10.15 Header beam layout.

Adjusting the loads for seasoned lumber (recall that $k = 0.5$ for seasoned lumber), the long term total load for deflection calculations is:

$$kw_{DL} + w_{LL} = (0.5)(372) + 375 = 561 \text{ lb/ft}$$

(Use this value for calculating deflections only.)

$$\text{Maximum moment } M = \frac{w_{TL}L^2}{8} = \frac{(747)(8)^2}{8} = 5976 \text{ ft-lb}$$

$$\text{Maximum shear and reaction } V_{TL} = \frac{w_{TL}L}{2} = \frac{(747)(8)}{2} = 2988 \text{ lb}$$

Bending:

$$F_b' = F_b C_D C_M C_t C_L C_F C_{fu} C_i C_r$$

$$C_D(DL + LL) = 1.0$$

$$C_M = 1.0$$

$$C_t = 1.0$$

$$C_F(F_b) = 1.0$$

$$C_{fu} = 1.0$$

$$C_i = 1.0$$

$$C_r = 1.0$$

From NDS Table 4A (Spruce-Pine-Fir, no. 1/no. 2)

$$F_b = 875 \text{ psi}$$

$$F_v = 135 \text{ psi}$$

$$F_{c\perp} = 425 \text{ psi}$$

$$E = 1.4 \times 10^6 \text{ psi}$$

$$E_{\min} = 0.51 \times 10^6 \text{psi}$$

Because the beam is unbraced laterally for its full length, the beam stability factor C_L will have to be calculated. See Section 3.4.3 for the C_L equations. For three 2 × 12′s:

$$d = 11.25 \text{ in.}$$

$$b = (3)(1.5 \text{ in.}) = 4.5 \text{ in.}$$

$$A = (3)(16.88 \text{ in.}^2) = 50.63 \text{ in.}^2$$

$$S_x = (3)(31.64 \text{ in.}^3) = 94.92 \text{ in.}^3$$

$$I_x = (3)(178 \text{ in.}^4) = 534 \text{ in.}^4$$

$$l_u = 8 \text{ ft} = 96 \text{ in. (span of header)}$$

$$\frac{l_u}{d} = \frac{96 \text{ in.}}{11.25 \text{ in.}} = 8.53$$

$$l_e = 1.63 l_u + 3d \text{ (Table 3.10 or NDS Table 3.3.3)}$$

$$= (1.63)(96 \text{ in.}) + (3)(11.25 \text{ in.}) = 191 \text{ in.}$$

$$R_B = \sqrt{\frac{\ell_e d}{b^2}} \leq 50 \text{ (NDS Equation 3.3-5)}$$

$$= \sqrt{\frac{(191)(11.25)}{(4.5)^2}} = \mathbf{10.3 < 50} \qquad \textbf{OK}$$

$$E'_{\min} = E_{\min} C_M C_t C_i = (0.51 \times 10^6)(1.0)(1.0)(1.0) = 0.51 \times 10^6$$

$$F_{bE} = \frac{1.20 E_{\min}}{R_B^2} = \frac{(1.20)(0.51 \times 10^6)}{(10.3)^2} = 5768 \text{ psi}$$

$$F_b^* = F_b C_D C_M C_t C_F C_{fu} C_i C_r = (875)(1.0)(1.0)(1.0)(1.0)(1.0)(1.0)(1.0) = 875 \text{ psi}$$

$$C_L = \frac{1 + F_{be}/F_b^*}{1.9} - \sqrt{\left(\frac{1 + F_{be}/F_b^*}{1.9}\right)^2 - \frac{F_{be}/F_b^*}{0.95}}$$

$$= \frac{1 + 5768/875}{1.9} - \sqrt{\left(\frac{1 + 5768/875}{1.9}\right)^2 - \frac{5768/875}{0.95}} = 0.991$$

The allowable bending stress is

$$F_b' = F_b^* C_L = (875)(0.991) = 867 \text{ psi}$$

The bending stress applied is

$$f_b = \frac{M}{S_x} = \frac{(5976)(12)}{94.92} = \mathbf{756 \text{ psi} < F_b'} \quad \textbf{OK}$$

Shear. The allowable shear stress is

$$F_v' = F_v C_D C_M C_t C_i = (135)(1.0)(1.0)(1.0)(1.0) = 135 \text{ psi}$$

The shear stress applied is

$$f_v = \frac{1.5V}{A} = \frac{2988}{50.63} = \mathbf{88.6 \text{ psi} < F_v'} \quad \textbf{OK}$$

Bearing. The allowable bearing stress perpendicular to the grain is

$$F_{c\perp}' = F_{c\perp} C_M C_t C_i = (425)(1.0)(1.0)(1.0) = 425 \text{ psi}$$

The required bearing length is

$$l_{b,\text{req'd}} = \frac{R}{F'_{c\perp}b}$$

$$= \frac{2988}{(425)(4.5)} = 1.6 \text{ in.}$$

The headers will bear directly on two jack studs (see Figure 10.16).
Deflection. The total load deflection is

$$E' = EC_M C_t C_i = \left(1.4\times 10^6\right)(1.0)(1.0)(1.0) = 1.4\times 10^6$$

$$\Delta_{kDL+LL} = \frac{5wL^4}{385EI} = \frac{(5)(561/12)(96)^4}{(384)(1.4\times 10^6)(534)} = 0.07 \text{ in.} < \frac{L}{240} = \frac{96}{240} = 0.4 \text{ in.} \quad \text{OK}$$

The live load deflection is

$$\Delta_{LL} = \frac{5wL^4}{385EI} = \frac{(5)(375/12)(96)^4}{(384)(1.4\times 10^6)(534)} = 0.046 \text{ in.} < \frac{L}{360} = \frac{96}{360} = 0.27 \text{ in.} \quad \text{OK}$$

Use three 2 × 12's for header H-1, bear on two 2 × 6 jack studs.

10.7.3.2 Design of header beam H-2

Header beam H-2 has a 16 ½-in.-deep girder framing into it. For practical framing considerations, the header should be at least as deep as the girder. For this example, a preengineered header beam will be selected with the following assumed properties:

$$F_b = 2600 \text{ psi}$$

$$F_v = 285 \text{ psi}$$

$$E = 1900 \text{ksi}$$

$$F_{c\perp} = 750 \text{ psi}$$

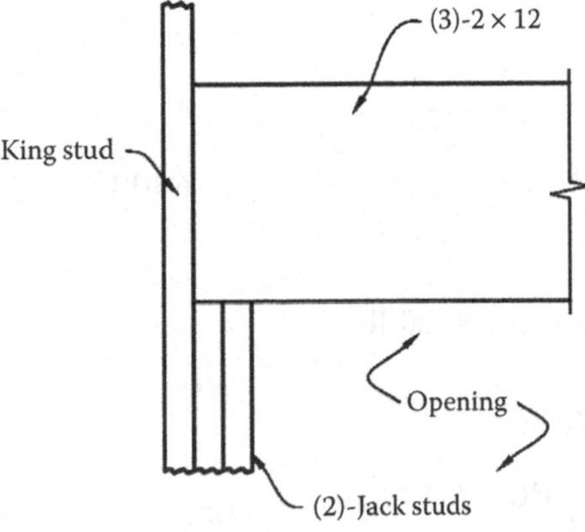

Figure 10.16 Header bearing.

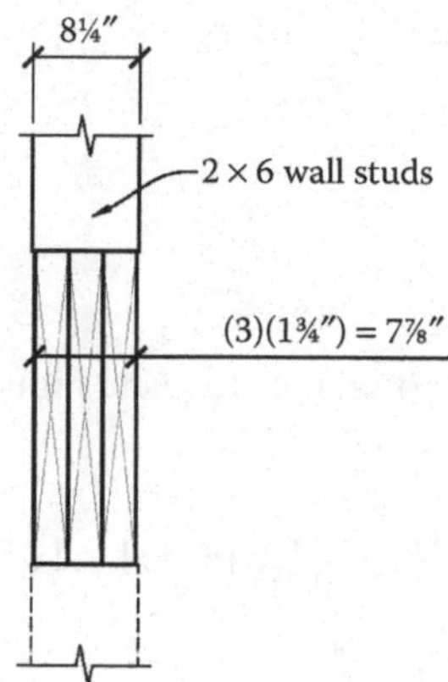

Figure 10.17 Header section.

For this example, three 1¾ × 16 in. Laminated veneer lumber (LVL) headers will be assumed. The design will show that one header would be adequate, but it is practical to use at least three members in a 2 × 6 stud wall so that a "void" is not created above the opening (Figure 10.17).

Section properties:

$$d = 16 \text{ in.}$$

$$b = (3)(1.75 \text{ in.}) = 5.25 \text{ in.}$$

$$A = (3)(1.75)(16) = 84 \text{ in.}^2$$

$$S_x = (3)\frac{bd^2}{6} = (3)\frac{(1.75)(16)^2}{6} = 224 \text{ in.}^3$$

$$I_x = (3)\frac{bd^3}{12} = (3)\frac{(1.75)(16)^3}{12} = 1792 \text{ in.}^4$$

Loads on header:

$$P_D = (40 \text{ psf})\left(\frac{12 \text{ ft}}{2}\right)(15 \text{ ft}) + 15 \text{ plf}\left(\frac{12 \text{ ft}}{2}\right) = 3690 \text{ lb}$$

$$P_L = (50 \text{ psf})\left(\frac{12 \text{ ft}}{2}\right)(15 \text{ ft}) = 4500 \text{ lb}$$

$$P_T = 3690 + 4500 = \mathbf{8190\ lb}$$

$$\text{Maximum moment } M = \frac{PL}{4} = \frac{(8190)(8)}{4} = 16{,}380 \text{ ft-lb}$$

Maximum shear and reaction $V_{TL} = \dfrac{p}{2} = \dfrac{8190}{2} = 4095\text{ lb}$

The bending stress applied is

$$f_b = \frac{M}{S_x} = \frac{(16{,}380)(12)}{224} = 878\text{ psi} < 2600\text{ psi} \quad \text{OK}$$

The shear stress applied is

$$f_v = \frac{1.5V}{A} = \frac{(1.5)(4095)}{84} = 74\text{ psi} < F'_v = 285\text{ psi} \quad \text{OK}$$

The required bearing length is

$$\begin{aligned} l_{b,\text{req'd}} &= \frac{R}{F'_{c\perp} b} \\ &= \frac{4095}{(750)(5.25)} = 1.04\text{ in.} \end{aligned}$$

The headers will bear directly on at least one jack stud.

Deflection. Because the dead load is more than one-half of the live load, the *L*/240 criteria will control. Therefore,

$$\Delta_{DL+LL} = \frac{PL^3}{48EI} = \frac{(8190)(96)^3}{(48)(1.9\times10^6)(1792)} = 0.05\text{ in.} < \frac{L}{240} = \frac{96}{240} = 0.4\text{ in.} \quad \text{OK}$$

Use three 1¾ in. × 16 in. LVLs for header H-2, bear on one 2 × 6 jack stud.

10.8 DESIGN OF A TYPICAL GROUND-FLOOR COLUMN

It is assumed that the ground-floor column in this building will be a wood column instead of the steel pipe column that is frequently used in residential structures. However, the wood column base should be protected from moisture at the slab-on-grade/footing level using a column base detail similar to that shown in Figure 10.18.

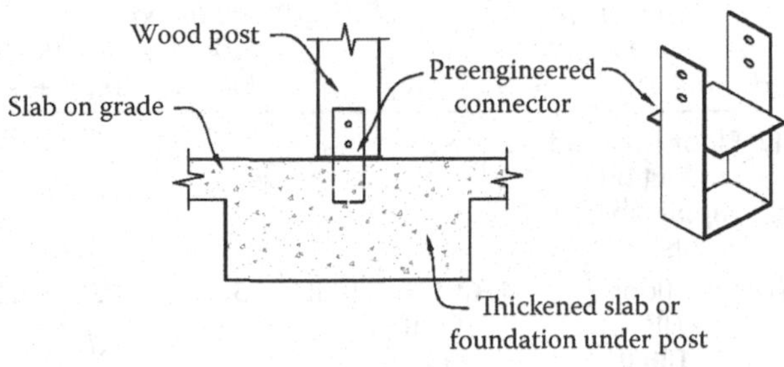

Figure 10.18 Column base support detail.

10.8.1 Determining the column axial load

We recall that this is a two-story building, but the roof tributary area for this column is zero because the column supports only the second floor.

$$\text{TA of typical interior column} = (15 \text{ ft})(12 \text{ ft}) = 180 \text{ ft}^2$$

From Chapter 2,

$$A_T = 180 \text{ ft}^2 \text{ (only the second-floor live load is reducible for this column)}$$

$$K_{LL} = 4 \text{ (interior column)}$$

$$K_{LL}A_T = (4)(180 \text{ ft}^2) = 720 \text{ ft}^2 > 400 \text{ ft}^2$$

Floor live load $L = 50$ psf < 100 psf.
Floor occupancy is not assembly occupancy.
Because all the conditions above are satisfied, live load reduction is permitted.

The governing IBC load combination from Chapter 2 for calculating the column axial loads is $D + 0.75L + 0.75(L_r$ or S or $R)$. The reduced or design floor live loads for the second and third floors are calculated using Table 10.8.

Using the loads calculated in Section 10.2, the column axial loads are calculated using Table 10.9. Therefore, the ground-floor column will be designed for a cumulative maximum axial compression load $P = 14.5$ kips.

It should be noted that the column load calculated previously assumes simply supported glulam girders. However, if the contractor chooses to use two 24-ft-long pieces for the girder, thus indicating two two-span continuous girders, the interior column load would be amplified due to the girder continuity effect. For a two-span continuous girder, this will result in a 25% increase in the second-floor load on the column. Therefore, the revised axial load on the column will be

$$P = (1.25)(14.5) = 18.1 \text{ kips}$$

Using Figure B.17, the allowable axial load at an unbraced length of 10.5 ft is approximately 25.0 kips with a load duration factor of 1.0. Therefore, use a 8 × 8 SPF No. 2 Column. The reader should note that the load capacity of a 6 × 6 column is approximately 11 kips and would not be adequate.

Table 10.8 Reduced or design floor live load calculation table

Member	*Levels supported*	*Summation of floor tributary area, A_T*	*K_{LL}*	*Unreduced floor live load, L_0*	*Live load reduction factor, $0.25 + 15/\sqrt{K_{LL}A_T}$*	*Design floor live load, L*
Second-floor column (i.e., column below roof)	Roof only	Floor live load reduction *not* applicable to roofs!	–	–	–	35 psf (snow load)
Ground-floor column (i.e., column below the second floor)	One floor + roof	One floor × 180 ft² **= 180 ft²**	4 $K_{LL}A_T = 720$ ft² >400 ft²; therefore, live load reduction allowed	50 psf	$0.25 + \frac{15}{\sqrt{(4)(180)}} = 0.809$	(0.809)(50) **= 40.5 psf** ≥ $0.50L_0$ = 25 psf

Table 10.9 Column load summation table (with floor live load reduction)

Level	*TA* (ft^2)	*Dead load, D (psf)*	*Live load, L_0 (S or L_r or R on the roof) (psf)*	*Design live load roof: S or L_r or R floor: L (psf)*	*Unfactored total load at each level, w_{s1} roof: D floor: D + L (psf)*	*Unfactored total load at each level, w_{s2} roof: D + 0.75S floor: D + 0.75L (psf)*	*Unfactored column axial load at each level, P = (TA) (w_{s1}) or (TA) (w_{s2}) (kips)*	Cumulative unfactored axial load, $2P_{D+L}$ (kips)	Cumulative unfactored axial load, $\Sigma P_{D+0.75L+0.75S}$ (kips)	*Maximum cumulative unfactored axial load, ΣP (kips)*
Roof	0	22	35	35	22	48.3	0	0	0	**0**
Second floor	180	40	50	40.5	80.5	70.4	14.5 or 12.7	14.5	12.7	**14.5**

10.9 DESIGN OF A TYPICAL EXTERIOR WALL STUD

The typical exterior wall stud is subjected to combined concentric axial compression load and bending due to the wind loads acting perpendicular to the face of the exterior wall. The ground-floor exterior wall stud, which is the more critically loaded than the second-floor stud, is designed in this section. The size obtained is also used for the second-floor stud.

10.9.1 Determining the gravity and lateral loads acting on a typical wall stud

Floor-to-floor height = 12 ft
TW of wall stud = 2 ft
TA per stud at the *roof* level = (30 ft/2 + 2 ft truss overhang) × (2 ft) = 34 ft²
TA per stud at the *second-floor* level = (15 ft/2) × (2 ft) = 15 ft²

From Chapter 2,

$$A_T = \text{floor TA} = 15\ \text{ft}^2\ \left(\text{only the second-floor live loadis reducible for this stud}\right)$$

$$K_{LL} = 2\ \left(\text{exterior column}\right)$$

$$K_{LL}A_T = \left(2\right)\left(15\ \text{ft}^2\right) = 30\ \text{ft}^2 < 400\ \text{ft}^2$$

Therefore, floor live load reduction is *not* permitted for this stud.

From Section 10.2, the gravity loads have been calculated as follows:

Roof dead load $D \approx 22$ psf
Roof snow load $S = 35$ psf
Roof live load $L_r = 20$ psf $< S$; therefore, the snow load controls
Second-floor dead load $D = 40$ psf
Second-floor live load $L = 50$ psf
Second-floor wall self-weight = 10 psf
Ground-floor wall self-weight = 10 psf

The lateral or horizontal unfactored wind pressure acting perpendicular to the face of the exterior wall represents the components and cladding wall pressures calculated in Section 10.5 as $0.6W = 22.6$ psf.

Gravity loads. The total *dead load* on the ground-floor studs is

$$P_D = P_{D(\text{roof})} + P_{D(\text{2ndfloor})} + P_{D(\text{wall})}$$

$$= (22\ \text{psf})\left(34\ \text{ft}^2\right) + (40\ \text{psf})\left(15\ \text{ft}^2\right)$$

$$+ (10\ \text{psf})(2\text{-ft stud spacing})\left(12\text{-ft wall height}\right)$$

$$= \mathbf{1588\ lb}$$

The total *snow load* on the ground-floor studs is

$$P_S = \left(35\ \text{psf}\right)\left(34\ \text{ft}^2\right) = \mathbf{1190\ lb}$$

The total *floor live load* on the ground-floor studs is

$$P_L = \left(50\ \text{psf}\right)\left(15\ \text{ft}^2\right) = \mathbf{750\ lb}$$

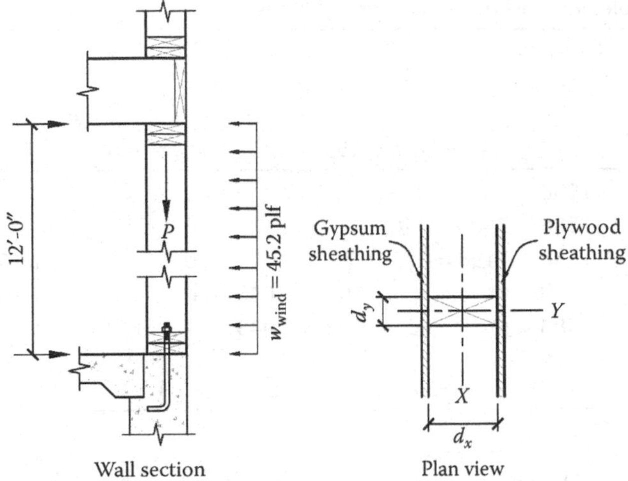

Figure 10.19 Stud wall section and plan view.

Lateral loads. The wind load acts perpendicular to the face of the stud wall, causing bending of the stud about the *x–x* (strong) axis (see Figure 10.19). The lateral wind load w_{wind} = (22.6 psf)(2-ft TW) = 45.2 plf. The maximum moment in the exterior wall stud due to wind load is calculated using the 12-ft floor-to-floor height (conservative for the ground-floor studs on the east and west faces of the building) as

$$M_w = \frac{wL^2}{8}$$

$$= \frac{(45.2 \text{ lb/ft})(12)^2}{8} = \mathbf{814 \text{ ft-lb}}$$

The most critical axial load combination and the most critical combined axial and bending load combination are determined using the normalized load method outlined in Chapter 3. The axial load *P* in the wall stud will be caused by the gravity loads *D*, *L*, and *S*, whereas the bending load and moment will be caused by the lateral wind load *W*. The applicable nonzero loads and load effect in the load combinations are

$$D = 1588 \text{ lb}$$

$$L = 750 \text{ lb}$$

$$S = 1190 \text{ lb}$$

$$0.6W = 814 \text{ ft-lb}$$

We can separate the load cases in Table 10.10 into two types of loads: pure axial load cases and combined load cases. Because not all the loads on this wall stud are pure axial loads only or bending loads only, the normalized load method discussed in Chapter 3 can then only be used to determine the most critical pure axial load case, not the most critical combined load case. The most critical combined axial plus bending load case would have to be determined by carrying out the design (or analysis) for all the combined load cases, with some of the load cases eliminated by inspection.

Table 10.10 Applicable and governing load combinations

Load combination	Axial load, P (lb)	Moment, M (ft-lb)	C_D	Normalized load and moment P/C_D	M/C_D
D	1588	0	0.9	1764	0
$D + L$	1588 + 750 = 2338	0	1.0	2388	0
$D + S$	1588 + 1190 = 2778	0	1.15	2416	0
$D + 0.75L + 0.75S$	1588 + (0.75)(750 + 1190) = **3043**	0	1.15	2646	0
$D + 0.75\,(0.6W) + 0.75L + 0.75S$	1588 + (0.75)(750 + 1190) = **3043**	(0.75)(814) = **611**	1.6	1902	382
$0.6D + 0.6W$	(0.6)(1588) = **953**	**814**	1.6	596	509

Pure axial load case. For the pure axial load cases, the load combination with the highest normalized load P/C_D from Table 10.10 is

$$D + 0.75(L + S) \quad \mathbf{P = 3043\ lb\ with\ C_D = 1.15}$$

Combined load case. For the combined load cases, the load combinations with the highest normalized load P/C_D and normalized moment M/C_D from Table 10.10 are

$$D + 0.75(L + S + 0.6W) \quad \mathbf{P = 3043\ lb} \quad \mathbf{M = 611\ ft\text{-}lb} \quad \mathbf{C_D = 1.6}$$

or

$$0.6D + 0.6W \quad \mathbf{P = 953\ lb} \quad \mathbf{M = 814\ ft\text{-}lb} \quad \mathbf{C_D = 1.6}$$

By inspection, it would appear as if the load combination $D + 0.75(L + S + 0.6W)$ will control for the combined load case, but this should be verified through analysis, and therefore, both of these combined load cases will have to be investigated. The figures in Appendix B will be utilized. A 2 × 6 wall stud will be assumed and Spruce-Pine-Fir lumber will be used.

For the pure axial load case, the applied axial load is $P = 3043$ lb. Using Figure B.44, the allowable axial load capacity is **4400 lb** at an unbraced length of 12 ft with no applied moment. Note that the load duration factor for this design aid is 1.0, so the use of this design aid is conservative because the actual load duration is 1.15. Note also that the lumber here is 2 × 6, SPF No. 1/No. 2.

For the first combined load case, the applied axial load is $P = 3043$ lb and the applied moment is $M = 611$ ft-lb. Using Figure B.27 ($C_D = 1.6$), the allowable axial load capacity is approximately **2900 lb** at an unbraced length of 12 ft with an applied moment of **611 ft-lb,** which is very close to the applied axial load of 3043 lb. Therefore, the stud is deemed to be adequate because the length of the stud is actually less than 12 ft. Note here that the lumber assumed is 2 × 6, SPF Select Structural. The reader should verify that the use of SPF No. 1/No. 2 is not adequate.

For the second combined load case, the applied axial load is $P = 953$ lb and the applied moment is $M = 814$ ft-lb. Using Figure B.27 ($C_D = 1.6$), the allowable axial load capacity is approximately **2400 lb** at an unbraced length of 12 ft with an applied moment of **814 ft-lb.** Note that the lumber assumed is 2 × 6, SPF Select Structural.

Therefore, use 2 × 6 at 24 in. o.c. SPF Select Structural wall studs.

10.10 DESIGN OF ROOF AND FLOOR SHEATHING

10.10.1 Gravity loads

10.10.1.1 Roof sheathing

From the floor load calculations in Section 10.2, the total loads acting on the roof sheathing are as follows:

Asphalt shingles	= 2.5 psf
Reroofing	= 2.5 psf
Plywood sheathing	= 1.2 psf
Dead load on sheathing, $D_{sheathing}$	= 6.2 psf
Snowload, S	= 35 psf

The total gravity load to roof sheathing = 6.2 + 35 = **41.2 psf**. Assuming a roof truss spacing of 24 in., the following is selected from IBC Table 2304.8(3):

$\frac{7}{16}$-in. CD–X (minimum thickness)
Strength axis perpendicular to the supports
Span rating = 24/16
Edge support not required (may be required for diaphragm strength)
Total load capacity = 50 psf > 41.2 psf, OK
Live load capacity = 40 psf > 35 psf, OK

10.10.1.2 Floor sheathing

From the floor load calculations in Section 10.2, the total loads acting on the floor sheathing alone are as follows:

Floor covering	= 4.0 psf
Plywood sheathing	= 3.2 psf
Partitions	= 20.0 psf
$DL_{sheathing}$	**= 27.2 psf**
LL	**= 50 psf**

The total gravity load to floor sheathing = 27.2 + 50 = **77.2 psf**. Assuming that the floor framing is spaced at 24 in. o.c., the following is selected from IBC Table 2304.8(3) or 2304.8(4):

$\frac{3}{4}$-in. Structural I sheathing (minimum thickness)
Note that Table 2304.7(4) indicates that a plywood panel with veneers made from group 1 wood species (i.e., Structural I) must be used if $\frac{3}{4}$ in. plywood thickness is specied. However, Table 2304.8(3) does not indicate the specie of wood used for the veneer.
Strength axis perpendicular to the supports
Span rating = 24 in.
Tongue-and-grooved joints
Total load capacity = 100 psf > 77.2 psf, OK

10.10.2 Lateral loads

Based on the lateral wind and seismic loads calculated in Sections 10.4 and 10.5 and Figures 10.2 and 10.4, the loads to the floor and roof diaphragms (Figure 10.20) are as summarized in Table 10.11. By inspection, wind loads will govern the diaphragm design in the transverse direction, and seismic loads will govern the longitudinal direction. From the gravity load design in Section 10.5, the minimum thickness is 7⁄16 in. for roof sheathing and 3⁄4 in. for the floor sheathing. From Table 6.3, the minimum required fastening is 8d nails spaced 6 in. o.c. at the supported edges and 12 in. o.c. on the intermediate members for both the roof and floor sheathing. These minimum parameters will be considered in the design of the diaphragm fasteners.

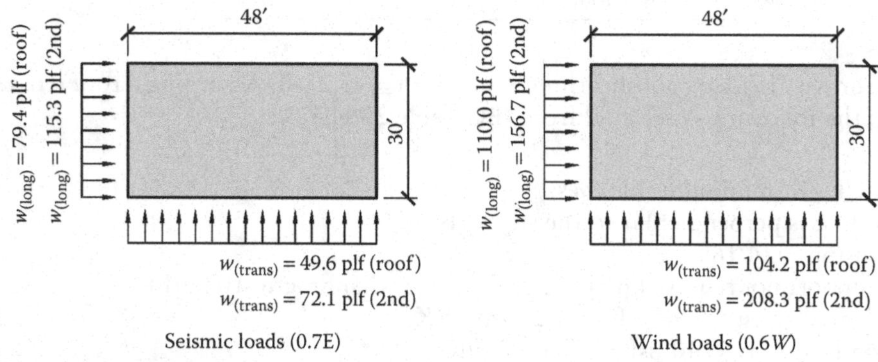

Figure 10.20 Lateral loads to the roof and floor diaphragms.

Table 10.11 Seismic and wind loads at each level

Direction	*Level*	*F (lb)*	*w*	*V*	v_d
Seismic loads (0.7E)					
Transverse	Roof	2380[a]	2380 lb/(48 ft) = 49.6 plf	(2380 lb)/2 = 1190 lb	1190 lb/(30 ft) = **39.7 plf**
	Second floor	3460	3460 lb/(48 ft) = 72.1 plf	(3460 lb)/2 = 1730 lb	1730 lb/(30 ft) = **57.7 plf**
Longitudinal	Roof	2380	2380 lb/(30 ft) = 79.4 plf	(2380 lb)/2 = 1190 lb	1190 lb/(48 ft) = **24.8 plf**
	Second floor	3460	3460 lb/(30 ft) = 115.3 plf	(3460 lb)/2 = 1730 lb	1730 lb/(48 ft) = **36.1 plf**
Wind loads (0.6W)					
Transverse	Roof	5000[b]	5000 lb/(48 ft) = 104.2 plf	(5000 lb)/2 = 2500 lb	2500 lb/(30 ft) = **83.3 plf**
	Second floor	10,000	10,000 lb/(48 ft) = 208.3 plf	(10,000 lb)/2 = 5000 lb	5000 lb/(30 ft) = **166.7 plf**
Longitudinal	Roof	3300	3300 lb/(30 ft) = 110 plf	(3300 lb)/2 = 1650 lb	1650 lb/(48 ft) = **34.4 plf**
	Second floor	4700	4700 lb/(30 ft) = 156.7 plf	(4700 lb)/2 = 2350 lb	2350 lb/(48 ft) = **49.0 plf**

[a] See Section 10.3.
[b] See Table 10.2.

10.10.2.1 Transverse direction (wind loads govern)

The applied diaphragm unit shears (see Table 10.11) are as follows:

$$v_{dr} = 83.3 \text{ plf (roof level)}$$

$$v_{d2} = 166.7 \text{ plf (second-floor level)}$$

For the *roof sheathing*, using load case 1 from the Special Design Provisions for Wind and Seismic (SDPWS) [3], the nominal capacity of a ⅜-in. CD–X Structural I panel with 8d nails into 2× framing members is 670 plf for an unblocked diaphragm, but it is valid only for framing of Douglas Fir-Larch or Southern Pine. The adjusted allowable shear is calculated as follows using the specific gravity adjustment factor (SGAF):

$$\begin{aligned} \text{SGAF} &= 1 - (0.5 - \text{G}) \le 1.0 \quad G = 0.42 \text{ (Spruce-Pine-Fir lumber is specified)} \\ &= 1 - (0.5 - 0.42) = 0.92 \end{aligned}$$

$$v_{\text{allowable}} = (0.92)(670 \text{ plf})/2.0 = \mathbf{308\ plf} > v_{dr} = 83.3 \text{ plf}$$

Therefore, the required fastening is 8d nails at 6 in. o.c. [edge nailing (EN)] and 12 in. o.c. [fielding nailing (FN)].

For the *floor sheathing*, using load case 1 from the SDPWS, the nominal capacity of a 15⁄32-in. CD–X Structural I panel with 10d nails into 2× framing members is 800 plf for an unblocked diaphragm. Adjusting this value for wind loads and Spruce-Pine-Fir lumber yields

$$v_{\text{allowable}} = (1.4)(0.92)(800 \text{ plf})/2.0 = \mathbf{368\ plf} > v_{dr} = 166.7 \text{ plf}$$

Note here that a comparison was made with 15⁄32-in. panels, which is conservative when a ¾-in. panel is used. Also, the minimum fastener penetration is 1½ in. (from IBC Table 2306.3.1). Because the length of a 10d nail is 3 in. (from NDS Table L4), the actual penetration is greater than the required penetration (3 in. – ¾ in. = 2¼ in. > 1½ in.). The required fastening is 10d nails at 6 in. on-center EN and 12 in. on-center FN).

10.10.2.2 Longitudinal direction (seismic loads govern)

The applied diaphragm unit shears (See Table 10.11) are as follows:

$$v_{dr} = 24.8 \text{ plf (roof level)}$$

$$v_{d2} = 36.1 \text{ plf (second-floor level)}$$

By inspection, the diaphragm shear in the transverse direction is much greater than that in the longitudinal direction. Because the diaphragm design did not require blocking in the transverse direction, it would be economical to select a diaphragm design that did not include blocking for the longitudinal direction.

For the *roof sheathing*, using load case 3 from the SDPWS, the nominal capacity of a ⅜-in. panel with 8d nails into 2× framing members is 360 plf for an unblocked diaphragm. Adjusting this value (see Section 6.5.1) for Spruce-Pine-Fir lumber yields

$$v_{\text{allowable}} = (0.92)(360 \text{ plf})/2.0 = \mathbf{166\ plf} > v_{dr} = 24.8 \text{ plf}$$

Therefore, the specified fastening used for the transverse direction is also adequate for the longitudinal direction.

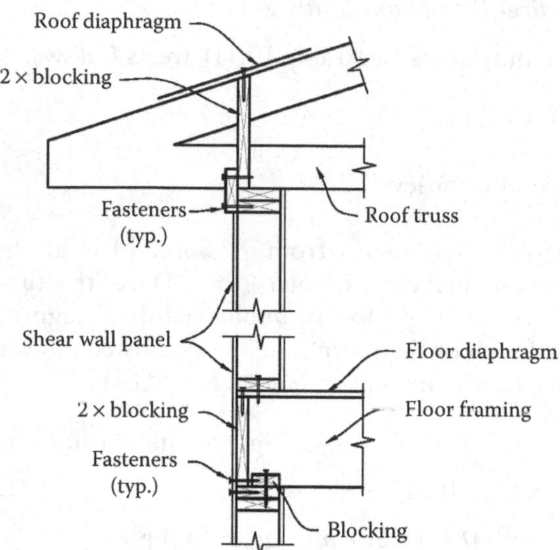

Figure 10.21 Diaphragms-to-shear wall connection.

For *the floor sheathing*, using load case 3 from the SDPWS, the ominal capacity of a $^{15}/_{32}$-in. panel with 10d nails into 2x framing members is 430 plf for an unblocked diaphragm. Adjusting this value for Spruce-Pine-Fir lumber yields

$$v_{\text{allowable}} = (0.92)(430 \text{ plf})/2 = \mathbf{197\ plf} > v_{dr} = 36.1 \text{ plf}$$

Therefore, the specified fastening used for the transverse direction is also adequate for the longitudinal direction.

The attachment of the horizontal diaphragm to the shear walls requires consideration. A variety of diaphragm-to-shear wall connections are possible (see Figure 10.21 for typical details). The most efficient connection would be one where the horizontal wood panels meet the vertical panels (Figure 10.21), where the boundary nails in the horizontal diaphragm transfer loads directly to the boundary nails in the vertical diaphragm. Other connection types often require more analysis and fasteners.

10.11 DESIGN OF WALL SHEATHING FOR LATERAL LOADS

From the SDPWS Table 3.2.1, a $^{7}/_{16}$-in. panelis selected and is adequate to resist a wind pressure perpendicular to the face of the wall sheathing. This panel can be oriented either parallel or perpendicular to the wall studs and allows the builder some latitude in selecting the most economical layout. Table 6-1 of the *Wood Structural Panels Supplement* indicates that the minimum fastening is 6d nails spaced 6 in. o.c. at the supported edges and 12 in. o.c. in the field of the plywood panel. These minimum parameters will be considered in the design of the shear wall fasteners.

The aspect ratio of the available shear walls must be calculated. The shortest available wall length is 8 ft and the total height of the walls is 12 ft. Therefore,

$$\frac{h}{w} = \frac{12 \text{ ft}}{8 \text{ ft}} = 1.5$$

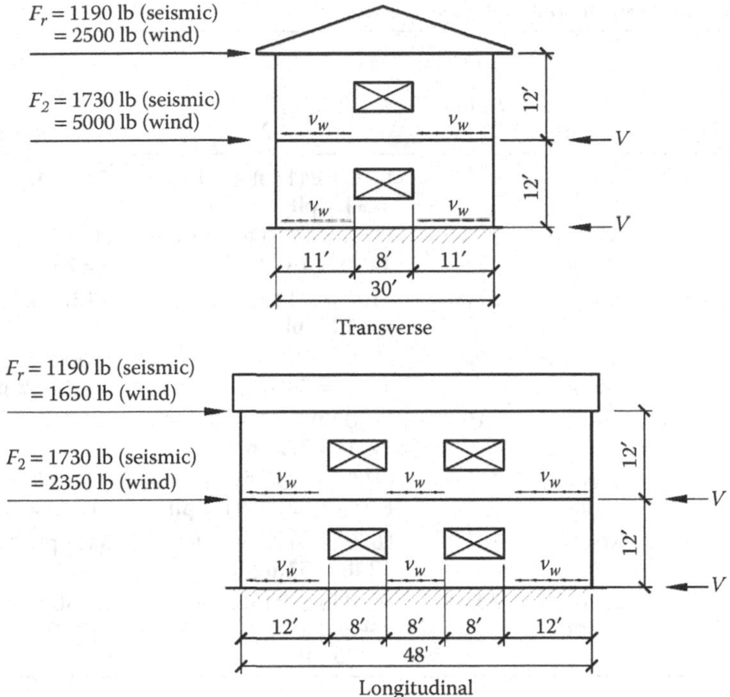

Figure 10.22 Unfactored lateral loads to shear walls (see Table 10.11).

The aspect ratio is less than 2.0, so no adjustment is necessary to the unit shear capacity.

The unit shear due to lateral loads will now be calculated. Note that the base shear V to each line of shear walls is cumulative at the ground level. With reference to Figure 10.22, the unit shear in the shear walls is as summarized in Table 10.12. By inspection, wind loads will govern the shear wall design in the transverse direction, and seismic loads will govern in the longitudinal direction

10.11.1 Transverse direction (wind governs)

The shear wall unit shears (See Table 10.12) are as follows:

$$v_{wr}\left(\text{roof level}\right) = 113.7 \text{ plf}$$

$$v_{w2}\left(\text{second-floor level}\right) = 340.9 \text{ plf}$$

From the SDPWS Table 4.3A, the nominal capacity of a $^{15}\!/_{32}$-in. panel with 8d nails into 2× framing members is 785 plf. This value is valid only for framing of Douglas Fir-Larch or Southern Pine. The species adjustment factor to be applied to the allowable shear is

$$\text{SGAF} = 1 - (0.5 - G) \leq 1.0 \qquad G = 0.42 \text{ (Spruce-Pine-Fir lumber is specified)}$$

$$= 1 - (0.5 - 0.42) = 0.92$$

Table 10.12 Summary of shear wall unit shears

Direction	*Load*	*Level*	*V (lb) (from Figure 10.22)*	$v_w = \frac{V}{\Sigma L_w}$	v_w *(for drag struts)*
Transverse	Seismic	Roof	1190	(1190 lb)/(11 ft + 11 ft) = **54.1 plf**	**54.1 plf**
		Second floor	1730	(1190 + 1730 lb)/(11 ft + 11 ft) = **132.8 plf**	(1730 lb)/(11 ft + 11 ft) = **78.7 plf**
	Wind	Roof	2500	(2500 lb)/(11 ft + 11 ft) = **113.7 plf**	**113.7 plf**
		Second floor	5000	(2500 + 5000 lb)/(11 ft + 11 ft) = **340.9 plf**	(5000 lb)/(11 ft + 11 ft) = **227.3 plf**
Longitudinal	Seismic	Roof	1190	(1190 lb)/(12 ft + 8 ft + 12 ft) = **37.2 plf**	**37.2 plf**
		Second floor	1730	(1190 + 1730 lb)/(12 ft + 8 ft + 12 ft) = **91.3 plf**	(1730 lb)/(12 ft + 8 ft + 12 ft) = **54.1 plf**
	Wind	Roof	1650	(1650 lb)/(12 ft + 8 ft + 12 ft) = **51.6 plf**	**51.6 plf**
		Second floor	2350	(1650 lb + 2350 lb)/(12 ft + 8 ft + 12 ft) = **125 plf**	(2350 lb)/(12 ft + 8 ft + 12 ft) = **73.4 plf**

The adjusted allowable unit shear using $\Omega = 2.0$ is

$$v_{\text{allowable}}\ (\text{wind}) = (0.92)(785\ \text{plf})/2.0 = \mathbf{359\ plf} > v_{wr} = 113.7\ \text{plf}$$

$$> v_{w2} = 340.9\ \text{plf} \qquad \text{OK}$$

Therefore, the required fastening for the shear walls between the second floor and the roof and between the ground floor and the second floor is 8d nails at 6 in. on-center EN and 12 in. on-center FN.

10.11.2 Longitudinal direction (seismic governs)

The shear wall unit shears are as follows:

$$V_{Er}\left(\text{roof level}\right) = 37.2\ \text{plf}$$

$$V_{E2}\left(\text{second-floor level}\right) = 91.3\ \text{plf}$$

The $\frac{15}{32}$-in. panel with 8d nails will be the minimum required for these walls; therefore, the nominal capacity of the shear walls is 560 plf. Adjusting this value for Spruce-Pine-Fir lumber framing yields

$$v_{\text{allowable (seismic)}} = (0.92)(560\,\text{plf})/2.0 = \mathbf{257\ plf} > v_{Er} = 37.2\,\text{plf}$$

$$> v_{E2} = 91.3\,\text{plf}$$

Therefore, the fastening specified for the shear wall in the transverse direction is adequate for the shear walls in the longitudinal direction.

10.12 OVERTURNING ANALYSIS OF SHEAR WALLS: SHEAR WALL CHORD FORCES

The shear walls (Figure 10.23) will now be analyzed for overturning. Through this analysis, the maximum compression and tension forces in the shear wall chords will be calculated. The first step will be to calculate the lateral and gravity loads to each shear wall. The lateral load to each shear wall is summarized in Tables 10.13 and 10.14.

The gravity loads to each shear wall will now be calculated. Recall the following from Section 10.2:

Roof dead load = 22 psf
Roof snow load = 35 psf
Floor dead load = 40 psf
Floor live load = 50 psf
Wall dead load =10 psf

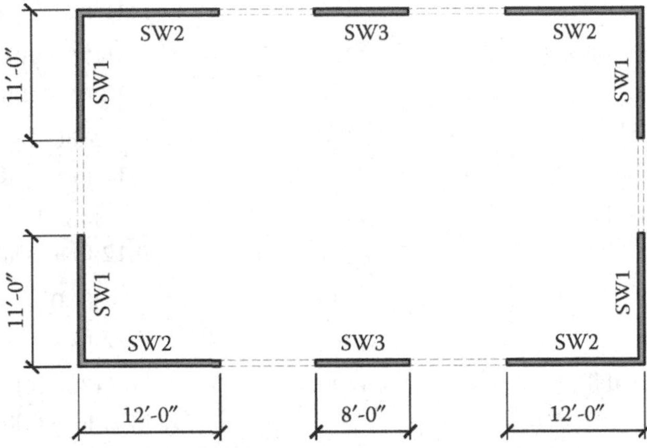

Figure 10.23 Shear wall layout.

Table 10.13 Summary of unfactored lateral forces on a building

Direction	*Load*	*Level*	*F (lb) (from Figures 10.2 and 10.4)*
Transverse	Seismic	Roof	2380
		Second floor	3460
	Wind	Roof	5000
		Second floor	10,000
Longitudinal	Seismic	Roof	2380
		Second floor	3460
	Wind	Roof	3300
		Second floor	4700

Table 10.14 Unfactored lateral forces on each shear wall

Wall	*Load*	*Level*	*F (lb)* *(from Figures 10.2 and 10.4)*	$F_w = \frac{F}{\Sigma L_w}(L_{ww})$
SW1	Seismic	Roof	2380	$\frac{(2380\text{ lb})}{(4)(11\text{ ft})}(11\text{ ft}) = \mathbf{595\text{ lb}}$
		Second floor	3460	$\frac{(3460\text{ lb})}{(4)(11\text{ ft})}(11\text{ ft}) = \mathbf{865\text{ lb}}$
	Wind	Roof	5000	$\frac{(5000\text{ lb})}{(4)(11\text{ ft})}(11\text{ ft}) = \mathbf{1250\text{ lb}}$
		Second floor	10,000	$\frac{(10{,}000\text{ lb})}{(4)(11\text{ ft})}(11\text{ ft}) = \mathbf{2500\text{ lb}}$
SW2	Seismic	Roof	2380	$\frac{(2380\text{ lb})}{(4)(12\text{ ft}) + (2)(8\text{ ft})}(12\text{ ft}) = \mathbf{447\text{ lb}}$
		Second floor	3460	$\frac{(3460\text{ lb})}{(4)(12\text{ ft}) + (2)(8\text{ ft})}(12\text{ ft}) = \mathbf{649\text{ lb}}$
	Wind	Roof	3300	$\frac{(3300\text{ lb})}{(4)(12\text{ ft}) + (2)(8\text{ ft})}(12\text{ ft}) = \mathbf{619\text{ lb}}$
		Second floor	4700	$\frac{(4700\text{ lb})}{(4)(12\text{ ft}) + (2)(8\text{ ft})}(12\text{ ft}) = \mathbf{882\text{ lb}}$
SW3	Seismic	Roof	2380	$\frac{(2380\text{ lb})}{(4)(12\text{ ft}) + (2)(8\text{ ft})}(8\text{ ft}) = \mathbf{298\text{ lb}}$
		Second floor	3460	$\frac{(3460\text{ lb})}{(4)(12\text{ ft}) + (2)(8\text{ ft})}(8\text{ ft}) = \mathbf{433\text{ lb}}$
	Wind	Roof	3300	$\frac{(3300\text{ lb})}{(4)(12\text{ ft}) + (2)(8\text{ ft})}(8\text{ ft}) = \mathbf{413\text{ lb}}$
		Second floor	4700	$\frac{(4700\text{ lb})}{(4)(12\text{ ft}) + (2)(8\text{ ft})}(8\text{ ft}) = \mathbf{588\text{ lb}}$

The calculation of gravity loads on shear walls is covered in Chapter 7. The summary of the loads on walls SW1, SW2, and SW3 is shown in Figures 10.24 through 10.26, respectively.

10.12.1 Maximum force in the tension chord

The maximum force that can occur in the tension chord force will now be calculated for each shear wall using the applicable IBC load combinations from Chapter 2. The reader should refer to Section 7.2 where the equations for the forces in the tension and compression chords of the shear wall are derived. It will be recalled that because the lateral wind or seismic loads can act in any direction, the summation of moments in the shear wall is taken about a point on the left- or right-end chord, whichever produces the greater chord forces.

SW1 wind loads (refer to Figure 10.24 for loads):

$$OM_2 = (1250)(12\text{ ft}) = 15{,}000\text{ ft-lb}$$

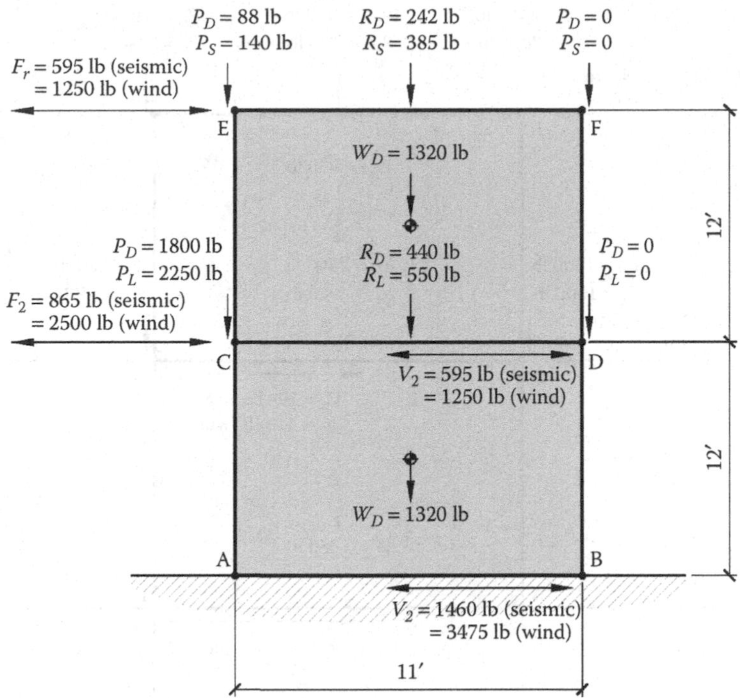

Figure 10.24 Free-body diagram of SW1.

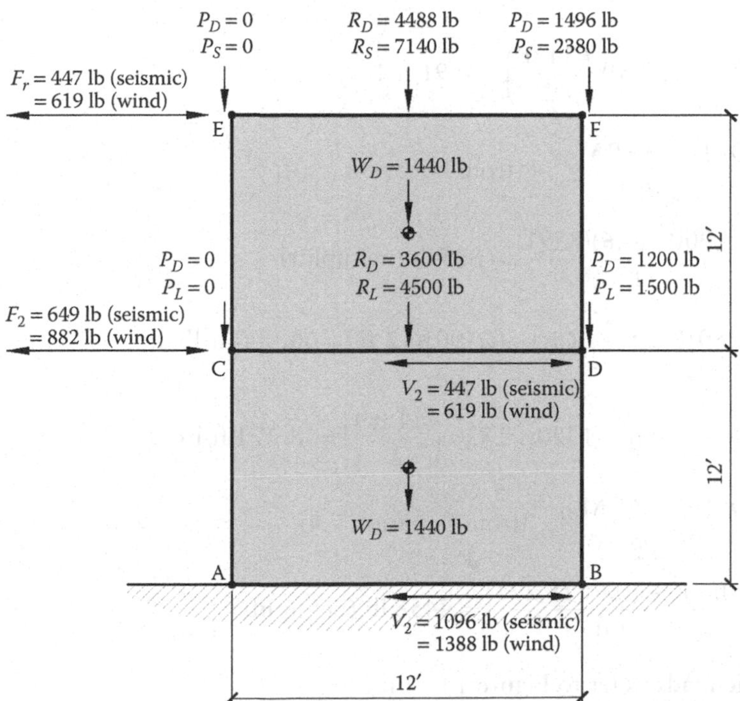

Figure 10.25 Free-body diagram of SW2.

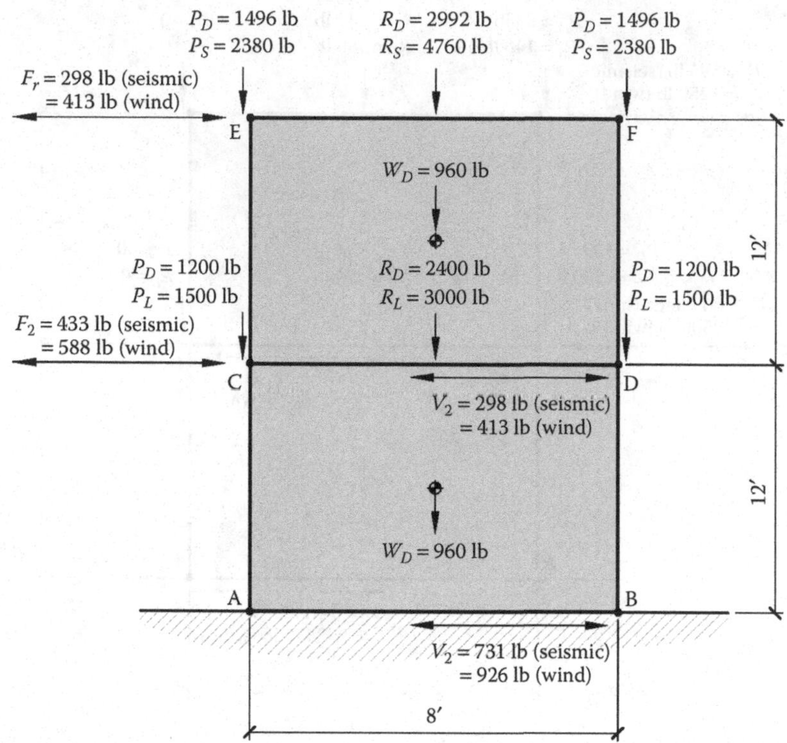

Figure 10.26 Free-body diagram of SW3.

$$\mathrm{RM}_{D2} = (242 + 1{,}320)\left(\frac{11 \text{ ft}}{2}\right) = 8591 \text{ ft-1b}$$

$$T_2 = \frac{\mathrm{OM}_2 - 0.6\mathrm{RM}_{D2}}{w} \quad \text{[from Equation (7.3)]}$$

$$= \frac{15{,}000 - (0.6)(8591)}{11 \text{ ft}} = \mathbf{895\ lb} \text{ (net uplift)}$$

$$\mathrm{OM}_1 = (1250)(12 \text{ ft} + 12 \text{ ft}) + (2500)(12 \text{ ft}) = 60{,}000 \text{ ft-lb}$$

$$\mathrm{RM}_{D1} = \left[(242 + 440 + 1320 + 1320)\left(\frac{11 \text{ ft}}{2}\right)\right] = 18{,}271 \text{ ft-lb}$$

$$T_1 = \frac{\mathrm{OM}_1 - 0.6\mathrm{RM}_{D1}}{w} \quad \text{[from Equation 7.4]}$$

$$= \frac{60{,}000 - (0.6)(18{,}271)}{11 \text{ ft}} = \mathbf{4458\ lb} \text{ (net uplift)}$$

SW1 seismic loads (refer to Figure 10.24 for loads):

$\rho = 1.0$ and $S_{DS} = 0.267$

$$\text{OM}_2 = (595)(12\text{ ft}) = 7140\text{ ft-lb}$$

$$T_2 = \frac{0.7\text{OM}_2 - 0.56\text{RM}_{D2}}{w} \quad \text{[from Equation (7.5)]}$$

$$= \frac{(0.7)(7140) - (0.56)(8591)}{11\text{ ft}} = \mathbf{17\ lb}\text{ (net uplift)}$$

$$\text{OM}_1 = \left[(595)(12\text{ ft} + 12\text{ ft})\right] + \left[(865)(12\text{ ft})\right] = 24{,}660\text{ ft-lb}$$

$$T_1 = \frac{0.7\text{OM}_1 - 0.56\text{RM}_{D1}}{w} \quad \text{[from Equation (7.6)]}$$

$$= \frac{(0.7)(24{,}660) - (0.56)(18{,}271)}{11\text{ ft}} = \mathbf{639\ lb}$$

SW2 wind loads (refer to Figure 10.25 for loads):

$$\text{OM}_2 = (619)(12\text{ ft}) = 7428\text{ ft-lb}$$

$$\text{RM}_{D2} = (4488 + 1440)\left(\frac{12\text{ ft}}{2}\right) = 35{,}568\text{ ft-lb}$$

$$T_2 = \frac{\text{OM}_2 - 0.6\text{RM}_{D2}}{w} \quad \text{[from Equation (7.3)]}$$

$$= \frac{7428 - (0.6)(35{,}568)}{12\text{ ft}} = \mathbf{-1159\ lb}\text{ (minus sign indicates no net uplift)}$$

$$\text{OM}_1 = (619)(12\text{ ft} + 12\text{ ft}) + (882)(12\text{ ft}) = 25{,}440\text{ ft-lb}$$

$$\text{RM}_{D1} = \left[(4488 + 3600 + 1440 + 1440)\left(\frac{12\text{ ft}}{2}\right)\right] = 65{,}808\text{ ft-lb}$$

$$T_1 = \frac{\text{OM}_1 - 0.6\text{RM}_{D1}}{w} \quad \text{[from Equation (7.4)]}$$

$$= \frac{25{,}440 - (0.6)(65{,}808)}{12\text{ ft}} = \mathbf{-1170\ lb}\text{ (minus sign indicates no net uplift)}$$

SW2 seismic loads (refer to Figure 10.25 for loads):

$\rho = 1.0$ and $S_{DS} = 0.267$

$$\text{OM}_2 = (447)(12\text{ ft}) = 5364\text{ ft-lb}$$

$$T_2 = \frac{0.7\text{OM}_2 - 0.56\text{RM}_{D2}}{w} \quad \text{[from Equation (7.5)]}$$

$$= \frac{(0.7)(5364) - (0.56)(35{,}568)}{12\text{ ft}} = \mathbf{-1346\ lb}\text{ (minus sign indicates no net uplift)}$$

$$OM_1 = \left[(447)(12\text{ ft} + 12\text{ ft})\right] + (649)(12\text{ ft}) = 18,516\text{ ft-lb}$$

$$T_1 = \frac{0.7OM_1 - 0.56RM_{D1}}{w} \quad \text{[from Equation (7.6)]}$$

$$= \frac{(0.7)(18,516) - (0.56)(65,808)}{12\text{ ft}} = -\mathbf{1991\ lb}\text{ (minus sign indicates no net uplift)}$$

SW3 wind loads (refer to Figure 10.26 for loads):

$$OM_2 = (413)(12\text{ ft}) = 4956\text{ ft-lb}$$

$$RM_{D2} = [(1496)(8\text{ ft})] + \left[(2992 + 960)\left(\frac{8\text{ ft}}{2}\right)\right] = 27,776\text{ ft-lb}$$

$$T_2 = \frac{OM_2 - 0.6RM_{D2}}{w} \quad \text{[from Equation (7.3)]}$$

$$= \frac{4956 - (0.6)(27,776)}{8\text{ ft}} = -\mathbf{1463\ lb}\text{ (minus sign indicates no net uplift)}$$

$$OM_1 = (413)(12\text{ ft} + 12\text{ ft}) + (588)(12\text{ ft}) = 16,968\text{ ft-lb}$$

$$RM_{D1} = [(1496 + 1200)(8\text{ ft})] + \left[(2992 + 2400 + 960 + 960)\left(\frac{8\text{ ft}}{2}\right)\right] = 50,816\text{ ft-lb}$$

$$T_1 = \frac{OM_1 - 0.6RM_{D1}}{w} \quad \text{[from Equation (7.4)]}$$

$$= \frac{(16,968) - (0.6)(50,816)}{8\text{ ft}} = -\mathbf{1690\ lb}\text{ (minus sign indicates no net uplift)}$$

SW3 seismic loads (refer to Figure 10.26 for loads):

$\rho = 1.0$ and $S_{DS} = 0.267$

$$OM_2 = (298)(12\text{ ft}) = 3576\text{ ft-lb}$$

$$T_2 = \frac{0.7OM_2 - 0.56RM_{D2}}{w} \quad \text{[from Equation (7.5)]}$$

$$= \frac{(0.7)(3576) - (0.56)(27,776)}{8\text{ ft}} = -\mathbf{1631\ lb}\text{ (minus sign indicates no net uplift)}$$

$$OM_1 = (298)(12\text{ ft} + 12\text{ ft}) + (433)(12\text{ ft}) = 12,348\text{ ft-lb}$$

$$T_1 = \frac{0.7OM_1 - 0.56RM_{D1}}{w} \quad \text{[from Equation (7.6)]}$$

$$= \frac{(0.7)(12{,}348) - (0.56)(50{,}816)}{8 \text{ ft}} = \mathbf{-2476\ lb} \text{ (minus sign indicates no net uplift)}$$

10.12.2 Maximum force in compression chord

The maximum compression chord force will now be calculated for each shear wall using the applicable load combinations from Chapter 2. Recall that in calculating the resisting moments (RM_{D1}, RM_{D2}, RM_{T1}, RM_{T2}) in Equations 7.7 through 7.10, it should be noted that the full P loads or header reactions (see Figure 7.8) and the tributary R and W (i.e., distributed) loads should be used because the compression chords only support distributed gravity loads that are tributary to it in addition to the concentrated header reactions or P loads (see Section 7.2 for further discussion).

SW1 wind loads (refer to Figure 10.24 for loads):

$$OM_2 = 15{,}000 \text{ ft-lb (from tension chord force calculations)}$$

$$RM_{T2} = (385)\left(\frac{2 \text{ ft}}{2}\right) + 140(11 \text{ ft}) = 1925 \text{ ft-1b}$$

$$RM_{D2} = (242 + 1320)\left(\frac{2 \text{ ft}}{2}\right) + 88(11 \text{ ft}) = 2530 \text{ ft-1b}$$

$$C_2 = \frac{0.75OM_2 - RM_{D2} + 0.75RM_{T2}}{w} \quad \text{[from Equation (7.7)]}$$

$$= \frac{(0.75)(15{,}000) + (2530) + (0.75)(1925)}{11 \text{ ft}} = \mathbf{1384\ lb}$$

$$OM_1 = 60{,}000 \text{ ft-lb (from tension chord force calculations)}$$

$$RM_{T1} = [(2{,}250 + 140)(11 \text{ ft})] + \left[(385 + 550)\left(\frac{2 \text{ ft}}{2}\right)\right] = 27{,}225 \text{ ft-lb}$$

$$RM_{D1} = [(1800 + 88)(11 \text{ ft})] + \left[(242 + 440 + 1320 + 1320)\left(\frac{2 \text{ ft}}{2}\right)\right] = 24{,}090 \text{ ft-lb}$$

$$C_1 = \frac{0.75OM_1 + RM_{D1} + 0.75RM_{T1}}{w} \quad \text{[from Equation (7.8)]}$$

$$= \frac{(0.75)(60{,}000) + (24{,}090) + (0.75)(27{,}225)}{11 \text{ ft}} = \mathbf{8137\ lb}$$

SW1 seismic loads (refer to Figure 10.24 for loads):

$\rho = 1.0$ and $S_{DS} = 0.267$

$$OM_2 = 7140 \text{ ft-lb (from tension chord force calculations)}$$

$$RM_{T2} = 1925 \text{ ft-lb (from wind loads)}$$

$$\text{RM}_{D2} = 2530 \text{ ft-lb (from wind loads)}$$

$$C_2 = \frac{0.525\text{OM}_2 + 1.03\text{RM}_{D2} + 0.75\text{RM}_{T2}}{w} \quad \text{[from Equation (7.9)]}$$

$$= \frac{(0.525)(7140) + (1.03)(2530) + (0.75)(1925)}{8 \text{ ft}} = \mathbf{975 \text{ lb}}$$

$$\text{OM}_1 = 24{,}660 \text{ ft-lb (from tension chord force calculations)}$$

$$\text{RM}_{T1} = 27{,}225 \text{ ft-lb (from wind loads)}$$

$$\text{RM}_{D1} = 24{,}090 \text{ ft-lb (from wind loads)}$$

$$C_1 = \frac{0.525\text{OM}_1 + 1.03\text{RM}_{D1} + 0.75\text{RM}_{T1}}{w} \quad \text{[from Equation (7.10)]}$$

$$= \frac{(0.525)(24{,}660) + (1.03)(24{,}090) + (0.75)(27{,}225)}{8 \text{ ft}} = \mathbf{7272 \text{ lb}}$$

SW2 wind loads (refer to Figure 10.25 for loads):

$$\text{OM}_2 = 7428 \text{ ft-lb (from tension chord force calculations)}$$

$$\text{RM}_{T2} = (2380)(12 \text{ ft}) + \left[(7140)\left(\frac{2 \text{ ft}}{2}\right)\right] = 35{,}700 \text{ ft-lb}$$

$$\text{RM}_{T2} = (1496)(12 \text{ ft}) + \left[(4488 + 1440)\left(\frac{2 \text{ ft}}{2}\right)\right] = 23{,}880 \text{ ft-lb}$$

$$C_2 = \frac{0.75\text{OM}_2 + \text{RM}_{D2} + 0.75\text{RM}_{T2}}{w} \quad \text{[from Equation (7.7)]}$$

$$= \frac{(0.75)(7428) + (23{,}880) + (0.75)(35{,}700)}{12 \text{ ft}} = \mathbf{4686 \text{ lb}}$$

$$\text{OM}_1 = 25{,}440 \text{ ft-lb (from tension chord force calculations)}$$

$$\text{RM}_{T1} = [(2380 + 1500)(12 \text{ ft})] + \left[(7140 + 4500)\left(\frac{2 \text{ ft}}{2}\right)\right] = 58{,}200 \text{ ft-lb}$$

$$\text{RM}_{D1} = [(1496 + 1200)(12 \text{ ft})] + \left[(4488 + 3600 + 1440 + 1440)\left(\frac{2 \text{ ft}}{2}\right)\right]$$

$$= 43{,}320 \text{ ft-lb}$$

$$C_1 = \frac{(0.75\text{OM}_1 + \text{RM}_{D1} + 0.75\text{RM}_{T1})}{w} \quad \text{[from Equation (7.8)]}$$

$$= \frac{(0.75)(25{,}440) + (43{,}320) + (0.75)(58{,}200)}{12 \text{ ft}} = \mathbf{8838 \text{ lb}}$$

SW2 seismic loads (refer to Figure 10.25 for loads):

$\rho = 1.0$ and $S_{DS} = 0.267$

$$\text{OM}_2 = 5364 \text{ ft-lb (from tension chord force calculations)}$$

$$\text{RM}_{T2} = 35{,}700 \text{ ft-lb (from wind loads)}$$

$$\text{RM}_{D2} = 23{,}880 \text{ ft-lb (from wind loads)}$$

$$C_2 = \frac{0.525\text{OM}_2 + 1.03\text{RM}_{D2} + 0.75\text{RM}_{T2}}{w} \quad \text{[from Equation (7.9)]}$$

$$= \frac{(0.525)(5364) + (1.03)(23{,}880) + (0.75)(35{,}700)}{12} = \mathbf{4516\ lb}$$

$$\text{OM}_1 = 18{,}516 \text{ ft-lb (from tension chord force calculations)}$$

$$\text{RM}_{T1} = 58{,}200 \text{ ft-lb (from wind loads)}$$

$$\text{RM}_{D1} = 43{,}320 \text{ ft-lb (from wind loads)}$$

$$C_1 = \frac{0.525\text{OM}_1 + 1.03\text{RM}_{D1} + 0.75\text{RM}_{T1}}{w} \quad \text{[from Equation (7.10)]}$$

$$= \frac{(0.525)(18{,}516) + (1.03)(43{,}320) + (0.75)(58{,}200)}{12 \text{ ft}} = 8166 \text{ lb}$$

SW3 wind loads (refer to Figure 10.26 for loads):

$$\text{OM}_2 = 4956 \text{ ft-lb (from tension chord force calculations)}$$

$$\text{RM}_{T2} = (2380)(8 \text{ ft}) + \left[(4760)\left(\frac{2 \text{ ft}}{2}\right)\right] = 23{,}800 \text{ ft-lb}$$

$$\text{RM}_{D2} = (1496)(8 \text{ ft}) + \left[(2992 + 960)\left(\frac{2 \text{ ft}}{2}\right)\right] = 15{,}920 \text{ ft-lb}$$

$$C_2 = \frac{0.75\text{OM}_2 + \text{RM}_{D2} + 0.75\text{RM}_{T2}}{w} \quad \text{[from Equation (7.7)]}$$

$$= \frac{(0.75)(4956) + (15{,}920) + (0.75)(23{,}800)}{8 \text{ ft}} = \mathbf{4686\ lb}$$

$$\text{OM}_1 = 16{,}968 \text{ ft-lb (from tension chord force calculations)}$$

$$\text{RM}_{T1} = [(2380 + 1500)(8 \text{ ft})] + \left[(4760 + 3000)\left(\frac{2 \text{ ft}}{2}\right)\right] = 38{,}800 \text{ ft-lb}$$

$$\text{RM}_{D1} = [(1496 + 1200)(8 \text{ ft}) + \left[(2992 + 2400 + 960 + 960)\left(\frac{2 \text{ ft}}{2}\right)\right] = 28{,}880 \text{ ft-lb}$$

$$C_1 = \frac{(0.75\text{OM}_1 + \text{RM}_{D1} + 0.75\text{RM}_{T1})}{w} \quad \text{[from Equation (7.8)]}$$

$$= \frac{(0.75)(16{,}968) + (28{,}880) + (0.75)(38{,}800)}{8 \text{ ft}} = \mathbf{8838\ lb}$$

SW3 seismic loads (refer to Figure 10.26 for loads):

$\rho = 1.0$ and $S_{DS} = 0.267$

$$\text{OM}_2 = 3576 \text{ ft-lb (from tension chord force calculations)}$$

$$\text{RM}_{T2} = 23{,}800 \text{ ft-lb (from wind loads)}$$

$$\text{RM}_{D2} = 15{,}920 \text{ ft-lb (from wind loads)}$$

$$C_2 = \frac{0.525\text{OM}_2 + 1.03\text{RM}_{D2} + 0.75\text{RM}_{T2}}{w} \quad \text{[from Equation (7.9)]}$$

$$= \frac{(0.525)(3576) + (1.03)(15{,}920) + (0.75)(23{,}800)}{8 \text{ ft}} = \mathbf{4516\ lb}$$

$$\text{OM}_1 = 12{,}348 \text{ ft-lb (from tension chord force calculations)}$$

$$\text{RM}_{T1} = 38{,}800 \text{ ft-lb (from wind loads)}$$

$$\text{RM}_{D1} = 28{,}880 \text{ ft-lb (from wind loads)}$$

$$C_1 = \frac{0.525\text{OM}_1 + 1.03\text{RM}_{D1} + 0.75\text{RM}_{T1}}{w} \quad \text{[from Equation (7.10)]}$$

$$= \frac{(0.525)(12{,}348) + (1.03)(28{,}880) + (0.75)(38{,}800)}{8 \text{ ft}} = \mathbf{8166\ lb}$$

The shear wall design forces are summarized in Table 10.15.

Table 10.15 Summary of shear wall chord design forces (lb)

		Level 2			*Ground level*		
Load	*Wall*	V_2	T_2	C_2	V_1	T_1	C_1
Seismic	SW1	595	17	975	1460	639	7272
	SW2	447	−1346	4516	1096	−1991	8166
	SW3	298	−1631	4516	731	−2476	8166
Wind	SW1	1250	**895**	1384	**3750**	**4458**	8137
	SW2	619	−1159	4686	1501	−1170	8838
	SW3	413	−1463	4686	1001	−1690	**8838**

Note: Negative values for T_1 and T_2 forces actually indicate net compressive forces; hold-down anchors would, theoretically, not be required. Bold values indicate controlling values that will be used for design. Note that the value of the wind uplift force of 710 lb in the shear wall chords calculated in Section 10.4 must be added to the net tension forces due to overturning from wind obtain the net uplift force on the MWFRS (see Section 10.4). For wall SW1 at the ground-floor level, the total net uplift force, T_1, will become 5168 lb (i.e., 4458 lb + 710 lb).

10.13 FORCES IN HORIZONTAL DIAPHRAGM CHORDS, DRAG STRUTS, AND LAP SPLICES

With reference to Figure 10.20, the diaphragm chord forces are summarized in Table 10.16. With reference to Figures 10.27 and 10.28, the drag strut forces are summarized in Table 10.17.

Table 10.16 Seismic and wind loads on horizontal diaphragm chords

Direction	*Level*	*w (plf) (from Figure 10.20)*	*L (ft)*	*b (ft)*	*M = (wL²/8) (ft-lb)*	*F_{dc} = M/b (lb)*
Seismic loads						
Transverse	Roof	49.6	48	30	$(49.6)(48^2)/8 =$ **14,285**	14,285/30 = **477**
	Second floor	72.1	48	30	$(72.1)(48^2)/8 =$ **20,765**	20,765/30 = **693**
Longitudinal	Roof	79.4	30	48	$(79.4)(30^2)/8 =$ **8933**	8933/48 = **187**
	Second floor	115.3	30	48	$(115.3)(30^2)/8 =$ **12,972**	12,972/48 = **271**
Wind loads						
Transverse	Roof	104.2	48	30	$(104.2)(48^2)/8 =$ **30,010**	30,010/30 = **1001**
	Second floor	208.3	48	30	$(208.3)(48^2)/8 =$ **59,990**	59,990/30 = **2000**
Longitudinal	Roof	110	30	48	$(110)(30^2)/8 =$ **12,375**	12,375/48 = **258**
	Second floor	156.7	30	48	$(156.7)(30^2)/8 =$ **17,629**	17,629/48 = **367**

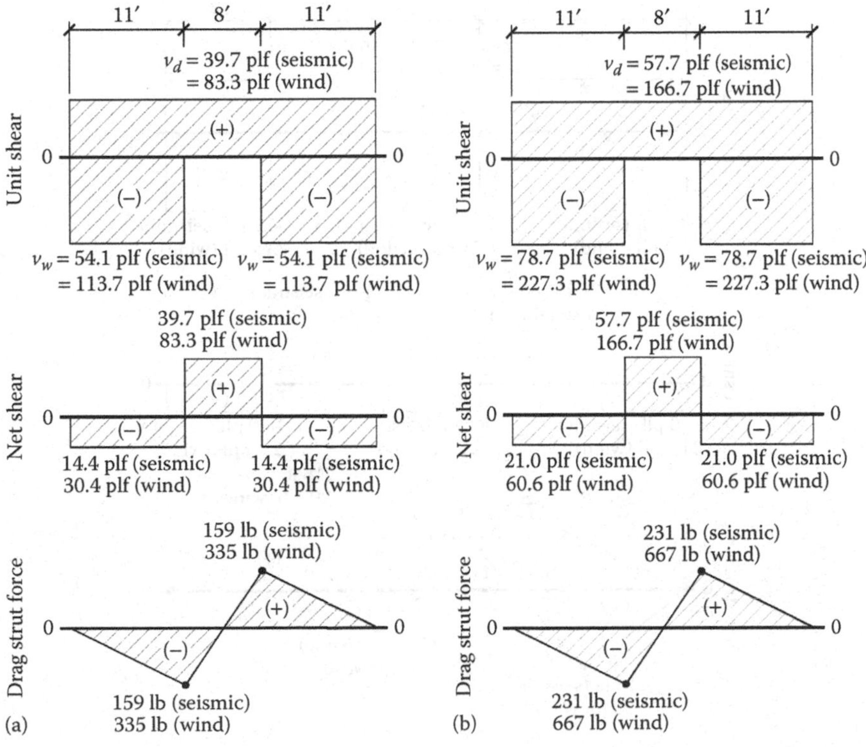

Figure 10.27 Unit shear, net shear, and drag strut force diagrams: (a) transverse—roof and (b) transverse—second floor.

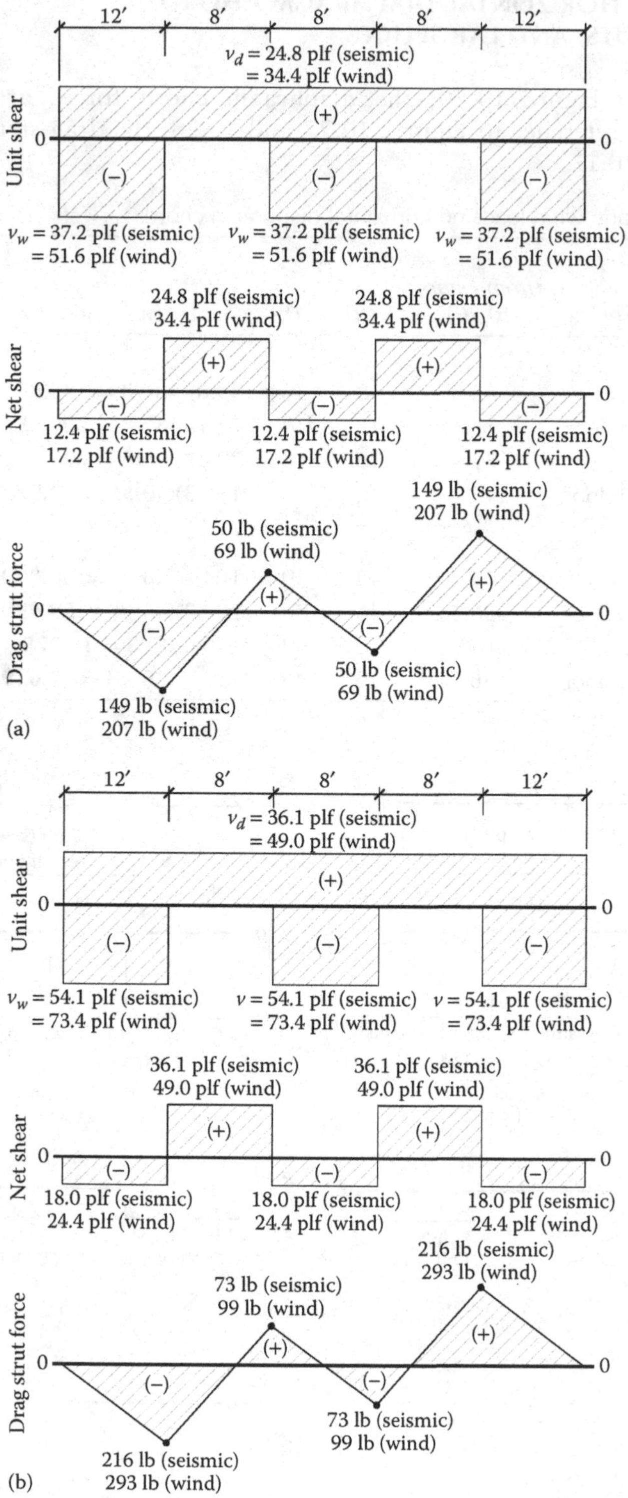

12′ 8′ 8′ 8′ 12′

v_d = 24.8 plf (seismic)
= 34.4 plf (wind)

Unit shear
0 (+) 0
(−) (−) (−)

v_w = 37.2 plf (seismic)
= 51.6 plf (wind)
v_w = 37.2 plf (seismic)
= 51.6 plf (wind)
v_w = 37.2 plf (seismic)
= 51.6 plf (wind)

24.8 plf (seismic)
34.4 plf (wind)
24.8 plf (seismic)
34.4 plf (wind)

Net shear
0 (+) (+) 0
(−) (−) (−)

12.4 plf (seismic)
17.2 plf (wind)
12.4 plf (seismic)
17.2 plf (wind)
12.4 plf (seismic)
17.2 plf (wind)

149 lb (seismic)
207 lb (wind)
50 lb (seismic)
69 lb (wind)

Drag strut force
0 (+) (+) 0
(−) (−)

50 lb (seismic)
69 lb (wind)
149 lb (seismic)
207 lb (wind)

(a)

12′ 8′ 8′ 8′ 12′

v_d = 36.1 plf (seismic)
= 49.0 plf (wind)

Unit shear
0 (+) 0
(−) (−) (−)

v_w = 54.1 plf (seismic)
= 73.4 plf (wind)
v = 54.1 plf (seismic)
= 73.4 plf (wind)
v = 54.1 plf (seismic)
= 73.4 plf (wind)

36.1 plf (seismic)
49.0 plf (wind)
36.1 plf (seismic)
49.0 plf (wind)

Net shear
0 (+) (+) 0
(−) (−) (−)

18.0 plf (seismic)
24.4 plf (wind)
18.0 plf (seismic)
24.4 plf (wind)
18.0 plf (seismic)
24.4 plf (wind)

216 lb (seismic)
293 lb (wind)
73 lb (seismic)
99 lb (wind)

Drag strut force
0 (+) (+) 0
(−) (−)

73 lb (seismic)
99 lb (wind)
216 lb (seismic)
293 lb (wind)

(b)

Figure 10.28 Unit shear, net shear, and drag strut force diagrams: (a) longitudinal—roof and (b) transverse—second floor.

Table 10.17 Seismic and wind loads on drag struts

Direction	*Level*	v_d *(plf) (from Table 10.11)*	v_w *(plf) (from Table 10.12)*	*Net shear (plf)* $v_w - v_d$	F_{ds} *(lb)*
Seismic loads					
Transverse	Roof	39.7	54.1	14.4	**159**
	Second floor	57.7	78.7	21.0	**231**
Longitudinal	Roof	24.8	37.2	12.4	**149**
	Second floor	36.1	54.1	18.0	**216**
Wind loads					
Transverse	Roof	83.3	113.7	30.4	**335**
	Second floor	166.7	227.3	60.6	**667**
Longitudinal	Roof	34.4	51.6	17.2	**207**
	Second floor	49.0	73.4	24.4	**293**

10.13.1 Design of chords, struts, and splices

The maximum force in the chord and drag struts is a diaphragm chord force of 2000 lb (see Table 10.16). It is common practice to use the maximum force in the design of all chords and lap splices.

10.13.1.1 Design for tension

The maximum applied tension force, $T_{\text{applied}} = 2000$ lb. For stress grade, assume Spruce-Pine-Fir stud grade. Because a 2 × 6 is the dimension lumber, use NDS-S Table 4A. From the table, we obtain the design stress values:

$$F_t = 350 \text{ psi}$$

$$C_D(\text{wind}) = 1.6$$

$$C_M = 1.0$$

$$C_t = 1.0$$

$$C_F(F_t) = 1.0$$

$$A_g = 8.25 \text{ in}^2 \left(\text{only one } 2\times 6 \text{ is effective in tension}\right)$$

The allowable tension stress is

$$f_t' = F_t C_D C_M C_t C_F C_i$$
$$= (350)(1.6)(1)(1)(1)(1) = 560 \text{ psi}$$

The allowable tension force in the top plates is

$$T_{\text{allowable}} = F_t' A_g = (560 \text{ psi})\left(8.25 \text{ in}^2\right) = \mathbf{4620 \text{ lb}} > T_{\text{applied}} = 2000 \text{ lb} \qquad \textbf{OK}$$

10.13.1.2 Design for compression

The top plate is fully braced about both axes of bending. It is braced by the stud wall for y–y or weak-axis bending, and it is braced by the roof or floor diaphragm for x–x or strong-axis bending. Therefore, the column stability factor $C_P = 1.0$. The maximum applied compression force $P_{max} = 2000$ lb. For stress grade, assume Spruce-Pine-Fir stud grade. Because a 2 × 6 is the dimension lumber, use NDS-S Table 4A for design values. From the table, the following design stress values are obtained:

$$F_c = 725 \text{ psi}$$

$$C_D(\text{wind}) = 1.6$$

$$C_M = 1.0$$

$$C_t = 1.0$$

$$C_F(F_c) = 1.0$$

$$C_i = 1.0 \text{ (assumed)}$$

$$C_P = 1.0$$

$$A_g = 16.5 \text{ in}^2 \text{ (for two } 2 \times 6\text{'s)}$$

The allowable compression stress is

$$F_c' = F_c C_D C_M C_t C_F C_i C_P$$
$$= (725)(1.6)(1.0)(1.0)(1.0)(1.0)(1.0) = 1160 \text{ psi}$$

The allowable compression force in the top plates is

$$P_{\text{allowable}} = F_c' A_g = (1160 \text{ psi})(16.5 \text{ in}^2) = \mathbf{19{,}140\ lb} > P_{\text{applied}} = 2000 \text{ lb} \quad \textbf{OK}$$

10.13.1.3 Lap splice

The IBC recommended the minimum length of a top plate lap splice is 4 ft (section 2308.5.3.2). If the length of a lap splice length is *more than twice* the minimum length of 4 ft (i.e., 8 ft or more), the nailed or bolted connection each side of the splice can be designed for *one-half* of the full chord force for double top plates and *one-third* of the full chord force for triple top plates. For lap splice lengths less than 8 ft, the splice is designed for the full chord or drag strut force. The IBC requires a minimum of 8–16d nails each side of a splice connection (Table 2304.10.1).

We will design both nailed and bolted lap splices.

1. Using a nailed connection and assuming a lap splice length of 6 ft, we will design the connection for the full chord force. The maximum applied tension or compression load on the top plate splice $T = 2000$ lb. Try 16d common nails.

$L = 3\frac{1}{2}$ in. (length of nail)

$D = 0.162$ in. (diameter of nail) (see NDS Table L4 for nail sizes)

Minimum spacing (see Table 8.1):

End distance = $15D$ = 2.43 in. ⇒ **2.5 in.**
Edge distance = $2.5D$ = 0.41 in. ⇒ **0.5 in.**
Center-to-center spacing = $15D$ = 2.43 in. ⇒ **2.5 in.**
Row spacing = $5D$ = 0.81 in. ⇒ **1 in.**

Allowable lateral resistance per nail:

$$Z' = ZC_DC_MC_tC_g\ C_\Delta C_{eg}C_{di}C_{tn}$$

where:

$C_D = 1.6$ (wind)
$C_M = 1.0$ (MC $\le$ 19%)
$C_t = 1.0$ ($T \le 100°$F)
$C_g, C_\Delta, C_{eg}\ C_{di}, C_{tn} = 1.0$

The tabulated single shear value for 16d nail (Spruce-Pine-Fir framing), Z = (120 lb) ($p/10D$).

$$Z = (120)[1.5\text{in.}/(10\times 0.162)] = \mathbf{111\ lb}\left(\text{see NDS Code Table 12N and footnote 3}\right)$$

The allowable shear per nail is

$$Z' = ZC_DC_MC_tC_gC_\Delta C_{eg}C_{di}C_{tn}$$

$$= (111)(1.6)(1)(1)(1)(1)(1)(1)(1) = 177 \text{ lb per nail}$$

$$\text{No. of nails req'd} = \frac{T}{Z'} = \frac{2000\text{ lb}}{177\text{ lb}} = 11.3 \Rightarrow \mathbf{12\ nails}$$

Using a 6 ft 0 in. lap splice length and two rows of nails (which gives six nails per row), and assuming a center-to-center spacing of 3 in., we obtain

$$\text{Minimum nail penetration } P \ge 6D = 6(0.162) = 0.972 \text{ in.}\quad \textbf{OK}$$

Use 12 16d nails in two rows *each side* of the splice (Figure 10.29).

2. *Bolted lap splice.* Nailed lap splices are more commonly used for top plates, but as an alternate, we will design a bolted lap splice in this section. Assuming a *lap splice length of 6 ft*, which is less than twice the minimum recommended lap splice length of 4 ft; therefore, the lap splice connection must be designed for the full chord force. The maximum applied tension or compression force on the top plate splice (see Tables 10.16 and 10.17) is

$$T = 2000 \text{ lb}$$

Assume $\frac{5}{8}$ in. diameter thru-bolts.
The length of the fastener is given as

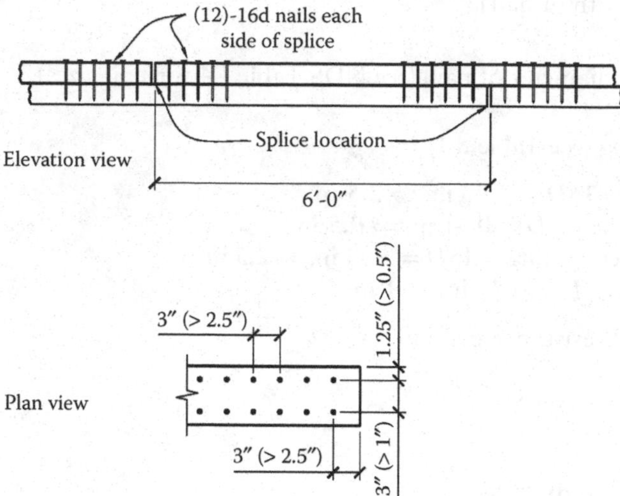

Figure 10.29 Lap splice detail.

$$L = 1.5 \text{ in. } (\text{side member}) + 1.5 \text{ in. } (\text{main member}) = 3 \text{ in.}$$

The minimum bolt spacing to achieve a geometry factor C_Δ of 1.0 is as follows:

End distance = $7D$ = (7)(⅝ in.) = 4.38 in. ⇒ **4.5 in.**

Edge distance: For $L/D = 3/0.625 = 4.8 < 6$

= $1.5D$ = 0.94 in.

Center-to-center spacing = $4D$ = 2.5 in.

Row spacing = $1.5D$ = 0.94 in.

Allowable lateral load per bolt:

$$Z' = ZC_DC_MC_tC_gC_\Delta C_{eg}C_{di}C_{tn}$$

where:

$C_D = 1.6$ (wind)

$C_M = 1.0$ (MC ≤ 19%)

$C_t = 1.0$ ($T \leq 100°$F)

C_Δ, C_e, C_{di}, $C_{tn} = 1.0$

C_g: $A_s = A_m$ = (1.5 in.) × (5.5 in.) = 8.25 in.^{2} ⇒ $A_s/A_m = 1.0$

Assume the number of fasteners in a row = 3 (this will be verified later).

Entering NDS Table 11.3.6A with A_s/A_m of 1.0, A_s of 8.25 in.^{2}, and three bolts in a row, the group factor, C_g, can be obtained by linear interpolation as

$$C_g = 0.97 + \left(\frac{0.99 - 0.97}{12 \text{ in.}^2 - 5 \text{ in.}^2}\right)(8.25 \text{ in.}^2 - 5 \text{ in.}^2) = 0.98$$

Because the lateral load is *parallel* to the grain in both the *side* and *main* members,

$$\mathbf{Z} = \mathbf{Z}_{\parallel} = 510 \text{ lb per bolt } (\text{See NDS Code Table 12A})$$

$$\mathbf{Z}' = \mathbf{Z}_{\parallel}C_DC_MC_tC_gC_\Delta C_{eg}C_{di}C_{tn}$$

$$= 510 \times 1.6 \times 1 \times 1 \times 0.98 \times 1 \times 1 \times 1 \times 1 = 799 \text{ lb per bolt}$$

The number of bolts required to resist the axial force in the top plate splice is T/Z' = 2000/799 = 2.5 < **3 bolts.**

Because the number of bolts required above is *less than or equal to* the number of bolts initially assumed in calculating C_g, this implies that the C_g value obtained previously is conservative, and the three bolts in a row initially assumed is adequate.

It should noted, however, that if the calculated number of bolts required had been *greater* than the number of bolts initially assumed, the C_g value would have had to be recalculated using the newly calculated required number of bolts; the process is repeated until the number of bolts required in the current cycle is *less than or equal to* the number of bolts used in the previous cycle.

Therefore, **use 3-⅝in. diameter bolts** in a single row each side of the lap splice.

10.13.2 Hold-down anchors

From Table 10.15, the maximum uplift at the second floor was 895 lb due to windloads. At the ground level, the maximum tension was 4458 lb also due to wind loads. Adding the 710 lb from the uplift on the MWFRS (see Section 10.4), the total net uplift forces become 1605 lb and 5168 lb at the second-floor and the ground-floor level, respectively. Although both connectors are typically selected from a variety of preengineered products, the connector at the second floor will be designed.

10.13.2.1 Hold-down strap

Assuming an 18-gage ASTM A653 grade 33 strap and 10d common nails, determine the strap length and nail layout.

$$L = 3 \text{ in. (length of nail)}$$

$$D = 0.148 \text{ in. (diameter of nail)}$$

Minimum spacing (see Table 8.1):

End distance = $15D$ = 2.22 in. ⇒ **2.5 in.**
Edge distance = $2.5D$ = 0.37 in. ⇒ **0.5 in.** (0.75 in. provided)
Center-to-center spacing = $15D$ = 2.22 in. ⇒ **2.5 in.**
Row spacing not applicable because there is only one row of nails
Minimum p = $10D$ = 1.48 in. (approximately 3 in. provided)

From NDS Table 12P, the nominal lateral design value is Z = 99 lb. The allowable shear per nail is

$$Z' = ZC_DC_MC_tC_gC_\Delta C_{eg}C_{di}C_{tn}$$

$$= (99)(1.6)(1)(1)(1)(1)(1)(1)(1) = 158.4 \text{ lb per nail}$$

$$\text{Number of nails required} = \frac{T}{Z'} = \frac{1605 \text{ lb}}{158.4 \text{ lb}} = 10.2 \Rightarrow \textbf{6 nails each side}$$

(A steel strap will be provided on both sides of the studs to avoid eccentric loading).

The required length of strap

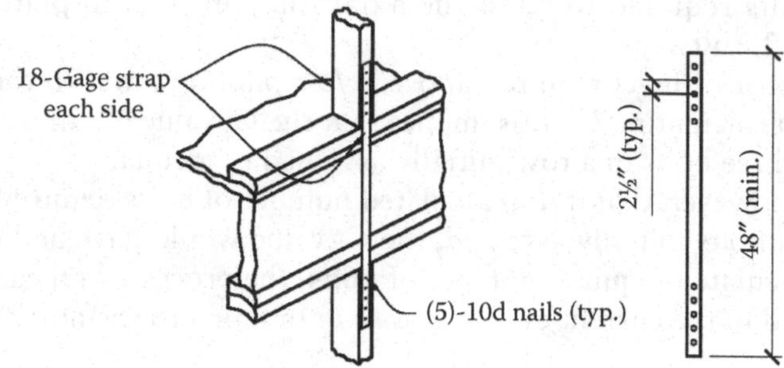

Figure 10.30 Hold-down strap detail.

$$L_{\text{min}} = (2)\left[(2.5 \text{ in. enddist.}) + (5 \text{ spaces})(2.5 \text{ in.}) + (2.5 \text{ in. end distance})\right]$$
$$+ (18 \text{ in. floor structure}) = \mathbf{53\ in.}$$

See Figure 10.30 for strap detail.

10.13.2.2 Hold-down anchor selection

A generic anchor selection table is shown in Figure 10.31. Comparing the maximum tension load of 4088 lb, the hold-down anchor HD4 is selected. With Spruce-Pine-Fir lumber and a load duration factor of 1.6 (160%), the capacity of this connector is 5200 lb. Note that this hold-down connector will be required at each end of the shear wall because the lateral loads can act in either direction. The effect of eccentricity on this connection will be discussed in Section 10.14 where the chords are designed.

10.13.3 Sill anchors

From Table 10.15, the maximum shear at the ground level is 3750 lb shear wall SW1. The anchor bolts will now be designed.

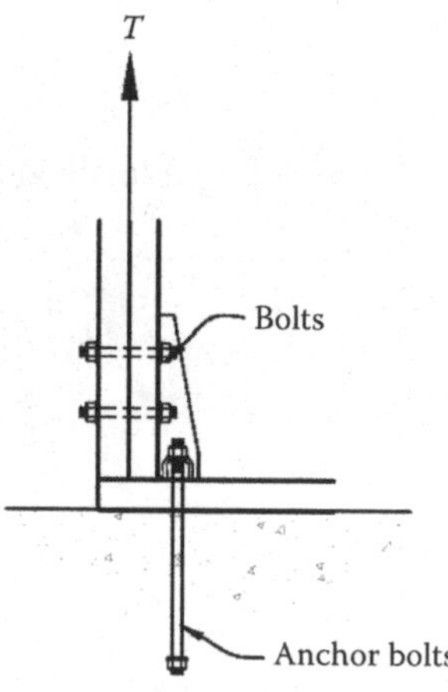

Model #	Fasteners		DF-L or S.P. Allowable tension load (lb) 160%		S.P.F. or H.F. Allowable tension load (lb) 160%	
	Anchor bolts	To studs (bolts)	Bolt length in wood		Bolt length in wood	
			1½″	3″	1½″	3″
HD1	(1)-⅝″	(2)-½″	2,900	4,200	2,200	3,600
HD4	(1)-⅞″	(2)-½″	3,500	6,100	2,800	4,800
xx	xx	xx	xx	xx	xx	xx

Figure 10.31 Hold-down anchor detail.

Assuming ½-in. anchor bolts embedded 7 in. into concrete with a minimum 28-day strength of 2500 psi, the allowable shear in the anchor bolts parallel to the grain of the sill plate is

$$Z' = ZC_DC_MC_tC_gC_\Delta C_{eg}C_{di}C_{tn}$$

where:

$$C_D = 1.6 \ (\text{wind or seismic})$$

$$C_M = 1.0 \ (MC \le 19\%)$$

$$C_t = 1.0 \ (T \le 100°F)$$

$$C_g, C_\Delta, C_e, C_{di}, C_{tn} = 1.0$$

Because the base shear is parallel to the grain of the sill plate, use

$Z_\parallel = 590$ lb (NDS Table 12E)

Minimum spacing (see Table 8.3):

End distance = $7D$ = **3.5 in.**
Edge distance = $1.5D$ = **0.75 in.**
Center-to-center spacing = $4D$ = **2.0 in.**
Row spacing is not applicable here, because there is only one row of bolts.

These distances will be maintained in the detailing; therefore, $C_\Delta = 1.0$.

The allowable shear per bolt is

$$Z' = ZC_DC_MC_tC_gC_\Delta C_{eg}C_{di}C_{tn}$$
$$= (590)(1.6)(1)(1)(1)(1)(1)(1)(1) = 944 \text{ lb per bolt}$$
$$\Rightarrow \text{number of sill anchor bolts required} = 3750 \text{ lb}/944 \text{ lb} = 3.97 \simeq 4 \text{ bolts}$$

Maximum bolt spacing = (11 ft wall length – 1 ft edge distance – 1 ft edge distance)/(4 bolts – 1) = 3 ft 0 in. This is less than the maximum spacing of 6 ft 0 in. permitted by the code.

Use ½-in.-diameter anchor bolts at 3 ft 0 in. o.c. See Figure 10.32.

10.14 SHEAR WALL CHORD DESIGN

The shear wall chords will now be designed. From Table 10.15, the maximum uplift at the ground level is 5168 lb and the maximum compression is 8838 lb. In addition to designing

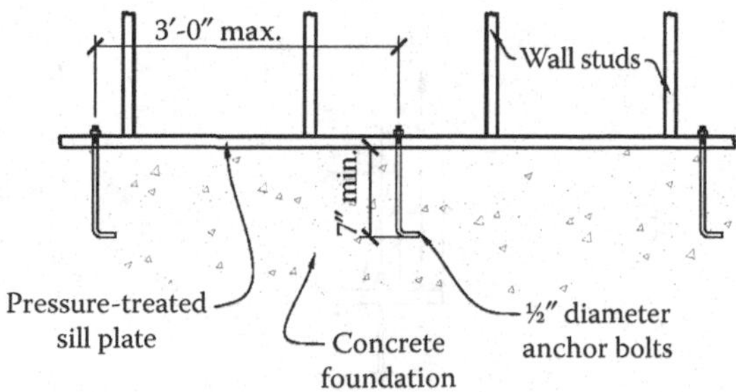

Figure 10.32 Anchor bolt details.

the shear wall chords for the maximum tension and compression loads, a connector needs to be provided to transfer the uplift load to the foundation. It is common practice to select a preengineered connector to resist this load (see Figure 10.33 for various hold-down anchors).

Note that if the connector is one-sided such that the load path is eccentric to the shear wall chord, the chord must be designed to resist the tension load plus the bending moment induced by the eccentric load (see Figure 10.34) using the methods discussed in Chapter 5. If the hold-down connector requires bolts through the chords, the net area of the chord should be used in the design. For this example, a hold-down anchor with a single anchor bolt placed 1.25 in. from the face of the inner stud and two ½-in. bolts through the chords was selected. Two 2 × 6 chords will be assumed. Note that multiple chords need to be nailed or bolted together in accordance with NDS Section 15.3.3 (nailed) or 15.3.4 (bolted).

The following data will be used for the design of the shear wall chords:

$$C_D(\text{wind}) = 1.6$$

$$C_M = 1.0$$

$$C_t = 1.0$$

$$C_F(F_t, F_b) = 1.3$$

$$C_F(F_c) = 1.1$$

$$C_{fu} = 1.15 \text{ (for the tension case)}$$

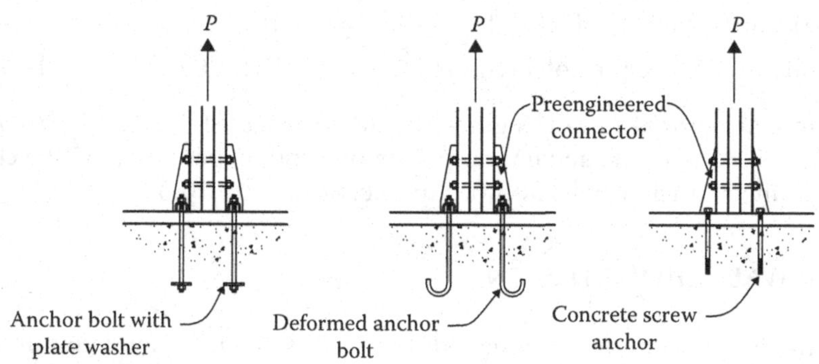

Figure 10.33 Hold-down anchor types.

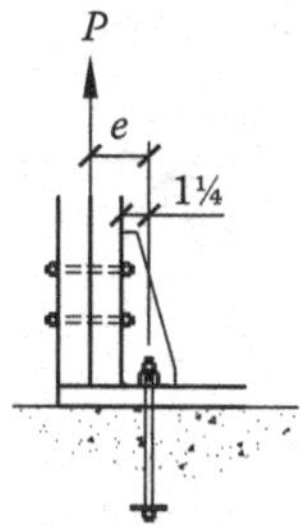

Figure 10.34 Eccentric hold-down anchor.

$$C_i = 1.0$$

$$C_r = 1.0$$

From NDS Table 4A (Spruce-Pine-Fir, Select Structural):

$$F_b = 1250 \text{ psi}$$

$$F_t = 700 \text{ psi}$$

$$F_c = 1400 \text{ psi}$$

$$E_{\min} = 0.55 \times 10^6 \text{psi}$$

For two 2 × 6′s:

$$d = 5.5 \text{ in.}$$

$$b = (2)(1.5 \text{ in.}) = 3 \text{ in.}$$

$$A_g = (2)(8.25 \text{ in.}^2) = 16.5 \text{ in.}^2$$

$$A_n = A_g - A_{\text{holes}} = (16.5) - (2)(1.5 \text{ in.})\left(\frac{1}{2} \text{ in.} + \frac{1}{8} \text{ in.}\right) = 14.6 \text{ in.}^2$$

$$S_y = (3)(31.64 \text{ in.}^3) = 94.92 \text{ in.}^3$$

$$l_u = 12 \text{ ft} = 144 \text{ in. (floor-to-floor height)}$$

10.14.1 Design for compression

Check the slenderness ratio:

$$\frac{\ell_e}{d} = \frac{(1.0)(144 \text{ in.})}{5.5 \text{ in.}} = \mathbf{26.1 < 50 \quad OK} \quad \text{(from Section 5.4, where } l_e = K_e l_u\text{)}$$

Column stability factor C_P:

$$c = 0.8 \text{ (sawn lumber)}$$

$$K_f = 1.0 \text{ (Section 5.4)}$$

$$E'_{\min} = E_{\min} C_M C_t C_i = (0.55 \times 10^6)(1.0)(1.0)(1.0) = 0.55 \times 10^6$$

$$F_{cE} = \frac{0.822 E'_{\min}}{\ell_e/d} = \frac{(0.822)(0.55 \times 10^6)}{(26.1)^2} = 659 \text{ psi}$$

$$\begin{aligned} F_c^* &= F_c C_D C_M C_t C_F C_i \\ &= (1400)(1.6)(1.0)(1.0)(1.0)(1.0)(1.1)(1.0) = 2464 \text{ psi} \end{aligned}$$

$$C_P = K_f\left[\frac{1+F_{cE}/F_c^*}{2c} - \sqrt{\left(\frac{1+F_{cE}/F_c^*}{2c}\right)^2 - \frac{F_{cE}/F_c^*}{2c}}\right] \quad \text{[Equation (5.37)]}$$

$$= (1.0)\left[\frac{1+659/2464}{(2)(0.8)} - \sqrt{\left[\frac{1+659/2464}{(2)(0.8)}\right]^2 - \frac{659/2464}{0.8}}\right] = 0.250$$

Check compression loads:

$$F_c' = F_c^* C_P = (2464\text{ psi})(0.250) = 616\text{ psi}$$

$$P_{\text{allow}} = \begin{cases} F_c' A_g = (616\text{ psi})(16.5\text{ in.}^2) = \mathbf{10{,}164\ lb > 8838\ lb} \ \ \text{(OK at midheight)} \\ F_c' A_n = (2464\text{ psi})(14.6\text{ in.}^2) = \mathbf{35{,}974\ lb > 8838\ lb} \ \ \text{(OK at base)} \end{cases}$$

10.14.2 Design for tension

The total tension force of 5168 lb (i.e., 4458 lb + 710 lb) is applied eccentrically to the shear wall chords (Table 10.18).

The allowable stresses are

$$F_t' = F_t C_D C_M C_t C_F C_i$$
$$= (700)(1.6)(1.0)(1.0)(1.0)(1.0)(1.3)(1.0) = 1456\text{ psi}$$

$$F_b' = F_b C_D C_M C_t C_F C_{fu} C_i C_r$$
$$= (1250)(1.6)(1.0)(1.0)(1.3)(1.15)(1.0)(1.0) = 2990\text{ psi}$$

The actual stresses are

$$f_t = \frac{T}{A_n} = \frac{5168}{14.6} = 354\text{ psi}$$

$$S_y = \frac{b_{\text{net}} d^2}{6}$$

The width of the chords is reduced by the presence of the bolts that connect the hold-down anchors to the chord member.

$$b_{\text{net}} = (5.5\text{ in.}) - \left(\frac{1}{2}\text{ in.} + \frac{1}{8}\text{ in.}\right) = 4.88\text{ in.} \quad \text{(effective width of chords)}$$

$$d = (2)(1.5\text{ in.}) = 3\text{ in.}$$

$$S_y = \frac{(4.88)(3\text{ in.})^2}{6} = 7.31\text{ in.}^3$$

$$f_b = \frac{M}{S_y} = \frac{14{,}212}{7.31} = 1944\text{ psi}$$

Table 10.18 Shear wall and hold-down schedule

Mark	Height (ft)	Length (ft)	Unit shear (lb/ft)	Shear wall chords	Compression force (lb)	Tension, hold-down (lb)	Plywood thickness and nailing	Sides	Level	Hold-down connector	Hold-down anchor bolt embedment	Sill anchor bolt
SW1	12	11	114/341	2-2 × 6 2-2 × 6	1384/8137	895/4458 (1605/5168)g	15/32″ Str.-1 C-DX, 8d nails @ 6″o.c. at edges, 12″ o.c. field	1	Second flr	18 ga.-53″ strap, (12)-10d nails	7″	1/2″ dia. @3′-0″ o.c.
								1	Ground floor	HD4		
SW2	12	12	51.6/125	2-2 × 6 2-2 × 6	4686/8838	No Uplift	15/32″ Str.-1 C-DX, 8d nails @ 6″ o.c. at edges, 12″ o.c. field	1	Second flr	18 ga.-53″ strap, (12)-10d nails	7″	1/2″ dia. @3′-0″ o.c.
								1	Ground floor	HD1		
SW3	12	8	51.6/125	2-2 × 6 2-2 × 6	4686/8838	No Uplift	15/32″ Str.-1 C-DX, 8d nails @ 6″ o.c. at edges, 12″ o.c. field	1	Second flr	18 ga.-53″ strap, (12)-10d nails		1/2″ dia. @3′-0″ o.c.
							15/32″ Str.-1 C-DX, 8d nails @ 6″ o.c. at edges, 12″ o.c. field	1	Ground floor	HD1	7″	

Notes: All plywood panel edges shall be blocked with 2× wood blocking; All plywood panels shall extend from the bottom of the sole or sill plate to the top of the top plates; Where oversized bolt holes in the sill plate are needed for ease of placement of the sill anchors, sleeve the bolt holes with metal sleeves and fill the gaps between the bolts and the inside face of the sleeve expansive cement to ensure snug tight condition; Plywood panels shall have a minimum span rating of 24/16; Locate the edge nails at an edge distance of not less than 3/8″ from panel edges and at equal distance from the panel edges and the edge of the wood framing; All shear wall anchor bolts shall have a minimum embedment of 7 in. into the concrete foundation wall; 1605 lb and 5168 lb uplift forces include 710 lb force due to wind uplift; $M = Te = (5168)(1.5 \text{ in.} + 1.25 \text{ in.}) = 14{,}212$ in.-lb.

Check the combined stresses:

$$\frac{f_t}{F_t'} + \frac{f_b}{F_b'} \leq 1.0$$

$$\frac{354}{1456} + \frac{1944}{2990} = 0.89 \leq 1.0 \quad \textbf{OK} \quad \textbf{for combined stresses}$$

Two 2 × 6 chords are structurally adequate for the worst-case loading conditions. The other chords could be designed, but it is common practice to provide the same chords and hold-down anchors at the ends of all shear walls for buildings of this size.

10.15 CONSTRUCTION DOCUMENTS

In a typical wood building project, the construction documents would include the structural, mechanical, electrical, and architectural drawings as well as the construction specifications that would enable the builder construct the building. For the building design case study, the drawings shown in Figures 10.35 through 10.38 are representatives of what would typically be shown on a construction set of structural drawings.

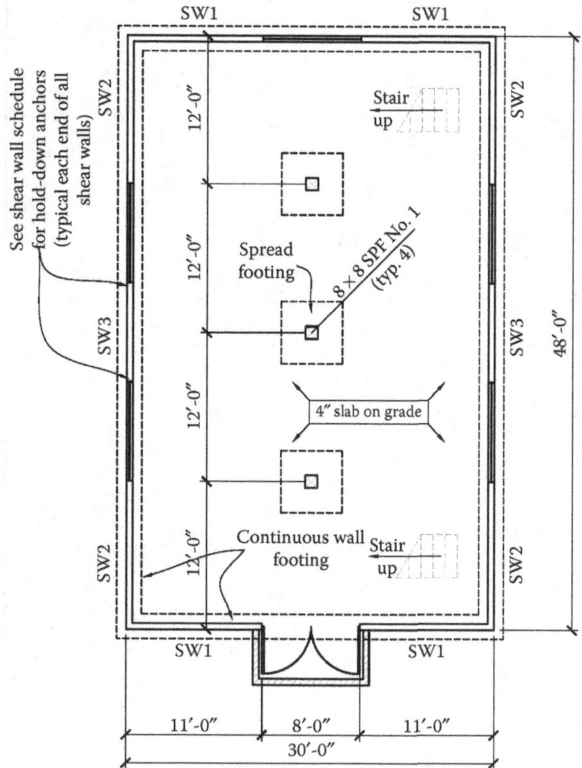

Figure 10.35 Ground-floor plan.

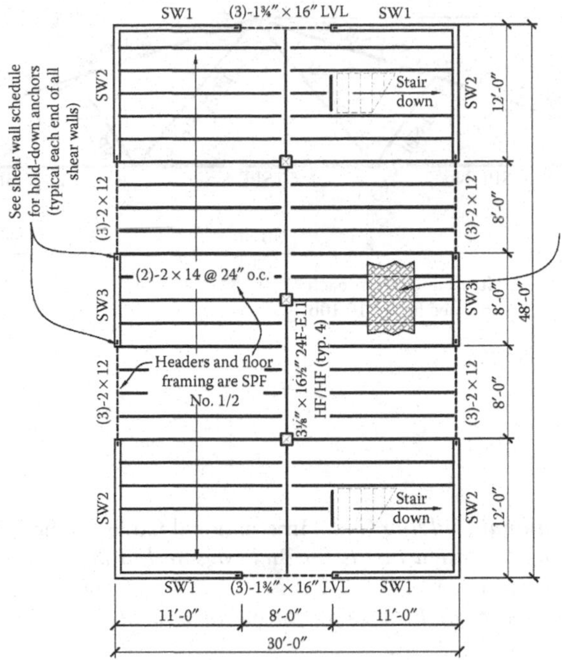

Figure 10.36 Second-floor framing plan.

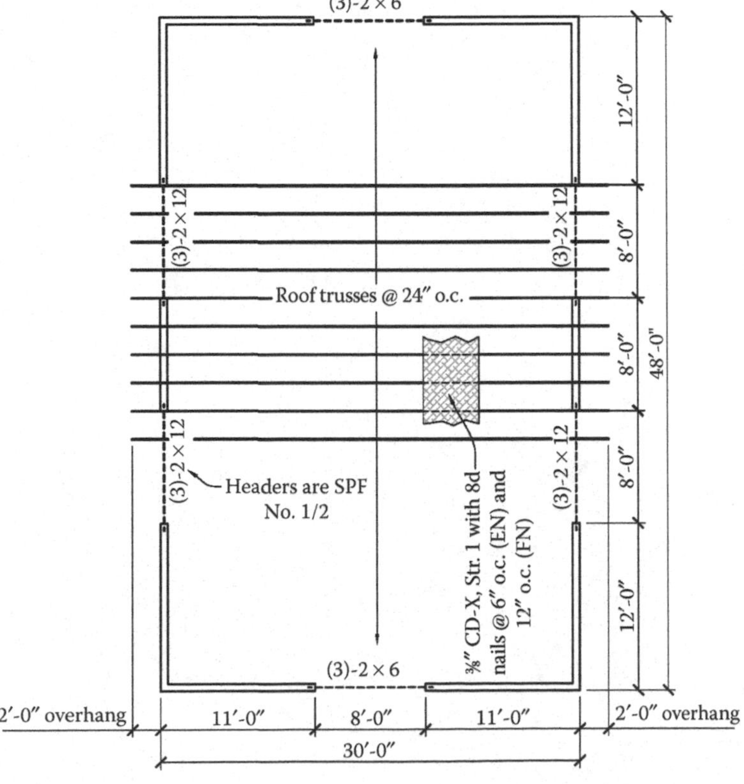

Figure 10.37 Roof framing plan.

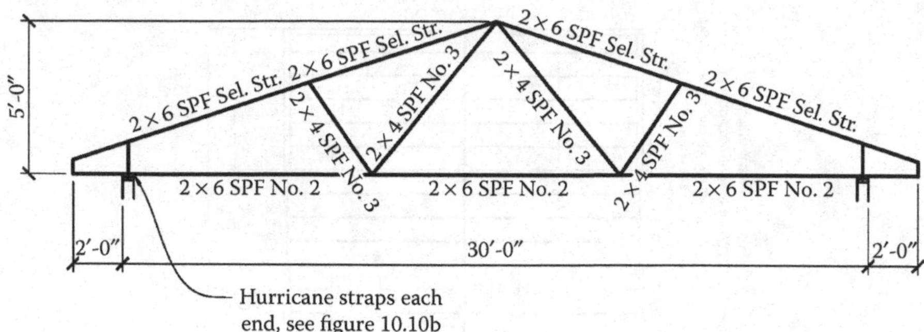

Figure 10.38 Roof truss members.

REFERENCES

1. ICC (2015), *International Building Code*, International Code Council, Washington, DC.
2. ASCE (2010), *Minimum Design Loads for Buildings and Other Structures*, American Society of Civil Engineers, Reston, VA.
3. AWC (2015), *Special Design Provisions for Wind and Seismic*, American Wood Council, Washington, DC.
4. AWC (2015), *National Design Specification Supplement: Design Values for Wood Construction*, American Wood Council, Leesburg, VA.

Appendix A
Weights of building materials

A.1 WOOD FRAMING

Weights shown are for Douglas Fir-Larch, $G = 0.50$, which is conservative for most structures.

	Weight (psf) for spacing			
Nominal size	*12 in. o.c.*	*16 in. o.c.*	*19.2 in. o.c.*	*24 in. o.c.*
2 × 4	1.2	0.9	0.8	0.6
2 × 6	1.8	1.4	1.2	0.9
2 × 8	2.4	1.8	1.5	1.2
2 × 10	3.1	2.3	1.9	1.5
2 × 12	3.7	2.8	2.3	1.9
2 × 14	4.4	3.3	2.7	2.2
2 × 16	5.0	3.8	3.1	2.5
3 × 4	1.9	1.5	1.2	1.0
3 × 6	3.0	2.3	1.9	1.5
3 × 8	4.0	3.0	2.5	2.0
3 × 10	5.1	3.8	3.2	2.6
3 × 12	6.1	4.6	3.9	3.1
3 × 14	7.2	5.4	4.5	3.6
3 × 16	8.3	6.2	5.2	4.2
4 × 4	2.7	2.0	1.7	1.4
4 × 6	4.2	3.2	2.7	2.1
4 × 8	5.5	4.2	3.5	2.8
4 × 10	7.1	5.3	4.4	3.6
4 × 12	8.6	6.4	5.4	4.3
4 × 14	10.1	7.6	6.3	5.1
4 × 16	11.6	8.7	7.3	5.8

A.2 PREENGINEERED FRAMING

A.2.1 Roof trusses

Span (ft.)	*Weight (psf of horizontal plan area)*
30	5.0
35	5.5
40	6.0
50	6.5
60	7.5
80	8.5
100	9.5
120	10.5
150	11.5

A.2.2 Floor joists (I-joists and open web)

Depth (in.)	*Weight (psf of horizontal plan area)*
9½	2.5
11⅞	3.5
14	4.0
16	4.5
18	5.0
20	5.5
24	6.0
26	6.3
28	6.6
30	7.0

A.3 OTHER MATERIALS

	Weight (psf)
Sheathing	
Plywood (per ⅛ in. thickness)	0.4
Gypsum sheathing (per 1 in. thickness)	5.0
Lumber sheathing (per 1 in. thickness)	2.5
Floors and floor finishes	
Asphalt mastic (per 1 in. thickness)	12.0
Ceramic or quarry tile (¾ in. thickness)	10.0
Concrete fill (per 1 in. thickness)	12.0
Gypsum fill (per 1 in. thickness)	6.0
Hardwood flooring (⅞ in. thickness)	4.0
Linoleum or asphalt tile (per ¼ in. thickness)	1.0
Slate (per 1 in. thickness)	15.0

(Continued)

	Weight (psf)
Softwood (3/4 in. thickness)	2.5
Terrazzo (per 1 in. thickness)	13.0
Vinyl tile (1/8 in. thickness)	1.4
Roofs	
Copper or tin	1.0
Felt	
Three-ply ready roofing	1.5
Three-ply with gravel	5.5
Five-ply	2.5
Five-ply with gravel	6.5
Insulation (per 1 in. thickness)	
Cellular glass	0.7
Expanded polystyrene	0.2
Fiberboard	1.5
Fibrous glass	1.1
Loose	0.5
Perlite	0.8
Rigid	1.5
Urethane foam with skin	0.5
Metal deck	3.0
Roll roofing	1.0
Shingles	
Asphalt (1/4 in.)	2.0
Clay tile	12.0
Clay tile with mortar	22.0
Slate (1/4 in.)	10.0
Wood	3.0
Ceilings	
Acoustical fiber tile	1.0
Channel-suspended system	
Steel	2.0
Wood furring	2.0
Plaster on wood lath	8.0
Walls and partitions	
Brick (4 in. nominal thickness)	40
Glass (per 1 in. thickness)	15
Glass block (4 in. thickness)	18
Glazed tile	18
Plaster	8.0
Stone (4 in. thickness)	55
Stucco (7/8 in. thickness)	10.0
Vinyl siding	1.0
Windows (glass, frame, sash)	8.0
Wood paneling	2.5

Appendix B
Design aids

B.1 NOTES ON DESIGN AIDS

B.1.1 Allowable uniform loads on floor joists

The allowable uniform load is calculated with respect to joist length for common 2× lumber sizes and wood species. The repetitive member factor applies and normal load duration and full lateral stability are assumed. Joists were analyzed for bending and deflection (creep factor $k = 1.0$). Shear is typically not a controlling factor and was not used; consequently, the allowable uniform load usually does not exceed 300 plf in the charts. For practical reasons, joist lengths exceeding 30 ft were not considered. Furthermore, joists in the deflection-controlled region should be investigated for vibration.

For the maximum span length charts, allowable live load deflections of $L/360$ and $L/480$ were used for floor framing members ($C_D = 1.0$) and allowable live load deflections of $L/360$ and $L/240$ were used assuming roof framing members ($C_D = 1.15$).

B.1.2 Allowable axial loads on columns

The allowable axial load is calculated with respect to the unbraced length for common square column sizes (4 × 4 through 8 × 8) using common wood species. Normal load duration was assumed. Values for Southern Pine in wet service conditions were included since Southern Pine is commonly used in exterior applications (such as wood-framed decks). All loading is based on no eccentricity.

B.1.3 Combined axial and bending loads on wall studs

A unity curve is plotted in accordance with NDS Equation 3.9.2 for wall studs supporting vertical gravity loads and lateral wind loads. The repetitive member factor applies and a load duration factor of $C_D = 1.6$ was assumed. The weak axis is assumed to be fully braced by wall sheathing.

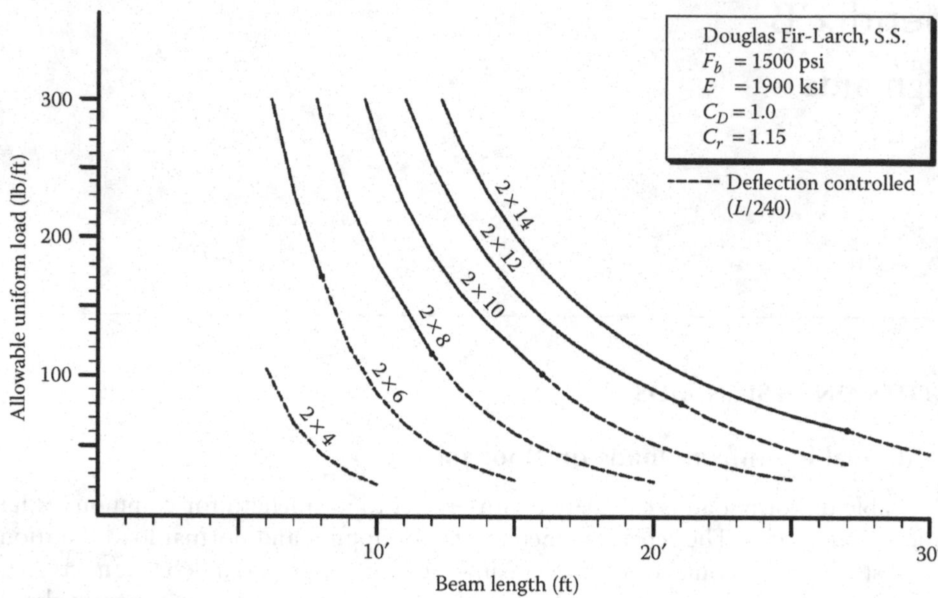

Figure B.1 Allowable uniform loads on floor joists (DF-L, Sel. Str.).

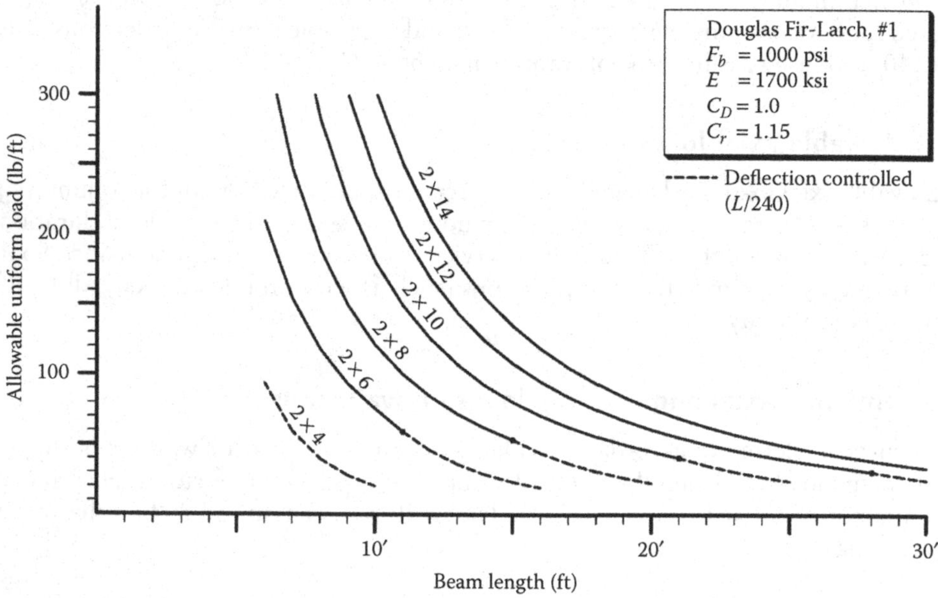

Figure B.2 Allowable uniform loads on floor joists (DF-L, No. 1).

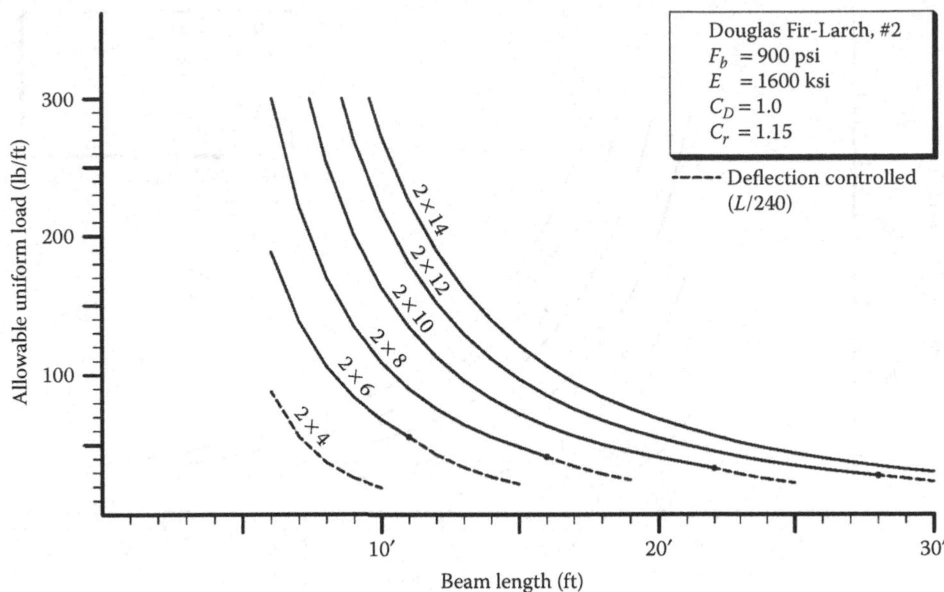

Figure B.3 Allowable uniform loads on floor joists (DF-L, No. 2).

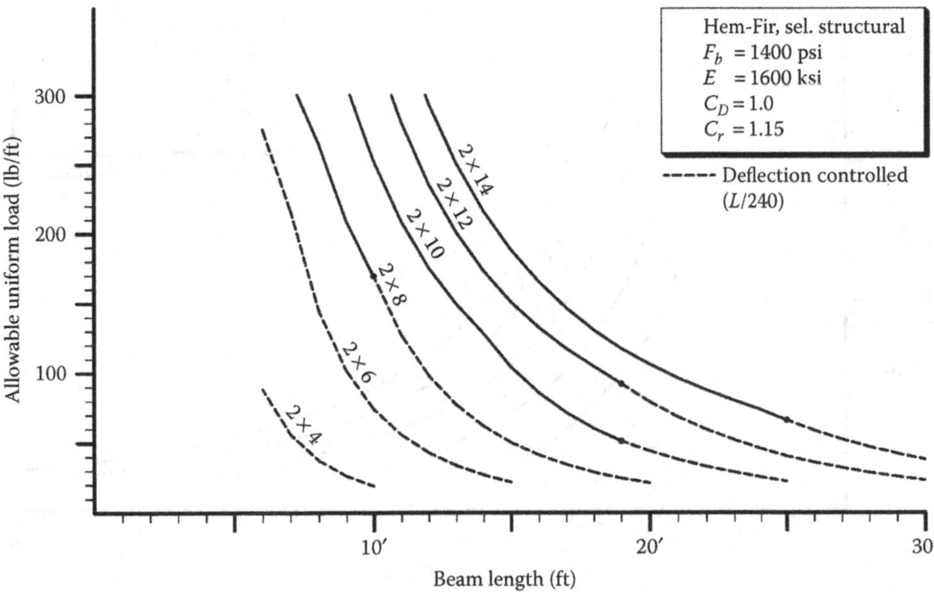

Figure B.4 Allowable uniform loads on floor joists (Hem-Fir, Sel. Str.).

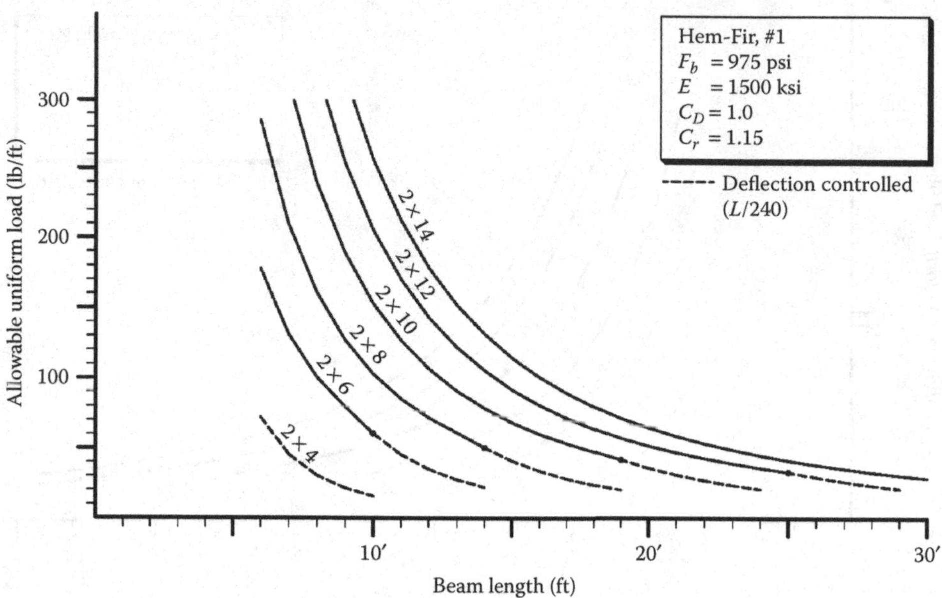

Figure B.5 Allowable uniform loads on floor joists (Hem-Fir, No. 1).

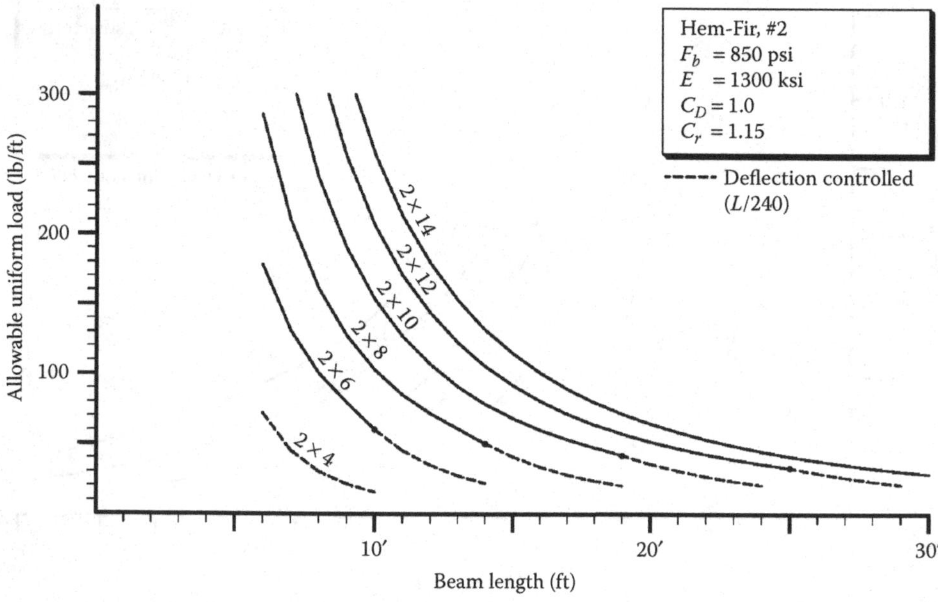

Figure B.6 Allowable uniform loads on floor joists (Hem-Fir, No. 2).

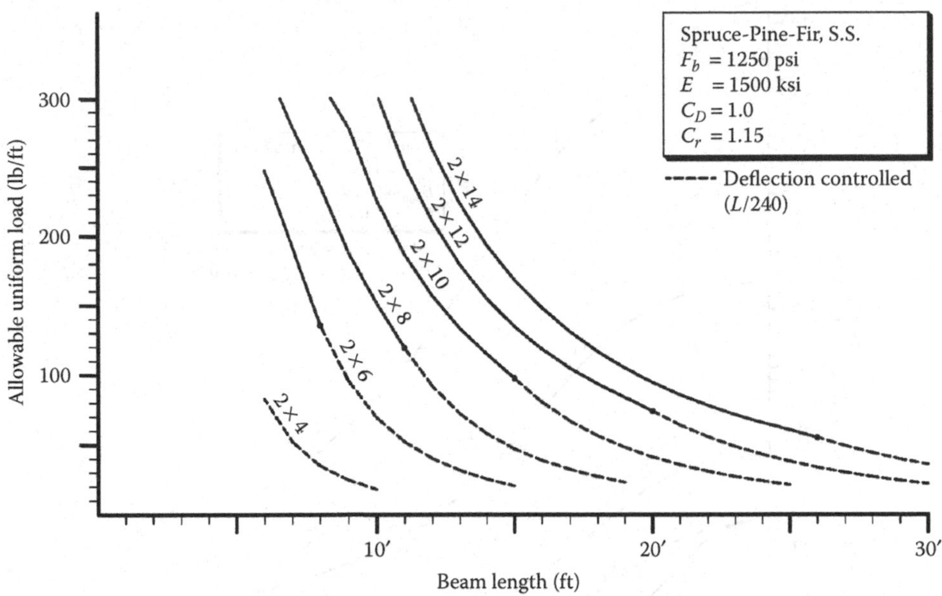

Figure B.7 Allowable uniform loads on floor joists (Spruce-Pine-Fir, Sel. Str.).

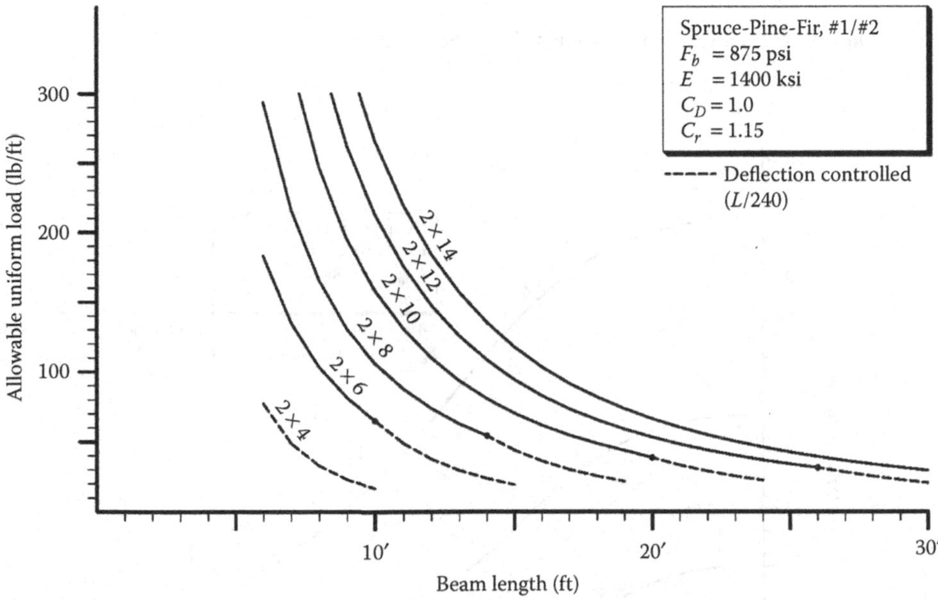

Figure B.8 Allowable uniform loads on floor joists (Spruce-Pine-Fir, No. 1/No. 2).

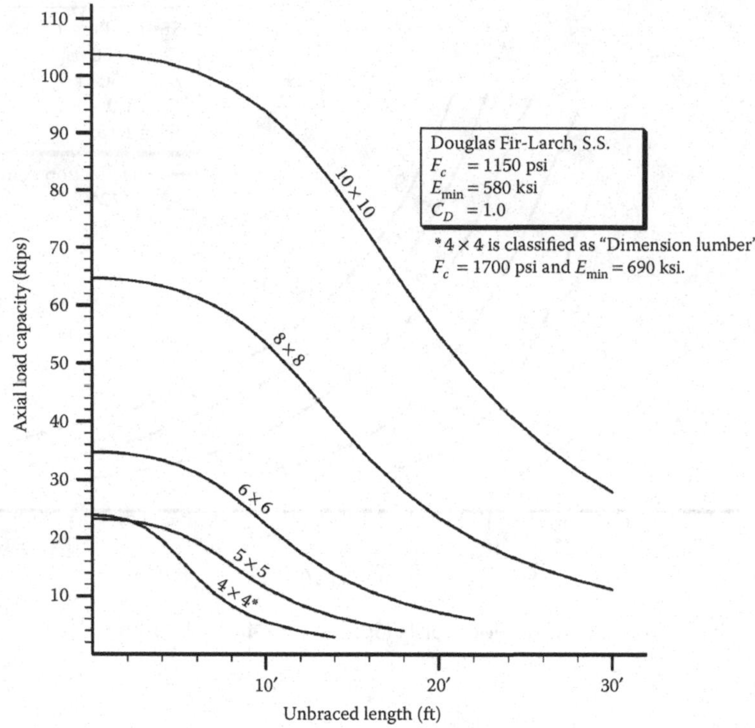

Figure B.9 Allowable axial loads on columns (DF-L, Sel. Str.).

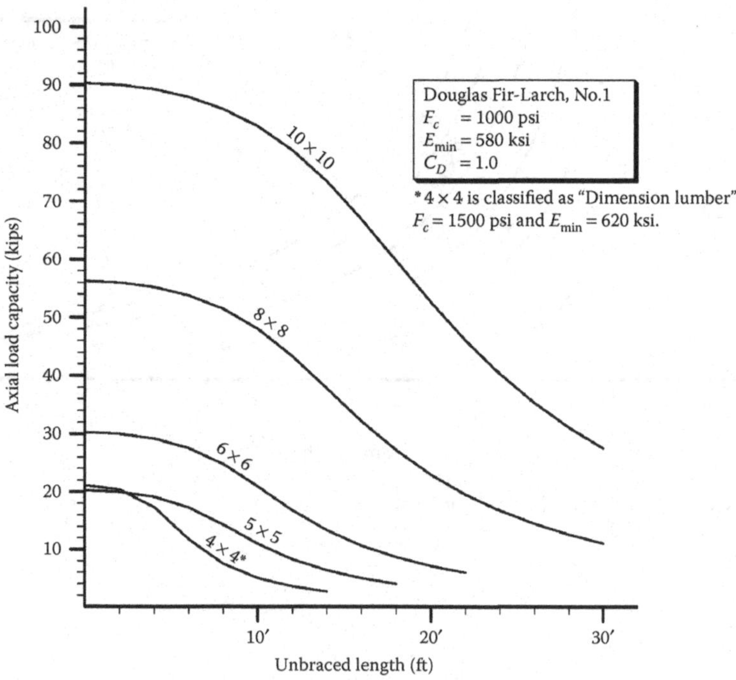

Figure B.10 Allowable axial loads on columns (DF-L, No. 1).

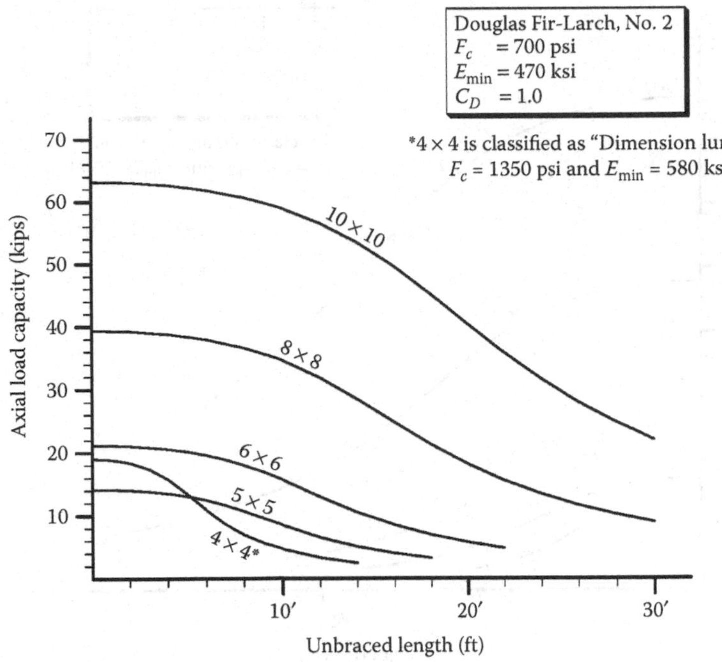

Figure B.11 Allowable axial loads on columns (DF-L, No. 2).

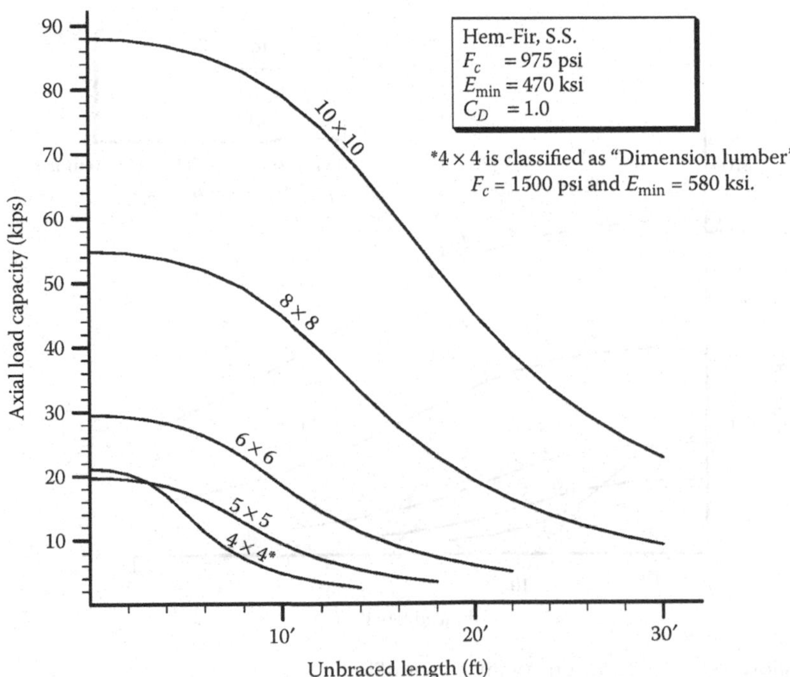

Figure B.12 Allowable axial loads on columns (Hem-Fir, Sel. Str.).

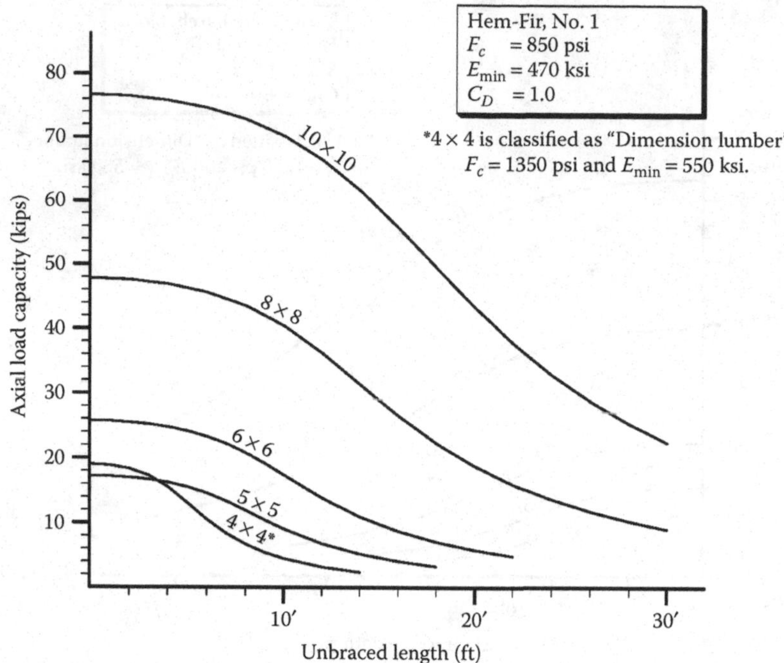

Figure B.13 Allowable axial loads on columns (Hem-Fir, No. 1).

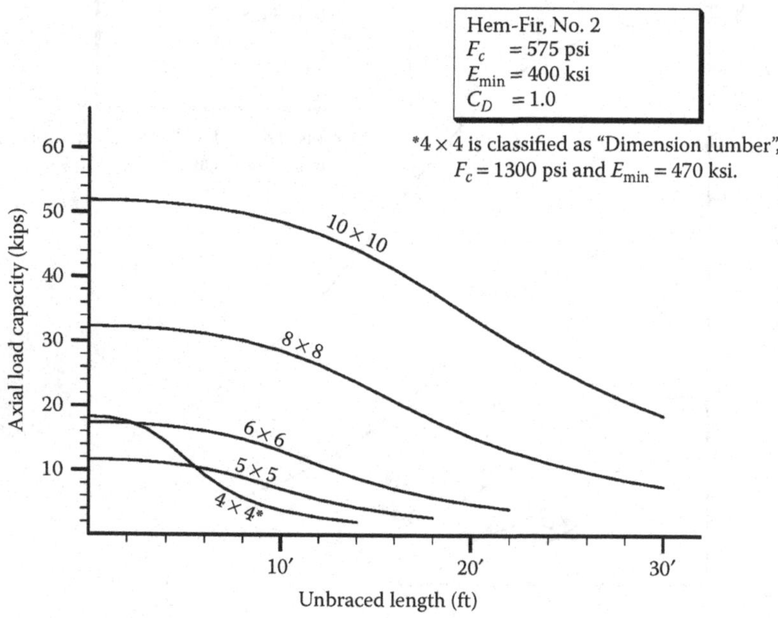

Figure B.14 Allowable axial loads on columns (Hem-Fir, No. 2).

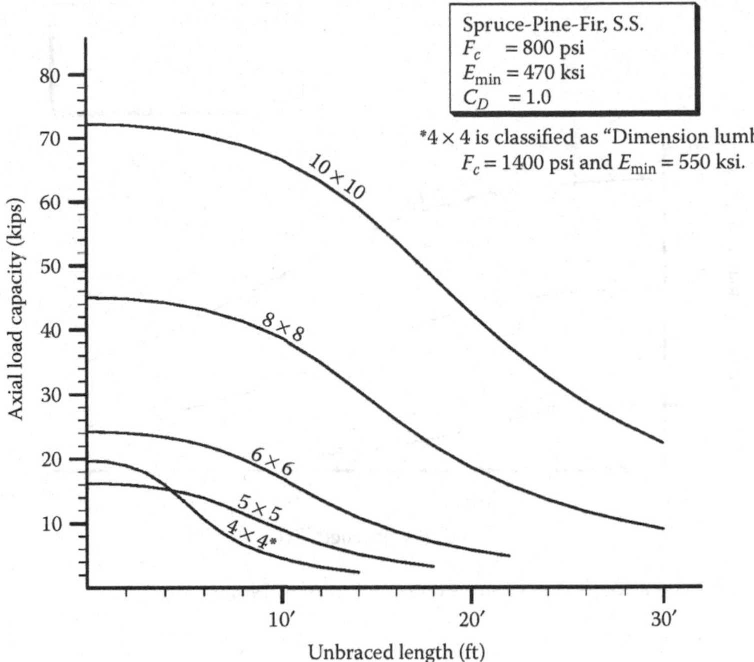

Figure B.15 Allowable axial loads on columns (Spruce-Pine-Fir, Sel. Str.).

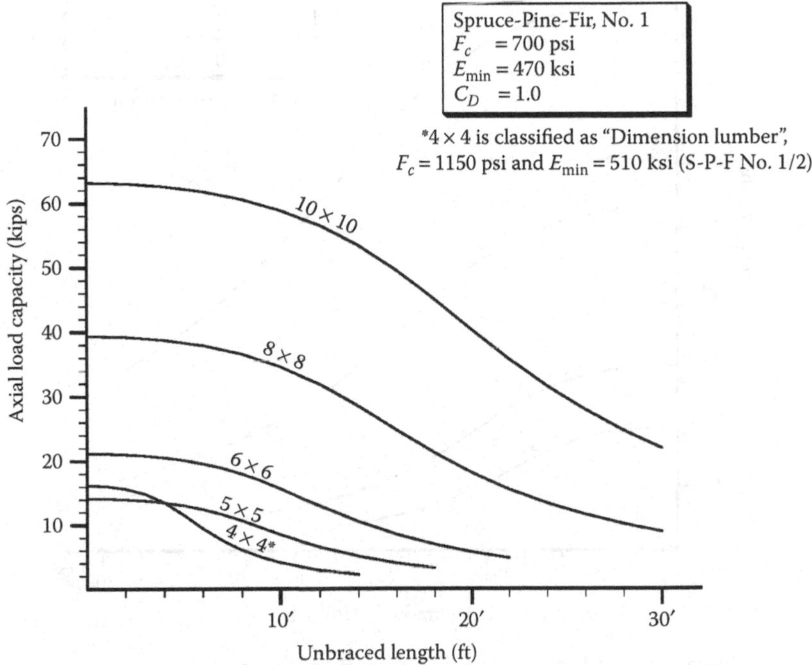

Figure B.16 Allowable axial loads on columns (Spruce-Pine-Fir, No. 1).

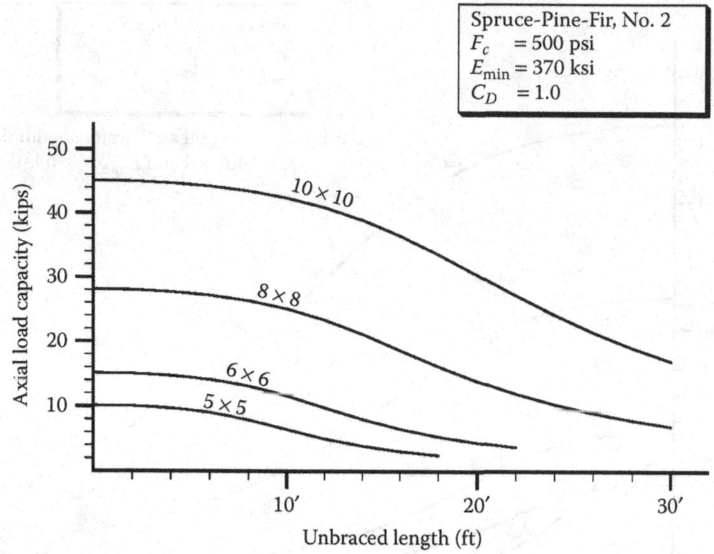

Figure B.17 Allowable axial loads on columns (Spruce-Pine-Fir, No. 2).

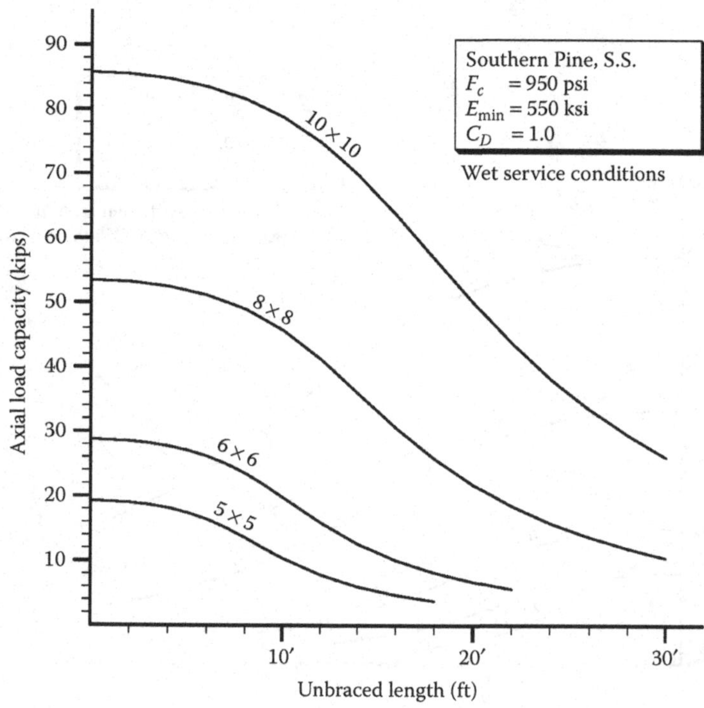

Figure B.18 Allowable axial loads on columns (Southern Pine, Sel. Str.).

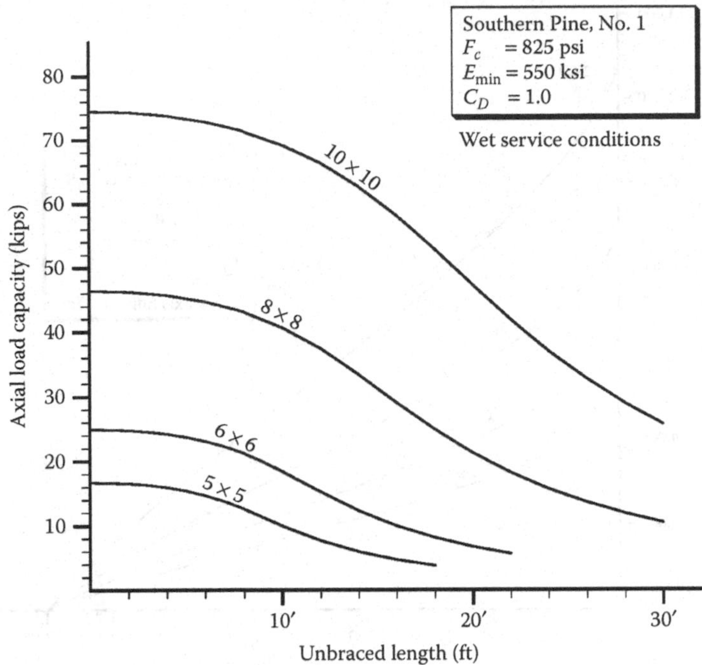

Figure B.19 Allowable axial loads on columns (Southern Pine, No. 1).

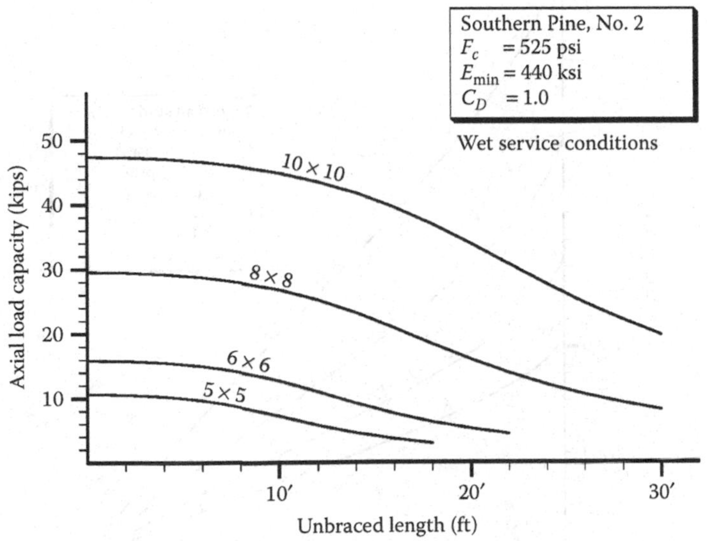

Figure B.20 Allowable axial loads on columns (Southern Pine, No. 2).

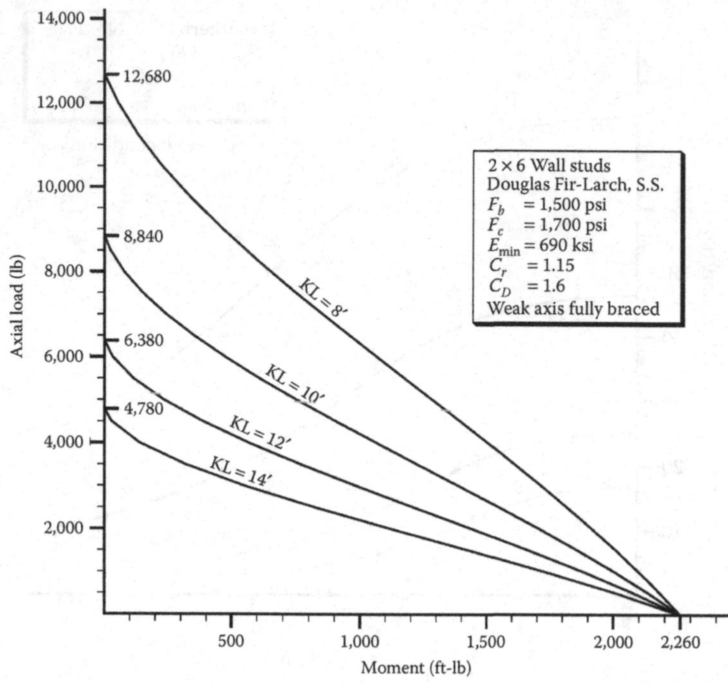

Figure B.21 Combined axial and bending loads on 2 × 6 wall studs (DF-L, Sel. Str., C_D = 1.6).

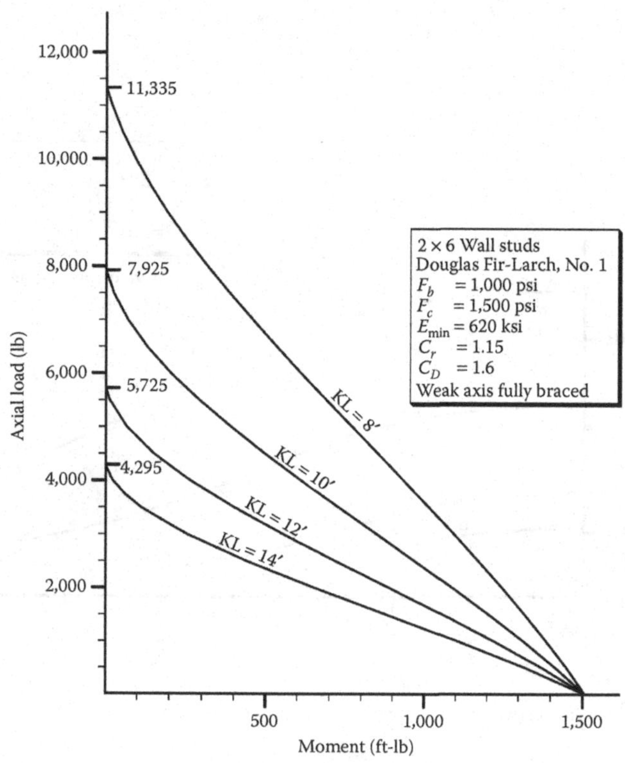

Figure B.22 Combined axial and bending loads on 2 × 6 wall studs (DF-L, No. 1, C_D = 1.6).

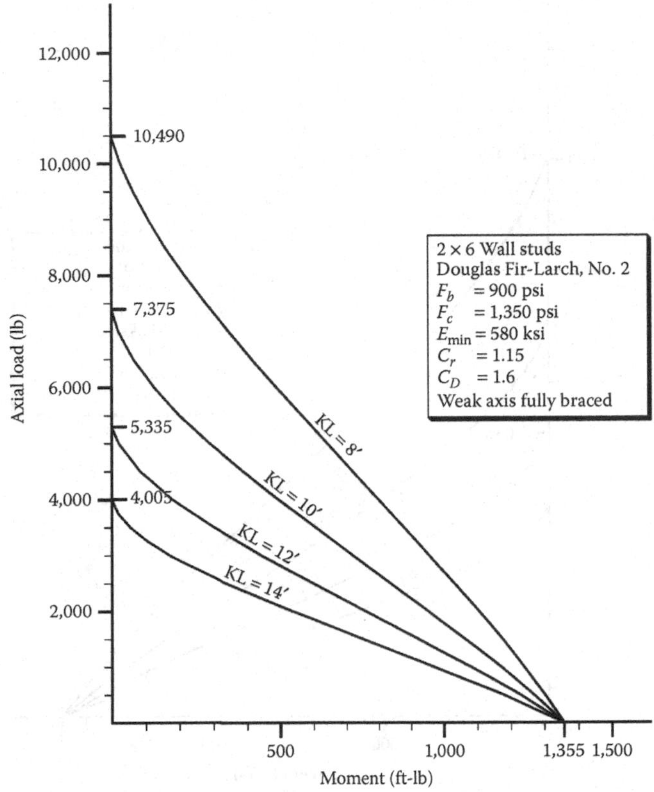

Figure B.23 Combined axial and bending loads on 2 × 6 wall studs (DF-L, No. 2, $C_D = 1.6$).

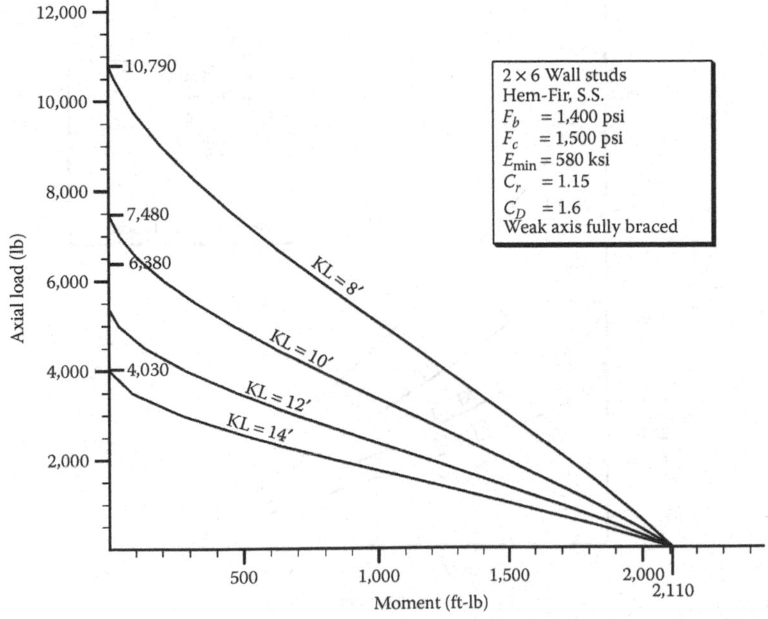

Figure B.24 Combined axial and bending loads on 2 × 6 wall studs (Hem-Fir, Sel. Str., $C_D = 1.6$).

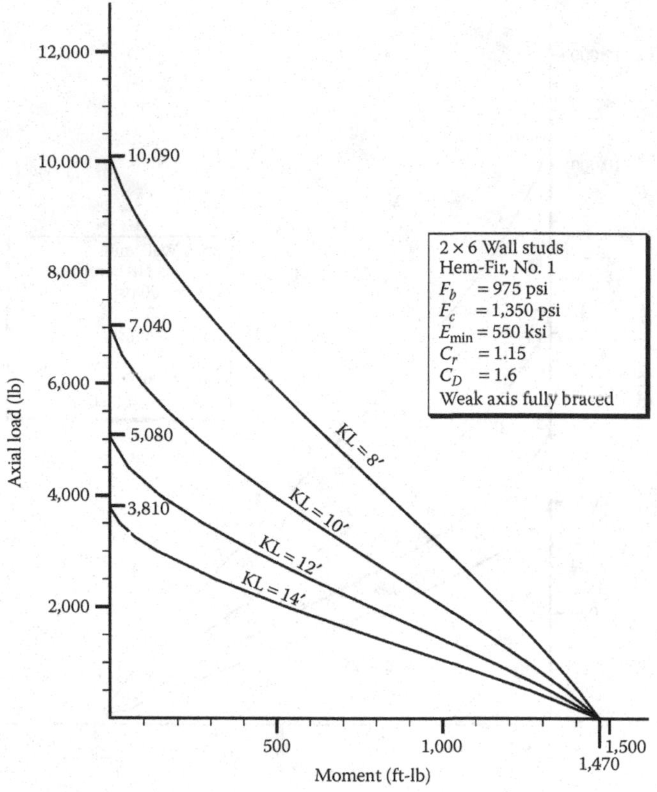

Figure B.25 Combined axial and bending loads on 2 × 6 wall studs (Hem-Fir, No. 1, C_D = 1.6).

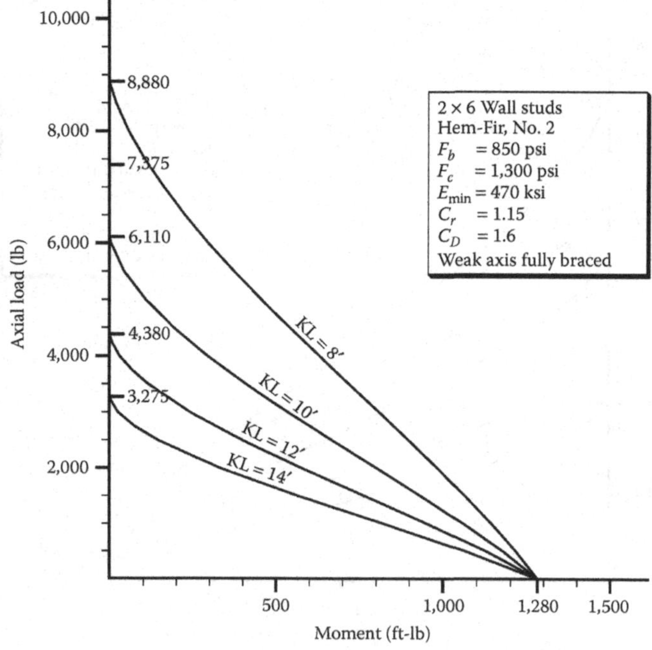

Figure B.26 Combined axial and bending loads on 2 × 6 wall studs (Hem-Fir, No. 2, C_D = 1.6).

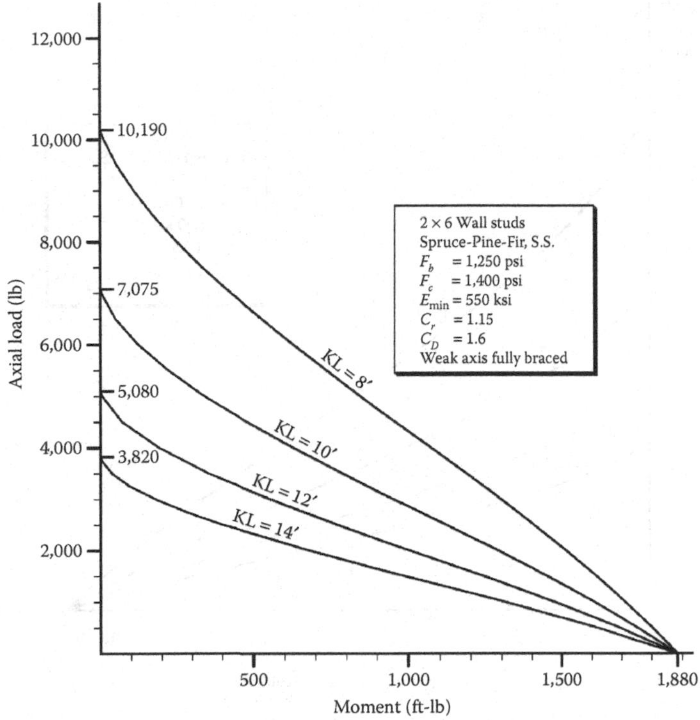

Figure B.27 Combined axial and bending loads on 2 × 6 wall studs (Spruce-Pine-Fir, Sel. Str., C_D = 1.6).

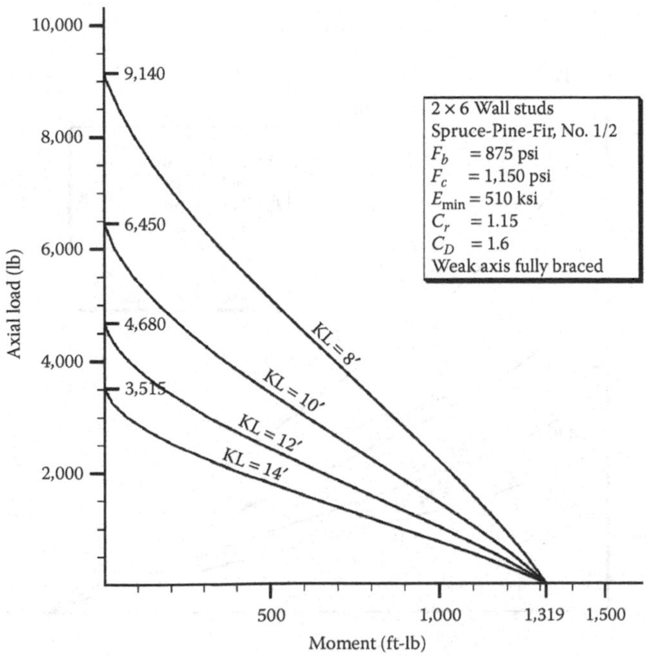

Figure B.28 Combined axial and bending loads on 2 × 6 wall studs (Spruce-Pine-Fir, No. 1/No. 2, C_D = 1.6).

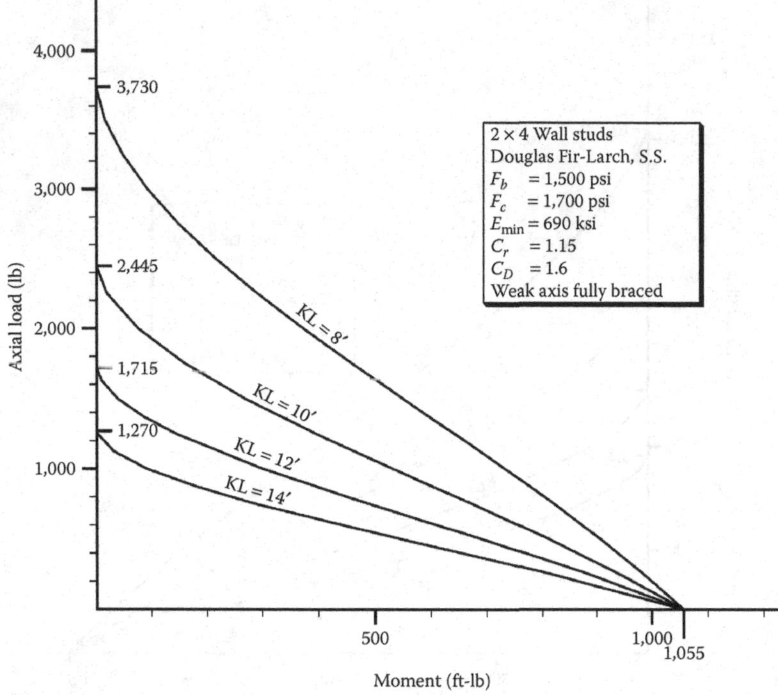

Figure B.29 Combined axial and bending loads on 2 × 4 wall studs (DF-L, Sel. Str., C_D = 1.6).

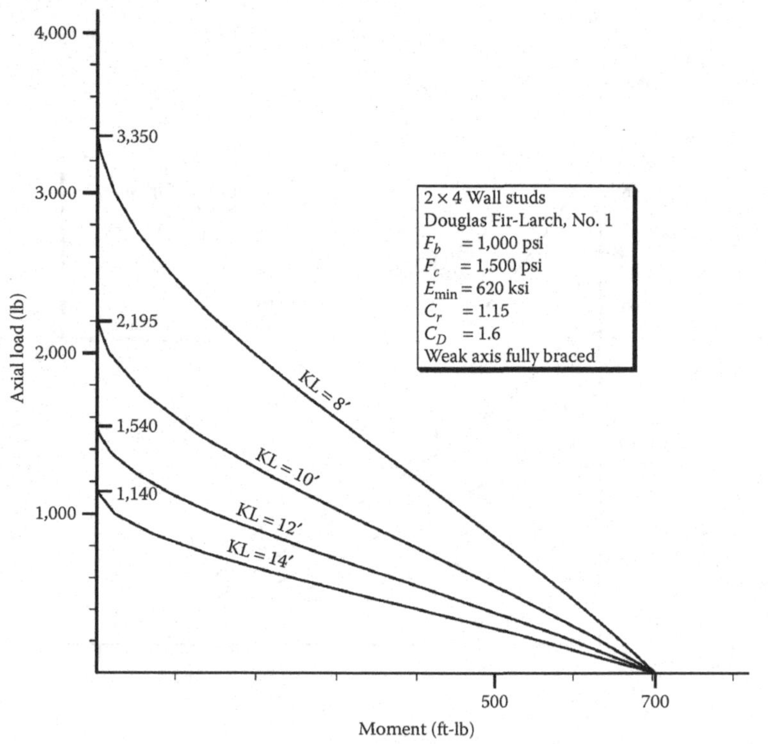

Figure B.30 Combined axial and bending loads on 2 × 4 wall studs (DFL, No. 1, C_D = 1.6).

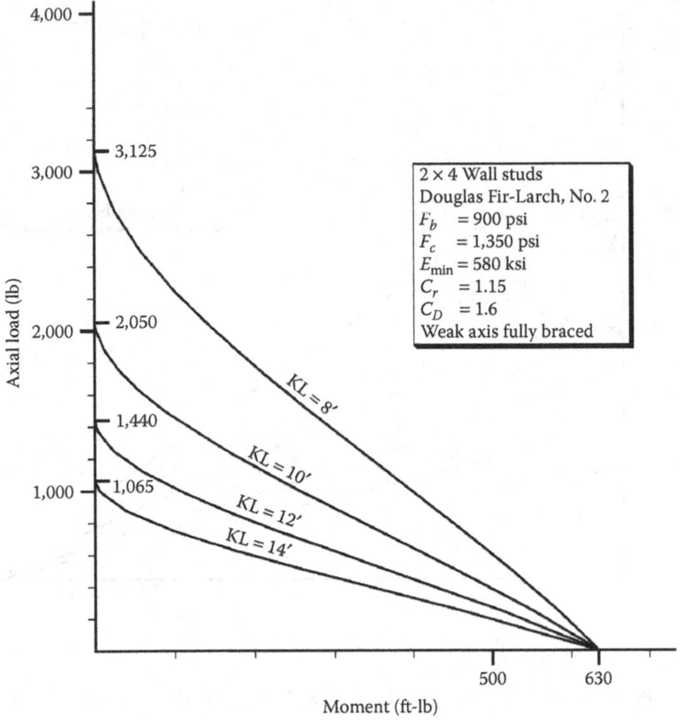

Figure B.31 Combined axial and bending loads on 2 × 4 wall studs (DF-L, No. 2, C_D = 1.6).

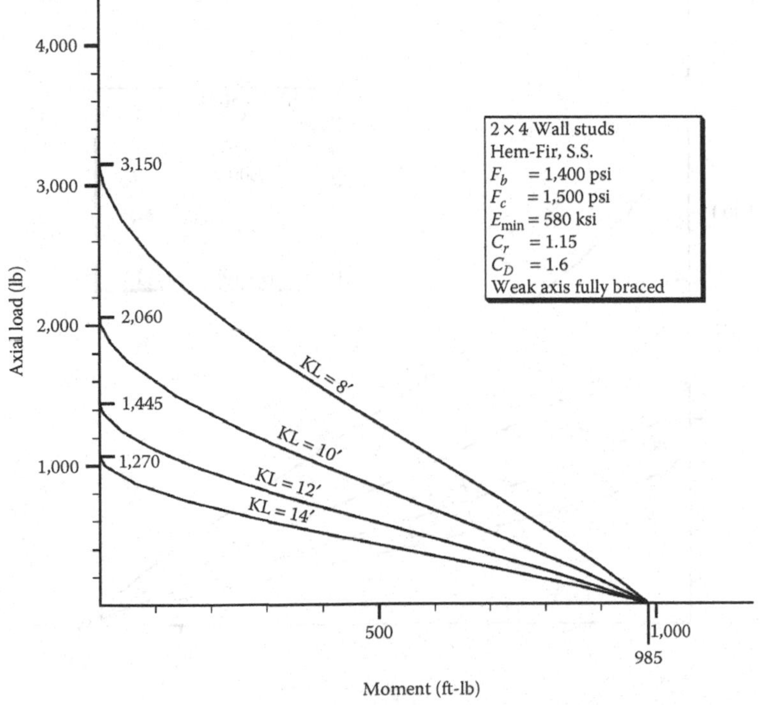

Figure B.32 Combined axial and bending loads on 2 × 4 wall studs (Hem-Fir, Sel. Str., C_D = 1.6).

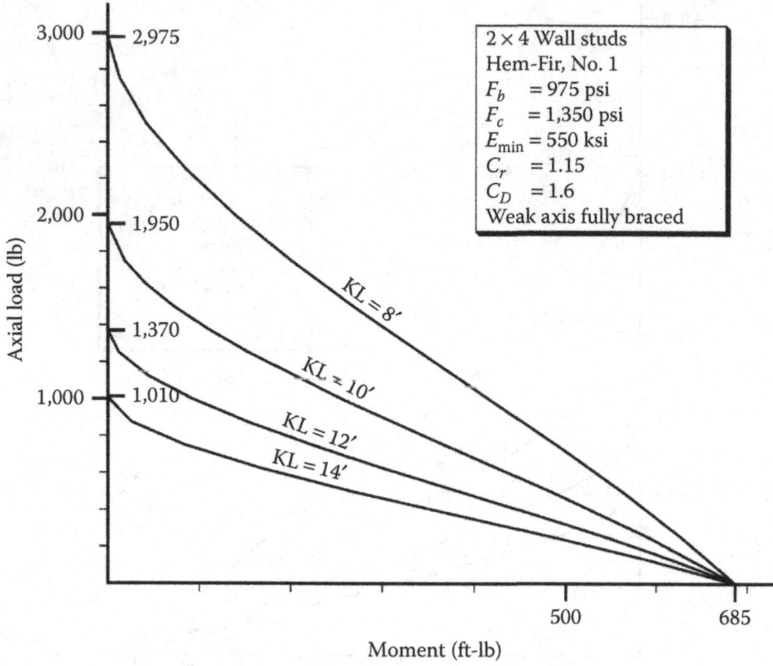

Figure B.33 Combined axial and bending loads on 2 × 4 wall studs (Hem-Fir, No. 1, C_D = 1.6).

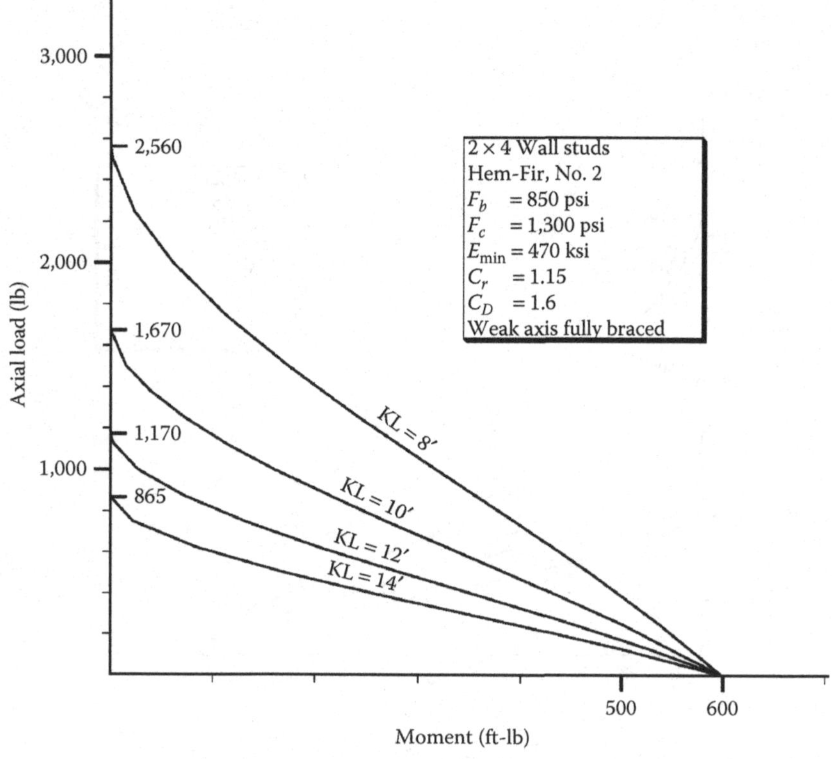

Figure B.34 Combined axial and bending loads on 2 × 4 wall studs (Hem-Fir, No. 2, C_D = 1.6).

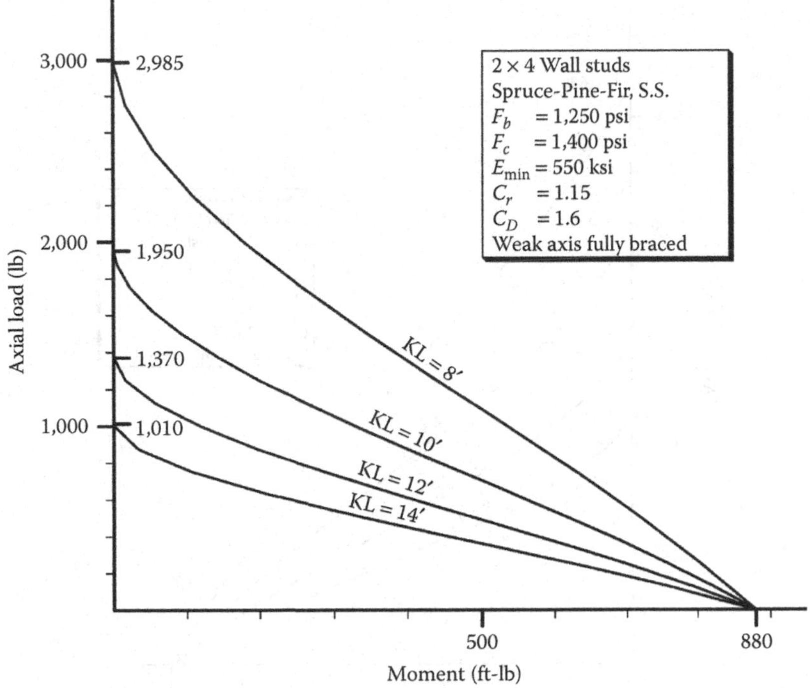

Figure B.35 Combined axial and bending loads on 2 × 4 wall studs (Spruce-Pine-Fir, Sel. Str., C_D = 1.6).

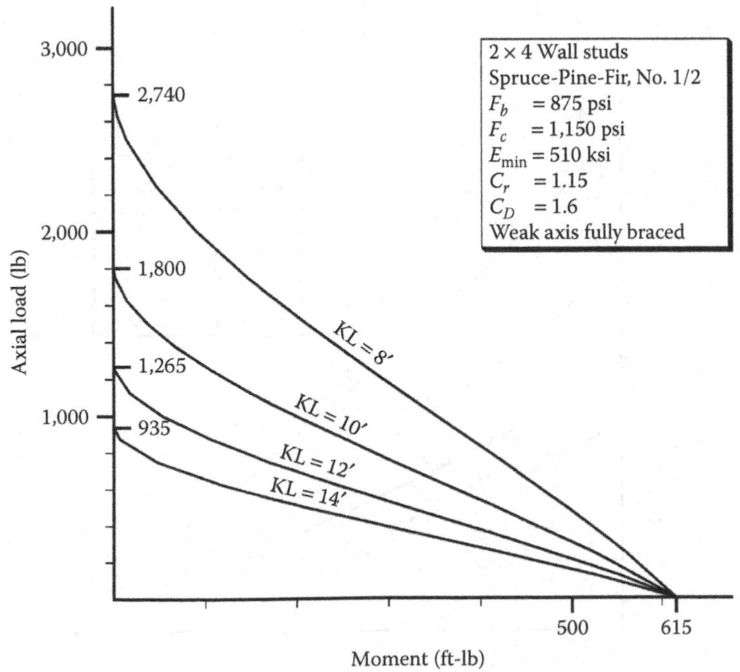

Figure B.36 Combined axial and bending loads on 2 × 4 wall studs (Spruce-Pine-Fir, No. 1/No. 2, C_D = 1.6).

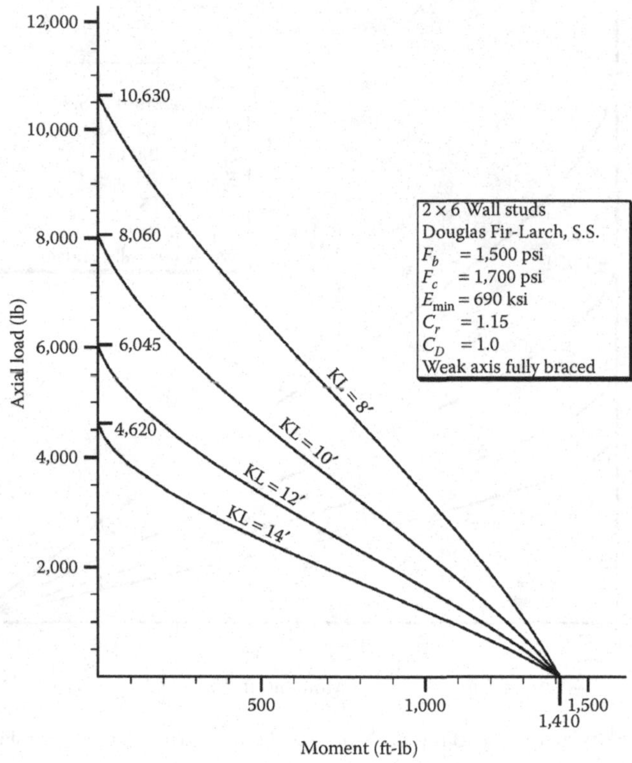

Figure B.37 Combined axial and bending loads on 2 × 6 wall studs (DFL, Sel. Str., C_D = 1.0).

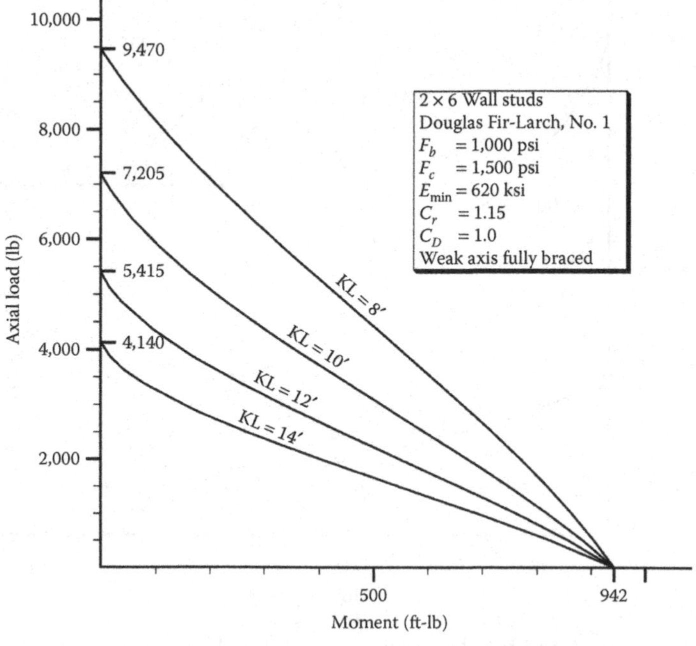

Figure B.38 Combined axial and bending loads on 2 × 6 wall studs (DF-L, No. 1, C_D = 1.0).

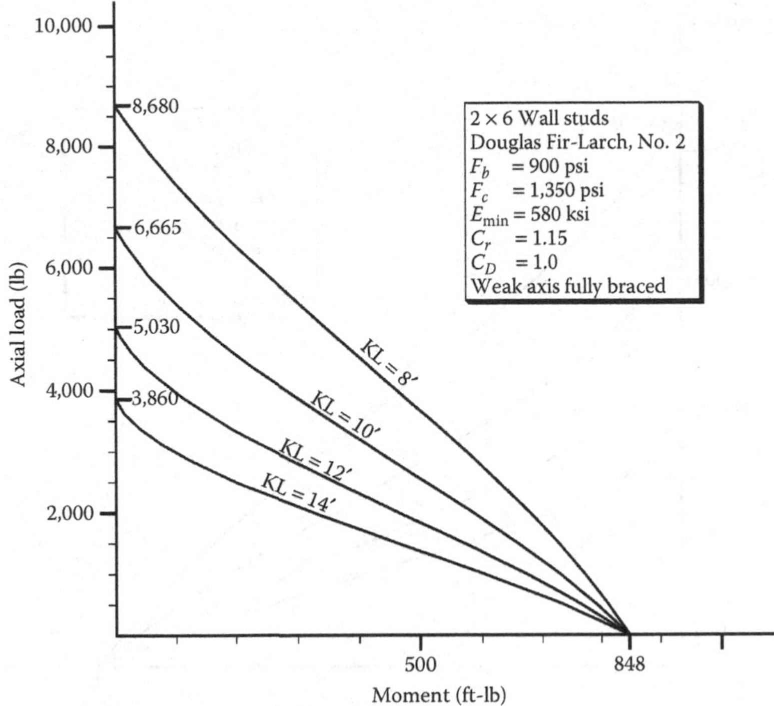

Figure B.39 Combined axial and bending loads on 2 × 6 wall studs (DF-L, No. 2, C_D = 1.0).

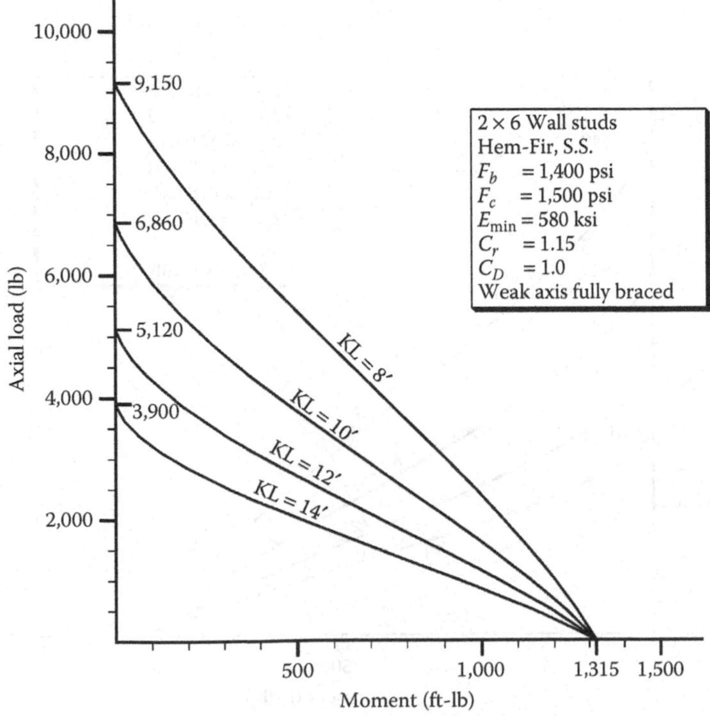

Figure B.40 Combined axial and bending loads on 2 × 6 wall studs (Hem-Fir, Sel. Str., C_D = 1.0).

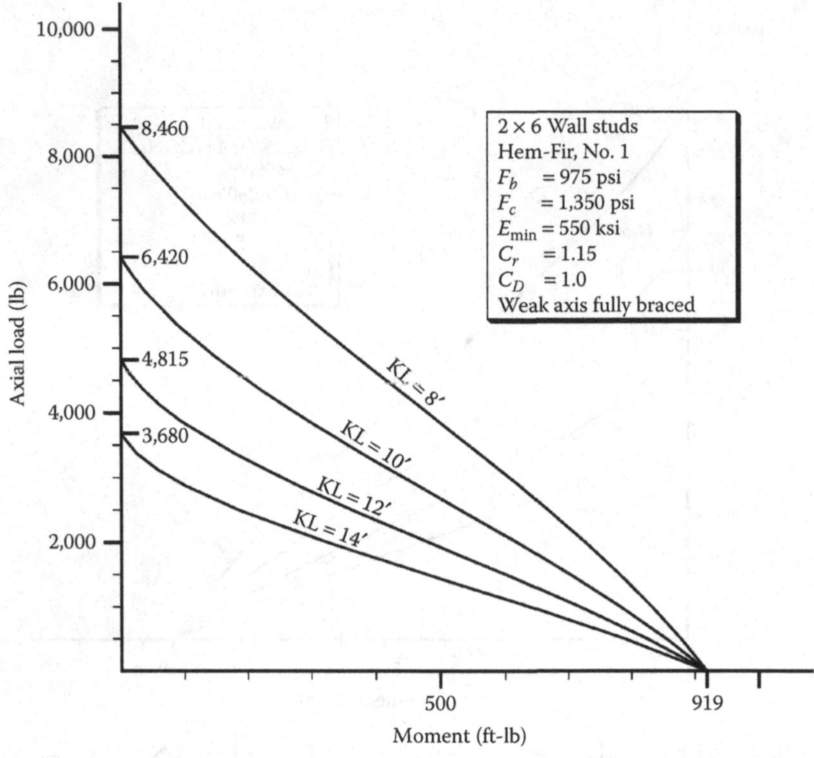

Figure B.41 Combined axial and bending loads on 2 × 6 wall studs (Hem-Fir, No. 1, C_D = 1.0).

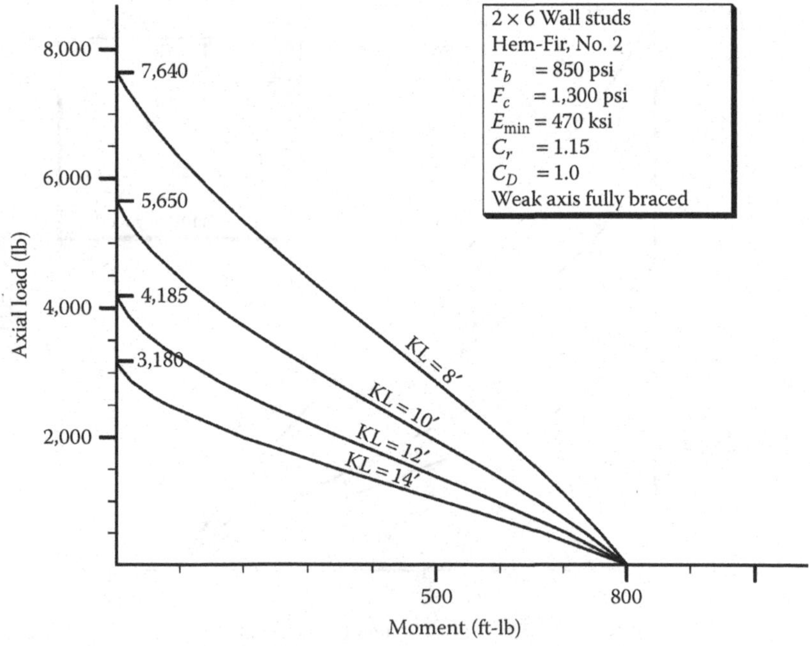

Figure B.42 Combined axial and bending loads on 2 × 6 wall studs (Hem-Fir, No. 2, C_D = 1.0).

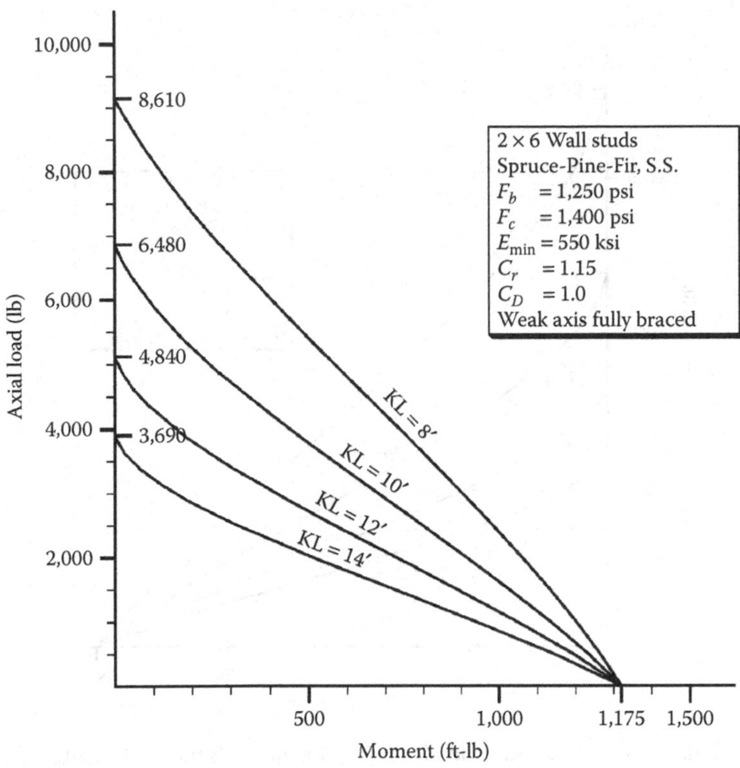

Figure B.43 Combined axial and bending loads on 2 × 6 wall studs (Spruce-Pine-Fir, Sel. Str., C_D = 1.0).

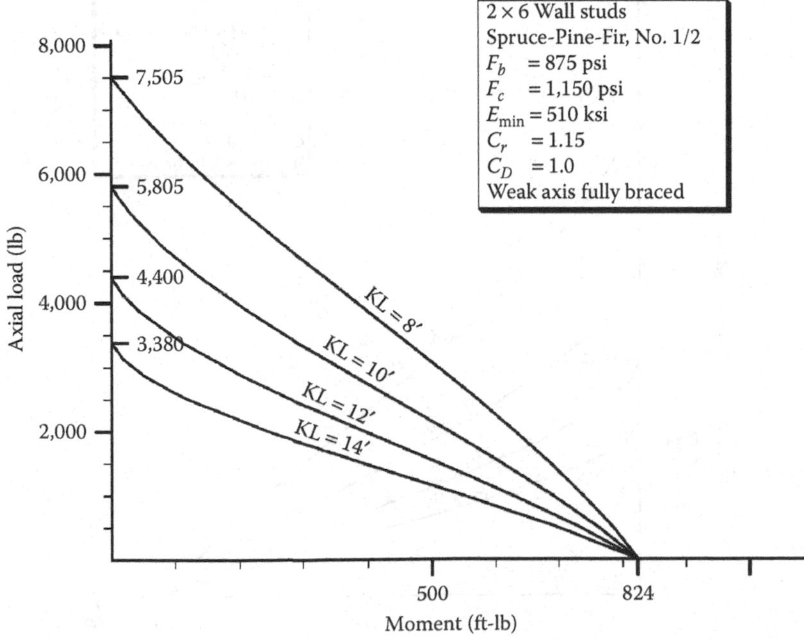

Figure B.44 Combined axial and bending loads on 2 × 6 wall studs (Spruce-Pine-Fir, No. 1/No. 2, C_D = 1.0).

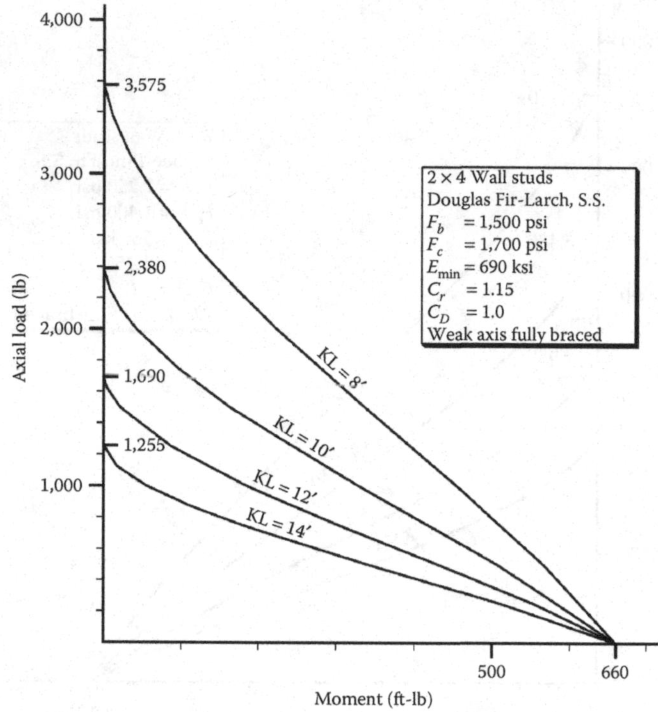

Figure B.45 Combined axial and bending loads on 2 × 4 wall studs (DF-L, Sel. Str., C_D = 1.0).

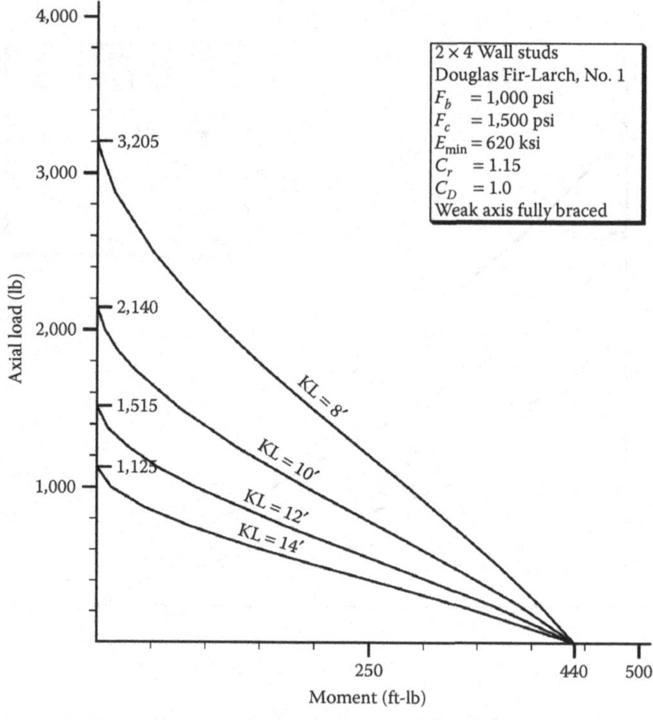

Figure B.46 Combined axial and bending loads on 2 × 4 wall studs (DF-L, No. 1, C_D = 1.0).

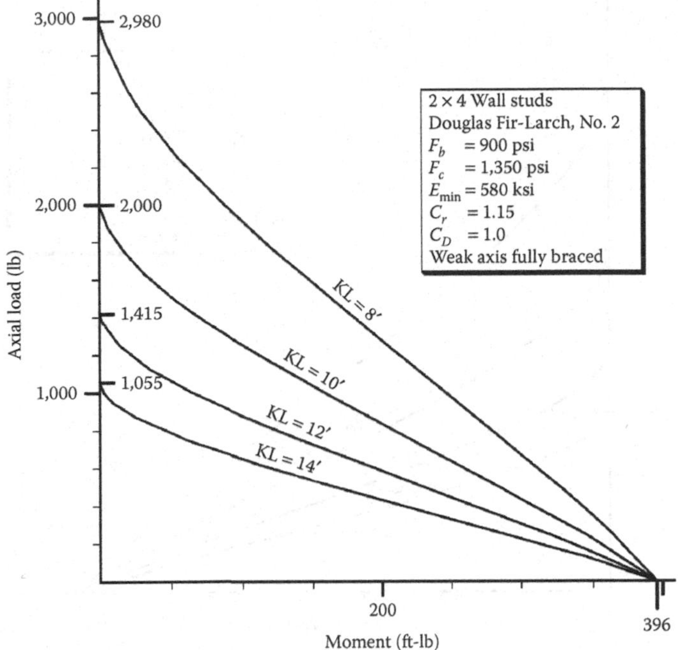

Figure B.47 Combined axial and bending loads on 2 × 4 wall studs (DF-L, No. 2, C_D = 1.0).

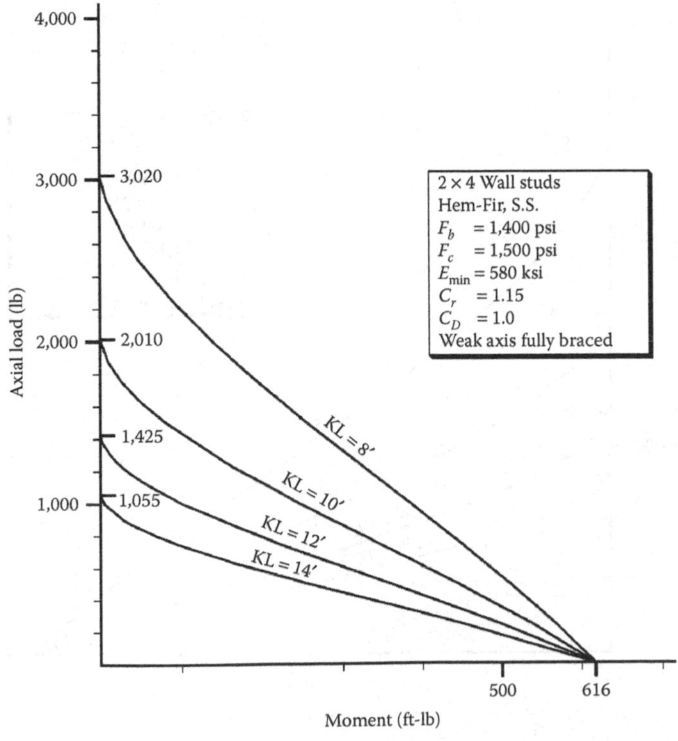

Figure B.48 Combined axial and bending loads on 2 × 4 wall studs (Hem-Fir, Sel. Str., C_D = 1.0).

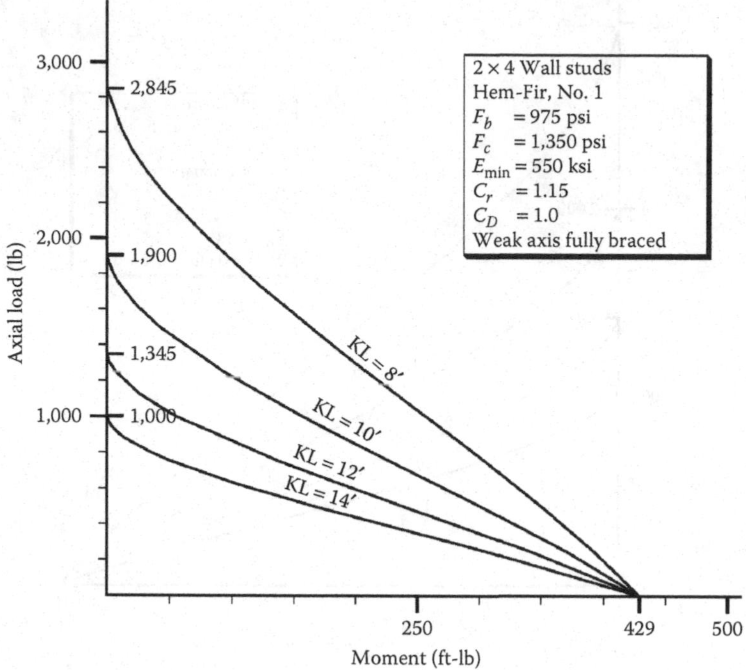

Figure B.49 Combined axial and bending loads on 2 × 4 wall studs (Hem-Fir, No. 1, C_D = 1.0).

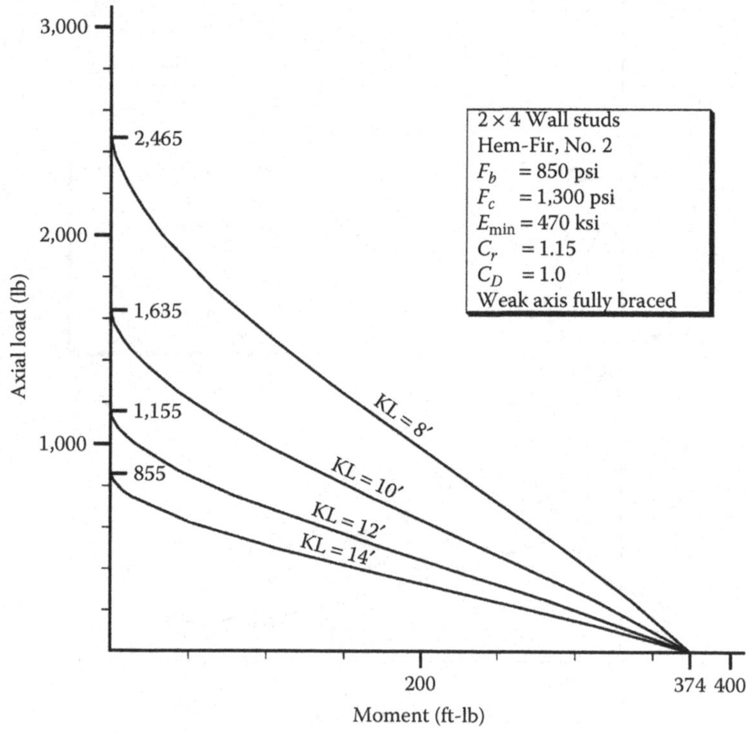

Figure B.50 Combined axial and bending loads on 2 × 4 wall studs (Hem-Fir, No. 2, C_D = 1.0).

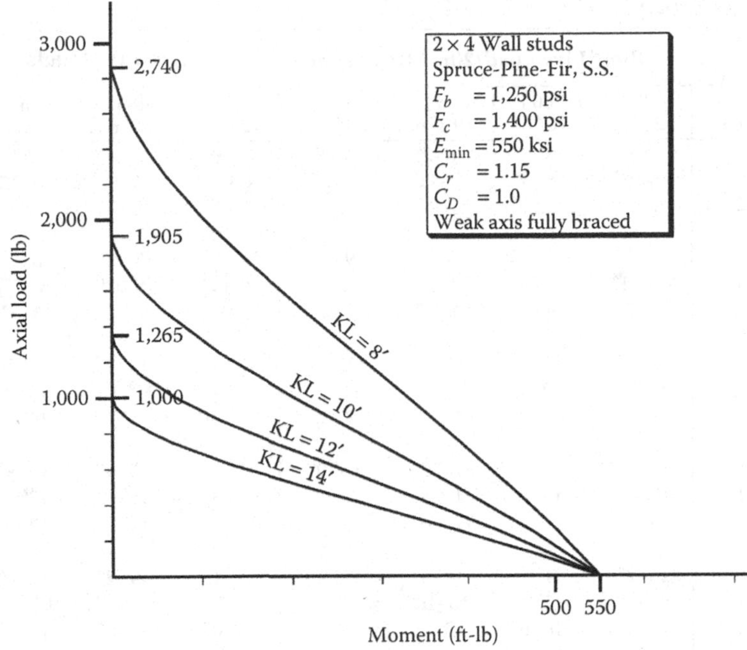

Figure B.51 Combined axial and bending loads on 2 × 4 wall studs (Spruce-Pine-Fir, Sel. Str., C_D = 1.0).

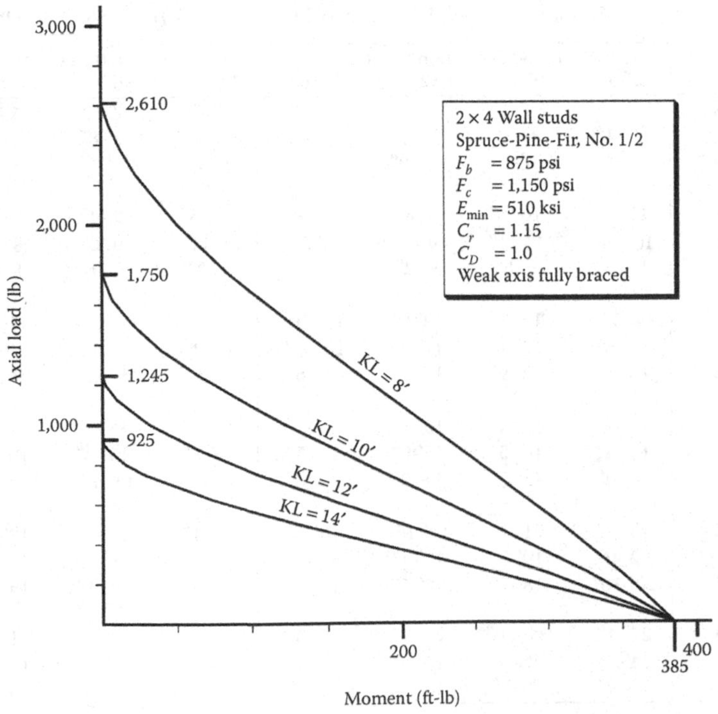

Figure B.52 Combined axial and bending loads on 2 × 4 wall studs (Spruce-Pine-Fir, No. 1/No. 2, C_D = 1.0).

Table B.1 Floor joist span tables

40psf Live Load/20psf Dead Load **$C_D = 1.0$, Douglas Fir-Larch**

Size	Grade	L/360 Deflection Limit 12″ o.c.	16″ o.c.	19.2″ o.c.	24″ o.c.	L/480 Deflection Limit 12″ o.c.	16″ o.c.	19.2″ o.c.	24″ o.c.
	SS	7′-2″	6′-6″	6′-2″	5′-8″	6′-6″	5′-11″	5′-7″	5′-2″
2×4	#1	6′-11″	6′-3″	5′-11″	5′-5″	6′-3″	5′-8″	5′-4″	5′-0″
	#2	6′-9″	6′-2″	5′-8″	5′-1″	6′-2″	5′-7″	5′-3″	4′-11″
	SS	11′-4″	10′-3″	9′-8″	9′-0″	10′-3″	9′-4″	8′-9″	8′-2″
2×6	#1	10′-11″	9′-8″	8′-10″	7′-11″	9′-11″	9′-0″	8′-5″	7′-10″
	#2	10′-7″	9′-2″	8′-4″	7′-6″	9′-8″	8′-10″	8′-3″	7′-6″
	SS	14′-11″	13′-7″	12′-9″	11′-10″	13′-7″	12′-4″	11′-7″	10′-9″
2×8	#1	14′-2″	12′-3″	11′-2″	10′-0″	13′-1″	11′-10″	11′-2″	10′-0″
	#2	13′-5″	11′-7″	10′-7″	9′-6″	12′-10″	11′-7″	10′-7″	9′-6″
	SS	19′-1″	17′-4″	16′-3″	15′-0″	17′-4″	15′-9″	14′-9″	13′-9″
2×10	#1	17′-4″	15′-0″	13′-8″	12′-3″	16′-8″	15′-0″	13′-8″	12′-3″
	#2	16′-5″	14′-2″	13′-0″	11′-7″	16′-4″	14′-2″	13′-0″	11′-7″
	SS	23′-2″	21′-1″	19′-5″	17′-4″	21′-1″	19′-2″	18′-0″	16′-8″
2×12	#1	20′-1″	17′-4″	15′-10″	14′-2″	20′-1″	17′-4″	15′-10″	14′-2″
	#2	19′-0″	16′-6″	15′-0″	13′-5″	19′-0″	16′-6″	15′-0″	13′-5″
	SS	27′-4″	23′-9″	21′-9″	19′-5″	24′-10″	22′-6″	21′-2″	19′-5″
2×14	#1	22′-5″	19′-5″	17′-9″	15′-10″	22′-5″	19′-5″	17′-9″	15′-10″
	#2	21′-3″	18′-5″	16′-10″	15′-0″	21′-3″	18′-5″	16′-10″	15′-0″

40psf Live Load/10psf Dead Load **$C_D = 1.0$, Douglas Fir-Larch**

Size	Grade	L/360 Deflection Limit 12″ o.c.	16″ o.c.	19.2″ o.c.	24″ o.c.	L/480 Deflection Limit 12″ o.c.	16″ o.c.	19.2″ o.c.	24″ o.c.
	SS	7′-2″	6′-6″	6′-2″	5′-8″	6′-6″	5′-11″	5′-7″	5′-2″
2×4	#1	6′-11″	6′-3″	5′-11″	5′-6″	6′-3″	5′-8″	5′-4″	5′-0″
	#2	6′-9″	6′-2″	5′-10″	5′-4″	6′-2″	5′-7″	5′-3″	4′-11″
	SS	11′-4″	10′-3″	9′-8″	9′-0″	10′-3″	9′-4″	8′-9″	8′-2″
2×6	#1	10′-11″	9′-11″	9′-4″	8′-8″	9′-11″	9′-0″	8′-5″	7′-10″
	#2	10′-8″	9′-8″	9′-2″	8′-2″	9′-8″	8′-10″	8′-3″	7′-8″
	SS	14′-11″	13′-7″	12′-9″	11′-10″	13′-7″	12′-4″	11′-7″	10′-9″
2×8	#1	14′-5″	13′-1″	12′-3″	10′-11″	13′-1″	11′-10″	11′-2″	10′-4″
	#2	14′-1″	12′-9″	11′-7″	10′-5″	12′-10″	11′-7″	10′-11″	10′-2″
	SS	19′-1″	17′-4″	16′-3″	15′-1″	17′-4″	15′-9″	14′-9″	13′-9″
2×10	#1	18′-4″	16′-5″	15′-0″	13′-5″	16′-8″	15′-2″	14′-3″	13′-3″
	#2	18′-0″	15′-7″	14′-2″	12′-8″	16′-4″	14′-10″	14′-0″	12′-8″
	SS	23′-2″	21′-1″	19′-10″	18′-5″	21′-1″	19′-2″	18′-0″	16′-8″
2×12	#1	22′-0″	19′-0″	17′-4″	15′-6″	20′-3″	18′-5″	17′-4″	15′-6″
	#2	20′-10″	18′-1″	16′-6″	14′-9″	19′-11″	18′-1″	16′-6″	14′-9″
	SS	27′-4″	24′-10″	23′-4″	21′-3″	24′-10″	22′-6″	21′-2″	19′-8″
2×14	#1	24′-7″	21′-3″	19′-5″	17′-4″	23′-11″	21′-3″	19′-5″	17′-4″
	#2	23′-4″	20′-2″	18′-5″	16′-6″	23′-4″	20′-2″	18′-5″	16′-6″

Table B.2 Floor joist span tables

80psf Live Load/15psf Dead Load **$C_D = 1.0$, Douglas Fir-Larch**

Size	Grade	L/360 Deflection Limit 12″ o.c.	16″ o.c.	19.2″ o.c.	24″ o.c.	L/480 Deflection Limit 12″ o.c.	16″ o.c.	19.2″ o.c.	24″ o.c.
	SS	5′-8″	5′-2″	4′-10″	4′-6″	5′-2″	4′-8″	4′-5″	4′-1″
2×4	#1	5′-6″	5′-0″	4′-8″	4′-3″	5′-0″	4′-6″	4′-3″	3′-11″
	#2	5′-4″	4′-11″	4′-6″	4′-1″	4′-11″	4′-5″	4′-2″	3′-10″
	SS	9′-0″	8′-2″	7′-8″	7′-1″	8′-2″	7′-5″	7′-0″	6′-5″
2×6	#1	8′-8″	7′-8″	7′-0″	6′-3″	7′-10″	7′-2″	6′-8″	6′-3″
	#2	8′-5″	7′-3″	6′-8″	5′-11″	7′-8″	7′-0″	6′-7″	5′-11″
	SS	11′-10″	10′-9″	10′-1″	9′-5″	10′-9″	9′-9″	9′-2″	8′-6″
2×8	#1	11′-3″	9′-9″	8′-11″	7′-11″	10′-4″	9′-5″	8′-10″	7′-11″
	#2	10′-8″	9′-3″	8′-5″	7′-6″	10′-2″	9′-3″	8′-5″	7′-6″
	SS	15′-1″	13′-9″	12′-11″	11′-11″	13′-9″	12′-6″	11′-9″	10′-11″
2×10	#1	13′-9″	11′-11″	10′-10″	9′-8″	13′-3″	11′-11″	10′-10″	9′-8″
	#2	13′-0″	11′-3″	10′-4″	9′-2″	13′-0″	11′-3″	10′-4″	9′-2″
	SS	18′-5″	16′-8″	15′-5″	13′-10″	16′-8″	15′-2″	14′-3″	13′-3″
2×12	#1	15′-11″	13′-10″	12′-7″	11′-3″	15′-11″	13′-10″	12′-7″	11′-3″
	#2	15′-1″	13′-1″	11′-11″	10′-8″	15′-1″	13′-1″	11′-11″	10′-8″
	SS	21′-8″	18′-11″	17′-3″	15′-5″	19′-8″	17′-11″	16′-10″	15′-5″
2×14	#1	17′-10″	15′-5″	14′-1″	12′-7″	17′-10″	15′-5″	14′-1″	12′-7″
	#2	16′-11″	14′-8″	13′-4″	11′-11″	16′-11″	14′-8″	13′-4″	11′-11″

100psf Live Load/15psf Dead Load **$C_D = 1.0$, Douglas Fir-Larch**

Size	Grade	L/360 Deflection Limit 12″ o.c.	16″ o.c.	19.2″ o.c.	24″ o.c.	L/480 Deflection Limit 12″ o.c.	16″ o.c.	19.2″ o.c.	24″ o.c.
	SS	5′-3″	4′-10″	4′-6″	4′-2″	4′-10″	4′-4″	4′-1″	3′-10″
2×4	#1	5′-1″	4′-7″	4′-4″	3′-10″	4′-7″	4′-2″	3′-11″	3′-8″
	#2	5′-0″	4′-6″	4′-1″	3′-8″	4′-6″	4′-1″	3′-10″	3′-7″
	SS	8′-4″	7′-7″	7′-1″	6′-7″	7′-7″	6′-10″	6′-5″	6′-0″
2×6	#1	8′-0″	7′-0″	6′-4″	5′-8″	7′-3″	6′-7″	6′-3″	5′-8″
	#2	7′-8″	6′-7″	6′-0″	5′-5″	7′-2″	6′-6″	6′-0″	5′-5″
	SS	11′-0″	10′-0″	9′-5″	8′-9″	10′-0″	9′-1″	8′-6″	7′-11″
2×8	#1	10′-3″	8′-10″	8′-1″	7′-3″	9′-7″	8′-9″	8′-1″	7′-3″
	#2	9′-8″	8′-5″	7′-8″	6′-10″	9′-5″	8′-5″	7′-8″	6′-10″
	SS	14′-0″	12′-9″	12′-0″	10′-10″	12′-9″	11′-7″	10′-11″	10′-1″
2×10	#1	12′-6″	10′-10″	9′-10″	8′-10″	12′-3″	10′-10″	9′-10″	8′-10″
	#2	11′-10″	10′-3″	9′-4″	8′-4″	11′-10″	10′-3″	9′-4″	8′-4″
	SS	17′-1″	15′-4″	14′-0″	12′-6″	15′-6″	14′-1″	13′-3″	12′-4″
2×12	#1	14′-6″	12′-6″	11′-5″	10′-3″	14′-6″	12′-6″	11′-5″	10′-3″
	#2	13′-9″	11′-11″	10′-10″	9′-8″	13′-9″	11′-11″	10′-10″	9′-8″
	SS	19′-10″	17′-2″	15′-8″	14′-0″	18′-3″	16′-7″	15′-7″	14′-0″
2×14	#1	16′-2″	14′-0″	12′-9″	11′-5″	16′-2″	14′-0″	12′-9″	11′-5″
	#2	15′-4″	13′-3″	12′-2″	10′-10″	15′-4″	13′-3″	12′-2″	10′-10″

Table B.3 Floor joist span tables

40psf Live Load/15psf Dead Load **C_D = 1.15, Douglas Fir-Larch**

Size	Grade	L/360 Deflection Limit 12″ o.c.	16″ o.c.	19.2″ o.c.	24″ o.c.	L/240 Deflection Limit 12″ o.c.	16″ o.c.	19.2″ o.c.	24″ o.c.
	SS	7′-2″	6′-6″	6′-2″	5′-8″	8′-3″	7′-6″	7′-0″	6′-6″
2×4	#1	6′-11″	6′-3″	5′-11″	5′-6″	7′-11″	7′-2″	5′-4″	6′-0″
	#2	6′-9″	6′-2″	5′-10″	5′-4″	7′-9″	7′-0″	5′-3″	5′-9″
	SS	11′-4″	10′-3″	9′-8″	9′-0″	12′-11″	11′-9″	8′-9″	10′-3″
2×6	#1	10′-11″	9′-11″	9′-4″	8′-8″	12′-6″	10′-10″	8′-5″	8′-10″
	#2	10′-8″	9′-8″	9′-2″	8′-5″	11′-10″	10′-3″	8′-3″	8′-5″
	SS	14′-11″	13′-7″	12′-9″	11′-10″	17′-1″	15′-6″	11′-7″	13′-7″
2×8	#1	14′-5″	13′-1″	12′-3″	11′-2″	15′-10″	13′-9″	11′-2″	11′-2″
	#2	14′-1″	12′-10″	11′-11″	10′-7″	15′-0″	13′-0″	10′-11″	10′-7″
	SS	19′-1″	17′-4″	16′-3″	15′-1″	21′-10″	19′-10″	14′-9″	16′-9″
2×10	#1	18′-4″	16′-8″	15′-4″	13′-8″	19′-5″	16′-9″	14′-3″	13′-8″
	#2	18′-0″	15′-11″	14′-6″	13′-0″	18′-5″	15′-11″	14′-0″	13′-0″
	SS	23′-2″	21′-1″	19′-10″	18′-5″	26′-7″	23′-10″	18′-0″	19′-6″
2×12	#1	22′-4″	19′-6″	17′-9″	15′-11″	22′-6″	19′-6″	17′-4″	15′-11″
	#2	21′-4″	18′-6″	16′-10″	15′-1″	21′-4″	18′-6″	16′-10″	15′-1″
	SS	27′-4″	24′-10″	23′-4″	21′-8″	30′-9″	26′-8″	21′-2″	21′-9″
2×14	#1	25′-1″	21′-9″	19′-10″	17′-9″	25′-1″	21′-9″	19′-10″	17′-9″
	#2	23′-10″	20′-8″	18′-10″	16′-10″	23′-10″	20′-8″	18′-10″	16′-10″

60psf Live Load/15psf Dead Load **C_D = 1.15, Douglas Fir-Larch**

Size	Grade	L/360 Deflection Limit 12″ o.c.	16″ o.c.	19.2″ o.c.	24″ o.c.	L/240 Deflection Limit 12″ o.c.	16″ o.c.	19.2″ o.c.	24″ o.c.
	SS	6′-3″	5′-8″	5′-4″	5′-0″	7′-2″	6′-6″	6′-2″	5′-8″
2×4	#1	6′-0″	5′-6″	5′-2″	4′-9″	6′-11″	6′-3″	4′-8″	5′-2″
	#2	5′-11″	5′-4″	5′-1″	4′-8″	6′-9″	6′-0″	4′-7″	4′-11″
	SS	9′-11″	9′-0″	8′-5″	7′-10″	11′-4″	10′-3″	7′-8″	9′-0″
2×6	#1	9′-6″	8′-8″	8′-2″	7′-7″	10′-9″	9′-3″	7′-5″	7′-7″
	#2	9′-4″	8′-6″	8′-0″	7′-2″	10′-2″	8′-9″	7′-3″	7′-2″
	SS	13′-0″	11′-10″	11′-2″	10′-4″	14′-11″	13′-7″	10′-1″	11′-9″
2×8	#1	12′-7″	11′-5″	10′-9″	9′-7″	12′-7″	11′-9″	9′-9″	9′-7″
	#2	12′-4″	11′-2″	10′-2″	9′-1″	12′-10″	11′-2″	9′-7″	9′-1″
	SS	16′-8″	15′-1″	14′-3″	13′-2″	19′-1″	17′-4″	12′-11″	14′-4″
2×10	#1	16′-0″	14′-4″	13′-1″	11′-9″	16′-7″	14′-4″	12′-5″	11′-9″
	#2	15′-9″	13′-7″	12′-5″	11′-1″	15′-9″	13′-7″	12′-2″	11′-1″
	SS	20′-3″	18′-5″	17′-4″	16′-1″	23′-2″	20′-5″	15′-9″	16′-8″
2×12	#1	19′-3″	16′-8″	15′-2″	13′-7″	19′-3″	16′-8″	15′-2″	13′-7″
	#2	18′-3″	15′-10″	14′-5″	12′-11″	18′-3″	15′-10″	14′-5″	12′-11″
	SS	23′-10″	21′-8″	20′-5″	18′-7″	26′-4″	22′-10″	18′-6″	18′-7″
2×14	#1	21′-6″	18′-7″	17′-0″	15′-2″	21′-6″	18′-7″	17′-0″	15′-2″
	#2	20′-5″	17′-8″	16′-1″	14′-5″	20′-5″	17′-8″	16′-1″	14′-5″

Table B.4 Floor joist span tables

40psf Live Load/20psf Dead Load **$C_D = 1.0$, Hem-Fir**

Size	Grade	*L*/360 Deflection Limit 12″ o.c.	16″ o.c.	19.2″ o.c.	24″ o.c.	*L*/480 Deflection Limit 12″ o.c.	16″ o.c.	19.2″ o.c.	24″ o.c.
	SS	6′-9″	6′-2″	5′-10″	5′-4″	6′-2″	5′-7″	5′-3″	4′-11″
2 × 4	#1	6′-8″	6′-0″	5′-8″	5′-3″	6′-0″	5′-6″	5′-2″	4′-9″
	#2	6′-4″	5′-9″	5′-5″	4′-11″	5′-9″	5′-3″	4′-11″	4′-7″
	SS	10′-8″	9′-8″	9′-2″	8′-6″	9′-8″	8′-10″	8′-3″	7′-8″
2 × 6	#1	10′-5″	9′-6″	8′-8″	7′-9″	9′-6″	8′-7″	8′-1″	7′-6″
	#2	10′-0″	8′-11″	8′-2″	7′-3″	9′-1″	8′-3″	7′-9″	7′-2″
	SS	14′-1″	12′-10″	12′-1″	11′-2″	12′-10″	11′-7″	10′-11″	10′-2″
2 × 8	#1	13′-9″	12′-1″	11′-0″	9′-10″	12′-6″	11′-5″	10′-8″	9′-10″
	#2	13′-1″	11′-4″	10′-4″	9′-3″	11′-11″	10′-10″	10′-2″	9′-3″
	SS	18′-0″	16′-4″	15′-5″	14′-3″	16′-4″	14′-10″	14′-0″	13′-0″
2 × 10	#1	17′-1″	14′-9″	13′-6″	12′-1″	16′-0″	14′-6″	13′-6″	12′-1″
	#2	15′-11″	13′-10″	12′-7″	11′-3″	15′-3″	13′-10″	12′-7″	11′-3″
	SS	21′-11″	19′-11″	18′-9″	16′-9″	19′-11″	18′-1″	17′-0″	15′-9″
2 × 12	#1	19′-10″	17′-2″	15′-8″	14′-0″	19′-6″	17′-2″	15′-8″	14′-0″
	#2	18′-6″	16′-0″	14′-7″	13′-1″	18′-6″	16′-0″	14′-7″	13′-1″
	SS	25′-9″	23′-0″	21′-0″	18′-9″	23′-5″	21′-3″	20′-0″	18′-7″
2 × 14	#1	22′-2″	19′-2″	17′-6″	15′-8″	22′-2″	19′-2″	17′-6″	15′-8″
	#2	20′-8″	17′-11″	16′-4″	14′-7″	20′-8″	17′-11″	16′-4″	14′-7″

40psf Live Load/10psf Dead Load **$C_D = 1.0$, Hem-Fir**

Size	Grade	*L*/360 Deflection Limit 12″ o.c.	16″ o.c.	19.2″ o.c.	24″ o.c.	*L*/480 Deflection Limit 12″ o.c.	16″ o.c.	19.2″ o.c.	24″ o.c.
	SS	6′-9″	6′-2″	5′-10″	5′-4″	6′-2″	5′-7″	5′-3″	4′-11″
2 × 4	#1	6′-8″	6′-0″	5′-8″	5′-3″	6′-0″	5′-6″	5′-2″	4′-9″
	#2	6′-4″	5′-9″	5′-5″	5′-0″	5′-9″	5′-3″	4′-11″	4′-7″
	SS	10′-8″	9′-8″	9′-2″	8′-6″	9′-8″	8′-10″	8′-3″	7′-8″
2 × 6	#1	10′-5″	9′-6″	8′-11″	8′-3″	9′-6″	8′-7″	8′-1″	7′-6″
	#2	10′-0″	9′-1″	8′-6″	7′-11″	9′-1″	8′-3″	7′-9″	7′-2″
	SS	14′-1″	12′-10″	12′-1″	11′-2″	12′-10″	11′-7″	10′-11″	10′-2″
2 × 8	#1	13′-9″	12′-6″	11′-9″	10′-10″	12′-6″	11′-5″	10′-8″	9′-11″
	#2	13′-2″	11′-11″	11′-3″	10′-1″	11′-11″	10′-10″	10′-2″	9′-6″
	SS	18′-0″	16′-4″	15′-5″	14′-3″	16′-4″	14′-10″	14′-0″	13′-0″
2 × 10	#1	17′-7″	16′-0″	14′-9″	13′-3″	16′-0″	14′-6″	13′-8″	12′-8″
	#2	16′-9″	15′-1″	13′-10″	12′-4″	15′-3″	13′-10″	13′-0″	12′-1″
	SS	21′-11″	19′-11″	18′-9″	17′-4″	19′-11″	18′-1″	17′-0″	15′-9″
2 × 12	#1	21′-5″	18′-10″	17′-2″	15′-4″	19′-6″	17′-8″	16′-8″	15′-4″
	#2	20′-3″	17′-7″	16′-0″	14′-4″	18′-7″	16′-10″	15′-10″	14′-4″
	SS	25′-9″	23′-5″	22′-1″	20′-6″	23′-5″	21′-3″	20′-0″	18′-7″
2 × 14	#1	24′-3″	21′-0″	19′-2″	17′-2″	22′-11″	20′-10″	19′-2″	17′-2″
	#2	22′-8″	19′-7″	17′-11″	16′-0″	21′-10″	19′-7″	17′-11″	16′-0″

Table B.5 Floor joist span tables

80psf Live Load/15psf Dead Load **$C_D = 1.0$, Hem-Fir**

Size	Grade	L/360 Deflection Limit 12″ o.c.	16″ o.c.	19.2″ o.c.	24″ o.c.	L/480 Deflection Limit 12″ o.c.	16″ o.c.	19.2″ o.c.	24″ o.c.
	SS	5′-4″	4′-11″	4′-7″	4′-3″	4′-11″	4′-5″	4′-2″	3′-10″
2×4	#1	5′-3″	4′-9″	4′-6″	4′-2″	4′-9″	4′-4″	4′-1″	3′-9″
	#2	5′-0″	4′-7″	4′-3″	3′-11″	4′-7″	4′-2″	3′-11″	3′-7″
	SS	8′-6″	7′-8″	7′-3″	6′-9″	7′-8″	7′-0″	6′-7″	6′-1″
2×6	#1	8′-3″	7′-6″	6′-11″	6′-2″	7′-6″	6′-10″	6′-5″	6′-0″
	#2	7′-11″	7′-1″	6′-5″	5′-9″	7′-2″	6′-6″	6′-2″	5′-8″
	SS	11′-2″	10′-2″	9′-7″	8′-10″	10′-2″	9′-3″	8′-8″	8′-1″
2×8	#1	10′-11″	9′-7″	8′-9″	7′-10″	9′-11″	9′-0″	8′-6″	7′-10″
	#2	10′-4″	9′-0″	8′-2″	7′-4″	9′-6″	8′-7″	8′-1″	7′-4″
	SS	14′-3″	13′-0″	12′-2″	11′-4″	13′-0″	11′-9″	11′-1″	10′-3″
2×10	#1	13′-7″	11′-9″	10′-9″	9′-7″	12′-8″	11′-6″	10′-9″	9′-7″
	#2	12′-8″	11′-0″	10′-0″	8′-11″	12′-1″	11′-0″	10′-0″	8′-11″
	SS	17′-4″	15′-9″	14′-10″	13′-4″	15′-9″	14′-4″	13′-6″	12′-6″
2×12	#1	15′-9″	13′-7″	12′-5″	11′-1″	15′-5″	13′-7″	12′-5″	11′-1″
	#2	14′-8″	12′-9″	11′-7″	10′-5″	14′-8″	12′-9″	11′-7″	10′-5″
	SS	20′-6″	18′-3″	16′-8″	14′-11″	18′-7″	16′-11″	15′-11″	14′-9″
2×14	#1	17′-7″	15′-3″	13′-11″	12′-5″	17′-7″	15′-3″	13′-11″	12′-5″
	#2	16′-5″	14′-3″	13′-0″	11′-7″	16′-5″	14′-3″	13′-0″	11′-7″

100psf Live Load/15psf Dead Load **$C_D = 1.0$, Hem-Fir**

Size	Grade	L/360 Deflection Limit 12″ o.c.	16″ o.c.	19.2″ o.c.	24″ o.c.	L/480 Deflection Limit 12″ o.c.	16″ o.c.	19.2″ o.c.	24″ o.c.
	SS	5′-0″	4′-6″	4′-3″	3′-11″	4′-6″	4′-1″	3′-10″	3′-7″
2×4	#1	4′-11″	4′-5″	4′-2″	3′-10″	4′-5″	4′-0″	3′-9″	3′-6″
	#2	4′-8″	4′-3″	4′-0″	3′-7″	4′-3″	3′-10″	3′-7″	3′-4″
	SS	7′-10″	7′-2″	6′-9″	6′-3″	7′-2″	6′-6″	6′-1″	5′-8″
2×6	#1	7′-8″	6′-11″	6′-3″	5′-7″	7′-0″	6′-4″	6′-0″	5′-6″
	#2	7′-4″	6′-5″	5′-10″	5′-3″	6′-8″	6′-1″	5′-8″	5′-3″
	SS	10′-4″	9′-5″	8′-10″	8′-3″	9′-5″	8′-7″	8′-1″	7′-6″
2×8	#1	10′-1″	8′-9″	8′-0″	7′-1″	9′-3″	8′-4″	7′-11″	7′-1″
	#2	9′-5″	8′-2″	7′-5″	6′-8″	8′-9″	8′-0″	7′-5″	6′-8″
	SS	13′-3″	12′-0″	11′-4″	10′-5″	12′-0″	10′-11″	10′-3″	9′-6″
2×10	#1	12′-4″	10′-8″	9′-9″	8′-8″	11′-9″	10′-8″	9′-9″	8′-8″
	#2	11′-6″	10′-0″	9′-1″	8′-1″	11′-3″	10′-0″	9′-1″	8′-1″
	SS	16′-1″	14′-8″	11′-7″	12′-1″	14′-8″	13′-4″	12′-6″	11′-7″
2×12	#1	14′-4″	12′-5″	11′-4″	10′-1″	14′-4″	12′-5″	11′-4″	10′-1″
	#2	13′-4″	11′-7″	10′-7″	9′-5″	13′-4″	11′-7″	10′-7″	9′-5″
	SS	19′-0″	16′-7″	15′-2″	13′-6″	17′-3″	15′-8″	14′-9″	13′-6″
2×14	#1	16′-0″	13′-10″	12′-8″	11′-3″	16′-0″	13′-10″	12′-8″	11′-3″
	#2	14′-11″	12′-11″	11′-9″	10′-6″	14′-11″	12′-11″	11′-9″	10′-6″

Table B.6 Floor joist span tables

40psf Live Load/15psf Dead Load **C_D = 1.15, Hem-Fir**

Size	Grade	L/360 Deflection Limit 12″ o.c.	16″ o.c.	19.2″ o.c.	24″ o.c.	L/240 Deflection Limit 12″ o.c.	16″ o.c.	19.2″ o.c.	24″ o.c.
	SS	6′-9″	6′-2″	5′-10″	5′-4″	7′-9″	7′-1″	6′-8″	6′-2″
2×4	#1	6′-8″	6′-0″	5′-8″	5′-3″	7′-7″	6′-11″	5′-2″	5′-11″
	#2	6′-4″	5′-9″	5′-5″	5′-0″	7′-3″	6′-7″	4′-11″	5′-7″
	SS	10′-8″	9′-8″	9′-2″	8′-6″	12′-3″	11′-1″	8′-3″	9′-8″
2×6	#1	10′-5″	9′-6″	8′-11″	8′-3″	12′-0″	10′-8″	8′-1″	8′-9″
	#2	10′-0″	9′-1″	8′-6″	7′-11″	11′-5″	10′-0″	7′-9″	8′-2″
	SS	14′-1″	12′-10″	12′-1″	11′-2″	16′-2″	14′-8″	10′-11″	12′-10″
2×8	#1	13′-9″	12′-6″	11′-9″	10′-11″	15′-8″	13′-7″	10′-8″	11′-1″
	#2	13′-2″	11′-11″	11′-3″	10′-4″	14′-7″	12′-8″	10′-2″	10′-4″
	SS	18′-0″	16′-4″	15′-5″	14′-3″	20′-7″	18′-9″	14′-0″	16′-2″
2×10	#1	17′-7″	16′-0″	15′-1″	13′-6″	19′-2″	16′-7″	13′-8″	13′-6″
	#2	16′-9″	15′-3″	14′-1″	12′-7″	17′-10″	15′-6″	13′-0″	12′-7″
	SS	21′-11″	19′-11″	18′-9″	17′-4″	25′-1″	22′-9″	17′-0″	18′-10″
2×12	#1	21′-5″	19′-3″	17′-6″	15′-8″	22′-2″	19′-3″	16′-8″	15′-8″
	#2	20′-5″	17′-11″	16′-4″	14′-8″	20′-9″	17′-11″	15′-10″	14′-8″
	SS	25′-9″	23′-5″	22′-1″	20′-6″	29′-6″	25′-9″	20′-0″	21′-0″
2×14	#1	24′-10″	21′-6″	19′-7″	17′-6″	24′-10″	21′-6″	19′-7″	17′-6″
	#2	23′-2″	20′-1″	18′-4″	16′-4″	23′-2″	20′-1″	18′-4″	16′-4″

60psf Live Load/15psf Dead Load **C_D = 1.15, Hem-Fir**

Size	Grade	L/360 Deflection Limit 12″ o.c.	16″ o.c.	19.2″ o.c.	24″ o.c.	L/240 Deflection Limit 12″ o.c.	16″ o.c.	19.2″ o.c.	24″ o.c.
	SS	5′-11″	5′-4″	5′-1″	4′-8″	6′-9″	6′-2″	5′-10″	5′-4″
2×4	#1	5′-10″	5′-3″	4′-11″	4′-7″	6′-8″	6′-0″	4′-6″	5′-1″
	#2	5′-6″	5′-0″	4′-9″	4′-4″	6′-4″	5′-9″	4′-3″	4′-9″
	SS	9′-4″	8′-6″	8′-0″	7′-5″	10′-8″	9′-8″	7′-3″	8′-6″
2×6	#1	9′-2″	8′-3″	7′-10″	7′-3″	10′-5″	9′-2″	7′-1″	7′-6″
	#2	8′-8″	7′-11″	7′-5″	6′-11″	9′-10″	8′-7″	6′-9″	7′-0″
	SS	12′-4″	11′-2″	10′-6″	9′-9″	14′-1″	12′-10″	9′-7″	11′-2″
2×8	#1	12′-1″	10′-11″	10′-3″	9′-6″	13′-5″	11′-7″	9′-4″	9′-6″
	#2	11′-6″	10′-5″	9′-10″	8′-10″	12′-6″	10′-10″	8′-11″	8′-10″
	SS	15′-9″	14′-3″	13′-5″	12′-6″	18′-0″	16′-4″	12′-2″	13′-10″
2×10	#1	15′-5″	14′-0″	12′-11″	11′-7″	16′-5″	14′-2″	11′-11″	11′-7″
	#2	14′-8″	13′-3″	12′-1″	10′-10″	15′-4″	13′-3″	11′-5″	10′-10″
	SS	19′-1″	17′-4″	16′-4″	15′-2″	21′-11″	19′-9″	14′-10″	16′-1″
2×12	#1	18′-9″	16′-5″	15′-0″	13′-5″	19′-0″	16′-5″	14′-6″	13′-5″
	#2	17′-9″	15′-4″	14′-0″	12′-6″	17′-9″	15′-4″	13′-10″	12′-6″
	SS	22′-6″	20′-6″	19′-3″	17′-10″	25′-5″	22′-0″	17′-6″	18′-0″
2×14	#1	21′-3″	18′-5″	16′-9″	15′-0″	21′-3″	18′-5″	16′-9″	15′-0″
	#2	19′-10″	17′-2″	15′-8″	14′-0″	19′-10″	17′-2″	15′-8″	14′-0″

Table B.7 Floor joist span tables

40psf Live Load/20psf Dead Load **C_D = 1.0, Spruce-Pine-Fir**

Size	Grade	L/360 Deflection Limit 12″ o.c.	16″ o.c.	19.2″ o.c.	24″ o.c.	L/240 Deflection Limit 12″ o.c.	16″ o.c.	19.2″ o.c.	24″ o.c.
	SS	6′-8″	6′-0″	5′-8″	5′-3″	7′-7″	6′-11″	6′-6″	6′-0″
2×4	#1/#2	6′-6″	5′-11″	5′-6″	5′-0″	7′-1″	6′-2″	5′-0″	5′-0″
	#3	5′-5″	4′-8″	4′-3″	3′-9″	5′-5″	4′-8″	4′-3″	3′-9″
	SS	10′-5″	9′-6″	8′-11″	8′-3″	12′-0″	10′-10″	8′-1″	8′-10″
2×6	#1/#2	10′-3″	9′-0″	8′-3″	7′-4″	10′-5″	9′-0″	7′-11″	7′-4″
	#3	7′-11″	6′-10″	6′-3″	5′-7″	7′-11″	6′-10″	6′-3″	5′-7″
	SS	13′-9″	12′-6″	11′-9″	10′-11″	15′-10″	13′-8″	10′-8″	11′-2″
2×8	#1/#2	13′-3″	11′-5″	10′-5″	9′-4″	13′-3″	11′-5″	10′-5″	9′-4″
	#3	10′-0″	8′-8″	7′-11″	7′-1″	10′-0″	8′-8″	7′-11″	7′-1″
	SS	17′-7″	16′-0″	15′-1″	13′-8″	19′-4″	16′-9″	13′-8″	13′-8″
2×10	#1/#2	16′-2″	14′-0″	12′-9″	11′-5″	16′-2″	14′-0″	12′-9″	11′-5″
	#3	12′-3″	10′-7″	9′-8″	8′-8″	12′-3″	10′-7″	9′-8″	8′-8″
	SS	21′-5″	19′-5″	17′-9″	15′-10″	22′-5″	19′-5″	16′-8″	15′-10″
2×12	#1/#2	18′-9″	16′-3″	14′-10″	13′-3″	18′-9″	16′-3″	14′-10″	13′-3″
	#3	14′-2″	12′-3″	11′-2″	10′-0″	14′-2″	12′-3″	11′-2″	10′-0″
	SS	25′-1″	21′-9″	19′-10″	17′-9″	25′-1″	21′-9″	19′-7″	17′-9″
2×14	#1/#2	21′-0″	18′-2″	16′-7″	14′-10″	21′-0″	18′-2″	16′-7″	14′-10″
	#3	15′-10″	13′-9″	12′-6″	11′-2″	15′-10″	13′-9″	12′-6″	11′-2″

40psf Live Load/10psf Dead Load **C_D = 1.0, Spruce-Pine-Fir**

Size	Grade	L/360 Deflection Limit 12″ o.c.	16″ o.c.	19.2″ o.c.	24″ o.c.	L/240 Deflection Limit 12″ o.c.	16″ o.c.	19.2″ o.c.	24″ o.c.
	SS	6′-8″	6′-0″	5′-8″	5′-3″	7′-7″	6′-11″	6′-6″	6′-0″
2×4	#1/#2	6′-6″	5′-11″	5′-6″	5′-2″	7′-5″	6′-9″	5′-0″	5′-6″
	#3	5′-11″	5′-1″	4′-8″	4′-2″	5′-11″	5′-1″	4′-8″	4′-2″
	SS	10′-5″	9′-6″	8′-11″	8′-3″	12′-0″	10′-10″	8′-1″	9′-6″
2×6	#1/#2	10′-3″	9′-3″	8′-9″	8′-1″	11′-5″	9′-11″	7′-11″	8′-1″
	#3	8′-8″	7′-6″	6′-10″	6′-1″	8′-8″	7′-6″	6′-10″	6′-1″
	SS	13′-9″	12′-6″	11′-9″	10′-11″	15′-10″	14′-4″	10′-8″	12′-3″
2×8	#1/#2	13′-6″	12′-3″	11′-5″	10′-3″	14′-6″	12′-7″	10′-6″	10′-3″
	#3	10′-11″	9′-6″	8′-8″	7′-9″	10′-11″	9′-6″	8′-8″	7′-9″
	SS	17′-7″	16′-0″	15′-1″	14′-0″	20′-2″	18′-4″	13′-8″	15′-0″
2×10	#1/#2	17′-2″	15′-4″	14′-0″	12′-6″	17′-9″	15′-4″	13′-4″	12′-6″
	#3	13′-5″	11′-7″	10′-7″	9′-5″	13′-5″	11′-7″	10′-7″	9′-5″
	SS	21′-5″	19′-6″	18′-4″	17′-0″	24′-6″	21′-3″	16′-8″	17′-4″
2×12	#1/#2	20′-7″	17′-10″	16′-3″	14′-6″	20′-7″	17′-10″	16′-3″	14′-6″
	#3	15′-6″	13′-5″	12′-3″	11′-0″	15′-6″	13′-5″	12′-3″	11′-0″
	SS	25′-3″	22′-11″	21′-7″	19′-5″	27′-6″	23′-9″	19′-7″	19′-5″
2×14	#1/#2	23′-0″	19′-11″	18′-2″	16′-3″	23′-0″	19′-11″	18′-2″	16′-3″
	#3	17′-4″	15′-0″	13′-9″	12′-3″	17′-4″	15′-0″	13′-9″	12′-3″

Table B.8 Floor joist span tables

100psf Live Load/15psf Dead Load **C_D = 1.0, Spruce-Pine-Fir**

Size	Grade	*L*/360 Deflection Limit 12″ o.c.	16″ o.c.	19.2″ o.c.	24″ o.c.	*L*/480 Deflection Limit 12″ o.c.	16″ o.c.	19.2″ o.c.	24″ o.c.
	SS	4′-11″	4′-5″	4′-2″	3′-10″	4′-5″	4′-0″	3′-9″	3′-6″
2×4	#1/#2	4′-9″	4′-4″	4′-1″	3′-7″	4′-4″	3′-11″	3′-8″	3′-5″
	#3	3′-10″	3′-4″	3′-1″	2′-9″	3′-10″	3′-4″	3′-1″	2′-9″
	SS	7′-8″	7′-0″	6′-7″	6′-1″	7′-0″	6′-4″	6′-0″	5′-6″
2×6	#1/#2	7′-6″	6′-6″	5′-11″	5′-4″	6′-10″	6′-2″	5′-10″	5′-4″
	#3	5′-8″	4′-11″	4′-6″	4′-0″	5′-8″	4′-11″	4′-6″	4′-0″
	SS	10′-2″	9′-3″	8′-8″	8′-1″	9′-3″	8′-4″	7′-11″	7′-4″
2×8	#1/#2	9′-7″	8′-3″	7′-6″	6′-9″	9′-0″	8′-2″	7′-6″	6′-9″
	#3	7′-3″	6′-3″	5′-8″	5′-1″	7′-3″	6′-3″	5′-8″	5′-1″
	SS	13′-0″	11′-9″	11′-0″	9′-10″	11′-9″	10′-8″	10′-1″	9′-4″
2×10	#1/#2	11′-8″	10′-1″	9′-3″	8′-3″	11′-6″	10′-1″	9′-3″	8′-3″
	#3	8′-10″	7′-8″	7′-0″	6′-3″	8′-10″	7′-8″	7′-0″	6′-3″
	SS	15′-9″	14′-0″	12′-10″	11′-5″	14′-4″	13′-0″	12′-3″	11′-4″
2×12	#1/#2	13′-7″	11′-9″	10′-8″	9′-7″	13′-7″	11′-9″	10′-8″	9′-7″
	#3	10′-3″	8′-10″	8′-1″	7′-3″	10′-3″	8′-10″	8′-1″	7′-3″
	SS	18′-1″	15′-8″	14′-4″	12′-9″	16′-11″	15′-4″	14′-4″	12′-9″
2×14	#1/#2	15′-2″	13′-1″	12′-0″	10′-8″	15′-2″	13′-1″	12′-0″	10′-8″
	#3	11′-5″	9′-11″	9′-0″	8′-1″	11′-5″	9′-11″	9′-0″	8′-1″

80psf Live Load/15psf Dead Load **C_D = 1.0, Spruce-Pine-Fir**

Size	Grade	*L*/360 Deflection Limit 12″ o.c.	16″ o.c.	19.2″ o.c.	24″ o.c.	*L*/480 Deflection Limit 12″ o.c.	16″ o.c.	19.2″ o.c.	24″ o.c.
	SS	5′-3″	4′-9″	4′-6″	4′-2″	4′-9″	4′-4″	4′-1″	3′-9″
2×4	#1/#2	5′-2″	4′-8″	4′-5″	4′-0″	4′-8″	4′-3″	4′-0″	3′-8″
	#3	4′-3″	3′-8″	3′-4″	3′-0″	4′-3″	3′-8″	3′-4″	3′-0″
	SS	8′-3″	7′-6″	7′-1″	6′-7″	7′-6″	6′-10″	6′-5″	6′-0″
2×6	#1/#2	8′-1″	7′-2″	6′-7″	5′-10″	7′-4″	6′-8″	6′-3″	5′-10″
	#3	6′-3″	5′-5″	4′-11″	4′-5″	6′-3″	5′-5″	4′-11″	4′-5″
	SS	10′-11″	9′-11″	9′-4″	8′-8″	9′-11″	9′-0″	8′-6″	7′-11″
2×8	#1/#2	10′-6″	9′-1″	8′-4″	7′-5″	9′-8″	8′-10″	8′-4″	7′-5″
	#3	7′-11″	6′-10″	6′-3″	5′-7″	7′-11″	6′-10″	6′-3″	5′-7″
	SS	14′-0″	12′-8″	11′-11″	10′-10″	12′-8″	11′-6″	10′-10″	10′-1″
2×10	#1/#2	12′-10″	11′-1″	10′-2″	9′-1″	12′-5″	11′-1″	10′-2″	9′-1″
	#3	9′-8″	8′-5″	7′-8″	6′-10″	9′-8″	8′-5″	7′-8″	6′-10″
	SS	17′-0″	15′-5″	14′-1″	12′-7″	15′-5″	14′-0″	13′-2″	12′-3″
2×12	#1/#2	14′-11″	12′-11″	11′-9″	10′-6″	14′-11″	12′-11″	11′-9″	10′-6″
	#3	11′-3″	9′-9″	8′-11″	7′-11″	11′-3″	9′-9″	8′-11″	7′-11″
	SS	19′-11″	17′-3″	15′-9″	14′-1″	18′-2″	16′-6″	15′-7″	14′-1″
2×14	#1/#2	16′-8″	14′-5″	13′-2″	11′-9″	16′-8″	14′-5″	13′-2″	11′-9″
	#3	12′-7″	10′-11″	9′-11″	8′-11″	12′-7″	10′-11″	9′-11″	8′-11″

Table B.9 Floor joist span tables

40psf Live Load/15psf Dead Load **C_D = 1.15, Spruce-Pine-Fir**

Size	Grade	*L*/360 Deflection Limit 12″ o.c.	16″ o.c.	19.2″ o.c.	24″ o.c.	*L*/240 Deflection Limit 12″ o.c.	16″ o.c.	19.2″ o.c.	24″ o.c.
	SS	6′-8″	6-0	5′-8″	5′-3″	7′-7″	6′-11″	6′-6″	6′-0″
2 × 4	#1/#2	6′-6″	5′-11″	5′-6″	5′-2″	7′-5″	6′-9″	5′-0″	5′-8″
	#3	6′-0″	5′-3″	4′-9″	4′-3″	6′-0″	5′-3″	4′-9″	4′-3″
	SS	10′-5″	9′-6″	8′-11″	8′-3″	12′-0″	10′-10″	8′-1″	9′-6″
2 × 6	#1/#2	10′-3″	9′-3″	8′-9″	8′-1″	11′-8″	10′-2″	7′-11″	8′-3″
	#3	8′-10″	7′-8″	7′-0″	6′-3″	8′-10″	7′-8″	7′-0″	6′-3″
	SS	13′-9″	12′-6″	11′-9″	10′-11″	15′-10″	14′-4″	10′-8″	12′-6″
2 × 8	#1/#2	13′-6″	12′-3″	11′-6″	10′-6″	14′-10″	12′-10″	10′-6″	10′-6″
	#3	11′-2″	9′-8″	8′-10″	7′-11″	11′-2″	9′-8″	8′-10″	7′-11″
	SS	17′-7″	16′-0″	15′-1″	14′-0″	20′-2″	18′-4″	13′-8″	15′-4″
2 × 10	#1/#2	17′-2″	15′-8″	14′-4″	12′-10″	18′-2″	15′-8″	13′-4″	12′-10″
	#3	13′-8″	11′-10″	10′-10″	9′-8″	13′-8″	11′-10″	10′-10″	9′-8″
	SS	21′-5″	19′-6″	18′-4″	17′-0″	24′-6″	21′-9″	16′-8″	17′-9″
2 × 12	#1/#2	20′-11″	18′-2″	16′-7″	14′-10″	21′-0″	18′-2″	16′-3″	14′-10″
	#3	15′-11″	13′-9″	12′-7″	11′-3″	15′-11″	13′-9″	12′-7″	11′-3″
	SS	25′-3″	22′-11″	21′-7″	19′-10″	28′-1″	24′-4″	19′-7″	19′-10″
2 × 14	#1/#2	23′-6″	20′-4″	18′-7″	16′-7″	23′-6″	20′-4″	18′-7″	16′-7″
	#3	17′-9″	15′-4″	14′-0″	12′-6″	17′-9″	15′-4″	14′-0″	12′-6″

60psf Live Load/15psf Dead Load **C_D = 1.15, Spruce-Pine-Fir**

Size	Grade	*L*/360 Deflection Limit 12″ o.c.	16″ o.c.	19.2″ o.c.	24″ o.c.	*L*/240 Deflection Limit 12″ o.c.	16″ o.c.	19.2″ o.c.	24″ o.c.
	SS	5′-10″	5′-3″	4′-11″	4′-7″	6′-8″	6′-0″	5′-8″	5′-3″
2 × 4	#1/#2	5′-8″	5′-2″	4′-10″	4′-6″	6′-6″	5′-11″	4′-5″	4′-10″
	#3	5′-2″	4′-6″	4′-1″	3′-8″	5′-2″	4′-6″	4′-1″	3′-8″
	SS	9′-2″	8′-3″	7′-10″	7′-3″	10′-5″	9′-6″	7′-1″	8′-3″
2 × 6	#1/#2	8′-11″	8′-1″	7′-7″	7′-1″	10′-0″	8′-8″	6′-11″	7′-1″
	#3	7′-7″	6′-6″	6′-0″	5′-4″	7′-7″	6′-6″	6′-0″	5′-4″
	SS	12′-1″	10′-11″	10′-3″	9′-7″	13′-9″	12′-6″	9′-4″	10′-9″
2 × 8	#1/#2	11′-9″	10′-8″	10′-0″	9′-0″	12′-8″	11′-0″	9′-2″	9′-0″
	#3	9′-7″	8′-4″	7′-7″	6′-9″	9′-7″	8′-4″	7′-7″	6′-9″
	SS	15′-5″	14′-0″	13′-2″	12′-2″	17′-7″	16′-0″	11′-11″	13′-1″
2 × 10	#1/#2	15′-0″	13′-5″	12′-3″	11′-0″	15′-6″	13′-5″	11′-8″	11′-0″
	#3	11′-9″	10′-2″	9′-3″	8′-3″	11′-9″	10′-2″	9′-3″	8′-3″
	SS	18′-9″	17′-0″	16′-0″	14′-10″	21′-5″	18′-8″	14′-6″	15′-2″
2 × 12	#1/#2	18′-0″	15′-7″	14′-3″	12′-9″	18′-0″	15′-7″	14′-2″	12′-9″
	#3	13′-7″	11′-9″	10′-9″	9′-7″	13′-7″	11′-9″	10′-9″	9′-7″
	SS	22′-1″	20′-0″	18′-10″	17′-0″	24′-1″	20′-10″	17′-1″	17′-0″
2 × 14	#1/#2	20′-1″	17′-5″	15′-11″	14′-3″	20′-1″	17′-5″	15′-11″	14′-3″
	#3	15′-2″	13′-2″	12′-0″	10′-9″	15′-2″	13′-2″	12′-0″	10′-9″

Length	Select structural 100%	Select structural 115%	Select structural 125%	#1/#2 100%	#1/#2 115%	#1/#2 125%	#3 100%	#3 115%	#3 125%
4′	19.55	21.53	22.70	17.34	19.11	20.17	15.77	17.42	18.41
5′	16.44	17.52	18.10	14.64	15.62	16.16	13.43	14.37	14.89
6′	13.19	13.70	13.97	11.78	12.26	12.50	10.89	11.35	11.59
7′	10.45	10.70	10.83	9.35	9.59	9.71	8.68	8.91	9.03
8′	8.34	8.48	8.55	7.48	7.61	7.67	6.96	7.08	7.15
9′	6.76	6.84	6.89	6.07	6.14	6.18	5.66	5.73	5.76
10′	5.57	5.62	5.65	5.00	5.05	5.07	4.67	4.71	4.73
11′	4.66	4.69	4.71	4.18	4.21	4.23	3.90	3.93	3.95
12′	3.95	3.97	3.98	3.55	3.57	3.58	3.31	3.33	3.34
13′	3.39	3.40	3.41	3.04	3.05	3.06	2.84	2.85	2.86
14′	2.93	2.95	2.95	2.63	2.65	2.65	2.46	2.47	2.48
15′									
16′									

Figure B.53 Allowable axial loads (kips) on Douglas Fir-Larch (4 × 4).

Length	Select structural 100%	Select structural 115%	Select structural 125%	#1 100%	#1 115%	#1 125%	#2 100%	#2 115%	#2 125%
4′	21.78	24.74	26.67	19.14	21.78	23.51	13.51	15.41	16.65
5′	20.73	23.28	24.89	18.37	20.73	22.23	13.07	14.80	15.92
6′	19.26	21.27	22.48	17.31	19.26	20.46	12.46	13.96	14.89
7′	17.41	18.82	19.61	15.94	17.41	18.26	11.66	12.86	13.57
8′	15.32	16.23	16.71	14.31	15.32	15.87	10.68	11.56	12.06
9′	13.26	13.82	14.11	12.60	13.26	13.60	9.58	10.19	10.51
10′	11.39	11.74	11.92	10.98	11.39	11.61	8.48	8.88	9.09
11′	9.80	10.02	10.14	9.53	9.80	9.94	7.45	7.71	7.85
12′	8.46	8.61	8.69	8.28	8.46	8.55	6.53	6.70	6.79
13′	7.35	7.46	7.51	7.23	7.35	7.42	5.73	5.85	5.92
14′	6.44	6.51	6.55	6.35	6.44	6.48	5.05	5.14	5.18
15′	5.67	5.72	5.75	5.60	5.67	5.70	4.48	4.54	4.57
16′	5.03	5.07	5.09	4.98	5.03	5.05	3.99	4.03	4.06

Figure B.54 Allowable axial loads (kips) on Douglas Fir-Larch (5 × 5).

Length	Select structural			#1			#2		
	100%	115%	125%	100%	115%	125%	100%	115%	125%
4′	33.36	38.09	41.19	29.19	33.36	36.11	20.54	23.50	25.46
5′	32.42	36.80	39.63	28.50	32.42	34.97	20.14	22.95	24.79
6′	31.13	35.02	37.47	27.56	31.13	33.41	19.59	22.21	23.89
7′	29.44	32.68	34.65	26.34	29.44	31.36	18.89	21.23	22.71
8′	27.33	29.83	31.29	24.80	27.33	28.83	17.99	20.00	21.22
9′	24.89	26.70	27.69	22.96	24.89	25.98	16.91	18.53	19.47
10′	22.31	23.55	24.21	20.92	22.31	23.07	15.66	16.89	17.59
11′	19.80	20.64	21.07	18.83	19.80	20.32	14.32	15.22	15.71
12′	17.50	18.06	18.36	16.82	17.50	17.85	12.96	13.60	13.94
13′	15.46	15.85	16.05	14.98	15.46	15.70	11.67	12.12	12.36
14′	13.68	13.96	14.10	13.35	13.68	13.85	10.47	10.80	10.97
15′	12.16	12.36	12.46	11.92	12.16	12.28	9.41	9.65	9.77
16′	10.85	11.00	11.08	10.67	10.85	10.94	8.46	8.64	8.73

Figure B.55 Allowable axial loads (kips) on Douglas Fir-Larch (6 × 6).

Length	Select structural			#1			#2		
	100%	115%	125%	100%	115%	125%	100%	115%	125%
4′	63.33	72.58	78.71	55.23	63.33	68.70	38.76	44.47	48.25
5′	62.50	71.45	77.36	54.61	62.50	67.70	38.40	43.97	47.66
6′	61.40	69.96	75.56	53.81	61.40	66.39	37.92	43.33	46.89
7′	60.00	68.04	73.23	52.79	60.00	64.70	37.33	42.52	45.91
8′	58.25	65.61	70.28	51.52	58.25	62.57	36.59	41.50	44.68
9′	56.09	62.63	66.67	49.96	56.09	59.95	35.69	40.26	43.17
10′	53.51	59.09	62.44	48.09	53.51	56.83	34.61	38.77	41.36
11′	50.53	55.10	57.74	45.89	50.53	53.27	33.34	37.01	39.25
12′	47.22	50.82	52.83	43.41	47.22	49.40	31.87	35.03	36.89
13′	43.73	46.49	47.98	40.69	43.73	45.42	30.23	32.86	34.35
14′	40.22	42.30	43.39	37.86	40.22	41.50	28.46	30.58	31.76
15′	36.82	38.38	39.19	35.01	36.82	37.78	26.62	28.30	29.21
16′	33.63	34.80	35.41	32.25	33.63	34.36	24.76	26.08	29.77

Figure B.56 Allowable axial loads (kips) on Douglas Fir-Larch (8 × 8).

Length	Select structural			#1/#2			#3		
	100%	115%	125%	100%	115%	125%	100%	115%	125%
4′	17.01	18.67	19.65	15.54	17.12	18.05	14.43	15.77	16.54
5′	14.15	15.02	15.49	13.08	13.94	14.41	11.81	12.48	12.83
6′	11.24	11.65	11.87	10.50	10.91	11.13	9.28	9.58	9.74
7′	8.86	9.06	9.17	8.32	8.53	8.63	7.26	7.41	7.49
8′	7.06	7.16	7.22	6.65	6.76	6.81	5.76	5.84	5.88
9′	5.71	5.77	5.81	5.39	5.45	5.49	4.65	4.70	4.72
10′	4.70	4.74	4.76	4.44	4.48	4.50	3.82	3.85	3.87
11′	3.93	3.95	3.97	3.71	3.74	3.75	3.19	3.21	3.22
12′	3.33	3.34	3.35	3.15	3.16	3.17	2.70	2.72	2.72
13′	2.85	2.86	2.87	2.70	2.71	2.72	2.32	2.32	2.33
14′	2.47	2.48	2.48	2.34	2.35	2.35	2.00	2.01	2.02
15′									
16′									

Figure B.57 Allowable axial loads (kips) on Hem-Fir (4 × 4).

Length	Select structural			#1			#2		
	100%	115%	125%	100%	115%	125%	100%	115%	125%
4′	18.40	20.88	22.49	16.21	18.44	19.90	11.12	12.69	13.72
5′	17.45	19.57	20.90	15.52	17.49	18.74	10.77	12.21	13.14
6′	16.13	17.77	18.74	14.56	16.16	17.14	10.29	11.55	12.33
7′	14.48	15.60	16.22	13.32	14.50	15.18	9.67	10.68	11.29
8′	12.66	13.37	13.74	11.89	12.68	13.11	8.89	9.65	10.08
9′	10.90	11.33	11.55	10.40	10.91	11.17	8.02	8.55	8.83
10′	9.33	9.60	9.73	9.02	9.34	9.50	7.12	7.47	7.66
11′	8.00	8.17	8.26	7.80	8.01	8.11	6.27	6.51	6.63
12′	6.90	7.01	7.07	6.76	6.90	6.97	5.51	5.67	5.75
13′	5.99	6.07	6.11	5.90	5.99	6.04	4.85	4.96	5.01
14′	5.24	5.29	5.32	5.17	5.24	5.27	4.28	4.36	4.40
15′	4.61	4.65	4.67	4.56	4.61	4.64	3.79	3.85	3.88
16′	4.08	4.12	4.13	4.05	4.09	4.10	3.38	3.42	3.44

Figure B.58 Allowable axial loads (kips) on Hem-Fir (5 × 5).

Length	Select structural			#1			#2		
	100%	115%	125%	100%	115%	125%	100%	115%	125%
4′	28.22	32.21	34.82	24.76	28.29	30.61	16.89	19.33	20.94
5′	27.38	31.05	33.42	24.14	27.44	29.59	16.57	18.90	20.42
6′	26.22	29.44	31.47	23.30	26.28	28.17	16.15	18.31	19.71
7′	24.69	27.35	28.95	22.19	24.74	26.32	15.60	17.55	18.78
8′	22.81	24.82	25.97	20.79	22.84	24.05	14.89	16.58	17.61
9′	20.66	22.08	22.85	19.15	20.68	21.55	14.04	15.42	16.23
10′	18.42	19.38	19.89	17.35	18.44	19.03	13.05	14.12	14.72
11′	16.28	16.92	17.25	15.54	16.29	16.69	11.97	12.77	13.19
12′	14.34	14.77	14.99	13.83	14.35	14.61	10.88	11.45	11.75
13′	12.63	12.93	13.09	12.28	12.64	12.82	9.82	10.22	10.44
14′	11.17	11.38	11.49	10.92	11.17	11.30	8.84	9.13	9.28
15′	9.91	10.06	10.14	9.73	9.91	10.01	7.95	8.16	8.27
16′	8.83	8.95	9.01	8.70	8.84	8.91	7.16	7.32	7.40

Figure B.59 Allowable axial loads (kips) on Hem-Fir (6 × 6).

Length	Select structural			#1			#3		
	100%	115%	125%	100%	115%	125%	100%	115%	125%
4′	53.64	61.46	66.64	46.90	53.77	58.32	31.86	36.55	39.67
5′	52.89	60.45	65.43	46.35	53.02	57.43	31.57	36.16	39.20
6′	51.91	59.11	63.82	45.62	52.04	56.24	31.20	35.65	38.59
7′	50.66	57.39	61.72	44.70	50.77	54.71	30.73	35.01	37.82
8′	49.08	55.20	59.07	43.55	49.19	52.79	30.15	34.22	36.86
9′	47.14	52.52	55.83	42.14	47.23	50.42	29.44	33.24	35.67
10′	44.82	49.36	52.05	40.45	44.90	47.61	28.60	32.07	34.25
11′	42.15	45.82	47.91	38.47	42.22	44.43	27.60	30.70	32.59
12′	39.23	42.08	43.65	36.24	39.29	41.01	26.45	29.13	30.72
13′	36.19	38.34	39.49	33.84	36.23	37.54	25.15	27.41	28.70
14′	33.16	34.77	35.61	31.36	33.20	34.18	23.75	25.58	26.61
15′	30.27	31.46	32.08	28.90	30.29	31.03	22.27	23.74	24.53
16′	27.58	28.47	28.93	26.54	27.60	28.15	20.77	21.93	22.54

Figure B.60 Allowable axial loads (kips) on Hem-Fir (8 × 8).

Length	Select structural 100%	Select structural 115%	Select structural 125%	#1/#2 100%	#1/#2 115%	#1/#2 125%	#3 100%	#3 115%	#3 125%
4′	15.95	17.53	18.46	13.54	14.99	15.86	8.26	9.31	9.98
5′	13.31	14.15	14.60	11.61	12.45	12.92	7.61	8.41	8.89
6′	10.61	11.01	11.22	9.47	9.89	10.12	6.73	7.26	7.55
7′	8.38	8.57	8.67	7.58	7.79	7.90	5.75	6.05	6.22
8′	6.68	6.78	6.84	6.09	6.21	6.26	4.82	4.99	5.08
9′	5.41	5.47	5.50	4.96	5.02	5.06	4.02	4.12	4.17
10′	4.45	4.49	4.51	4.09	4.13	4.15	3.38	3.44	3.47
11′	3.72	3.75	3.76	3.43	3.45	3.47	2.86	2.90	2.92
12′	3.15	3.17	3.18	2.91	2.92	2.93	2.44	2.47	2.48
13′	2.70	2.71	2.72	2.49	2.51	2.51	2.10	2.12	2.13
14′	2.34	2.35	2.35	2.16	2.17	2.18	1.83	1.84	1.85
15′									
16′									

Figure B.61 Allowable axial loads (kips) on Spruce-Pine-Fir (4 × 4).

Length	Select structural 100%	Select structural 115%	Select structural 125%	#1 100%	#1 115%	#1 125%	#2 100%	#2 115%	#2 125%
4′	15.32	17.44	18.83	13.51	15.41	16.65	9.70	11.07	11.98
5′	14.72	16.61	17.82	13.07	14.80	15.92	9.42	10.69	11.51
6′	13.89	15.46	16.43	12.46	13.96	14.89	9.04	10.16	10.87
7′	12.80	14.00	14.69	11.66	12.86	13.57	8.53	9.46	10.03
8′	11.52	12.34	12.80	10.68	11.56	12.06	7.90	8.62	9.03
9′	10.16	10.70	10.98	9.58	10.19	10.51	7.18	7.69	7.98
10′	8.86	9.20	9.38	8.48	8.88	9.09	6.43	6.78	6.96
11′	7.70	7.92	8.04	7.45	7.71	7.85	5.69	5.93	6.05
12′	6.70	6.84	6.92	6.53	6.70	6.79	5.02	5.18	5.26
13′	5.85	5.95	6.00	5.73	5.85	5.92	4.43	4.54	4.60
14′	5.14	5.21	5.25	5.05	5.14	5.18	3.92	4.00	4.04
15′	4.54	4.59	4.62	4.48	4.54	4.57	3.48	3.54	3.57
16′	4.03	4.07	4.09	3.99	4.03	4.06	3.11	3.15	3.17

Figure B.62 Allowable axial loads (kips) on Spruce-Pine-Fir (5 × 5).

Length	Select structural			#1			#2		
	100%	115%	125%	100%	115%	125%	100%	115%	125%
4′	23.36	26.70	28.91	20.54	23.50	25.46	14.72	16.85	18.26
5′	22.82	25.96	28.01	20.14	22.95	24.79	14.46	16.50	17.84
6′	22.08	24.95	26.78	19.59	22.21	23.89	14.12	16.03	17.27
7′	21.12	23.62	25.17	18.89	21.23	22.71	13.67	15.42	16.52
8′	19.91	21.95	23.18	17.99	20.00	21.22	13.11	14.64	15.58
9′	18.45	20.03	20.92	16.91	18.53	19.47	12.42	13.70	14.46
10′	16.84	17.98	18.61	15.66	16.89	17.59	11.62	12.63	13.21
11′	15.18	15.98	16.40	14.32	15.22	15.71	10.73	11.49	11.91
12′	13.58	14.13	14.42	12.96	13.60	13.94	9.81	10.36	10.66
13′	12.10	12.49	12.69	11.67	12.12	12.36	8.90	9.30	9.51
14′	10.79	11.07	11.21	10.47	10.80	10.97	8.04	8.33	8.48
15′	9.64	9.84	9.94	9.41	9.65	9.77	7.25	7.47	7.58
16′	8.63	8.78	8.86	8.46	8.64	8.73	6.55	6.71	6.79

Figure B.63 Allowable axial loads (kips) on Spruce-Pine-Fir (6 × 6).

Length	Select structural			#1			#2		
	100%	115%	125%	100%	115%	125%	100%	115%	125%
4′	44.20	50.68	54.98	38.76	44.47	48.25	27.73	31.82	34.53
5′	43.71	50.02	54.19	38.40	43.97	47.66	27.49	31.50	34.16
6′	43.07	49.16	53.16	37.92	43.33	46.89	27.19	31.09	33.66
7′	42.27	48.06	51.83	37.33	42.52	45.91	26.81	30.58	33.04
8′	41.27	46.68	50.15	36.59	41.50	44.68	26.35	29.94	32.27
9′	40.05	44.98	48.09	35.69	40.26	43.17	25.78	29.16	31.32
10′	38.57	42.95	45.64	34.61	38.77	41.36	25.10	28.22	30.18
11′	36.85	40.60	42.83	33.34	37.01	39.25	24.30	27.11	28.85
12′	34.89	37.99	39.77	31.87	35.03	36.89	23.38	25.85	27.33
13′	32.74	35.22	36.60	30.23	32.86	34.35	22.33	24.44	25.66
14′	30.49	32.43	33.47	28.46	30.58	31.76	21.18	22.93	23.92
15′	28.23	29.71	30.50	26.62	28.30	29.21	19.95	21.37	22.15
16′	26.02	27.16	27.75	24.76	26.08	26.77	18.69	19.83	20.43

Figure B.64 Allowable axial loads (kips) on Spruce-Pine-Fir (8 × 8).

Index

Note: Page numbers followed by f and t refer to figures and tables, respectively.

Printed in the United States
by Baker & Taylor Publisher Services

Printed in the United States
by Baker & Taylor Publisher Services